Representative
Elements

					18 8A
13 3A	14 4A	15 5A	16 6A	17 7A	2 **He** Helium 4.003
5 **B** Boron 10.811	6 **C** Carbon 12.011	7 **N** Nitrogen 14.007	8 **O** Oxygen 15.999	9 **F** Fluorine 18.998	10 **Ne** Neon 20.180
13 **Al** Aluminum 26.982	14 **Si** Silicon 28.086	15 **P** Phosphorus 30.974	16 **S** Sulfur 32.066	17 **Cl** Chlorine 35.453	18 **Ar** Argon 39.948

10	11 1B	12 2B						
28 **Ni** Nickel 58.693	29 **Cu** Copper 63.546	30 **Zn** Zinc 65.39	31 **Ga** Gallium 69.723	32 **Ge** Germanium 72.61	33 **As** Arsenic 74.922	34 **Se** Selenium 78.96	35 **Br** Bromine 79.904	36 **Kr** Krypton 83.80
46 **Pd** Palladium 106.42	47 **Ag** Silver 107.868	48 **Cd** Cadmium 112.411	49 **In** Indium 114.82	50 **Sn** Tin 118.710	51 **Sb** Antimony 121.757	52 **Te** Tellurium 127.60	53 **I** Iodine 126.904	54 **Xe** Xenon 131.29
78 **Pt** Platinum 195.08	79 **Au** Gold 196.967	80 **Hg** Mercury 200.59	81 **Tl** Thallium 204.383	82 **Pb** Lead 207.2	83 **Bi** Bismuth 208.980	84 **Po** Polonium (209)	85 **At** Astatine (210)	86 **Rn** Radon (222)
110 **Uun** (269)	111 **Uuu** (272)	112 **Uub** (277)						

Metals ← → Nonmetals

63 **Eu** Europium 151.965	64 **Gd** Gadolinium 157.25	65 **Tb** Terbium 158.925	66 **Dy** Dysprosium 162.50	67 **Ho** Holmium 164.930	68 **Er** Erbium 167.26	69 **Tm** Thulium 168.934	70 **Yb** Ytterbium 173.04	71 **Lu** Lutetium 174.967
95 **Am** Americium (243)	96 **Cm** Curium (247)	97 **Bk** Berkelium (247)	98 **Cf** Californium (251)	99 **Es** Einsteinium (252)	100 **Fm** Fermium (257)	101 **Md** Mendelevium (258)	102 **No** Nobelium (259)	103 **Lr** Lawrencium (260)

F.V.

An Introduction to
PHYSICAL SCIENCE

An Introduction to
PHYSICAL SCIENCE

NINTH EDITION

James T. Shipman
Ohio University

Jerry D. Wilson
Lander University

Aaron W. Todd
Middle Tennessee State University

Houghton Mifflin Company   **Boston**   **New York**

TO

Students who desire to advance their comprehension and knowledge of science.
—*J. T. S.*

Marianne Stepanian, for her outstanding editorial and diplomatic ability.
—*J. D. W.*

Roger and Chris Todd, my two fine sons.
—*A. W. T.*

Editor-in-Chief: Kathi Prancan
Associate Editor: Marianne Stepanian
Production/Design Coordinators: Jodi O'Rourke and Jill Haber
Senior Manufacturing Coordinator: Priscilla J. Bailey
Executive Marketing Manager: Andy Fisher

Cover design by Diana Coe / ko Design Studio
Cover photograph © Jim Erickson/The Stock Market

Printed in the U.S.A.

Library of Congress Catalog Card Number: 99-71934

ISBN: 0-395-95570-X

123456789-DW-03 02 01 00 99

Contents

11 THE CHEMICAL ELEMENTS 265

12 CHEMICAL BONDING 293

13 CHEMICAL REACTIONS 321

14 ORGANIC CHEMISTRY 349

15 THE SOLAR SYSTEM 380

16   PLACE AND TIME   421

17   THE MOON   443

18   THE UNIVERSE   466

19   THE ATMOSPHERE   500

20   ATMOSPHERIC EFFECTS   530

Preface

Science and technology are the driving forces of change in our world today. They have created a revolution in all aspects of our lives including communication, transportation, medical care, and education. Because the world is rapidly being transformed, it is important that today's students be motivated to advance their comprehension and knowledge of science. Equipped with this knowledge, they can thrive in our changing world and make informed decisions that ultimately affect their lives and the lives of others.

The primary goal of the ninth edition of *An Introduction to Physical Science* is in keeping with that of previous editions: to stimulate students' interest in and knowledge of the physical sciences. Additionally, we continue to present the content in such a way that it helps students develop the critical reasoning and problem-solving skills that are needed to cope in our ever-changing modern technological world. This revision builds on the successes of previous editions in advancing students' scientific knowledge and awareness.

An Introduction to Physical Science, Ninth Edition, is intended for the first-year college nonscience major. The five divisions of physical science—physics, chemistry, astronomy, meteorology, and geology—are covered. Each division of physical science is defined and explained in the context of real-world examples. The textbook is readily adaptable to either a one- or a two-semester course.

Approach

One of the outstanding features of this textbook continues to be its emphasis on fundamental concepts. We build on the basic concepts as we progress through the chapters. For example, Chapter 1, which introduces the concepts of measurement, is followed by chapters on the basic topics of physics: motion, force, energy, heat, wave motion, electricity, magnetism, and modern physics. This foundation in physics is useful in developing the principles of chemistry, astronomy, meteorology, and geology that follow.

Physical concepts in the five divisions, or disciplines, of physical science are made *accessible* by developing them in a logical rather than a chronological fashion. Concepts are made *interesting* by discussing them in the context of everyday experience. Real-world examples throughout the textbook enhance the students' understanding of the world around them.

Although the approach remains constant, in order to make the presentation of concepts as clear and logical as possible, the ninth edition of *An Introduction to Physical Science* has undergone a major revision. By shortening the overall length, simplifying the math coverage, refining the pedagogy, and streamlining the art and photo program we have made the book more manageable for teachers and students alike.

Organizational Updates in the Ninth Edition

In accordance with our goal of reducing the overall length, we have reduced the number of chapters in the textbook. Two previous chapters—Moles, Solutions, and Gases, and Isostasy and Diastrophism—have been eliminated. A briefer, simpler discussion of moles and Avogadro's number is now included in Chapter 13 (Chemical Reactions). A streamlined discussion of solutions is now in Chapter 11 (The Chemical Elements) and Chapter 13. Much of the material on gases, especially the mathematics, has been deleted, and the remaining material has been moved to Chapter 5 (Temperature and Heat). Crustal deformation and mountain building from the old Isostasy and Diastrophism chapter are now included in Chapter 22 (Structural Geology).

We have also done some rewriting and reorganizing of the physics portion of the book. In Chapter 1 (Measurement), the sequence of rewritten sections in now: 1.2, Introduction to the Scientific Method; 1.3, Standard Units and Systems of Units; 1.4, More on the Metric System; and 1.6, Significant Figures.

In Chapter 3 (Force and Motion), Newton's laws are now covered sequentially, followed by the law of gravitation. Also, momentum is discussed without impulse. And, in Chapter 5 (Temperature and Heat), the reordered section topics are: 5.4, Heat Transfer; 5.5, Phases of Matter; 5.6, The Kinetic Theory of Gases; and 5.7, Thermodynamics. Section 5.6 has been moved forward from former Chapter 14 and simplified.

To follow the sequence of most physical science courses more closely, meteorology is now covered before geology. Therefore, The Atmosphere (formerly Chapter 25) and Atmospheric Effects (formerly Chapter 26) are now Chapters 19 and 20. The four geology chapters that follow have been completely rewritten and reorganized to bring the coverage more up-to-date.

Each chapter has been carefully revised with an eye toward tightening the presentation and clarifying the material. Among numerous examples, in Chapter 9 (Atomic Physics), the section on the quantum mechanical model of the atom has been shortened and simplified so that the nonscience student is not overwhelmed by quantum numbers and subshell electronic configuration. Similarly, Chapter 15 (The Solar System), the longest chapter in the eighth edition, has been reduced, while still covering the latest information on the extrasolar planets.

Simplified Math Coverage

Each discipline is treated both descriptively and quantitatively. To make the ninth edition more user-friendly for students who are not mathematically inclined, we introduce concepts to be treated mathematically as follows: First the concept is defined, as briefly as possible, using words. The definition is then presented as an equation in word form. Finally, the concept is expressed in symbol notation. Mathematical assistance is provided for students who may need it, but the relative emphasis, whether descriptive or quantitative, is left to the discretion of the instructor. To those who wish to emphasize the descriptive approach in teaching physical science, we recommend using only the Review Questions and Applying Your Knowledge questions at the end of each chapter and omitting the Exercises.

We have, as much as possible, used terms instead of symbols when discussing scientific concepts in the textbook. This eliminates the need for students to remember what concept the symbol represents.

Outstanding Pedagogical Features

- Each section begins with *Learning Goals* that provide the student with a focus for the text that follows.

- *A Spotlight On* is a new feature that uses figures, photos, or flowcharts to visually summarize, at a glance, a section of the chapter or the entire chapter. Thirteen of these features appear throughout the textbook.

- A significant number of *Highlights* have been updated, and nearly ten are new to this edition, which include such up-to-date topics as El Niño and La Niña, tsunamis, and the disappearance of dinosaurs.

- At the end of many sections, thought-provoking *Relevance Questions* tie the material just covered to students' everyday lives and develop their critical reasoning skills.

- All worked-out *Examples* within the chapter give step-by-step solutions to problems and are followed by a *Confidence Exercise* that gives the student immediate practice in solving that specific type of problem. Answers to Confidence Exercises appear at the end of the chapter.

- *Review Questions* at the end of each chapter contain both short-answer and multiple-choice questions that are classified by chapter section, allowing students to quiz their knowledge of material just covered. Answers to the multiple-choice questions are provided.

- *Applying Your Knowledge* questions have been added to the end of each chapter. These questions involve practical applications of material covered in the chapter and everyday topics relevant to the subject matter.

- All end-of-chapter *Exercises* are paired. Each odd-numbered exercise has an even-numbered exercise that is similar in content to provide further practice. Answers are provided for the odd-numbered exercises only, allowing instructors to assign the even-numbered ones as homework.

Design, Photo, and Illustration Program

The ninth edition features an entirely new design that continues to adhere to our long-standing goal of presenting the material in an appealing, nonthreatening way. The palette and design were developed carefully, to be inviting yet appropriate for the college level.

The illustration and photo programs were completely overhauled for this revision. Over one-third of the illustrations are revised or new, and many new photos were chosen to better depict how current events and everyday life are tied to physical science. Numerous figures and photos have been deleted in order to tighten the presentation and focus attention on the concepts being presented.

Complete Instructional Package

Web Site.   New to this edition, the material for the web site was developed by the textbook authors and by Clyde D. Baker of Ohio University. It features information for both students and instructors.

- **Students** can access, on a chapter-by-chapter basis, study goals, a discussion of each text section, sample solved problems, and solved paired exercises. Additionally, there are two on-line quizzes per chapter featuring both multiple-choice and short-answer questions. Students can take a quiz on-line and receive immediate feedback on their results.

- **Instructors** will also visit this site frequently, as it includes teaching tips for use in both the classroom and the laboratory. To support the textbook, the web site features teaching suggestions, a list of potential in-class demonstrations, and answers to all end-of-chapter material. Additionally, a comprehensive list of teaching aids is included. Information pertaining to the *Laboratory Guide* includes a brief explanation of the experiments plus caution notes, sample data, answers to questions, and additional or alternate questions.

Laboratory Guide.   Written by James T. Shipman and Clyde D. Baker, the Laboratory Guide contains 53 experiments that are arranged to correlate with the textbook presentation. Each experiment includes an introduction, learning objectives, a list of required apparatus, a detailed procedure for collecting data, and questions. Safety is stressed throughout the guide.

Test Bank.   Prepared by the textbook authors, this ancillary is available to adopters and contains more than 2,000 questions in completion, multiple-choice, and short-answer formats. The *Test Bank* is also available electronically for Windows and Macintosh computers. The computerized format allows instructors to produce chapter tests, midterms, and final exams easily and with excellent graphics capability. The instructor can also edit existing questions and add new ones as desired.

Transparencies.   This package contains a set of more than 130 full-color transparencies—approximately 25 more than in the last edition—depicting figures and tables from the ninth edition.

Acknowledgments

We wish to thank our colleagues and students for the many contributions they made to this textbook through correspondence, questionnaires, and classroom use of the text material. We would also like to thank the following reviewers for their suggestions and comments: Leletemeskel Asfaw, Alcorn State University; Clyde D. Baker, Ohio University; Claude Bolze, Tulsa Community College; Terry Bradfield, Northeastern Oklahoma State University; Jack G. Couch, Bloomsburg University; Rahul Mehta, University of Central Arkansas; Neil E. Miller, Lander University; Duane L. Sea, Bemidji State University; David Shiner, University of North Texas; Leonard M. Thomas, Phillips University; Lynn P. Thomson, Ricks College; Norman R. Russell, Jefferson Community College; David L. Wagner, Edinboro University; Linda A. Wilson, Middle Tennessee State University.

We express our deep appreciation to Mary Falcon, Development Editor, for her excellent assistance in updating the geology chapters for this edition. We are grateful to those individuals and organizations who contributed photographs, illustrations, and other information used in the text. We are also indebted to the Houghton Mifflin staff and several others for their dedicated and conscientious efforts in the production of *An Introduction to Physical Science,* Ninth Edition. We especially wish to thank Kathi Prancan, Editor-in-Chief; Marianne Stepanian, Associate Editor; Peggy J. Flanagan, Project Editor; Jill Haber, Senior Production Design Coordinator; Jodi O'Rourke, Associate Production Design Coordinator; Ron Kosciak, Designer; Penny Peters, Art Editor; Martha Shethar, Photo Researcher; and Jim Madru, Copyeditor. Finally, we acknowledge the contributions of Genevieve and Sudie Shipman, Sandy Wilson, and Clara Todd.

We welcome comments from students and instructors of physical science and invite you to send us your impressions and suggestions.

MEASUREMENT

1

When you can measure what you are speaking about and express it in numbers, you know something about it; but when you cannot measure it, when you cannot express it in numbers, your knowledge is of a meager and unsatisfactory kind.

Lord Kelvin (1824–1907)

A first step in understanding our environment is to find out about the physical world through measurements. Over the centuries, humans have developed increasingly sophisticated methods of measurement, and scientists make use of the most advanced of these.

We are continually making measurements in our daily life. Each day we plan our work, play, and rest schedules as a function of time. With watches and clocks we measure the time for events to take place. Every 10 years we take the census and determine (measure) the population. We count money, calories, and the minutes, hours, days, and years of our lives.

Some of us keep precise records of food and drugs we take into our bodies because of illness. Many lives depend on accurate measurements being made by the medical doctor, laboratory technician, and pharmacist in the diagnosis and treatment of disease.

Measurements of the energy released by earthquakes and on samples of ocean sediments have helped geologists formulate a picture of Earth's internal structure and a theory for continental drift.

Meteorologists measure the many elements (temperature, pressure, humidity, precipitation, wind) that make up the weather. This information is relayed to millions by the communications media.

Photo: Measurements tell us about our environment, as illustrated here in the operation of an automobile.

At one time it was thought that all things could be measured with exact certainty. However, as we measured smaller and smaller objects, it became evident that the very act of measuring distorted the measurement. This uncertainty in making measurements of the very small is discussed in more detail in Chapter 9.

The preceding examples show the relevance of measurement and underscore our need to know the related concepts. Understanding measurement is the first step in understanding our physical environment. ■

1.1 The Senses

LEARNING GOAL*

▼ Explain the importance and limitations of our senses in making measurements.

Our environment stimulates our senses, either directly or indirectly. The five senses (sight, hearing, touch, taste, and smell) make it possible for us to know the environment. Therefore, the senses are a good starting point in studying and understanding the physical world.

Most information about our environment comes through sight. Hearing ranks second to sight in supplying the brain with information about the external world. Touch, taste, and smell, although important, rank well below sight and hearing in providing environmental information.

All the senses have limitations. For example, the unaided eye cannot see the vast majority of stars and galaxies and cannot immediately distinguish the visible stars of our galaxy from the planets of our solar system, all of which appear as points of light. (The planets, however, move relative to the stars. The word *planet* comes from the Greek word meaning "wanderer.") The limitations of the senses can be reduced by using measuring instruments.

Not only do the senses have limitations, but they also can be deceived, thus providing false information about our environment. For example, perceived sight information may not always be a true representation of the facts because the brain can be fooled.

There are many well-known optical illusions, such as those in ● Fig. 1.1. Some people may be quite con-

*Learning Goals will be listed at the beginning of each section. Students should be able to achieve these goals after reading and studying the section.

vinced that what they see in such drawings actually exists as they perceive it. However, we can generally eliminate deception by using measuring instruments. For example, in Fig. 1.1a, lines *a* and *b* can be measured with a ruler to see how their lengths compare.

Indeed, we extend our ability to measure and learn about our environment with various instruments. But these too have their limitations. We commonly measure temperature with a thermometer, but a reading can only be estimated to 0.1 of a degree on most thermometers. Even the most precise instruments have their limitations.

1.2 Introduction to the Scientific Method

LEARNING GOALS

▼ Distinguish among a scientific law, a hypothesis, and a theory.

▼ Describe the scientific method and its application.

Experiment and explanation are at the heart of scientific research. An **experiment** is an observation of natural phenomena carried out in a controlled manner so that the results can be duplicated. For example, in Chapter 12 the investigation to determine what happens to mass during a chemical reaction is discussed. These experiments were done in a controlled manner by using closed containers so that no substances could enter or leave the vessels. Very accurate balances were used to find the masses before and after each of many experiments on various chemical reactions.

After a series of experiments, perhaps a regularity or relationship can be detected in the results. If so, then we have a scientific **law**—a concise statement, in words or a mathematical equation, about a fundamental relationship or regularity of nature. For example, in every one of the previously mentioned experiments concerning mass and chemical reactions, it was found that the mass was the same after the reaction as it was before the reaction. Thus the law of conservation of mass was discovered: *During a chemical reaction, no detectable gain or loss of mass occurs.* Note that the law does not *explain* this behavior of nature; it just *states* the generalized experimental finding. And analogous to legal laws, a scientific law may have to be repealed or changed if inconsistencies are found.

FIGURE 1.1   Some Optical Illusions
We can be deceived by what we see. Answer the questions under the drawings.

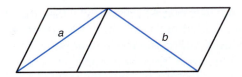

(a) Is the diagonal line *b* longer than the diagonal line *a*?

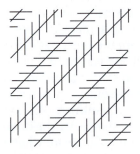

(b) Are the diagonal lines parallel?

(c) Going down?

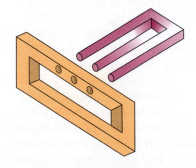

(d) Is something dimensionally wrong here?

Of course, we are usually not content just to know how nature operates; in addition, we want an explanation of the behavior. Explanations allow us to organize knowledge and make predictions about related phenomena. We call a tentative explanation of some regularity in nature (or a tentative answer of any kind) a **hypothesis.** For example, to explain the law of conservation of mass, an early scientist hypothesized that matter consists of atoms that only rearrange themselves during a chemical reaction (see Section 12.3).

For a hypothesis to be useful, it must suggest new experiments, the results of which serve as a test of the validity of the hypothesis. For example, if the *atomic hypothesis* were correct, another regularity would be found if certain experiments were done. The predicted regularity was found, and the hypothesis was supported. (If the experimental results had not been as predicted, then the hypothesis would need to be modified to fit both the new and the original results, or an entirely new hypothesis would have to be proposed.)

If a hypothesis successfully passes many tests over a long period of time and proves useful in knitting together a large body of scientific work, it takes on the status of a **theory**—a tested explanation of a broad segment of basic natural phenomena. For example, the original atomic hypothesis has withstood testing for almost 200 years now, and this basic idea (or *concept*, or *model*) of nature has been essentially proven and is important in all other areas of science. So it is now referred to as the *atomic theory.*

As new scientific knowledge and insights are gained, it is expected that modification (or even replacement) of laws, hypotheses, and theories will occur often. And this is welcomed, because that is how science makes progress and why it is said to be self-correcting.

This process of investigating nature is known as the **scientific method,** which holds that no concept or model of nature is valid unless the predictions are in agreement with experimental results. That is, all hypotheses—tentative answers—should be based on as much relevant data as possible and then should be tested and verified. If a hypothesis does not withstand rigorous testing, it must be modified and retested or rejected and replaced by a new hypothesis. An attitude of curiosity, objectivity, rationality, and willingness to go where the evidence leads is associated with use of the scientific method. Note carefully that the scientific method not only is used in scientific work but also is applicable in many areas of our daily lives.

RELEVANCE QUESTION: One morning you turn the key in your car's ignition and the car doesn't start. How would the scientific method come into play in this predicament?

1.3 Standard Units and Systems of Units

LEARNING GOALS

▼ List some standard units for various systems of units.

▼ Describe the metric system units of length, mass, and time.

To describe nature, we make measurements, and we express these measurements in terms of the magnitudes of units. Units allow us to describe things in a concrete way, that is, numerically. (See the chapter-opening quotation.) Suppose that you were given the following directions to find the campus when you first arrived in town: "Keep going on this street for a few blocks, turn left at a traffic light, go quite a ways, and you're there." Certainly some units or numbers would have helped.

A great many objects and phenomena may be described in terms of the *fundamental* physical quantities of length, mass, and time. (*Fundamental* because they are the most basic quantities or properties of which we can think.) In fact, the topics of mechanics—the study of motion and force—studied in the first few chapters of this textbook require *only* these physical quantities. Another fundamental quantity, electric charge, will be discussed in Chapter 8 (Electricity and Magnetism) when it is needed. For now, let's focus on the units of length, mass, and time.

To measure these fundamental quantities, we compare them with a reference, or standard, that is taken to be a standard unit. That is, a **standard unit** is a fixed and reproducible value for the purpose of taking accurate measurements. Traditionally, a government or international body establishes a standard unit.

A group of standard units and their combinations is called a **system of units.** Two major systems of units are in use today—the **metric system** and the **British system.** The latter is used widely in the United States but is gradually being phased out in favor of the metric system, which is used throughout most of the world (● Fig. 1.2). A brief history of the metric system is given in Table 1.1.

FIGURE 1.2 A Mostly Metric World

The sign warns that the metric system is in use. Note the differences in the magnitudes of the speed limit—better not go 90 mi/h!

Length

The description of space might refer to a location or the size of an object (amount of space occupied). To measure these, we use the fundamental quantity of **length,** which is defined as the measurement of space in any direction.

Space has three dimensions, each of which can be measured by length. The three dimensions can be seen easily by considering a rectangular object (● Fig. 1.3). It has length, width, and height, but each of these dimensions is actually a length.

The standard unit of length in the metric system is the **meter** (m), from the Greek *metron,* "to measure." It was defined originally as one ten-millionth of the distance from the Earth's equator to the geographic

TABLE 1.1 A Brief, Chronological History of the Metric System

1670	Gabriel Moulton, a French mathematician, proposes a measurement system based on a physical quantity of nature and not on human anatomy.
1790	The French Academy of Science recommends the adoption of a system with a unit of length equal to one ten-millionth of the distance on a meridian between Earth's North Pole and equator.
1870	A French conference is set up to work out standards for a unified metric system.
1875	The Treaty of the Meter is signed by 17 nations, including the United States. This establishes a permanent body with the authority to set standards.
1893	The United States officially adopts the metric system standards as bases for weights and measures (but continues to use British units).
1975	The Metric Conversion Act is enacted by Congress. It states, "The policy of the United States shall be to coordinate and plan the increasing use of the metric system in the United States and to establish a United States Metric Board to coordinate the voluntary conversion to the metric system." (No mandatory requirements are made.)

FIGURE 1.3 Space Has Three Dimensions

The bathtub has dimensions of length (l), width (w), and height (h), but all are actually measurements of length.

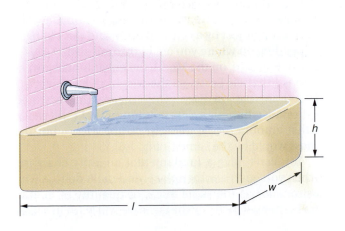

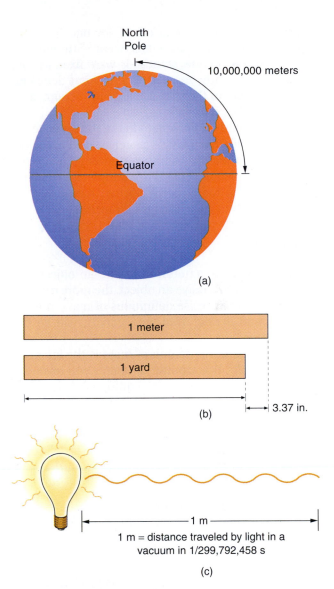

FIGURE 1.4 The Meter

(a) The meter was defined originally so that the distance from the North Pole to the equator would be 10 million meters. (b) The meter is a little longer than a yard (not to scale). (c) The meter is now defined by the distance light travels in a vacuum in a very short time.

North Pole (● Fig. 1.4a). A portion of the meridian was measured, and the unit was first adopted in France in the 1790s. The meter is slightly longer than a yard, as illustrated in Fig. 1.4b.

From 1889 to 1960, the standard meter was referenced to a platinum-iridium bar kept at the International Bureau of Weights and Measures in Paris,

France. However, the stability of the bar was questioned (for example, variations occur with temperature changes), so new standards were fixed in 1960 and, most recently, in 1983. The current definition links the meter to the speed of light in vacuum, as illustrated in Fig. 1.4c.

The standard unit of length in the British system is the *foot,* which historically was referenced to the human foot. Other early units commonly were referenced to parts of the body. For example, the *hand* is a unit that even today is used when measuring the heights of horses (one hand is four inches).

Mass

Mass refers to the amount of matter an object contains. The more massive an object, the more matter it contains. (More precise definitions of mass in terms of force and acceleration, and gravity, will be discussed later.)

The standard metric unit of mass is the **kilogram** (kg). Originally, this amount of matter was related to length and was defined as the amount of water in a cubic container 0.10 m on a side (at 4°C, where water has its maximum density). However, for convenience, the mass standard was referenced to a material standard (an object). Currently, the kilogram is defined to be the mass of a cylinder of platinum-iridium kept at

FIGURE 1.5 Kilogram Standard

Prototype kilogram number 20 is the U.S. standard unit of mass. The prototype is a platinum-iridium cylinder 39 mm in diameter and 39 mm high.

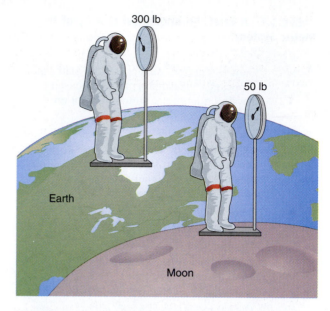

FIGURE 1.6 Mass Is the Fundamental Quantity

The weight of an astronaut on the Moon is $\frac{1}{6}$ of what it is on Earth. However, the mass is the same.

the International Bureau of Weights and Measures. The U.S. prototype is kept at the National Institute of Standards and Technology (NIST) in Washington, D.C. (● Fig. 1.5).

The unit of mass in the British system is the *slug,* which you've probably never heard of (and which will only be mentioned, not used, in this textbook). This is so because a quantity of matter in the British system is expressed in terms of weight on the surface of Earth and in units of pounds. (The British system is sometimes said to be a gravitational system.) This is unfortunate because weight is not a *fundamental* quantity, and this often gives rise to confusion. Certainly a fundamental quantity should be the same and not change. However, weight is the gravitational attraction on an object by a celestial body, and this changes depending on where you are in the universe.

For example, on the Moon, the gravitational attraction is $\frac{1}{6}$ that on Earth, so the weight of an object is $\frac{1}{6}$ less than that on Earth. This means that a suited astronaut who weighs 300 pounds on Earth will weigh $\frac{1}{6}$ that amount, or 50 pounds, on the Moon, but the mass would be the same (● Fig. 1.6).

You can't have a fundamental quantity changing when you go to another place, and mass doesn't. The astronaut has the same mass, or quantity of matter, no matter where he or she is. As we will learn in a later

FIGURE 1.7 Time and Events

Events mark intervals of time. Here, at the New York City Marathon, after starting out (beginning event), a runner crosses the finish line (end event) in a time interval of 2 hours, 12 minutes, and 38 seconds.

chapter, mass and weight are related, but they are not the same, as the preceding example shows. For now, keep in mind that *mass, not weight, is the fundamental quantity.*

Time

Each of us has an idea of what time is, but when asked to define it, we probably would have difficulty. Some terms that are often used in referring to time are *duration, period,* and *interval.* A common definition is that **time** is the continuous, forward flowing of events.

Without events or happenings of some sort, there would be no perceived time (● Fig. 1.7). The mind has no innate awareness of time, merely an awareness of events taking place in time. In other words, we do not perceive time as such, only events that mark locations in time, similar to marks on a meterstick indicating length intervals.

Note that time has only one direction—forward. Time has never been observed to run backwards. This would be like watching a film run backwards in a projector.

Time and space seem to be linked together. In fact, time is sometimes added as a fourth dimension to the three dimensions of space. If something exists in space, it also must exist in time. But, for our discussion, we will regard space and time as separate quantities.

Fortunately, the standard unit of time is the same in both the metric and British systems—the **second** (s). The second originally was defined in terms of observations of the Sun, as a certain fraction of a solar day, but now the second is referenced to the invariable properties of the atom (● Fig. 1.8). Changes occur in atoms regardless of outside influences, and time is kept by an "atomic" clock, as shown in Fig. 1.8c. Perhaps it doesn't look like much of a clock, but it keeps time accurately to ±1 second in 30,000 years.

RELEVANCE QUESTION: What are some examples of your daily uses of the standard units of length, mass, and time?

1.4 More on the Metric System

LEARNING GOALS

▼ Obtain a better understanding of the metric system and the SI.

▼ Become familiar with common metric prefixes.

The standard units for length, mass, and time in the metric system give rise to an acronym for it—the **mks system.** The letters *mks* stand for *m*eter, *k*ilogram, and *s*econd. These are also standard units for a modernized version of the metric system, called the International System of Units (abbreviated **SI,** from the French, *Le Système International d'Unités*). The SI was established in 1960 to make comprehension and the exchange of ideas among the people of different nations as simple as possible. It now contains seven base units: the meter (m); the kilogram (kg); the second (s); the ampere (A), to measure the flow of electric charge; the kelvin (K), to measure temperature; the mole (mol), to measure the amount of a substance; and the candela (cd), to measure luminous intensity. A definition of each of these units is given in Appendix I. But, as stated earlier, we will be concerned with only the first three of these for several chapters.

One of the greatest advantages of the metric system is that it is a decimal (base-10) system. The British system is a duodecimal (base-12) system, as in 12 inches to the foot. The base 10 allows easy expression and conversion to larger and smaller units. A series of **metric prefixes** is used to express the multiples of 10, but you will need to know only a few of the most common ones. These are

mega- (M) 10^6 or 1,000,000 (million, meg′a, as in *mega*phone)

kilo- (k) 10^3 or 1000 (thousand, kil′o, as in *kilo*watt)

centi- (c) 10^{-2} or $\frac{1}{100}$ = 0.01 (hundredth, sen′ti, as in *senti*mental)

milli- (m) 10^{-3} or $\frac{1}{1000}$ = 0.001 (thousandth, mil′li, as in *mili*tary)

(See Appendix I for a complete listing of metric prefixes.)

Examples of using these prefixes are

1 meter is equal to 100 centimeters (cm) or 1000 millimeters (mm).

1 kilogram is equal to 1000 grams (g).

1 millisecond (ms) is equal to 0.001 second (s).

1 megabyte (Mb) is equal to a million bytes.

You are familiar with another base-10 system—our currency. A cent is $\frac{1}{100}$ of a dollar, or a centidollar. Tax assessments and school bond levies are given in mills. Although not as common as a cent, a mill is $\frac{1}{1000}$ of a dollar, or a millidollar.

When more applicable, smaller units than those in the mks system may be used. Although the mks system is the *standard* system, we sometimes use the smaller *cgs system*, where the *cgs* stands for *c*entimeter, *g*ram, and *s*econd. For comparison, the units for length, mass, and time for the various systems are listed in Table 1.2.

As will be learned by using factors of 10, the decimal metric system makes it much simpler than the British system to convert from one unit to another. For example, it is easy to see that 1 kilometer is 1000 meters, whereas in the British system 1 mile is 5280 feet.

FIGURE 1.8 A Second of Time

(a) The second was defined originally in terms of a fraction of the average solar day. (b) It is defined currently in terms of the frequency of radiation from the cesium atom. (c) The "atomic" clock is the primary frequency or time standard at the National Institute of Standards and Technology (NIST). This device keeps time with an accuracy of about three millionths of a second per year.

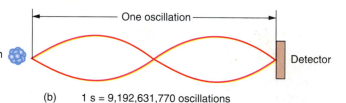

(b)   1 s = 9,192,631,770 oscillations

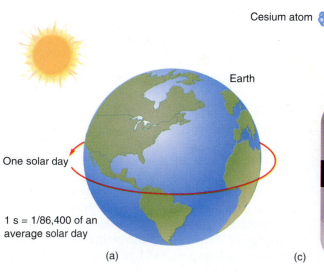

(a)

(c)

A SPOTLIGHT ON: Fundamental Quantities

Many objects and phenomena may be described in terms of the fundamental physical quantities of

| **Length** | **Mass** | **Time** |

The study of force and motion require only these quantities.

Another fundamental quantity that describes important phenomena, the electric charge, will be studied in Chapter 8 (Electricity and Magnetism).

The force associated with electric charges causes the hair to stand on end.

TABLE 1.2   Units of Length, Mass, and Time for the Metric and British Systems of Measurement

| Fundamental Quantity | Metric | | British |
	mks or SI	cgs	
Length	meter (m)	centimeter (cm)	foot (ft)
Mass	kilogram (kg)	gram (g)	slug
Time	second (s)	second (s)	second (s)

Memorizing the various conversions, such as 12 inches in 1 foot, 3 feet in 1 yard, and 5280 feet in 1 mile, makes the British system unwieldy compared with the metric system. Ever wondered why there are 5280 feet in a mile? Check out the chapter Highlight to get an insight on how the British system developed.

Some nonstandard metric units are in use. One of the most common of these is a unit of fluid volume or capacity. Recall that the kilogram was defined originally as the mass of a cube of water 0.10 m, or 10 cm, on a side. This is illustrated in ● Fig. 1.9. This volume was defined to be a **liter** (L). Hence a liter has a volume of 10 cm × 10 cm × 10 cm = 1000 cm^3 (cubic centimeters, sometimes abbreviated as cc). Since 1 liter contains 1000 milliliters (mL), it follows that 1 cm^3 is the same volume as 1 mL.

Also, because a liter or 1000 cm^3 of water has a mass of 1 kilogram or 1000 grams, it follows that 1 cm^3 of water has a mass of 1 gram. These relationships are shown in Fig. 1.9.

You may be wondering why a nonstandard volume such as the liter is used when the standard metric volume would be that using the standard length, that is, a cube 1 m on a side. This is a rather large volume, but it, too, is used to define a unit of mass. The mass of a quantity of water in a cubic container 1 m on a side (1 m^3) is taken to be a *metric ton* (or *tonne*). This is a relatively large mass. One cubic meter contains 1000 L (Can you show this?), so 1 m^3 of water = 1000 kg = 1 metric ton.

The liter is now used commonly for soft drinks and other liquids, taking the place of the quart. A liter is a little larger than a quart, 1 L = 1.06 qt (● Fig. 1.10).

RELEVANCE QUESTION: Name one standard unit of measurement that is related to purchases you make at (a) the grocery store and (b) the drug store.

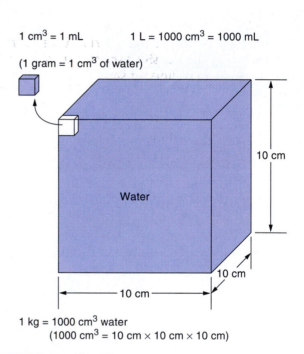

1 cm^3 = 1 mL   1 L = 1000 cm^3 = 1000 mL

(1 gram = 1 cm^3 of water)

10 cm

Water

10 cm

10 cm

1 kg = 1000 cm^3 water
(1000 cm^3 = 10 cm × 10 cm × 10 cm)

FIGURE 1.9   The Kilogram

The kilogram was related originally to length. The mass of the quantity of water in a cubic container 10 cm on a side was taken to be 1 kg. As a result, 1 cm^3 of water has a mass of 1 g. The volume of the container was defined to be one liter (L), and 1 cm^3 = 1 mL.

1.5 Derived Units and Conversion Factors

LEARNING GOALS

▼ Explain derived units, in particular, density.

▼ Show how quantities are converted to different units using conversion factors.

So how do we generally describe the mechanics of nature using *only* the three basic units of length, mass, and time? This is done by using **derived units,** which are multiples or combinations of units. The various derived units will become evident to you during the course of your study. Some examples of derived units follow.

Derived Quantity	Unit
Area (length)^2	m^2, cm^2, ft^2, etc.
Volume (length)^3	m^3, cm^3, ft^3, etc.
Speed (length/time)	m/s, cm/s, ft/s, etc.

Let's focus on a particular quantity with derived units—density, which involves mass and volume.

Density refers to how compact a substance is. In more formal language, **density** is the amount of mass located in a definite volume, or simply the mass per unit volume.

$$\text{density} = \frac{\text{mass}}{\text{volume}} = \frac{\text{mass}}{(\text{length})^3}$$

In equation form, density (ρ, Greek rho) is written

$$\rho = \frac{m}{V} \qquad \textbf{1.1}$$

(density = mass/volume)

Thus something with a mass of 20 kg that occupies a volume of 5.0 m^3 has a density of 20 kg/5.0 m^3 = 4.0 kg/m^3. In this example we use standard units, and the mass is measured in kilograms and the volume in cubic meters.

The idea of how density expresses the compactness of matter is illustrated in ● Fig 1.11a. Also, if mass is distributed uniformly throughout a volume, then the density of the matter will be constant. As illustrated in Fig. 1.11b, if you have a uniform substance, such as water, its density remains the same no matter how much of the substance you have.

Notice that from the definitions of kilogram and liter, the density of water is exactly 1.00 g/cm^3, or 1000 kg/m^3. If density is expressed in units of grams per cubic centimeter, we can easily compare the density of a substance with that of water. For example, a rock with a density of 3.3 g/cm^3 is 3.3 times as dense as water. Iron has an average density of 7.9 g/cm^3, and Earth as a whole has an average density of 5.5 g/cm^3. The planet Saturn with a density of 0.68 g/cm^3 is, then, less dense than water.

Densities of liquids such as blood and alcohol can be measured by means of a *hydrometer*, which is a weighted glass bulb that floats in the liquid (● Fig. 1.12). The higher the bulb floats, the greater is the density of the liquid.

When a medical technologist checks a sample of urine, one test he or she performs is for density. For a

FIGURE 1.10 Liter and Quart

(a) The liter of drink on the right contains a little more liquid than a quart of milk.

(b) A quart is equivalent to 946 mL, or 1 L = 1.06 qt.

(a)

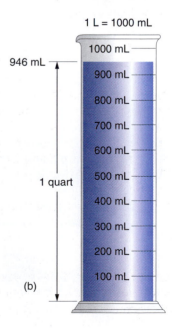

1 L = 1000 mL

946 mL

1000 mL

900 mL

800 mL

700 mL

600 mL

500 mL

1 quart

400 mL

300 mL

200 mL

100 mL

(b)

HIGHLIGHT: Who Put 5280 Feet in a Mile?

It seems strange to have 5280 sub-units making up our common length of a mile. The reason for this is historical and illustrates the difficulty with British units.

The word *mile* comes from the Latin *mille passus,* literally meaning "one thousand paces" (*mille,* as in *millennium*). This unit was introduced to Britain by the Romans. Each *passus* contained 5 *pes,* the Roman foot, and a *mille passus* was 5000 ft. Thus the original mile was 5000 feet, even though the Roman foot was slightly shorter than our current foot. (The length of the foot has varied with different cultures, as might be expected when referencing an anatomic unit.)

An old English unit of length was the furlong, which is said to have originated as the average length of a furrow a team of oxen could plow before resting—a "furrow long." The furlong was used in measuring land acreage and is still used today in horse racing. About A.D. 1500, the "Old London" mile was defined to be eight furlongs. At that time, the furlong, in terms of a longer (German) foot, contained 625 feet. So, by this definition, the mile again equaled (8 × 625 ft =) 5000 ft.

So where did the other 280 feet come in? Well, in 1593, Queen Elizabeth I established by statute, or decree, that the furlong would be defined in terms of the shorter British foot, which

made the furlong 660 feet. Hence the 8-furlong mile became (8 × 660 ft =) 5280 ft.

Our common mile is called a *statute mile* because it was established by statute. There is also a *nautical mile,* which, as the name implies, is used in marine (and air transportation) measurements. Its definition is somewhat similar to that of the meter, being defined as the length of one minute of arc (latitude) on a meridian. An international committee in 1929 established the nautical mile to be 6076 ft, or 1.852 km. A *knot,* a unit of speed, is a nautical mile per hour.

---

healthy person, urine has a density of 1.015 to 1.030 g/cm^3; that is, it consists mostly of water and dissolved salts. When the density is greater or less than this normal range, the urine may have an excess or deficiency of dissolved salts, perhaps caused by an illness.

(a)

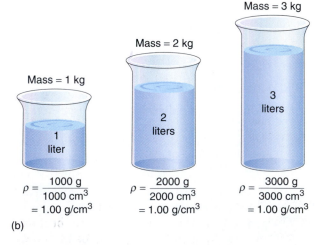

Mass = 1 kg

$\rho = \dfrac{1000 \text{ g}}{1000 \text{ cm}^3}$
$= 1.00 \text{ g/cm}^3$

Mass = 2 kg

$\rho = \dfrac{2000 \text{ g}}{2000 \text{ cm}^3}$
$= 1.00 \text{ g/cm}^3$

Mass = 3 kg

$\rho = \dfrac{3000 \text{ g}}{3000 \text{ cm}^3}$
$= 1.00 \text{ g/cm}^3$

(b)

FIGURE 1.11   Mass, Volume, and Density

(a) Both the weight and the pillow have the same mass, but they have very different volumes. Hence they have different densities. The weight is much denser. (b) Whether you have one, two, or three liters of water (with masses of one, two, and three kilograms, respectively), the density of the water is the same. The density of the water in each of the three containers is 1.00 g/cm^3, or 1000 kg/m^3.

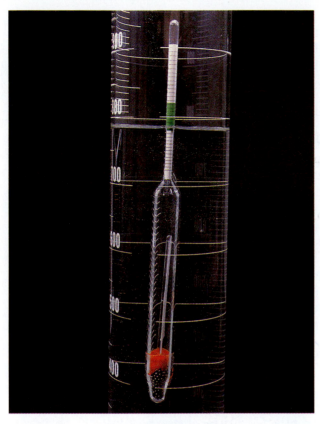

FIGURE 1.12 Measuring Liquid Density

A hydrometer is used to measure the density of a liquid. The denser the liquid, the higher the hydrometer floats, and the density can be read from the calibrated stem.

A hydrometer also is used to test the antifreeze in a car radiator. The hydrometer is calibrated directly in degrees rather than actual density. The closer the density is to 1.00 g/cm^3, the closer the antifreeze and water solution is to being pure water. When the density corresponds to 1.00 g/cm^3, the hydrometer will read a temperature of 0°C, or 32°F, the freezing point of water. The further the mixture is from being pure water, the lower the temperature reading will be.

Finally, when a combination of units becomes complicated, we frequently give it a name of its own. Consider the following examples, which are discussed in later chapters:

$$\text{newton (N)} = \text{kg} \times \text{m/s}^2$$

$$\text{joule (J)} = \text{kg} \times \text{m}^2/\text{s}^2$$

$$\text{watt (W)} = \text{kg} \times \text{m}^2/\text{s}^3$$

There are many more, but the point is that it is easier to talk about watts than $\text{kg} \times \text{m}^2/\text{s}^3$. As you might guess, the preceding unit names are in honor of early scientists. The unit abbreviations for such names are capitalized, but the unit names are not.

Conversion Factors

Often we want to convert from one system of units to another in order to make comparisons. For instance, we frequently want to make comparisons between the metric and the British systems. To do this, we use a **conversion factor,** which relates one unit to another. Many of these conversion factors are listed on the inside back cover. For instance,

$$1 \text{ in.} = 2.54 \text{ cm}$$

Although commonly written in equation form, this is really an equivalence statement; that is, 1 in. has an *equivalent* length of 2.54 cm. (To be a true equation, the expression must have the same units or dimensions on both sides.) However, in the process of expressing a quantity in different units, we use the conversion relationship in factor form:

$$\frac{1 \text{ in.}}{2.54 \text{ cm}} \quad \text{or} \quad \frac{2.54 \text{ cm}}{1 \text{ in.}}$$

For example, suppose you are 5 ft 5 in., or 65.0 in., tall, and you want to express your height in centimeters. Then

$$65.0 \text{ in.} \times \frac{2.54 \text{ cm}}{1 \text{ in.}} = 165 \text{ cm}$$

Notice how the units cancel and that we are left with the cm unit on both sides of the equation.

The steps may be summarized as follows:

Steps for Converting One Unit to Another

Step 1. **Use a conversion factor, that is, a ratio that may be obtained from an equivalence statement. (Often these factors or statements must be looked up in a table; see the inside back cover of this book.)**

Step 2. **Choose the appropriate form of conversion factor (or factors) so that the unwanted units cancel.**

Step 3. **Check to see that the units cancel and that you have the desired unit. Then perform the multiplication or division of the numerical quantities.**

Here is an example done in stepwise form.

Conversion Factors: One-Step Conversion

The length of a football field is 100 yards. In constructing a football field in Europe, the specifications have to be given in metric units. How long is a football field in meters?

SOLUTION

One might represent this problem by an equivalence statement:

$$100 \text{ yd} = ? \text{ m}$$

In other words, 100 yd is equivalent to how many meters?

Step 1

We could readily change 100 yd to 3000 ft to make the conversion, but there is a convenient, direct equivalence statement given inside the back cover of the textbook under Length:

$$1 \text{ yd} = 0.914 \text{ m}$$

We may form the ratios

$$\frac{1 \text{ yd}}{0.914 \text{ m}} \quad \text{or} \quad \frac{0.914 \text{ m}}{\text{yd}} \quad \text{(conversion factors)}$$

Notice how the number 1 is commonly left out of the denominator of the conversion factor; that is, we write 0.914 m/yd, instead of 0.914 m/1 yd.

Step 2

The second form of conversion factor, 0.914 m/yd, would allow the yd unit to be canceled. (The yd is the unwanted unit in the denominator of the ratio.)

Step 3

Checking this unit cancellation and performing the operation:

$$100 \text{ yd} \times \frac{0.914 \text{ m}}{\text{yd}} = 91.4 \text{ m}$$

CONFIDENCE EXERCISE 1.1

In a football game, you often hear the expression: "First and 10" (yd). How would you express this in meters to a friend from Europe?

Often we are concerned with converting from the metric unit of speed, which is meters per second, to the more familiar miles per hour. The necessary equivalence statement is

$$1 \text{ m/s} = 2.24 \text{ mi/h}$$

In this case the units are compound units, but the procedure is the same. For example, converting 15 m/s to mi/h,

$$15 \text{ m/s} \times \frac{2.24 \text{ mi/h}}{\text{m/s}} = 34 \text{ mi/h}$$

As we expand our use of the metric system in the United States, conversions and the ability to make them will become increasingly important (● Fig. 1.13).

In some instances, more than one conversion factor may be used. Here's an example to illustrate this. We will not write out the steps explicitly as in the preceding example.

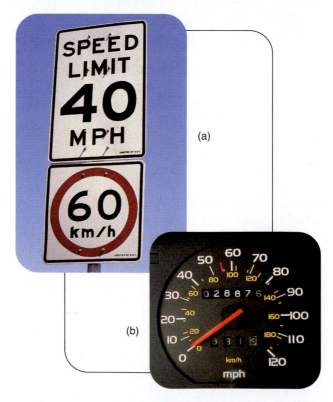

FIGURE 1.13   Unit Conversions

(a) A highway sign shows the speed limit in both British and metric units. (b) The speedometers of most automobiles today are calibrated in mi/h (mph) and km/h.

EXAMPLE 1.2

Conversion Factors: Multistep Conversion

A computer printer has a width of 18 in. What is its width in meters?

SOLUTION

$$18 \text{ in.} = ? \text{ m}$$

Suppose you didn't know or couldn't look up the conversion factor for inches to meters but remembered that 1 in. = 2.54 cm (which is a good length equivalence statement to remember between the British and metric systems). Then, using this and another well-known equivalence statement (1 m = 100 cm), we'll do the multiple conversion as follows:

$$\text{inches} \longrightarrow \text{centimeters} \longrightarrow \text{meters}$$

$$18 \text{ in.} \; \frac{2.54 \text{ cm}}{\text{in.}} \; \frac{1 \text{ m}}{100 \text{ cm}} = 0.46 \text{ m}$$

Note that the units cancel correctly.

Let's check this directly with the equivalence statement 1 m = 39.4 in.

$$18 \text{ in.} \; \frac{1 \text{ m}}{39.4 \text{ in.}} = 0.46 \text{ m}$$

CONFIDENCE EXERCISE 1.2

How many seconds are there in a day? (Use multiple conversion factors, starting with 24 h/day.)

RELEVANCE QUESTION: *Can you describe* walking *using only one fundamental quantity or standard unit? Why or why not?*

1.6 Significant Figures

LEARNING GOALS

▼ Explain why we use significant figures.

▼ Know how to use significant figures in simple calculations.

When working with quantities, we often use hand calculators to do mathematical operations. Suppose you divided 6.8 cm by 1.67 cm and got the result as shown

in ● Fig. 1.14. Would you report 4.0718563? We hope not, because your instructor might get upset.

The reporting problem is solved by what we call **significant figures** (or *significant digits*), which is a method of expressing measured numbers properly. This involves the accuracy of measurement and mathematical operations (see Appendix VII).

Notice in our example that 6.8 cm has two figures or digits, and 1.67 has three. These figures are significant because they indicate a magnitude read from some measurement instrument. In general, more digits in a measurement implies more accuracy, or the greater fineness of the scale of the measurement instrument. That is, the smaller the scale, or more divisions, the more numbers you can read, resulting in a better measurement. The 1.67 cm reading is more accurate because it has one more digit than 6.8 cm.

However, a mathematical operation cannot give you a better "reading," or more significant figures. So, as a general rule:

When multiplying or dividing quantities, report only as many significant figures in the result as there are in the quantity with the least number of significant figures.

Applying this rule to the example in Fig. 1.14, the result of the division should have two significant figures. Hence we round off the result:

$$6.8 \text{ cm}/1.67 \text{ cm} = 4.1$$

Limiting term has 2 s.f.   4.0718563 is rounded to 4.1, 2 s.f.

We generally follow this procedure in the text examples, as you should in your calculations. (See Appendix VII for the significant figure rule for addition and subtraction.)

As we have seen, it is necessary to round off numbers to obtain the appropriate number of significant figures. Use the following rules to do this.*

Rules for Rounding Off Numbers

1. **If the first digit to be dropped is less than 5, leave the preceding digit (the one to the left) unchanged.**

---

*In a calculation with more than one step, only the final answer should be rounded. Such multistep calculations are easily done on calculators. However, in examples in this textbook, an exercise may be worked in several steps for instructional clarity and the answer rounded each time.

FIGURE 1.14 Significant Figures and Insignificant Figures

The division operation of 6.8/1.67 on a calculator with a floating decimal point gives many figures. However, most of these figures are insignificant, and the result should be rounded to the proper number of significant figures, which is 2. See text for description.

2. If the first digit to be dropped is 5 or greater, increase the preceding digit by 1.

(Notice that in this method, five digits—0, 1, 2, 3, and 4—are less than 5 and would be left unchanged, and five digits—5, 6, 7, 8, and 9—are 5 or greater and so are rounded up.)

For example, if 2.348 were to be rounded to two significant figures, by rule 1 the 3 is left unchanged because the first digit to be dropped (the 4) is less than 5, and we have 2.3. But, in rounding 2.451 to two significant figures, by rule 2 we would have 2.5 because the first digit to be dropped (the 5) is 5 or greater.

EXAMPLE 1.3

Rounding Off Numbers

Round off each of the following:

(a) 26.142 to three significant figures.*

The 4 is the first digit to be removed and is less than 5. Then, by rule 1,

$$26.142 \longrightarrow 26.1$$

(b) 10.063 to three significant figures.

The 6 is the first digit to be removed. (Here, the zeros on each side of the decimal point are significant be-

*See Appendix VII for rules on determining the number of significant figures.

cause they have nonzero digits on both sides or are "captive" zeros.) Then, by rule 2,

$$10.063 \longrightarrow 10.1$$

(c) 0.09970 to two significant figures.

In this case the first nondigit to be removed is the 7. (The zeros to the immediate left and right of the decimal point are not significant but only locate the decimal point. They are called "leading" zeros.) Since 7 is greater than 5, by rule 2,

$$0.0997 \longrightarrow 0.10$$

(d) The result of the product of the measured numbers 5.0×356. Performing the multiplication,

$$5.0 \times 356 = 1780$$

Since the result should have only two significant figures as limited by the 5.0, we round off

$$1780 \longrightarrow 1800$$

CONFIDENCE EXERCISE 1.3

The result of multiplying 2.55 times 3.14 on a calculator is 8.007. How should you report the result?

Zeros often cause a problem with significant figures. It may seem that 1800 in the preceding example has four significant figures rather than two. This problem may be eliminated by using powers-of-10 notation, as described in the next section. Using this notation, we would write,

$$5.0 \times 356 = 1780 \boxed{\text{rounded to}} \!\!> 1.8 \times 10^3$$

and the two significant figures of the result are evident.

1.7 Powers-of-10 (Scientific) Notation

LEARNING GOALS

▼ Write numbers in powers-of-10 notation.

▼ Perform mathematical operations with numbers in this notation.

In physical science, many numbers are very big or very small. To express them, we frequently use the powers-of-10 (scientific) notation. When the number 10 is squared or cubed, we get

$$10^2 = 10 \times 10 = 100$$

$$10^3 = 10 \times 10 \times 10 = 1000$$

You can see that the number of zeros is just equal to the power of 10. As an example, 10^{23} is a 1 followed by 23 zeros.

Negative powers of 10 also can be used. For example,

$$10^{-2} = \frac{1}{10^2} = \frac{1}{100} = 0.01$$

We see that if a number has a negative exponent, we shift the decimal place to the left once for each power of 10. For example, one centimeter (cm) is 1/100 m or 10^{-2} m, which is 0.01 m.

We also can multiply numbers by powers of 10. Table 1.3 shows examples of various large and small numbers expressed in powers-of-10 notation.

We can represent a number in powers-of-10 notation in many different ways—all correct. For example, the distance from Earth to the Sun is 93 million miles. This value can be represented as 93,000,000 miles, or 93×10^6 miles, or 9.3×10^7 miles, or 0.93×10^8 miles.

Any of the given representations of 93 million miles is correct, although 9.3×10^7 is preferred. (In expressing powers-of-10 notation, it is customary to have one digit to the left of the decimal point. This is called *conventional*, or *standard, form*.)

Thus it can be seen that the exponent, or power of 10, changes when the decimal point of the prefix number is shifted. General rules for this are as follows:

Rules for Using Powers-of-10 Notation

1. **The exponent, or power of 10, is *increased* by 1 for every place the decimal point is shifted to the *left*.**
2. **The exponent, or power of 10, is *decreased* by 1 for every place the decimal point is shifted to the *right*.**

This is simply a way of saying that if the coefficient (prefix number) gets smaller, the exponent gets correspondingly larger, and vice versa. Overall, the number is the same.

TABLE 1.3 Numbers Expressed in Powers-of-10 Notation

Number	Powers-of-10 Notation*
247	2.47×10^2
186,000	1.86×10^5
4,705,000	4.705×10^6
9,000,000,000	9×10^9
30,000,000,000	3×10^{10}
602,200,000,000,000,000,000,000	6.022×10^{23}
0.025	2.5×10^{-2}
0.0000408	4.08×10^{-5}
0.00000010	1.0×10^{-7}
0.00000000000000000016	1.6×10^{-19}

*Note: The exponent (power of 10) is increased by 1 for every place the decimal point is shifted to the left, and decreased by 1 for every place the decimal point is shifted to the right.

EXAMPLE 1.4

Expressing Numbers in Powers-of-10 Notation

Express the following numbers in powers-of-10 notation in conventional form:

(a) 360,000   (c) 0.0694

(b) 246.7   (d) 0.000011

SOLUTION

Applying the preceding rules:

(a) $360,000 = 3.6 \times 10^5$ (shift to left, rule 1)
          5 4 3 2 1

(b) $246.7 = 2.467 \times 10^2$ (shift to left, rule 1)
        2 1

(c) $0.0694 = 6.94 \times 10^{-2}$ (shift to right, rule 2)
      1 2

(d) $0.000011 = 1.1 \times 10^{-5}$ (shift to right, rule 2)
      1 2 3 4 5

(Check out the notations in Table 1.3 and see if they are correct.)

CONFIDENCE EXERCISE 1.4

(a) Express 76,500 and 0.0345 in conventional powers-of-10 notation.

(b) Express the following numbers as a series of digits: 4.6×10^3 and 2.8×10^{-3}.

Arithmetic procedures of multiplication and division, as well as addition and subtraction, can be done in powers-of-10 notation. (See Appendix VI.)

Powers-of-10 notation is useful in expressing the results of mathematical operations with the proper number of significant figures, as discussed in the preceding section. For example, consider the calculator operation

$$325 \times 45 = 14,625$$

To express the result with two significant figures by our general rule in Section 1.6, we write

$$325 \times 45 = 1.5 \times 10^4$$

EXAMPLE 1.5

Using Powers-of-10 Notation to Express Calculation Results

Perform the following mathematical operation on a calculator and express the result properly using significant figures and scientific notation:

$$\frac{0.0024}{8.05} = ?$$

SOLUTION

Doing this operation on a calculator gives

$$\frac{0.0024}{8.05} = 0.000298136$$

(*Note:* The number of digits in the result may vary with different calculators.)

The number 0.0024 has two significant figures, since the zeros to the left simply locate the decimal point, and 8.05 has three significant figures. In the latter number the zero is significant because it is a "captive" zero.* Then we should write

*See Appendix VII for rules on determining the number of significant figures.

$$\frac{0.0024}{8.05} = 3.0 \times 10^{-4}$$

CONFIDENCE EXERCISE 1.5

Write the numbers 6000 and 0.0020 in powers-of-10 notation with two significant figures, and then (a) multiply them together and (b) divide the first by the second.

RELEVANCE QUESTION: Express the national debt in powers-of-10 notation. The debt is increasing daily, but assume 5.50 trillion dollars. What is your share (in dollars) of the debt if the U.S. population is about 270 million? (Solve in scientific notation.)

1.8 Approach to Problem Solving

LEARNING GOALS

▼ Establish an approach to problem solving and gain a firm foundation therein.

▼ Eliminate any math anxiety that might exist.

One of the major difficulties many students have in science is solving problems, particularly word problems. There is no set way to work a problem; it may be possible to use more than one approach to obtain a correct solution. Even so, a general procedure is helpful in most cases. The following procedure will guide you in solving a problem by giving you steps to follow as you analyze it. (See also Appendix II.)

Step 1

Read the problem, and identify the chapter principle that applies to it. Write down the given quantities using symbol representation. (Be sure to include units.) Make a rough sketch if applicable.

Step 2

Determine what is wanted, and write it down. (This is important. You have to know what is wanted before you can find it.) Then check to see that the units of the given quantities are appropriate. If they are not, use appropriate conversion factors. In general, all quantities should be expressed in the same system of units. Your answer will then be in units of that system.

Step 3

Survey the chapter equations, and determine the one that relates what is given to what is wanted. (In some instances, two equations may be necessary.) Perform the mathematical operation, and express your answer with the appropriate units and number of significant figures.

Let's apply these steps to an example. In general, we will show an *expanded* procedure, as follows, when a new concept is introduced, in order to help you understand its application.

EXAMPLE 1.6

Applying Problem-Solving Procedures to Find the Distance Earth Travels Around the Sun

Earth goes around the Sun in a nearly circular orbit with a radius of 93 million miles. How many miles does Earth travel in making one revolution about the Sun?

SOLUTION

Step 1

Here, the given quantity is the radius of Earth's (approximately) circular orbit, and we have

Given: $r = 93$ million miles $= 9.3 \times 10^7$ mi

(Using powers-of-10 notation, 93,000,000 mi $= 9.3 \times 10^7$ mi; see Section 1.7.)

Step 2

The distance (length) or circumference of Earth's orbit in miles is wanted, so

Wanted: $d = ?$ (distance)

(The unit of the given quantity is in miles, which is OK. Our answer will then come out in miles.)

Step 3

If you survey the Important Equations on the next page, you will find for the circumference of a circle,

$$c = 2\pi r \qquad \textbf{1.2}$$

The circumference of a circle with the given radius of Earth's orbit is how far Earth travels in one revolution, so

$$d = c = 2\pi r = 2\,(3.14)\,(9.3 \times 10^7\,\text{mi})$$
$$= 58 \times 10^7\,\text{mi} = 5.8 \times 10^8\,\text{mi}$$

Here the value of π is known (or may be looked up in a table of constants, or you may have π on your calculator). Notice that we simply multiply the number prefix of the power of 10 (the 9.3 in the mi term) by 2π, express the answer with two significant figures (since there are two significant figures in 93 million miles), and then put the power of 10 in standard form with one digit to the left of the decimal point.

CONFIDENCE EXERCISE 1.6

Earth has a radius of 6.4×10^3 km. What is Earth's surface area in square meters? (Consider Earth to be a sphere. The surface area of a sphere is given by $A = 4\pi r^2$.)

Important Terms

Important terms will be listed at the end of each chapter for review. After reading and studying the chapter, you should be able to define and/or explain each important term.

experiment (1.2)	British system	SI	density
law	length	metric prefixes	conversion factor
hypothesis	meter	mega-	significant figures (1.6)
theory	mass	kilo-	powers-of-10 (scientific)
scientific method	kilogram	centi-	notation (1.7)
standard unit (1.3)	time	milli-	
system of units	second	liter	
metric system	mks system (1.4)	derived units (1.5)	

Important Equations

Density: $\rho = \dfrac{m}{V}$ $\left(\text{density} = \dfrac{\text{mass}}{\text{volume}}\right)$

Circumference of a Circle: $c = 2\pi r$ $(\pi = 3.14159\ldots)$

Review Questions

1.1 The Senses

1. Which sense ranks second in providing knowledge about our environment?
   (a) touch (b) smell (c) sight (d) hearing

2. The loss of which sense would deprive you of the most information about the environment?
   (a) touch (b) smell (c) sight (d) hearing

3. Do all measurements ultimately depend on our senses? Explain.

4. State three limitations of our senses in obtaining accurate information.

5. Answer the questions accompanying ● Fig. 1.15.

1.2 Introduction to the Scientific Method

6. A scientific hypothesis
   (a) is a physical law.
   (b) is always accepted.
   (c) must be substantiated by the scientific method.
   (d) is none of the preceding.

7. A scientific law
   (a) is a hypothesis.
   (b) is the result of a theory that has been tested once or twice.
   (c) is the same as a concept.
   (d) must consistently describe a regularity in nature.

8. How are phenomena, hypotheses, and theories related?

9. Describe what is meant by the scientific method.

1.3 Standard Units and Systems of Units

10. The standard unit of mass in the SI is which of the following?
    (a) slug (b) kilogram (c) gram (d) pound

11. Which of the following is *not* a fundamental quantity?
    (a) weight (b) mass (c) length (d) time

12. Compare how the following were originally defined and how they are currently defined:
    (a) meter (b) kilogram (c) second

13. The standard of which fundamental quantity is still based on an object?

14. Which is more basic, the standard unit of length or of time? Explain.

15. Which standard unit is the same in all systems of units?

1.4 More on the Metric System

16. Which one of the following is larger?
    (a) a second   (c) a kilosecond
    (b) a centisecond   (d) a millisecond

17. Which one of the following volumes is used in defining the liter?
    (a) 1 mL (b) 0.001 m^3 (c) 1 m^3 (d) 100 cm^3

FIGURE 1.15 Seeing Is Believing

See Review Question 5.

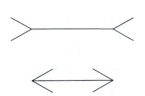

(a) Is the upper horizontal line longer?

(b) Are the men different heights?

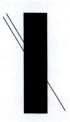

(c) With which of the upper lines does the line on the right connect?

18. What is the mks system?

19. How does the cgs system compare with the mks system?

20. Distinguish between a cubic centimeter (cm^3) and a milliliter (mL).

21. Using the SI standard unit of length to construct a cube, what would be the mass of the water needed to fill the cube?

1.5 Derived Units and Conversion Factors

22. Which of the following is true: A derived quantity
    (a) only applies to the British gravitational system.
    (b) is a mathematical derivation.
    (c) describes nonfundamental quantities.
    (d) is a combination of fundamental quantities.

23. The expression 1 in. = 2.54 cm is
    (a) an equation.
    (b) an equivalence statement.
    (c) a conversion factor.
    (d) a conversion statement.

24. In the following pairs, which metric unit is smaller than its British counterpart on a one-to-one basis?
    (a) meter-yard   (c) liter-quart
    (b) kilogram-pound   (d) kilometer-mile

25. Distinguish between fundamental quantities and derived quantities.

26. Define *density* in terms of fundamental quantities.

27. State the standard units for measuring density in the mks system and the SI.

28. (a) Which is more dense, a kilogram of iron or a kilogram of feathers?
    (b) Which has more mass?

29. How is the antifreeze in a car radiator tested?

1.6 Significant Figures

30. In rounding a number, if the first digit to be dropped were a 5, the preceding digit would be
    (a) increased by 1.
    (b) decreased by 1.
    (c) left unchanged.

31. What is the general purpose in using significant figures?

32. In doing a mathematical operation on a calculator, should you report the result exactly as shown on the calculator display? Explain. How about when you do the calculation by hand?

1.7 Powers-of-10 (Scientific) Notation

33. For a number expressed in scientific notation, when the decimal point is shifted to the right, does the exponent in the power of 10
    (a) increase?
    (b) decrease?
    (c) sometimes increase, sometimes decrease?

34. Why would you write Avogadro's number (6.02×10^{23}) in powers-of-10 notation?

35. 1×10^7 is how many times larger than 1×10^5?

36. 1×10^{-7} is how many times larger than 1×10^{-10}?

1.8 Approach to Problem Solving

37. State the steps of the procedure recommended for problem solving, and give a logical reason for each step.

Applying Your Knowledge

1. Why is a wrong hypothesis considered better than no hypothesis?

2. In general, common metric units are larger than their British counterparts. For example, a meter is a little longer than a yard. Give other examples, as well as any notable exceptions.

3. If you were going to create your own system of units, what would you choose for the standard units of the fundamental quantities?

4. In originally defining the kilogram as a volume of water, it was specified that the water be at its maximum density, which occurs at about 4°C. Why is this, and what would be the effect if another temperature were used? (*Hint:* See the first Highlight in Chapter 5.)

5. Suppose you could buy a quart or a liter of a soft drink at the same price. Which would you choose and why?

Exercises

1.4 More on the Metric System

1. What is the volume of a liter in (a) m^3 and (b) mm^3?
   Answer: (a) $0.001 \ m^3$ (b) $1,000,000 \ mm^3$

2. Show that one cubic meter contains 1000 L.

3. Water is sold in a 2.0-L bottle. What is the mass, in kilograms and grams, of the water in such a full bottle?
   Answer: 2.0 kg, 2000 g

5. A quantity of water with a mass of 0.085 kg is poured into a graduated cylinder. If the cylinder is graduated in mL, what would be the volume reading of the water?
   Answer: 85 mL

4. A rectangular container measuring 20 cm × 20 cm × 30 cm is filled with water. What is the mass of this volume of water in kilograms and grams?

6. A chemist measures out 350 cm^3 of water. What is the mass of this volume of water in kilograms?

1.5 Derived Units and Conversion Factors

7. Compute the density in g/cm^3 of a piece of metal that has a mass of 0.500 kg and a volume of 63 cm^3.
   Answer: 7.9 g/cm^3 (the density of iron)

9. What is the mass of a cube of ice with a side length of 3.0 cm? (ρ_{ice} = 0.92 g/cm^3, less than that of water, so ice floats.)
   Answer: 25 g

11. Compute the height in centimeters and meters of a person who is 6.00 ft tall.   Answer: 183 cm, 1.83 m

13. In Fig. 1.13a, is the conversion exact? Justify your answer.   Answer: no, 40 mi/h is 64 km/h

15. A car travels through a school zone at a speed of 25 mi/h. What is this speed in km/h?   Answer: 40 km/h

8. What is the volume of a piece of iron (ρ = 7.9 g/cm^3) that has a mass of 0.55 kg?

10. A seawater aquarium contains 1.25 m^3 of water. What is the mass of the seawater? ($\rho_{seawater}$ = 1030 kg/m^3, greater than fresh water, 1000 kg/m^3, because of dissolved salts.)

12. Compute the height in feet and inches of a woman who is 157 cm tall.

14. If we changed our speed limit signs to metric, what would probably replace a 60 mi/h sign?

16. Which is traveling faster, a car going 90 km/h or one going 60 mi/h? Justify your answer.

1.6 Significant Figures
1.7 Powers-of-10 (Scientific) Notation

17. Round off the following numbers to three figures, and express them in standard powers-of-10 notation.
    (a) 0.009992   (c) 0.010559
    (b) 6487.33   (d) 87,645
    Answer: (a) 9.99×10^{-3} (b) 6.49×10^3 (c) 1.06×10^{-2} (d) 8.76×10^4

19. Perform the operations on the measured quantities, and express the answer properly.
    $$(3.2 \text{ m} \times 1.04 \text{ m})/0.015 \text{ m}$$
    Answer: 2.2×10^2 m

21. Write the following numbers as a series of digits.
    (a) 7.3×10^4 (c) 0.399×10^3
    (b) 3.25×10^{-4} (d) 0.234×10^{-2}
    Answer: (a) 73,000   (b) 0.000325   (c) 399   (d) 0.00234

23. Write the following quantities in standard powers-of-10 notation instead of with metric prefixes.
    (a) 255 Ms   (c) 65 μg
    (b) 607 km   (d) 0.18 mL
    Answer: (a) 2.55×10^8 s   (b) 6.07×10^5 m   (c) 6.5×10^{-5} g   (d) 1.8×10^{-4} L

18. Round off the following numbers to two figures, and express them in standard powers-of-10 notation.
    (a) 105.25   (c) 9438
    (b) 0.00208   (d) 0.000344

20. The calculator result of multiplying $2.15 \times \pi$ is shown in ● Fig. 1.16. Round off the result to the proper number of significant figures.

22. Round the following numbers to three figures, and express them in standard powers-of-10 notation (one digit to the left of the decimal).
    (a) 4,256,000   (c) 0.01020
    (b) 2783   (d) 0.00006279

24. Fill in the blanks with the correct power of 10.
    (a) 0.25 megaton = 2.5 × _____ tons
    (b) 99 mg = 9.9 × _____ grams
    (c) 150 μL = 1.50 × _____ L
    (d) 300 kilobucks = 3.00 × _____ bucks

FIGURE 1.16   How Many Significant Figures?
See Exercise 20.

6.75442 42

1.8   Approach to Problem Solving

25. The thickness of a textbook, not including the covers, is measured to be 3.10 cm. If the last page is numbered 778, what is the average thickness per sheet?

$\qquad$ Answer: 7.97×10^{-3} cm/sheet

26. A textbook is 3.50 cm thick, including the covers. In a bookstore display, how many of these books could be placed upright on a shelf 1.50 m long?

27. Compute the area in square inches of a pizza with a diameter of (a) 7.00 in., and (b) 14.0 in. (*Hint:* The area of a circle is given by $A = \pi r^2$, where r is the radius.) (c) The area of the 14.0-in. pizza is how many times larger than that of the 7.00-in. pizza?

$\qquad$ Answer: (a) 38.5 in.^{2} (b) 154 in.^{2} (c) 4.00 times

28. Suppose the prices of the pizzas in Exercise 27 were $4.95 and $13.95, respectively. Which size is the best buy? (Check the price values of your own pizza parlor with a similar calculation. You may save some money.)

Solutions to Confidence Exercises

1.1 10 yd (0.914 m/yd) = 9.1 m ("First and 9.1")

1.2 1 day $\left(\dfrac{24 \text{ h}}{\text{day}}\right)\left(\dfrac{60 \text{ min}}{\text{h}}\right)\left(\dfrac{60 \text{ s}}{\text{min}}\right) = 86{,}400$ s

1.3 8.01 (to three significant figures)

1.4 (a) 7.65×10^4 and 3.45×10^{-2}, (b) 4600 and 0.0028

1.5 (a) $(6.0 \times 10^3)(2.0 \times 10^{-3}) = 12$
   (b) $6.0 \times 10^3/2.0 \times 10^{-3} = 3.0 \times 10^6$

1.6 $A = 4\pi r^2 = 4\pi(6.4 \times 10^6 \text{ m})^2 = 5.1 \times 10^{14} \text{ m}^2$

Answers to Multiple-Choice Review Questions

1. d   6. c   10. b   16. c   22. d   24. d   33. b
2. c   7. d   11. a   17. b   23. b   30. a

MOTION

2 •••••••••

Give me matter and motion, and I will construct the universe.

René Descartes (1596–1650)

Motion is everywhere. We walk to class. We drive to the store. Birds fly. The wind blows the trees. The rivers flow. Even the continents drift. In the larger environment, Earth rotates on its axis, the Moon revolves around Earth, Earth revolves around the Sun, the Sun moves in the galaxy, and the galaxies move with respect to one another.

This chapter focuses on the description of motion, with definitions and discussion of terms such as *speed, velocity,* and *acceleration.* We will study these concepts without considering the forces involved, reserving that discussion for Chapter 3.

Two basic kinds of motion are straight-line motion and circular motion. We experience examples of these each day. For instance, we know that driving around a curve feels different from driving in a straight line. Understanding acceleration is the key to understanding these basic kinds of motion. ▪

Photo: Motion in action at the Paraolympics in Barcelona, Spain.

2.1 **Defining Motion**

▼ Define motion.

The term **position** refers to the location of an object. To designate the position of an object, we must give or imply a reference point. For example, the entrance to campus is 1.6 km (1 mi) from the intersection with the traffic light. The book is on the table. Atlanta is in Georgia. The Cartesian coordinates of the point on the graph are $(x, y) = (2.0\ \text{cm}, 3.0\ \text{cm})$.

If an object changes its position, we say that motion has occurred. When an object is undergoing a continuous change in position, we say the object is moving or is in **motion.**

Consider an automobile traveling on a straight highway. The motion of the automobile may or may not be at a constant rate. In either case, the motion is described by using the fundamental units of length and time.

That length and time describe motion is evident in running. For example, as shown in ● Fig. 2.1, the cheetah runs a certain length in the shortest possible time. Combining length and time to give the *time rate of change of position* is the basis of describing motion

FIGURE 2.1   Motion

We describe motion in terms of time and distance. Here, a running cheetah appears to be trying to run a distance in the shortest time possible. The cheetah is the fastest of all land animals, capable of attaining speeds of up to 113 km/h, or 70 mi/h. (The slowest land creature is the snail, 0.05 km/h, or 0.03 mi/h.)

in terms of speed and velocity, as discussed in the following section.

RELEVANCE QUESTION: *In coming to physical science class from your previous location, how would you describe your motion?*

2.2 **Speed and Velocity**

▼ Differentiate between *scalars* and *vectors, distance* and *displacement,* and *speed* and *velocity.*

The terms *speed* and *velocity* are often used interchangeably. In physical science, however, they have distinct meanings. One is a scalar quantity and one is a vector quantity. Let's distinguish between scalars and vectors now, because other terms will fall into these categories during the course of our study. The distinction is simple. A **scalar** quantity is one that has only magnitude (number plus the unit of measurement). For example, you may be traveling in a car at 90 km/h (about 55 mi/h). This figure is your speed, which is a scalar quantity—it has magnitude only.

A **vector** quantity, on the other hand, is one that has magnitude *and* direction. For example, suppose you are traveling 90 km/h *north*. This quantity describes your velocity, which is a vector quantity—magnitude *plus* direction. By including direction, a vector quantity gives us more information than does a scalar quantity. No direction is associated with a scalar quantity.

Vector quantities may be represented by arrows (● Fig. 2.2). The length of a vector arrow is proportional to the magnitude and may be drawn to scale. The arrowhead indicates the direction of the vector. Notice in the figure for the cars that a negative vector, that is, a negative velocity of the car $(-v_\text{c})$, is equal in magnitude, but opposite in direction, to a positive velocity of the car $(+v_\text{c})$. (As for numbers, the $+$ sign is often omitted as being understood.) The velocity vector for the man, v_m, is shorter than those for the cars because he is moving more slowly.

Now let's look more closely at speed and velocity as used in the description of motion. The **average speed** *of an object is the total distance traveled divided by the time spent in traveling the total distance.* **Distance** *is*

FIGURE 2.2 Vectors

Vector quantities may be represented by arrows. The length of a vector arrow is proportional to the magnitude (size or value) of a quantity, and the arrowhead indicates the direction. Notice that $-v_c$ is equal in magnitude and opposite in direction to $+v_c$. (Here the v arrows represent velocities, which will be discussed shortly.) The v_m arrow represents the velocity of the walking man. It is shorter than the other velocity arrows because the man moves more slowly.

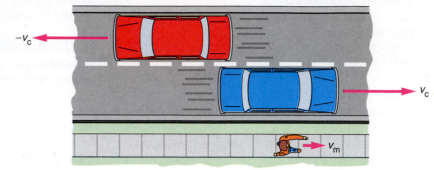

the actual path length that is traveled. In equation form, we have

$$\frac{\text{average}}{\text{speed}} = \frac{\text{distance traveled}}{\text{time to travel distance}}$$

$$\bar{v} = \frac{d}{t} \qquad \textbf{2.1}$$

where the bar over the symbol (read "vee-bar") indicates that it is an average value.

Note that length d and time t are *intervals.* They are sometimes written Δd and Δt to indicate explicitly that they are intervals. The Δ (Greek delta) means "change in" or "difference in"; for example, $\Delta t = t - t_o$, where t_o and t are the original and final times (on the clock), respectively. (If $t_o = 0$, then $\Delta t = t$.) Notice that speed has the standard units of m/s or ft/s (length/time). Other common units: mi/h and km/h.

Taken over a time interval, speed is an average. This is somewhat analogous to an average class grade. As such, average speed gives only a general description of motion. During a long time interval—say, that of making a trip by car—you speed up, slow down, and even stop. The average speed, however, is a single value that represents the average rate of motion for the entire trip.

The description of motion usually can be made more specific by taking a smaller time interval—for example, a few seconds or even an instant. The speed of an object at any instant of time may be quite different from its average speed, and it gives a more accurate description of the motion. In this case, we have an instantaneous speed.

The **instantaneous speed** of an object is its speed at that instant of time (Δt being extremely small).

A common example of nearly instantaneous speed is the speed registered on an automobile speedometer (● Fig. 2.3). This value is the speed at which the automobile is traveling right then, or instantaneously.

Now let's look at describing motion with velocity. Velocity is similar to speed, *but* a direction is involved. **Average velocity** is the displacement divided by the total travel time, where **displacement** is the straight-line distance between the initial and final positions, with direction toward the final position—a vector quantity (● Fig. 2.4). For straight-line motion in one direction, speed and velocity are very similar. Their magnitudes are the same because the lengths of the distance and the displacement are the same. The distinction between them in this case is that a direction must be specified for the velocity.

As you might guess, there is also **instantaneous velocity,** which is the velocity at any instant of time. For example, a car's instantaneous speedometer reading plus the direction it is traveling at that instant give its instantaneous velocity. Of course, the speed and direction of the car may and usually do change. This

FIGURE 2.3 Instantaneous Speed

The speed indicated on an automobile speedometer is a practical example of instantaneous speed—the speed of the car at a particular instant.

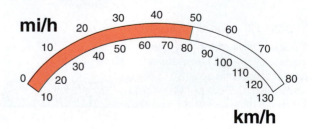

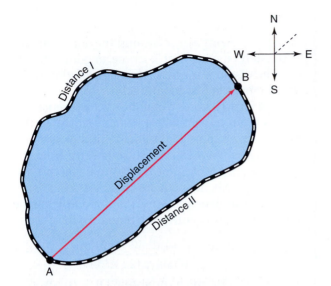

FIGURE 2.4   Displacement and Distance

Displacement is a vector quantity and is the straight-line distance between two points, plus direction. If the points A and B are 100 km apart, then in traveling directly from A to B, the displacement would be 100 km NE (northeast). Distance is a scalar quantity and is the actual path length traveled. Different routes may have different distances. Two are shown in the figure.

motion is then accelerated motion, which is discussed in the following section.

If the velocity is *constant,* or *uniform,* then we don't have to worry about changes. Suppose an airplane is flying at a constant speed of 320 km/h (200 mi/h) directly eastward. Then the airplane has a constant velocity and flies in a straight line. (Why?)

Also, for this special case, you should be able to convince yourself that the instantaneous velocity and average velocity are the same. By analogy, think about everyone in your class getting the same (constant) test score. How do the class average and individual scores compare?

A car traveling with a constant velocity is illustrated in ● Fig. 2.5. Here are some examples using speed and velocity.

EXAMPLE 2.1

Finding Speed and Velocity

Describe the motion (speed and velocity) of the car in Fig. 2.5.

SOLUTION

Step 1

The data are taken from the figure.

Given:

$d = 80$ m and $t = 4.0$ s
(The car travels 80 m in 4.0 s.)

Step 2

Wanted:

Speed and velocity. The units of the data are standard.

Step 3

Calculate the speed using Eq. 2.1:

$$\bar{v} = \frac{d}{t} = \frac{80 \text{ m}}{4.0 \text{ s}} = 20 \text{ m/s}$$

Notice that the car has a constant (uniform) speed and travels 20 m each second. Assuming that the motion is in one direction (straight-line motion), then the car's velocity is also constant and is 20 m/s *in the direction of the motion.*

CONFIDENCE EXERCISE 2.1

How far would the car in Example 2.1 travel in 10 s?

The preceding example was worked in stepwise fashion, as suggested in the approach to problem solving in Chapter 1. This will be done initially in the first chapters as a reminder. Examples generally will be worked directly unless a stepwise solution is considered helpful.

FIGURE 2.5   Constant Velocity

The car travels equal distances in equal periods of time in straight-line motion. With a constant speed and constant direction, the velocity of the car is constant, or uniform.

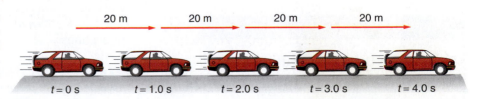

EXAMPLE 2.2

Finding the Time for Sunlight to Reach Earth

The speed of light in space is 3.00×10^8 m/s. How many seconds does it take light from the Sun to reach Earth if the distance from the Sun to Earth is 1.50×10^8 km?

SOLUTION

Given:

$$v = 3.00 \times 10^8 \text{ m/s}$$

and

$$d = 1.50 \times 10^8 \text{ km} \left(\frac{10^3 \text{ m}}{\text{km}}\right) = 1.50 \times 10^{11} \text{ m}$$

If we rearrange Eq. 2.1 for the unknown t, we have $t = d/v$, where the bar over the v is omitted because the speed is constant and $\bar{v} = v$. (The average value is equal to the constant value.)

$$t = \frac{d}{v} = \frac{1.50 \times 10^{11} \text{ m}}{3.00 \times 10^8 \text{ m/s}}$$
$$= 5.00 \times 10^2 \text{ s} = 500 \text{ s}$$

From this example we see that although light travels very fast, it still takes 500 seconds, or 8.33 minutes, to reach Earth after leaving the Sun (● Fig. 2.6). Here again we are working with a constant speed and velocity (straight-line motion).

CONFIDENCE EXERCISE 2.2

A communications satellite is in a circular orbit at an altitude of 3.56×10^4 km. How many seconds does it take a signal from the satellite to reach a TV-receiving station? (Radio signals travel at the speed of light, 3.00×10^8 m/s.)

EXAMPLE 2.3

Finding the Orbital Speed of Earth

What is the average speed in miles per hour of Earth for one revolution about the Sun?

SOLUTION

Our planet revolves once around the Sun in an approximately circular orbit in a time, or period, of one year. The distance it travels in one revolution is the circumference of this circular orbit.

Recall from Chapter 1 that the circumference of a circle is given by $2\pi r$, where r is the radius. In this case, r is taken to be 93.0 million miles (9.30×10^7 mi), which is the mean distance of Earth from the Sun (see Fig. 2.6). The time it takes to travel this distance is one year, or about 365 days.

Putting this data into Eq. 2.1 and multiplying the time by 24 h/day to convert to hours, we have

$$\bar{v} = \frac{d}{t} = \frac{2\pi r}{t} = \frac{2(3.14)(9.30 \times 10^7 \text{ mi})}{365 \text{ days } (24 \text{ h/day})}$$
$$= 6.67 \times 10^4 \text{ mi/h (or about } 3.0 \times 10^4 \text{ m/s)}$$

CONFIDENCE EXERCISE 2.3

What is the average speed in mi/h of a person at the equator as a result of Earth's rotation? (Take the radius of Earth to be $R_E = 4000$ mi.)

How about that! The solution to Example 2.3 shows that Earth, *and all of us*, are traveling through space at a speed of 66,700 mi/h (or 18.5 mi/s). Pretty fast! Even though this speed is exceedingly high and relatively constant, the velocity is continuously changing because the direction of the motion is continuously changing. We generally don't sense this great speed because of the small relative motions (apparent motions) of the stars and because our atmosphere moves with us. Think about how you know you are in motion when riding in a perfectly smooth-riding and quiet car. You see trees and other fixed objects "moving" in relative motion.

Also, we do not sense any change in velocity if the change is too small. We usually can sense changes in motion if they are appreciable. Think about being in the smooth-riding car and being blindfolded. You would be able to tell if the car suddenly went faster, slowed down, or went around a sharp curve, all of which are changes in velocity. A change in velocity is called an *acceleration*, which is the topic of the following sections.

RELEVANCE QUESTION: In describing your everyday movements from place to place, are you more often stating speeds or velocities?

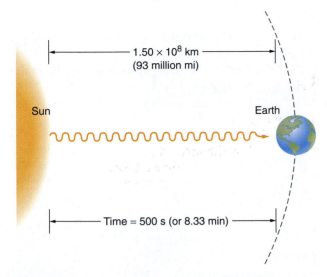

FIGURE 2.6 Traveling at the Speed of Light
Although light travels 3.00×10^8 m/s, it still takes over 8 minutes for light from the Sun to reach Earth. (See Example 2.2.)

2.3 Acceleration

LEARNING GOALS

▼ Explain how acceleration and velocity are related.

▼ Define *g*, the acceleration due to gravity.

When you drive down a straight interstate highway and suddenly increase your speed—say, from 20 m/s to 30 m/s—you feel as though you are being forced back against the seat. When driving fast on a circular cloverleaf, you feel forced to the outside of the circle. These experiences result from changes in velocity.

You can change the velocity of an object in three ways. You can (1) increase its magnitude, (2) decrease its magnitude, and/or (3) change the direction of the velocity vector. When any of these changes occurs, we say that the object is accelerating. The faster the change in the velocity, the greater the acceleration.

Acceleration is defined as *the time rate of change of velocity*. If we take the symbol Δ (delta) to mean "change in," the equation for average acceleration can be written as

$$\frac{\text{average}}{\text{acceleration}} = \frac{\text{change in velocity}}{\text{time for change to occur}}$$

$$\bar{a} = \frac{\Delta v}{t} = \frac{v_f - v_o}{t} \qquad \textbf{2.2}$$

Note that the change in velocity (Δv) is just the final velocity v_f minus the original velocity v_o. Keep in mind that t is also an interval, $\Delta t = t - t_o = t$, with t_o taken to be zero.

The units of acceleration in the SI are meters per second per second (m/s)/s, or meters per second squared (m/s^2). These units may be confusing at first. Keep in mind that an acceleration is a measure of a *change* in velocity during a given time period.

Consider a constant acceleration of 9.8 m/s^2. This value means that the velocity changes by 9.8 m/s *each* second. Thus for straight-line motion, as the number of seconds increases, the velocity goes from 0 to 9.8 m/s during the first second, to 19.6 m/s (9.8 m/s + 9.8 m/s) during the second second, to 29.4 m/s (19.6 m/s + 9.8 m/s) during the third second, and so forth, adding 9.8 m/s each second. This sequence is illustrated in ● Fig. 2.7 for an object that falls with a constant acceleration of 9.8 m/s^2. Notice that since the velocity increases, the distance traveled by the falling object each second also increases.

We will limit our discussion to such constant accelerations. For a constant acceleration, we may write

FIGURE 2.7 Constant Acceleration
For an object with a constant downward acceleration of 9.8 m/s^2, its velocity increases 9.8 m/s each second. The increasing lengths of the velocity arrows indicate the increasing velocity of the ball. (*Note:* The distances of fall are not to scale.)

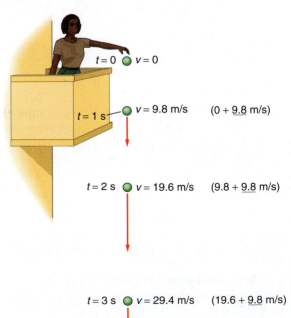

$t = 0$ ● $v = 0$

$t = 1$ s ● $v = 9.8$ m/s $(0 + \underline{9.8}$ m/s$)$

$t = 2$ s ● $v = 19.6$ m/s $(9.8 + \underline{9.8}$ m/s$)$

$t = 3$ s ● $v = 29.4$ m/s $(19.6 + \underline{9.8}$ m/s$)$

Acceleration (Speed increases)

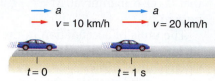

a $v = 10$ km/h a $v = 20$ km/h

$t = 0$ $t = 1$ s

Deceleration (Speed decreases)

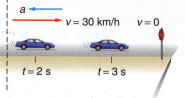

a $v = 30$ km/h $v = 0$

$t = 2$ s $t = 3$ s

FIGURE 2.8   Acceleration and Deceleration

When the acceleration is in the same direction as the velocity of an object in straight-line motion, its speed increases.

If the acceleration is in the opposite direction of the velocity (beginning at the dashed line), there is a deceleration, and the speed decreases.

$\bar{a} = a$, or drop the overbar, since the average acceleration is equal to the constant value. We may rearrange Eq. 2.2 to give an expression for *the final velocity of an object in terms of its original velocity, time, and constant acceleration.*

$$v_f - v_o = at$$

or $$v_f = v_o + at \qquad \textbf{2.3}$$

This equation is useful for working problems in which the quantities a, v_o, and t are all known and we wish to find v_f. If the original velocity $v_o = 0$, then

$$v_f = at \qquad \textbf{2.3a}$$

It should be noted that the use of Eq. 2.3 requires a constant acceleration—that is, one for which the velocity changes uniformly.

EXAMPLE 2.4

Finding Acceleration

A race car starting from rest accelerates uniformly along a straight track, reaching a speed of 90 km/h in 7.0 s. What is the magnitude of the acceleration of the car in m/s^2?

SOLUTION

Let's work this example stepwise for clarity.

Step 1

Given:

$v_o = 0$; $v_f = 90$ km/h; $t = 7.0$ s
(Notice that the car being initially at rest tells you that $v_o = 0$.)

Step 2

Wanted:

a (acceleration in m/s^2)
Since the acceleration is wanted in m/s^2, we convert 90 km/h to m/s using the conversion factor from the inside back cover:

$$v_f = 90 \; \text{km/h} \left(\frac{0.278 \; \text{m/s}}{\text{km/h}} \right) = 25 \; \text{m/s}$$

Step 3

The acceleration may be calculated using Eq. 2.2,

$$a = \frac{v_f - v_o}{t} = \frac{25 \; \text{m/s} - 0}{7.0 \; \text{s}} = 3.6 \; \text{m/s}^2$$

CONFIDENCE EXERCISE 2.4

If the car in the preceding example continues to accelerate at the same rate for three more seconds, what would be the magnitude of its velocity in m/s at the end of this time?

Because velocity is a vector quantity, acceleration is also a vector quantity. For an object in straight-line motion, the acceleration (vector) may be in the same direction as the velocity (vector), as in Example 2.4, *or* the acceleration may be in the opposite direction of the velocity (● Fig. 2.8). In the first instance, the acceleration causes the object to speed up, and the velocity increases. If the velocity and acceleration are in opposite directions, then the acceleration slows down the object, which is sometimes called a *deceleration.**

*Note that a deceleration is not necessarily a *negative* acceleration ($-a$). If the motion is in the negative direction ($-v$), the motion and the acceleration are in the *same* direction, and the object speeds up.

Let's consider an example. What is the common name for the gas pedal of a car? Right, the *accelerator.* When you push down on the accelerator, you speed up (increase the magnitude of the velocity). But when you let up on the accelerator, you slow down, or decelerate. Putting on the brakes will give an even greater deceleration. (Maybe we should call the brake pedal the "decelerator.")

In general, we will consider only constant, or uniform, accelerations. One very special constant acceleration is associated with the acceleration of falling objects. The **acceleration due to gravity** at Earth's surface is directed downward and is denoted by the letter g. Its magnitude in SI units is

$$g = 9.80 \text{ m/s}^2$$

This value corresponds to 980 cm/s^2, or about 32 ft/s^2.

The acceleration due to gravity varies slightly depending on such factors as how far you are from the equator and your altitude. However, the variations are very small, and for our purposes, we will take g to be the same everywhere on Earth's surface.

The Italian physicist Galileo Galilei (1564–1642) was one of the first scientists to assert that all objects fall with the same acceleration. Of course, this assertion assumes that frictional effects are negligible. We can state Galileo's principle as follows:

If frictional effects can be neglected, every freely falling object near Earth's surface accelerates downward at the same rate, regardless of the mass of the object.

One can illustrate this experimentally by dropping a small mass, such as a coin, and a larger mass, such as a ball, at the same time from the same height. They will hit the floor, as best as can be judged, at the same time (negligible air resistance). Legend has it that Galileo himself performed such experiments. (See the chapter Highlight.)

The effect of air resistance or friction can be demonstrated by dropping a piece of tissue paper and a coin. The air friction will prevent the tissue paper from falling as fast as the coin. If the tissue paper is wadded up into a very small ball to minimize air friction, its acceleration will be closer to that of the coin.

On the Moon there is no atmosphere, so there is no air resistance. In 1971, astronaut David Scott dropped a feather and a hammer at the same time while on the lunar surface (● Fig. 2.9). They both hit the surface of

FIGURE 2.9   No Air Resistance
Astronaut David Scott demonstrated that a feather and a hammer fall at the same rate on the Moon. There is no atmosphere on the Moon and hence no resistance, so the feather and the hammer fall side by side with the same acceleration.

the Moon at the same time because neither the feather nor the hammer was slowed by air resistance. This experiment shows that Galileo's assertion applies on the Moon as well as on Earth. Of course, on the Moon all objects fall at a slower rate than do objects on Earth's surface because the acceleration due to gravity on the Moon is only $\frac{1}{6}$ the value on Earth.

The velocity of a freely falling object on Earth increases 9.80 m/s each second, or increases uniformly with time. But how about the distance covered each second? Distance covered is not uniform, because the object speeds up. *The distance a dropped object travels downward with time* can be computed from the equation*

$$d = \tfrac{1}{2}gt^2 \qquad\qquad \textbf{2.4}$$

*In general, the distance an object travels when it starts at an original velocity v_o and is accelerated at an acceleration a is

$$d = v_\text{o}t + \tfrac{1}{2}at^2$$

HIGHLIGHT: Galileo and the Leaning Tower of Pisa

One of Galileo's greatest contributions to science was his emphasis on experimentation, a basic part of the scientific method. However, a question exists as to whether or not he actually carried out a now-famous experiment. There is a popular story that Galileo dropped stones or cannonballs of different masses from the top of the Tower of Pisa to determine experimentally whether objects fall with the same acceleration (Fig. 1).

Galileo did indeed question Aristotle's view that objects fell because of their "earthiness"; and the heavier or more earthy an object, the faster it would fall in seeking its "natural" place at the center of Earth. His ideas are evident in the following excerpts from his writings.*

How ridiculous is this opinion of Aristotle is clearer than light. Who ever would believe, for example, that . . . if two stones were flung at the same moment from a high tower, one stone twice the size of the other, . . . that when

*From L. Cooper, Aristotle, Galileo, and the Tower of Pisa (Ithaca, N.Y.: Cornell University Press, 1935).

the smaller was half-way down the larger had already reached the ground?

And:

Aristotle says that "an iron ball of one hundred pounds falling a height of one hundred cubits reaches the ground before a one-pound ball has fallen a single cubit." I say that they arrive at the same time.

Although Galileo (Fig. 2) mentions a *high tower*, the Tower of Pisa is not mentioned in his writings, and there is no independent record of such an experiment. Fact or fiction? No one really knows. What we do know is that all freely falling objects near Earth's surface fall with the same acceleration.

As an interesting sidelight, the Tower of Pisa has been in the news again recently. It was closed to the public in 1990 because of the possibility of it toppling. The tower started leaning

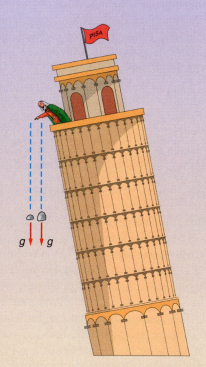

FIGURE 1 Free-fall.   All freely falling objects near Earth's surface have the same constant acceleration. Galileo is alleged to have shown this by dropping cannonballs or stones of different masses from the Leaning Tower of Pisa. Over short distances, the air resistance can be neglected, so objects dropped at the same time will strike the ground together.

EXAMPLE 2.5

How Far a Dropped Object Falls

A ball is dropped from atop a tall building (● Fig. 2.10, p. 34). How far does the ball fall in 1.50 s?

SOLUTION

With $t = 1.50$ s, the distance is given by Eq. 2.4, where it is known that $g = 9.80$ m/s^2.

$$d = \tfrac{1}{2} g t^2 = \tfrac{1}{2}(9.80 \text{ m/s}^2)(1.50 \text{ s})^2 = 11.0 \text{ m}$$

CONFIDENCE EXERCISE 2.5

What is the speed of the ball 1.50 s after being dropped?

Distances and velocities for a falling object are listed in ● Fig. 2.10. The figure is not drawn to scale because of the distance increases. Note that the distance increase is proportional to time squared (t^2, Eq. 2.4) and the velocity increase is proportional to time (t, Eq. 2.3a). From the figure we see that at the end of the first second the object has fallen 4.90 m. At the

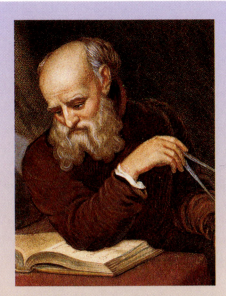

because of more construction, so another 230 tons of lead were added (Fig. 3).

At the time of this writing, plans call for soil to be removed from beneath the high side of the tower to balance the natural settling. Moreover, plastic-coated steel cables will be attached to the tower so as to support the structure should there be any sudden shifts. It is hoped this will correct the tilt by about one-half of a degree—enough to save the tower and still have a "leaning" attraction.

FIGURE 2   (left) **Galileo Galilei (1564–1642).**   The motion of objects was one of Galileo's many scientific inquiries.

FIGURE 3   (below) **Don't Let It Fall!**   Lead counterweights have been placed on the base of the Tower of Pisa to help correct the leaning.

before its completion in 1350, because of soft subsoil. It is currently leaning almost 6 degrees from the vertical.

Cement was injected into the base in the 1930s, but the leaning accelerated. During 1993–1995, a concrete ring was built at the base, and 620 tons of lead counterweights were placed on it. This corrected the lean somewhat, but there was a sudden overnight shift

end of the second second, the total distance fallen is 19.6 m. At the end of the third second, the object has fallen 44.1 m, which is about the height of a 10- or 11-story building, and it is falling at a speed of 29.4 m/s, which is about 65 mi/h—pretty fast!

If we throw an object straight upward, it slows down. (The velocity decreases 9.80 m/s each second.) In this case the velocity and acceleration are in opposite directions, and there is a deceleration (● Fig. 2.11). The object slows down (its velocity decreases) until it stops instantaneously at its maximum height. Then it starts to fall downward as though it were dropped from that height. The travel time upward is the same as the travel time downward to the original starting position.

Also, the object returns with the same speed it had initially. For example, if an object is thrown upward with an initial speed of 29.4 m/s, it will return with a velocity of 29.4 m/s downward. You should be able to conclude that it would travel 3 s upward, to a maximum height of 44.1 m, and return to its starting point in another 3 s. (*Hint:* See Fig. 2.10.)

RELEVANCE QUESTION: *While driving, do you accelerate only when you step on the accelerator? Explain.*

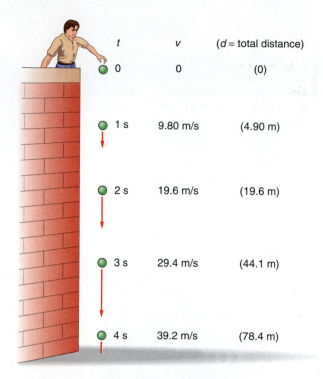

t	v	(d = total distance)
0	0	(0)
1 s	9.80 m/s	(4.90 m)
2 s	19.6 m/s	(19.6 m)
3 s	29.4 m/s	(44.1 m)
4 s	39.2 m/s	(78.4 m)

FIGURE 2.10   Time, Velocity, and Distance

Listed in the figure are the magnitudes of the velocities and distances traveled by a freely falling object at the end of the first few seconds. The magnitude of the velocity increases by 9.80 m/s each second. When dropped from rest, the distance an object falls in a particular time is computed by using the equation $d = \frac{1}{2}gt^2$. (The distances of the fall are obviously not to scale.)

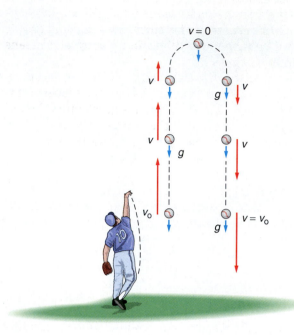

FIGURE 2.11   Up and Down

An object projected straight upward slows down because the acceleration due to gravity is in the opposite direction of the velocity, and the object stops ($v = 0$) for an instant at its maximum height. It then accelerates downward and returns to the starting point with a velocity equal and opposite to the initial velocity (v_o). (The downward path is displaced to the right in the figure for clarity.)

2.4 Acceleration in Uniform Circular Motion

LEARNING GOALS

▼ Describe centripetal acceleration.

▼ Explain why centripetal acceleration is necessary for circular motion.

An object in uniform circular motion has a constant speed. For example, a car going around a circular track at a uniform rate of 90 km/h (about 55 mi/h) has a constant speed. However, the velocity of the car is *not* constant, because its velocity is continuously changing direction. Since there is a change in velocity, there is an acceleration.

This acceleration cannot be in the direction of the instantaneous motion or velocity, or else the object would speed up and the motion would not be uniform. So in what direction is the acceleration? Because it causes a change in direction that keeps the object in a circular path, the acceleration is actually perpendicular, or at a right angle, to the velocity vector.

Consider a car traveling in uniform circular motion, as illustrated in ● Fig. 2.12. At any point, the instantaneous velocity is tangential to the curve (at an angle of 90° to a radial line at that point). After a short time, the velocity vector has changed (in direction). The change in velocity (Δv) is given by a vector triangle, as illustrated in the figure.

This change is an average over a time interval Δt, but notice how the Δv vector generally points inward toward the center of the circle. For instantaneous measurement this generalization is true, so for an object in uniform circular motion, the acceleration is toward the center of the circle. This acceleration is called **centripetal acceleration** (*centripetal* means "center seeking").

Even though it is traveling at a constant speed, an object in uniform circular motion must have an inward acceleration. For a car, this acceleration is supplied by friction on the tires. When a car hits an icy spot on a curved road, it slides outward because the centripetal acceleration is not great enough to keep it in a circular path.

In general, whenever an object moves in a circle of radius r with a constant speed v, *the magnitude of the centripetal acceleration a_c is given by the equation*

$$a_c = \frac{v^2}{r} \qquad \textbf{2.5}$$

(centripetal acceleration)

Note that the centripetal acceleration increases as the square of the speed—double the speed and the acceleration increases by a factor of 4. Also, the smaller the radius, the greater the centripetal acceleration needed to keep an object in circular motion for a given speed.

FIGURE 2.12   Centripetal Acceleration

A car traveling with a constant speed on a circular track is accelerating because its velocity is changing (direction change). The acceleration is toward the center of the circular path and is called *centripetal* ("center-seeking") *acceleration*.

EXAMPLE 2.6

Finding Centripetal Acceleration

Determine the magnitude of the acceleration of a car going 12 m/s (about 27 mi/h) on a circular cloverleaf with a radius of 50 m (● Fig. 2.13).

SOLUTION

The centripetal acceleration is given by Eq. 2.5:

$$a_c = \frac{v^2}{r} = \frac{(12 \text{ m/s})^2}{50 \text{ m}}$$
$$= 2.9 \text{ m/s}^2$$

CONFIDENCE EXERCISE 2.6

Using the result of Example 2.3, compute the centripetal acceleration in m/s^2 of Earth in its nearly circular orbit about the Sun ($r = 1.5 \times 10^{11}$ m).

The value of 2.9 m/s^2 in Example 2.6 is about 30% of the acceleration due to gravity, $g = 9.8$ m/s^2, and is a fairly large acceleration. We experience the effects of such an acceleration when riding in a car going in a circular path.

RELEVANCE QUESTION: *On Earth, we are continuously experiencing a centripetal acceleration. Explain this statement. Give another example of a centripetal acceleration you frequently undergo in your daily activities.*

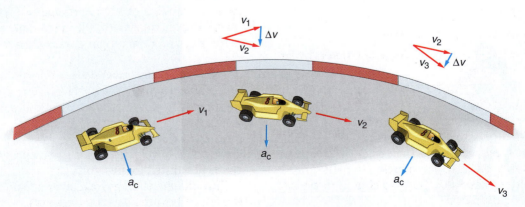

FIGURE 2.13 Frictional Centripetal Acceleration

The inward acceleration necessary for a vehicle to go around a level curve is supplied by friction on the tires. On a banked curve, some of the acceleration is supplied by a component of the acceleration due to gravity.

2.5 Projectile Motion

LEARNING GOAL

▼ Describe projectile motion, and explain some of its effects.

An object thrown horizontally will fall at the same rate as an object that is dropped. The velocity in the horizontal direction does not affect the velocity and acceleration in the vertical direction. This is illustrated in ● Fig. 2.14a.

The multiflash photo in the figure shows a dropped ball and one projected horizontally at the same time. Notice how the balls fall vertically together as the projected ball moves to the right.

Neglecting air resistance, a horizontally projected object essentially travels in a horizontal direction with a constant velocity (no acceleration in that direction) while falling vertically under the influence of gravity. The resulting path is a curved arc, as shown in Fig. 2.14. Occasionally, a sports announcer claims that a hard-throwing quarterback can throw the football so many yards "on a line," meaning a straight line. This statement, of course, must be false. All objects thrown horizontally begin falling as soon as they leave the hand.

If an object is projected at an angle θ (Greek theta) to the horizontal, it will follow a symmetric curved path, as illustrated in ● Fig. 2.15, where air resistance is again neglected. The curved path is essentially the result of the combined motions in the vertical and

FIGURE 2.14 Same Vertical Motions

(a) If a ball is thrown horizontally and another ball is dropped simultaneously from the same height, both will hit the ground at the same time (if air resistance is neglected) because the vertical motions are the same. (Diagram not to scale.) (b) A multiflash photograph of two balls—one projected horizontally at the same time the other ball was dropped. Note from the horizontal lines that they fall vertically at the same rate.

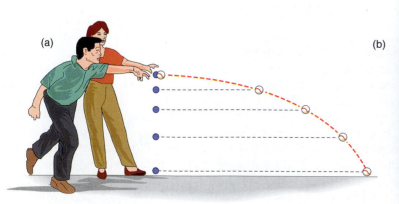

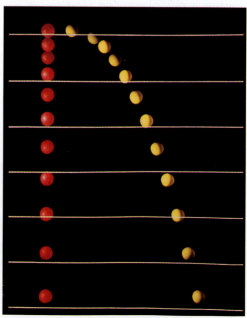

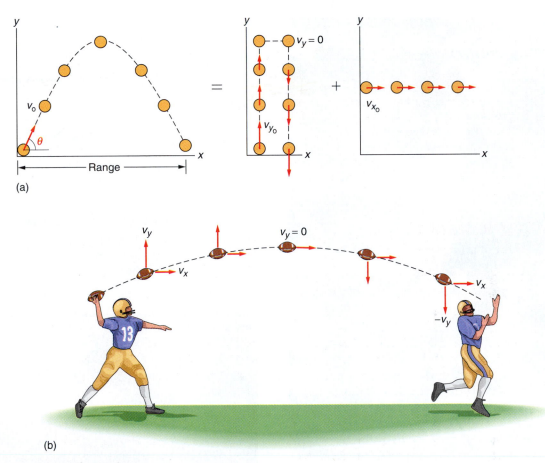

FIGURE 2.15   Projectile Motion

(a) The curved path of a projectile is a result of the combined motions in the vertical and horizontal directions. (b) Neglecting air resistance, the projected football has the same horizontal velocity (v_x), but the vertical velocity (v_y) changes like that of an object thrown upward.

horizontal directions. The projectile goes up and down vertically, while at the same time traveling horizontally with a constant velocity.

Now you can understand why basketball players, when jumping to score, seem to be suspended momentarily, or "hang" in midair (● Fig. 2.16). Notice that for the vertical motion near the maximum height, the velocity is quite small—going to zero and then increasing from zero (see Fig. 2.11). During this time, the combination of the slow vertical motion and the constant horizontal motion gives the illusion that the player is hanging in the air.

If a ball or other object is thrown at various angles, the path that it takes will depend on the angle at which it is thrown. Neglecting air resistance, each path will resemble one of those in ● Fig. 2.17. As shown, the range or horizontal distance the object travels is maximum when the object is projected at an angle of 45° relative to level ground. Notice in the figure that for a given initial speed, projections at complementary angles—for example, 30° and 60°—have the same range as long as there is no air resistance.

With little or no air friction, projectiles have symmetric paths. However, when a ball or object is thrown or hit hard, air friction comes into effect. In this case the projectile path resembles one of those shown in ● Fig. 2.18 and is no longer symmetric. Air friction reduces the velocity of the projectile, particularly in the horizontal direction. As a result, the maximum range occurs at an angle less than 45°.

Athletes such as football quarterbacks and baseball players are aware of the best angle at which to throw in order to get the maximum distance. A good golfing drive also depends on the angle at which the

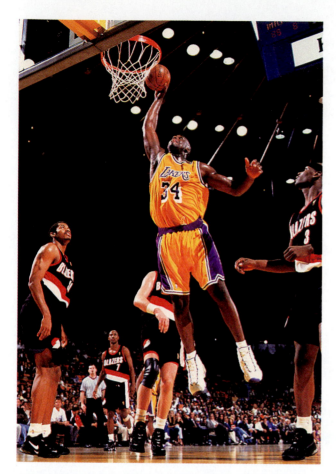

FIGURE 2.16   Hanging In There

Shaquille O'Neal seems to "hang" in the air at the top of his "projectile" path. (See text for description.)

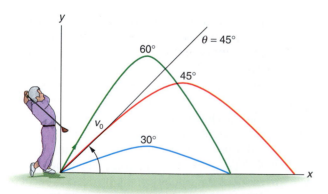

FIGURE 2.17 Maximum Range

A projectile's maximum range on a horizontal plane is achieved with a projection angle of 45°. This assumes no air resistance.

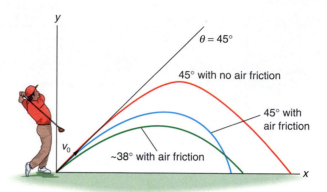

FIGURE 2.18   Effects of Air Resistance on Projectiles

Long football passes, hard-hit baseballs, and driven golf balls follow trajectories similar to those shown here. Frictional air resistance reduces the range.

FIGURE 2.19   Going for the Distance

With air resistance, the athlete hurls the javelin at an angle of less than 45° for maximum range.

ball is hit. Of course, in most of these instances there are other considerations, such as spin. The angle of throw is also a consideration in track and field events such as discus and javelin throwing. ● Figure 2.19 shows an athlete hurling a javelin at an angle of less than 45° in order to achieve the maximum distance.

RELEVANCE QUESTION: *Does the ideal projection angle for a golf drive change depending on whether there is a wind blowing from the fairway toward the tee or down the fairway in the direction of the drive? Explain.*

Important Terms

position (2.1)	average speed	displacement	acceleration due to
motion	distance	instantaneous velocity	gravity
scalar (2.2)	instantaneous speed	acceleration (2.3)	centripetal
vector	average velocity		acceleration (2.4)

Important Equations

Average Speed: $\bar{v} = \dfrac{d}{t}$

Distance Traveled by an Object Starting from Rest:
$$d = \tfrac{1}{2}at^2 \quad \text{(constant acceleration)}$$

Distance Traveled by a Dropped Object: $d = \tfrac{1}{2}gt^2$

Acceleration due to gravity: $g = 9.80 \text{ m/s}^2 = 32 \text{ ft/s}^2$

Acceleration: $\bar{a} = \dfrac{\Delta v}{t} = \dfrac{v_\mathrm{f} - v_\mathrm{o}}{t}$

or $\qquad v_\mathrm{f} = v_\mathrm{o} + at \qquad$ (constant acceleration)

Centripetal Acceleration: $a_\mathrm{c} = \dfrac{v^2}{r}$

Review Questions

2.1 Defining Motion

1. Which one of the following describes an object in motion?
   (a) A period of time has passed.
   (b) Its position is known.
   (c) It is continuously changing position.
   (d) It has reached its final position.

2. What is needed to designate the position of an object? Analyze each of the examples given at the beginning of Section 2.1 in terms of this.

3. How is position described on a Cartesian graph, and what is necessary to do this?

4. What is meant by the *time rate of change of position*?

2.2 Speed and Velocity

5. Which one of the following is always true about the magnitude of a displacement?
   (a) greater than the distance traveled
   (b) equal to the distance traveled
   (c) less than the distance traveled
   (d) less than or equal to the distance traveled

6. Which one of the following does *not* have units of m/s?
   (a) average speed
   (b) average acceleration
   (c) average velocity
   (d) instantaneous velocity

7. What does the speedometer of an automobile read?
   (a) instantaneous speed
   (b) average speed
   (c) instantaneous velocity
   (d) average velocity

8. Explain the difference between scalar and vector quantities.

9. What is the difference between distance and displacement? How are these quantities associated with speed and velocity?

10. A jogger travels two blocks directly north.
    (a) How do the jogger's average speed and the average magnitude of the average velocity compare?
    (b) If the return trip follows the same path, how do the average speed and the magnitude of the average instantaneous velocity compare for the total trip?

2.3 Acceleration

11. Acceleration is the time rate of change of what?
    (a) distance   (c) displacement
    (b) speed   (d) velocity

12. Which one of the following is true for a deceleration?
    (a) The velocity remains constant.
    (b) The acceleration is negative.
    (c) The acceleration is in the direction opposite the motion.
    (d) The acceleration is zero.

13. What changes when there is an acceleration? Give an example.

14. Can an object simultaneously have an instantaneous velocity of 9.8 m/s in one direction and an acceleration of 9.8 m/s^2 in the same or opposite direction? Explain.

15. When Galileo dropped two objects simultaneously from a high place, they hit the ground at almost exactly the same time. Why was there a slight difference in when they hit?

16. A ball is dropped. What is its initial speed? What is its initial acceleration?

2.4 Acceleration in Uniform Circular Motion

17. If the speed of an object in uniform circular motion is doubled and the radial distance remains constant, then the centripetal acceleration increases by what factor? (a) 2   (b) 3   (c) 4   (d) 6

18. Can a car be moving at a constant rate of 60 km/h and still be accelerating? Explain.

19. What does the term *centripetal* mean?

20. What would happen to an object in uniform circular motion if the centripetal acceleration (a) were reduced and (b) went to zero?

21. Are we accelerating due to Earth's spinning on its axis? Can you sense this motion? Explain.

22. What is the direction of the acceleration vector of a person on the spinning Earth if the person is (a) at the equator and (b) at some other latitude?

2.5 Projectile Motion

23. Neglecting air resistance, which of the following is true for a ball thrown at an angle θ to the horizontal?
    (a) a constant velocity in the x direction
    (b) a constant acceleration in the $-y$ direction
    (c) a changing velocity in the $+y$ direction
    (d) all of the preceding

24. In the absence of air resistance, a projection at an angle of 33° above the horizontal will have the same range as a projection of which one of the following angles?
    (a) 45°   (b) 57°   (c) 66°   (d) 180° − 33° = 147°

25. For projectile motion, what quantities are constant? (Neglect air resistance.)

26. On what does the range of a projectile depend?

27. Can a baseball player throw a fastball on a straight, horizontal line? Why or why not?

28. Does a basketball player driving in and jumping for a "slam dunk" really hang in the air at the maximum height of the jump? Explain.

29. Does a football quarterback throw a long pass at a greater or smaller angle than a short pass?

30. Taking into account air resistance or friction, how would you throw a ball to get the maximum range and why?

Applying Your Knowledge

1. Do highway speed limit signs refer to average speeds or instantaneous speeds? Explain.

2. If we are moving at high speed through space as Earth revolves about the Sun, why don't we generally sense the motion?

3. The gas pedal of a car is commonly referred to as the "accelerator." Would this term be appropriate for (a) the steering wheel, (b) the brakes, and, for a stick-shift car, (c) the clutch, and (d) the gears? Explain.

4. Roadways around curves are sometimes banked or at an incline toward the center of the curve. Why is this?

Exercises

2.2 Speed and Velocity

1. At a track meet, a runner runs the 100-m dash in 15 s. What was the runner's average speed?   Answer: 6.7 m/s

2. A jogger jogs around a circular track with a diameter of 300 m in 13 minutes. What was the jogger's average speed in m/s?

3. The Moon is about 3.87×10^5 km from Earth, and an astronaut on the Moon communicates by using radio waves, which travel at the speed of light (3.00×10^8 m/s). How many seconds does it take the astronaut's signal to reach Earth?   Answer: 1.29 s

4. A group of college students anxious to get to Florida on a spring break drive the 630-mi trip with only minimum stops. They compute their average speed for the trip to be 55.3 mi/h. How many hours did the trip take?

5. A student drives the 100-mi trip back to campus after spring break and travels with an average speed of 55 mi/h for 1 hour and 30 minutes.
   (a) What distance was traveled during this time?
   (b) Traffic gets heavier, and the last part of the trip takes another half hour. What was the average speed during this leg of the trip?
   (c) Find the average speed for the total trip.
   Answer: (a) 83 mi (b) 34 mi/h (c) 50 mi/h

6. (a) A car is driven 180 km in 2.0 h. What is the average speed of the car?
   (b) The return trip over the same route takes 3.0 h because of heavy traffic. What is the average return speed?
   (c) The total distance traveled for the entire trip is 360 km and the total time is 5.0 h. What is the average speed for the entire trip?

7. An airplane flying directly eastward at a constant rate travels 300 km in 2.0 h.
   (a) What is the average velocity of the plane?
   (b) What is its instantaneous velocity?
   Answer: (a) 150 km/h, east (b) same

8. A race car travels northward on a straight, level track at a constant speed, traveling 0.750 km in 20.0 s. The return trip over the same track is made in 25.0 s.
   (a) What is the average velocity of the car in m/s for the first leg of the run?
   (b) What is the average velocity for the total trip?

2.3 Acceleration

9. A sprinter starting from rest on a straight and level track is able to achieve a speed of 12 m/s in a time of 4.0 s. What is the sprinter's average acceleration?
   Answer: 3.0 m/s^2

10. Modern oil tankers weigh over a half-million tons and have lengths of up to a quarter of a mile. Such massive ships require a distance of 5.0 km (about 3.0 mi) and a time of 20 min to come to a stop from a top speed of 30 km/h.
    (a) What is the magnitude of such a ship's average acceleration in m/s^2 in coming to a stop?
    (b) What is the magnitude of the ship's average velocity in m/s? Comment on the potential of a tanker running aground.

11. A motorboat starting from rest travels in a straight line on a lake.
    (a) If the boat achieves a speed of 12 m/s in 10 s, what is the boat's average acceleration?
    (b) Then, in 5.0 more seconds, the boat's speed is 18 m/s. What is the boat's average acceleration for the total time?
    Answer: (a) 1.2 m/s^2 in direction of motion (b) same

12. (Here are some convenient British system conversions.) A car travels on a straight, level road.
    (a) Starting from rest, the car is going 44 ft/s (30 mi/h) at the end of 5.0 s. What is the car's average acceleration in ft/s^2?
    (b) In 4.0 more seconds, the car is going 88 ft/s (60 mi/h). What is the car's average acceleration for this time period?
    (c) The car then slows down to 66 ft/s (45 mi/h) in 3.0 s. What is the average acceleration for this time period?
    (d) What is the overall average acceleration for the total time?

13. A drag racer accelerates from rest at a constant rate of 2.0 m/s^2.
    (a) How fast will the racer be going at the end of 5.0 s?
    (b) How far does the racer travel in this time?
    Answer: (a) 10 m/s (b) 25 m

14. A rocket accelerating at a constant rate of 4.0 m/s^2 in a straight line is traveling with a speed of 60 m/s after a 12-s interval.
    (a) Did the rocket initially blast off from its pad at the beginning of this time interval?
    (b) If not, describe the initial situation.

15. A student drops an object out of the window of the top floor of a high-rise dormitory.
    (a) Neglecting air resistance, how fast is the object traveling when it strikes the ground at the end of 3.0 s? First use standard British units and then express the speed in mi/h for a familiar comparison. Can things dropped from high places be dangerous to people below?
    (b) How far, in m, does the object fall during the 3.0 s? Comment on how many floors the dormitory probably has.
        Answer: (a) 96 ft/s (65 mi/h)   (b) 44 m (14 floors)

16. A girl standing on a bridge drops a stone.
    (a) If it takes 2.50 s for the stone to hit the water below, what is the girl's distance above the water?
    (b) With what speed did the stone hit the water?

2.4 Acceleration in Uniform Circular Motion

17. A person drives a car around a circular cloverleaf with a radius of 70 m at a uniform speed of 10 m/s.
    (a) What is the acceleration of the car?
    (b) Compare this answer with the acceleration due to gravity as a percentage. Would you be able to sense the car's acceleration if you were riding in it?
        Answer: (a) 1.4 m/s², toward center   (b) 14%, yes

18. A race car goes around a level, circular track with a diameter of 1.00 km at a constant speed of 90.0 km/h. What is the car's centripetal acceleration in m/s²?

2.5 Projectile Motion (Neglect air resistance)

19. If you drop an object from a height of 1.5 m, it will hit the ground in 0.55 s. If you throw a baseball horizontally with an initial speed of 30 m/s from the same height, how long will it take the ball to hit the ground?
        Answer: 0.55 s

20. An airplane flies west at a speed of 200 m/s (448 mi/h). A package of supplies is dropped from the plane to some stranded campers.
    (a) What is the package's initial velocity? Give both speed and direction.
    (b) What are the magnitude and direction of the package's acceleration?
    (c) If the plane is at an altitude of 1000 m, how far from the campers should the package be dropped to land near them?

Solutions to Confidence Exercises

2.1 $d = vt = (20 \text{ m/s})(10 \text{ s}) = 2.0 \times 10^2 \text{ m}$

2.2 $t = \dfrac{d}{v} = \dfrac{3.56 \times 10^7 \text{ m}}{3.00 \times 10^8 \text{ m/s}} = 0.119 \text{ s}$

(Now you know why people on TV pause when asked a question on satellite interviews.)

2.3 $\bar{v} = \dfrac{d}{t} = \dfrac{2\pi R_E}{t} = \dfrac{2\pi(4.00 \times 10^3 \text{ mi})}{24.0 \text{ h}} = 1.05 \times 10^3 \text{ mi/h}$

Pretty fast—over 1000 mi/h.

2.4 $v_f = v_o + at = 0 + (3.6 \text{ m/s}^2)(10 \text{ s}) = 36 \text{ m/s}$

2.5 $v = gt = (9.80 \text{ m/s}^2)(1.50 \text{ s}) = 14.7 \text{ m/s}$

2.6 $a_c = \dfrac{v^2}{r} = \dfrac{(3.0 \times 10^4 \text{ m/s})^2}{1.5 \times 10^{11} \text{ m}} = 6.0 \times 10^{-3} \text{ m/s}^2$

Answers to Multiple-Choice Review Questions

1. c   6. b   11. d   17. c   24. b
5. d   7. a   12. c   23. d

FORCE AND MOTION

3

The whole burden of philosophy seems to consist of this—from the phenomena of motions to investigate the forces of nature, and from these forces to explain the other phenomena.

Sir Isaac Newton (1642–1727)

What is force? What is it that sets an object in motion? What stops it? These are basic questions dealt with in this chapter on force and motion. Our discussion will include Newton's three laws of motion, his law of universal gravitation, and the laws of conservation of linear and angular momentum.

Galileo (1564–1642) was one of the first scientists to experiment on moving objects. But it remained for Sir Isaac Newton, who was born the year Galileo died, actually to formulate the laws of motion and to explain the phenomena of moving objects on Earth and the motions of the planets and other celestial bodies.

Newton was only 25 years old when he formulated most of his discoveries in mathematics and mechanics. His book *Mathematical Principles of Natural Philosophy* (commonly referred to as the *Principia*), published in Latin in 1687 when he was 45, is considered by many to be the most important publication in the history of physics. Certainly it established Newton as one of the greatest scientists of all time. (See the chapter's first Highlight.) ■

Photo: Force and motion in action. Forces are applied through collisions, and motion, or a change in motion, results.

3.1 Newton's First Law of Motion

LEARNING GOALS

▼ Explain Newton's first law of motion.

▼ Describe and give examples of the concept of inertia.

Long before the time of Galileo and Newton, scientists had asked themselves: "What is the natural state of motion?" The early Greek scientist Aristotle had presented a theory, which prevailed for almost 20 centuries after his death. According to this theory, an object required a force in order to be kept in motion. That is, the natural state of an object was one of rest, with the exception of celestial bodies, which were naturally in motion. It was easy to observe that moving objects tended to slow down and come to rest, so a natural state of being at rest must have seemed logical to Aristotle.

Galileo studied motion using a ball rolling onto a level surface from an inclined plane. The smoother he made the surface, the farther the ball would roll (● Fig. 3.1). He reasoned that if he could make a very long surface perfectly smooth, there would be nothing to stop the ball, so it would continue indefinitely or until something did stop it. Thus, contrary to Aristotle, Galileo concluded that objects could *naturally* remain in motion rather than come to rest.

Newton also recognized this phenomenon and incorporated Galileo's result in his **first law of motion.** This law can be stated as follows:

An object will remain at rest or in uniform motion in a straight line unless acted on by an external, unbalanced force.

Uniform motion in a straight line means that the velocity is constant. An object at rest has a constant velocity of zero.

Because of the ever-present forces of friction and gravity on Earth, it is difficult to observe an object in a state of constant velocity. However, in free space where there is no friction and negligible gravitational attraction, an object initially in motion maintains a constant velocity. For example, after being projected on its way, an interplanetary spacecraft approximates this condition quite well between planets. On going out of the solar system, as some *Pioneer* and *Voyager* spacecrafts have done, a spacecraft travels with a con-

stant velocity until an external, unbalanced force alters this velocity.

Let's focus on what is meant by an *external, unbalanced force.* Newton's first law indicates that if an object is to speed up, slow down, and/or change direction (for example, as in circular motion), then an unbalanced force is required. But first, what is a force? We all have an intuitive feeling for the concept of a force. A force is simply a push or pull. In terms of the first law, we might say that a **force** is a quantity that is capable of producing motion or a change in motion—that is, a change in velocity, or an acceleration (Chapter 2).

This statement does not necessarily mean that a force produces a change in motion, only that the *capability* is there. Here is where the term *unbalanced* comes in. In the tug-of-war shown in ● Fig. 3.2a, forces are being applied, but there is no motion. The forces in this case are balanced—that is, equal (in magnitude) and opposite (in direction)—and in effect cancel each other out. Motion occurs when the applied forces are unbalanced—that is, when they are not equal and opposite, as in Fig. 3.2b.

Since forces have direction as well as magnitude, they are *vector* quantities. Forces may act on an object in different directions, but in order for a change in motion or an acceleration to occur, there must be an unbalanced or *net* force.

An *external* force is an applied force—that is, one applied on or to the object or system. There are also internal forces. For example, suppose the object is an automobile and you are a passenger traveling inside.

FIGURE 3.1   Motion Without Resistance

If the level surface could be made perfectly smooth, how far would the ball travel?

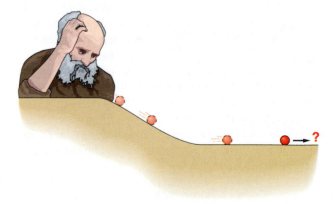

$F_1 = F_2$

(a)

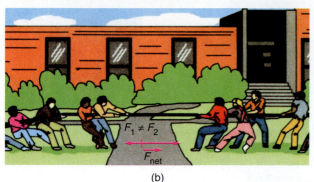

$F_1 \neq F_2$

F_{net}

(b)

FIGURE 3.2   Balanced and Unbalanced Forces

(a) When two applied forces are equal (in magnitude) and opposite (in direction), they are said to be *balanced,* and there is no net force and no motion if the system is initially at rest. (b) When F_2 is greater than F_1, there is an unbalanced, or net, force, and motion occurs.

You can push (apply a force) on the floor or the dashboard, but this has no effect on the car's velocity because this push is an internal force.

Motion and Inertia

Galileo also introduced another concept. It appeared that objects had a behavior or a property of maintaining a state of motion; that is, there was a resistance to changes in motion. Similarly, if an object was at rest, it seemed to "want" to remain at rest. Galileo called this property *inertia,* and we say that **inertia** is the natural tendency of an object to remain in a state of rest or in uniform motion in a straight line.

Newton went one step further and related the concept of inertia to something that could be measured—mass; that is,

mass is a measure of inertia

The greater the mass of an object, the greater is its inertia, and vice versa.

As an example of the relationship of mass and inertia, suppose you horizontally push two different people on swings initially at rest, one a very large man and the other a small child (● Fig. 3.3a). You'd quickly find that it was more difficult to get the adult moving; that is, there would be a noticeable difference in the resistance to motion between the man and the child.

Also, once you got them swinging and then tried to stop the motions, you'd notice a difference in the resistance to a change in motion again. Being more

FIGURE 3.3   Mass and Inertia

(a) An external, applied force is necessary to put an object in motion, and the acceleration is inversely proportional to the mass or inertia of the object. (b) Because of inertia, the paper strip can be removed from beneath the stack of quarters without toppling it by giving the paper a quick jerk.

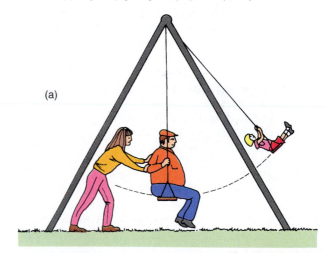

(a)

(b)

HIGHLIGHT: Isaac Newton

Isaac Newton and Albert Einstein are usually considered to be the two greatest scientists in history. Newton's laws of motion were some of many contributions he made to a variety of subjects in physics. Newton was born on Christmas day in 1642 in the village of Woolsthorpe in Lincolnshire, England (Fig. 1a). He showed no particular genius in his early schooling, but fortunately, a teacher encouraged him to pursue his education, and in 1661 he entered Trinity College at Cambridge.

Four years later he received his degree. He planned to continue studying for a master's degree, but an epidemic of the bubonic plague broke out, and the university was closed. Newton returned to Woolsthorpe, and in the next two years he laid the groundwork for many of his contributions in physics, mathematics, and astronomy. In Newton's own words, "I was in the prime of my age for invention, and minded mathematics and philosophy [science] more than any time since."

Over the next twenty years Newton was very productive, and at the age of forty-five he published his famous treatise, *Philosophiae Naturalis Principia Mathematica* [*Mathematical Principles of Natural Philosophy*],* or *Principia* for short (Fig. 1b). In this book he set forth his laws of motion, together with the theory of gravitation. The publication of the *Principia* was financed by a friend, Edmund Halley, who used Newton's theories to predict the return of the comet that bears his name.

*Physics was once called *natural philosophy*.

(a)

FIGURE 1   The Man and the Book

(a) Sir Isaac Newton (1642–1727).

(b) The title page from the *Principia*. Can you read the Roman numerals at the bottom of the page that give the year of publication?

(b)

Newton was reportedly a shy man, but he often got into disputes about his theories and achievements. His dispute with Gottfried Leibniz about who first developed calculus is famous. Newton was elected to Parliament and later appointed Master of the Mint, where he supervised the task of recoining the English currency. He was knighted by Queen Anne in 1705.

Sir Isaac Newton died in 1727 at the age of eighty-five and was buried with honor in Westminster Abbey. Some insight about this austere bachelor and giant of science may be gleaned from the following excerpts from his writings.

I seem to have been only a boy playing on the seashore and diverting myself in now and then finding a smoother pebble or a prettier shell than ordinary, whilst the great ocean of truth lay all undiscovered before me.

If I have been able to see farther than some, it is because I have stood on the shoulders of giants.

About Newton, the poet Alexander Pope wrote:

Nature and Nature's laws lay hid in night. God said, Let Newton be! and all was light.

massive, the man has greater inertia and a greater resistance to a change in motion.

Another example of inertia is shown in Fig. 3.3b. The stack of coins has inertia and resists being moved when at rest. If the paper is jerked quickly, the inertia of the coins will prevent them from toppling. The small external force is not sufficient to overcome the inertia of the coins. You may have pulled a magazine from the bottom of a stack with a similar action.

Because of the relationship between motion and inertia, Newton's first law is sometimes called the *law of inertia*. The law may be used to describe some of the observed effects in everyday life. For example, suppose you were in the front seat of a car traveling at a high speed down a straight road and the driver suddenly put on the brakes for an emergency stop. What would happen, according to Newton's first law, if you were not wearing a seat belt?

The friction on the seat of your pants would not be enough to change your motion appreciably, so you'd keep on moving while the car was coming to a stop. The next external, unbalanced force you'd experience would not be pleasant. Newton's first law makes a good case for buckling up.

RELEVANCE QUESTION: *Give an example of an everyday situation where inertia would be (a) an advantage and (b) a disadvantage.*

3.2 Newton's Second Law of Motion

LEARNING GOALS

▼ Describe the relationships in Newton's second law of motion, and apply the law to simple situations.

▼ Differentiate between mass and weight using Newton's second law.

In our initial study of motion (Chapter 2), acceleration was defined as the time rate of the change of velocity ($\Delta v/\Delta t$). Nothing was said about what *causes* acceleration, only that a change in velocity was required. So what causes an acceleration? Newton's first law answers the question: If an external, unbalanced force is required to produce a change in velocity, then we see that an external, unbalanced force causes an acceleration.

Newton was aware of this result, but he went further and also related acceleration to inertia or mass. Because inertia is the tendency not to have a *change* in motion, a reasonable assumption is that the greater the inertia or mass of an object, the smaller is the change in motion or velocity when a force is applied. Such insight was typical of Newton in his many contributions to science. Hence we have this summary:

1. The acceleration produced by an unbalanced force acting on an object (or mass) is directly proportional to the magnitude of the force ($a \propto F$) and in the direction of the force ($\propto$ is a proportionality sign). In other words, the greater the unbalanced force, the greater is the acceleration.

2. The acceleration of an object being acted on by an unbalanced force is inversely proportional to the mass of the object ($a \propto 1/m$); that is, for a given unbalanced force, the greater the mass of an object, the smaller is the acceleration.

Combining these effects of force and mass on acceleration, we have a statement of **Newton's second law of motion,** which may be expressed simply and conveniently as

$$\textbf{acceleration} \propto \frac{\textbf{unbalanced force}}{\textbf{mass}}$$

or
$$a \propto \frac{F}{m}$$

These relationships are illustrated in ● Fig. 3.4. Notice that if you double the force acting on a mass, the acceleration doubles (direct proportion, $a \propto F$). If, however, the same force is applied to twice as much mass, the acceleration is one-half as much (inverse proportion, $a \propto 1/m$).

When the appropriate units are used, we can write in equation form

$$a = \frac{F}{m}$$

Newton's second law, as commonly written, is

$$F = ma \qquad \textbf{3.1}$$
(Newton's second law)

Notice from Fig. 3.4 that the mass m is the *total* mass of the system, or all the mass that is accelerated. A system may be two or more separate masses, as will be shown in a later example. Also, F is the net, or unbalanced, force, which may be the vector sum of two or more forces. Unless otherwise stated, a general reference to force means an *unbalanced* force. The acceleration a is in the direction of this force.

In the metric system (SI), the unit of force is the **newton** (abbreviated N). This is a derived unit, and the standard unit equivalent may be seen from Eq. 3.1 by putting in standard units: force = mass × acceleration = kg × m/s^2 = kg-m/s^2 = N.

In the British system, the unit of force is the *pound* (lb). This unit also has derived units of mass times acceleration (ft/s^2). The unit of mass is the *slug*, which is rarely used, and we will not employ it in this textbook. Recall that the British system is a gravitational or force

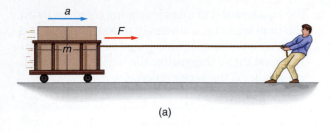

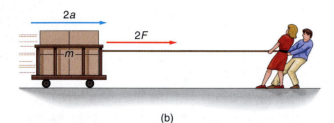

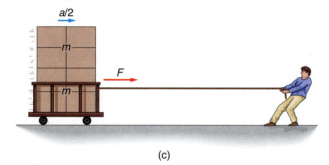

FIGURE 3.4 Force, Mass, and Acceleration

(a) An unbalanced force *F* acting on a mass *m* produces an acceleration *a*. (b) If the mass remains the same and the force is doubled, the acceleration is doubled. (c) If the mass is doubled and the force remains the same, the acceleration is reduced by one-half.

system, and things are weighed in pounds (force). As will be seen shortly, **weight** is a force and is expressed in newtons in the metric system and in pounds in the British system. Equivalent weight units in the two systems are shown in ● Fig. 3.5. If you are familiar with the story that Newton gained insight by observing (or being struck by) a falling apple while meditating on the concept of gravity, you can easily remember that an average-sized apple weighs about one newton.

Here's an example using $F = ma$ that illustrates that *F* is the unbalanced, or net, force and *m* is the total mass.

EXAMPLE 3.1

Finding Acceleration with Two Applied Forces

Forces are applied to blocks connected by a string and resting on a frictionless surface, as illustrated in ● Fig. 3.6. If the mass of each block is 1.0 kg and the mass of the string is negligible, what is the acceleration of the system?

SOLUTION

Step 1

Given:

$m_1 = 1.0$ kg $F_1 = -5.0$ N (left, negative direction)
$m_2 = 1.0$ kg $F_2 = 8.0$ N (right, positive direction)

Step 2

Wanted: *a* (acceleration)

(The units are standard in the metric system.)

Step 3

The acceleration may be calculated using Eq. 3.1, $F = ma$, or $a = F/m$. Note, however, that *F* is the unbalanced (net) force, which in this case is the vector sum of the forces. Effectively, F_1 cancels part of F_2, and there is a net force $F_{net} = 8.0$ N $- 5.0$ N in the direction of F_2. The total mass of the system being accelerated is $m = m_1 + m_2$. Hence we have

$$a = \frac{F}{m} = \frac{F_{net}}{m_1 + m_2} = \frac{8.0 \text{ N} - 5.0 \text{ N}}{1.0 \text{ kg} + 1.0 \text{ kg}}$$
$$= 1.5 \text{ m/s}^2$$

in the direction of the net force (right).

Question

What would be the case if the surface were not frictionless and the mass of the string were not negligible? Right. You'd have another force opposing the motion in the direction of F_1, and another mass m_3 would contribute to the total mass.

CONFIDENCE EXERCISE 3.1

Given the same conditions as in Example 3.1, suppose $F_1 = 9.0$ N and $F_2 = 6.0$ N in magnitude. What would be the acceleration of the system in this case?

FIGURE 3.5   About a Newton

An average-sized apple weighs about 1 newton
(or 0.225 lb).

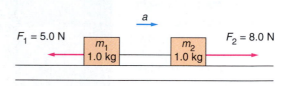

FIGURE 3.6   F = ma

Net force and total mass. (See Example 3.1.)

It should be noted how general an expression Newton's second law is. It can be used to analyze many situations. A dynamic example is **centripetal force.** Recall from Chapter 2 that the centripetal acceleration for uniform circular motion was given by $a_c = v^2/r$ (Eq. 2.5). If you want to know the magnitude of the centripetal force that supplies such an acceleration, you can use Newton's second law, $F = ma_c = mv^2/r$.

Another application of the second law follows.

Mass and Weight

This is a good place to make a firm distinction between mass and weight. These quantities were generally defined previously: *Mass* refers to the amount of matter an object contains (it is also a measure of inertia), and *weight* is related to the force of gravity (that is, the gravitational force acting on an object). However, the quantities are related, and Newton's second law clearly shows this relationship.

On the surface of Earth, where the acceleration due to gravity is relatively constant ($g = 9.80$ m/s^2), the weight force w on an object with a mass m is given by

$$w = mg \qquad 3.2$$

weight = mass × acceleration due to gravity

(special case of $F = ma$)

Notice that this equation is a special case of $F = ma$, where different symbols have been used to distinguish this special case.

Now you can see easily why mass is the fundamental quantity. Generally, a physical object has the same amount of matter, and so it has a constant mass. But the weight of an object may differ, depending on the

value of g. For example, on the surface of the Moon the acceleration due to gravity (g_M) is one-sixth what it is on the surface of Earth [$g_M = g/6 = (9.8$ m/s$^2)/6 = 1.6$ m/s^2]. Thus an object will have the same mass on the Moon as on Earth, but its weight will be different.

EXAMPLE 3.2

Computing Weight

What is the weight of a 1.0 kg mass on (a) Earth, and (b) the Moon?

SOLUTION

(a) Using Eq. 3.2 and Earth's $g = 9.8$ m/s^2, we get

$$w = mg = (1.0\text{ kg})(9.8\text{ m/s}^2) = 9.8\text{ N}$$

(b) On the Moon, where $g_M = g/6 = 1.6$ m/s^2, we have

$$w = mg_M = (1.0\text{ kg})(1.6\text{ m/s}^2) = 1.6\text{ N}$$

Notice that the mass is the same in both cases, but the weights are different (because of different g's).

CONFIDENCE EXERCISE 3.2

On the surface of Mars, the acceleration due to gravity is 0.39 times that on Earth. What would a kilogram weigh in newtons on Mars?

When we have an unknown mass or object, we "weigh" it on a scale. The scale may be calibrated in mass units (kilograms or grams) or in weight units (newtons or pounds; see ● Fig. 3.7).

FIGURE 3.7 Mass and Weight

A mass of 1.0 kg is suspended on a scale calibrated in newtons, which shows the weight to be 9.8 N. This is equivalent to a weight of 2.2 lb.

EXAMPLE 3.3

Finding the Force

What is the magnitude of the upward tension force in a string supporting a 2.0-kg mass, as illustrated in ● Fig. 3.8a?

SOLUTION

Since the mass isn't going anywhere, its acceleration is zero, which means that the net force acting on it is zero ($F = ma = 0$). The downward force (weight) must then be balanced by an upward force supplied by the string.

You can get a feel for this situation if you think about holding a 2.0-kg (4.4-lb) mass stationary. You have to supply an upward force to balance the downward weight force (Fig. 3.8b). Then the magnitude of the weight force is

$$w = mg$$

$$= (2.0 \text{ kg}) (9.8 \text{ m/s}^2)$$

$$= 20 \text{ N} \quad (19.6 \text{ N rounded off})$$

and the tension force in the string has the same magnitude in the opposite direction.

CONFIDENCE EXERCISE 3.3

Suppose the string in Example 3.3 had a tensile strength of 30 N (the maximum force it could sustain without breaking). What is the maximum mass that could be supported from the string?

Finally, an object in free-fall has an unbalanced force acting on it of $F = w = mg$ (downward).* So why do objects in free-fall all descend at the same rate, as

stated in the last chapter? Even Aristotle thought a heavy object would fall faster than a lighter one. Newton's laws explain.

Suppose two objects were in free-fall, one with twice the mass of the other, as illustrated in ● Fig. 3.9. According to Newton's second law, the more massive

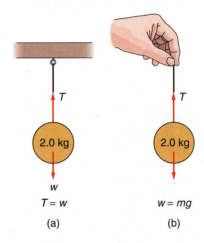

FIGURE 3.8 Balanced Forces

The tension force in the string can be experienced by holding the mass. (See Example 3.3.)

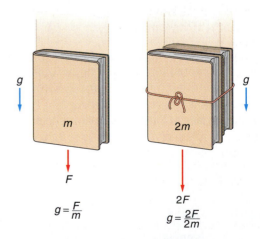

FIGURE 3.9 Acceleration Due to Gravity

The acceleration due to gravity is independent of the mass of a freely falling object. Thus the acceleration is the same for all such falling objects. An object with twice the mass of another has twice the gravitational force acting on it, but it also has twice the inertia and so falls at the same rate.

*An object falling solely under the influence of gravity (no air resistance) is said to be in *free-fall.*

object would have twice the weight, or gravitational force of attraction. However, by Newton's first law, the more massive object has twice the inertia, so it needs twice the force to fall at the same rate.

RELEVANCE QUESTION: *What is your weight in newtons and your mass in kilograms? (In Europe, you would weigh yourself in kilograms, or "kilos.")*

3.3 Newton's Third Law of Motion

LEARNING GOALS

▼ Explain Newton's third law of motion.

▼ Identify the third-law force pair in practical applications.

Newton formulated a third law of motion that is of great physical significance. We commonly think of single forces. However, Newton recognized that it is impossible to have a single force. He observed that in the application of a force, there is always a mutual reaction, and so forces always occur in pairs. As Newton illustrated: "If you press a stone with your finger, the finger is also pressed upon by the stone." Another example might be for a person riding in a car that is braked to a quick stop: "If you push against your seat belt, your seat belt pushes against you" (Newton's third law in action).

Newton's third law of motion is sometimes called the law of action and reaction. **Newton's third law** states:

For every action there is an equal and opposite reaction.

In this statement, the words *action* and *reaction* refer to forces. A more precise statement would be the following:

Whenever one object exerts a force on a second object, the second object exerts an equal (in magnitude) and opposite (in direction) force on the first object.

Expressed in equation form, Newton's third law may be written

action = opposite reaction

$$F_1 = -F_2 \qquad \textbf{3.3}$$

where F_1 = force exerted on object 1 by object 2
F_2 = force exerted on object 2 by object 1

The minus sign indicates that F_2 is in the opposite direction of F_1.

Jet propulsion is an example of Newton's third law. In the case of a rocket (● Fig. 3.10a), the exhaust gas is accelerated from the rocket, and the rocket accelerates in the opposite direction. A common misconception is that the exhaust gas pushes against the launch pad to accelerate the rocket. This is nonsense.

If this misconception were true, there would be no space travel, since there is nothing to push against in space. The correct explanation is one of action (gas going out the back) and reaction (rocket propelled forward). The gas (or gas particles) exerts a force on the rocket, and the rocket exerts a force on the gas. The equal and opposite actions of Newton's third law should be evident in Fig. 3.10b.

Let's take a look at the third law in terms of the second law ($F = ma$). Writing Eq. 3.3 in this form, we have

$$F_1 = -F_2$$

or

$$m_1 a_1 = -m_2 a_2$$

From this we can see for the equal and opposite force pair that if m_2 is much greater than m_1, then a_1 is much larger than a_2.

As an example, consider dropping a book on the floor. As the book falls, it has a force acting on it (Earth's gravity) that causes it to accelerate. What is the equal and opposite force? The equal and opposite force is the force of the book's gravitational pull on Earth. Technically, Earth accelerates upward to meet the book! However, since the planet's mass is so huge compared with the book's, Earth's acceleration is minuscule and cannot be measured.

An important distinction to keep in mind is: Newton's third law relates two equal and opposite forces that act on two separate objects. Newton's second law concerns how forces acting on a single object can cause an acceleration. If two forces acting on a single object are equal and opposite, then there will be no acceleration, but these are not the third law force pair.

Let's take a look at one more example that illustrates the application of Newton's laws. Imagine yourself as a passenger in a car traveling down a straight road and entering a circular curve at a constant speed (● Fig. 3.11). As you know, there must be a centripetal force to provide the centripetal acceleration necessary to negotiate the curve (Section 2.4). This is supplied by friction on the tires, and the magnitude of

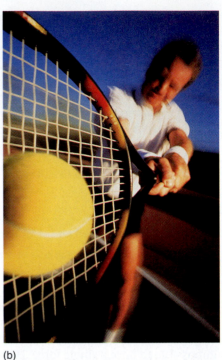

FIGURE 3.10 **Newton's Third Law in Action**

(a) The rockets and exhaust gases exert equal and opposite forces on each other and so are accelerated in opposite directions. (b) The equal and opposite forces should be obvious here.

(a)   (b)

this frictional force is given by $f = ma_c = mv^2/r$. Should the frictional force not be great enough, say, if you hit an icy spot (reduced friction), the car would slide outward because the centripetal force would not be great enough to keep the car in a circular path.

FIGURE 3.11 **Newton's First Law in Action**

When a car initially traveling along a straight road goes around a sharp curve or corner, the people in it tend to continue in the original direction, according to Newton's first law. This gives the sensation of being thrown outward as the car rounds the curve. The door, seat belt, and/or friction on the seat of the pants supplies the necessary centripetal force and acceleration needed to go around the curve with the car.

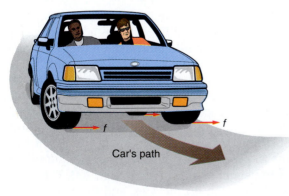

Car's path

You have no doubt experienced a lack of centripetal force when going around a curve in a car and having a feeling of being "thrown" outward. From what we've learned about force and motion, you might think that there is an outward force acting on you—a centrifugal (center-fleeing) force. However, such a force doesn't exist. Riding in the car before entering the curve, you tend to go in a straight line, in accord with Newton's first law. As the car makes the turn, you continue to maintain your straight-line motion, until the car turns in front of you. You push against the door (action), and the door pushes against you (reaction, third law).

You may feel that you are being thrown outward toward the door, but actually, the door is coming toward you because the car is turning, and when the door gets to you, it exerts a force on you that supplies the necessary centripetal force you need to go around the curve with the car. The friction on the seat of your pants is not enough to do this, but the force could be supplied by a seat belt instead of the door if you are buckled up.

RELEVANCE QUESTION: *What are the two forces that normally act on you when you are standing on Earth? Are these a third-law force pair? Explain.*

3.4 **Newton's Law of Gravitation**

LEARNING GOALS

▼ State and apply Newton's law of gravitation.

▼ Explain on what the acceleration due to gravity depends.

Now that we have a basic understanding of forces, let's take a look at a common fundamental force of nature—gravity. It is fundamental because we do not know what causes it and can only describe it. Gravity is associated with the fundamental quantity of mass and describes the mutual attraction of mass particles. Again, Newton's insight may be seen.

The law describing the gravitational force of attraction between two particles was formulated by Newton from his studies on planetary motion. Known as the **law of universal gravitation,** it may be stated as follows:

Every particle in the universe attracts every other particle with a force that is directly proportional to the product of their masses and inversely proportional to the square of the distance between them.

Suppose the masses of two particles are designated as m_1 and m_2, and the distance between them is r (● Fig. 3.12a). We may write this statement of the law in symbol form as

$$F \propto \frac{m_1 m_2}{r^2}$$

where $\propto$ is a proportionality sign. Note in the figure that F_1 and F_2 are equal and opposite—a mutual interaction *and* a third-law force pair.

An object is made up of a lot of point particles, so the gravitational force on an object is the vector sum of all the particle forces. This computation can be quite complicated, but one simple and convenient case is a homogeneous* sphere. In this case the net force acts as though all the mass were concentrated at the center of the sphere, and the separation distance for the mutual interaction with another object is measured from there. If we assume that Earth, other planets, and the Sun are spheres with uniform mass distri-

Homogeneous means that the mass particles are distributed uniformly throughout the object.

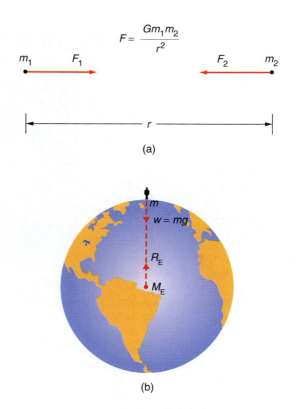

FIGURE 3.12   Newton's Law of Gravitation
(a) Two particles attract each other gravitationally, and the magnitude of the force is given by Newton's law of gravitation. The forces are equal and opposite—Newton's third law. (b) For a homogeneous, or uniform, sphere, the force acts as if all the mass of the sphere were concentrated as a particle at its center. Using such an approximation, your weight, or the gravitational attraction of Earth on your mass, is computed as though all the mass of Earth were concentrated at its center, and the distance between the masses is the radius of Earth.

butions, we can apply the law of gravitation to such bodies (Fig. 3.12b).

Newton's law of universal gravitation may be written in equation form by putting in an appropriate constant (of proportionality):

$$F = \frac{Gm_1 m_2}{r^2} \qquad \textbf{3.4}$$

(Newton's law of gravitation)

where G is called the *universal gravitational constant* and has a value of $G = 6.67 \times 10^{-11}$ N-m²/kg².

HIGHLIGHT: The "Fourth Law of Motion": The Automobile Air Bag

A major automobile safety feature is the air bag. As learned earlier, seat belts restrain you so you don't follow along with Newton's first law when the car comes to a sudden stop. But where does the air bag come in, and what is its principle?

When a car has a head-on collision with another vehicle or hits an immovable object such as a tree, it stops almost instantaneously. Even with seat belts, the impact of a head-on collision could be such that seat belts might not restrain you completely, and injuries could occur.

Enter the air bag. This balloon-like bag inflates automatically on hard impact and cushions the driver (Fig. 1). Passenger-side air bags are becoming more common, and back-seat air bags are available.

The air bag tends to "cushion" or increase the contact time in stopping a person, thereby reducing the impact force (as compared to hitting the dashboard or steering column). Also, the impact force is spread over a large general area and not applied to certain parts of the body as in the case of seat belts.

Being inquisitive, you might wonder what causes an air bag to inflate and what inflates it. Keep in mind that this must occur in a fraction of a second to do any good. (How much time would there be between the initial collision contact and a driver hitting the steering wheel column?) The air bag's inflation is initiated by an electronic sensing unit. This unit contains sensors that detect rapid decelerations, such as those in a

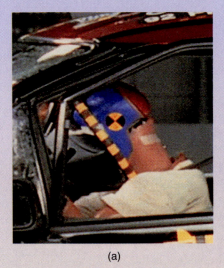
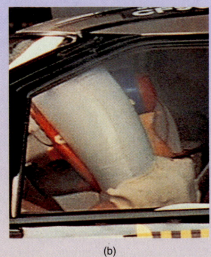

(a)   (b)

FIGURE 1   Life-Saving Cushioning
(a) The effects on a crash dummy in a car not equipped with air bags. (b) An inflated air bag provides a much safer and softer landing.

high-impact collision. The sensors have threshold settings so that normal hard braking does not activate them.

Sensing an impact, a control unit sends an electric current to an igniter in the air bag system that sets off a chemical explosion. The gases (mostly nitrogen) rapidly inflate the thin nylon bag. The total process of sensing to complete inflation takes about 25 thousandths of a second (0.025 s). Pretty fast, and a good thing too!

The sensing unit is equipped with its own electrical power source because, in a front-end collision, a car's battery and alternator are among the first things to go. The currently installed automobile air bags offer protection for only front-end collisions, in which the car's occupants are thrown forward (excuse me—continue to travel forward—Newton's first law).

Recent Concerns

A recent concern about air bags is the injuries and deaths resulting from their deployment. An air bag is not a soft, fluffy pillow. When activated, it comes out of the dashboard at speeds of up to 320 km/h (200 mi/h) and could hit a person close by with enough force to cause severe injury and even death. Adults are advised to sit at least 25 cm (10 in.) from the air bag cover. This allows a margin of safety from the 5- to 8-cm (2- to 3-in.) "risk zone." Seats

Newton used his law of gravitation and his second law of motion to show that gravity supplied the necessary centripetal acceleration and force on the Moon for it to move in its nearly circular orbit about Earth. However, he did not know or experimentally measure the value of G.

This value was measured some 70 years after Newton's death by an English scientist, Henry Cavendish, who used a very delicate balance to measure the force between two masses. The value of G is very small, as can be seen. Table 3.1 on p. 56 summarizes what G is and is not. (The units of G are such that when it is used

should be adjusted to allow for the proper safety distance.

Probably a more serious concern is associated with children. Children may get close to the dashboard if they are not buckled in or not buckled in securely so that they can see (Fig. 2a). Another bad situation is using a rear-facing child seat in the front passenger seat. An inflating air bag could have serious effects (Fig. 2b). Thus it is recommended that (1) *children 12 years old and younger ride in the back seat (buckled up, of course), and* (2) *never put a rear-facing child restraint in front of an air bag.*

Sometimes it may be impossible to follow these safety rules, and to avoid the possibility of air bag injury, it is now possible to have a cutoff switch installed in an automobile to turn off one or both the front air bags. Application for this must be made to and approved by the appropriate government safety agency. Air bag deactivation may be authorized for the following reasons:

- A rear-facing child restraint must be placed in the front seat because the car either has no back seat or it has one that is too small.

- A child 12 years old or younger must ride in the front seat because of a medical condition that requires frequent monitoring.

- An individual who drives (or rides in the front seat) has a medical condition that would make it safer to have the air bag(s) turned off.

- A driver must sit within a few inches of the air bag (typically because of extremely short stature, 4 feet 6 inches or less).

Specific problems may exist, but air bags save many lives. All new passenger cars must now have dual air bags, and manufacturers are beginning to install air bags that inflate with less force, so as to reduce the possibility of injuries. Even if your car is equipped with air bags, however, *always* remember to buckle up. (Maybe we should make that Newton's "fourth law of motion.")

FIGURE 2   Some No-No's

(a) Children not buckled in or not buckled in securely so that they can see over the dashboard. **No!** (b) Using a rear-facing child seat in the passenger front seat. **No!**

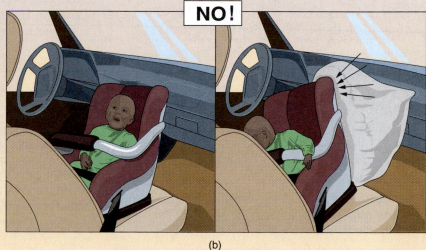

(a)   (b)

in Eq. 3.3, the gravitational force is in newtons. Try it and see.)

When we drop something, the force of gravity is made evident by the acceleration of the falling object. Yet if there is a gravitational force of attraction between every two objects, why don't you feel this at-traction between yourself and this textbook? (No pun intended.) Indeed, there is a force of attraction between you and this textbook, but it is so small that you don't notice it. Gravitational forces can be very small, as the following example illustrates.

TABLE 3.1 WHAT IS *G*?

$G = 6.67 \times 10^{-11}$ N-m^2/kg^2

G is a very small quantity.

G is a universal constant. It is thought to be the same throughout the universe, as is the law of gravitation.

G is not *g* ("little" *g* is the acceleration due to gravity).

G is not a force.

EXAMPLE 3.4

Applying Newton's Law of Gravitation

Two objects with masses of 1.0 kg and 2.0 kg are 1.0 m apart (● Fig. 3.13). What is the magnitude of the gravitational force between the masses?

SOLUTION

The magnitude of the force is given by Eq. 3.4:

$$F = \frac{Gm_1 m_2}{r^2}$$

$$= \frac{(6.67 \times 10^{-11} \text{ N-m}^2/\text{kg}^2)(1.0 \text{ kg})(2.0 \text{ kg})}{(1.0 \text{ m})^2}$$

$$= 1.3 \times 10^{-10} \text{ N}$$

This number is very, very small. A grain of sand would weigh more. For an appreciable gravitational force to exist between two masses, at least one of the masses must be relatively large (see Fig. 3.13).

CONFIDENCE EXERCISE 3.4

If the distance between the two masses in Fig. 3.13 were doubled, by what factor would the mutual gravitational force change? (*Hint:* Use a ratio.)

Newton's law of gravitation gives us insight as to why the **acceleration due to gravity** at Earth's surface is constant and why this acceleration does not depend on the mass of an object. (Recall that all objects in free-fall have the same acceleration, or fall at the same rate.) For a mass on Earth's surface, the distance between the masses is essentially the radius of Earth.

The force of gravity on an object of mass *m* is then

$$F = \frac{GmM_E}{R_E^2}$$

where M_E and R_E are the mass and radius of Earth, respectively. This force is just the object's weight, so we can write

$$w = mg = \frac{GmM_E}{R_E^2}$$

Canceling the *m*'s, we have an expression for *g*:

$$g = \frac{GM_E}{R_E^2} \qquad \text{3.5}$$

(acceleration due to gravity at Earth's surface)

Notice that since the mass of the object *m* cancels out of the equation, it does not appear in the expression for *g*. Thus *g* is independent of the mass of the object or is the same for all objects. Hence all objects in free-fall have the same acceleration (resulting from the force of gravity).

Also, from Eq. 3.5 we can see why *g* is relatively constant near Earth's surface. The term *G* is a constant, and in general, so are the mass and radius of Earth (M_E and R_E). Hence, to a good approximation, *g* is a constant near Earth's surface or over the normal

FIGURE 3.13 The Amount of Mass Makes a Difference

A 1.0-kg mass and a 2.0-kg mass separated by a distance of 1.0 m have a negligible mutual gravitational attraction (about 10^{-10} N). However, because Earth's mass is quite large, the masses are attracted to Earth with forces of 9.8 N and 19.6 N, respectively. These forces are the weights of the masses.

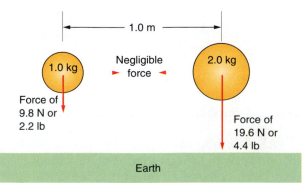

A SPOTLIGHT ON: Newton's Laws of Motion

Newton's First Law

(Law of Inertia)

An object will remain at rest or in uniform motion in a straight line unless acted upon by an external unbalanced force.

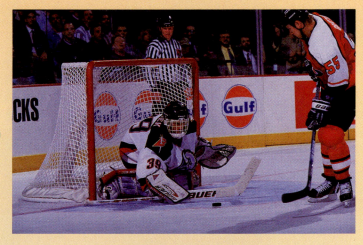

Newton's Second Law

$F = ma$ (cause and effect)

An unbalanced force acting on an object produces an acceleration, or an acceleration is evidence of an unbalanced force.

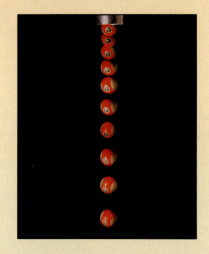

Newton's Third Law

For every action, there is an equal and opposite reaction.

(Forces always occur in pairs and act on *different* objects.)

heights of falling objects. The force of gravity on an object (Eq. 3.4) and g do decrease with increasing height or altitude (increasing r). However, at an altitude of 1.6 km (1.0 mi), the acceleration due to gravity, and hence an object's weight ($w = mg$), is only 0.05% less than g at Earth's surface. At an altitude of 160 km (100 mi), g is still 95% of its surface value.

The gravitational force between two masses is said to be an infinite-range force; that is, the only way to get it to approach zero is to separate the masses by a distance approaching infinity ($r \rightarrow \infty$). No matter what the altitude is or how far an object is from Earth, there is still a force resulting from gravity. In general, we may write Eq. 3.5 as

$$g = \frac{GM}{r^2} \qquad \textbf{3.6}$$

(acceleration due to gravity)

where M is the mass of any spherical uniform object, and r is the distance from its center (such as the center of Earth).

This equation can be used to find the acceleration due to gravity on the Moon and other planets (assuming uniform mass distribution). For example, if the mass and radius of the Moon are used, then we get the acceleration due to gravity on the surface of the Moon, g_M. As noted earlier, this acceleration turns out to be one-sixth that on Earth ($g_M = g/6$), basically because the mass of the Moon is much less than that of Earth. Hence things weigh only one-sixth as much on the Moon as on Earth. If you want to lose weight fast, go to the Moon (● Fig. 3.14).

Another interesting thing to note about Eq. 3.5 is that it provides a way to compute the mass of Earth. Knowing the values of g, G, and R_E, we can find M_E easily. All the quantities were known once G was determined in 1797. Scientists were then able to calculate Earth's mass, which turns out to be about 6.0×10^{24} kg. (Try the calculation yourself. You can do it now, using Eq. 3.5 and R_E from inside the back cover.)

While on the subject of gravity, let's take a look at orbiting satellites, which are quite common today. When a satellite is put into orbit, it must be given a sufficient tangential velocity, which depends on the altitude of its orbit. To help understand this idea, imagine throwing an object from a *very* tall building (● Fig. 3.15). If not thrown hard enough, the object would fall to Earth. However, with the proper velocity, the object would go into a circular orbit. Essentially, it

FIGURE 3.14   Strong Man

An astronaut on the Moon carries equipment that on Earth may weigh 300 lb (a mass of about 136 kg). On the Moon, where the acceleration due to gravity is one-sixth that on Earth, the mass of the equipment is the same, but it has a weight of only about 50 lb. The astronaut also experiences a weight reduction. For example, a 180-lb (82-kg) astronaut on Earth would weigh 30 lb on the Moon (but still have a mass of 82 kg).

would "fall" around Earth. (Newton himself imagined a similar situation of firing cannonballs from the top of a high mountain.)

Gravity supplies the centripetal force or acceleration that keeps satellites in orbit ($F_c = ma_c = mv^2/r$). This law applies also to Earth's only natural satellite, the Moon. With astronauts in space, we now hear the terms *zero g* and *weightlessness* (● Fig. 3.16).

These terms are not really true descriptions. Gravity certainly acts on an astronaut in an orbiting spacecraft. Otherwise, the astronaut (and spacecraft) would not remain in orbit. Because gravity is acting, the astronaut, by definition, has weight.

The reason an astronaut floats in the spacecraft and feels "weightless" is that the spacecraft and the

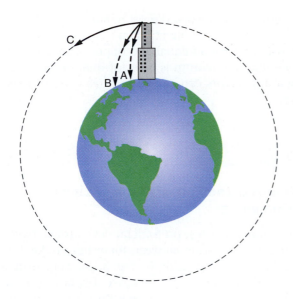

FIGURE 3.15 Earth Orbit

An imaginary situation of projecting objects from a very tall building (atmosphere negligible). Ordinarily, horizontally projected objects A and B would fall back to Earth. However, an object projected with a greater and just the right horizontal speed would "fall around" Earth, as illustrated by path C. The Moon "falls around" Earth in a similar manner.

FIGURE 3.16 Floating Around

Astronauts in the space shuttle in orbit around Earth are said to "float" around because of "zero *g*" or "weightlessness." Actually, the gravitational attraction of objects keeps them in orbit with the spacecraft. Here, astronaut pilot Susan L. Still enters data into an onboard computer.

astronaut are both "falling" toward Earth. Imagine yourself in a freely falling elevator standing on a scale. The scale would read zero because it is falling as fast as you are. However, you are not weightless and *g* is not zero, as you would find out on reaching the bottom of the elevator shaft.

RELEVANCE QUESTION: *When might you have a "weightless" feeling on a roller coaster ride?*

3.5 Momentum

LEARNING GOALS

▼ Define linear momentum.

▼ Define torque and angular momentum, and state their relationship.

▼ Explain the conditions for the conservations of linear and angular momenta, and give examples.

Another important quantity in the description of motion is *momentum*. We use this term commonly; for example, the team has a lot of momentum—or lost its momentum. But let's see what momentum means scientifically. There are two types: linear and angular.

Linear Momentum

Stopping a speeding bullet is difficult because it has a high velocity. Stopping a slowly moving oil tanker is difficult because it has a large mass. In general, the product of mass and velocity is called **linear momentum.**

> linear momentum = mass × velocity
>
> or in symbols,
>
> $$p = mv \qquad \textbf{3.7}$$
>
> where v is the instantaneous velocity.

Both mass and velocity are involved in momentum. A small car and a large truck both traveling at 50 km/h may have the same velocity, but the truck has more momentum because it has a much larger mass. If we have a system of masses, we find the linear momentum of the system by adding the linear momentum vectors of all the individual masses.

The linear momentum of a system is important because if there are no external unbalanced forces, it is conserved; it does not change with time. In other words, *with no unbalanced forces, no acceleration occurs, so there is no change in velocity or change in momentum*. This property makes it extremely important in analyzing the motion of various systems. The **law of conservation of linear momentum** can be stated as follows:

The total linear momentum of an isolated system remains the same if there is no external, unbalanced force acting on the system.

> Total final momentum = total initial momentum
>
> or $\qquad P_f = P_i$ $\qquad\qquad$ **3.8**
>
> where $\qquad P = p_1 + p_2 + p_3 + \cdots$
> **(sum of individual momentum vectors)**

Suppose you are standing in a boat near the shore; you and the boat are the system (● Fig. 3.17). Let the boat be stationary, so the total linear momentum of the system is zero. (No motion, no linear momentum.) If you jumped toward the shore, you would notice something immediately—namely, the boat would move in the opposite direction. The boat moves because the force you exerted in jumping is an *internal* force, so the total linear momentum of the

system is conserved and must remain zero. You have momentum in one direction, and to cancel this, the boat must have an equal and opposite momentum. Note that momentum is a vector quantity, and momentum vectors can add to zero (as can force vectors).

EXAMPLE 3.5

Applying the Conservation of Linear Momentum

Two masses at rest on a frictionless surface have a compressed spring between them but are held together by a light string (● Fig. 3.18). The string is burned, and the masses fly apart. If v_1 has a velocity of 1.8 m/s (to the right), what is the velocity of m_2?

SOLUTION

Let's solve this example stepwise for clarity.

Step 1

Given: $\quad m_1 = 1.0$ kg $\qquad v_1 = 1.8$ m/s

$\qquad\qquad m_2 = 2.0$ kg (masses from figure)

Step 2

Wanted: $\quad v_2$ (velocity of m_2)

Step 3

Reasoning: The total momentum of the system is initially zero ($P_i = 0$) before the string is burned. (The reason for releasing the masses by this manner is so that no external forces are applied, as there might be if the masses were held together with hands.)

Note that the spring is part of the system and so applies an *internal* force to each of the masses. Hence the linear momentum is conserved. After leaving the spring, the moving masses have nonzero momenta, and the total momentum of the system is $P_f = p_1 + p_2$. (Don't forget, momentum is a vector.)

So, with the total linear momentum being conserved (no unbalanced *external* forces acting), using Eq. 3.8, we may write

$$P_f = P_i$$

or $\qquad\qquad\qquad p_1 + p_2 = 0$

and $\qquad\qquad\qquad p_1 = -p_2$

FIGURE 3.17 Conservation of Linear Momentum

Initially at rest, the total momentum of the system (man and boat) is zero. When the man jumps toward the shore (an internal force), the boat moves in the opposite direction to conserve linear momentum.

Man jumps forward

Boat moves backward

which tells us the momenta are equal and opposite. Then, in terms of mv,

$$m_1v_1 = -m_2v_2$$

And solving for v_2,

$$v_2 = -\frac{m_1v_1}{m_2} = -\frac{(1.0 \text{ kg})(1.8 \text{ m/s})}{2.0 \text{ kg}} = -0.90 \text{ m/s}$$

or to the left in Fig. 3.18, since v_1 was taken to be positive to the right.

CONFIDENCE EXERCISE 3.5

Suppose you were not given the values of the masses but only that $m_1 = m$ and $m_2 = 3m$. What could you say about the velocities in this case?

We looked at jet propulsion or rockets in terms of Newton's third law. This phenomenon also can be explained in terms of linear momentum. The burning of the rocket fuel gives energy by which *internal* work is done, and hence internal forces act. As a result, the exhaust gas goes out the back of the rocket with momentum in that direction, and the rocket goes in the opposite direction to conserve linear momentum.

FIGURE 3.18   An Internal Force and Conservation of Linear Momentum

By burning the string so as not to apply an external force, the compressed spring supplies an internal force to the system. See Example 3.5.

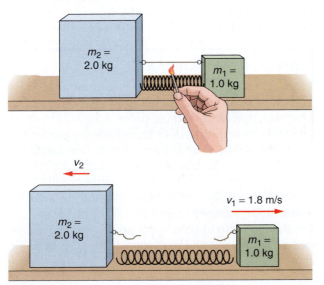

Here the exhaust gas molecules have small masses and large velocities, whereas the rocket has a large mass and a relatively small velocity.

Also, there are many, many gas molecules. You can and probably have demonstrated this rocket effect by blowing up a balloon and letting it go. Without a guidance system, the balloon zigzags wildly, but the air comes out the back and the balloon is "jet" propelled.

Angular Momentum

Another important quantity that Newton found to be conserved is angular momentum. Angular momentum arises when objects go in paths around a fixed point. Consider a planet going around the Sun in an elliptical orbit, as illustrated in ● Fig. 3.19. The **angular momentum** (L) is given by

$$L = mvr \qquad\qquad \textbf{3.9}$$

where m = mass
$\quad v$ = velocity
$\quad r$ = distance of object from center of motion (in this case, the Sun)

The linear momentum of a system can be changed by the introduction of an external unbalanced force. Similarly, the angular momentum of a system can be changed by an external unbalanced torque. A **torque** is a twisting effect caused by one or more forces. For example, in ● Fig. 3.20, a torque on a steering wheel is caused by two equal and opposite forces acting on different parts of the steering wheel. Notice these

FIGURE 3.19   Angular Momentum

The angular momentum of a planet going around the Sun is given at different points in the orbit by mv_1r_1 and mv_2r_2. Angular momentum is conserved in this case, and $mv_1r_1 = mv_2r_2$. As the planet comes closer to the Sun, the radial distance r decreases, so the speed v must increase. Similarly, the speed decreases when r increases. Thus a planet moves fastest when it is closest to the Sun and slowest when farthest from the Sun.

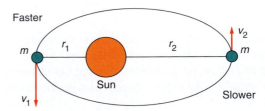

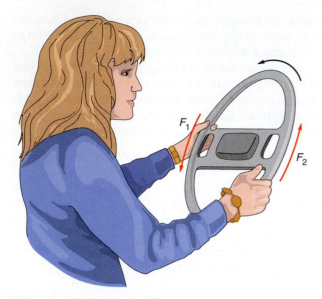

FIGURE 3.20   Torque

A torque is a twisting action that produces rotational motion or a change in rotational motion. This is analogous to a force producing linear motion or a change in linear motion. The forces F_1 and F_2 supply the torque.

forces at a distance (r) from the center of motion or axis of rotation. (If $r = 0$, then the angular momentum is zero, Eq. 3.9.) The forces in Fig. 3.20 cause a twist, or torque, and cause the steering wheel to turn. In general, a torque tends to produce a rotational motion.

There is also a conservation law for angular momentum. The **law of conservation of angular momentum** states:

The angular momentum of an object remains constant if there is no external, unbalanced torque acting on it.

That is, the magnitudes of the angular momenta are equal at times 1 and 2:

$$L_1 = L_2$$

or
$$m_1 v_1 r_1 = m_2 v_2 r_2 \qquad \textbf{3.10}$$

where the 1 and 2 subscripts denote the angular momentum of the object at general times 1 and 2.

In our example of a planet, the angular momentum mvr remains the same because the gravitational at-

traction is internal to the system. As the planet gets closer to the Sun, r decreases, so the speed v increases. For this reason, a planet moves faster when it is closer to the Sun (see Fig. 3.19). (Although approximated as circles, planet orbits are slightly elliptical, so planets do move at different speeds in different parts of their orbits.) The same is true for comets in elliptical orbits about the Sun.

EXAMPLE 3.6

Applying the Conservation of Angular Momentum

A comet at its farthest point from the Sun is 900 million miles away and traveling at 6000 mi/h. What is its speed at its closest point to the Sun, which is 30 million miles away?

SOLUTION

We know v_2, r_2, and r_1, so we can calculate v_1 as follows (Eq. 3.10):

$$mv_1 r_1 = mv_2 r_2$$

or
$$v_1 r_1 = v_2 r_2$$

or
$$v_1 = \frac{v_2 r_2}{r_1}$$

$$= \frac{(6.0 \times 10^3 \text{ mi/h})(900 \times 10^6 \text{ mi})}{30 \times 10^6 \text{ mi}}$$

$$= 1.8 \times 10^5 \text{ mi/h or } 180{,}000 \text{ mi/h}$$

Thus we see that the comet moves much faster when it is close to the Sun than it does when it is far away from the Sun.

CONFIDENCE EXERCISE 3.6

Earth's orbit about the Sun is not quite circular. At its closest approach, our planet is 1.47×10^8 km from the Sun, and at its farthest point, 1.52×10^8 km. At which of these points does Earth have the greater orbital speed and by what factor? (*Hint:* Use a ratio.)

Another example of the conservation of angular momentum is demonstrated in ● Fig. 3.21. Ice skaters use the principle to spin faster on the ice. The skater extends both arms and perhaps one leg and obtains a slow rotation; then the arms and leg are drawn inward, and rapid angular velocity is gained because of the decrease in the radial distance of the mass.

Angular momentum also plays an important role for helicopters. The law of conservation of angular momentum would cause the body of a helicopter with only a single rotor to rotate. For conservation of angular momentum, the body of the helicopter would have to rotate in the direction opposite that of the rotor. To prevent rotation, large helicopters have two oppositely rotating rotors (● Fig. 3.22a). Smaller heli-copters, however, have small antitorque rotors on the tail (Fig. 3.22b). These are like small airplane pro-pellers that provide a torque to counteract the rota-tion of the helicopter body.

RELEVANCE QUESTION: *When you open a door, is there a torque involved? Why is it harder to push open a door if you mistakenly push on the side closer to the hinges?*

FIGURE 3.21 Conservation of Angular Momentum

(a) The skater starts his spin with the arms out-stretched. (b) When the arms are drawn inward and the average radial distance of the mass de-creases, the angular velocity (spin) increases to conserve angular momentum.

(a)   (b)

FIGURE 3.22 Conservation of Angular Momentum in Action

(a) Large helicopters have two overhead rotors that rotate in opposite directions so as to balance the angular momentum. (b) Small helicopters with one overhead rotor have an "antitorque" tail rotor to bal-ance the angular momentum and prevent the rota-tion of the helicopter body.

(a)   (b)

Important Terms

Newton's first law of
  motion (3.1)
force
inertia
mass
Newton's second law of
  motion (3.2)

newton
weight
centripetal force
Newton's third law of
  motion (3.3)
Newton's law of universal
  gravitation (3.4)

G
acceleration due to
  gravity (g)
linear momentum (3.5)
law of conservation of
  linear momentum
angular momentum

torque
law of conservation of
  angular momentum

Important Equations

Newton's Second Law: $F = ma$
  for weight: $w = mg$

Newton's Third Law: $F_1 = -F_2$

Newton's Law of Gravitation: $F = \dfrac{Gm_1m_2}{r^2}$

(gravitational constant: $G = 6.67 \times 10^{-11}$ N-m^2/kg^2)

Acceleration Due to Gravity: $g = \dfrac{GM}{r^2}$

Linear Momentum: $p = mv$

Conservation of Linear Momentum:
$$P_f = P_i$$
where
$$P = p_1 + p_2 + p_3 + \cdots$$

Angular Momentum: $L = mvr$

Conservation of Angular Momentum: $L_1 = L_2$
$$m_1v_1r_1 = m_2v_2r_2$$

Review Questions

3.1 Newton's First Law of Motion

1. The natural state of motion is
   (a) at a constant velocity.
   (b) in a circular orbit.
   (c) accelerated.
   (d) in free-fall.

2. An unbalanced force is necessary for an object to be
   (a) at rest.
   (b) in motion with a constant velocity.
   (c) accelerated.
   (d) all of the preceding.

FIGURE 3.23 Make It Tight
See Review Question 5.

3. What is meant when we say that a person has a lot of inertia?

4. An old party trick is to pull a tablecloth out from under dishes and glasses on a table. Explain how this is done without disturbing the dishes and glasses.

5. To tighten the loose head on a hammer, the base of the handle is sometimes struck on a hard surface (● Fig. 3.23). Explain the physics behind this.

6. Explain the principle of automobile seat belts in terms of Newton's first law.

3.2 Newton's Second Law of Motion

7. A change in velocity
   (a) results from inertia.
   (b) requires an unbalanced force.
   (c) results from a zero net force.
   (d) is the natural state of motion.

8. The acceleration of an object is
   (a) inversely proportional to its mass.
   (b) directly proportional to the applied force.
   (c) resisted by inertia.
   (d) all of the preceding.

9. Describe the relationship between (a) force and acceleration, and (b) mass and acceleration.

10. (a) Can an object be at rest if forces are being applied to it? Explain.
    (b) If no forces are acting on an object, can the object be in motion? Explain.

11. What is the unbalanced force acting on a car moving with a constant velocity of 25 m/s (56 mi/h)?

12. Is Newton's first law consistent with his second law? Explain.

13. A 10-lb rock and a 1-lb rock are dropped simultaneously from the same height.
    (a) Some say that because the 10-lb rock has 10 times as much force acting on it, it should reach the ground first. Do you agree?
    (b) Describe the situation if the rocks were dropped by an astronaut on the Moon.

3.3 Newton's Third Law of Motion

14. The force pair of Newton's third law
    (a) never produce an acceleration.
    (b) act on different objects.
    (c) cancel each other.
    (d) only exist for internal forces.

15. Two books, one on top of the other, lie on a table. How many forces act on the bottom book?

    (a) 1   (b) 2   (c) 3   (d) 4

16. When a rocket blasts off, is it the fiery exhaust gases "pushing against" the launch pad that cause it to lift off? Explain.

17. If there is an equal and opposite reaction for every force, how can an object be accelerated when the vector sum of the forces is zero?

18. If an object is accelerating, does Newton's third law apply? Explain.

19. Explain the kick of a rifle or shotgun in terms of Newton's third law. Do the masses of the gun and the bullet or shot make a difference?

3.4 Newton's Law of Gravitation

20. The acceleration due to gravity
    (a) is a universal constant.
    (b) is a fundamental property.
    (c) decreases with increasing altitude.
    (d) is different for different objects in free-fall.

21. The constant G
    (a) is a very small quantity.
    (b) is a force.
    (c) is the same as g.
    (d) decreases with altitude.

22. In the equation for Newton's law of gravitation, what does the r stand for?

23. The gravitational force is said to have an infinite range. What does this mean?

24. How can the acceleration due to gravity be taken to be constant near Earth's surface, when g varies with altitude?

25. (a) Are astronauts "weightless" in a spacecraft orbiting Earth? Explain.
    (b) Is "zero g" possible? Explain.

3.5 Momentum

26. A change in linear momentum requires
    (a) a change in velocity.
    (b) an unbalanced force.
    (c) an acceleration.
    (d) all of the preceding.

27. Angular momentum is conserved in the absence of
    (a) inertia.
    (b) gravity.
    (c) a net torque.
    (d) linear momentum.

28. What is meant when we say that an athletic team has a lot of momentum or loses its momentum?

29. Explain how the conservation of linear momentum follows directly from Newton's first law of motion.

30. In Example 3.5, there are external forces acting on the block.
    (a) What are these?
    (b) Why is the total linear momentum still conserved?

31. A key on a string is whirled around your forefinger. Describe what happens as the string wraps around your finger and explain why.

32. After diving from a platform, divers tuck in and spin before straightening out to cleave the water. How does tucking accomplish spinning?

33. Would you take a ride on a helicopter with only one rotor? If you did, what would happen after liftoff?

Applying Your Knowledge

1. Why is it easier to tear a paper towel off of a full roll than a used (smaller) roll?

2. Astronauts walking on the Moon are seen "bounding" rather than walking normally. Why is this?

3. Suppose a hole could be drilled through the center of Earth to the other side. If an object were dropped down the hole, what would happen?

4. In a washing machine, water is extracted from clothes by rapid spinning, as illustrated in ● Fig. 3.24. Explain the physics behind this process.

5. Someone suggested that a new, super space station should include a basketball court so the astronauts could exercise. Describe how a basketball game might look if it were played in a space station orbiting Earth.

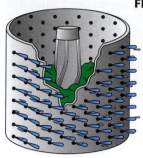

FIGURE 3.24 Get the Water Out

See Applying Your Knowledge Question 4.

Exercises

3.2 Newton's Second Law of Motion

1. Determine the force necessary to give an object with a mass of 4.0 kg an acceleration of 6.0 m/s^2.

   Answer: 24 N

2. A force of 2.1 N is exerted on a 7.0-g rifle bullet. What is the bullet's acceleration?

3. A 1000-kg automobile is pulled by a horizontal tow line with a net force of 850 N. What is the acceleration of the auto?

   Answer: 0.85 m/s^2

4. A constant net force of 1500 N gives a rocket an acceleration of 2.5 m/s^2. What is the mass of the rocket?

5. What is the weight in newtons of a 6.0-kg package of nails?

   Answer: 59 N

6. What is the force in newtons acting on a 6.0-kg package of nails that falls off a roof and is on its way to the ground?

7. (a) What is the weight in newtons of a 140-lb person?
   (b) What is your weight in newtons? Answer: (a) 623 N

8. What is the weight in newtons of a 500-g package of breakfast cereal?

9. A vertical fishing line supports a 5.0-kg fish when it is held up by the proud angler for a picture. What is the tension in the line?

   Answer: 49 N

10. A 4.0-kg block is suspended by a string. Using another piece of string, a 2.0-kg block is suspended from the bottom of the 4.0-kg block.
    (a) What is the tension in the upper string?
    (b) What is the tension in the lower string?

11. A 2.0-kg block sits fixed on top of a 3.0-kg block on a level, frictionless surface.
    (a) If a horizontal force of 10.0 N is applied to the bottom block, what is the acceleration of the system?
    (b) If there were a constant friction force of 4.0 N between the bottom block and the surface, what would be the acceleration in this case?

    Answer: (a) 2.0 m/s^2 (b) 1.2 m/s^2 (both in the direction of applied force)

12. A constant force of 12 N is applied to a 4.0-kg block sitting on a level, frictionless surface. A short time later, a 2.0-kg block is dropped onto the moving 4.0-kg block. By what percent does the acceleration change?

3.3 Newton's Law of Gravitation

13. Two 1.5-kg physical science textbooks on a bookshelf are 0.35 m apart. What is the magnitude of the gravitational attraction between the books?

    Answer: (a) 1.2 × 10^{-9} N

14. (a) What is the force of gravity between two 1000-kg cars separated by a distance of 50 m on an interstate highway? (b) How does this force compare with the weight of a car?

15. How would the force of gravity between two masses be affected if the separation distance between them were (a) tripled and (b) decreased by one-half?

    Answer: (a) $F_2 = F_1/9$ (b) $F_2 = 4F_1$

16. The separation distance between two 1.0-kg masses is (a) decreased by one-third and (b) increased by a factor of 5. How is the mutual gravitational force affected in each case?

17. Compute the acceleration due to gravity at Earth's surface. (Use $M_E = 5.96 \times 10^{24}$ kg and $R_E = 6.37 \times 10^6$ m.)

    Answer: 9.80 m/s^2

18. Show that the acceleration due to gravity on the surface of the Moon is about one-sixth that on Earth's surface. (*Hint:* Use $M_M = 7.35 \times 10^{22}$ kg, $R_M = 1.74 \times 10^6$ m, and the Earth data in Exercise 17 in a ratio.)

19. What is the acceleration due to gravity at an altitude of 1000 km (620 mi) above Earth's surface ? (*Hint:* Use $R_E = 6.37 \times 10^6$ m and form a ratio.)

Answer: $g = (0.747)g_E = 7.32$ m/s^2

20. At what altitude is the acceleration due to gravity one-half that on Earth's surface?

21. (a) Determine the weight of a person on the Moon whose weight on Earth is 180 lb.
    (b) What would be your weight on the Moon?

Answer: (a) 30 lb

22. An astronaut on the Moon places a package on a scale and finds its weight to be 18 N.
    (a) What would be the weight of the package on Earth?
    (b) What is the mass of the package on the Moon?
    (c) What is the package's mass on Earth?

3.5 Momentum

23. Calculate the linear momentum of a truck with a mass of 13,000 kg that is traveling 20 m/s eastward.

Answer: 2.6×10^5 kg-m/s, east

24. A small car with a mass of 900 kg travels northward at 30 m/s. Does the car have more or less momentum than the truck in Exercise 23, and how much more or less? (Is direction a factor in this exercise?)

25. Two ice skaters stand together as illustrated in ● Fig. 3.25a. They "push off" and travel directly away from each other, the boy with a speed of 0.50 m/s. If the boy weighs 735 N and the girl 490 N, what is the girl's speed after pushing off? (Consider the ice to be frictionless.)

Answer: 0.75 m/s

26. Suppose that for the couple in Fig. 3.25 you were told that the girl's mass was three-fourths that of the boy. What would be the girl's speed in this case?

27. A comet goes around the Sun in an elliptical orbit. At its farthest point, 600 million miles from the Sun, it is traveling with a speed of 15,000 mi/h. How fast is it traveling at its closest approach to the Sun at a distance of 100 million miles?

Answer: 90,000 mi/h

28. An asteroid in an elliptical orbit about the Sun travels 1.2×10^6 m/s at perihelion (point of closest approach) at a distance of 2.0×10^8 km from the Sun. How fast is it traveling at aphelion (farthest point), which is 8.0×10^8 km from the Sun?

FIGURE 3.25 Pushing Off

See Exercises 25 and 26.

(a) Before

(b) After

Solutions to Confidence Exercises

3.1 $a = \dfrac{F_{net}}{m_1 + m_2} = \dfrac{6.0\,\text{N} - 9.0\,\text{N}}{2.0\,\text{kg}} = -1.5$ m/s^2

(in direction opposite to that in Example 3.1)

3.2 $w_M = mg_M = m(0.39)g_E = (1.0\,\text{kg})(0.39)(9.8\,\text{m/s}^2) = 3.8$ N

3.3 $m = \dfrac{w}{g} = \dfrac{30\,\text{N}}{9.8\,\text{m/s}^2} = 3.1$ kg

3.4 $\dfrac{F_2}{F_1} = \dfrac{r_1^2}{r_2^2}$ or $\dfrac{F_2}{F_1} = \left(\dfrac{r_1}{r_2}\right)^2$ and with $r_2 = 2r_1$ or $\dfrac{r_1}{r_2} = \dfrac{1}{2}$, then

$$F_2 = \left(\dfrac{1}{2}\right)^2 F_1 = \dfrac{F_1}{4} \text{ (reduced to } \tfrac{1}{4} \text{ original value)}$$

3.5 $\dfrac{m_1}{m_2} = \dfrac{m}{3m} = \dfrac{1}{3}$ and $v_2 = -\left(\dfrac{m_1}{m_2}\right)v_1 = -\left(\dfrac{1}{3}\right)(1.8 \text{ m/s})$

$$= -0.60 \text{ m/s}$$

3.6 Greater speed at closer distance.

$mv_1r_1 = mv_2$, and with r_1 the closer distance,

$$v_1 = \left(\dfrac{r_1}{r_2}\right)v_2 = \left(\dfrac{1.52 \times 10^8 \text{ km}}{1.47 \times 10^8 \text{ km}}\right)v_2 = (1.03)v_2, \text{ or 3\% greater}$$

Answers to Multiple-Choice Review Questions

1. a   7. b   14. b   20. c   26. d
2. c   8. d   15. c   21. a   27. c

WORK AND ENERGY

4 ● ● ● ● ● ● ● ● ●

I like work; it fascinates me.
I can sit and look at it for hours.

Jerome K. Jerome (1859–1927)

The terms *work* and *energy* are used commonly and have general meanings for most people. For example, work is done in order to accomplish some task or job. When work is done, energy is expended. Hence work and energy are related. After a day's work, a person is usually tired, and rest and food are needed to regain one's energy.

The technical or mechanical meaning of work is quite different from the common meaning. A student standing and holding an overloaded book bag is technically doing no work, yet he or she will feel tired after a time. Isn't this definition of work incorrect, then? As will be learned in this chapter, work involves force *and* motion.

Energy, while one of the cornerstones of science, is more difficult to define. Matter and energy make up the universe. Matter is easily understood—we can touch it and feel it. Energy is not actually tangible; it is a concept. We are only aware of it when it is being used or transformed, for example, when it is used to do work. For this reason, energy is sometimes described as stored work.

Our main source of energy is the Sun, which radiates enormous amounts of energy into space each day. Only a small portion of this energy is received by Earth, where it goes into sustaining plant and animal life. On Earth, we find

Photo: Loading a truck. Work is done and energy is expended.

energy in various forms, including chemical, electrical, nuclear, and gravitational energies. These forms, as we will learn, can be classified more generally as either kinetic or potential energy. ■

4.1 Work

Mechanically, work involves force and motion. One can apply a force all day long, but if there is no motion, there is technically no work. We define the work done by a constant force F as follows:

The **work** done by a constant force F acting on an object is the product of the magnitude of the force (or component of force) and the parallel distance d through which the object moves while the force is applied.

In equation form,

> work = force × parallel distance
>
> $$W = Fd \qquad \textbf{4.1}$$

In this form, it is easy to see that work involves motion. If $d = 0$, the object has not moved, and no work is done.

Figures 4.1, 4.2, and 4.3 illustrate this concept of work and the notion of a component of force and parallel distance. In ● Fig. 4.1, a force is being applied to the wall, but no work is done because the wall doesn't move. ● Figure 4.2 shows an object being moved through a distance d by an applied force F. Note that the force and the directed distance are parallel to each other, and the force F acts through the parallel distance d. The work is then the product of the force and distance, $W = Fd$.

As shown in ● Fig. 4.3, when the force and distance are not parallel to one another, only a component or part of the force acts through a parallel distance. In pushing a lawn mower at an angle to the horizontal, only the component of the force parallel to the level lawn (horizontal component F_h) moves through a parallel distance and does work ($W = F_h d$). The vertical component of the force (F_v) does no work because this part of the force does not act through a distance. It only tends to push the lawn mower against the ground.

FIGURE 4.1   No Work Done
A force is applied to the wall, but no work is done because there is no movement ($d = 0$).

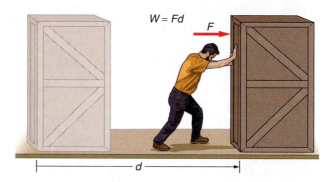

FIGURE 4.2   Work Being Done
An applied force F acts through a parallel distance d. The amount of work equals the force times the distance the object moves (while the force is applied) when the force and the displacement are in the same direction.

An important property of work is that it is a scalar quantity. Both force and parallel distance (actually, displacement) have directions associated with them, but work does not. Work is expressed only as a magnitude (a number with proper units). There is no direction associated with it.

Because work is the product of force and distance, the units of work are those of force times length ($W = Fd$). The SI unit of work is thus the newton-meter (N-m, force × length). This unit combination is given the special name of **joule** (abbreviated J and pro-

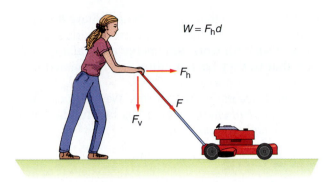

$$W = F_h d$$

F_h

F

F_v

FIGURE 4.3 Work and No Work

Only the horizontal component F_h does work because only it is in the direction of the motion. The vertical component F_v does no work because $d = 0$ in that direction.

nounced "jool") in honor of the early English scientist James Prescott Joule.

One joule is the amount of work done by a force of 1 N acting through a distance of 1 m. Similarly, the unit of force times length in the British system is the pound-foot. For some reason, however, the units are commonly listed in the reverse order, and we express work in **foot-pound** (ft-lb) units. A force of 1 lb acting through a distance of 1 ft does 1 ft-lb of work. The units of work are summarized in Table 4.1.

When doing work, we apply a force, and we feel the other part of Newton's third law force pair acting against us. Therefore, we sometimes say that we are doing work *against* something—for example, work against gravity or friction. When we lift something, we must apply a force to overcome the force of gravity (as expressed by an object's weight $w = mg$), so we do work *against* gravity. This work is given by $W = Fd = wh = mgh$, where h is the height to which the object was lifted.

Similarly, we do work *against* friction. Friction always opposes motion. Hence, to move something on

a surface in a real situation, we must apply a force. In doing so, work is done against friction. As illustrated in ● Fig. 4.4a, if an object is moved with a constant velocity, the applied force F is equal and opposite to the frictional force f (zero net force). The work done by the applied force against friction is $W = Fd = fd$.

When we walk, we must have friction between our feet and the floor; otherwise, we would slip. However, in this case, we do not work against friction (Fig. 4.4b) because the frictional force prevents the foot from moving (slipping). Hence no motion of the foot, no

FIGURE 4.4 Work and No Work Done Against Friction

(a) The mass is moved to the right at a constant velocity by a force F, which is equal and opposite to the frictional force f. (b) When walking, there is friction between your feet and the floor. This is a static case, and the frictional force prevents the foot from moving or slipping. No motion, no work.

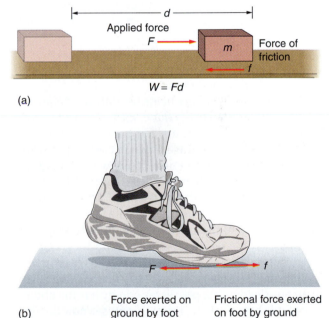

d

Applied force

F

m Force of friction

f

$$W = Fd$$

(a)

(b)

Force exerted on ground by foot

F f

Frictional force exerted on foot by ground

TABLE 4.1 Work Units (Energy Units)

System	Force × Distance Units $W = F \times d$	Special or Common Name
SI	newton × meter (N-m)	joule (J)
British	pound × foot	foot-pound (ft-lb)

work. Of course, other (muscle) forces do work, because, when walking, you are in motion. It is interesting to note in Fig. 4.4b, however, that to move (walk) forward, we actually exert a backward force on the floor.

RELEVANCE QUESTION: *How is work done when you walk up stairs?*

4.2 Kinetic Energy and Potential Energy

LEARNING GOALS

▼ Explain the relationship between work and energy.

▼ Define and distinguish kinetic and potential energies.

When work is done on an object, what happens? A force acting on an object changes its speed. When work is done against gravity, an object's height is changed, and when done against friction, heat is produced. In all these examples, some physical quantity changes when work is done.

The concept of energy helps us unify all the possible changes when work is done. Basically, when work is done, there is a change in energy, and the amount of work done is equal to the change in energy. But what is energy? You may find energy somewhat difficult to define, because it is abstract. Like force, it is a concept—easier to describe in terms of what it can do rather than what it is.

Energy, one of the most fundamental concepts in science, may be described as a quantity possessed by an object or system (a group of objects). A common definition of **energy** *is the ability to do work;* that is, an object or system that possesses energy has the ability or capability to do work. From this we see how the notion of energy as stored work arises. We say that when work is done *by* a system, the amount of energy of the system decreases. Conversely, when work is done *on* a system, the system gains energy. (However, it should be kept in mind that *all* the energy possessed by an object may not be available to do work.)

Hence we see that work is the process by which energy is transferred from one object to another. An object with energy can do work on another object and give it energy. This being the case, it should not surprise you to learn that work and energy have the same units. In the SI, energy is measured in joules, as is work. Also, both work and energy are scalar quantities; that is, they have no direction associated with them.

Energy occurs in many forms. We will focus here on *mechanical energy*, which has two fundamental forms: *kinetic* and *potential*.

Kinetic Energy

As noted previously, when a net force acts on, or work is done on, an object, the object's speed or velocity changes. This can be seen by using equations from Chapter 2, where we had $d = \frac{1}{2}at^2$ and $v = at$. Work is given by $W = Fd$, and since $F = ma$, we have

$$W = Fd = mad$$
$$= ma(\tfrac{1}{2}at^2) = \tfrac{1}{2}m(at)^2$$
$$= \tfrac{1}{2}mv^2$$

This amount of work is now motional energy, and it is defined as kinetic energy.

Kinetic energy *is the energy an object possesses because of its motion, or simply stated, it is the energy of motion.* The amount of kinetic energy an object has when traveling with a velocity v is given by[*]

$$\text{kinetic energy} = \tfrac{1}{2} \times \text{mass} \times (\text{velocity})^2$$
$$E_k = \tfrac{1}{2}mv^2 \qquad \textbf{4.2}$$

As an example of the relationship between work and kinetic energy, consider a pitcher throwing a baseball (● Fig. 4.5). The amount of work required to accelerate a baseball from rest to a speed v is equal to the baseball's kinetic energy, $\frac{1}{2}mv^2$.

Suppose work is done on a moving object. Since the object is moving, it already has some kinetic energy, and *the work done goes into changing the kinetic energy.* Hence we may write

$$\text{work} = \text{change in kinetic energy}$$
$$W = \Delta E_k = E_{k_2} - E_{k_1} = \tfrac{1}{2}mv_2^2 - \tfrac{1}{2}mv_1^2 \qquad \textbf{4.3}$$

Note that if an object is initially at rest ($v_1 = 0$), then the change in kinetic energy is equal to the kinetic en-

---

[*]Although velocity is a vector, the product of $v \times v$, or v^2, gives a scalar, so kinetic energy is a scalar quantity. Either instantaneous velocity or speed may be used to determine kinetic energy.

v ⟶ ◯

FIGURE 4.5   Work and Energy

The work necessary to increase the velocity of a mass is equal to the increase in kinetic energy—here, that of the thrown ball. (It is assumed that no energy is lost.)

ergy of the object. Also keep in mind that to find the change in kinetic energy, you must calculate the kinetic energy for each velocity and then subtract—*not* find the change in velocities to compute the change in kinetic energy.

EXAMPLE 4.1

Finding the Change in Kinetic Energy

A 1.0-kg ball is fired from a cannon. What is the change in the ball's kinetic energy when it accelerates from 4.0 m/s to 8.0 m/s?

SOLUTION

We are given

$$m = 1.0 \text{ kg} \quad \text{and} \quad v_1 = 4.0 \text{ m/s}$$
$$v_2 = 8.0 \text{ m/s}$$

Equation 4.3 can be used directly to compute the change in kinetic energy. Note that the kinetic energy is calculated for each speed.

$$\Delta E_k = E_{k_2} - E_{k_1} = \tfrac{1}{2}mv_2^2 - \tfrac{1}{2}mv_1^2$$
$$= \tfrac{1}{2}(1.0 \text{ kg})(8.0 \text{ m/s})^2 - \tfrac{1}{2}(1.0 \text{ kg})(4.0 \text{ m/s})^2$$
$$= 32 \text{ J} - 8.0 \text{ J} = 24 \text{ J}$$

CONFIDENCE EXERCISE 4.1

Suppose in working the preceding example that a student first subtracts the velocities and says $\Delta E_k = \tfrac{1}{2}m(v_2 - v_1)^2$. Would this give the same (correct) answer? Explain.

Thus work is done to get an object moving, and the object then has kinetic energy. Suppose you wanted to stop a moving object such as an automobile. Work must be done here too, and the amount of work needed to stop the automobile is equal to its change in kinetic energy. The work is generally supplied by brake friction.

In bringing an automobile to a stop, we are sometimes concerned about the braking distance, that is, the distance the car travels after the brakes are applied. The work done to stop a moving car is equal to the braking force times the braking distance ($W = fd$). As was noted, the required work is equal to the kinetic energy of the car ($fd = \tfrac{1}{2}mv^2$). Assuming the braking force to be constant, we see that the braking distance is directly proportional to the square of the velocity ($d \propto v^2$).

The squaring of the velocity makes a big difference in the braking distances for different speeds. For example, suppose you doubled your speed; that is, $v_2 = 2v_1$. Then, with $d_1 \propto v_1^2$ and $d_2 \propto v_2^2 = (2v_1)^2 = 4v_1^2$, we see that $d_2 = 4d_1$ and that the braking distance is increased by a factor of 4.

This braking distance concept explains why school zones have relatively low speed limits—commonly 32 km/h (20 mi/h). The braking distance of a car traveling at this speed is about 8.0 m (26 ft). For a car traveling 64 km/h (40 mi/h), the braking distance is four times this, $4 \times 8.0 \text{ m} = 32 \text{ m}$ (105 ft) (● Fig. 4.6). The driver's reaction time is also a consideration. As this simple calculation shows, if a driver exceeds the speed limit in a school zone, he or she may not be able to stop for a child darting into the street. Remember v^2 the next time you are driving through a school zone.

Potential Energy

An object doesn't have to be in motion to have energy. It also may have energy by virtue of where it is, which is described as potential energy. **Potential energy** *is the energy an object has because of its position or*

FIGURE 4.6 Energy and Braking Distance

Considering a constant braking force, if the braking distance of a car traveling 32 km/h is 8.0 m, then for a car traveling twice as fast, or 64 km/h, the braking distance is *four* times greater, or 32 m.

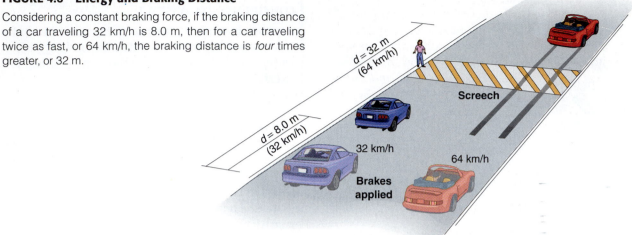

location, or simply, the energy of position. Work is done in changing the position of an object; hence there is a change in energy.

For example, if we lift an object at a (slow) constant velocity, there is no net force on it because it is not accelerating. The weight mg of the object acts downward, and we push up with an equal and opposite force. The distance parallel to the applied upward force is the height h we lift the object (● Fig. 4.7). Thus the *work done against gravity* is, in equation form,

$$\text{work} = \text{weight} \times \text{height}$$
$$W = mgh \qquad \textbf{4.4}$$
$$(W = Fd)$$

Suppose you lift a 1.0-kg book to a height of 1.0 m to a tabletop. The amount of work done in lifting the book is then

$$W = mgh$$
$$= (1.0 \text{ kg})(9.8 \text{ m/s}^2)(1.0 \text{ m}) = 9.8 \text{ J}$$

With work being done, the energy of the book changes (increases), and the book on the table has energy and the ability to do work because of its height or position. This energy is called **gravitational potential energy.** If the book were allowed to fall back to the floor, it could do work; that is, it could crush something.

As another example, the water behind a dam has potential energy because of its position. We use this

FIGURE 4.7 Work Against Gravity

In lifting the weights to a height h, an upward lifting force F equal to the total weight mg is applied. The work done in lifting the weights is then mgh. While standing there with the weights overhead, are you doing any work?

gravitational potential energy to generate electrical energy. Also, when you walk up stairs to a laboratory, you do work. Thus, on the upper floor, you have more gravitational potential energy than a person on the lower floor. (Call down and tell the person so.)

The gravitational potential energy E_p is equal to the work done, and this is equal to the weight of the object times the height (Eq. 4.4). That is,

> gravitational potential energy = weight × height
>
> $$E_p = mgh \qquad \textbf{4.5}$$

When work is done by or against gravity, the potential energy changes, and

$$\text{work} = \text{change in potential energy}$$
$$= E_{p2} - E_{p1}$$
$$= mgh_2 - mgh_1$$
$$= mg(h_2 - h_1)$$
$$= mg\Delta h$$

As with the kinetic energy of an object with a particular velocity, an object has a potential energy for each particular height or position. Work is done when there is a *change* in position, so keep in mind that the *h* in Eq. 4.5 is really a height *difference* (Δh). Note that the *h* in Eq. 4.4 is really a height difference with $h_1 = 0$.

Because the potential energy depends only on the initial and final positions or the difference in height, we say that the potential energy is independent of path. When work is done only against gravity, the force necessary to lift an object from the floor to a tabletop height is the force needed to overcome gravity or the downward weight force (● Fig. 4.8). Therefore, the applied force used to *lift* the object must be upward, no matter what path is taken to arrive at the tabletop. (Notice in Fig. 4.8 that *h* is really a height difference.)

The value of the gravitational potential energy at a particular position depends on the reference point, that is, the reference or zero point from which the height is measured. Near the surface of Earth, where the acceleration due to gravity (*g*) is relatively constant, the designation of the zero position or height is arbitrary. Any point will do. Using an arbitrary zero point is like using a point other than the zero mark on a meterstick to measure length (● Fig. 4.9). This may give rise to negative positions, such as the minus (−) positions on a Cartesian graph.

Heights (actually, displacements or directed lengths) may be positive or negative relative to the zero reference point. However, notice that the height *difference* or change in the potential energy between

FIGURE 4.8   Independent of Path

The change in potential energy in placing a mass on the table is independent of the path taken and depends only on the initial and final positions, or the difference in height. The mass was initially on the floor and finally on the table, so the change in potential energy is the same no matter how it got there. (It is assumed that no energy is lost to friction.)

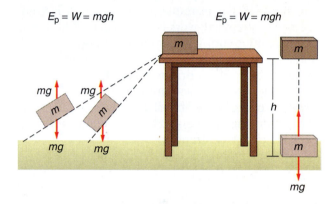

FIGURE 4.9   Reference Point

The reference point for measuring heights is arbitrary. For example, the zero reference point may be that on a Cartesian axis (left) or at one end of the meterstick (right). For positions below the chosen zero reference point on the Cartesian *y* axis, the potential energy is negative because of negative displacement. Referring to the zero end of the meterstick, the potential energy values of these positions would be positive. The important point is that the energy *differences* are the same for any reference.

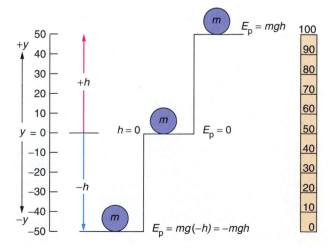

two positions is the same in any case. For example, notice in Fig. 4.9 that the top ball is at a height of $h = y_2 - y_1 = 50$ cm $- 0$ cm $= 50$ cm, according to the scale on the left, or $h = 100$ cm $- 50$ cm $= 50$ cm according to the meterstick. Basically, you can't change a length or height by changing scales.

A negative $(-)$ h gives a negative potential energy. A negative potential energy is somewhat analogous to being in a hole or a well shaft, since we usually designate $h = 0$ at Earth's surface. Negative energy "wells" will be important in understanding atomic theory in Chapter 9.

There are other types of potential energy besides gravitational. For example, when a spring is compressed or stretched, work is done (against the spring force), and the spring has potential energy as a result of the change in length (position). Also, work is done when a bowstring is drawn back. The bow and the bowstring bend and acquire potential energy. This potential energy is capable of doing work on an arrow, thus producing motion and kinetic energy. Note how work is a process of transferring energy.

RELEVANCE QUESTION: *Both a 0.20-kg ball traveling at 30 m/s and a 0.10-kg ball traveling at 60 m/s have the same momentum (Ch. 3). Would they both impart the same energy to your hand if you caught them? Explain.*

4.3 **Conservation of Energy**

LEARNING GOALS

▼ Explain the difference between the conservation of *total* energy and the conservation of *mechanical* energy.

▼ Give examples of the conservation of energy.

As we have seen, energy changes from one form to another, and it does so without a net loss or net gain. That is to say, energy is *conserved*—the amount remains constant. The study of energy transformations has led to one of the most basic scientific principles—the law of conservation of energy. Although the meaning is the same, the law of conservation of energy (or, simply, the conservation of energy) can be stated in many different ways. For example, *energy can neither be created nor destroyed;* or *in changing from one form to another, energy is always conserved.*

Our formal definition of the **conservation of total energy** will be

The total energy of an isolated system remains constant.

Thus, although energy may be changed from one form to another, energy is not lost from the system. A *system* is something enclosed within boundaries, which may be real or imaginary, and *isolated* means that nothing from the outside affects the system (and vice versa).

For example, the students in a classroom might be considered a system. They may move around in the room, but if no one leaves or enters, then the number of students is conserved ("law of conservation of students"). We sometimes say that the total energy of the universe is conserved. This is true, since the universe is the largest system we can think of, and since all the energy in the universe is in it somewhere in some form.

In equation form, we may write the conservation of energy as

$$(\text{total energy})_{\text{time 1}} = (\text{total energy})_{\text{time 2}}$$

That is to say, the total energy does not change with time.

To simplify the understanding of the conservation of energy, we often use *ideal* systems in which the energy is only in two forms—kinetic and potential. In this case, we talk about the **conservation of mechanical energy,** which may be written in equation form as

$$(E_k + E_p)_1 = (E_k + E_p)_2$$
$$(\tfrac{1}{2}mv^2 + mgh)_1 = (\tfrac{1}{2}mv^2 + mgh)_2 \quad \textbf{4.6}$$

where the subscripts indicate the energy at different times, that is, initial and final. Here we assume that no energy is lost in the form of heat because of frictional effects (or any other cause). It is an ideal but instructive situation in helping to understand the conservation of energy.

EXAMPLE 4.2

Finding Kinetic and Potential Energies

A 0.10-kg stone is dropped from a height of 10.0 m. What will be the kinetic and potential energies of the stone at the heights indicated in ● Fig. 4.10? (Neglect air resistance.)

SOLUTION

With no frictional losses, by the conservation of mechanical energy, the total energy ($E_T = E_k + E_p$) will be the same at all heights above the ground. When the stone is released, the total energy is all potential energy ($E_T = E_p$), since $v = 0$ and $E_k = 0$.

At any height h, the potential energy will be $E_p = mgh$. Thus the potential energies at the heights of 10 m, 7.0 m, 3.0 m, and 0 m are

$h = 10$ m: $E_p = mgh = (0.10 \text{ kg})(9.8 \text{ m/s}^2)(10.0 \text{ m})$
$= 9.8 \text{ J}$

$h = 7.0$ m: $E_p = mgh = (0.10 \text{ kg})(9.8 \text{ m/s}^2)(7.0 \text{ m})$
$= 6.9 \text{ J}$

$h = 3.0$ m: $E_p = mgh = (0.10 \text{ kg})(9.8 \text{ m/s}^2)(3.0 \text{ m})$
$= 2.9 \text{ J}$

$h = 0$ m: $E_p = mgh = (0.10 \text{ kg})(9.8 \text{ m/s}^2)(0 \text{ m})$
$= 0 \text{ J}$

Because the total mechanical energy is conserved or constant, the kinetic energy (E_k) at any point can be found from the equation $E_T = E_k + E_p$. That is to say, $E_k = E_T - E_p = 9.8 \text{ J} - E_p$, since we know that all the energy is potential ($E_T = E_p = 9.8$ J) at the 10 m height. (Check to see if the equation gives $E_k = 0$ at the release height.) A summary of the results is given in Table 4.2.

CONFIDENCE EXERCISE 4.2

Find the kinetic energy of the stone in the preceding example when it has fallen 5.0 m.

TABLE 4.2 Energy Summary for Example 4.2

Height (m)	E_T (J)	E_p (J)	E_k (J)	v (m/s)
10.0	9.8	9.8	0	0
7.0	9.8	6.9	2.9	7.7
3.0	9.8	2.9	6.9	12
0	9.8	0	9.8	14

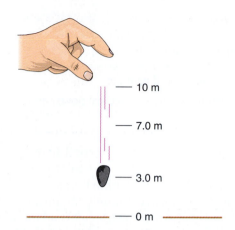

FIGURE 4.10   Changing Kinetic and Potential Energies
See Example 4.2.

Notice from Table 4.2 that as the stone falls the potential energy becomes less (decreasing h), and the kinetic energy becomes greater (increasing v); that is, potential energy is converted into kinetic energy. Just before hitting the ground ($h = 0$), all the energy is kinetic, and the velocity is a maximum. We can use this relationship to compute the magnitude of the velocity, or the speed, of a falling object released from rest. The potential energy lost is $mg\Delta h$, where Δh is the change or decrease in height measured *from the release point down*. This is converted into kinetic energy, $\frac{1}{2}mv^2$. By the conservation of mechanical energy, these quantities are equal, so we may write

$$\tfrac{1}{2}mv^2 = mg\Delta h$$

Then, canceling the m's and solving for v, we have for *the speed of a dropped object after having fallen a distance Δh from the point of release*

$$v = \sqrt{2g\Delta h} \qquad \textbf{4.7}$$

This equation was used to compute the v's in Table 4.2. For example, at a height of 3.0 m, $\Delta h = 10.0$ m − 3.0 m = 7.0 m (a decrease of 7.0 m), and

$$v = \sqrt{2g\Delta h} = \sqrt{2(9.8 \text{ m/s}^2)(7.0 \text{ m})} = 12 \text{ m/s}$$

Because of relatively recent developments, we now consider mass to be a form of energy and more correctly speak of the conservation of mass-energy. (More about this in a Chapter 9 Highlight.)

4.4 Power

LEARNING GOALS

▼ Define power and its units.

▼ Distinguish between electrical power and electrical energy.

When a family moves into a second-floor apartment, a lot of work must be done to carry their belongings up the stairs. In fact, each time the steps are climbed, the movers must carry not only the furniture, boxes, and so on but also their own weights up the stairs.

If the movers do all the work in 3 hours, they will not have worked as rapidly as if the job had been done in 2 hours. The same amount of work will have been done in each case, but there's something different— the *rate* at which the work is done.

To express how fast work is done, we use the concept of power. **Power** *is the time rate of doing work.* This is calculated by dividing the work done by the time required to do the work. In equation form, we have

$$\text{power} = \frac{\text{work}}{\text{time}}$$

or
$$P = \frac{W}{t} \qquad \textbf{4.8}$$

Because work is the product of force and distance ($W = Fd$), power also may be written in terms of these quantities:

$$P = \frac{W}{t} = \frac{Fd}{t} \qquad \textbf{4.9}$$

In the SI, work is measured in joules, so power has the units of joule/second (J/s). This unit is given the special name of **watt** (W), after James Watt, a Scottish engineer who developed an improved steam engine, and 1 W = 1 J/s (● Fig. 4.11).

We rate our light bulbs in watts—for example, a 100-W bulb. This means that such a bulb uses 100 joules of electrical energy each second (100 W = 100 J/s). You have been introduced to several SI units in a short time. These are summarized in Table 4.3.

One should be careful not to confuse the meaning of the letter W. In the equation $P = W/t$, the W stands for work. In the statement $P = 25$ W, the W stands for

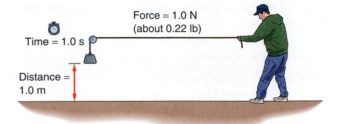

FIGURE 4.11 The Watt

In applying a force of 1.0 N to raise a mass a distance of 1.0 m, the amount of work done is 1.0 J. If this work is done in a time of 1.0 s, then the power, or time rate of doing work, is 1.0 W. ($P = W/t = 1.0$ J/1.0 s = 1.0 W.)

watts. Notice that in equations the letters are in italic, whereas unit letters are regular (roman) type.

In the British system, where the unit of work is the foot-pound, the units of power are foot-pounds per second (ft-lb/s). However, a larger unit, the **horsepower** (hp), is commonly used to rate the power of motors and engines, and

$$1 \text{ horsepower (hp)} = 550 \text{ ft-lb/s} = 746 \text{ W}$$

The horsepower unit was originated by James Watt, after whom the SI unit of power is named. In the 1700s, horses were used in coal mines to bring coal to the surface and to power water pumps. In trying to sell his improved steam engine to replace horses, Watt cleverly rated the engines in horsepower so as to compare the rates at which work could be done by an engine and by an average horse.

The greater the power of an engine or motor, the faster it can do work; that is, it can do more work in a given time. For instance, a 2-hp motor can do twice as much work as a 1-hp motor in the same amount of time, or a 2-hp motor can do the same amount of work as a 1-hp motor in half the time.

TABLE 4.3 SI Units of Force, Work, Energy, and Power

Quantity	Unit	Symbol	Equivalent Units
Force	newton	N	kg-m/s^2
Work	joule	J	N-m
Energy	joule	J	N-m
Power	watt	W	J/s

The following example shows how power is calculated.

EXAMPLE 4.3

Calculating Power

A constant force of 150 N is used to push a student's stalled motorcycle 10 m along a flat road in 20 s. Calculate the power in watts.

SOLUTION

Listing the given data and what we want to find in symbol form:

Given: $F = 150$ N   **Find:** P (power)

   $d = 10$ m

   $t = 20$ s

We note that the units are basic and all SI, so we can use Eq. 4.9 with the work expressed explicitly as Fd:

$$P = \frac{W}{t} = \frac{Fd}{t} = \frac{(150 \text{ N})(10 \text{ m})}{20 \text{ s}} = 75 \text{ W}$$

Note that the units are consistent, N-m/s = J/s = W. The given units are all SI, so we know that the answer will have the SI unit of power—the watt.

CONFIDENCE EXERCISE 4.3

A student expends 7.5 W of power in lifting a textbook 0.50 m in 1.0 s with a constant velocity. (a) How much work is done, and (b) how much does the book weigh (in newtons)?

As we have seen, work produces a change in energy. Thus *power may be thought of as energy produced or consumed divided by the time taken,* and we may write

$$\text{power} = \frac{\text{energy produced or consumed}}{\text{time taken}}$$

or $P = \dfrac{E}{t}$ **4.10**

Rearranging this equation, we see that

$$E = Pt$$

This equation is useful in computing the amount of electrical energy consumed in the home. In particular, since energy is power times time ($P \times t$), it has units of watt-second (W-s), or joule (J). To have a bigger unit, we can use a kilowatt (kW) and an hour (h), which would give a **kilowatt-hour** (kWh).

Keeping in mind that a kWh is an energy unit, when you pay the power company for electricity, in what units are you charged—that is, what do you pay for? If you check an electric bill, you will find that the bill is for so many kilowatt-hours (kWh). Hence we really pay the power company for the amount of energy consumed, which is used to do work (● Fig. 4.12). The following example illustrates how the energy consumed may be calculated when the power rating is known. Also, Table 4.4 gives some typical power requirements for a few common household appliances.

FIGURE 4.12   Energy Consumption

Electrical energy is consumed as the motor of the grinder does work and turns the grinding wheel. Notice the flying sparks and the fact that the operator wisely is wearing a face shield rather than just goggles, as the sign in the background suggests. An electric *power* company is really charging us for *energy* in units of kilowatt-hours (kWh).

TABLE 4.4 Typical Power Requirements of Some Household Appliances

Appliance	Power (W)
Air conditioner	
Room	1500
Central	5000
Coffee maker	1650
Dishwasher	1200
Hot-water heater	4500
Microwave oven	1250
Refrigerator	500
Stove	
Range-top	6000
Oven	4500
Television (color)	100

EXAMPLE 4.4

Computing Energy Consumed

A 1.0-hp electric motor runs for 10 hours. How much energy is consumed, in kWh?

SOLUTION

We have

Given: $P = 1.0$ hp   **Find:** E (energy in kWh)

        $t = 10$ h

We note that the time is in hours, which is what we want. However, the power needs to be converted to kW. Since 1 hp = 746 W, we have

$$1.0 \text{ hp} = 746 \text{ W} (1 \text{ kW}/1000 \text{ W}) = 0.746 \text{ kW}$$

$$= 0.75 \text{ kW (rounding off)}$$

Then, using Eq. 4.10,

$$E = Pt = (0.75 \text{ kW})(10 \text{ h}) = 7.5 \text{ kWh}$$

This is the electrical energy consumed when the motor is running (doing work). We often complain about our electric bills. In the United States, the cost of electricity ranges from about 7¢ to 14¢ per kWh, depending on location. Thus, running the motor for 10 hours at a rate of 10¢/kWh costs 75¢. That's pretty cheap for 10 hours of work output. (Electrical energy is discussed further in Chapter 8.)

CONFIDENCE EXERCISE 4.4
A household uses 2.00 kW of power each day for one month (30 days). If the charge for electricity is 8¢ per kWh, how much is the electric bill for the month?

RELEVANCE QUESTION: *What are three common ways to "save" electricity to reduce electric bills?*

4.5 Forms and Sources of Energy

LEARNING GOALS

▼ Identify some common forms of energy.

▼ Compare the major sources of energy and the main sectors of energy consumption.

▼ List some "alternative" energy sources, and explain their pros and cons.

We commonly talk about various *forms* of energy— for example, chemical energy and electrical energy. Many forms of energy exist; however, the main unifying concept is the conservation of energy. We cannot create or destroy energy, but we can change it from one form to another.

If we consider the conservation of energy to its fullest, we have to account for all the energy. Consider a swinging pendulum. The kinetic and potential energies of the pendulum bob change at each point in the swing. Ideally, the *sum* of the kinetic and potential energies—the total mechanical energy—will remain constant at each point in the swing, and the pendulum would swing indefinitely. But in actuality we know that the pendulum will eventually come to a stop. Then, we might ask, where did the energy go? Of course, friction is involved. In most practical situations, the kinetic and potential energies of objects eventually end up as heat. *Thermal,* or *heat, energy* will be studied at some length in Chapter 5, but for now let's say that heat is transferred energy that becomes associated with kinetic and potential energies on a molecular level.

We have already studied *gravitational potential energy.* We use the gravitational potential energy of water to generate electricity in hydroelectric plants. Electricity may be described in terms of electrical force and *electrical energy* (Chapter 8). This energy is associated with the motions of electric charges—that is, with electric currents. It is electrical energy that runs numerous appliances and machines that do work for us.

Electrical forces hold or bond atoms and molecules together, and there is potential energy in these bonds. When fuel is burned (a chemical reaction), a rearrangement of the electrons and atoms in the fuel occurs, and energy is released. We refer to this as *chemical energy.* Our main fuels—wood, coal, petroleum, and natural gas—are indirectly the result of the Sun's energy. This *radiant energy,* or light from the Sun, is electromagnetic radiation. When electrically charged particles are accelerated, they "radiate" electromagnetic waves (Chapter 6). Visible light, radio waves, TV waves, and microwaves are examples of electromagnetic waves.

A more recent entry into the energy sweepstakes is *nuclear energy.* Nuclear energy is the source of the Sun's energy. Fundamental nuclear forces are involved, and the rearrangement of nuclear particles to form different nuclei results in the release of energy as some of the mass of the nuclei is converted to energy. Thus we now consider mass to be a form of energy (Chapter 10).

We have learned to control one type of nuclear process that releases energy (called *fission*) and use it to generate electricity. This process is different from the process that releases the Sun's energy (called *fusion*). The Sun gives off vast amounts of energy in burning huge amounts of nuclear fuel (hydrogen); however, the Sun should be with us for at least another couple of billion years. Meanwhile, on Earth, we have not yet learned to control nuclear fusion. Research goes on, but it is estimated that nuclear energy from this source will not be available until well into the twenty-first century. (See Chapter 10.)

As we go about our daily lives, each of us is constantly using and giving off energy in the form of body heat. The source of this energy is food (● Fig. 4.13). An average adult radiates heat energy at about the same rate as a 100-W light bulb. This heat radiation explains why a crowded room can soon get hot. In winter, extra clothing helps keep our body heat from escaping. In the summer, the evaporation of perspiration helps remove heat and cool our bodies.

The commercial sources of energy on a national scale are mainly coal, oil (petroleum), and natural gas. Nuclear and hydroelectric energies were once the only other significant commercial sources, but note how renewable sources have made an appearance. ● Figure 4.14a shows the percentage of energy sup-

FIGURE 4.13   Refueling

The source of human energy is food. This fuel is necessary to obtain energy that is used in performing tasks, used in body functions, given off as heat, or stored for later use.

plied by each of these resources over the last several decades. Over one-half of our oil consumption comes from imported oil. The United States does have large reserves of coal, but there are some pollution problems with this resource. (See Chapter 20.) Figure 4.14b shows the comparative energy sources for electrical generation.

Perhaps you're wondering where all this energy goes and who consumes it. ● Figure 4.15 gives a general breakdown of use by sector. How do you think the sector breakdown would have looked for the other years in Fig. 4.14a?

All these forms of energy go into satisfying a growing demand. Although the United States has less than 5% of the world's population, it accounts for approximately 25% of the world's annual energy consumption of fossil fuels—namely, coal, oil, and natural gas. With increasing world population and development in third world countries, there is an ever-increasing demand for energy. Where will it come from?

Of course, fossil fuels and nuclear processes will continue to be used, but increasing use gives rise to pollution and environmental concerns (Chapter 20). Research is being done on so-called alternative fuels and energy sources, which would be nonpolluting supplements to our energy supply. In closing this

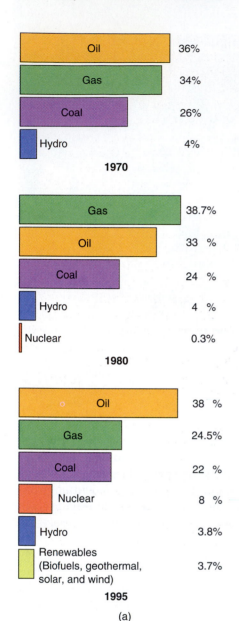

1970	
Oil	36%
Gas	34%
Coal	26%
Hydro	4%

1980	
Gas	38.7%
Oil	33 %
Coal	24 %
Hydro	4 %
Nuclear	0.3%

1995	
Oil	38 %
Gas	24.5%
Coal	22 %
Nuclear	8 %
Hydro	3.8%
Renewables (Biofuels, geothermal, solar, and wind)	3.7%

(a)

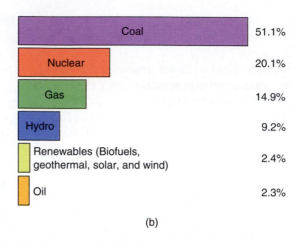

(b)	
Coal	51.1%
Nuclear	20.1%
Gas	14.9%
Hydro	9.2%
Renewables (Biofuels, geothermal, solar, and wind)	2.4%
Oil	2.3%

FIGURE 4.14 Comparative Fuel Consumption

(a) The bar graphs show the approximate relative percentages of fuel consumption in the United States for various years. Notice how the relative use of nuclear energy has increased, and notice the changes in gas and oil. (b) For electrical generation, the relative fuel usage on average is quite different from the general fuel consumption.

Now this frozen gas-water combination is the focus of research and exploration. Methane hydrate occupies as much as 50% of the space between sediment particles in samples obtained by exploratory drilling. It has been estimated that the energy locked in methane hydrates amounts to twice the global reserves of conventional sources (coal, oil, and natural gas). You may be reading more about this energy source in the future. For now, there are many problems to solve, such as finding and drilling into deposits of methane hydrates and separating the methane from the water. Back to the more conventional *alternative fuels.*

Hydropower is used widely to produce electricity (● Fig. 4.16). We would like to increase this production because falling water is nonpolluting. However, most of the best sites for dams have already been de-

FIGURE 4.15 Energy Consumption by Sectors

The bar graph shows the relative consumption of energy by various sectors of the economy.

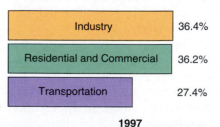

1997	
Industry	36.4%
Residential and Commercial	36.2%
Transportation	27.4%

chapter, let's consider a few of these. But, before looking at some common alternative fuels, let's consider a relatively new and potential *alternative* form of a common fuel—a frozen substance called *methane hydrate,* which is described as "ice that burns."

Found under the ocean floors and below polar regions, methane hydrate is a crystalline form of natural gas and water. (Methane is the major constituent of natural gas.) Methane hydrate resembles ice, but it burns if ignited. Until recently, it was looked upon as a nuisance because it sometimes plugged natural gas lines in polar regions.

FIGURE 4.16 Hoover Dam

The potential energy of dammed water can be used to do work. Here, Lake Mead stretches into the distance behind Hoover Dam, the major dam on the Colorado River.

veloped. Also, the damming of rivers usually results in the loss of agricultural land and alters ecosystems. (Another hydropower consideration is discussed in the chapter Highlight.)

Because of our large agriculture capacities, we can produce large amounts of corn, from which ethanol (an alcohol) can be made. A mixture of gasoline and ethanol, called "gasohol," can be used to run cars. Ethanol has been advertised as reducing air pollution when mixed and burned with gasoline. Actually, some pollutants are reduced, but others are added or increased. Also, there is the disposal of waste by-products from the ethanol production to consider, and unfortunately, twice as much fossil energy is used in ethanol production than the ethanol produces.

If you drive from Los Angeles to Palm Springs, California, in the desert you will suddenly come upon acres and acres of windmills (● Fig. 4.17). *Wind power* has been used for centuries. Windmills for pumping water were once common on American farms. There have been significant advances in wind technology, and the wind turbines shown in the figure generate electricity directly. The wind is free and nonpolluting. However, the limited availability of sites with sufficient wind (at least 20 km/h) limits widespread development of wind power. And the wind does not blow continuously.

The Sun is our major source of energy, and we can put this *solar power* to more use. Solar heating and cooling systems are used in some homes and businesses. Also, there are technologies that focus on concentrating solar radiation for energy production. The

FIGURE 4.17 Wind Energy

Wind turbines in Tehachapi, California, generate electricity using the desert wind.

HIGHLIGHT: Tidal Power

If you have been to the beach, you probably have observed that the ocean water periodically rises and falls in the form of tides (see Chapter 17). This is a change in gravitational potential energy, and it is free, so to speak. In certain bays and river estuaries, tidal rises of several meters occur. Can this tidal power be harnessed? The answer is yes. Hydroelectric power from dammed rivers is quite common, but the lesser-known tidal power does make a contribution. Although a rather small one, it is an intriguing subject.

The idea is simple. Tidewater flows into a dam-controlled basin on a bay or estuary (where the mouth of a river meets the sea tide) during high tide and is discharged during low tide to generate intermittent power. One such plant is located in France on the estuary of the Rance River near Saint-Malo in Brittany on the English Channel. This was the world's first large-scale tidal plant (Fig. 1). Smaller plants exist in Nova Scotia and in Russia on the Bering Sea.

The Rance River dam reservoir is 4 km inland from the mouth of the river and is filled by tides ranging up to 13 m. The power station is equipped with reversible propeller turbines so that electricity may be generated with either direction of water flow. The idea is illustrated in Fig. 2.

During each high tide, water flows through the dam and the generators. At maximum high tide, the dam gates are closed, and water is trapped behind. This is then released at low tide for another phase of electrical generation. However, the times of the tides change regularly, and water outflow may not

FIGURE 1 Tidal Power
During high tide, water flows through this station on the Rance River in France and electricity is generated. At maximum high tide, the dam gates are closed and water is trapped behind. The water is released at low tide for another phase of electrical generation.

occur at times of peak power demands. To help with this, pumped storage may be used. Water is pumped behind the dam for release at peak demand times.

However, tidal power is not totally "free," and there can be environmental impacts. The construction of the generating dam can be quite costly. Also, once the dam is built, navigation is blocked (unless locks are installed), fish migration is impeded and fish are killed passing through the turbines,

and the tidal environment is changed downstream.

New technology addresses these issues by proposed offshore tidal generators. An impoundment enclosure sits on the ocean floor. During high tide, power is generated, and the enclosure is filled. The tide goes out, and the water in the full enclosure may then be used to generate electricity—all done offshore to minimize environmental impacts.

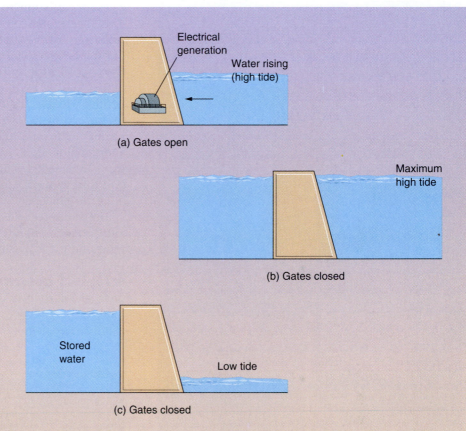

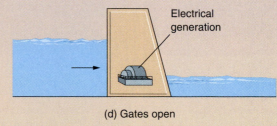

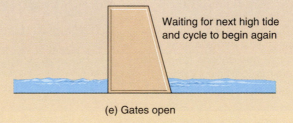

FIGURE 2 Coming and Going

(a–e) Electricity is generated as the rising water of high tide goes through the gates. The gates are closed, and the stored water is used to generate water at low tide. The cycle begins again with the next high tide.

most environmentally promising solar application is the photovoltaic cell (or photocell, for short). These cells convert sunlight directly into electricity (● Fig. 4.18). The light meter used in photography is a photocell. A problem with photocells has been efficiency, but this is over 20% now with advanced technology. Even so, electricity from photocells costs approximately 30¢ per kWh, which is not economically competitive with electricity produced from fossil fuels (on the order of 8¢ to 10¢ per kWh). Photocell arrays could be put on the roofs of buildings to reduce the need for additional land. However, electrical backup systems would be needed, since the photocells could be used only during the daylight hours. Also, clouds would reduce the efficiency.

Hopefully, we will one day learn to mimic the Sun and produce energy by nuclear fusion. You'll learn more about this in Chapter 10.

RELEVANCE QUESTION: *What is the energy source (or sources) that supplies the electricity you use?*

(a)

FIGURE 4.18 Solar Energy
Photocells convert solar energy directly into electrical energy. This is convenient for remote applications (a) or for places where electrical outlets are not convenient (b).

(b)

Important Terms

work (4.1)	kinetic energy	conservation of total	power (4.4)
joule	potential energy	energy (4.3)	watt
foot-pound	gravitational potential	conservation of	horsepower
energy (4.2)	energy	mechanical energy	kilowatt-hour

Important Equations

Work: $W = Fd$

Kinetic Energy: $E_k = \frac{1}{2}mv^2$

and work: $W = \Delta E_k = E_{k_2} - E_{k_1}$
$$= \frac{1}{2}mv_2^2 - \frac{1}{2}mv_1^2$$

Potential Energy (Gravitational): $E_p = mgh$

Conservation of Mechanical Energy:

$$(E_k + E_p)_1 = (E_k + E_p)_2$$
or $\left(\frac{1}{2}mv^2 + mgh\right)_1 = \left(\frac{1}{2}mv^2 + mgh\right)_2$

Speed and Height (from Rest): $v = \sqrt{2g\Delta h}$

Power: $P = \dfrac{W}{t} = \dfrac{Fd}{t}$

$\left(\text{or } P = \dfrac{E}{t} \quad \text{and} \quad E = Pt\right)$

Review Questions

4.1 Work

1. What does work require?
   (a) a net force
   (b) an object changing position
   (c) motion
   (d) all of the preceding

2. What is the SI unit of work?
   (a) ft-lb   (b) joule   (c) newton   (d) watt

3. What is required for work to be done?

4. Do all forces do work? Explain.

5. (a) What are the units of work?
   (b) Show that in terms of fundamental units, those of work are kg-m^2/s^2.

6. A weight lifter holds 900 N (about 200 lb) over his head. Is he doing any work on the weights? Did he do any work on the weights? Explain.

4.2 Kinetic Energy and Potential Energy

7. Car B is traveling twice as fast as car A, but car A has three times as much mass as car B. Which has the greater kinetic energy?
   (a) car A
   (b) car B
   (c) Both have the same kinetic energy.

8. Two identical cars, A and B, start from the same point and travel to the top of a hill by different routes. If the distance traveled by car A is 1.5 times that of car B, which car has the greater change in potential energy?
   (a) car A
   (b) car B
   (c) Both have the same change.

9. Define energy in general terms.

10. Explain how the braking distance of an automobile is related to its velocity.

11. A book sits on a library shelf 1.5 m above the floor. One friend tells you that the book's total mechanical energy is zero, and another says it is not. Who is correct? Explain.

12. (a) A car traveling at a constant speed on a level road rolls up an incline until it stops. Assuming no frictional losses, comment on how far up the hill the car would roll.
    (b) Suppose the car rolled back down the hill. Again assuming no frictional losses, comment on the speed of the car.

13. An object is said to have a negative potential energy. Since it is preferable not to work with negative numbers, can you change the value without moving the object? Explain.

14. Why do pole vaulters run so fast before vaulting?

4.3 Conservation of Energy

15. Which of the following is always conserved?
    (a) power
    (b) mechanical energy
    (c) kinetic energy
    (d) total energy of the universe

16. An object in a conservative system has kinetic energy and gravitational potential energy. If the speed of the object doubles, then the height of the object must decrease by how much?
    (a) twofold
    (b) threefold
    (c) fourfold
    (d) by half as much

17. Distinguish between total energy and mechanical energy.

18. What is meant by a system?

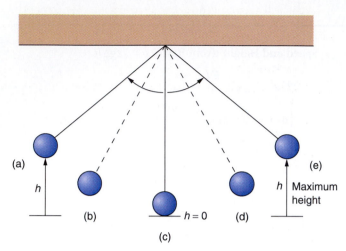

FIGURE 4.19 The Simple Pendulum and Energy

See Review Question 20.

19. When is total energy conserved?

20. A simple pendulum as shown in ● Fig. 4.19 oscillates back and forth. Use the letter designations in the figure to identify the pendulum's position(s) for the following conditions. (There may be more than one answer. Consider the pendulum to be ideal with no energy losses.)
    (a) Position(s) of instantaneous rest _____
    (b) Position(s) of maximum velocity _____
    (c) Position(s) of maximum E_k _____
    (d) Position(s) of maximum E_p _____
    (e) Position(s) of minimum E_k _____
    (f) Position(s) of minimum E_p _____
    (g) Position(s) after which E_k increases _____
    (h) Position(s) after which E_p increases _____
    (i) Position(s) after which E_k decreases _____
    (j) Position(s) after which E_p decreases _____

21. Two students throw identical snowballs as shown in ● Fig. 4.20, with both snowballs having the same initial speed v_o. Which ball has the greater speed on striking the level ground at the bottom of the slope? Justify your answer using energy considerations.

22. A mass suspended on a spring is pulled down and released. It oscillates up and down as illustrated in ● Fig. 4.21. Assuming the total energy to be conserved, use the letter designations to identify the spring's position(s) as listed in Question 20. (There may be more than one answer.)

4.4 Power

23. Power is the time rate of
    (a) work being done.
    (b) energy being expended.
    (c) both of the preceding.

24. If one motor has three times as much power as another, then the second motor can do which of the following?
    (a) the same work in three times the time
    (b) the same work in the same time
    (c) the same work in one-third the time

25. (a) What is the SI unit of power?
    (b) Show that in terms of fundamental units, the units of power are kg-m^2/s^3.

FIGURE 4.20   Away They Go!

See Review Question 21.

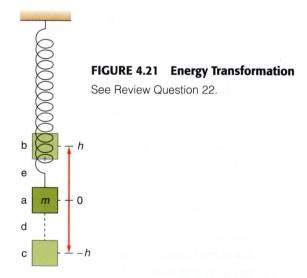

FIGURE 4.21   Energy Transformation

See Review Question 22.

26. Persons A and B do the same job, but B takes longer. Who does the greater amount of work? Who is more "powerful"?

27. What does a greater power rating mean in terms of (a) the amount of work that can be done in a given time, and (b) how fast a given amount of work can be done?

28. What is the unit for which we are charged by the electric company? Are we paying for energy or power?

29. Which common household appliances consume the most electrical energy and why? (See Table 4.4.)

4.5 Forms and Sources of Energy

30. Which fuel has the greatest consumption in the United States?
(a) natural gas   (b) oil   (c) coal   (d) nuclear

31. Which sector of the economy consumes the most energy?
(a) residential and commercial
(b) industry
(c) transportation

32. On average, how much energy do you radiate each second?

33. List five different general forms of energy (other than kinetic and potential energies).

34. Comparing the graphs in Fig. 4.14, what do they tell you about nuclear energy?

Applying Your Knowledge

1. A fellow student tells you that she has both zero kinetic energy and zero potential energy. Is this possible? Explain.

2. Some factory workers are paid by the hour. Others may be paid on a piecework basis (paid according to the number of pieces or items they process or produce). Is there a power consideration involved in either of these methods of payment? Explain.

3. You are at a track meet and watch the high jump and the long jump events. Why do the jumpers run so fast before jumping?

4. With which of our five senses can we detect energy?

5. Discuss some of the problems, limitations, and advantages of each of the fossil fuel energy sources.

Exercises

4.1 Work

1. A worker pushes horizontally on a large crate with a force of 300 N, and the crate is moved 5.0 m. How much work was done?   Answer: 1.5×10^3 J

3. A 5.0-"kilo" bag of sugar is on a counter. How much work is required to put the bag on a shelf a distance of 0.45 m above the counter?   Answer: 22 J

5. A man pushes a lawn mower on a level lawn with a force of 200 N. If 40% of this force is directed downward, how much work is done by the man in pushing the mower 6.0 m?   Answer: 7.2×10^2 J

2. While rearranging a dorm room, a student does 300 J of work in moving a desk 2.0 m. What was the magnitude of the applied horizontal force?

4. How much work is required to lift a 4.0-kg concrete block a height of 2.0 m?

6. If the man pushes the mower with the force directed at an angle of 45° below the horizontal, the horizontal and vertical components of the force are equal. What is the work done in this case? (*Hint:* Draw a vector triangle and use the Pythagorean theorem.)

4.2 Kinetic Energy and Potential Energy

7. (a) What is the kinetic energy in joules of a 1000-kg automobile traveling at 90 km/h?
(b) How much work would have to be done to bring a 1000-kg automobile traveling at 90 km/h to a stop?   Answer: (a) 3.1×10^5 J   (b) same

8. A 60-kg student traveling in a car with a constant velocity has a kinetic energy of 1.2×10^4 J. What is the speedometer reading of the car in km/h?

9. What is the kinetic energy of a 20-kg dog that is running with a speed of 9.0 m/s (about 20 mi/h)?

    Answer: 8.1×10^2 J

10. Which has more kinetic energy: a 0.0020-kg bullet traveling at 400 m/s or a 6.4×10^4-metric-ton ocean liner traveling at 10 m/s (20 knots)? Justify your answer.

11. A student wishes to calculate the work done in changing the speed of a 1.0-kg object from 2.0 m/s to 6.0 m/s. She first finds the difference in the speeds (4.0 m/s) and then uses this for v in the kinetic energy equation $E = \frac{1}{2}mv^2$, to compute an *incorrect* answer of 8.0 J. Why is the answer incorrect, and what is the correct answer?

    Answer: 16 J

12. How much work is done in slowing a 1000-kg automobile from 90 km/h to 30 km/h?

13. By what factor is the kinetic energy of an object increased when its speed increases (a) from 1.0 m/s to 3.0 m/s, and (b) from 2.0 m/s to 8.0 m/s?

    Answer: (a) 9 (b) 16

14. (a) By what factor is the kinetic energy increased when the speed of a car is increased from 30 km/h to 60 km/h?

    (b) What is the factor of decrease in slowing from 60 km/h to 30 km/h?

15. What is the potential energy of a 3.00-kg object at the bottom of a well 10.0 m deep as measured from ground level? Explain the sign of the answer. Answer: -294 J

16. How much work is required to lift a 3.00-kg object from the bottom of a 10.0-m-deep well?

4.3 Conservation of Energy

17. An object is dropped from a height of 10.0 m. At what height will its kinetic energy and potential energy be equal?

    Answer: 5.0 m

18. A 1.0-kg rock is dropped from a height of 6.0 m. At what height is the rock's kinetic energy twice its potential energy?

19. A sled and rider with a combined weight of 50 kg are at rest on the top of a hill 10 m high.

    (a) What is their total energy at the top of the hill?

    (b) Assuming there is no friction, what is the total energy on sliding halfway down the hill?

    Answer: (a) 4.9×10^3 J (b) same

20. A 25.0-kg child starting from rest slides down a water slide with a vertical height of 20.0 m. What is the child's speed (a) halfway down the slide's vertical distance, and (b) three-quarters of the way down? (Neglect friction.)

21. A 100-kg rollercoaster car starts from rest on the top of a 30-m-high track. It rolls down to a 10-m-high dip, then back up to a 15-m crest. What are the speeds of the car at (a) the dip and (b) the crest?

    Answer: (a) 20 m/s (b) 17 m/s

22. A 1.00-kg ball is dropped from the top of a 50.0-m building. What are the kinetic and potential energies of the ball at (a) $t = 1.00$ s and (b) $t = 2.00$ s? (*Hint:* See Chapter 2.)

4.4 Power

23. If the man in Exercise 5 pushes the lawn mower 6.0 m in 30 s, how much power does he expend?

    Answer: 24 W

24. If the man in Exercise 5 expends 18 W of power in pushing the mower 6.0 m, how much time is spent in pushing the mower this distance?

25. A student weighing 500 N climbs a stairway (vertical height of 4.0 m) in 25 s.

    (a) How much work was done?

    (b) What was the power output of the student?

    Answer: (a) 2.0×10^3 J (b) 80 W

26. A 154-lb student races up stairs with a vertical height of 6.0 m in 5.0 s to get to a class on the second floor. How much power in watts does the student expend in doing work against gravity?

27. What is the horsepower necessary to carry a 20-lb bag of books to a height of 22 ft in 40 s?   Answer: 0.020 hp

28. A 3.0-hp motor is run for 1.0 h. What is the work output in ft-lb?

29. On a particular day the following appliances are used for the times indicated: coffee maker, 40 min, and microwave oven, 12 min. Using the power requirements given in Table 4.4, how much does it cost to use the appliances at an electrical cost of 8¢ per kWh?

    Answer: $0.11

30. A microwave oven has a power requirement of 1250 W. A frozen dinner requires 4.0 min to heat on full power.

    (a) How much electrical energy is used in kWh to prepare the dinner?

    (b) If the rate of electricity is 10¢ per kWh, how much does it cost to heat the dinner? (See Table 4.4.)

Solutions to Confidence Exercises

4.1 No. $\Delta v = v_2 - v_1 = 8.0\,\text{m/s} - 4.0\,\text{m/s} = 4.0\,\text{m/s}$, and $\frac{1}{2}m(\Delta v)^2 = \frac{1}{2}(1.0\,\text{kg})(4.0\,\text{m/s}^2) = 8.0\,\text{J}$. Wrong!
$(v_2 - v_1)^2 \neq (v_2^2 - v_1^2)$. Must compute E_k for each speed.

4.2 Since the stone has fallen halfway, half of the original E_p is converted to E_k, or 4.9 J.

4.3 (a) $W = Pt = (7.5\,\text{W})(1.0\,\text{s}) = 7.5\,\text{W-s (J)}$
(b) $F = W/d = 7.5\,\text{J}/0.50\,\text{m} = 15\,\text{N}$

4.4 30 days = 720 h
$E = Pt = (2.00\,\text{kW})(720\,\text{h}) = 1440\,\text{kWh}$ ($0.08/kWh)
$= \$115.20$

Answers to Multiple-Choice Review Questions

1. d   7. b   15. d   23. c   30. b
2. b   8. c   16. c   24. a   31. b

TEMPERATURE AND HEAT

5

If you can't stand the heat, stay out of the kitchen.

Harry S. Truman (1884–1972)

The heating effects produced by fire, the sensation experienced when holding a piece of ice, and the warmth produced when hands are rubbed together are well known. Explaining what takes place in each of these cases, however, is not easy. Both *heat* and *temperature* are terms commonly used when referring to hotness or coldness. However, heat and temperature are not the same thing. They have different and distinct meanings, as we will see.

The concepts of heat and temperature play an important part in our daily lives. We like our coffee hot and our ice cream cold. The temperature of our living and working quarters must be carefully adjusted to our bodies' heat demands. The daily temperature reading is perhaps the most important part of a weather report. How cold or how hot it will be affects the clothes we wear and the plans we make.

We discuss in detail in Chapter 19 how the Sun provides heat to our Earth. The heat balance between various parts of Earth and its atmosphere gives rise to wind, rain, and other weather phenomena. The thermal pollution of rivers that can be caused by hot water from power plants is a cause for concern because of its effect on the ecology of these rivers. On a cosmic scale, the temperature of various stars gives clues to their ages and the origin of the universe.

Photo: Molten steel being poured in a steel mill.

What is temperature? What is heat? How is heat transferred? The answers to these questions are examined in this chapter. Knowing the answers, we can explain many phenomena occurring around us. ■

5.1 Temperature

LEARNING GOALS

▼ Define temperature.

▼ Describe the common temperature scales, and convert temperature readings from one scale to another.

Temperature tells us whether something is hot or cold. In fact, we can say that temperature is a *relative* measure of hotness or coldness. For example, if the water in one cup has a higher temperature than the water in another cup, then we know that the water in the first cup is hotter. But it would be colder than a cup of water with an even higher temperature. Thus hot and cold are *relative* terms, that is, comparisons.

On the molecular level, we find that temperature depends on the kinetic energy of the molecules of a substance. The molecules of all substances are in constant motion. This observation is true even for solids in which the molecules are held together by intermolecular forces, which are sometimes likened to springs. The molecules move back and forth about their equilibrium positions.

In general, the greater the temperature of a substance, the greater is the motion of its molecules. On this basis, we say that **temperature** is a measure of the average kinetic energy of the molecules of a substance.

Humans have temperature perception in the sense of touch. However, this perception is unreliable and may vary a great deal for different people. Our sense of touch doesn't allow us to measure temperature accurately or quantitatively. The quantitative measurement of temperature is accomplished through the use of a thermometer. A *thermometer* is an instrument that utilizes the physical properties of materials for the purpose of accurately determining temperature. The temperature-dependent property most commonly used to measure temperature is *thermal expansion*. Nearly all substances expand with increasing temperature and contract with decreasing temperature.

The change in the linear dimensions or volume of a substance is quite small, but the effects of thermal expansion can be made evident by using special arrangements. For example, a bimetallic strip is made of pieces of different metals bonded together (● Fig. 5.1). When it is heated, one metal expands more than the other, and the strip bends toward the metal with the smaller thermal expansion. As illustrated in Fig. 5.1, the strip can be calibrated with a scale to measure temperature. Bimetallic strips in the form of a coil or helix are used in dial-type thermometers and thermostats.

The most common type of thermometer is the liquid-in-glass thermometer, with which you are probably familiar. It consists of a glass bulb on a glass stem with a capillary bore and sealed upper end. A liquid in the bulb (usually mercury, or alcohol colored with a red dye to make it visible) expands on heating, and a column of liquid is forced up the capillary tube. The glass also expands, but the liquid expands much more.

Thermometers are calibrated so that numerical values can be assigned to different temperatures. The calibration requires two reference or fixed points and a choice of unit. By analogy, think of constructing a stick to measure length. You need two marks or reference points, and then you divide the interval between the marks into sections or units. For example, you might use 100 units between the reference marks to calibrate the length of a meter in centimeters.

Two common reference points for a temperature scale are the ice and steam points of water. The *ice point* is the temperature of a mixture of ice and water at normal atmospheric pressure. The *steam point* is the temperature at which pure water boils at normal atmospheric pressure. Common names for the ice and steam points are freezing and boiling points, respectively.

Two common temperature scales are the Fahrenheit and Celsius scales (● Fig. 5.2). The **Fahrenheit scale** has an ice point of 32° (read "32 degrees") and a steam point of 212°. The interval between the ice and steam points is evenly divided into 180 units. Each unit is called a *degree*. Thus a *degree Fahrenheit*, abbreviated °F, is $\frac{1}{180}$ of the temperature change between the ice point and the steam point.

The **Celsius scale** is based on an ice point of 0° and a steam point of 100°. There are 100 equal units or divisions between these points. A *degree Celsius*, abbreviated °C, is thus $\frac{1}{100}$ of the temperature change

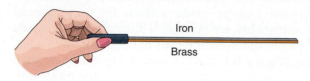

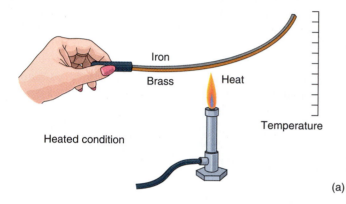

Initial temperature condition

Heated condition

Heat

Temperature

(a)

FIGURE 5.1 Bimetallic Strip and Thermal Expansion

(a) (above; above right) Because of different degrees of thermal expansion, a bimetallic strip of two different metals bends when heated. The degree of deflection of the strip is proportional to the temperature, and a calibrated scale could be added for temperature readings. Scale is shown but not calibrated. (b) (right) Bimetallic coils are used not only in oven thermometers but also in refrigerator-freezer thermometers, as shown here. Note that the indicator arrow is attached directly to the coil.

(b)

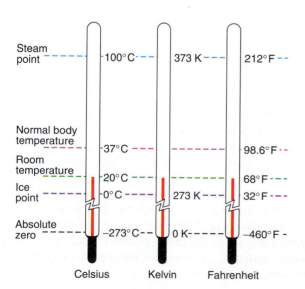

Steam point — 100°C — — 373 K — — 212°F

Normal body temperature — 37°C — — 98.6°F

Room temperature — 20°C — — 68°F

Ice point — 0°C — — 273 K — — 32°F

Absolute zero — −273°C — — 0 K — — −460°F

Celsius   Kelvin   Fahrenheit

FIGURE 5.2 Temperature Scales

The common temperature scales are the Fahrenheit and Celsius scales. They have 180- and 100-degree intervals, respectively, between their ice and steam points. A third scale, the Kelvin (absolute) scale, is used primarily in scientific work and takes zero as the lower limit of temperature—absolute zero (0 K). The unit or interval on the absolute Kelvin scale is the kelvin (K).

between the ice point and the steam point. So a degree Celsius is 1.8 times (almost twice) as large as a degree Fahrenheit (100 degrees Celsius and 180 degrees Fahrenheit for the same temperature interval). The Celsius temperature scale is used predominantly in metric countries and hence throughout most of the world.

There is no known upper limit of temperature; however, there is a lower limit. The lower limit occurs at about $-273°C$ or $-460°F$, and it is called *absolute zero*. Another temperature scale, called the **Kelvin scale,** has its zero temperature at this absolute limit (see Fig. 5.2).* It is sometimes called the *absolute temperature scale*. The unit of the Kelvin scale is the **kelvin,** abbreviated K (*not* °K), and it has the same size as a degree Celsius. Notice that since the Kelvin scale has absolute zero as its lowest reading, it can have no negative temperatures.

Because the kelvin and degree Celsius are equal intervals, we can easily convert from the Celsius scale to the Kelvin scale. We simply add 273 to the Celsius temperature. In equation form, we have

$$T_K = T_C + 273 \quad \text{(Celsius } T_C \text{ to Kelvin } T_K) \quad \textbf{5.1}$$

As examples, a temperature of 0°C equals 273 K, and a Celsius temperature of 27°C is equal to 300 K ($T_K = T_C + 273 = 27 + 273 = 300$ K).

Converting from Fahrenheit to Celsius, and vice versa, is also quite easy. The equations for these conversions are

$$T_F = \tfrac{9}{5}T_C + 32 \quad \text{(Celsius } T_C \text{ to Fahrenheit } T_F) \quad \textbf{5.2}$$
$$\text{(or} \quad T_F = 1.8T_C + 32)$$
$$\text{and} \quad T_C = \tfrac{5}{9}(T_F - 32) \quad \text{(Fahrenheit } T_F \text{ to Celsius } T_C) \quad \textbf{5.2a}$$
$$\left(\text{or} \quad T_C = \frac{T_F - 32}{1.8}\right)$$

Equations 5.2 and 5.2a are the same, merely different arrangements.

As examples, try converting 100°C and 32°F to their equivalent temperatures on the other scales. (You already know the answers.) Remember that on these scales you can have negative temperatures (as opposed to no negative values on the Kelvin scale).

*Named after Lord Kelvin (William Thomson, 1824–1907), the British physicist who developed it.

EXAMPLE 5.1

Converting Temperatures on Different Scales

One winter day the temperature is 5°F outside. What is the equivalent temperature on the Celsius scale?

SOLUTION

With $T_F = 5°F$, and using Eq. 5.2a to find T_C, we have
$$T_C = \tfrac{5}{9}(T_F - 32) = \tfrac{5}{9}(5 - 32)$$
$$= \tfrac{5}{9}(-27) = -15°C$$

CONFIDENCE EXERCISE 5.1

On a winter day the temperature is $-15°C$. What is the equivalent temperature on the Fahrenheit scale? (You know the answer, but use the appropriate equation to show this.)

RELEVANCE QUESTION: *The temperature perception of the sense of touch cannot be used to measure temperature quantitatively, and also is limited. What is the major limitation?*

5.2 Heat

LEARNING GOALS

▼ Define heat.
▼ List the common units of heat.

We commonly say that heat is a form of energy. However, we can be more descriptive. The molecules of a substance may vibrate back and forth, rotate, or move from place to place. Hence they have kinetic energy. As stated previously, the average kinetic energy of the molecules of a substance is related to its temperature. For example, if an object has a high temperature, the average kinetic energy of the molecules is relatively high. (The molecules move relatively fast.)

Potential energy also exists on the molecular level. This is associated with the vibrations and/or rotational modes of the atoms within molecules and the molecular bonds, which simplistically may be thought of as "springs." [Think of the potential energy of a mass (atom) oscillating on a spring.] The total energy (kinetic plus potential) contained within an object is called its *internal energy*.

We say that heat is transferred from one body to another and that heat "flows" from a region of higher temperature to a region of lower temperature. Actually, **heat** is energy that is transferred from one object to another as a result of a temperature difference. In other words, heat is energy in transit. When heat is added to a body, its internal energy increases. Some of the transferred energy may go into the kinetic energy of the molecules, which is manifested as a temperature increase, and some may go into the potential energy part of the internal energy.

Since heat is energy, it has the SI unit of joule (J). However, a common and traditional unit for measuring heat energy is the calorie. A **calorie** (cal) is defined as the amount of heat necessary to raise one gram of pure water by one Celsius degree at normal atmospheric pressure. In terms of the SI energy unit,

$$1 \text{ cal} = 4.186 \text{ J} (\approx 4.2 \text{ J})$$

Heat measurements commonly have been made in calories rather than joules. We are now in a transition period going from calories to joules, so both are used in this book.

The calorie that we have defined is not the same as the one used when discussing diets and nutrition. This is the kilocalorie (kcal), and 1 kcal is equal to 1000 cal. A **kilocalorie** is the amount of heat necessary to raise the temperature of one kilogram of water one Celsius degree. A food calorie (Cal) is equal to one kilocalorie and is commonly written with a capital C to avoid confusion. We sometimes refer to a "big" (kilogram) Calorie and a "little" (gram) calorie for distinction.

$$1 \text{ food Calorie} = 1000 \text{ calories } (1 \text{ kcal})$$

$$1 \text{ food Calorie} = 4186 \text{ joules } (\approx 4.2 \text{ kJ})$$

Food Calories indicate the amount of energy produced when a given amount of the particular food is completely burned.

The unit of heat in the British system is the British thermal unit, or Btu. One **Btu** is the amount of heat required to raise one pound of water one Fahrenheit degree at normal atmospheric pressure. Air conditioners and heating systems are commonly rated in Btu's. These ratings are actually the Btu's removed or supplied per hour. Some relationships among Btu, kcal, joules, and kWh are given in Table 5.1.

TABLE 5.1 Relationships Among Some Common Energy Units

1 cal = 4.2 J
1 kcal = 4.2 kJ = 3.97 Btu = 0.00116 kWh
1 Btu = 1055 J = 0.25 kcal = 0.00029 kWh
1 kWh = 3.6×10^6 J = 860 kcal = 3413 Btu

As we have seen in the measurement of temperature, one effect of heating a material is expansion. As a general rule, almost all matter—solids, liquids, and gases—expands when heated and contracts when cooled. *The most important exception to this rule is water.* If water is frozen, it expands; that is, ice at 0°C occupies a larger volume than the same mass of water at 0°C. This leads to some interesting environmental effects, as discussed in the chapter's first Highlight.

The change in length or volume of a substance due to heat and temperature changes is a major factor in the design and construction of items ranging from steel bridges and automobiles to watches and dental cements. The cracks in a highway are designed so that in summer the concrete will not buckle as it expands due to the heat. Expansion joints are designed into bridges for the same reason (● Fig. 5.3). The Golden Gate Bridge across San Francisco Bay varies in length about 1 m between summer and winter.

Heat-expansion characteristics are used to control such things as the flow of water in car radiators and the flow of heat in homes through the operation of thermostats.

RELEVANCE QUESTION: *We talk about the summer "heat." What does this mean? How does air conditioning fit in?*

5.3 Specific Heat and Latent Heat

LEARNING GOALS

▼ Explain how specific heat relates heat and temperature.

▼ Distinguish between specific heat and latent heat.

Specific Heat

Heat and temperature, although different, are intimately related. When heat is added to a substance,

FIGURE 5.3   Thermal Expansion Joints

Expansion joints are built into bridges and connecting road-ways to allow for the expansion and contraction of the steel girders caused by the addition and removal of heat. If the girders were allowed to come into contact when expanding, serious damage could result.

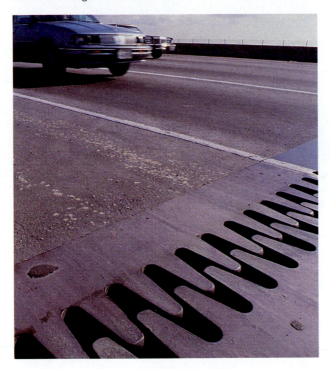

The specific heat of a substance is the amount of heat necessary to raise the temperature of one kilogram of the substance one Celsius degree.

By definition, one kilocalorie is the amount of heat that raises the temperature of one kilogram of water one Celsius degree, so it follows that water has a specific heat of 1.00 kcal/kg-C°. For ice and steam, the specific heats are nearly the same, 0.50 kcal/kg-C°. Other substances require different amounts of heat to raise the temperature of one kilogram of the substance by one degree. That is, a specific heat is *specific* for a particular substance. The specific heats of a few common substances are given in Table 5.2.

The SI units of specific heat are J/kg-C°, but kcal is sometimes used for energy. We will work in both kcal and J. (The latter is generally preferred, but the larger kcal unit makes the specific heat values smaller and more manageable mathematically, particularly for water. See Table 5.2.) The greater the specific heat of a substance, the greater is the amount of heat required to raise the temperature of a unit of mass. Put another way, the greater the specific heat of a substance, the greater is its capacity for heat (given equal masses and temperature change). In fact, the full technical name for specific heat is *specific heat capacity.*

the temperature generally increases. For example, suppose you added equal amounts of heat to equal masses of iron and aluminum. How do you think their temperatures would change? You might be surprised to find that if the temperature of the iron increased by 100 C°, the corresponding temperature change in the aluminum would be only 48 C°. You would have to add more than twice the amount of heat to the aluminum to get the same temperature change as for an equal mass of iron.*

This result reflects the fact that the internal forces of the materials are different (different intermolecular "springs," so to speak). In aluminum, more of the energy goes into internal potential energy than into kinetic energy, which is manifested as temperature.

We express this difference in terms of **specific heat.**

*A particular temperature, such as $T = 48°C$, is written °C (48 degrees Celsius), whereas a temperature interval or difference, such as $\Delta T = T_2 - T_1 = 100°C - 52°C = 48$ C°, is written C° (48 Celsius degrees).

TABLE 5.2   Specific Heats of Some Common Substances

Substance	Specific Heat (20°C)	
	kcal/kg-C°	J/kg-C°
Air (0°C, 1 atm)	0.24	1000
Alcohol (ethyl)	0.60	2510
Aluminum	0.22	920
Copper	0.092	385
Glass	0.20	670
Human body (average)	0.83	3470
Ice	0.50	2100
Iron	0.105	440
Mercury	0.033	138
Silver	0.056	230
Soil (average)	0.25	1050
Steam (at 1 atm)	0.50	2100
Water	1.000	4186
Wood (average)	0.40	1700

HIGHLIGHT: Freezing from the Top Down

As a general rule, a substance expands when heated and contracts when cooled. An important exception to this rule is water. A volume of water does contract when cooled up to a point—about 4°C. When cooled from 4°C to 0°C, a volume of water *expands*. This behavior is illustrated in the graph in Fig. 1. Another way of looking at this is in terms of density ($\rho = m/V$, Chapter 1). As a volume of water is cooled to 4°C, its density increases (volume decreases), and from 4°C to 0°C, the density decreases (volume increases). Hence we say that water has its maximum density at 4°C (actually 3.98°C).

The reason for this rather unique behavior is molecular structure. When water freezes, the water molecules go together in an open hexagonal (six-sided) structure, as is evident in snowflakes (Fig. 2). The open space in the molecular structure explains why ice is less dense than water and floats. Again, this is a rather unique feature. The solids of most substances are denser than their liquids.

The fact that water has its maximum density at 4°C accounts for why lakes freeze from the top down. Most of the cooling takes place at the open surface. As the temperature of the top layer of water is cooled toward 4°C, the cooler water at the surface is denser than the water below and sinks to the bottom. This takes place until 4°C is reached. Below 4°C, however, the surface layer is less dense than the water below and remains on top, where it freezes at 0°C.

Thus, because of water's very unusual feature in density versus temperature, lakes freeze from the top down. If the water thermally contracted and the density increased all the way to 0°C, the coldest layer would be on the bottom, and freezing would begin there first. Think of what this would mean to aquatic life, let alone ice skating.

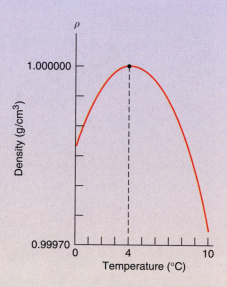

FIGURE 1   Strange Behavior

Like most substances, the volume of a quantity of water decreases with decreasing temperature—but only down to 4°C. Below this temperature, the volume increases slightly. With a minimum volume at 4°C, the density of water is a maximum at this temperature and decreases with lower temperatures.

FIGURE 2   Structure of Ice

(a) An illustration of the open hexagonal (six-sided) molecular structure of ice. (From Ebbing, Darrell D., *General Chemistry*, Sixth Edition. Copyright © 1999 by Houghton Mifflin Company. Used with permission.)
(b) This hexagonal pattern is evident in snowflakes.

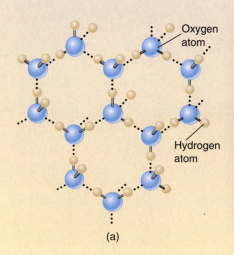

(a)

(b)

(a)

(b)

FIGURE 5.4   Specific Heat Helps
(a) The high specific heat of water causes it to have a moderating affect on climate, as in Hawaii. (b) Lacking water, a desert experiences extreme temperatures—hot during the day and cold at night.

Water has one of the highest specific heats and so can store more heat energy for a given temperature change. Because of this, water is used in solar energy applications. Solar energy is collected during the day and is used to heat water, which can store more energy than most liquids without getting overly hot. At night, the warm water may be pumped through a home to heat it. The high specific heat of water also has a moderating effect on climates (● Fig. 5.4).

The specific heat, or the amount of heat necessary to change the temperature of a given amount of a substance, depends on three factors: the mass (m), the specific heat (designated by c), and the temperature change (ΔT). In equation form, we write

amount of heat to change temperature = mass × specific heat × temperature change

or

$$H = mc\Delta T \qquad 5.3$$

This equation applies to a substance that does not undergo a phase change (e.g., changing from a solid to a liquid) when the heat is added. When heat is added to (or removed from) a substance that is chang-

ing phase, the temperature does not change, and a different equation must be used. This will be presented shortly.

EXAMPLE 5.2

Using Specific Heat

How much heat in kcal does it take to heat 80 kg of bathwater from 12°C (about 54°F) to 42°C (about 108°F)?

SOLUTION

Step 1

Given: $m = 80$ kg

$\Delta T = 42°C - 12°C = 30$ C°

$c = 1.00$ kcal/kg-C° (for water, known)

Step 2

Wanted: H (heat)

(The units are consistent, and the answer will come out in kilocalories. If the specific heat c were expressed in units of J/kg-C°, the answer would be in joules.)

Step 3

The amount of heat required may be computed directly from Eq. 5.3,

$$H = mc\Delta T$$

$$= (80 \text{ kg})(1.00 \text{ kcal/kg-C}°)(30 \text{ C}°)$$

$$= 2.4 \times 10^3 \text{ kcal}$$

Let's get an idea of how much it costs to electrically heat the bathwater. Each kilocalorie corresponds to 0.00116 kWh (Table 5.1), so this amount of heat in kWh is

$$H = 2.4 \times 10^3 \text{ kcal} \; \frac{0.00116 \text{ kWh}}{\text{kcal}} = 2.8 \text{ kWh}$$

At 10¢ per kWh, the cost of the electricity to heat the bathwater is 2.8 kWh × 10¢/kWh = 28¢. For four people each taking a similar bath each day for one month (30 days), it would cost 4 × 30 × $0.28 = $33.60 to heat the water.

CONFIDENCE EXERCISE 5.2

A liter of water at room temperature (20°C) is placed in a refrigerator with a temperature of 5°C. How much heat in kcal must be removed from the water for it to reach the refrigerator temperature?

Latent Heat

Substances in our environment are usually classified as solids, liquids, or gases. These forms are called *phases of matter.* When heat is added to (or removed from) a substance, it may undergo a change of phase. For example, when water is heated sufficiently, it changes to steam, or when enough heat is removed, water changes to ice.

As we know, water changes to steam at a temperature of 100°C (or 212°F) under normal atmospheric pressure. If we keep adding heat to a quantity of water at 100°C, it continues to boil as the liquid is converted to gas, but the temperature remains constant. Here is a case of adding heat to a substance without a resulting temperature change. Where, then, does the energy go?

On a molecular level, when a substance goes from a liquid to a gas, work must be done to break the intermolecular bonds and to separate the molecules. The

molecules of a gas are farther apart than the molecules in a liquid, relatively speaking. Hence, during a phase change, the heat energy must go into the work of separating the molecules and not into increasing the molecular kinetic energy, which would increase the temperature. (Phase changes are discussed in more detail in Section 5.5.) The heat associated with a phase change is called *latent heat* (*latent* means "hidden").

Referring to ● Fig. 5.5, let's go through the process of heating a substance and changing phases from solid to liquid to gas. In the lower left-hand corner, the substance is represented in the solid phase. As heat is added, the temperature rises. When point *A* is reached, adding more heat does not change the temperature. Instead, the heat energy goes into changing the solid into a liquid. The amount of heat necessary to change one kilogram of a solid into a liquid at the same temperature is called the **latent heat of fusion** of the substance. In Fig. 5.5 this heat is simply the amount of heat necessary to go from *A* to *B*.

When point *B* has been reached, the substance is all liquid. The temperature of the substance at which this change from solid to liquid takes place is known as the *melting point.* After point *B*, further heating again causes a rise in temperature. The temperature continues to rise as heat is added until point *C* is reached.

From *B* to *C* the substance is in the liquid phase. When point *C* is reached, adding more heat does not change the temperature. The added heat now goes into changing the liquid into a gas. The amount of heat necessary to change one kilogram of a liquid into a gas is called the **latent heat of vaporization** of the substance. In Fig. 5.5 this heat is simply the amount of heat necessary to go from *C* to *D*.

When point *D* has been reached, the substance is all in the gas phase. The temperature of the substance at which this change from liquid to gas phase occurs is known as the *boiling point.* After point *D* is reached, further heating again causes a rise in the temperature.

In some instances a substance can change directly from the solid to the gaseous phase. This change is called *sublimation.* Examples are dry ice (solid carbon dioxide, CO_2), mothballs, and solid air fresheners. (The reverse process of changing directly from the gaseous to the solid phase is called *deposition.*)

From Fig. 5.5 we get a better understanding of the difference between temperature and heat. We see that the temperature of a body rises as heat is added only when the substance is not undergoing a change in phase.

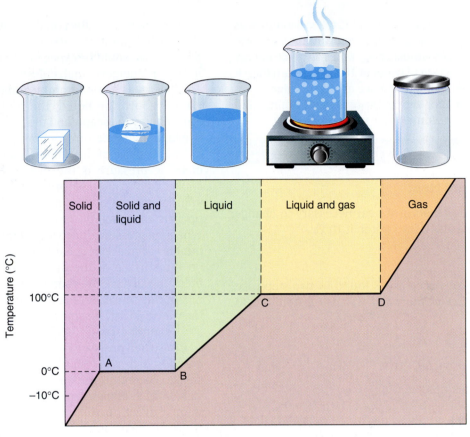

FIGURE 5.5   Graph of Temperature Versus Heat for Water

The solid, liquid, and gas phases are ice, water, and steam, as shown over the sloping graph lines. Notice that during a phase change (A–B and C–D) heat is added, but the temperature does not change. Also, the two phases exist together.

(From Ebbing, Darrell D., and R. A. D. Wentworth, *Introductory Chemistry*, Second Edition. Copyright © 1998 by Houghton Mifflin Company. Used with permission.)

When heat is added to ice at 0°C, the ice melts without changing its temperature. The more ice there is, the more heat is needed to melt it. In general, *the heat required to change a solid into a liquid at the melting point can be found by multiplying the mass of the substance by its heat of fusion.* Thus we can write

$$\text{heat needed to melt a substance} = \text{mass} \times \text{latent heat of fusion}$$

or $H = mL_f$ **5.4**

Similarly, at the boiling point, *the amount of heat necessary to change a liquid into a gas* can be written as

$$\text{heat needed to boil a substance} = \text{mass} \times \text{latent heat of vaporization}$$

or $H = mL_v$ **5.5**

For water, the latent heats are

$$L_f = 80 \text{ kcal/kg} = 3.35 \times 10^5 \text{ J/kg}$$

$$L_v = 540 \text{ kcal/kg} = 2.26 \times 10^6 \text{ J/kg}$$

These values are very large numbers compared with the specific heat. Notice that it takes 80 times more heat energy to melt 1 kg of ice at 0°C than to raise the temperature of 1 kg of water by 1 C°. Similarly, changing 1 kg of water to steam at 100°C takes 540 times as much energy as raising its temperature by 1 C°. Also notice that it takes almost seven times as much energy to change a kilogram of water at 100°C to steam than it does to change a kilogram of ice at 0°C to water.

See Table 5.3 for the latent heats and boiling and melting points of some other substances.

EXAMPLE 5.3

Using Latent Heat

Calculate the amount of heat necessary to change 0.20 kg of ice at 0°C into water at 10°C.

SOLUTION

The total heat necessary is found in two steps. Ice melts at 0°C, and the water warms up from 0°C to 10°C.

$$H = H_{\text{melt ice}} + H_{\text{change }T}$$
$$= mL_f + mc\Delta T$$
$$= (0.20 \text{ kg})(80 \text{ kcal/kg}) +$$
$$(0.20 \text{ kg})(1.00 \text{ kcal/kg-C°})(10°C - 0°C)$$
$$= 18 \text{ kcal}$$

CONFIDENCE EXERCISE 5.3

How much heat must be removed from 0.20 kg of water at 10°C to form ice at 0°C? (Show your calculations.)

Pressure has an effect on the boiling point of water. The boiling point of water increases with increasing pressure, as would be expected. *Boiling* is the process by which energetic molecules escape from a liquid. This energy is gained from heating. If the pressure is greater above the liquid, the molecules must have more energy to escape, and the liquid has to be heated to a higher temperature for boiling to take place.

The increase of the boiling point of water with increasing pressure is the principle of the pressure cooker (● Fig. 5.6). Normally, when a heated liquid approaches the boiling point in an open container, pockets of energetic molecules form gas bubbles. When the pressure due to the molecular activity in the bubbles is great enough, or greater than the pressure on the surface of the liquid, the bubbles rise and break the surface. We then say that the liquid is boiling. In this sense, boiling is a cooling mechanism for the water. Energy is removed, and the water's temperature cannot exceed 100°C.

In a sealed pressure cooker, the pressure above the liquid is increased, causing the boiling point to increase. The extra pressure is regulated by a pressure valve, which allows vapor to escape. (There is also a safety valve in the lid in case the pressure valve gets stuck.) Hence the contents of the cooker boil at a temperature greater than 100°C, and the cooking time is less.

At mountain altitudes, the boiling point of water may be several degrees less than at sea level. For example, at the top of Pike's Peak (elevation 4300 m, or 14,000 ft), the atmospheric pressure is reduced to the point where water boils at about 86°C rather than at

TABLE 5.3 Temperatures of Phase Changes and Latent Heats for Some Substances (atmospheric pressure)

Substance	Latent Heat of Fusion, L_f (kcal/kg)	Melting Point	Latent Heat of Vaporization, L_v (kcal/kg)	Boiling Point
Alcohol, ethyl	25	−114°C	204	78°C
Helium*	——	——	377	−269°C
Lead	5.9	328°C	207	1744°C
Mercury	2.8	−39°C	65	357°C
Nitrogen	6.1	−210°C	48	−196°C
Water	80	0°C	540	100°C

*Not a solid at 1 atm pressure; melting point −272°C at 26 atm.

FIGURE 5.6   The Pressure Cooker

Because of the increased pressure in the cooker, the boiling point of water is raised, and food cooks faster at the higher temperature.

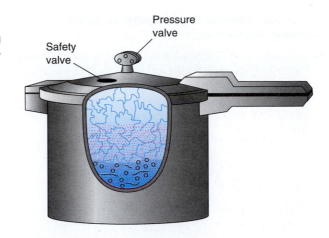

100°C. Pressure cookers come in handy at high altitudes—if you want to eat on time.

Finally, let's consider a type of phase change that is important to our personal lives. *Evaporation* is a relatively slow phase change of a liquid to a gas, and it is a major cooling mechanism of our bodies. When hot, we perspire, and the evaporation of perspiration has a cooling effect because energy is lost. This cooling effect is quite noticeable on the bare skin when one gets out of a bath or shower or has a rubdown with alcohol.

The comforting evaporation of perspiration is promoted by moving air. When we are sweaty and standing in front of a blowing electric fan, we often say that the air is cool. But the air is the same temperature as the other air in the room. The motion of the air promotes evaporation by carrying away molecules (and energy) and provides a cooling effect.

On the other hand, air can hold only so much moisture at a given temperature. The amount of moisture in the air is commonly expressed in terms of *relative humidity* (see Chapter 19). When it is quite humid, there is little evaporation of perspiration, and we feel hot.

RELEVANCE QUESTION: When we are hot, we perspire. What is the cooling mechanism of our bodies in terms of latent heat?

5.4 Heat Transfer

LEARNING GOALS

▼ Describe three methods of heat transfer.

▼ Give examples of these methods.

Because heat is energy in transit, how the transfer is done is an important consideration. Heat transfer is accomplished by three methods: conduction, convection, and radiation.

Conduction *is the transfer of heat by molecular collisions.* The kinetic energy of molecules is transferred from one molecule to another through collision. How well a substance conducts heat depends on the molecular bonding. Solids are generally the best thermal conductors, especially metals.

In addition to undergoing molecular collisions, metals contain a large number of "free" electrons (not permanently bound) that can move around. These electrons contribute significantly to heat transfer, or thermal conductivity. The *thermal conductivity* of a substance is a measure of its ability to conduct heat. As shown in Table 5.4, metals have relatively high thermal conductivities.

Liquids and gases are, in general, relatively poor thermal conductors. Liquids are better than gases because their molecules are closer together and collide more often. Gases are poor conductors because their molecules are relatively far apart, and so conductive collisions do not occur as often. Substances that are poor thermal conductors are sometimes referred to as *thermal insulators.*

We make cooking pots and pans out of metals so that heat will be readily conducted to the foods inside. Pot holders, on the other hand, are made out of cloth, a poor thermal conductor, for obvious reasons (● Fig. 5.7). Many solids, such as cloth, wood, and plastic foam (Styrofoam), are porous and have large numbers of air (gas) spaces that add to their poor conductivity. For example, thermal underwear and Styrofoam coolers depend on this property, as does fiberglass insulation used in the walls and attics of our homes.

Convection *is the transfer of heat by the movement of a substance, or mass, from one position to*

TABLE 5.4 Thermal Conductivities of Some Common Substances

Substance	W/C°-m*
Silver	425
Copper	390
Iron	80
Brick	3.5
Floor tile	0.7
Water	0.6
Glass	0.4
Wood	0.2
Cotton	0.08
Styrofoam	0.033
Air	0.026
Vacuum	0

*Note that W/C° = (J/s)/C°, the rate of heat flow per temperature difference $(\Delta H/\Delta t)/\Delta T$, where W represents the watt. The length unit (m) comes from considering the dimensions (area and thickness) of the conductor.

FIGURE 5.7   Thermal Insulator

Pot holders are made of cloth, a poor thermal conductor (or good thermal insulator), thus preventing heat from being quickly conducted to the hand and causing a burn. Pots and pans, on the other hand, are made of metals so as to promote heat conduction.

another. The movement of heated air or water is an example.

Most homes are heated by convection (movement of hot air). The air is heated at the furnace and circulated throughout the house by way of metal ducts. When the air temperature has dropped, it passes through a cold-air return on its way back to the furnace to be reheated and recirculated (● Fig. 5.8).

The warm-air vents in a room are usually in the floor (under windows). Cold-air return ducts are in the floor too, but on opposite sides of a room. The warm air entering the room rises (being "lighter," or less dense, and therefore buoyant). As a result, cooler air is forced toward the floor, and convection cycles that promote even heating are set up in the room. Heat is distributed in Earth's atmosphere (see Chapter 19) in a manner similar to the transfer of heat by convection currents set up in a room.

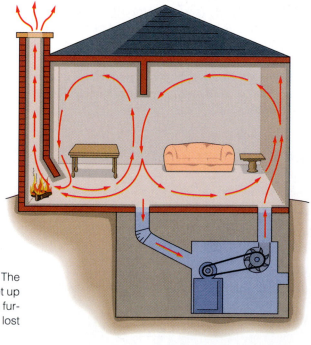

FIGURE 5.8   Convection Cycles

In a forced-air heating system, warm air is blown into a room. The warm air rises, the cold air descends, and a convection cycle is set up that promotes heat distribution. Some of the cold air returns to the furnace for heating and recycling. Notice that a great deal of heat is lost up the chimney of a fireplace.

The transfer of heat by convection and conduction requires a material medium for the process to take place. The heat we get from the Sun is transmitted through the vacuum of space by radiation. **Radiation** *is the process of transferring energy by means of electromagnetic waves.* Electromagnetic waves carry energy and can travel through a vacuum.

Another example of heat transfer by radiation occurs in an open fire or a fireplace. We can readily feel the warmth of the fire on exposed skin. Yet air is a poor conductor; moreover, the air warmed by the fire is rising (up the chimney in a fireplace). Therefore, the only mechanism for appreciable heat transfer is radiation.

In general, we find that dark objects are good absorbers of radiation, whereas light-colored objects are poor absorbers and good reflectors. For this reason, we commonly wear light-colored clothing in the summer so as to be cooler. In the winter, we generally wear dark-colored clothes to take advantage of the absorption of the reduced solar radiation.

An application using all three methods of heat transfer is the thermos bottle, which is used to keep liquids either hot or cold (● Fig. 5.9). Actually, knowledge of the methods of heat transfer is used to *prevent* the transfer of heat in this case. The sealed, double-walled glass bottle is partially evacuated. Glass is a relatively poor conductor, and any heat conducted

through a wall (from the outside in or the inside out) will find the partial vacuum an even greater thermal insulator. This also reduces heat transfer from one glass wall to the other by convection. Finally, the inner surface of the glass bottle is silvered to prevent heat transfer through the glass by radiation. Thus a hot or cold drink in the bottle remains hot or cold for some time.

RELEVANCE QUESTION: *Your automobile has a "radiator" as part of its cooling system. Is radiation the main method of heat transfer?*

5.5 Phases of Matter

LEARNING GOALS

▼ Identify the phases of matter.

▼ Compare the molecular arrangements of the phases.

As we saw in Section 5.3, the addition of heat can cause a substance to change phase. The three common **phases of matter** are solid, liquid, and gas. All substances exist in each phase at some temperature and pressure. At normal room temperature and atmospheric pressure, a substance will be in one of the three phases. For instance, at room temperature, oxygen is a gas, water is a liquid, and copper is a solid.

The principal distinguishing features of solids, liquids, and gases can be understood if we look at the phases from a *molecular* point of view. Most substances are made up of very small particles called *molecules,* chemical combinations of atoms (see Section 11.3). For example, two hydrogen atoms attached to an oxygen atom form a water molecule, H_2O.

A **solid** has a definite shape and volume. In a *crystalline* solid, the molecules are arranged in a particular repeating pattern. This orderly arrangement of molecules is called a *lattice.* ● Figure 5.10 illustrates a lattice structure in two dimensions. The molecules (represented by small circles in the figure) are bound to each other by electrical forces.

Upon heating, the molecules gain kinetic energy and vibrate about their positions in the lattice. The more heat that is added, the stronger the vibrations become. The molecules move farther apart, and as shown diagrammatically in Fig. 5.10, the solid expands.

When the melting point of a solid is reached, additional energy (the heat of fusion, 80 kcal/kg for water) breaks apart the bonds that hold the molecules in

FIGURE 5.9   A Vacuum Bottle

A vacuum thermos bottle is designed to keep hot foods hot and cold foods cold. The bottle usually has a stopper and protective case, which are not shown. (See text for description.)

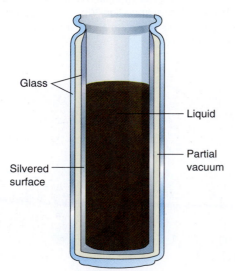

Glass

Liquid

Partial vacuum

Silvered surface

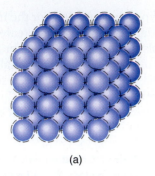

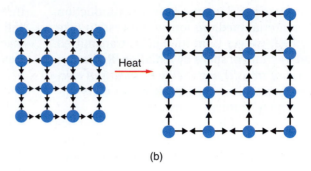

Heat

(a)   (b)

FIGURE 5.10   Crystalline Lattice Expansion

(a) In a crystalline solid, the molecules are arranged in a particular repeating pattern. This orderly array is called a *lattice.* (b) A schematic diagram of a solid crystal lattice in two dimensions (left). Heating causes the molecules to vibrate with greater amplitudes in the lattice, thereby increasing the volume of the solid (right). The arrows represent the molecular bonds, and the drawing is obviously not to scale.

place. As bonds break, holes are produced in the lattice, and nearby molecules can move toward the holes. As more and more holes are produced, the lattice becomes significantly distorted.

Solids that lack an ordered molecular structure are said to be *amorphous.* Examples are glass and asphalt. They do not melt at definite temperatures but gradually soften when heated.

● Figure 5.11 illustrates an arrangement of the molecules in a liquid. A liquid has only a slight, if any, lattice structure. Molecules are relatively free to move. A **liquid** is an arrangement of molecules that may move and assume the shape of the container. Hence a liquid has a definite volume but no definite shape.

When a liquid is heated, the individual molecules gain kinetic energy. The result is that the liquid expands. When the boiling point is reached, the heat energy is sufficient to break the molecules completely apart from each other. The heat of vaporization is the heat per kilogram necessary to free the molecules completely from each other. Because the electrical forces holding the different molecules together are quite strong, the heat of vaporization is fairly large. When the molecules are completely free from each other, the gaseous phase is reached.

A **gas** is made up of fast-moving molecules that exert little or no force on one another except when they collide. The distance between molecules in a gas is quite large compared with the size of the molecules (● Fig. 5.12). As a result, a quantity of gas assumes the size and shape of its container—that is, a gas has no definite shape or volume. Its pressure, volume, and temperature are closely related, as will be discussed in the next section.

Continued heating of a gas causes the molecules to move faster and faster. Eventually, the molecules and atoms are ripped apart by collisions with one another. Inside hot stars, such as our Sun, atoms and molecules do not exist, and another phase of matter, called a plasma (no relationship to blood plasma), occurs. A **plasma** is a hot gas of electrically charged particles.

FIGURE 5.11   Liquids and Molecules

This illustration depicts the arrangement of molecules in a liquid. The molecules are packed closely together and form only a slight lattice structure. Some surface molecules may acquire enough energy through collisions to break free of the liquid. This is called *evaporation.* When the liquid is heated and surface molecules break free from the boiling liquid, this is called *vaporization.*

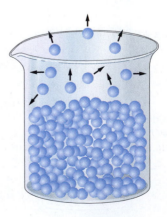

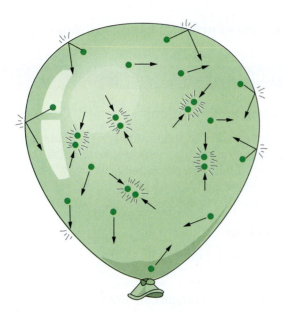

FIGURE 5.12   Gases and Molecules

Gas molecules, on average, are relatively far apart. They move randomly at high speeds, colliding with each other and the walls of the container. The average force of their collisions with the container walls causes pressure on the walls.

On Earth, we have plasmas in gas discharge tubes, such as fluorescent and neon lamps.

5.6 The Kinetic Theory of Gases

LEARNING GOALS

▼ Describe how the temperature and molecular speed of a gas are related.

▼ Solve exercises involving pressure, temperature, volume, and number of molecules of a gas.

Unlike solids and liquids, gases take up the entire volume and shape of any enclosing container and are easily compressed. Gas pressure rises as the temperature increases, and gases *diffuse* (travel) slowly into the air when their containers are opened. These observations and others lead to a model called the *kinetic theory of gases.*

The **kinetic theory** describes a gas as consisting of molecules moving independently in all directions at high speeds—the higher the temperature, the higher is the average speed. The molecules collide with each

other and the walls of the container. The distance between molecules is, on average, large compared with the size of the molecules themselves. See Fig. 5.12 for an illustration of this model. Theoretically, an *ideal gas* (or *perfect gas*) is one in which the molecules are point particles (have no size at all) and interact only by collision. A *real gas* behaves like an ideal gas unless it is under so much pressure that the space between its molecules becomes small relative to the size of the molecules or unless the temperature drops to the point at which attractions among the molecules can be significant.

Each collision with a wall exerts only a tiny force on the wall. However, the frequent collisions of billions of gas molecules with the wall exert a steady average force per unit area, or pressure, on the wall. **Pressure** is defined as the force per unit area: $p = F/A$. ● Figure 5.13 illustrates how these three quantities are related.

The SI unit of pressure is N/m^2, which is called a *pascal* (Pa), in honor of Blaise Pascal, a seventeenth-century Frenchman who was one of the first scientists to develop the concept of pressure. However, a more common unit of pressure used when dealing with gases is the *atmosphere* (atm), where 1 atm is the atmospheric pressure at sea level and 0°C (1.01×10^5 Pa or 14.7 lb/in^2).

Pressure and Number of Molecules

To see how pressure (p), volume (V), Kelvin temperature (T), and number of molecules (N) are related, let's examine the effect of each on pressure when the other two are held constant.

FIGURE 5.13   Force and Pressure

When holding a tack as shown, the thumb and the finger experience equal forces (Newton's third law). However, at the sharp end of the tack, the area is smaller and the pressure is greater, and it can hurt if you apply enough force. (Ouch!)

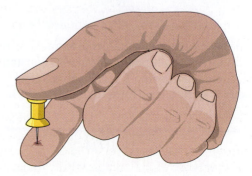

If the temperature and volume (T and V) are held constant for a gas, pressure is directly proportional to the number of gas molecules present: $p \propto N$ (● Fig. 5.14). It is logical that the greater the number of molecules, the greater is the number of collisions with the sides of the container.

Pressure and Temperature

If the volume and number of molecules (V and N) are held constant for a gas, pressure is directly proportional to the Kelvin temperature: $p \propto T$ (● Fig. 5.15). As T increases, the molecules move faster and strike the container walls harder and more frequently. No wonder the pressure increases!

Be aware that many accidental deaths have resulted from lack of knowledge of how pressure builds up in a closed container when it is heated. An explosion can result. A discarded spray can is a good example of such a dangerous container.

Pressure and Volume

If the number of molecules and the Kelvin temperature (N and T) are held constant for a gas, pressure and volume are found to be inversely proportional: $p \propto 1/V$ (● Fig. 5.16). As the volume decreases, the molecules do not have as far to travel and have a smaller surface area to hit. It is logical that they exert more pressure (more force per unit area) than they did before. This relationship was found in 1662 by Robert Boyle (Section 11.2) and is called *Boyle's law*.

The Ideal Gas Law

Summarizing the factors affecting the pressure of a confined gas, we find that pressure (p) is directly proportional to the number of molecules (N) and the Kelvin temperature (T) and inversely proportional to the volume (V); that is,

$$p \propto \frac{NT}{V} \qquad \textbf{5.6}$$

This proportion can be used to make a useful equation for a given amount of gas (n is constant). In this case, the relationship $p \propto T/V$ applies at any time, and we may write a ratio form of the **ideal gas law** as

$$\frac{p_2}{p_1} = \left(\frac{V_1}{V_2}\right)\left(\frac{T_2}{T_1}\right) \qquad \textbf{5.7}$$

(N, number of molecules, constant)

where the subscripts indicate conditions at different times.

EXAMPLE 5.4

Changing Conditions

A closed, rigid container holds a particular amount of hydrogen gas that behaves like an ideal gas. If the gas is initially at a pressure of 1.80×10^6 Pa at room temperature (20°C), what will be the pressure if the gas is heated to 40°C? (See Fig. 5.15.)

SOLUTION

Let's work this in steps for clarity.

Step 1

Given: $V_1 = V_2$ (rigid container)

$p_1 = 1.80 \times 10^6$ Pa (or N/m^2)

$T_1 = 20°C + 273 = 293$ K

$T_2 = 40°C + 273 = 313$ K

Note that the temperatures were converted to kelvins. This is a *must* when using the ideal gas law.

Step 2

Wanted: p_2 (new pressure)

Step 3

Equation 5.7 may be used directly. Since $V_1 = V_2$, the volumes cancel, and we have

$$p_2 = \left(\frac{T_2}{T_1}\right)p_1 = \left(\frac{313 \text{ K}}{293 \text{ K}}\right)(1.80 \times 10^6 \text{ Pa}) = 1.92 \times 10^6 \text{ Pa}$$

and the pressure increases, as would be expected, since $p \propto T$.

CONFIDENCE EXERCISE 5.4

Suppose it was observed that the pressure of the gas in Example 5.4 doubled. What would be the final temperature in kelvins in this case?

Some applications of the ideal gas law are given in the chapter's second Highlight.

RELEVANCE QUESTION: *An inflated balloon is put in the freezer compartment of a refrigerator. What happens to the balloon, and why? (Explain in terms of what you learned in this section.)*

FIGURE 5.14   Pressure and Molecules

In both containers, the temperature and volume are constant. However, in the container in (b) there are twice as many molecules as in the container in (a). This causes the pressure to be twice as great, as indicated on the gauge. (More molecules, more collisions, and greater pressure.)

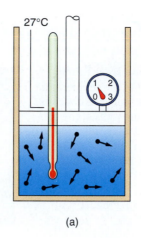

(a)

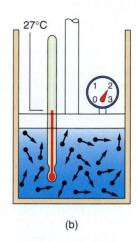

(b)

FIGURE 5.15   Pressure and Kelvin Temperature

In both containers, the number of molecules and the volume are constant. However, the gas in (b) has been heated to twice the Kelvin temperature of that in (a), that is, 600 K (327°C) versus 300 K (27°C). This causes the pressure to be twice as great, as shown on the gauge. (Higher temperature, more kinetic energy, more collisions, and greater pressure.)

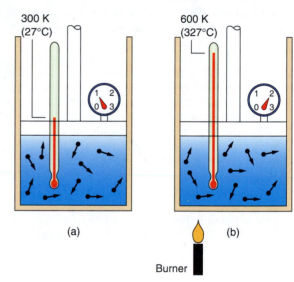

(a)

Burner

(b)

FIGURE 5.16   Pressure and Volume

In both containers, the temperature and the number of molecules are constant. However, the container in (b) has only half the volume of the container in (a). This causes the pressure to be twice as great, as shown on the gauge. (Same average kinetic energy, but less distance to travel, on average, in a smaller volume, so more collisions.)

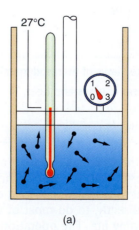

(a)

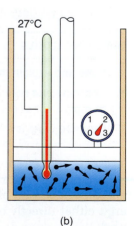

(b)

HIGHLIGHT: Hot Gases: Aerosol Cans and Popcorn

Suppose a trapped gas (constant volume) is heated. What happens? As the temperature of the gas increases, its molecules become more active and collide with the walls of the container more frequently, and the pressure increases. This may be seen from the ideal gas law in the form $p \propto NT/V$ (Eq. 5.6), or with the number of molecules (N) and volume (V) constant, the pressure (p) will change in direct proportion to the change in Kelvin temperature (T); that is,

$$p \propto T$$

Continued heating and temperature increase may cause the gas pressure to build up to the point that the container is ruptured or explodes. This could be a dangerous situation for an aerosol can, and warnings to this effect are printed on can labels (Fig. 1a).

A more beneficial case of a hot gas explosion is for popcorn. When heated, moisture inside the popcorn kernel is vaporized and trapped therein. Continued heating raises the gas (steam) pressure until it becomes great enough to rupture the kernel (Fig. 1b). The "explosion" causes the cornstarch inside to expand up to 40 times its original size. (Butter and salt anyone?)

Recall that we had another example of using hot gases in the Chapter 4 Highlight on air bags. Here, a chemical explosion occurred, and the rapidly expanding hot gas inflated the air bag. Could you analyze this situation in terms of the ideal gas law?

(a)

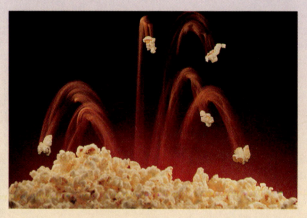

(b)

FIGURE 1   Hot Gases

Hot gases can be dangerous in some situations (a) but beneficial in others (b).

5.7 Thermodynamics

LEARNING GOALS

▼ State and explain the three laws of thermodynamics.

▼ Compare and contrast heat engines and heat pumps.

▼ Describe what is meant by *entropy*.

Thermodynamics means the dynamics of heat and deals with the production of heat, the flow of heat, and the conversion of heat to work. We use heat energy, either directly or indirectly, to do most of the work that is done in everyday life. The operation of heat engines, such as internal-combustion engines, and of refrigerators is based on the laws of thermodynamics. These laws are important because they state the relationships between heat energy, work, and the directions that thermodynamic processes may occur.

First Law of Thermodynamics

Because one aspect of thermodynamics is concerned with energy transfer, accounting for the energy involved in a thermodynamic process is important. This accounting is done by the principle of the conservation of energy, which, as you may recall, states that energy can neither be created nor destroyed. The first law of thermodynamics is simply the principle of conservation of energy applied to the thermodynamic processes.

Suppose that some heat (H) is added to a system. Where does it go? One possibility is that it goes into increasing the system's internal energy (ΔE_i). Another possibility is that it could result in work (W) being done by the system. Or both possibilities could occur. Thus *heat added to a closed system goes into the internal energy of the system and/or doing work.*

For example, consider heating an inflated balloon. As heat is added to the system (the balloon and the air inside), the temperature increases, and the system expands. The temperature of the air inside the balloon increases because some of the heat goes into the internal energy of the air. The gas expands and does work in expanding the balloon (work done by the system).

The **first law of thermodynamics** expresses this and other such energy balances and may be written in general as follows:

heat added to (or removed from) a system	= change in internal energy of the system	+ work done by (or on) the system
or	$H = \Delta E_i + W$	**5.8**

For this equation, a positive value of heat ($+H$) means that heat is *added to* the system, and a positive value of work ($+W$) means that work is *done by* the system. (These were the cases in the preceding heated balloon example.) Negative values indicate the opposite conditions.

Here are a few simple examples to illustrate the sign convention of the first law. Suppose heat is added to a gas in a rigid container. No expansion can take place (rigid container), and no work is done ($W = 0$). By the first law, then, $H = \Delta E_i$, and all the heat goes into the internal energy of the system and increases the temperature of the gas.

Say work is done on a perfectly insulated system; for example, a gas in a piston-cylinder arrangement is compressed. *Perfectly insulated* means that no heat can escape from the system, or $H = 0$. Then, from the first law, $H = 0 = \Delta E_i - W$ (negative W, work done *on the* system). So we have $\Delta E_i = W$, and ΔE_i is positive, which means that the internal energy of the gas increases.

Finally, suppose a gas under pressure expands without heat being added. (Air rapidly escaping from an automobile tire is an example.) Here, $H = 0$ and work is positive, $+W$, so we have $H = 0 = \Delta E_i + W$ and

$\Delta E_i = -W$. This means that the internal energy of the gas is decreased (work done in expanding against the atmosphere at the expense of internal energy), so the temperature of the gas decreases. This is why the valve stem of a tire feels cool when a lot of air escapes rapidly.

Heat Engines and Thermal Efficiency

Another good example of the first law is the heat engine. A **heat engine** is a device that converts heat into work. Many types of heat engines exist: gasoline engines on lawn mowers and in cars, diesel engines in trucks, and steam engines in old-time locomotives. They all operate on the same principle. Heat input—for example, from the combustion of fuel—goes into doing useful work, but some of the input energy is lost or wasted.

In thermodynamics, we are not concerned with the components of an engine but rather its general operation. We may represent a heat engine schematically as illustrated in ● Fig. 5.17.

A heat engine operates between a high-temperature reservoir and a low-temperature reservoir. These reservoirs are systems from which heat may be readily absorbed and to which heat may be readily expelled. In the process, the engine uses some of the input energy to do work. Notice that the widths of the heat and work paths in the figure are in keeping with the conservation of energy:

$$\text{heat in} = \text{work} + \text{heat out}$$

or

$$\text{work} = \text{heat in} - \text{heat out}$$

or, in symbols,

$$W = H_{\text{hot}} - H_{\text{cold}} \qquad \textbf{5.9}$$

The conversion of heat into work is expressed in terms of **thermal efficiency.** *Similar to mechanical efficiency, it is a ratio of the work output and the energy input. That is,*

thermal efficiency $= \dfrac{\text{work output}}{\text{heat input}} \times 100\%$	
or $\qquad \varepsilon_{\text{th}} = \dfrac{W}{H_{\text{hot}}} \times 100\%$	**5.10**
or, using Eq. 5.9, $\qquad \varepsilon_{\text{th}} = \dfrac{H_{\text{hot}} - H_{\text{cold}}}{H_{\text{hot}}} \times 100\%$	**5.10a**

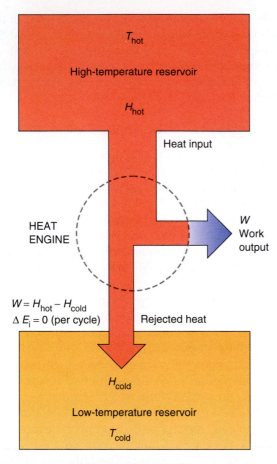

FIGURE 5.17   Schematic Diagram of a Heat Engine

A heat engine takes heat, H_{hot}, from a high-temperature reservoir, converts some to useful work, W, and rejects the remainder, H_{cold}, to a low-temperature reservoir.

For example, if a heat engine absorbs 1000 J each cycle and rejects 400 J, it does 600 J of work. Then the engine has an efficiency of 600 J/1000 J = 0.60, or 60%. This efficiency is quite high. The efficiency of an automobile is on the order of 15%; that is, 85% of the energy from fuel combustion is wasted or goes into doing nonessential work not associated with moving the car—for example, running a tape or CD player.

Second Law of Thermodynamics

As we have seen, the first law of thermodynamics is concerned with the conservation of energy. As long as the energy check sheet is balanced, the first law is satisfied. Suppose we had a hot body at 100°C and a colder body at room temperature (20°C), and when placed in contact, heat flowed from the colder body to the hotter body. The energy is easily accounted for,

and the first law is satisfied. But something is physically wrong. As we know, heat does not spontaneously flow from a colder body to a hotter body (● Fig. 5.18). (This would be like heat flowing up a "temperature hill," analogous to a ball spontaneously rolling *up* a hill rather than down.)

Something more than the first law is required to describe a thermodynamic process. As you might have guessed, we have to know the direction of the process, that is, whether or not something actually occurs. The **second law of thermodynamics** tells us what can or cannot happen thermodynamically. This law can be stated in several ways depending on the situation. A common statement of the second law as applied to our preceding example of heat flow is

It is impossible for heat to flow spontaneously from a colder body to a hotter body.

Another statement of the second law applies to heat engines. Suppose that a heat engine operated so that

FIGURE 5.18   Heat Flow

(a) When objects are in thermal contact, heat flows spontaneously from a hotter object to a colder object until they are at the same temperature, or come to thermal equilibrium. (b) Heat never flows spontaneously from a colder object to a hotter one. That is, a cold object never gets colder when placed in thermal contact with a warm object.

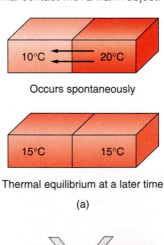

Occurs spontaneously

Thermal equilibrium at a later time

(a)

Does *not* occur spontaneously

(b)

all the heat input was converted into work. Such an engine doesn't violate the first law, but by Eq. 5.10 it would have a thermal efficiency of 1.0, or 100%, which has never been observed. Then the second law as applied to heat engines may be stated as

No heat engine operating in a cycle can convert thermal energy completely into work.

Another way of saying this is that no heat engine operating in a cycle can have 100% efficiency.

An engine must lose some heat. If this were not the case, with 100% thermal efficiency we could put the work output back into the engine as heat input and it would run perpetually. Since the efficiency cannot be as great as 100%, what is the maximum efficiency that can be obtained?

French engineer Sadi Carnot (1796–1832) studied this question and came up with the answer. He showed that the thermal efficiency of a heat engine is limited by the operating temperatures (temperatures of the hot and cold reservoirs; see Fig. 5.17). This **ideal** or **Carnot efficiency** (ε_C) is given by

$$\varepsilon_C = \frac{T_{hot} - T_{cold}}{T_{hot}} \times 100\%$$

$$= 1 - \frac{T_{cold}}{T_{hot}} \times 100\% \qquad \textbf{5.11}$$

where T_{hot} and T_{cold} are the *absolute* temperatures of the high-temperature reservoir and the low-temperature reservoir, respectively. *Don't forget* that the temperatures must be expressed as absolute temperatures on the Kelvin scale.

The actual efficiency of a heat engine will always be less than its ideal Carnot efficiency. The ideal Carnot efficiency sets an upper limit, but this limit can never be achieved.

EXAMPLE 5.5

Finding the Carnot Efficiency

What is the Carnot efficiency of a coal-fired steam power plant operating between temperatures of 300°C and 100°C? (Express as a percentage.)

SOLUTION

First, use Eq. 5.1 ($T_K = T_c + 273$) to convert the temperatures to their Kelvin equivalents:

$$T_{hot} = 300 + 273 = 573 \text{ K}$$

and
$$T_{cold} = 100 + 273 = 373 \text{ K}$$

Then, by Eq. 5.11,

$$\varepsilon_C = 1 - \frac{T_{cold}}{T_{hot}} \times 100\%$$

$$= 1 - \frac{373 \text{ K}}{573 \text{ K}} \times 100\% = 35\%$$

When other losses are taken into account, an actual efficiency of 32% or so is typical for a power plant.

CONFIDENCE EXERCISE 5.5

Suppose you could raise the power plant's high operating temperature (T_{hot}) by 50°C or lower the low operating temperature by 50°C. Which would you choose in order to increase the Carnot efficiency?

Notice from Eq. 5.11 that the second law forbids a cold reservoir of $T_{cold} = 0$ K. If absolute zero could be used, then we could have an ideal efficiency of 100%, which would violate the second law. Actually, a temperature of absolute zero cannot be attained; that is, thermodynamically,

It is impossible to attain a temperature of absolute zero.

This is sometimes called the **third law of thermodynamics.**

Scientists have tried to reach absolute zero and have achieved a low temperature of about 20 nK (nanokelvins), or 20×10^{-9} K, but they have never reached absolute zero.

The third law has still never been violated experimentally. It becomes more difficult to lower the temperature of a material (pump heat from it) the closer the temperature gets to absolute zero. The difficulty increases with each step, to the point at which an infinite amount of work would be required to reach the very bottom of the temperature scale.

Heat Pumps

As we have seen, the second law states that heat will not flow *spontaneously* from a colder body to a hotter body, or up a "temperature hill," so to speak. The analogy of a ball spontaneously rolling up a regular hill was used. Of course, we can get a ball to roll up a hill by applying a force and doing work on it. Similarly, we can get heat to flow up the "temperature hill" by doing work. This is the principle of a heat pump.

A **heat pump** is a device that uses work input to transfer heat from a low-temperature reservoir to a high-temperature reservoir (● Fig. 5.19). Work input is required to "pump" heat from a low-temperature reservoir to a high-temperature reservoir. Essentially, a heat pump is the reverse of a heat engine.

Refrigerators and air conditioners are examples of heat pumps. Heat is transferred from the inside volume of a refrigerator to the outside by the compressor doing work on a gas (and the expenditure of electrical energy). The heat transferred to the room (high-temperature reservoir) is much greater than the electrical energy used to perform the work. Similarly, an air conditioner transfers heat from the inside of a home or car to the outside.

The heat pumps used for home heating and cooling have a descriptive name. In the summer, they operate as air conditioners. In the winter, heat is extracted from the outside air or from a water source and pumped inside the home for heating. Heat pumps are used extensively in the South, where the climate is mild. In places with very cold winter months, an auxiliary heating unit (usually an electric heater) must supply extra heat when needed.

A heat pump is generally more expensive than a regular furnace, but it has long-term advantages. It has no associated fuel costs (other than the cost of the electricity to supply the work input), because it takes heat from the air or from water in a reservoir such as a well or a system of underground coils.

Entropy

You may have heard the term *entropy* and wondered what it means. Thermodynamically speaking, **entropy** is a mathematical quantity. Its change tells whether or not a process can take place naturally. Hence it is associated with the second law.

FIGURE 5.19 Schematic Diagram of a Heat Pump
Work input is required to pump heat from a low-temperature reservoir to a high-temperature reservoir. This may be an air conditioner that pumps heat to the outside in the summer or a household heat pump that can act as an air conditioner in the summer. In the winter, a heat pump extracts heat from outside air or in-ground water and pumps it indoors.

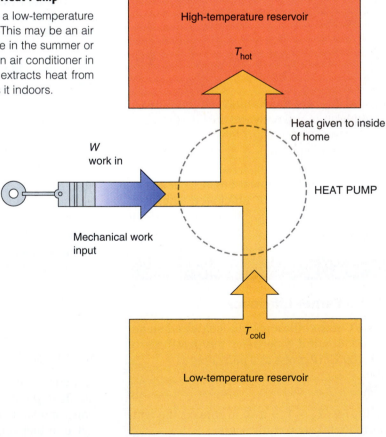

High-temperature reservoir

T_{hot}

Heat given to inside of home

HEAT PUMP

W
work in

Mechanical work input

T_{cold}

Low-temperature reservoir

In terms of entropy, the second law of thermodynamics can be stated as

The entropy of an isolated system never decreases.

To help understand this idea, we sometimes say that entropy is a measure of the disorder of a system. When heat is added to an object, its entropy increases because the added energy increases the disordered motion of the molecules. As a natural process takes place, the disorder increases. For example, when a solid melts, the molecules are freer to move in a random motion in the liquid phase than in the solid phase. Similarly, when evaporation takes place, the result is greater disorder and increased entropy.

Systems that are left to themselves tend to become more and more disordered—never the reverse. By analogy, a student's dormitory room or room at home naturally becomes disordered—never the reverse. Of course, the room can be cleaned and items put in order, and the entropy *of the room system* decreases. But to put things back in order, someone must expend energy, with a greater entropy increase than the room's entropy decrease. We sometimes say that *the total entropy of the universe increases in every natural process.*

Here's another long-term implication of the second law of thermodynamics: Heat naturally flows from a region of higher temperature to one of lower temperature. In terms of order, heat energy is more "orderly" when it is concentrated. When transferred naturally to a region of lower temperature, it is "spread out" or more "disorderly," and the entropy increases. Hence the universe—the stars and the galaxies—eventually should cool down to a final common temperature when the entropy of the universe has reached a maximum. This possible fate, billions of years from now, is sometimes referred to as the "heat death" of the universe.

RELEVANCE QUESTION: In terms of the first law of thermodynamics, why does exercise help keep a person from gaining weight?

Important Terms

temperature (5.1)	specific heat (5.3)	liquid	thermal efficiency
Fahrenheit scale	latent heat of fusion	gas	second law of
Celsius scale	latent heat of	plasma	thermodynamics
Kelvin scale	vaporization	kinetic theory (5.6)	Carnot (ideal) efficiency
kelvin	conduction (5.4)	pressure	third law of
heat (5.2)	convection	ideal gas law	thermodynamics
calorie	radiation	first law of	heat pump
kilocalorie	phases of matter (5.5)	thermodynamics (5.7)	entropy
Btu	solid	heat engine	

Important Equations

Kelvin-Celsius Conversion: $T_K = T_C + 273$

Fahrenheit-Celsius Conversion:

$T_F = \frac{9}{5}T_C + 32$ (or $T_F = 1.8T_C + 32$)

$T_C = \frac{5}{9}(T_F - 32)$ [or $T_C = (T_F - 32)/1.8$]

Specific Heat: $H = mc\Delta T$

$c_{water} = 1.00$ kcal/kg-C°

$c_{ice} = c_{steam} = 0.50$ kcal/kg-C°

Latent Heat: $H = mL$

(water) $\begin{array}{l} L_f = 80 \text{ kcal/kg} \\ L_v = 540 \text{ kcal/kg} \end{array}$

Changing Conditions for a Gas:

$$p \propto \frac{NT}{V}$$

Ideal Gas Law:

$$\frac{p_2}{p_1} = \left(\frac{V_1}{V_2}\right)\left(\frac{T_2}{T_1}\right)$$

First Law of Thermodynamics: $H = \Delta E_i + W$

Thermal Efficiency:

$$\varepsilon_{th} = \frac{W}{H_{hot}} \times 100\% = \frac{H_{hot} - H_{cold}}{H_{hot}} \times 100\%$$

$$= 1 - \frac{H_{cold}}{H_{hot}} \times 100\%$$

Carnot (Ideal) Efficiency: $\varepsilon_C = \dfrac{T_{hot} - T_{cold}}{T_{hot}} \times 100\%$

$$= 1 - \frac{T_{cold}}{T_{hot}} \times 100\%$$

Review Questions

5.1 Temperature

1. Temperature is
   (a) the same as heat.
   (b) always measured with a liquid-in-glass thermometer.
   (c) commonly expressed in kelvins.
   (d) a measure of the average kinetic energy of the molecules of a substance.

2. Which temperature scale has the smallest degree interval?
   (a) Celsius
   (b) Fahrenheit
   (c) Kelvin
   (d) absolute

3. (a) Why is temperature a *relative* measurement?
   (b) On what property does temperature measurement commonly depend, and how is it applied?

4. What are the ice and steam points on the Fahrenheit, Celsius, and Kelvin scales?

5. A thermostat of the type used in furnace (and heat pump) control is shown in ● Fig. 5.20. The glass vial tilts back and forth so that electrical contacts are made via the mercury (an electrically conducting liquid metal), and the furnace is turned off and on at a set temperature. Explain why the vial tilts back and forth.

5.2 Heat

6. Which of the following best describes heat?
   (a) energy in transit
   (b) the same as temperature
   (c) the internal energy of a substance
   (d) unrelated to energy

7. Which of the following is the largest unit of heat energy?
   (a) kilocalorie   (b) calorie   (c) joule   (d) Btu

8. Explain what is meant by the statement: Heat is energy in transit.

9. What is the difference between a calorie and a Calorie?

10. What is the British system unit of heat, and how does it compare in magnitude with metric units?

11. Why do sidewalks have joints built into them?

5.3 Specific Heat and Latent Heat

12. From the equation $H = mc\Delta T$, we see that the units of specific heat are which of the following?
    (a) J/kg-C°
    (b) kcal-kg/J
    (c) kcal/kg
    (d) kcal-kg/C°

FIGURE 5.20   An Exposed View of a Thermostat

See Review Question 5.

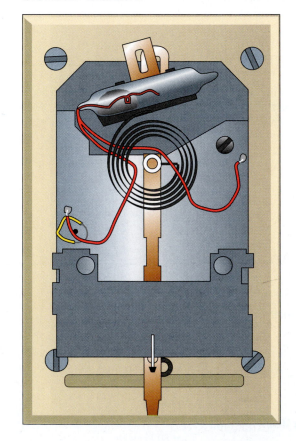

13. Sublimation refers to a phase change of which of the following?
    (a) solid to liquid
    (b) solid to gas
    (c) gas to liquid
    (d) none of these

14. What is *specific* about specific heat?

15. Water is a common heat-storage medium in solar heating. Why?

16. Why does it take a long time for lakes to freeze when the air temperature is below freezing?

17. When does the addition of heat *not* result in a temperature increase?

18. Compare the SI units of specific heat and latent heat, and explain any differences.

5.4 Heat Transfer

19. Much of the thermal conductivity in metals is contributed by which of the following?
    (a) radiation
    (b) electrons
    (c) entropy
    (d) latent heat

20. Which of the following methods of heat transfer generally involve mass movement?
    (a) conduction
    (b) convection
    (c) radiation

21. What are several examples of good thermal conductors and good thermal insulators? In general, what makes a substance a conductor or an insulator?

22. Why does a vinyl tile floor feel colder to the bare feet than a rug, even though they are both at the same temperature?

23. Explain how thermos bottles keep things both hot and cold.

24. (a) Is an open fireplace an efficient way of heating a room? Explain.
    (b) Thermal underwear is knitted with large holes in it (● Fig. 5.21). Doesn't this defeat the purpose of thermal underwear? Explain.

5.5 Phases of Matter

25. In which of the following is intermolecular bonding greatest?
    (a) solids   (c) gases
    (b) liquids   (d) plasmas

26. What determines the phase of a substance?

27. Give descriptions of a solid, a liquid, and a gas in terms of shape and volume.

5.6 The Kinetic Theory of Gases

28. Pressure is defined as which of the following?
    (a) force
    (b) force times area
    (c) area divided by force
    (d) force divided by area

29. When using the ideal gas law, the temperature must be in which of the following units?
    (a) °C   (b) °F   (c) K

30. How does the kinetic theory describe a gas?

31. What is meant by an *ideal* gas? When does a real gas not approximate an ideal gas?

32. What causes gas pressure?

33. In terms of the kinetic theory, explain why a basketball stays inflated.

5.7 Thermodynamics

34. When heat is added to a system, it goes into which of the following?
    (a) doing work
    (b) adding to the internal energy
    (c) doing work and/or increasing the internal energy

FIGURE 5.21   Holey Heat Transfer

See Review Question 24.

35. The direction of a natural process is indicated by which of the following?
    (a) conservation of energy
    (b) change in entropy
    (c) thermal efficiency
    (d) specific heat

36. A balloon is put in a refrigerator freezer and it shrinks. How does the first law of thermodynamics apply to this case?

37. What do the first and second laws of thermodynamics tell you about a thermodynamic process?

38. Distinguish between thermal efficiency and Carnot (ideal) efficiency. Expressed as a percentage, efficiency is a percentage of what?

39. Why is approximately two-thirds of the heat generated by an electrical generating plant wasted? What happens to the waste heat?

40. Compare and explain the following statements in terms of the laws of thermodynamics.
    (a) Energy can be neither created nor destroyed.
    (b) Entropy can be created but not destroyed.

Applying Your Knowledge

1. On a cold day when someone leaves an outside door open, it is often said that the person is letting in cold air. Is this correct? Explain.

2. When eating a pizza or a hot pie, the crust may be only warm, but you may burn your mouth when eating the topping or filling. Why is this?

3. When you freeze ice cubes in a tray, there is a decrease in entropy because there is more order in the crystalline lattice of the ice. Are you violating the second law? Explain.

4. Why should you never burn trash that contains an aerosol can?

5. Do any of the following statements have any general association to the laws of thermodynamics? (a) You can't get something for nothing. (b) You can't even break even. (c) You can't sink that low.

Exercises

(Assume temperatures to be exact numbers in the Exercises.)

5.1 Temperature

1. While in Europe, a tourist hears on the radio that the temperature that day will reach a high of 15°C. What is this temperature on the Fahrenheit scale?

    Answer: 59°F

2. A recipe calls for a cake to be baked at 180°C. On what temperature (Fahrenheit) should the oven be set?

3. Normal room temperature is about 68°F. What is the equivalent temperature on the Celsius scale?

    Answer: 20°C

4. Normal body temperature is 98.6°F. What is the equivalent temperature on the Celsius scale?

5. Researchers in the Antarctic measure the temperature to be −40°F. What is this temperature on (a) the Celsius scale and (b) the Kelvin scale?

    Answer: (a) −40°C   (b) 233 K

6. The temperature of outer space is about 3 K. What is this temperature on (a) the Celsius scale and (b) the Fahrenheit scale?

5.2 Heat

7. A skier comes down a slope from a height of 50 m, reaching a speed of 10 m/s at the bottom. If the skier's mass is 60 kg, how much heat in kilocalories was lost to friction? (*Hint:* See E_K and E_p in Chapter 4.)

    Answer: 6.2 kcal

8. A ball dropped from a height of 2.5 m strikes the ground with a speed of 6.6 m/s. What percentage of the ball's potential energy was lost because of air resistance?

9. A college student produces about 100 kcal of heat per hour on the average. What is the rate of energy production in watts?

    Answer: 116 W

10. How many kilocalories of heat does a 1200-W hair dryer produce each second?

11. A piece of pie contains 100 Cal of food energy. How would the piece of pie be rated in Btu? Answer: 397 Btu

12. A small window air conditioner is rated at 12,000 Btu. How many kilocalories does the unit remove in 2.0 h? (*Hint:* Recall the "Btu" rating is actually Btu per hour.)

5.3 Specific Heat and Latent Heat

13. How much heat in kcal must be added to 0.50 kg of water at room temperature (20°C) to raise its temperature to 30°C?   Answer: 5.0 kcal

14. How much heat in joules is required to raise the temperature of 1.0 L of water from 0°C to 100°C? (*Hint:* Recall the original definition of the liter.)

15. (a) How much energy is required to heat 1.0 kg of water from room temperature (20°C) to its boiling point? (Assume no energy loss.)
    (b) Assuming electrical energy was used, how much would this cost at 12¢ per kWh?
                              Answer: (a) 80 kcal   (b) 1.1¢

16. Equal amounts of heat are added to equal masses of aluminum and copper at the same initial temperature. Which metal will have the higher final temperature and how many times greater will that temperature change be than the temperature change of the other metal?

17. How much heat is required to change 500 g of ice at −10°C to water at 20°C?   Answer: 52.5 kcal

18. A quantity of steam (200 g) at 110°C is condensed and the resulting water is frozen into ice at 0°C. How much heat was removed?

5.6 The Kinetic Theory of Gases

19. A sample of neon gas has its volume tripled and its temperature held constant. What will be the new pressure relative to the initial pressure?   Answer: $p_2 = p_1/3$

20. A fire breaks out and increases the Kelvin temperature of a cylinder of compressed gas by a factor of 1.2. What is the final pressure of the gas relative to its initial pressure?

21. A cylinder of gas is at room temperature (20°C). The air conditioner breaks down, and the temperature rises to 40°C. What is the new pressure of the gas relative to its initial pressure?   Answer: $p_2 = 1.07p_1$

22. A cylinder of gas at room temperature has a pressure p_1. To what temperature in degrees Celsius would the temperature have to be increased for the pressure to be $1.5p_1$?

23. A quantity of gas in a piston cylinder has a volume of 0.500 m^3 and a pressure of 200 Pa. The piston compresses the gas to 0.150 m^3 in an isothermal (constant-temperature) process. What is the final pressure of the gas?   Answer: 667 Pa

24. If the gas in Exercise 23 is initially at room temperature (20°C) and is heated in an isobaric (constant-pressure) process, what will be the temperature of the gas in degrees Celsius when it has expanded to a volume of 0.600 m^3?

25. A quantity of (ideal) gas in an industrial machine has a pressure and volume of 2.6×10^6 Pa and 0.40 m^3, respectively, at a temperature of 100°C. The gas expands to 0.80 m^3, and the pressure is reduced to 1.3×10^6 Pa. What is the final temperature of the gas in kelvins?
                              Answer: 373 K

26. Suppose the gas in Exercise 25 was cooled and its temperature fell to 50°C while the volume was reduced to 0.20 m^3. What would be the final pressure of the gas?

5.7 Thermodynamics

27. A heat engine takes 300 kcal of heat and rejects 180 kcal each cycle. What is the engine's thermal efficiency?
                              Answer: 40%

28. A heat engine has a thermal efficiency of 25%. How much work in joules is obtained from a 100-kcal heat input per cycle?

29. A coal-fired power plant has operating temperatures of $T_{hot} = 270°C$ and $T_{cold} = 100°C$. What can you say about the efficiency of the power plant?
                  Answer: ε_{th} must be less than 31%

30. A proposed nuclear electrical generation plant would have $T_{hot} = 350°C$ and $T_{cold} = 100°C$, with a thermal efficiency of 42%. Would you support this project and buy stock in the electric company? Justify your answer.

Solutions to Confidence Exercises

5.1 $T_F = \frac{9}{5}T_C + 32 = \frac{9}{5}(-15) + 32 = 5°F$

5.2 $H = mc\Delta T = (1.0\text{ kg})(1.0\text{ kcal/kg-C°})(5°C - 20°C) = -15$ kcal

5.3 $H_1 = mc\Delta T = (0.20\text{ kg})(1.0\text{ kcal/kg-C°})(0°C - 10°C) = -2.0\text{ kcal}$
$H_2 = mL_f = (0.20\text{ kg})(-80\text{ kcal/kg}) = -16\text{ kcal}$ (minus because removed)
$H_T = H_1 + H_2 = (-2.0\text{ kcal}) + (-16\text{ kcal}) = -18\text{ kcal}$

5.4 $p_2 = 2p_1$, and $T_2 = (p_2/p_1)T_1 = (2)(293\text{ K}) = 586\text{ K}$

5.5 Lower the lower temperature. (Try the calculation each way to prove this.)

Answers to Multiple-Choice Review Questions

1. d   6. a   12. a   19. b   25. a   29. c   35. b
2. b   7. a   13. b   20. b   28. d   34. c

WAVES

6

When it came night, the white waves paced to and fro in the moonlight, and the wind brought the sound of the great sea's voice to those on shore, and they felt that they could then be interpreters.

Stephen Crane (1871–1900)

Since we began studying energy, much has been said about its forms, its relationship to work, and its conservation. Many questions have been raised and answered; other interesting questions remain. For example, how is energy changed from one form to another? How is energy transferred to the ball in a baseball game? How do our bodies obtain energy? How do we get energy from the Sun?

A partial answer to these questions may be found in terms of particle collisions and wave motion, two important ways of transferring energy (● Fig. 6.1). If a particle applies a force to another particle through a distance, then a transfer of energy has taken place through particle collision; one particle will gain energy, and the other will lose energy.

If we are to believe the conservation of energy (and we know of no instance where the principle has failed), then the increase in energy of one particle will be equal to the decrease in energy of the other, assuming that we neglect heat losses. In all cases, we find that a transfer of energy takes place whenever two or more particles collide with one another.

Energy is transmitted not only by particle collision but also by wave motion. When matter is disturbed, energy emanates from the disturbance. This propagation of energy is known as a **wave.** For example, a stone dropped

Photo: Ride the wave! A surfer rides the wave of a breaking surf.

FIGURE 6.1 Energy Transfer

Some examples of transferring energy. Energy is transferred by collisions and wave motion.

ergy transmissions that require a medium. The neighboring particles of the medium act on one another through molecular bonding to transfer the disturbance. Electromagnetic waves, on the other hand, can be transferred without a medium. These disturbances, such as radio waves, infrared radiation, visible light, and X-rays, propagate through the vacuum of space.

The data from which we learn about the planets and stars come to us by means of light and other electromagnetic radiation. Strangely enough, light, which we consider to be a wave, has some of the properties of particles. In Chapter 9, where quantum mechanics is discussed, we will study the dual nature of light.

Our eyes and ears are two wave-detecting devices that link us to our environment. Study of wave motion is relevant to understanding our physical environment, and knowledge of wave motion is essential to understanding many scientific principles. ■

in a pond of water will disturb the water, and a wave (energy) will move outward from the disturbance. Only energy is transferred, not matter (the water), as can be noted by observing a floating fishing bobber that goes up and down with the water.

A similar situation can occur in a solid. For example, during an earthquake, a disturbance takes place because of a slippage or other cause, and this disturbance is transmitted through Earth as waves. Again, this disturbance is a transfer of energy, not matter.

A transfer of energy may take place with or without a medium. Sound waves in the air and waves on stretched strings and steel wires are examples of en-

6.1 Wave Properties

LEARNING GOALS

▼ Distinguish between longitudinal and transverse waves.

▼ Identify the terms used to describe waves.

A disturbance generating a wave may be a simple pulse or shock, such as a book hitting the floor. A disturbance also may be periodic—that is, repeated again and again at regular intervals. A vibrating guitar string or a whistle sets up periodic waves, and the waves are continuous (● Fig. 6.2).

FIGURE 6.2 Vibrations

A couple of examples of vibrational disturbances that produce (sound) waves.

Steel or nylon string

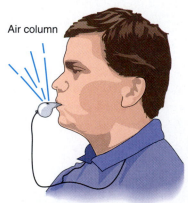

Air column

In general, waves may be classified into two categories, longitudinal and transverse, based on particle motion and wave direction. In a **longitudinal wave,** *the particle motion and the wave velocity are parallel to each other.* For example, consider a stretched spring, as illustrated in ● Fig. 6.3. When several coils at one end are compressed and released, the disturbance is propagated along the length of the spring with a certain wave velocity. Note that the displacements of the spring "particles" and the wave velocity vector are parallel to each other. The directions of the "particle" oscillations can be seen by tying a small piece of ribbon to the spring. The ribbon will oscillate similarly to the way the particles of the spring oscillate.

The single disturbance in the spring is an example of a longitudinal wave pulse. Periodic longitudinal waves are common. As we will learn shortly, sound is a longitudinal wave.

In a **transverse wave,** *the particle motion is perpendicular to the direction of the wave velocity.* A transverse wave may be generated by shaking one end of a stretched cord up and down or side to side (● Fig. 6.4). Notice how the cord "particles" oscillate perpendicularly (that is, at an angle of 90°) to the direction of the wave velocity vector. The particle motion may again be demonstrated by tying a small piece of ribbon to the cord.

Another example of a transverse wave is light. Recall that visible light and other electromagnetic radiation can propagate through space without a medium. What oscillates in this case are the electric and magnetic fields. As we will find in Section 6.2, these fields oscillate perpendicularly to the direction of wave motion. Hence all electromagnetic radiations are transverse waves.

Certain wave characteristics are used in describing periodic wave motion (● Fig. 6.5). The wave velocity describes the speed and direction of the wave motion. The **wavelength** (λ, Greek lambda) is the distance from any point on the wave to the adjacent point with

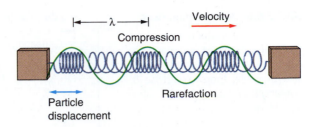

FIGURE 6.3   Longitudinal Wave

In a longitudinal wave, as illustrated here for a stretched spring, the wave velocity (vector), or direction of the wave propagation, is parallel to the (spring) particle displacements (back and forth).

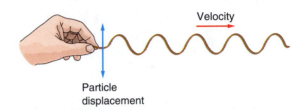

FIGURE 6.4   Transverse Wave

In a transverse wave, as illustrated here for a stretched cord, the wave velocity (vector), or the direction of the wave propagation, is perpendicular to the (cord) particle displacements (up and down).

similar oscillation, that is, the distance of one complete "wave," or where it starts to repeat itself. For example, the wavelength distance may be measured from one wave crest to an adjacent crest (or from one wave trough to the next wave trough).

The **amplitude** of a wave refers to the maximum displacement of any part of the wave (or wave particle) from its equilibrium position. The amplitude of the wave does not affect the wave speed. The energy transmitted by a wave is directly proportional to the square of its amplitude. (See amplitude in Fig. 6.5.)

FIGURE 6.5   Wave Description

Some terms used to describe wave characteristics. (See text for description.)

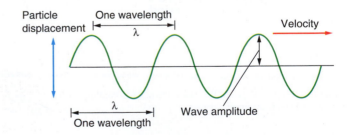

FIGURE 6.6 Wave Comparison

Wave (a) has a frequency of 4 Hz, which means that 4 wavelengths pass by a point in 1 second. With a wave of 8 Hz, there would be 8 wavelengths passing by in 1 second. Hence the wavelength of wave (a) is twice as long as the wavelength of wave (b). The period (*T*) is the time it takes one wavelength to pass by, so the period of wave (a) is also twice as long as that of wave (b). The relationship between the frequency and period is *f* = 1/*T*.

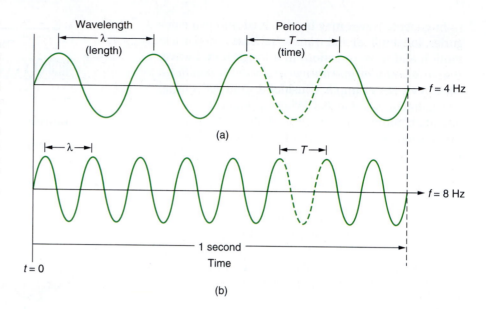

The oscillations of a wave may be characterized in terms of frequency. The wave **frequency** (*f*) is the number of oscillations or cycles that occur during a given period of time, usually a second. Frequency is the number of cycles per second, but this unit is given the name **hertz** (Hz).* One hertz is one cycle per second. For example, as illustrated in ● Fig. 6.6a, if four complete wavelengths pass a given point in 1 second, the frequency of the wave is four cycles per second, or 4 Hz. The more "wigglies" or wavelengths per time period, the greater is the frequency. A wave with a frequency of 8 Hz is illustrated in Fig. 6.6b.

Another quantity used to characterize a wave is the **period** (*T*). This is the time it takes for the wave to travel a distance of one wavelength (Fig. 6.6a). Looking at the dashed line, we see that a particle in the medium makes one complete oscillation in a time of one period.

The frequency and period are inversely proportional. Expressing this in equation form, we may write

$$\text{frequency} = \frac{1}{\text{period}}$$

or

$$f = \frac{1}{T} \qquad \textbf{6.1}$$

*After Heinrich Hertz (1857–1894), a German scientist and an early investigator of electromagnetic waves.

Examine the units. Frequency is cycles per second, and the period is the seconds per cycle.

Suppose a wave has a frequency of *f* = 4 Hz (Fig. 6.6a). Then four wavelengths pass by a point in 1 second, and one wavelength passes in $\frac{1}{4}$ second (*T* = 1/*f* = $\frac{1}{4}$ s). That is, the period, or time for one cycle, is $\frac{1}{4}$ second. What would be the period for the wave in Fig. 6.6b? (Right, $\frac{1}{8}$ s.)

Another simple relationship for wave characteristics relates the **wave speed** (*v*) to the wavelength and period (or frequency). Because speed is the distance divided by time, and a wave moves one wavelength in a time of one period, we may write

$$v = \frac{\lambda}{T} \qquad \textbf{6.2}$$

or, by Eq. 6.1,

$$v = \lambda f \qquad \textbf{6.3}$$

where *v* = wave speed measured in meters per second or other speed units

λ = wavelength, measured in length units

T = period of the wave, usually measured in seconds

f = frequency of the wave, measured in Hz (cycles/s)

EXAMPLE 6.1

Calculating Wavelengths

Consider sound waves with a speed of 344 m/s and frequencies of (a) 20 Hz and (b) 20 kHz. Find the wavelength of each of these sound waves.

SOLUTION

Step 1

Given: $v = 344$ m/s

(a) $f = 20$ Hz
(b) $f = 20$ kHz $= 20 \times 10^3$ Hz

(Notice that kHz was converted to Hz, the standard unit, and recall that Hz = 1/s.)

Step 2

Wanted: λ (wavelength)

Step 3

We rearrange Eq. 6.3, and solve for λ:

$$v = \lambda f$$

(a) $\qquad \lambda = \dfrac{v}{f}$

$$= \frac{344 \text{ m/s}}{20 \text{ Hz}} = \frac{344 \text{ m/s}}{20 \text{ 1/s}} = 17 \text{ m}$$

(b) $\qquad \lambda = \dfrac{v}{f} = \dfrac{344 \text{ m/s}}{20 \times 10^3 \text{ Hz}}$

$$= 17 \times 10^{-3} \text{ m} = 0.017 \text{ m}$$

CONFIDENCE EXERCISE 6.1

A sound wave has a speed of 344 m/s and a wavelength of 0.500 m. What is the frequency of the wave?

As we will see in Section 6.3, the frequencies of 20 and 20,000 Hz given in Example 6.1 define the general range of audible sound wave frequencies. Thus the wavelengths of sound cover the range from about 1.7 cm for the highest-frequency sound we can hear up to about 17 m for the lowest-frequency sound we can hear. In British units, sound waves range from wavelengths of approximately $\frac{1}{2}$ in. up to about 50 ft.

RELEVANCE QUESTION: Why do the cones in a stereo speaker vibrate? When do they vibrate more?

6.2 **Electromagnetic Waves**

LEARNING GOALS

▼ Identify various waves in the electromagnetic spectrum.

▼ List the speed of light.

When charged particles such as electrons are accelerated, energy is radiated away from them in the form of electromagnetic waves. **Electromagnetic waves** consist of vibrating electric and magnetic fields, which are more thoroughly studied in Chapter 8. They are vector fields. In electromagnetic waves, the field energy radiates outward at the speed of light, which is 3.00×10^8 m/s in vacuum.

An illustration of an electromagnetic wave is shown in ● Fig. 6.7. The wave is traveling in the x direction. The electric (E) and magnetic (B) field vectors are at angles of 90° to one another, and the velocity vector of the wave is at an angle of 90° to both the field vectors.

Charged particles are accelerated in many different ways to produce electromagnetic waves of various

FIGURE 6.7 Electromagnetic Wave

An illustration of the vector components of an electromagnetic wave. The wave consists of two force fields [electric (E) and magnetic (B)] oscillating perpendicularly to each other and to the direction of wave propagation (velocity vector).

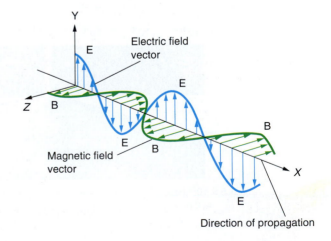

frequencies. Waves with relatively low frequencies, or long wavelengths, are known as *radio waves* and are produced primarily by causing electrons to oscillate, or vibrate, in an antenna. The frequency of oscillation is controlled by the physical dimensions and other properties of the driving circuit.

Electromagnetic waves with frequencies greater than radio waves and less than visible light are produced by molecular excitation. In such cases, radiation occurs from the collision of molecules in hot gases and solids. Because the molecules contain charged particles that are greatly accelerated as the molecules vibrate, the particles will radiate electromagnetic waves ranging from 10^{11} Hz to about 4.3×10^{14} Hz. This portion of the electromagnetic spectrum includes the *microwave* and *infrared* regions (● Fig. 6.8).

As the temperature of gases and solids is increased to higher and higher values, the atoms making up the molecules become more excited, and electromagnetic radiation in the *visible* and *ultraviolet* regions of the spectrum is emitted. Notice that the small portion of the spectrum visible to the human eye (Fig. 6.8) lies between the infrared (IR) and ultraviolet (UV) regions. It is the ultraviolet radiation in sunlight that tans and burns our skin.

Still more energy applied to the atom will generate waves of higher frequencies, called *X-rays*, which range from 3.0×10^{17} Hz to 3.0×10^{19} Hz. If sufficient energy to disturb the nucleus is applied to the atom,

radiation known as *gamma rays* is emitted, as discussed in Section 10.3.

The term *light* is commonly used for electromagnetic radiations in or near the visible region; for example, we say ultraviolet light. Only the frequency (or wavelength) distinguishes visible electromagnetic radiation from the other portions of the spectrum. Our human eyes are only sensitive to certain frequencies or wavelengths, but other instruments can detect other portions of the spectrum. For example, a radio receiver can detect radio waves.

Radio waves are *not* sound waves. If so, you'd "hear" many stations at once. Radio waves are electromagnetic waves that are detected and distinguished by frequency and then amplified by the circuits of the radio receiver. The signal is then demodulated—that is, the audio signal is separated from the radiofrequency carrier. The audio signal is amplified and then applied to the speaker system that produces sound waves.

Electromagnetic radiation consists of transverse waves. These waves can travel through a vacuum. For instance, radiation from the Sun travels through the vacuum of space before arriving at Earth. All electromagnetic waves travel at the same speed in a vacuum. This speed, called the **speed of light,** is designated by the letter c and has a value of

$$c = 3.00 \times 10^8 \text{ m/s}$$

or

$$c = 1.86 \times 10^5 \text{ mi/s}$$

FIGURE 6.8   Electromagnetic (EM) Spectrum

Different frequency (or wavelength) regions are given names.

Notice how the visible region forms only a very small part of the EM spectrum (much smaller than shown in the figure).

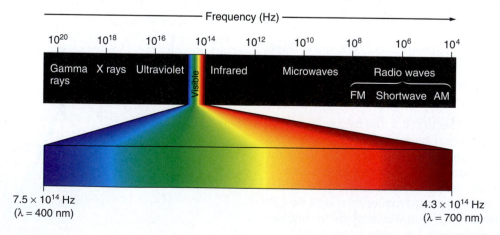

7.5 × 10¹⁴ Hz
(λ = 400 nm)

4.3 × 10¹⁴ Hz
(λ = 700 nm)

(To a good approximation, this value is also the speed of light in air.)

We can use Eq. 6.3 with $v = c$, that is, $c = \lambda f$, to find the wavelength of light or any electromagnetic radiation in a vacuum and to a good approximation in air. For example, let's calculate the wavelength of a typical radio wave.

EXAMPLE 6.2

Computing the Wavelength of a Radio Wave

What is the wavelength of the radio waves produced by a station with an assigned frequency of 600 kHz?

SOLUTION

First, we convert the frequency from kilohertz (kHz) to hertz. You can do this directly, since *kilo-* is 10^3.

$$f = 600 \text{ kHz} = 600 \times 10^3 \text{ Hz} = 6.00 \times 10^5 \text{ Hz}$$

Then, since radio waves are electromagnetic waves, we use Eq. 6.3 with the speed of light, $c = \lambda f$. Rearranging the equation,

$$\lambda = \frac{c}{f} = \frac{3.00 \times 10^8 \text{ m/s}}{6.00 \times 10^5 \text{ Hz}}$$

$$= 0.500 \times 10^3 \text{ m} = 500 \text{ m}$$

CONFIDENCE EXERCISE 6.2

The station in this example is an AM station, which generally uses kHz frequencies. FM stations have MHz frequencies. What is the wavelength of an FM station with an assigned frequency of 90.0 MHz?

Thus we see that the wavelengths of AM radio waves are quite long compared to FM radio waves, which have shorter wavelengths because the frequencies are higher. Visible light, with frequencies on the order of 10^{14} Hz, has relatively short wavelengths, as can be seen by the approximation

$$\lambda = \frac{c}{f} \approx \frac{10^8 \text{ m/s}}{10^{14} \text{ Hz}} = 10^{-6} \text{ m}$$

So the wavelength of visible light is on the order of one-millionth of a meter. To avoid using negative powers of 10, we commonly express the wavelengths in a smaller unit called the *nanometer* (nm, with 1 nm $= 10^{-9}$ m).

Using the values given in Fig. 6.8, you should be able to show that the approximate wavelength range for the visible region is between 4×10^{-7} and 7×10^{-7} m. This range corresponds to a span of 400 to 700 nm.

For visible light, different frequencies are perceived by the eye as different colors, and the brightness depends on the energy of the wave.

RELEVANCE QUESTION: Name a health hazard that arises from the ultraviolet radiation received from the Sun. How can we diminish the hazardous effects?

6.3 Sound Waves

LEARNING GOALS

▼ Define sound.

▼ Describe the sound spectrum and the decibel scale.

Technically, **sound** is defined as the propagation of longitudinal waves through matter. Sound waves involve longitudinal particle displacements in any kind of matter—solid, liquid, or gas.

We are most familiar with sound waves in air, which affect our sense of hearing. However, sound also travels in liquids and solids. When you are swimming underwater and someone clicks two rocks together, you can hear this disturbance. Also, we can hear sound through thin (but solid) walls.

The wave motion of sound depends on the elasticity of the medium. A longitudinal disturbance produces varying pressures and stresses in the medium. For example, consider a vibrating tuning fork, as illustrated in ● Fig. 6.9.

As an end of the fork moves outward, it compresses the air in front of it, and a *compression* is propagated outward. When the fork end moves back, it produces a region of decreased air pressure and density called a *rarefaction*. With the continual vibration, a series of high- and low-pressure regions travels outward, forming a longitudinal sound wave. The waveform may be displayed electronically on an oscilloscope, as shown in ● Fig. 6.10.

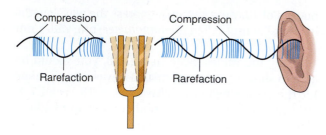

FIGURE 6.9 Sound Waves

Sound waves consist of a series of compressions (high-pressure regions) and rarefactions (low-pressure regions), as illustrated here for waves in air from a vibrating tuning fork. Notice how these regions can be described by a waveform.

FIGURE 6.10 Waveform

The waveform of a tone from a tuning fork can be displayed on an oscilloscope by using a microphone to convert the sound wave into an electrical signal.

Sound waves may have different frequencies and so form a spectrum similar to the electromagnetic spectrum (● Fig. 6.11). However, the **sound spectrum** has much lower frequencies and is much simpler, with only three frequency regions. These regions are referenced to the *audible range of human hearing*, which is about 20 Hz to 20 kHz (20,000 Hz) and defines the *audible region* of the spectrum.

Below it is the *infrasonic region*, and above it is the *ultrasonic region*. (Note the analogy to infrared and ultraviolet light.) The sound spectrum has an upper limit of about a billion hertz (or 1 GHz, gigahertz) because of the elastic limitations of materials.

Waves in the infrasonic region, which we cannot hear, are found in nature. Earthquakes produce infrasonic waves, and we study earthquakes using these waves (see Chapter 22). Infrasound is also associated with wind and weather patterns. Elephants and cattle have hearing response in the infrasonic region and may get advance warnings of earthquakes and weather disturbances. Aircraft, automobiles, and other rapidly moving objects produce infrasound.

The audible region is of immense importance to us in terms of hearing. Sound is sometimes defined as those disturbances perceived by the human ear. As can be seen from Fig. 6.11, this definition would omit a majority of the sound spectrum. Indeed, ultrasound, which we cannot hear, has many practical applications, some of which are discussed shortly.

We hear sound because the propagating disturbance causes the eardrum to vibrate, and sensations are transmitted to the auditory nerve through the fluid and bones of the inner ear. The characteristics associated with human hearing are physiologic and can differ from their physical counterparts.

For example, *loudness* is a relative term. One sound may be louder than another, and as you might guess, this property is associated with the energy of the wave. The measurable physical quantity is **intensity** (I), which is the rate of energy transfer through a

FIGURE 6.11 Sound Spectrum

The sound spectrum consists of three regions: the infrasonic region ($f < 20$ Hz), the audible region (20 Hz $< f >$ 20 kHz), and the ultrasonic region ($f > 20$ kHz).

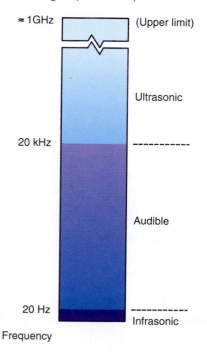

given area. Intensity may be given as so many joules per second (J/s) through a square meter (m²). But recall that a joule per second is a watt (W), so intensity has the units of W/m².

The loudness or intensity of sound decreases the farther one is from the source. As the sound is propagated outward, it is "spread" over a greater area and so has less energy per unit area. This characteristic is illustrated for a point source in ● Fig. 6.12. In this case the intensity is inversely proportional to the square of the distance from the source ($I \propto 1/r^2$); that is, it is an inverse-square relationship. This relationship is analogous to painting a larger room with the same amount of paint. The paint (energy) must be spread thinner and so is less "intense."

The minimum sound intensity that can be detected by the human ear (called the *threshold of hearing*) is about 10^{-12} W/m². At a much greater intensity of about 1 W/m², sound becomes painful to the ear. Because of the wide range, intensity is commonly measured on a logarithmic scale, which conveniently compresses the scale. The sound-intensity level is measured on a decibel (dB) scale, as illustrated in ● Fig. 6.13.

A **decibel** is one-tenth of a bel (B), a unit named in honor of Alexander Graham Bell, the inventor of the telephone. Because the decibel scale is not linear with intensity, when the sound intensity is doubled, the dB level is not doubled. Instead, the intensity level increases by only 3 dB. In other words, a sound with an intensity level of 63 dB has twice the intensity of a sound with an intensity level of 60 dB.

TABLE 6.1 Decibel Differences and Sound Intensity*

Decibel Difference (ΔdB)	Intensity
3 dB	Doubles
10 dB	Increases by a factor of 10
20 dB	Increases by a factor of 100
30 dB	Increases by a factor of 1000
.	
.	
.	

*Similar decreases in intensity occur for −ΔdB.

Comparisons are conveniently made on the dB scale in terms of decibel differences and factors of 10, as shown in Table 6.1.

Exposure to loud sounds or noise can be detrimental to one's hearing, as discussed in the chapter Highlight.

Loudness is related to intensity, but it is subjective, and estimates differ from person to person. Also, the ear does not respond equally to all frequencies. For example, two sounds with different frequencies but the same intensity level are judged by the ear to have different loudnesses.

The frequency of a sound wave may be physically measured, whereas *pitch* is the *perceived* highness or lowness of a sound. For example, a soprano has a high-pitched voice compared with a baritone. Pitch is

FIGURE 6.12 Sound Intensity

An illustration of the inverse-square law for a point source ($I \propto 1/r^2$). Notice that when the distance from the source doubles—for example, from 1 m to 2 m—the intensity decreases to one-fourth in value because the sound must pass through four times the area.

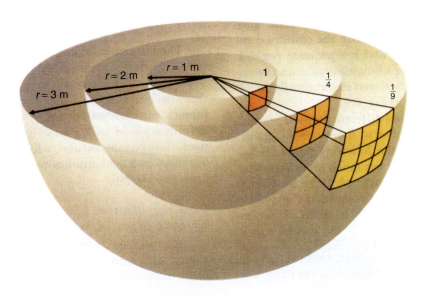

FIGURE 6.13   Sound-Intensity Levels

The decibel scale of sound-intensity levels with examples of typical sources.

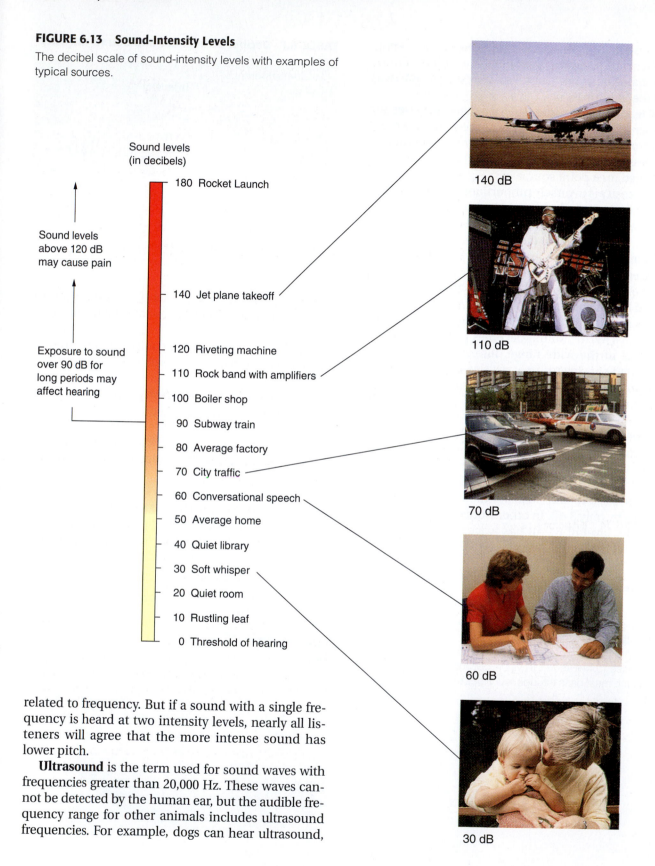

Sound levels
(in decibels)

180	Rocket Launch
140	Jet plane takeoff
120	Riveting machine
110	Rock band with amplifiers
100	Boiler shop
90	Subway train
80	Average factory
70	City traffic
60	Conversational speech
50	Average home
40	Quiet library
30	Soft whisper
20	Quiet room
10	Rustling leaf
0	Threshold of hearing

Sound levels
above 120 dB
may cause pain

Exposure to sound
over 90 dB for
long periods may
affect hearing

140 dB

110 dB

70 dB

60 dB

30 dB

related to frequency. But if a sound with a single frequency is heard at two intensity levels, nearly all listeners will agree that the more intense sound has lower pitch.

Ultrasound is the term used for sound waves with frequencies greater than 20,000 Hz. These waves cannot be detected by the human ear, but the audible frequency range for other animals includes ultrasound frequencies. For example, dogs can hear ultrasound, and ultrasonic whistles used to call dogs don't disturb humans. Ultrasonic whistles are used on cars to alert deer to oncoming traffic so that they won't leap across the road in front of cars. (There is some question, however, about the effectiveness of these deer whistles.) Bats use ultrasonic sonar for their night navigation and to catch insects (● Fig. 6.14).

An important use of ultrasound is in examining parts of the body. Thus ultrasound is an alternative to X-rays, which may be harmful. The ultrasonic waves allow different tissues such as organs and bone to be "seen" or distinguished by bouncing waves off the object examined. The waves are detected, analyzed, and stored in a computer. An *echogram*, such as the one of an unborn fetus shown in ● Fig. 6.15, is then reconstructed. X-rays might harm the fetus and cause birth defects, but ultrasonic waves have less energetic vibrations and have given no evidence of harming a fetus.

Ultrasound also can be used as a cleaning technique. Minute foreign particles can be removed from objects placed in a liquid bath through which ultrasound is passed. The wavelength of ultrasound is on the same order of magnitude as the particle size, and the wave vibrations can get into small crevices and "scrub" particles free. Thus ultrasound is especially useful in cleaning objects with hard-to-reach reces-

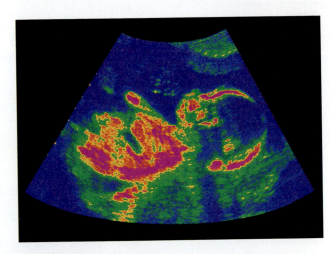

FIGURE 6.15   Echogram

A fetal echogram, or sonogram, in which the outline of a baby at 21 weeks can be clearly seen. There is no evidence that ultrasound harms a fetus, as X-rays can.

ses, such as rings and other jewelry. Ultrasonic cleaning baths for false teeth are also commercially available.

The **speed of sound** in a particular medium depends on the makeup of the material. In general, the speed of sound in air at 20° C is

$$v_{\text{sound}} = 344 \text{ m/s} \quad (770 \text{ mi/h})$$

or approximately $\frac{1}{3}$ km/s or $\frac{1}{5}$ mi/s.

The speed of sound increases with increasing temperature. For example, it is 331 m/s at 0°C, and 344 m/s at 20°C. Note that the speed of sound in air is much less than the speed of light.

The relatively slow speed of sound in air may be observed at a baseball game. A spectator may see a batter hit the ball but hear the "crack" of the bat slightly later if he or she is sitting far from home plate. Similarly, we see a lightning flash almost instantaneously, but the resulting thunder comes rumbling along afterwards at about $\frac{1}{5}$ mi/s.

By counting the seconds between seeing a lightning flash and hearing the thunder, you can estimate the distance you are from the lightning or the storm's center (where the lightning usually occurs). For example, if 5 seconds elapsed, the storm center would be at a distance of approximately $\frac{1}{3}$ km/s × 5 s (= 1.6 km), or $\frac{1}{5}$ mi/s × 5 s (= 1.0 mi).

FIGURE 6.14   Ultrasonic Sonar

Bats use the reflections of ultrasound for navigation and to locate food. The emitted sound waves (blue) are reflected, and the echoes (red) allow the bat to locate the wall and an insect.

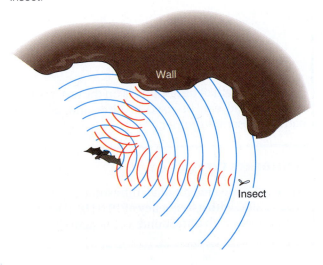

Wall

Insect

HIGHLIGHT: Noise Exposure Limits

Sounds with intensities of 120 dB and higher can be painfully loud to the ear. Brief exposures to even higher sound intensity levels can rupture eardrums and cause permanent hearing loss. However, long exposure to relatively lower sound (noise) levels also can cause hearing problems. (*Noise* is defined as unwanted sound.) Such exposures may be an occupational hazard, and in some jobs, ear protectors must be worn (Fig. 1). You may have experienced a temporary hearing loss after being exposed to a loud band for a long time or a loud bang for a short time. Ear protectors are available at hardware stores and are now commonly worn when mowing grass or using a chain saw.

Federal standards now set permissible noise exposure limits for occupational loudness. These limits are listed in Table 1. Notice that a person can work on a subway train (90 dB, Fig. 6.13) for 8 hours, but a person should only play in (or listen to) an amplified rock band (110 dB) continuously for a half hour.

FIGURE 1 Sound-Intensity Safety
An airport cargo worker wears ear protectors to prevent hearing damage from the high sound-intensity levels of jet plane engines.

TABLE 1 Permissible Noise Exposure Limits

Max. Duration per Day (h)	Sound Level Intensity (dB)
8	90
6	92
4	95
3	97
2	100
$1\frac{1}{2}$	102
1	105
$\frac{1}{2}$	110
$\frac{1}{4}$ or less	115

In general, as the density of the medium increases, the speed of sound increases. The speed of sound in water is about 4 times faster than in air, and in general, the speed of sound in solids is about 15 times faster than in air.

Using the speed of sound and the frequency, we can easily compute the wavelength of a sound wave.

EXAMPLE 6.3

Computing the Wavelength of Ultrasound

What is the wavelength of a sound wave in air at 20°C with a frequency of 22 MHz?

SOLUTION

The speed of sound at 20°C is $v_{\text{sound}} = 344$ m/s (given previously), and the frequency is $f = 22$ MHz $= 22 \times 10^6$ Hz, which is in the ultrasonic region.

To find the wavelength λ, we use the wave equation (Eq. 6.3) with the speed of sound:

$$\lambda = \frac{v_{\text{sound}}}{f}$$

$$\lambda = \frac{344 \text{ m/s}}{22 \times 10^6 \text{ Hz}} = 16 \times 10^{-6} \text{ m}$$

This wavelength of ultrasound is on the order of particle size and can be used in ultrasonic cleaning baths as described earlier.

CONFIDENCE EXERCISE 6.3

What is the wavelength of an infrasonic sound wave in air at 20°C with a frequency of 10.0 Hz? (How does this compare with ultrasound wavelengths?)

RELEVANCE QUESTION: *What are the decibel levels of four sounds you commonly hear around campus?*

6.4 The Doppler Effect

▼ Explain and give examples of the Doppler effect.

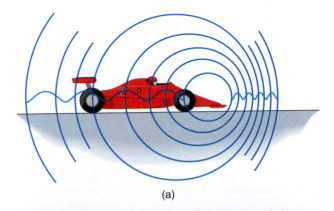

(a)

When we watch a race and a racing car with a loud engine approaches, we hear a higher-than-usual sound frequency. When the car passes by, the frequency suddenly shifts lower, and a low-pitched "whoom" sound is heard. Similar frequency changes may be heard when a large truck passes by. The apparent change in the frequency of the moving source is called the **Doppler effect.***

The reason for the observed change in frequency (and wavelength) of a sound is illustrated in ● Fig. 6.16. As a moving sound source approaches an observer, the waves are "bunched up" in front of the source. With the waves closer together (shorter wavelength), an observer perceives a higher frequency. Behind the source the waves are spread out, and with an increase in wavelength, a lower frequency is heard ($f = v/\lambda$). If the source is stationary and the observer moves toward and passes the source, the shifts in frequency are also observed. Hence the Doppler effect depends on the *relative* motion of the source and the observer.

Waves propagate outward in front of a source as long as the speed of sound is greater than the speed of the source (● Fig. 6.17). However, as the speed of the source approaches the speed of sound in a medium, the waves begin to bunch up closer and closer. When the speed of the source exceeds the speed of sound in the medium, a V-shaped bow wave is formed. Such a wave is readily observed for a motorboat traveling faster than the wave speed in water.

In air, when a jet aircraft travels at a supersonic speed (a speed greater than the speed of sound in air), the bow wave is in the form of a conical shock wave that trails out and downward from the aircraft. When this high-pressure, compressed wave front passes over an observer, he or she hears a sonic boom. But the bow wave travels with the supersonic aircraft and

*After Christian Doppler (1803–1853), the Austrian physicist who first described the effect.

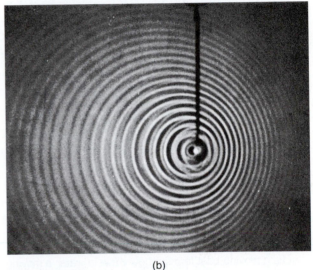

(b)

FIGURE 6.16   Doppler Effect

(a) Because of the motion of a sound source, illustrated here as a racing car, sound waves are "bunched up" in front and "spread out" in back. This results in shorter wavelengths or higher frequencies in front of the source and longer wavelengths or lower frequencies behind the source. (b) The Doppler effect in water waves in a ripple tank. The source of the disturbance is moving to the right.

does not occur only at the instant the aircraft "breaks the sound barrier" (first exceeds the speed of sound).

The Doppler effect is a general effect that occurs for all kinds of waves, such as water waves, sound waves, and light (electromagnetic) waves. In the Doppler effect for visible light, the frequency is shifted toward the blue end of the spectrum when the light source (such as a star) is approaching. (Blue light has a shorter wavelength, or higher frequency.) In this case we say a Doppler *blue shift* has occurred.

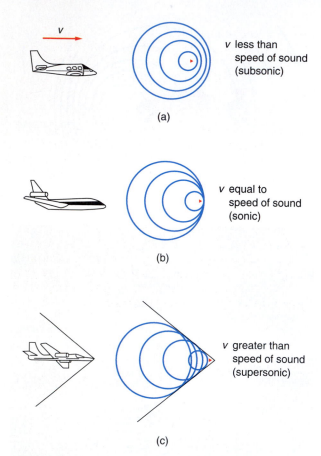

FIGURE 6.17 Bow Waves and Sonic Boom

Just as a moving boat forms a bow wave in water, a moving aircraft forms a bow wave in air. The sound waves bunch up in front of the airplane for increasing sonic speeds [(a) to (b)]. A plane traveling at supersonic speeds forms a high-pressure shock wave in air (c) that is heard as a sonic boom when passing over an observer. (Actually, shock waves are formed from both the nose and tail of the aircraft. Only one is shown here for simplicity.)

When a stellar light source moves away from us and the frequency is shifted toward the red (longer wavelength) end of the spectrum, we say that a Doppler **redshift** has occurred. The magnitude of the frequency shift is related to the speed of the source. The rotations of the planets and stars can be established by looking at the Doppler shifts from opposite sides—one is receding (redshift) and the other approaching (blue shift). Also, Doppler shifts of light from stars in our galaxy, the Milky Way, indicate that the galaxy is rotating.

Light from other galaxies shows redshifts, which indicate that they are moving away from us according to the Doppler effect. By modern interpretations, however, this is not a Doppler shift but a shift in the wavelength of light influenced by the expansion of the universe. The wavelength of light expands along with the universe, giving a *cosmological redshift* (see Chapter 18).

The Doppler shift of electromagnetic waves also is applied here on Earth. You may have experienced one such application. Radar (radio waves), which police use in determining the speed of moving vehicles, utilizes the Doppler effect.

6.5 Standing Waves and Resonance

LEARNING GOALS

▼ Explain standing waves and frequencies.
▼ Describe what is meant by resonance.

Many of us have shaken one end of a stretched cord or rope and have observed wave patterns that seem to "stand" along the rope when we shake it just right. We refer to these waveforms as **standing waves** (● Fig. 6.18).

They are caused by the interference of waves traveling down and back along the rope. When two waves meet, they interfere, and the combined waveform of the superimposed waves is the sum of the waveforms or particle displacements of the medium.

Note that the string in Fig. 6.18 vibrates at only *particular* frequencies (as evidenced by the number of loops or half wavelengths). These frequencies are referred to as the *characteristic*, or *natural*, *frequencies* of the stretched string.

When a stretched string or an object is acted on by a periodic driving force with a frequency equal to one of the natural frequencies, the oscillations have large amplitudes. This phenomenon is called **resonance,** and in this case there is maximum energy transfer to the system.

A common example of driving a system in resonance is pushing a swing. A swing is essentially a pendulum with only one natural frequency, which depends on the rope length. When a swing is pushed periodically with a period of $T = 1/f$, energy is transferred to the swing, and its amplitude gets larger

a natural frequency of the bridge and result in resonance and large oscillations that could cause structural damage. In 1850, in France, about 500 soldiers marching over a suspension bridge caused a resonant vibration that rose to such a level that the bridge broke, and over 200 of the soldiers were drowned.

Electrical resonance also occurs. When we tune a radio or TV receiver, turning the dial adjusts the natural frequency of an alternating electric circuit to the assigned broadcast frequency of a particular station.

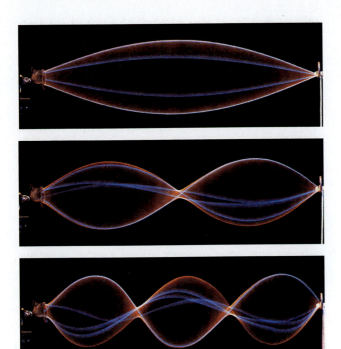

FIGURE 6.18   Standing Waves

Actual standing waves in a stretched rubber string. Standing waves are formed only when the string is vibrated at particular frequencies. Notice the differences in wavelengths.

FIGURE 6.19   Resonance

(a) When one tuning fork is activated, the other tuning fork of the same frequency will be driven in resonance and start to vibrate. (b) Unwanted resonance. The famous Tacoma Narrows Bridge in the state of Washington, which collapsed in 1940 after wind drove the bridge into resonance vibrations.

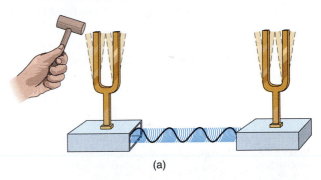

(a)

(higher swings). If the swing is not pushed at its natural frequency, the pushing force may be applied as the swing approaches or after it has reached its maximum amplitude and is swinging back. In either case, the swing is not driven in resonance.

A stretched string has many natural frequencies—not just one, as a pendulum has. When the string is shaken, standing waves are set up. But when the driving frequency corresponds to one of the natural frequencies, the amplitude at the antinodes will be larger because of resonance.

There are many examples of resonance. The structures of the throat and nasal cavities give the human voice a particular tone due to resonances. Tuning forks of the same frequency can be made to resonate, as illustrated in ● Fig. 6.19a.

A steel bridge or any elastic structure is capable of vibrating at natural frequencies, sometimes with dire consequences (Fig. 6.19b). Soldiers marching in columns across bridges are told to break step and not march at a periodic cadence that might correspond to

(b)

This results in a maximum energy transfer to the circuit for this frequency, and other frequencies or stations are excluded.

Musical instruments use standing waves and resonance to produce different tones. Standing waves are formed on strings fixed at both ends on stringed instruments such as the guitar, violin, and piano. When a stringed instrument is tuned, a string is tightened or loosened, which adjusts the tension and the wave speed in the string. This adjustment changes the frequency because the length of the string is fixed ($\lambda f = v$).

A vibrating string does not produce a great disturbance in air, but the body of a stringed instrument such as a violin acts as a sounding board and amplifies the sound. Thus the body of such an instrument acts as a resonance cavity, and sound comes out through holes in the top surface.

Similarly, standing waves are set up in wind instruments in air columns. Organ pipes have fixed lengths similar to fixed strings, so only a certain number of wavelengths can be fitted in. However, the length of an air column and the frequency or tone can be varied in some instruments, such as a trombone or trumpet, by varying the length of the column.

The *quality* (or timbre) of a sound depends on the waveform or the number of waveforms present. For example, you can sing the same note as a famous singer, but a different combination of waveforms gives the singer's voice a different quality, perhaps a pleasing "richness." It is the quality of our voices that gives them different sounds.

RELEVANCE QUESTION: *Why is the sound of your voice different from other people's voices? Explain using two factors that determine the sound of your voice.*

Important Terms

wave	hertz	sound (6.3)	Doppler effect (6.4)
longitudinal wave (6.1)	period	sound spectrum	redshift
transverse wave	wave speed	intensity	standing waves (6.5)
wavelength	electromagnetic	decibel	resonance
amplitude	waves (6.2)	ultrasound	
frequency	speed of light	speed of sound	

Important Equations

Frequency-Period Relationship: $f = \dfrac{1}{T}$

Wave Speed: $v = \dfrac{\lambda}{T} = \lambda f$

Speed of Light: $c = 3.00 \times 10^8$ m/s

$$(= 1.86 \times 10^5 \text{ mi/s})$$

Review Questions

6.1 Wave Properties

1. If a piece of ribbon were tied to a stretched string carrying a transverse wave, how would the ribbon be observed to oscillate?
   (a) perpendicular to wave direction
   (b) parallel to wave direction
   (c) neither a nor b
   (d) both a and b

2. The energy of a wave is related to its
   (a) amplitude.
   (b) frequency.
   (c) wavelength.
   (d) period.

3. What is the difference between a longitudinal wave and a transverse wave? Give an example of each.

4. What are the SI units of (a) wavelength, (b) frequency, and (c) period?

5. What determines the amplitude of a wave?

6.2 Electromagnetic Waves

6. Which of the following is true for electromagnetic waves?
   (a) have different speeds in vacuum for different frequencies
   (b) are transverse waves
   (c) require a medium for propagation
   (d) none of the preceding

7. Which one of the following regions has frequencies just higher than the visible region in the EM frequency spectrum?
   (a) radio wave
   (b) ultraviolet
   (c) microwave
   (d) infrared

8. List seven types of electromagnetic waves, in order of increasing frequency.

9. Which end (blue or red) of the visible spectrum has the longer wavelength? Which has the higher frequency?

10. Are radio waves sound waves? Explain.

11. What is the range of the wavelengths of visible light? How do these wavelengths compare to those of audible sound?

6.3 Sound Waves

12. Sound waves propagate in which of the following?
    (a) solids   (b) liquids   (c) gases   (d) all of these

13. What is the upper frequency limit of the audible range of human hearing?
    (a) 20 kHz
    (b) 2,000 Hz
    (c) 2 kHz
    (d) 2,000,000 Hz

14. Referring to Fig. 6.12, how many squares would the sound waves spread over for $r = 4$ m? The intensity would decrease to what fraction in value?

15. What is the chief physical property that describes (a) pitch, (b) loudness, and (c) quality?

16. Can ultrasound be heard? Give some examples of applications of ultrasound.

17. Why does the music from a marching band in a spread-out formation on a football field sometimes sound discordant?

18. Does doubling the decibels of a sound level double the intensity? Explain.

6.4 The Doppler Effect

19. A moving observer approaches a stationary sound source. What does the observer hear?
    (a) an increase in frequency
    (b) a decrease in frequency
    (c) the same frequency as the source

20. If an astronomical light source is moving away from us, what is observed?
    (a) a blue shift
    (b) a shift toward longer wavelengths
    (c) a shift toward higher frequencies
    (d) a sonic boom

21. How is the wavelength of sound affected when (a) a sound source moves toward a stationary observer, and (b) an observer moves away from a stationary sound source?

22. What would be the situation for sound to have a Doppler (a) blue shift and (b) redshift?

23. Compare the crack of a whip with a sonic boom.

24. Radar and sonar are based on similar principles. Sonar (which stands for *so*und *nav*igation and *r*anging) uses ultrasound, and radar (which stands for *ra*dio *d*etecting *a*nd *r*anging) uses radio waves. Explain the principles of detecting and ranging in these applications.

6.5 Standing Waves and Resonance

25. What is the effect if a system is driven in resonance? Is a particular frequency required?

26. What determines the pitch or frequency of a string on a violin? How does a violinist get a variety of notes from one string?

Applying Your Knowledge

1. (Here's an old one.) If a tree falls in the forest and no one is around to hear it, is there sound?

2. If an astronaut on the Moon dropped a hammer, would there be sound? Explain. (*Follow-up:* How do astronauts communicate with each other and mission control?)

3. We cannot see ultraviolet (UV) radiation, which can be dangerous to the eyes. However, we can detect some UV radiations (without instruments). How is this done?

4. If a jet pilot is flying faster than the speed of sound, will he or she be able to hear sound?

5. When one sings in the shower, the tones sound full and rich. Why is this?

Exercises

6.1 Wave Properties

1. A periodic wave has a period of 0.25 s. What is the wave frequency?
   Answer: 4.0 Hz

2. What is the period of the wave motion for a wave with a frequency of 10 kHz?

3. Waves moving on a lake have a speed of 2.0 m/s and a distance of 1.5 m between adjacent crests. (a) What is the frequency of the waves? (b) Find the period of the wave motion.
   Answer: (a) 1.3 Hz (b) 0.77 s

4. A sound wave has a frequency of 2000 Hz. What is the distance between crests or compressions of the wave? (Take the speed of sound to be 344 m/s.)

6.2 Electromagnetic Waves

5. Compute the wavelength of the radio waves from an AM station operating at a frequency of 650 kHz.
   Answer: 4.62×10^2 m

6. Compute the wavelength, in nm, of an X-ray with a frequency of 10^{18} Hz.

7. What is the frequency of blue light that has a wavelength of 420 nm?
   Answer: 7.14×10^{14} Hz

8. An electromagnetic wave has a wavelength of 6.00×10^{-6} m. In what region of the EM spectrum is this radiation?

6.3 Sound Waves

9. Compute the wavelength in air of ultrasound with a frequency of 50 kHz if the speed of sound is 344 m/s.
   Answer: 6.9×10^{-3} m

10. What are the wavelength limits of the audible range of the sound spectrum? (Use the speed of sound in air.)

11. During a thunderstorm, 4.0 s elapses between observing a lightning flash and hearing the resulting thunder. Approximately how far away in km and mi was the lightning flash?
    Answer: 1.3 km or 0.80 mi

12. Picnickers see a lightning flash and hear the resulting thunder 9.0 s later. If the storm is traveling at a rate of 12 km/h, how long, in minutes, do the picnickers have before the storm arrives at their location?

13. A subway train has a sound intensity level of 90 dB, and a rock band has a sound intensity level of about 110 dB. How many times greater is the sound intensity of the band than the train?
    Answer: 100 times

14. A loudspeaker has an output of 80 dB. If the volume of the sound is turned up so that the output intensity is 10,000 times greater, what is the new sound intensity level?

15. A new jackhammer has a 1/10 reduction in sound level intensity from that of the older model that was rated at 118 dB. What is the sound intensity level of the new hammer?
    Answer: 108 dB

16. A rock band's music has an average sound intensity level of 123 dB. What would the average level be if the intensity were (a) reduced by one-half, and (b) reduced by a factor of $\frac{1}{100}$?

Solutions to Confidence Exercises

6.1 $f = \dfrac{v}{\lambda} = \dfrac{344 \text{ m/s}}{0.500 \text{ m}} = 688$ Hz

6.3 $\lambda = \dfrac{v}{f} = \dfrac{344 \text{ m/s}}{10.0 \text{ Hz}} = 34.4$ m (infrasound λ much longer)

6.2 $\lambda = \dfrac{v}{f} = \dfrac{3.00 \times 10^8 \text{ m/s}}{9.00 \times 10^7 \text{ Hz}} = 0.333 \times 10^1 \text{ m} = 3.33$ m

Answers to Multiple-Choice Review Questions

1. a   6. b   12. d   19. a
2. a   7. b   13. a   20. b

WAVE EFFECTS AND OPTICS

7

So when the sun in bed,
Curtain'd with cloudy red,
Pillows his chin upon an orient wave.

John Milton (1608–1674)

The effects of waves—particularly sound and light waves—are all around us. We are aware of many of these effects, but they are such a part of our experience that we take them for granted and rarely try to analyze them. For example, if you speak loudly, a person around the corner of an outside doorway can hear you, indicating that sound waves "bend" around corners. Light waves do not appear to do so. In other words, you can be heard in the next room but not seen.

When light waves strike a soap bubble or an oil slick, we commonly see brilliant displays of colors. Similarly, we sometimes see a colorful rainbow. These phenomena can be described and explained through the effects and interactions of light (electromagnetic) waves.

Mirrors and lenses are commonplace. We look into mirrors daily, and many of us wear lenses (eyeglasses). Here, the descriptions involve reflection and refraction. We will discuss these two effects and then describe the basic principles of mirrors and lenses. From this discussion you will gain an understanding of many optical devices such as the human eye, slide projectors, and eyeglasses.

Photo: Because of refraction, the pencil appears to be almost severed.

There are many wave effects, two of which—the Doppler effect and resonance—have already been discussed (Chapter 6). In this chapter we will consider some other important wave phenomena that affect us all the time. ■

7.1 Reflection

LEARNING GOALS

▼ Explain the law of reflection.

▼ Distinguish between regular (specular) and irregular (diffuse) reflections.

Light waves travel through space in a straight line and will continue to do so unless diverted from their original direction. A change in direction takes place when light strikes and rebounds from a surface or the boundary between two media. A change in direction by this method is called **reflection.**

Reflection may be thought of as light "bouncing off" a surface. However, it is really much more complicated and involves the absorption and emission of complex atomic vibrations of the reflecting medium. To describe reflection simply, we consider the reflection of light rays in a manner that ignores the wave nature of light. A **ray** is a straight line that represents the motion of light. A *beam* of light may be represented by a group of parallel rays.

An incident light ray is reflected from a surface in a particular way. As illustrated in ● Fig. 7.1, the angles of the incident and reflected rays (θ_i and θ_r) are measured relative to the *normal,* a line perpendicular to the reflecting surface. The **law of reflection** states:

The angle of reflection θ_r is equal to the angle of incidence θ_i.

Also, the reflected and incident rays are in the same plane.

The reflection from very smooth (mirror) surfaces is called **regular** (or **specular**) **reflection** (● Fig. 7.2). In regular reflection, incident parallel rays are parallel on reflection. However, rays reflected from relatively rough surfaces are not parallel, and this is called **irregular** (or **diffuse**) **reflection.** The reflection from the pages of this book is diffuse. The law stated above applies to each type of reflection, but the rough surface causes the light rays to be reflected in different directions.

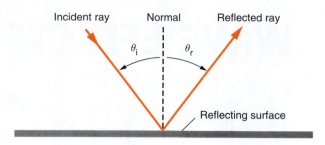

FIGURE 7.1 Law of Reflection
The angle of incidence θ_i equals the angle of reflection θ_r relative to the normal (a line perpendicular to the reflecting surface). The rays and normal line lie in the same plane.

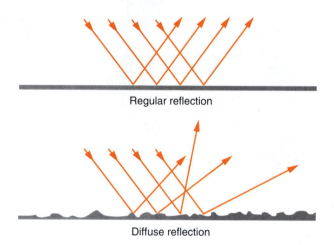

FIGURE 7.2 Reflection
A smooth (mirror) surface produces regular (or specular) reflection. A rough surface produces irregular (or diffuse) reflection. Both reflections follow the law of reflection.

Rays may be used to determine the image formed by a mirror. A *ray diagram* for determining the apparent location of an image formed by a plane mirror is shown in ● Fig. 7.3. The image is located by drawing two rays emitted by the object and applying the law of reflection. Where the rays intersect or appear to intersect locates the image. Notice that for a plane mirror the image is located behind or "inside" the mirror at the same distance as the object is in front of the mirror.

● Figure 7.4 shows a ray diagram for the light rays involved when a person sees a complete or head-to-toe image. Applying the law of reflection reveals that you can see your total image in a plane mirror that is

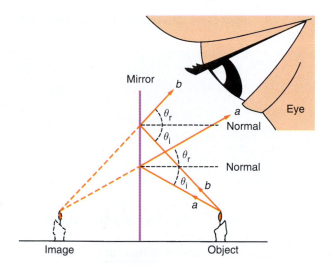

FIGURE 7.3 Ray Diagrams
By tracing the reflected rays, you can locate the mirror image where the rays intersect or appear to intersect behind the mirror.

FIGURE 7.5 Natural Reflection
Beautiful reflections are often seen in nature, such as the one shown here on a still lake. Is the picture really right-side up? Without the tree limbs to the left, would you be able to tell?

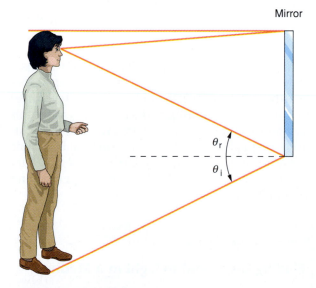

FIGURE 7.4 Complete Figure
For a person to see his or her complete figure in a plane mirror, the minimum height of the mirror must be one-half the height of the person, as we see by ray tracing.

only half your height. Also, your distance from the mirror is not a factor.

It is the reflection of light that allows us to see things. Look around you. What you see in general is light reflected from the walls, ceiling, floor, and other objects. We often see beautiful reflections in nature, as shown in ● Fig. 7.5.

RELEVANCE QUESTION: When inside a lighted room, why can you see your reflection in a windowpane when it is dark outside but not during the day?

7.2 Refraction and Dispersion

LEARNING GOALS

▼ Describe refraction, and explain how light is dispersed.

▼ Describe how the boundary of a transparent medium can be used as a mirror and how this is applied to fiber optics.

Refraction

When light strikes a transparent medium, some light is reflected and some is transmitted. This is illustrated in ● Fig. 7.6 for a beam of light incident on the surface of a body of water. On investigation, we find that the transmitted light has changed direction because the speed of light changes in going from one medium to another. The deviation of light from its original path because of a speed change is called **refraction.** You

FIGURE 7.6   Refraction in Action
A beam of light is refracted—that is, its direction is changed—on entering the water.

have probably observed refraction effects for an object in a glass of water. For example, a fork or pencil in a glass of water will appear to be displaced and perhaps severed (see the chapter-opening photo).

The directions of the incident and refracted rays are expressed in terms of the angle of incidence θ_1 and the angle of refraction θ_2, which are measured relative to the normal to the surface boundary of the medium (● Fig. 7.7). The different speeds in the different media are expressed in terms of the ratio of the speeds.

FIGURE 7.7   Refraction
When light enters a transparent medium at an angle, it is deviated, or refracted, from its original path. As illustrated here, when light passes from air into a denser medium such as glass, the rays are refracted, or "bent," toward the normal ($\theta_2 < \theta_1$—that is, the angle of refraction is less than the angle of incidence). Some of the light is also reflected from the surface, as indicated by the dashed ray.

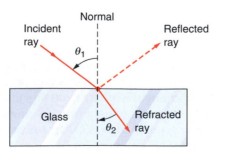

TABLE 7.1   Indexes of Refraction of Some Common Substances

Substance	n
Water	1.33
Crown glass	1.52
Diamond	2.42
Air (0°C, 1 atm)	1.00029
Vacuum	1.00000

This ratio is known as the **index of refraction** n:

$$\frac{\text{index of}}{\text{refraction}} = \frac{\text{speed of light in vacuum}}{\text{speed of light in medium}}$$

or

$$n = \frac{c}{c_m} \qquad \text{7.1}$$

The index of refraction is a pure number, because c and c_m are measured in the same units. The indexes of refraction of some common substances are given in Table 7.1. Notice that the index of refraction for air is close to that for a vacuum.

When light passes obliquely ($\theta_1 > 0°$) into a denser medium—for example, from air into water or glass—the light rays are refracted or bent toward the normal ($\theta_2 < \theta_1$). It is the slowing of the light that causes this deviation. Complex processes are involved, but intuitively, we might expect the passage of light by atomic absorption and emission through the denser medium to take longer. For example, the speed of light in water is about 75% of that in air or a vacuum.

EXAMPLE 7.1

Finding the Speed of Light in a Medium

What is the speed of light in water?

SOLUTION

Step 1
Given:
Nothing directly; therefore, quantities are known or available from tables.

Step 2
Wanted:
Speed of light in medium (here, water).

This may be found from Eq. 7.1. We rearrange this equation to $c_m = c/n$, so we have to know c and n. The speed of light c is known, $c = 3.00 \times 10^8$ m/s, and from Table 7.1, $n = 1.33$ for water.

Step 3

Doing the calculation, we have

$$c_m = \frac{c}{n} = \frac{3.00 \times 10^8 \text{ m/s}}{1.33}$$

$$= 2.26 \times 10^8 \text{ m/s}$$

This is about 75% of the speed of light in a vacuum (c), as we see by the ratio

$$\frac{c_m}{c} = \frac{1}{n} = \frac{1}{1.33} = 0.752 \ (= 75.2\%)$$

CONFIDENCE EXERCISE 7.1

According to Eq. 7.1, what is the speed of light in (a) a vacuum and (b) air, and how do they compare?

To help understand how light is bent or refracted when it passes into another medium, consider a band marching across a field and entering a wet, muddy region obliquely (at an angle), as illustrated in ● Fig. 7.8a. As the marchers enter the muddy region, they keep marching at the same frequency (cadence). But slipping in the muddy earth, they don't cover as much ground and are slowed down (smaller wave speed).

The marchers in the same row on solid ground continue on with the same stride, and as a result, the direction of the marching column is changed as it enters the muddy region. This change in direction with change in marching speed is also seen when a marching band turns a corner and the inner members mark time.

We may think of wave fronts as analogous to marching rows (Fig. 7.8b). In the case of light, the wave frequency (cadence) remains the same, but the wave speed is reduced, as is the wavelength ($c_m = \lambda_m f$). The wavelength may be thought of as the distance covered in each step (shorter when slipping).

Several refraction effects are commonly observed. The index of refraction of a gas varies with its density, which varies with temperature. At night, starlight passes through the atmosphere, which in turn, has temperature density variations and turbulence. As a re-

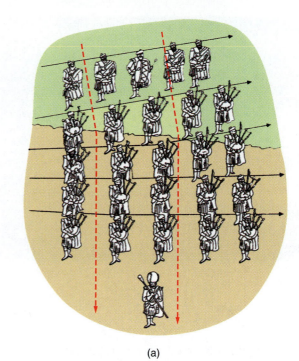

(a)

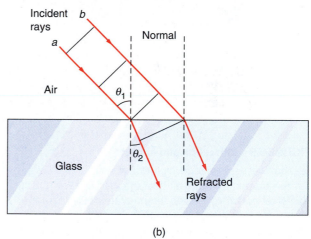

(b)

FIGURE 7.8 Refraction Analogy

(a) Marching obliquely into a muddy field causes a band column to change direction. The cadence (or frequency) remains the same, but the marchers in the mud slip and travel shorter distances (shorter wavelengths). This is analogous to the refraction of a wavefront (b).

sult, the refraction causes the star's image to appear to move and vary in brightness—the "twinkling" of stars.

A couple of other effects are shown in ● Fig. 7.9. You probably have seen the "wet spot" mirage on a road on a hot day (Fig. 7.9a). You travel toward the "wet spot" but never reach it. As illustrated, this is so

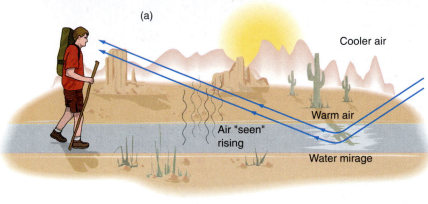

FIGURE 7.9 Refraction and Mirages

(a) The "wet spot" mirage is produced when light from the sky is refracted by warm air near the road surface (the index of refraction varies with temperature). Such refraction also enables us to perceive heated air rising. Although we cannot see air, the refraction in the turbulent updrafts causes variations in the light passing through. (b) Try to catch the fish. We tend to think of light as traveling in straight lines, but because light is refracted, the fish is not where it appears to be.

because of refraction in the hot air near the road surface, and the "wet spot" is really a view of the sky via refracted skylight. The variation in the density of the rising hot air causes refractive variations that allow us to "see" hot air rising from the road surface. (You can't see air.)

Have you ever tried to catch a fish under water and missed? Figure 7.9b shows you why. We tend to think

of our line of sight as a straight line, but light bends due to refraction at the air-water surface. The fish is not where you think, unless you take refraction into account.

Finally, take a look at the opening photo for Chapter 8. Notice how the Sun is apparently flattened. This is a refractive effect. Light coming from the top and bottom portions is refracted differently as it passes through different atmospheric densities near the horizon. Light from the sides of the Sun is refracted the same, so there is no apparent difference.

Reflection by Refraction

When light goes from a denser medium into a less dense medium—for example, from water into air—the ray is refracted and bent away from the normal. We may see this refraction by tracing, in reverse, the ray of light going from glass into air in Fig. 7.7. This type of refraction is shown in ● Fig. 7.10.

But notice that an interesting thing happens as the angle of incidence becomes larger. The refracted ray is bent farther from the normal, and at a particular critical angle θ_c the refracted ray is along the boundary of the two media. For angles greater than θ_c the light is reflected and none is refracted. This phenomenon is called **total internal reflection.** Refraction and total internal reflection are illustrated in ● Fig. 7.11a. With total internal reflection, a prism can be used as a mirror.

Internal reflection is used to enhance the *brilliance* of diamonds. In the so-called brilliant cut, a diamond is cut so that entering light is totally reflected (Fig. 7.11b). The light emerging from the upper portion gives the diamond its beautiful sparkle.

Another example of total internal reflection occurs when a fountain of water is illuminated from below.

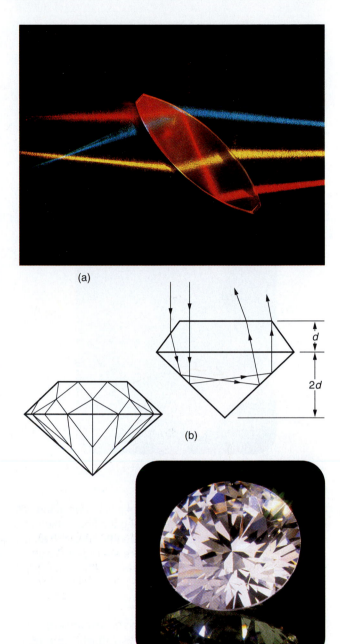

(a)

(b)

FIGURE 7.10 Internal Reflection

When light goes from a denser medium into a less dense medium, such as from water into air, as illustrated here, it is refracted away from the normal. At a certain critical angle θ_c, the angle of refraction is 90°. For incidence greater than the critical angle, the light is reflected internally.

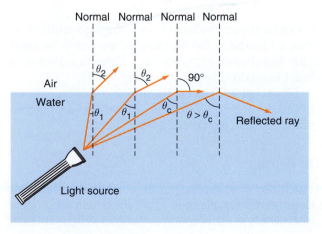

FIGURE 7.11 Refraction and Internal Reflection

(a) Beams of colored light from the left are incident on a piece of glass. As the light passes from air into the glass, the beams are refracted. The incident angle of the red beam at the glass–air interface exceeds the critical angle θ_c, and the beam is internally reflected (twice). (b) Refraction and internal reflection give rise to the "brilliance" of a diamond. In the so-called brilliant cut, a diamond is cut with a certain number of faces, or "facets," along with the correct depth to give the proper refraction and internal reflection.

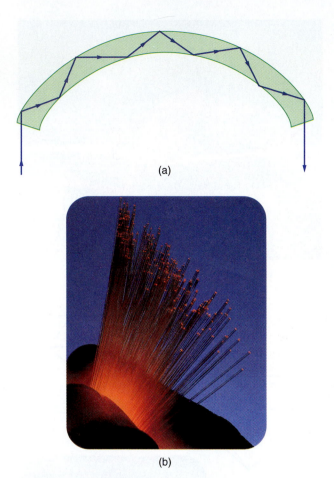

(a)

(b)

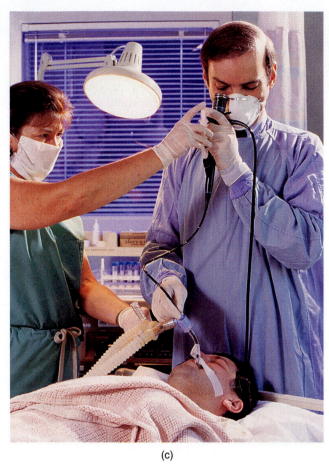

(c)

FIGURE 7.12   Fiber Optics

(a) When light is incident on the end of a fiber above the critical angle, it is internally reflected along the fiber, which acts as a "light pipe." (b) A fiber-optics bundle held between a person's fingers. The ends of the fibers are lit up because of transmission of light by multiple internal reflections. (c) A fiber-optic application. Endoscopes are used to view various internal body parts. Light can be reflected down some fibers and back through others, allowing a view of otherwise inaccessible places.

The light is totally reflected within the streams of water, providing a spectacular effect. Similarly, light can travel along transparent plastic tubes called "light pipes." When the incident angle for light in the tube is greater than the critical angle, the light undergoes a series of internal reflections down the tube (● Fig. 7.12a).

Light also can travel along thin fibers, and bundles of such fibers are used in the relatively new field of *fiber optics* (Fig. 7.12b). You probably have seen fiber optics used in decorative lamps. An important use of the flexible fiber bundle is to pipe light to hard-to-reach places. Light also may be transmitted down one set of fibers and reflected back through another so that an image of the illuminated area may be seen. This illuminated area may be a person's stomach or heart in medical applications (Fig. 7.12c).

Fiber optics is used in telephone communications. In this application, electronic signals in wires are replaced by light (optical) signals in fibers.

Dispersion

The index of refraction for a material actually varies slightly with wavelength. This means that when light is refracted, the different wavelengths of light are bent

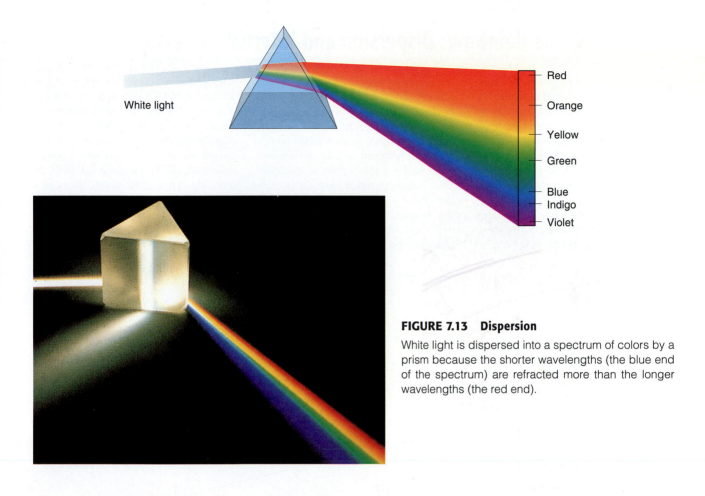

FIGURE 7.13   Dispersion
White light is dispersed into a spectrum of colors by a prism because the shorter wavelengths (the blue end of the spectrum) are refracted more than the longer wavelengths (the red end).

at slightly different angles. This phenomenon is called **dispersion.** When white light (light containing all wavelengths of the visible spectrum) passes through a glass prism, as shown in ● Fig. 7.13, the light rays are refracted on entering the glass. With different wavelengths refracted at slightly different angles, the light is dispersed into a spectrum of colors (wavelengths).

That the amount of refraction is a function of wavelength can be seen by combining the two equations $n = c/c_m$ and $c_m = \lambda f$. Substituting the second into the first for c_m, we have $n = c/\lambda f$, and the index of refraction n varies inversely with wavelength λ. This means that shorter wavelengths have greater indexes of refraction and are deviated from their path by the greatest amount. Blue light has a shorter wavelength than red light; hence blue light is refracted more than red light, as shown in Fig. 7.13.

As a common example, a diamond is said to have "fire" due to colorful dispersion. This is in addition to having brilliance due to internal reflection. A natural phenomenon involving dispersion and internal reflection is discussed in the chapter's first Highlight.

The fact that light can be separated into its component frequencies or wavelengths provides an important investigative tool. Basically, this is done as illustrated in ● Fig. 7.14a (page 150). Light from a source goes through a narrow slit, and when the light is passed through a prism, the respective wavelengths are separated into "line" images of the slit. The line images, representing definite wavelengths, appear as bright lines. A scale is added for measurement of the wavelengths of the lines in the line spectrum, and the instrument is called a *spectrometer.* The line spectra of various elements are shown in Fig. 7.14b.*

---

*In a spectrometer, a diffraction grating (Section 7.4), which produces sharper lines, is commonly used instead of a prism.

HIGHLIGHT: The Rainbow: Dispersion and Internal Reflection

A beautiful atmospheric phenomenon often seen after rain is the rainbow. The colorful arc across the sky is the result of several optical effects: refraction, internal reflection, and dispersion. The conditions must be just right, however. As we all know, a rainbow is seen after a rain but not after *every* rain.

Following a rain, the air contains many tiny water droplets. Sunlight incident on the droplets produces a rainbow. But whether a rainbow is visible or not depends on the relative positions of the Sun and the observer. As you may have noticed, the Sun is generally behind you when you see a rainbow.

To understand the formation and observation of a rainbow, consider what happens when sunlight is incident on a water droplet. On entering the droplet, the light is refracted and dispersed into component colors as it travels in the water (Fig. 1a). If the light strikes the water-air interface of the droplet above the critical angle, it is internally reflected, and the component colors emerge from the droplet at slightly different angles. Because of the conditions for refraction and internal reflection, the component colors lie in a narrow range of 40° to 42° for an observer on the ground.

Thus you see the display of colors only when the Sun is positioned so that the dispersed light is reflected to you

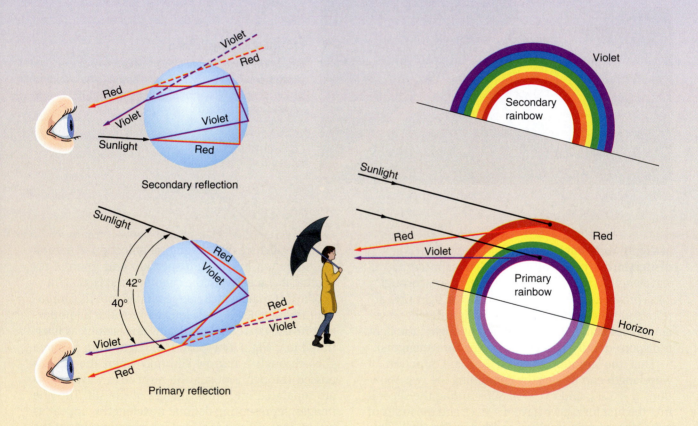

Secondary reflection

Primary reflection

through these angles. With this condition satisfied and an abundance of water droplets in the air, you see the colorful arc of a *primary rainbow* with colors running vertically upward from violet to red (Fig. 1a and 1c).

Occasionally, conditions are such that you see sunlight that has undergone two internal reflections in water droplets. The result is a vertical inversion of colors in a higher, fainter, and less frequently seen *secondary rainbow*

(Fig. 1b and 1c). Note the bright region below the primary rainbow. Light from the rainbows combines to form this illuminated region.

The arc length of a rainbow that you see depends on the altitude (angle above the horizon) of the Sun. As the altitude of the Sun increases, you see less of the rainbow. On the ground you cannot see a (primary) rainbow if the altitude of the Sun is greater than 42°. The rainbow is below the horizon in

this case. However, if your elevation is increased, you see more of the rainbow arc. For instance, airplane passengers commonly view a completely circular rainbow, similar to the miniature one that can be seen in the mist produced by a lawn sprayer.

FIGURE 1   Rainbow Formation

(a) Sunlight may be internally reflected once or twice in water droplets. (b) The dispersion of sunlight in the droplets produces a separation of colors, and an observer may see an arc or bow of colors in a particular annular region. (c) Both the Sun and the observer must be properly positioned for the observer to see a rainbow. In the photo, the observer was positioned so as to see both the primary and secondary rainbows.

(c)

FIGURE 7.14   Line Spectra

(a) A line spectrum is generated when light from a heated source passes through a slit to produce a sharp beam and then through a prism, which disperses the beam into line images of the slit. (b) Line spectra of various elements in the visible region. Notice that each spectrum is unique or characteristic of that element.

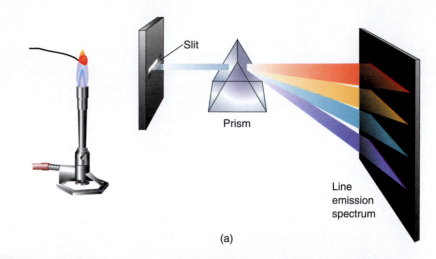

(a)

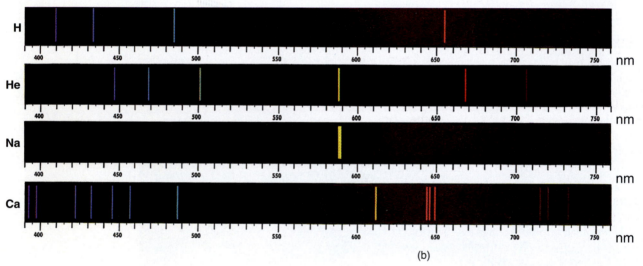

(b)

Notice in Fig. 7.14b that the spectra are different, or characteristic. Every substance, when sufficiently heated, gives off light of characteristic frequencies. Using spectrometers, the spectra can be studied and substances identified. Astronomers, chemists, physicists, and other scientists have acquired much basic information from the study of light. In fact, the element helium was first identified in the spectrum of sunlight—hence the name helium, from *helios,* Greek for "Sun."

RELEVANCE QUESTION: *When looking into a swimming pool that has a logo (for example, your college's) or some picture painted on the bottom, the logo or picture often looks somewhat blurry or distorted. Why is this?*

7.3 Polarization

LEARNING GOALS

▼ Describe the phenomenon of polarization.

▼ Identify some of the effects and applications of polarization.

The nature of light waves gives rise to an interesting and practical optical phenomenon. Recall that light waves are transverse electromagnetic waves with electric and magnetic field vectors oscillating perpendicular to the direction of propagation (see Fig. 6.6). The atoms of a light source generally emit light waves that are randomly oriented, and a beam of light has transverse field vectors in all directions. When we view a beam of light from the front, the transverse field vectors may be represented as shown in ● Fig. 7.15a. (Only electric field vectors are represented for simplicity.)

In the figure, the field vectors are in planes perpendicular to the direction of propagation. Such light is said to be *unpolarized*—that is, the field vectors are randomly oriented. **Polarization** refers to the preferential orientation of the field vectors. If the vectors have some partial preferential orientation, the light is *partially polarized* (Fig. 7.15b). If the vectors are in a single plane, the light is **linearly polarized** (Fig. 7.15c).

A light wave may be polarized by several means. A common method uses Polaroid film. Polaroid film has a polarization direction associated with the long molecular chains of the polymer film. The *polarizer* allows only the components in a specific plane to pass, as illustrated in ● Fig. 7.16a. The other field vectors are absorbed and do not pass through the polarizer.

The human eye cannot detect that light is polarized, so an *analyzer,* perhaps another polarizing sheet, is needed. If a second polarizer is placed in front of the first polarizer, as illustrated in Fig. 7.15b, then little (theoretically no) light is transmitted, and the sheets appear dark. When the polarization directions of the sheets are at 90°, the Polaroid films are said to be "crossed" (Fig. 7.16c).

The polarization of light is experimental proof that light is a transverse wave. Longitudinal waves, such as sound, cannot be polarized.

FIGURE 7.16   Polarized Light
(a) Light is linearly polarized when it passes through a polarizer. (The lines on the polarizer indicate the direction of polarization.) The polarized light passes through the analyzer if it is similarly oriented. (b) When the polarization direction of the analyzer is perpendicular to that of the polarizer ("crossed Polaroids"), little or no light is transmitted. (c) A photo showing the condition of (b), "crossed Polaroids," using polarizing sunglass lenses.

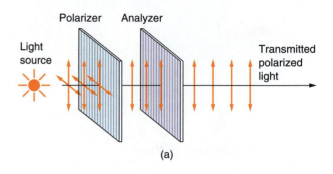

(a)

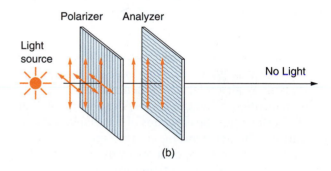

(b)

FIGURE 7.15   Polarization
(a) When the electric field vectors are randomly oriented, as viewed in the direction of propagation, the light is unpolarized. (b) With preferential orientation, the light is partially polarized. (c) If the field vectors lie in a plane, the light is linearly polarized.

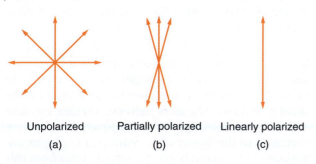

| Unpolarized | Partially polarized | Linearly polarized |
| (a) | (b) | (c) |

(c)

(a)

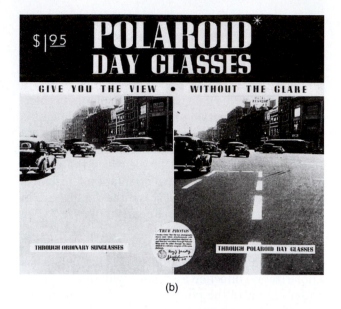

(b)

FIGURE 7.17   Polarizing Sunglasses
(a) Light reflected from a surface is partially polarized in the horizontal direction of the plane of the surface. By orienting the polarizing direction of the sunglasses vertically, the horizontally polarized component of the reflected light is blocked, thereby reducing the intensity or glare. (b) Old-time glare-reduction advertisement from the Polaroid Corporation Archives. Notice the vintage of the cars and the price of the glasses.

Although we cannot detect it with our eyes, sky light is partially polarized as a result of atmospheric scattering by air molecules. When unpolarized sunlight is incident on air molecules, the electric field of the light wave sets the electrons of the molecules into vibration. The accelerating charges emit radiation, similar to the vibrating charges in the antenna of a broadcast station. The radiated, or "scattered," sky light has polarized components, as may be observed with an analyzer. (The best direction to observe the polarization depends on the location of the Sun. At sunset and sunrise, the best direction is directly overhead.) It is believed that some insects, such as bees, use polarized sky light to determine navigational directions relative to the Sun.

A common application of polarization is in polarizing sunglasses. The lenses of these glasses are polarizing sheets oriented so that the polarization direction is vertical. When sunlight is reflected from a surface, such as water or a road, the light is partially polarized in the horizontal direction. The reflections increase the intensity, which an observer sees as glare. Polarizing sunglasses allow only the vertical component of the light to pass; the horizontal component is blocked out, which reduces the glare (● Fig. 7.17).

Another common, but not so well known, application of polarized light is discussed in the chapter's second Highlight.

RELEVANCE QUESTION: *You wish to buy a second pair of polarizing sunglasses. How can you check to make certain the new glasses are polarizing?*

7.4 **Diffraction and Interference**

LEARNING GOALS

▼ Describe the phenomena of diffraction and interference.

▼ Identify some of the effects and applications of these phenomena.

In our study of reflection and refraction, we located a mirror image by drawing straight-line rays; that is, we solved the problem by using geometric methods. The study of optics using these methods is known as *geometric optics*. However, for the topics in this section, we return to the theory of wave motion to give explanations of observed effects. A study of optics from this wave nature point of view is known as *physical optics* (sometimes called *wave optics*).

Diffraction

Water waves passing through slits are shown in ● Fig. 7.18. Notice how the waves are bent, or deviated, around the corners of the slit as they pass through. All waves—sound, light, and so on—show this type of bending as they go through relatively small slits or pass by the corners of objects. The deviation of waves in such cases is referred to as **diffraction.**

FIGURE 7.18   Diffraction

(a) The diffraction or bending of the water waves passing through a slit can be seen in a ripple tank. Notice the circular bending of the waves. (b) When the slit is made smaller or the wavelength (distance between wave crests) is larger compared with the size of the opening, the diffraction becomes greater.

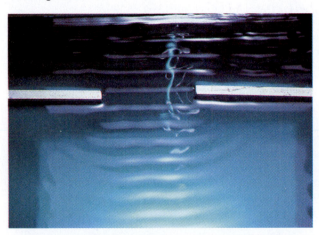

(a)

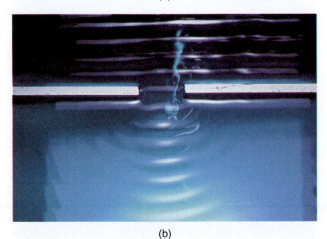

(b)

You will note in Fig. 7.18 that there are different degrees of bending or diffraction. The degree of diffraction depends on the wavelength of the wave and the size of the opening or object. In general, the larger the wavelength compared to the size of the opening or object, the greater is the diffraction.

We know from Chapter 6 that audible sound waves have wavelengths of centimeters to meters, whereas visible light waves have wavelengths of about 10^{-6} m (a millionth of a meter). Ordinary objects (and slits) have dimensions of centimeters to meters. Thus the wavelengths of sound are larger than or about the same size as objects, and diffraction readily occurs for sound. For example, you can talk through a doorway into another room in which persons are standing around the corner on each side, and they can hear you.

However, the dimensions of ordinary objects or slits are much greater than the wavelengths of visible light, so the diffraction of light is not commonly observed. For instance, if you shine a beam of light at an object, there will be a shadow zone behind the object with very sharp boundaries.

Some light diffraction does occur at corners, but this goes largely unnoticed because it is difficult to see. On very close inspection, you find that the shadow boundary is blurred or fuzzy, and there is actually a pattern of bright and dark regions (● Fig. 7.19). This is an indication that some diffraction has occurred.

Think about this: When you sit in a lecture room or movie theater, sound is easily diffracted around the people in front of you, but light is not. What does this say about wavelengths and size of people?

As another example, radio waves are electromagnetic waves of very long wavelengths—in some cases, hundreds of meters long. In this case, ordinary objects and slits are much smaller than the wavelength, so radio waves are easily diffracted around buildings, trees, and so on, making radio reception generally quite efficient. However, you may have noticed a difference in the reception of the AM and FM bands, which have different frequencies and wavelengths.

The wavelengths of the AM band range from about 180 m to 570 m, whereas the wavelengths of the FM band range from 2.8 m to 3.4 m. Hence the longer AM waves are easily diffracted around buildings and so on, whereas FM waves may not be. As a result, the AM reception may be better than FM reception in some areas.

HIGHLIGHT: Liquid Crystal Displays (LCDs)

When a crystalline solid melts, the resulting liquid generally has no orderly arrangement of atoms or molecules. However, some organic compounds have an intermediate state in which the liquid still retains some orderly molecular arrangement—hence the name liquid crystal (LC).

Some liquid crystals are transparent and have an interesting property. When an electrical voltage is applied, the liquid crystal becomes opaque. The applied voltage upsets the orderly arrangement of the molecules, and light is scattered, thus making the LC opaque.

Another property of some liquid crystals is how they affect linearly polarized light—they "twist" or rotate the polarization direction 90°. However, this does not occur if a voltage is applied, causing molecular disorder, as previously mentioned.

A common application of these properties is in liquid crystal displays (LCDs), which are found on wristwatches, calculators, and small TV and laptop computer screens. How LCDs work is illustrated in Fig. 1.

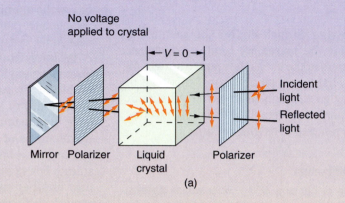

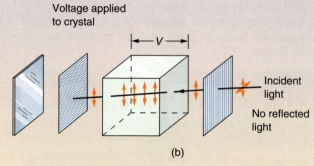

FIGURE 1 Liquid Crystal Display (LCD)

(a) An illustration of how a liquid crystal "twists" the light polarization through 90°. The light passes through the other polarizer and is reflected back and out the crystal with another twist. (b) When a voltage is applied to the crystal, there is no twisting, and light does not pass through the second polarizer. In this case, the light is absorbed, and the crystal appears dark.

FIGURE 7.19 Diffraction Pattern

Using special lighting, diffraction patterns can be seen clearly in the opening of a razor blade.

Interference

When two or more waves meet, they are said to interfere. For example, water waves on a lake or pond commonly interfere with each other. The resultant waveform of the interfering waves is a combination of the individual waves. Specifically, the waveform is given by the *principle of superposition,* which states: At any time, the combined waveform of two or more interfering waves is given by the sum of the displacements of the individual waves at each point in the medium.

The displacement of the combined waveform of two waves at any point is given by $y = y_1 + y_2$, where the directions are indicated by plus and minus signs.

Trace the incident light in the top diagram of Fig. 1. Unpolarized light is linearly polarized by the first polarizer. The LC then rotates the polarization direction, and the polarized light passes through the second polarizer (which is "crossed" with the first) and is then reflected by the mirror. On the reverse path, the polarization direction rotation in the LC allows the light to emerge from the LCD, which would appear bright or white.

However, if a voltage is applied to the LC so that it loses its rotational property, then light is not passed by the second polarizer. With no reflected light, the display would appear dark.

So, by applying voltages to segments of numeral and letter displays, dark regions can be formed on a white background (Fig. 2). The white background is the reflected, polarized light. This can be demonstrated by using an analyzer as shown in Fig. 2.

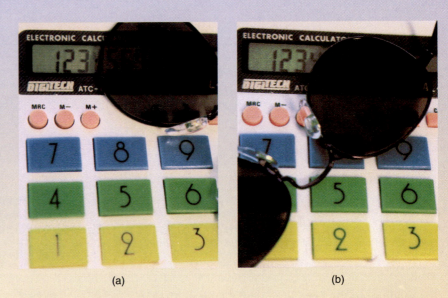

(a) (b)

FIGURE 2   LCDs and Polarization

(a) The light from the bright regions of an LCD is polarized, as can be shown by using polarizing sunglasses as an analyzer. (b) Notice that the glasses have been rotated 90°.

The waveform of the interfering waves changes with time, and after interfering, the waves pass on with their original forms.

While interfering, it is possible that the wave pulses reinforce one another, causing the amplitude of the combined waveform to be greater than either pulse. This is called **constructive interference.** On the other hand, if two pulses tend to cancel each other when they overlap or interfere (one pulse has a negative displacement), the amplitude of the combined waveform is smaller than that of either pulse. This is called **destructive interference.**

Special cases of constructive and destructive interference are shown in ● Fig. 7.20 for pulses with the same amplitude A. When the interfering pulses are exactly in phase (crest coincides with crest when overlapped), the amplitude of the combined waveform is twice that of either individual pulse ($y = A + A = 2A$), and this is referred to as *total constructive interference* (Fig. 7.20a). However, if two pulses are completely out of phase (crest coincides with trough when overlapped), then the waveforms disappear; that is, the amplitude of the combined waveform is zero ($y = A - A = 0$), and this is referred to as *total destructive interference* (Fig. 7.20b).

The word *destructive* is misleading. Do not get the idea that the energy of the pulses is destroyed as is the waveform. The propagating energy is still there in

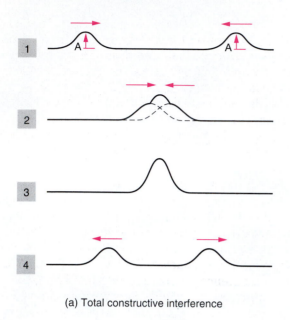

1

2

3

4

(a) Total constructive interference

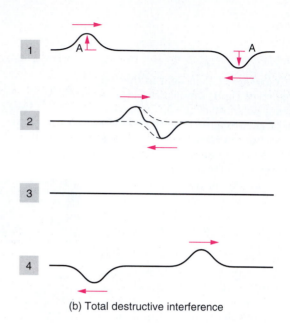

1

2

3

4

(b) Total destructive interference

FIGURE 7.20 **Interference**

(a) When wave pulses of equal amplitude and in phase meet and overlap, there is total constructive interference. At the instant of the overlap, the amplitude of the combined pulse is twice that of the individual ones ($A + A = 2A$). (b) When two wave pulses of equal amplitude are out of phase, the waveform disappears when waves exactly overlap; that is, the combined amplitude of the waveform is zero ($A - A = 0$).

the medium in the form of potential energy. After interfering, the individual pulses continue on with their original waveforms.

The colorful displays seen in oil films and soap bubbles can be explained by interference. Consider light waves incident on a thin film of oil on the surface of water or on a wet road. Part of the light is reflected at the air-oil surface, and part is transmitted. Part of the light in the oil film is then reflected at the oil-water surface (● Fig. 7.21).

The two reflected waves may be in phase, totally out of phase, or somewhere in between. In Fig. 7.21

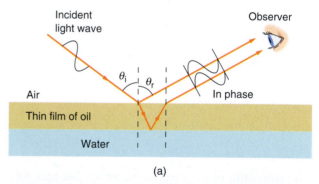

(a)

FIGURE 7.21 **Thin-Film Interference**

(a) When reflected rays from the top and bottom surfaces of an oil film are in phase, constructive interference occurs, and an observer sees only the color of light for a certain angle and film thickness. If the reflected rays are out of phase, destructive interference occurs, which means that light is transmitted at the oil-water interface rather than reflected, and this area appears dark. (b) Because the oil film thickness varies, a colorful display is seen for different wavelengths of light.

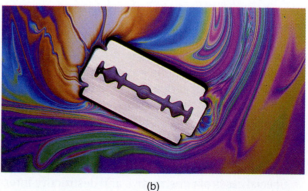

(b)

the waves are shown in phase, but this result will occur only for certain angles of observation, wavelengths of light (colors), and thicknesses of oil film. At certain angles and oil thicknesses, only one wavelength of light shows constructive interference. The other visible wavelengths interfere destructively; thus these wavelengths are transmitted and not reflected.

Hence different wavelengths interfere constructively for different oil-film thicknesses, and an array of colors is seen. In soap bubbles, where the thickness of the soap film moves and changes with time, so does the array of colors.

Diffraction also can give rise to interference. This interference can arise from the bending of light around the corners of a single slit. But an instructive technique employs two narrow double slits that can be considered point sources, as illustrated in ● Fig. 7.22.

When the slits are illuminated with monochromatic light (light of only one wavelength), the diffracted light through the slits spreads out and interferes constructively and destructively at different points where crest meets crest and crest meets trough, respectively. By placing a screen a distance from the slits, an observer can see an interference pattern of alternate bright and dark fringes.

A double-slit experiment was done in 1801 by the English scientist Thomas Young. It demonstrated the wave nature of light. Such an experiment allows the computation of the wavelength of the light from the geometry of the experiment.

This double-slit experiment may be extended. The intensity of the lines becomes less when the light has to pass through a number of narrow slits, but this produces sharp lines that are useful in the analysis of light sources and other applications. A *diffraction grating* consists of many narrow, parallel lines spaced very close together. If light is transmitted through a grating, it is called a *transmission grating*. Such gratings are made by using a laser to etch fine lines on a photosensitive material. The interference of waves passing through such a diffraction grating produces an interference pattern, as shown in ● Fig. 7.23a.

There are also *reflection gratings* (reflecting lines), which are made by etching lines on a thin film of aluminum deposited on a flat surface. The narrow grooves of a compact disk (CD) act as a reflection diffraction grating, producing colorful displays (Fig. 7.23b).

Diffraction gratings are actually more effective than prisms (Fig. 7.14) for separating the component wavelengths of light from stars and other light sources.

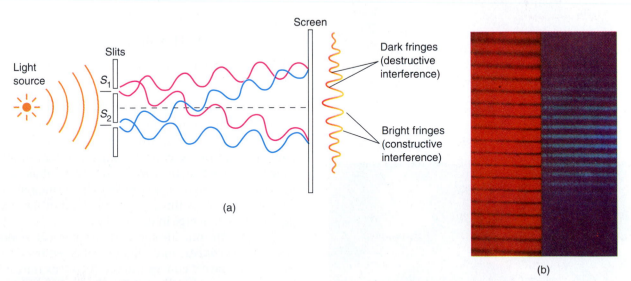

(a)

(b)

FIGURE 7.22   Double-Slit Interference

(a) Light waves interfere as they pass through two narrow slits that act as point sources, giving rise to regions of constructive interference, or bright fringes, and regions of destructive inter-ference, or dark fringes. (b) Actual diffraction patterns of different colors of light through different sized slits.

FIGURE 7.23   Diffraction Grating Interference

(a) The many slits of a diffraction grating produce very sharp interference patterns compared with those of only two slits. The photo shows the colorful separation of colors (wavelengths) of white light passing through a transmission grating. (b) Diffraction is now readily observed. The grooves of a compact disk (CD) form a diffraction grating, and the incident light is separated into a spectrum of colors.

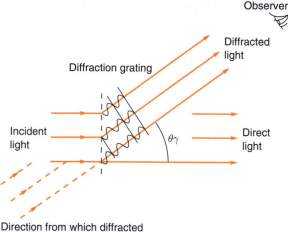

Observer

Diffracted light

Diffraction grating

Incident light

$\theta\gamma$

Direct light

Direction from which diffracted light appears to be coming

(a)

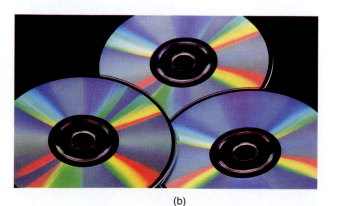

(b)

7.5 Spherical Mirrors

LEARNING GOALS

▼ Distinguish between converging and diverging spherical mirrors.

▼ Describe image formation, and contrast real and virtual images.

Spherical surfaces can be used to make practical mirrors. The geometry of a spherical mirror is shown in ● Fig. 7.24. A spherical mirror is a section of a sphere of radius R. A line drawn through the center of curvature C perpendicular to the mirror surface is called the *principal axis*. The point where the principal axis meets the mirror surface is called the *vertex* (V in the figure).

Another important point in spherical mirror geometry is the *focal point F*. The distance from the vertex to the focal point is called the **focal length** f. (What is "focal" about the focal point and focal length will become evident shortly.) For a spherical mirror, the focal length is one-half the value of the radius of curvature of the spherical surface. Expressed in symbols, the *focal length of a spherical mirror* is

$$f = \frac{R}{2} \qquad \textbf{7.2}$$

where f = the focal length

R = the radius of curvature for spherical mirror

The inside surface of a spherical section is said to be *concave* (as though looking into a recess or cave), and if light is reflected from this surface, we have a **concave mirror.** A concave mirror is commonly called a **converging mirror,** for the reason illustrated in ● Fig. 7.25a. Light rays parallel to the principal axis converge and pass through the focal point. Off-axis parallel rays converge in the focal plane.

Similarly, the outside surface of a spherical section is said to be *convex*, and if light is reflected from this surface, we have a **convex mirror.** A convex mirror is commonly called a **diverging mirror.** Parallel rays along the principal axis are reflected so that they *appear* to diverge from the focal point (Fig. 7.25b).

Considering reverse ray tracing, light rays coming to the mirror from the surroundings are made paral-

FIGURE 7.24 Spherical Mirror Geometry

A spherical mirror is a section of a sphere with a center of curvature C. The focal point F is halfway between C and the vertex V. The distance from V to F is called the *focal length f.* the distance from V to C is the radius of curvature R (the radius of the sphere). Hence $R = 2f$ or $f = R/2$.

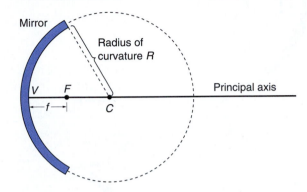

FIGURE 7.25 Spherical Mirrors

(a) Rays parallel to the principal axis of a concave spherical mirror converge at the focal point. Rays not parallel to the principal axis converge in the focal plane so as to form extended images. (b) Rays parallel to the axis of a convex, or diverging, spherical mirror are reflected so as to appear to diverge from the focal point inside the mirror. (c) The divergent property of a diverging mirror is used to give an expanded field of view, as shown here. [Consider reverse ray tracing in (b).]

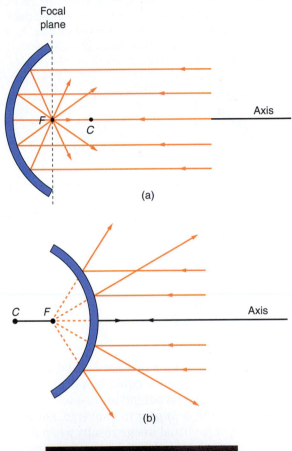

lel, and an expanded field of view is seen in the diverging mirror. Diverging mirrors are used to monitor store aisles and to give truck drivers a wide rear view of traffic (Fig. 7.25c).

Ray Diagrams

The images formed by spherical mirrors may be found by using ray diagrams. We commonly use an arrow as the object, and we determine the location and size of the image by drawing two rays:

1. Draw a ray parallel to the principal axis that is reflected through the focal point.
2. Draw a ray through the center of curvature C that is perpendicular to the mirror surface and is reflected back along the incident path.

The intersection of these rays locates the position of the image.

These rays are shown in ● Fig. 7.26 for an object at various positions in front of a concave mirror. The characteristics of an image are described as being real or virtual, upright (erect) or inverted, and larger or smaller than the object. A **real image** is one for which the light rays converge so that an image can be formed on a screen. A **virtual image** is one for which the light rays diverge and cannot be formed on a screen. Both real and virtual images are formed by a concave (converging) spherical mirror, depending on the object distance.

FIGURE 7.26 Ray Diagrams

(a) A ray diagram for an object beyond the center of curvature C for a concave spherical mirror shows where the image is formed. (b) A ray diagram for a concave mirror for an object located between F and C. Both in (a) and (b), the images are real and thus could be seen on a screen placed at the image distances. Notice how the image moves out and grows larger as the object moves toward the mirror. (c) A ray diagram for a concave mirror with the object inside the focal point F. In this case, the image is virtual and formed behind, or "inside," the mirror. The object and image heights, h_o and h_i, are shown for each case. These heights are related by the magnitude of the magnification factor M, where $h_i = Mh_o$.

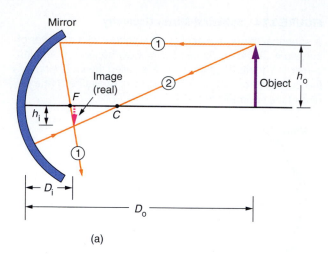

(a)

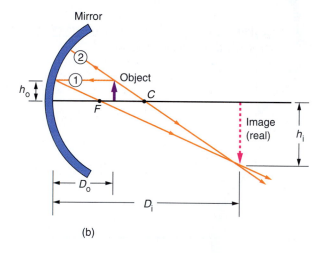

(b)

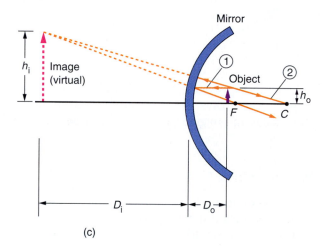

(c)

Notice that the real images are formed in front of the mirror where a screen could be positioned. Virtual images are formed behind or "inside" the mirror where the light rays *appear* to converge. For a converging mirror, a virtual image results when the object is inside the focal point. For a diverging mirror, a virtual image always results wherever the object is located. (What type of image is formed by a plane mirror?)

A convex mirror also may be treated by ray diagrams. The rays of a ray diagram are drawn by using the law of reflection, but they are extended through the focal point and the center of curvature inside or behind the mirror, as shown in ● Fig. 7.27. A virtual image is formed where these extended rays intersect. As we can gather from the figure, even though the object distance may vary, *the image of a convex mirror is always virtual, upright, and smaller than the object.*

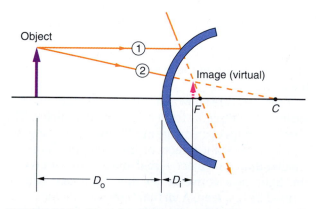

FIGURE 7.27 Diverging Mirror Ray Diagram

A ray diagram for a convex spherical mirror with an object in front of the mirror. A convex mirror always forms a reduced, virtual image.

Spherical Mirror Equation

Our ray diagrams have to be drawn to scale if we are to find the locations and sizes of the images. However, the location of the image may be found more quickly by using the **spherical mirror equation**:

$$\frac{1}{D_o} + \frac{1}{D_i} = \frac{1}{f} \qquad \text{7.3}$$

where D_o = object distance from the vertex of the mirror
  D_i = image distance from the vertex of the mirror
  f = focal length

Another useful form of the spherical mirror equation is

$$D_i = \frac{D_o f}{D_o - f} \qquad \text{7.3a}$$

The values of the object distance D_o and the focal length f are considered to be positive for a concave mirror. (For a convex mirror, f is given a negative value.) Then if D_i is positive, the image is real or formed in front of the mirror. If D_i is negative, the image is virtual and behind the mirror. The sign convention for the distances applies to both concave and convex mirrors. See Table 7.2.

The image size or height (h_i) is generally different from the height of the object (h_o). (See Fig. 7.26c.) The ratio of these heights gives what is called the *magnification factor M*. That is, $h_i/h_o = M$, or $h_i = Mh_o$ (image height = magnification factor × object height). So, if M is greater than 1, the image is taller than the object, and if M is less than 1, the image is reduced or shorter

than the object. (See Fig. 7.26a.) What would be the case for $M = 1$?

In terms of the parameters from a ray diagram, the **magnification factor** is given by

$$M = -\frac{D_i}{D_o} \qquad \text{7.4}$$

(magnification factor)

The minus sign in the equation tells whether the image is upright or inverted. For example, if D_i is positive, M will be negative, which indicates that the image is inverted. But if M is positive (negative D_i), then the image is upright. This sign convention is listed in Table 7.2 and illustrated in the following examples. Even when we use the mirror equation, initially drawing a ray diagram is helpful, so that the general image characteristics are known in advance.

EXAMPLE 7.2

Using the Spherical Mirror Equation

An object is placed 20 cm in front of a concave mirror with a focal length of 5 cm.

(a) At what distance in cm from the mirror will the image be formed?

(b) What are the image characteristics?

SOLUTION

(When working with fractions in this and the following examples, we will ignore significant figures in the data for clarity and report 2 or 3 figures in the answers.)

Step 1

Given: D_o = 20 cm
  f = 5 cm

Step 2

Wanted:

(a) D_i (image distance in cm)

(b) Image characteristics: real or virtual, upright or inverted, and magnification

Step 3

(a) We use Eq. 7.3 to illustrate the calculation with the equation in this form. Rearranging the equation, we obtain

TABLE 7.2 Sign Convention for Spherical Mirrors

Concave (converging) mirror: f positive
Convex (diverging) mirror: f negative
D_o always positive

D_i	Image	M	Image
+	Real	+	Upright
−	Virtual	−	Inverted

$$\frac{1}{D_i} = \frac{1}{f} - \frac{1}{D_o}$$

Then

$$\frac{1}{D_i} = \frac{1}{5} - \frac{1}{20}$$

$$= \frac{4}{20} - \frac{1}{20} = \frac{3}{20} \quad \text{(using a common denominator)}$$

and $\quad D_i = \dfrac{20}{3} \text{ cm} = 6.7 \text{ cm}$

(The centimeter units were omitted in the computation of the fractions for clarity.) Notice in the last step that the fraction was inverted to find D_i. *Don't forget to do this.*

(b) Because D_i is positive, we know that the image is real; that is, it is formed 6.7 cm *in front of* the mirror. To find the other characteristics, we use the magnification equation (Eq. 7.4):

$$M = -\frac{D_i}{D_o} = -\frac{\dfrac{20}{3} \text{ cm}}{20 \text{ cm}} = -\frac{1}{3}$$

Thus, because M is negative, the image is inverted. The numerical value of M indicates that the image is smaller than the object. A magnification factor of $\frac{1}{3}$ indicates that the image is one-third as tall as the object. For example, if the object were 30 cm tall, the image would be $h_i = Mh_o = \frac{1}{3}(30 \text{ cm}) = 10 \text{ cm}$ tall. Note that this example generally corresponds to the ray diagram in Fig. 7.26a.

CONFIDENCE EXERCISE 7.2

If the object is moved to 10 cm in front of the mirror in the preceding example, will the image have greater magnification or be reduced?

EXAMPLE 7.3

Using the Spherical Mirror Equation (Again)

Consider a concave mirror with a radius of curvature of 50 cm. An object is located 20 cm in front of the mirror.

(a) Where will its image be?

(b) Will the image be upright or inverted, larger or smaller?

SOLUTION

(a) We are given $D_o = 20$ cm and $R = 50$ cm. From the latter, we know that the focal length is (Eq. 7.2)

$$f = \frac{R}{2} = \frac{50 \text{ cm}}{2} = 25 \text{ cm}$$

To find D_i, let's use Eq. 7.3a. This alternative form of the spherical mirror equation is convenient because you solve directly for D_i.

$$D_i = \frac{D_o f}{D_o - f}$$

Then

$$D_i = \frac{(20 \text{ cm})(25 \text{ cm})}{20 \text{ cm} - 25 \text{ cm}}$$

$$= -100 \text{ cm}$$

The minus sign indicates that the image is virtual, and it is located 100 cm behind or "inside" the mirror.

(b) Using the magnification equation, we get

$$M = -\frac{D_i}{D_o} = -\frac{(-100 \text{ cm})}{20 \text{ cm}} = +5$$

Because M is positive, the image is upright, and it is five times larger than the object. This example corresponds to the ray diagram in Fig. 7.26c with the object inside the focal point. *A virtual, upright image is always formed for a concave mirror when the object is inside the focal point.*

CONFIDENCE EXERCISE 7.3

Suppose the object in Example 7.3 is 25 cm from the mirror. Where will the image be located in this case?

Example 7.3 corresponds to the case of a makeup mirror, which is a concave mirror of relatively long focal length. The mirror gives moderate magnification so that facial features can be seen better (● Fig. 7.28).

When we use the spherical mirror equation for convex mirrors, everything is the same as before, *except* that the focal length of a convex mirror is taken to

FIGURE 7.28 Magnification

Concave makeup mirrors give moderate magnification so that facial features can be seen better.

be negative. The following example illustrates the use of the mirror equation for this case.

EXAMPLE 7.4

Finding the Magnification of a Diverging Mirror

What is the magnification of an image in a convex mirror for an object 6 m from the mirror if the focal length of the mirror is 2 m? (Consider an image in a store mirror, as in Fig. 7.25c.)

SOLUTION

We are given $f = -2$ m (negative, convex mirror, Table 7.2) and $D_o = 6$ m. To find the magnification factor M, we need to know both D_o and D_i. Using Eq. 7.3a to find D_i, we have

$$D_i = \frac{D_o f}{D_o - f} = \frac{(6\text{ m})(-2\text{ m})}{6\text{ m} - (-2\text{ m})}$$

$$= -\frac{12\text{ m}}{8} = -1.5\text{ m}$$

where the minus tells us that the image is virtual (which we already know). Then

$$M = -\frac{D_i}{D_o} = -\frac{(-1.5\text{ m})}{6\text{ m}} = +\frac{1}{4}$$

and the image is smaller than the object—only one-fourth as tall. The positive value of M indicates that the image is upright.

RELEVANCE QUESTION: If you look into the front side of a shiny spoon, you see an inverted image of yourself. If you look into the back side of the spoon, your image is upright. Why is this?

7.6 Lenses

LEARNING GOALS

▼ Distinguish between converging and diverging spherical lenses.

▼ Describe image formation, and contrast real and virtual images.

A lens consists of material such as a transparent piece of glass or plastic that refracts light waves to give an image of an object. Lenses are extremely useful and are found in eyeglasses, telescopes, magnifying glasses, cameras, and many other optical devices.

In general, there are two main classes of lenses. A **converging**, or **convex, lens** is thicker at the center than at the edge. A **diverging**, or **concave, lens** is thicker at the edges. These two classes and some of the possible shapes for each are illustrated in ● Fig. 7.29. In general, we will investigate the spherical biconvex and biconcave lenses at the left of each group

FIGURE 7.29 Lenses

Different types of converging and diverging lenses. Notice that converging lenses are thicker at the centers that at the edges, whereas diverging mirrors are thinner at the center.

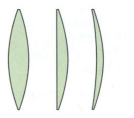

Converging, or convex, lenses; greatest thickness at center

Diverging, or concave, lenses; greatest thickness at edge

FIGURE 7.30 Lens Focal Points

For a converging spherical lens, rays parallel to the principal axis and incident on the lens converge at the focal point on the opposite side of the lens. Rays parallel to the axis of a diverging lens appear to diverge from a focal point on the incident side of the lens.

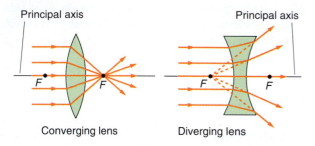

Converging lens   Diverging lens

FIGURE 7.31 Ray Diagrams

(a) A ray diagram for a converging lens with the object outside the focal point. The image is real and inverted and can be seen on a screen placed at the image distance, as shown in the photo. (b) A ray diagram for a converging lens with the object inside the focal point. In this case, a virtual image is formed on the object side of the lens.

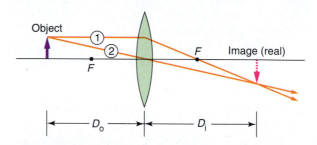

in the figure (*bi-* because they have similar spherical surfaces on each side).

Light passing through a lens is refracted twice—once at each surface. The lenses most commonly used are known as *thin lenses*. Thus, when constructing ray diagrams, we can neglect the thickness of the lens and assume that the two surfaces that cause refraction are in essentially the same plane.

The principal axis for a lens goes through the center of the lens and is labeled in ● Fig. 7.30. Rays coming in parallel to the principal axis are refracted toward the principal axis by a converging lens. For a converging lens, the rays are focused at point *F*, the focal point. For a diverging lens, the rays are refracted away from the principal axis and appear to emanate from the focal point.

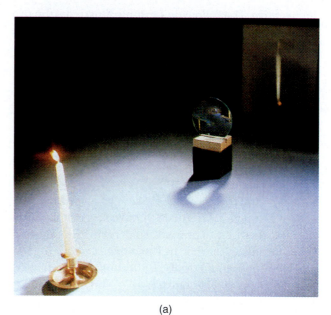

(a)

Ray Diagrams

We can show how lenses refract light to form images by a graphic procedure similar to what we did with mirrors.

1. The first ray is drawn parallel to the principal axis and then refracted by the lens along a line drawn through a focal point of the lens.
2. The second ray is drawn through the center of the lens and without a change in direction.

The image of the tip of the arrow occurs where these two rays meet.

Examples of this procedure are shown in ● Fig. 7.31. Notice that only the focal points for the respective surfaces are shown, which are all that are needed.

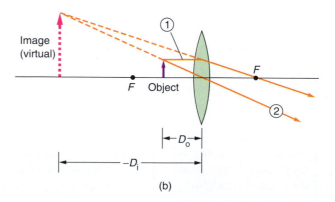

(b)

The lenses do have radii of curvature, but for spherical lenses $f \neq R/2$, in contrast to $f = R/2$ for spherical mirrors. The characteristics of the images formed by a converging or convex lens change just the way those of a converging mirror change as the object is brought toward the mirror from a distance. Beyond the focal point, an inverted, real image is formed, which becomes larger as the object approaches the focal point.

The magnification becomes greater than 1 when the object distance is less than $2f$. Once inside the focal point of a converging mirror, an object always forms a virtual image. For lenses, a *real image* is formed on the opposite side of a lens from the object and can be seen on a screen (Fig. 7.31a). A *virtual image* is formed on the object side of the lens (Fig. 7.31b).

For a diverging or concave lens, the image is always upright and smaller than the object. When looking through a concave lens, we see images as shown in ● Fig. 7.32.

Thin-Lens Equation

Thin-lens problems can be solved analytically with a single equation as long as the proper signs are used. The **thin-lens equation** is

$$\frac{1}{D_o} + \frac{1}{D_i} = \frac{1}{f} \qquad 7.5$$

or

$$D_i = \frac{D_o f}{D_o - f} \qquad 7.5a$$

Note that the thin-lens equation and the spherical mirror equation (Eq. 7.3) are mathematically identical. The lens equation also may be expressed in rearranged form as in Eq. 7.3a.

The sign convention for spherical lenses is similar to that for spherical mirrors (Table 7.2) inasmuch as for a converging (convex) lens f is positive (+) and for a diverging (concave) lens f is negative (−).

In analyzing lenses, we usually want to know the image distance D_i, which is the distance of the image from the lens (see Fig. 7.31). The object distance, or the distance of the object from the lens, is labeled D_o.

If D_i comes out to be positive, then the image is real and formed on the opposite side of the lens from the object, and it can be brought to focus on a screen placed there (Fig. 7.31b). If D_i is negative, the image

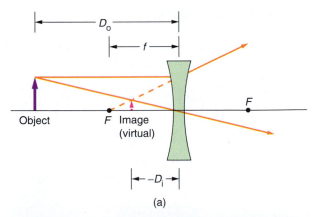

(a)

(b)

FIGURE 7.32 Diverging Lens
(a) A ray diagram for a diverging lens. A virtual image is formed on the object side of the lens. Diverging lenses form only virtual images. (b) Like a diverging mirror, a diverging (concave) lens gives an expanded field of view.

will be virtual (on the same side of the lens as the object) and can be viewed only by looking through the lens (Fig. 7.32).

The *magnification factor* of a lens is given by the same equation we had for mirrors:

$$M = -\frac{D_i}{D_o} \qquad 7.6$$

The sign convention for the magnification factor for lenses is also the same as for spherical mirrors. If M is positive, the image is upright, and if M is negative, the image is inverted.

The following example illustrates the use of the lens equation and the sign convention.

Finding Convex Lens Characteristics

An object is placed 6.0 cm in front of a convex lens with a focal length of 10 cm. Calculate the image position.

SOLUTION

We have $D_o = 6.0$ cm and $f = 10$ cm. Then, using the convenient form of the lens equation (Eq. 7.5a),

$$D_i = \frac{D_i f}{D_i - f} = \frac{(6.0 \text{ cm})(10 \text{ cm})}{6.0 \text{ cm} - 10 \text{ cm}} = -\frac{60 \text{ cm}}{4.0}$$

or $\qquad D_i = -15$ cm

Because D_i is negative, the image is virtual and located on the same side of the lens as the object.

Such a convex lens acts like a simple magnifying glass. The object to be viewed and magnified must be located inside the focal point of the lens. The magnification factor is given by

$$M = -\frac{D_i}{D_o} = -\frac{(-15 \text{ cm})}{6 \text{ cm}} = +2.5$$

CONFIDENCE EXERCISE 7.5

A concave lens with a focal length of -10 cm is used to view an object 25 cm from the lens. (a) Where will the image be located? (b) What are its characteristics?

The Human Eye

The human eye contains a convex lens, together with other refractive media in which most of the light refraction occurs. Even so, we can learn a great deal about the optics of the eye by considering only the focusing action of the lens. As illustrated in ● Fig. 7.33, the lens focuses the light entering the eye on the *retina*. The photoreceptors of the retina, called *rods* and *cones,* are connected to the optic nerve, which sends signals to the brain. The rods are more sensitive

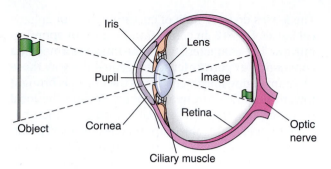

FIGURE 7.33 The Human Eye

The lens of the human eye forms an image on the retina, which contains rod and cone cells. The rods are more sensitive than the cones and are responsible for light and dark "twilight" vision; the cones are responsible for color vision.

than the cones and are responsible for light and dark "twilight" vision; the cones are responsible for color vision.

Notice in Fig. 7.33 that an image focused on the retina in normal vision is upside down. Why, then, don't we see things upside down? For some reason, the brain interprets images properly oriented.

Because the distance between the lens and retina does not vary, we have a situation in which D_i is constant. Because D_o varies for different objects, the focal length of the lens of the eye must vary. The lens is called the *crystalline lens* and consists of glassy fibers. By action of the attached ciliary muscles, the shape and focal length of the lens vary as the lens is made thinner and thicker. The optical adjustment of the eye is truly amazing. Objects can be seen quickly at distances that range from a few centimeters (the near point) to infinity (the far point).

By speaking of the "normal" eye, we imply that visual defects do exist in some eyes. That this is the case is readily apparent from the number of people who wear glasses or contact lenses. The eyes of many people have problems seeing objects at certain distances. These folks have one of the two most common visual defects: nearsightedness and farsightedness.

Nearsightedness is the condition of being able to see nearby objects clearly but not distant objects. This is so because for some reason the distant image is focused in front of the retina (● Fig. 7.34a). Glasses with diverging lenses that move the image back can be used to correct this defect.

Farsightedness is the condition of being able to see distant objects clearly but not nearby objects. The im-

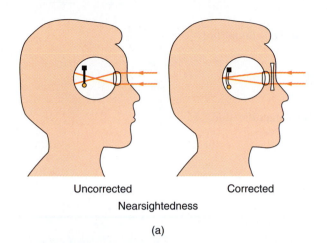

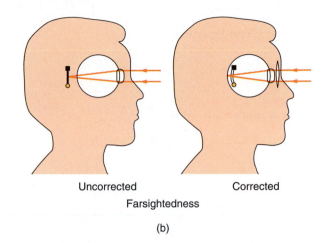

Uncorrected   Corrected   Uncorrected   Corrected

Nearsightedness   Farsightedness

(a)   (b)

FIGURE 7.34 Vision Defects

Two common vision defects arise because the image is not focused on the retina. (a) Nearsightedness occurs when the image is formed in front of the retina. The condition is corrected by wearing glasses with diverging lenses. (b) Farsightedness occurs when the image is formed behind the retina. This condition is corrected by wearing glasses with converging lenses.

ages of such objects are focused behind the retina (Fig. 7.34b). The near point is the position closest to the eye at which objects can be seen clearly. (Bring your finger toward your nose. The position where the tip of your finger goes out of focus is your near point.) For farsighted people, the near point is not at the normal position but at some point farther from the eye.

Children can see sharp images of objects as close as 10 cm (4 in.) to their eyes. The crystalline lens of the normal young adult eye can be deformed to produce sharp images of objects as close as 12 to 15 cm (5 to 6

in.). However, at about the age of 40, the near point normally moves beyond 25 cm (10 in.). You may have noticed people over 40 holding reading material at some distance from their eyes in order to see it clearly. When the print is too small or the arm too short, reading glasses with converging lenses are the solution (Fig. 7.34b). The recession of the near point with age is not considered an abnormal defect of vision. It proceeds at about the same rate in all normal eyes. You, too, may need reading glasses someday.

Important Terms

reflection (7.1)	refraction (7.2)	constructive interference	virtual image
ray	index of refraction	destructive interference	spherical mirror equation
law of reflection	total internal reflection	focal length (7.5)	magnification factor
regular (specular) reflection	dispersion	concave (converging) mirror	convex (converging) lens (7.5)
irregular (diffuse) reflection	polarization (7.3)	convex (diverging) mirror	concave (diverging) lens
	linearly polarized light	real image	thin-lens equation
	diffraction (7.4)		

Important Equations

Index of Refraction: $n = \dfrac{c}{c_\mathrm{m}}$

Condition for Diffraction: $\lambda \geq d$

Spherical Mirror Radius and Focal Length Equation: $f = \dfrac{R}{2}$

Spherical Mirror Equation: $\dfrac{1}{D_o} + \dfrac{1}{D_i} = \dfrac{1}{f}$

or $\qquad\qquad\qquad D_i = \dfrac{D_o f}{D_o - f}$

Mirror Magnification: $M = -\dfrac{D_i}{D_o}$

$\qquad$ (or $h_i = Mh_o$)

Thin-Lens Equation: $\dfrac{1}{D_o} + \dfrac{1}{D_i} = \dfrac{1}{f}$

or $\qquad\qquad\qquad D_i = \dfrac{D_o f}{D_o - f}$

Lens Magnification: $M = -\dfrac{D_i}{D_o}$

$\qquad$ (or $h_i = Mh_o$)

Review Questions

7.1 Reflection

1. What can be said about the angles of incidence and reflection?
   (a) can never be equal
   (b) add to 90°
   (c) are not related
   (d) are measured from a line perpendicular to the reflecting surface

2. To what does the law of reflection apply?
   (a) regular reflection
   (b) irregular reflection
   (c) diffuse reflection
   (d) all of these

3. Explain how the law of reflection applies to diffuse reflection, and give an example of this type of reflection.

4. A sign on the back of a tractor-trailer rig reads, "If you can't see my mirror, then I can't see you." Why is this statement correct?

7.2 Refraction and Dispersion

5. What is the case when the angle of refraction is greater than the angle of incidence?
   (a) The critical angle is exceeded.
   (b) The first medium is less dense.
   (c) The second medium has a smaller index of refraction.
   (d) The speed of light is less in the second medium.

6. Which of the following is true about dispersion?
   (a) occurs above a critical angle
   (b) is the principle of internal reflection
   (c) is responsible in part for a rainbow
   (d) occurs for diffuse reflection

7. What is refraction, and what causes it?

8. Explain how the refraction of light on entering a denser medium is analogous to the situation of a marching band.

9. On entering a certain medium from air, light is bent or refracted away from the normal. What does this tell you about the medium?

10. For any substance, the index of refraction is always greater than 1. Why is this?

11. What is dispersion? Name a scientific instrument that gives valuable information using the dispersion of light.

12. Explain why diamonds have brilliance and "fire."

7.3 Polarization

13. What happens if the polarization directions of two polarizing sheets are at an angle of 45° to each other?
    (a) No light gets through.
    (b) There is maximum transmission.
    (c) Maximum transmission is reduced.
    (d) None of these.

14. What is needed for the human eye to detect polarized light?

15. Give two examples of the practical use of polarization.

16. How could you use polarization to distinguish between transverse and longitudinal waves?

17. While looking through two polarizing sheets, one of the sheets is rotated 90°. Would there be any change in the observation? Explain.

7.4 Diffraction and Interference

18. Which is true for diffraction?
    (a) occurs best when a slit width is less than the wavelength of a wave
    (b) depends on refraction
    (c) is caused by interference
    (d) does not occur for light

19. When does total destructive interference occur?
    (a) when waves are in phase
    (b) at the same time as total constructive interference
    (c) when the waves have equal amplitudes and are completely out of phase
    (d) when total internal reflection occurs

20. A slit opening in a water tank is 0.50 cm wide. For what wavelength(s) will the diffraction of water waves passing through the slit produce an appreciable shadow zone?

21. Why do sound waves bend around everyday objects, whereas the bending of light rays is not generally observed?

22. Describe the interference of two wave pulses with different amplitudes if they are (a) in phase and (b) completely out of phase.

7.5 Spherical Mirrors

23. Which is true for a concave mirror?
    (a) has a radius of curvature equal to $2f$
    (b) is a diverging mirror
    (c) forms only virtual images
    (d) forms only magnified images

24. Which is true for a real image?
    (a) is always magnified
    (b) is formed by converging light rays
    (c) is formed behind a mirror
    (d) occurs for only $D_i = D_o$

25. What are the relationships among the center of curvature, the focal point, and the vertex for spherical mirrors?

26. Distinguish between real images and virtual images for spherical mirrors.

27. Explain when real and virtual images are formed by (a) a concave mirror and (b) a convex mirror.

7.6 Lenses

28. A biconvex lens is which of the following?
    (a) a converging lens
    (b) thicker at the edge than at the center
    (c) a lens that forms virtual images for $D_o > f$
    (d) a Fresnel lens

29. Which is true for a virtual image?
    (a) is always formed by a convex lens
    (b) can be formed on a screen
    (c) is formed on the object side of a lens
    (d) cannot be formed by a concave lens

30. Explain when real and virtual images are formed by (a) a convex lens and (b) a concave lens.

31. Why are slides put into a slide projector upside down, and what is done in focusing a slide image on a screen?

32. A biconvex lens is a magnifying glass. It is looked through with the object just inside the focal point on the opposite side of the lens. Explain why a magnified image is seen.

Applying Your Knowledge

1. If the Moon's spherical surface gave regular reflection, how would the full moon look?

2. How would a fish see the world above the surface of its water environment?

3. While looking through two polarizing sheets, one of the sheets is rotated 360°. Describe what would be observed.

4. If you walk toward a plane mirror at a given speed, how does your image approach you?

5. (a) Is a satellite TV dish a spherical mirror? Where is the receiver placed?
   (b) Some dishes are made of open-wire mesh. How can this be a reflecting mirror?

6. On most automobile passenger-side rearview mirrors, a warning is printed such as, "Objects in mirror are closer than they appear." Why is this, and what makes the difference? (*Hint:* The mirrors are convex mirrors.)

Exercises

7.1 Reflection

1. Light is incident on a plane mirror at an angle of 30° relative to the normal. What is the angle of reflection?

   Answer: 30°

2. Light is incident on a plane mirror at an angle of 30° relative to its surface. What is the angle of reflection?

3. Show that for a person to see his or her complete (head-to-toe) image in a plane mirror, the mirror must have a length (height) of at least one-half of a person's height (see Fig. 7.4). Does the person's distance from the mirror make a difference? Explain.
   Answer: Bisecting triangles in the figure give one-half height. Same for any distance.

4. How much longer must the minimum length of a plane mirror be for a 6-ft 4-in. man to see his complete head-to-toe image than for a 5-ft 2-in. woman?

7.2 Refraction and Dispersion

5. What is the speed of light in a diamond?
   Answer: 1.24×10^8 m/s

6. The speed of light in a particular type of glass is 1.60×10^8 m/s. What is the index of refraction of the glass?

7. What percentage of the speed of light in vacuum is the speed of light in crown glass?
   Answer: 65.8%

8. The speed of light in a certain transparent material is 41.3% of the speed of light in vacuum. What is the index of refraction of the material? (Can you identify the material?)

7.5 Spherical Mirrors

(Assume significant figures to 0.1 cm.)

9. Sketch a ray diagram for a concave mirror with an object at $D_o = R$, and describe the image characteristics.
   Answer: Real, inverted, and same size

10. Sketch ray diagrams for a concave mirror with objects at (a) $D_o > R$, (b) $R > D_o > f$, and (c) $D_o < f$. Describe how the image changes as the object is moved toward the mirror.

11. A woman holds a makeup mirror with a radius of curvature of 120 cm a distance of 20 cm from her face. What is the magnification of the observed image? Answer: 1.5

12. A 3.0-cm lighted candle is placed 20 cm from a concave spherical mirror with a radius of curvature of 30 cm. (a) Where should a screen be placed in order to see the candle's image clearly? (b) What is the minimum height of the screen in order to see the candle's complete image?

13. An object is placed 15 cm from a convex spherical mirror with a focal length of 10 cm. Where is the image located, and what are its characteristics?
    Answer: $D_i = 6.0$ cm, virtual, upright, and $M = 0.40$

14. A reflecting, spherical Christmas tree ornament has a diameter of 8.0 cm. If a child looks at the ornament from a distance of 20 cm, describe the image she sees.

7.6 Lenses

15. Sketch a ray diagram for a spherical convex lens with an object at $D_o = 2f$, and describe the image characteristics.
    Answer: Real, inverted, and same size

16. Sketch ray diagrams for a spherical convex lens with objects at (a) $D_o > 2f$, (b) $2f > D_o > f$, and (c) $D_o < f$. Describe how the image changes as the object is moved closer to the lens.

17. An object with a height of 5.0 cm is placed 45 cm in front of a converging lens with a focal length of 20 cm. Where is the image formed, and what is its height? How is the image oriented? Answer: $D_i = 36$ cm, $h_i = 4.0$ cm, inverted

18. A slide projector has a convex lens with a focal length of 8.00 cm. A slide is placed 8.20 cm from the lens. (a) How far from the lens should a screen be placed so that the image of the slide is in focus? (b) If the slide film is 1 in. $\times$ 1 in., what is the minimum size screen needed in order to see the total projected image?

19. A particular convex lens has a focal length of 15 cm. If an object is placed at the focal point, what do the thin-lens and magnification equations yield in this case?
    Answer: $D_i = \infty$, $M = \infty$ (mathematical blow-up, cross-over point for real and virtual images)

20. A simple magnifying glass (convex lens) has a focal length of 8.0 cm. It is positioned so that an object is 6.0 cm from the lens. What is the magnification of the observed image?

21. A spherical concave lens has a focal length of 20 cm. If an object is placed 10 cm from the lens, where is the image formed and what are its characteristics?
    Answer: $D_i = 6.7$ cm, virtual, upright, and reduced ($M = 0.67$)

22. A student is given a spherical concave lens with a focal length of 20 cm. He thinks that if he puts an object 5.0 cm on either side of one of its focal points that the image magnification will be the same. Is this true? Justify your answer.

Solutions to Confidence Exercises

7.1 (a) $c_m = \dfrac{c}{n} = \dfrac{c}{1.000} = c$ (or 3.00×10^8 m/s)

(b) $c_m = \dfrac{c}{n} = \dfrac{3.00 \times 10^8 \text{ m/s}}{1.00029} = 2.999 \times 10^8$ m/s

(significant figures ignored)

Essentially the same for normal calculations.

7.2 $D_i = \dfrac{D_o f}{D_o - f} = \dfrac{(10 \text{ cm})(5 \text{ cm})}{10 \text{ cm} - 5 \text{ cm}} = 10$ cm

$M = -\dfrac{D_i}{D_o} = -\dfrac{10 \text{ cm}}{10 \text{ cm}} = -1,$

so inverted image is the same size as the object.

7.3 $D_i = \dfrac{D_o f}{D_o - f} = \dfrac{(25 \text{ cm})(25 \text{ cm})}{25 \text{ cm} - 25 \text{ cm}} \rightarrow \infty$

For an object located at the focal point, we say that the image is formed at infinity. This is the point where the image switches from being real to virtual. (Draw a ray diagram for this case. The reflected rays are parallel and never meet—or *may* meet at infinity.)

7.4 The focal length f is always negative for a convex mirror. Expressing this explicitly in Eq. 7.3a,

$D_i = \dfrac{D_o f}{D_o - f} = \dfrac{D_o (-f)}{D_o - (-f)} = -\left(\dfrac{D_o f}{D_o + f} \right)$

so D_i is always negative, and the image is always virtual.

7.5 (a) $D_i = \dfrac{D_o f}{D_o - f} = \dfrac{(25 \text{ cm})(-10 \text{ cm})}{25 \text{ cm} - (-10 \text{ cm})} = -7.1$ cm

(b) $M = -\dfrac{D_i}{D_o} = -\dfrac{(-7.1 \text{ cm})}{25 \text{ cm}} = 0.28$

Virtual, upright, and reduced.

Answers to Multiple-Choice Review Questions

1. d 5. c 13. c 19. c 24. b 29. c
2. d 6. c 18. a 23. a 28. a

ELECTRICITY AND MAGNETISM

8

Like charges repel, and unlike charges attract each other, with a force that varies inversely with the square of the distance between them. . . . Frictional forces, wind forces, chemical bonds, viscosity, magnetism, the forces that make the wheels of industry go round—all these are nothing but Coulomb's law. . . .

J. R. Zacharias (1905–1986)

Ours is indeed an electrical society. Think of how your life might be without electricity. We get some idea during an occasional power outage or when the electricity is off. Yet, when asked to define electricity, many people have difficulty. The terms *electric current* and *charge* come to mind, but what are these?

You may recall from Chapter 1 that *electric charge* was mentioned as a fundamental quantity. That is, we really don't know what it is, so our chief concern is what it does—the description of electrical phenomena.

As you will learn in this chapter, we know that electric charge is associated with certain particles and that there are interacting forces. With a force, we can go on to the motion of electric charges (current) and then to electrical energy and power. By applying these principles, we have the benefits of electricity at our disposal. Electricity runs motors, heats our homes, gives us lighting, powers our televisions and stereos, and on and on.

But the *electric force* is even more basic than electricity. It keeps atoms and molecules together—even the ones that make up our bodies. It may be said that the electric force holds matter together, whereas the gravitational force (Chapter 3) holds our solar system and galaxies together.

Photo: Electrical transmission lines transport electrical energy over long distances.

Closely associated with electricity is *magnetism*. In fact, we refer to *electromagnetism* because these phenomena are basically inseparable. For example, without magnetism, we would not be able to generate electrical power. As children (and perhaps as adults), most of us have been fascinated with the properties of small magnets. Have you ever wondered what causes magnets to attract and repel?

This chapter introduces the basic properties of electricity and magnetism—exciting topics, examples of which are everywhere around you. ◼

8.1 Electric Charge and Current

LEARNING GOALS

▼ Describe electric charge and current.

▼ State the law of charges and Coulomb's law.

Electric charge is a fundamental quantity. The property of electric charge is associated with certain subatomic particles, and experimental evidence leads us to the conclusion that there are two types of charges. They are designated as *positive* (+) and *negative* (−) for distinction. All matter, according to modern theory, is made up of small particles called *atoms,* which are composed in part of negatively charged particles called **electrons,** positively charged particles called **protons,** and neutral particles called *neutrons* that have no electric charge and are slightly more massive than protons. Table 8.1 summarizes the fundamental properties of these atomic particles, which are discussed in more detail in Chapters 9 and 10.

As the table indicates, all three particles have certain masses, but only electrons and protons possess electric charges. The magnitudes of the electric charges on the electron and the proton are equal, but their natures are different, as expressed by the plus and minus signs. When we have the same number of electrons and protons, the *total* charge is zero (same number of positive and negative charges of equal magnitude), and we have an electrically *neutral* situation.

The unit of electric charge is called the **coulomb** (C), after Charles Coulomb (1736–1806), a French scientist who studied electrical effects. Electric charge is usually designated by the letter *q*. A +*q* indicates that an object has an excess number of positive charges, or fewer electrons than protons. A −*q* indicates an excess of negative charge, or more electrons than protons.

When charge flows, or is in motion, we say we have an electric current. **Current** is defined as the time rate of flow of electric charge.

$$\text{current} = \frac{\text{charge}}{\text{time}}$$

or
$$I = \frac{q}{t} \qquad \textbf{8.1}$$

where I = electric current, measured in amperes

 q = electric charge flowing past a given point, measured in coulombs

 t = time for the charge to move past the point, measured in seconds

Current is measured in units of amperes, named after André Ampère (1775–1836), another early French investigator of electricity. One **ampere** (A) is equal to a flow of one coulomb of charge per second.

From Eq. 8.1 we can write

$$q = It \qquad \textbf{8.1a}$$

or 1 coulomb = 1 ampere × 1 second. Hence we see that the *coulomb* is defined as the amount of charge that flows past a given point in 1 second when the current is 1 ampere.

Early theories considered electrical phenomena to be due to some type of fluid in materials, which is

TABLE 8.1   Some Properties of Atomic Particles

Particle	Symbol	Mass	Charge
Electron	e^-	9.109×10^{-31} kg	-1.60×10^{-19} C
Proton	p^+	1.673×10^{-27} kg	$+1.60 \times 10^{-19}$ C
Neutron	n	1.675×10^{-27} kg	0

probably why we sometimes say a current "flows," when actually it is electric charge that flows. Electrical *conductors* are materials in which an electric charge flows readily. Metals are good conductors. We use metal wires widely to conduct electric currents. This conduction is due primarily to the outer, loosely bound electrons of the atoms. (Recall from Chapter 5 that electrons also contribute significantly to thermal conduction.)

Materials in which electrons are more tightly bound do not conduct electricity very well, and they are referred to as electrical *insulators*. Examples are wood, glass, and plastics. We coat our electric cords with rubber or plastic so that we can handle them safely. Materials that are neither good conductors nor good insulators are called *semiconductors*; graphite (carbon) is an example.

In the definition of current in Eq. 8.1, we speak of an amount of charge flowing past a given point. This is *not* a flow of charge in a manner similar to fluid flow. In a metal wire, for example, the electrons move randomly and chaotically. Some go in one direction past a point, and others go in the opposite direction. However, with a current, more electrons go in one direction than the other, so we are really talking about a *net* charge q in the equations (analogous to a *net* force as in Chapter 3).

EXAMPLE 8.1

Finding the Amount of Electric Charge

A wire carries a current of 0.50 A for 2 minutes. (a) How much (net) charge goes past a point in the wire in this time? (b) How many electrons make up this amount of charge?

SOLUTION

Step 1

Given: $I = 0.50$ A
$t = 2.0$ min $= 120$ s

Step 2

Wanted:

(a) q (charge)

(b) n (number of electrons)

Checking units, we note that the time is given in minutes. This was converted directly to seconds when written in Step 1.

Step 3

(a) To find q, we use Eq. 8.1a:

$$q = It = (0.50 \text{ A})(120 \text{ s}) = 60 \text{ C}$$

(b) A charge q is made up of a number (n) of charged particles. For example,

$$q_1 = 1e = 1(1.6 \times 10^{-19} \text{ C}) = 1.6 \times 10^{-19} \text{ C}$$

$$q_2 = 2e = 2(1.6 \times 10^{-19} \text{ C}) = 3.2 \times 10^{-19} \text{ C}$$

$$q_3 = 3e = 3(1.6 \times 10^{-19} \text{ C}) = 4.8 \times 10^{-19} \text{ C}$$

and so on, such that we may write in general, for any charge q,

$$q = ne$$

where e is the electronic charge. Then, solving for n using the result from (a) with the $-$ sign for electrons neglected,

$$n = \frac{q}{e} = \frac{60 \text{ C}}{1.6 \times 10^{-19} \text{ C/electron}}$$

$$= 3.8 \times 10^{20} \text{ electrons}$$

Quite a few electrons!

CONFIDENCE EXERCISE 8.1

If 5.0×10^{19} electrons go by a certain point in a wire in two-thirds of a minute, what is the current in the wire?

Electric Force

An electric force exists between any two charged particles. On investigation, it is found that the mutual forces on the particles may be either attractive or repulsive, depending on the types of charges ($+$ or $-$). In fact, we know there are two different types of charge because of the different force interactions. (Recall from Chapter 3 that for gravitation there is only one type of mass, and the force interaction between masses is always attractive.) The attraction and repulsion between different types of charges are described by the **law of charges:**

Like charges repel, and unlike charges attract.

In other words, two negative charges (charged particles) or two positive charges experience repulsive electric forces—forces equal and opposite (Newton's third law), whereas a positive charge and a negative charge experience attractive forces (forces toward each other).

The law of charges gives the direction of an electric force, but what about its magnitude? In other words, how strong is the electric force between charged particles or bodies? Charles Coulomb derived a relationship for the magnitude of the electric force between two charged bodies, which is appropriately known as **Coulomb's law:**

The force of attraction or repulsion between two charged bodies is directly proportional to the product of the two charges and inversely proportional to the square of the distance between them.

This law can be written in equation form as

$$F = \frac{kq_1q_2}{r^2} \qquad \textbf{8.2}$$

where F = force of attraction or repulsion
q_1 = magnitude of first charge
q_2 = magnitude of second charge
r = distance between charges

The k is a proportionality constant with the value of

$$k = 9.0 \times 10^9 \ \frac{\text{N-m}^2}{\text{C}^2}$$

Notice that Coulomb's law is similar in form to Newton's law of universal gravitation (Chapter 3, $F = Gm_1m_2/r^2$). Both forces depend on the square of the separation distance. One obvious difference between them is that Coulomb's law depends on charge, whereas Newton's law depends on mass.

Two other important differences exist. One is that Coulomb's law can give rise to either an attractive or a repulsive force, depending on whether the two charges are different or the same (law of charges). The force of gravitation, on the other hand, is *always* attractive.

The other important difference is that the electric forces are comparatively much stronger than the gravitational forces. For example, an electron and a proton are attracted to each other both electrically and gravitationally. However, when we deal with such charged particles, the gravitational forces are so relatively weak that they can be ignored; we consider only the electric forces of attraction and repulsion. (See Exercise 6.)

An object with an excess of electrons is said to be *negatively charged,* and an object with a deficiency of electrons is said to be *positively charged.* A negative charge can be placed on a rubber rod by stroking the rod with fur. (Electrons are transferred from the fur to the rod by friction.) In ● Fig. 8.1a, a rubber rod that has been rubbed with fur and given a net charge is shown suspended by a thin thread that allows the rod to swing freely. The charge on the rod is negative.

FIGURE 8.1   Repulsive and Attractive Electrical Forces

(a) Two negatively charged objects repel one another. (b) Two positively charged objects repel one another. (c) A negatively charged object and a positively charged object attract one another.

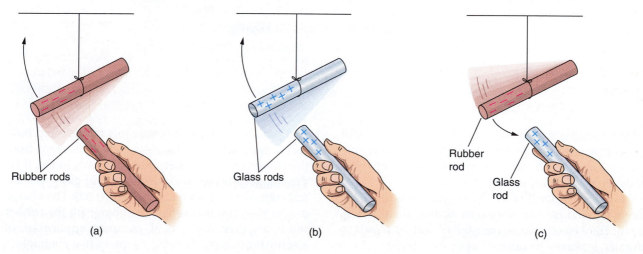

(a) Rubber rods

(b) Glass rods

(c) Rubber rod   Glass rod

HIGHLIGHT: Photocopiers

One of the important principles of electrostatics—that unlike charges attract—is the basis of photocopying machines. The process is known as *xerography* (coined from the Greek words *xeros,* meaning "dry," and *graphein,* meaning "to write") and refers to the operation by which almost any printed material can be copied.

In transfer xerography, a photoconductive plate, cylinder, or belt is electrostatically charged (Fig. 1a). A photoconductor, such as selenium, allows charge to leak away when exposed to light. The material to be copied is placed face down on the photocopying machine, and a projected image of the page falls on the charged plate. The illuminated portions become conducting and discharge, leaving a charged electrostatic image of the dark regions or print of the page (Fig. 1b).

The photoconductor copy then comes into contact with a negatively charged powder called *toner,* or *dry ink.* The toner is attracted to the paper, and heating causes it to be permanently fused to the paper. All this takes place very quickly—and out comes your copy.

Although the inner workings of the machine are fairly complicated, the main principle is simply the attraction of unlike charges. Perhaps you have noticed the static charge on photocopies when they exit a machine.

(a) Surface of a plate or cylinder coated with a photoconductive metal is electrically charged as it passes under wires.

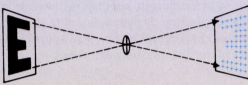

(b) Original document is projected through a lens. Plus marks represent latent image retaining positive charge. Charge drained in areas exposed to light.

(c) Negatively charged powder (toner or "dry ink") is applied to the latent image, which now becomes visible.

(d) Positively charged paper attracts dry ink from plate, forming direct positive image.

(e) Image is fused into the surface of the paper or other material by heat for permanency.

FIGURE 1   Electrostatic Copying

The steps involved in electrostatic copying. (See the Highlight text for description.)

When a similar rubber rod that has been stroked with fur and negatively charged is brought close to the suspended rod, it will swing away; that is, the charged rods repel one another (like charges repel).

The same procedure using two glass rods that have been stroked by silk will show similar results (Fig. 8.1b). Here electrons are transferred from the rods to the silk, leaving a positive charge on each rod.

Although the experiments show repulsion in both cases, the charge on the glass rods is different from the charge on the rubber rods. As shown in Fig. 8.1c, the charges on the stroked rubber and glass rods attract one another (unlike charges attract). The charge on the glass rod is positive; the charge on the rubber rod is negative. A practical, common application of electrostatics is given in the chapter's first Highlight.

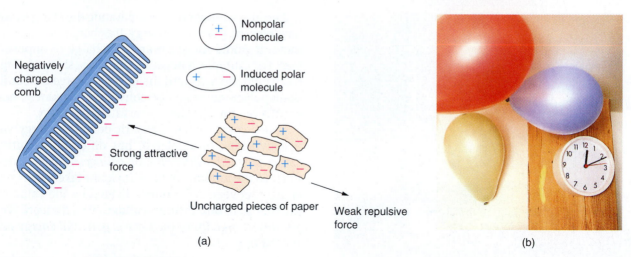

FIGURE 8.2 Polarization of Charge

(a) When a negatively charged comb is brought near small pieces of paper, the molecules are polarized with definite regions of charge, giving rise to a net attractive force. As a re-sult, the bits of paper are attached to and cling to the comb. (b) Charged balloons cling to the ceiling and wall because of attractive forces resulting from molecular polarization.

Static charge also can be a problem. After walking across a carpet, you have probably been annoyingly zapped by a spark when you reach for a doorknob. You were charged by friction in crossing the carpet, and when reaching for the door, the electric force was strong enough to cause the air to ionize and conduct charge to the metal doorknob. This occurs best on a dry day. With high humidity (a lot of moisture in the air), there is a thin film of moisture on objects, and charge is conducted away before it can build up. Even so, such sparks are undesirable when working around flammable materials—for example, in an operating room with explosive gases or around gasoline.

From Coulomb's law (Eq. 8.2), we see that as two charges get closer together, the force of attraction or repulsion increases. This effect can give rise to electric forces, as illustrated in ● Fig. 8.2. When a negatively charged rubber comb is brought near small pieces of paper, the charges in the paper molecules are acted on by electric forces—positive charges attracted, negative charges repelled—and the result is an effective separation of charge. The molecules are then said to be *polarized;* that is, they possess definite regions of charge.

Because the positive-charge regions are closer to the comb than the negative-charge regions, the attractive forces are stronger than the repulsive forces. Thus a net attraction exists between the comb and the pieces of paper. Small bits of paper may be picked up by the comb, which indicates that the attractive electric force is greater than the paper's weight (the gravitational force on it).

Keep in mind, however, that overall the paper is uncharged; that is, it is electrically neutral. Only molecular regions within the paper are charged. This procedure is termed *charging by induction.*

Now you know why a balloon will stick to a ceiling or wall after being rubbed on a person's hair or clothing. The balloon is charged by the frictional rubbing, which causes a transfer of charge. When the balloon is placed on a wall, the charge on the surface of the balloon induces regions of charge in the molecules of the wall material, attracting the balloon to the wall (Fig. 8.2b).

Another demonstration of electric force is shown in ● Fig. 8.3. When a charged rubber rod is brought close to a thin stream of water, the water is attracted toward the rod, and the stream is bent. Water molecules have a permanent separation of charge or regions of different charges. They are called *polar molecules* (Section 12.5).

RELEVANCE QUESTION: *What is the cause of "static cling" in clothes, and what is an economical way to get rid of it?*

FIGURE 8.3   Bending Water

A charged rod brought close to a small stream of water will bend the stream because of polarization of the water molecules.

8.2 Voltage and Electrical Power

LEARNING GOALS

▼ Define voltage, and state how Ohm's law relates it to current and resistance.

▼ Explain electrical power, and identify the electrical terms needed to describe it.

The effects produced by moving charges give rise to what we generally call *electricity*. For charges to move, they must be acted on by other positive or negative charges.

Consider the situation shown in ● Fig. 8.4. We start out with some unseparated charges and then begin to separate them. It takes very little work to pull the first negative charge to the left and the first positive charge to the right. When the next negative charge is moved to the left, it is repelled by the negative charge already there, so more work is needed. Similarly, it takes more work to move the second positive charge to the right. As we separate more and more charges, it takes more and more work.

Because work is done in separating the charges, we have **electric potential energy.** If a charge were free to move, it would move toward the charge of opposite sign. For example, a negative charge, as shown in Fig. 8.4, would move toward the positive charges. Electric potential energy would be converted into kinetic energy, as required by the conservation of energy.

Instead of speaking of electric potential energy, we usually speak of a related, but different, quantity called *potential difference,* or *voltage.* Voltage is defined as the amount of work it would take to move a charge between two points, divided by the value of the charge. In other words, **voltage** *(V) is the work (W) per unit charge (q) or the electric potential energy per unit charge.*

$$\text{Voltage} = \frac{\text{work}}{\text{charge}}$$

or
$$V = \frac{W}{q} \qquad \textbf{8.3}$$

The **volt** (V) is the unit of voltage and is equal to one joule per coulomb. Voltage is caused by a separation of charge. When work is done in separating the charges, then we have electric potential energy, which may be used to set up a current.

When there is a current, it meets with some opposition because of collisions within the conducting material. This opposition to the flow of charge is called **resistance** (R). The unit of resistance is the **ohm** (Ω, Greek letter omega). A simple relationship

FIGURE 8.4   Electric Potential Energy

Work must be done to separate positive and negative charges. The work is done against the attractive electric force. When separated, the charges have electric potential energy and would move if free to do so.

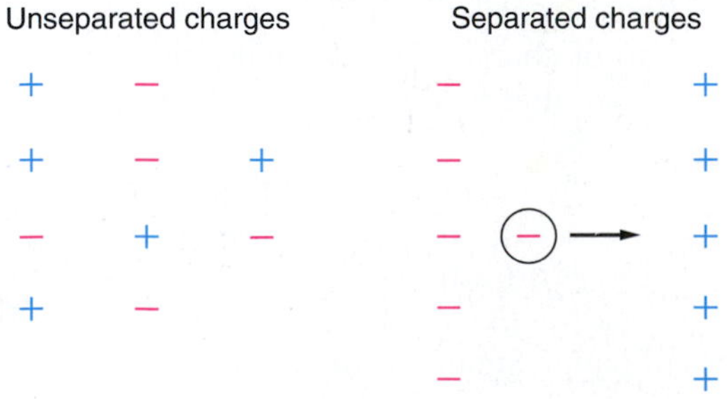

Unseparated charges	Separated charges

involving voltage, current, and resistance was formulated by Georg Ohm (1787–1854), a German physicist, and applies to many materials. This is called Ohm's law, and in equation form it is written as

Voltage = current × resistance

or $\qquad V = IR \qquad$ **8.4**

From this equation, we see that one ohm is one volt per ampere $(R = V/I)$.

An example of a simple electric circuit is illustrated in ● Fig. 8.5a, together with a circuit diagram. The water circuit analogy shown in Fig. 8.5b may help you better understand the components of the electric circuit. The battery provides the voltage to drive the circuit through chemical activity (chemical energy). This is analogous to the pump driving the water circuit. When the switch is closed (the valve is opened in the water circuit), there is a current in the circuit. Electrons move away from the negative terminal of the battery toward the positive terminal.

The light bulb in the circuit offers resistance, and work is done in lighting it, with electrical energy being converted to heat and radiant energy. The waterwheel in the water circuit provides analogous resistance to the water flow and uses gravitational potential energy to do work. Notice that there is a voltage or potential difference (drop) across the bulb, similar to the gravitational potential difference across the waterwheel. The components of an electric circuit are represented by symbols in a circuit diagram, as shown in the figure.

The switch in the circuit allows the path of the electrons to be open or closed. When the switch is open, there is not a complete path or circuit through which charge can flow, and there is no current. (This is called an *open* circuit.) When the switch is closed, the circuit is completed, and there is a current. (The circuit is then said to be *closed.*) A sustained electric current requires a closed path or circuit.

Notice in the circuit diagram in Fig. 8.5a that the conventional current (I) is in the opposite direction around the circuit to that of the electron flow. Even though we know that electron charges are flowing in the circuit, it is customary to designate the *conventional current I* in the direction that positive charges would flow. This is a historical leftover. Ben Franklin once advanced a fluid theory of electricity. All bodies supposedly contained a certain normal amount of this mysterious fluid, a surplus or deficit of which

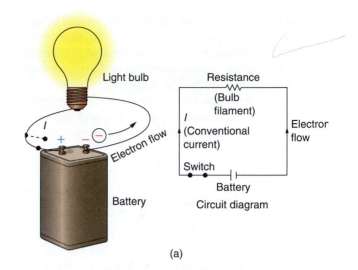

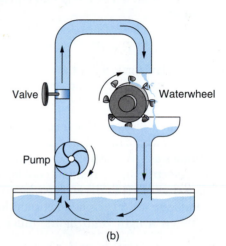

(b)

FIGURE 8.5   Simple Electric Circuit and Water Analogy

(a) A simple electric circuit in which a battery supplies the voltage and the light bulb the resistance. When the switch is closed, electrons flow from the negative terminal of the battery toward the positive terminal. Electrical energy is expended in heating the bulb filament. A circuit diagram with the component symbols is at the right. (b) In the water "circuit," the pump is analogous to the battery, the valve is analogous to the switch, and the waterwheel is analogous to the light bulb in furnishing resistance. Energy is expended or work is done in turning the waterwheel. (See the text for a more detailed description.)

gave rise to electrical properties. With an excess resulting from a fluid flow, a body was positively "excited."

This theory gave rise to the later idea that it was the positive charges that flowed or moved. (Electrons were unknown at the time.) In any case, we still designate the current direction in the conventional sense or in the direction that the positive charges would flow in the circuit—away from the positive terminal of the battery and toward the negative terminal.

When current exists in a circuit, work is done to overcome the resistance of the circuit, and power is expended. Recall that one definition of power (P) is

$$P = \frac{W}{t}$$

From Eq. 8.3, $W = qV$, and substituting for W, we get

$$P = \frac{q}{t}V$$

We now recognize that $q/t = I$ to get an equation for **electrical power:**

$$P = IV \qquad \text{8.5}$$

If we use Ohm's law for V, we get $P = I(IR)$, or

$$P = I^2R \qquad \text{8.6}$$

The power that is dissipated in an electric circuit is frequently in the form of heat. This heat is called *joule heat* or I^2R losses as given by Eq. 8.6. This heating effect is used in electric stoves, heaters, cooking ranges, hair dryers, and so on. Hair dryers have heating coils of low resistance so as to get a large current for large I^2R losses. When a light bulb lights, much of the power goes to produce heat as well as light. The unit of power is the watt, and light bulbs are rated in watts (● Fig. 8.6).

EXAMPLE 8.2

Finding Current and Resistance

Find the current and resistance of a 60-W, 120-V light bulb in operation.

SOLUTION

Step 1
Given: $P = 60$ W   (power)
$\qquad\quad$ $V = 120$ V   (voltage)

Step 2
Wanted: I (current)
$\qquad\quad$ R (resistance)

The units are standard. Notice that the electrical units we commonly use are metric units.

Step 3
The current is obtained using Eq. 8.5, $P = IV$. Rearranging,

$$I = \frac{P}{V} = \frac{60 \text{ W}}{120 \text{ V}} = 0.50 \text{ A}$$

We can rearrange Eq. 8.4 (Ohm's law) to solve for resistance:

$$R = \frac{V}{I} = \frac{120 \text{ V}}{0.50 \text{ A}} = 240 \ \Omega$$

Notice we also could solve for R from Eq. 8.6. Rearranging this equation, we have

$$R = \frac{P}{I^2} = \frac{60 \text{ W}}{(0.50 \text{ A})^2} = 240 \ \Omega$$

CONFIDENCE EXERCISE 8.2

A coffeemaker draws 10 A of current operating at 120 V. How much electrical energy is used by the coffeemaker each second?

FIGURE 8.6   Wattage (Power) Ratings

(a) A 60-W light bulb dissipates 60 J of electrical energy each second. (b) The curling iron uses 13 W at 120 V. Given the wattage and voltage ratings, you can find the current drawn by an appliance by using $I = P/V$.

(a)

(b)

RELEVANCE QUESTION: The heating element of a coffee pot is required to heat up quickly (joule heat). Should the element have a large resistance, or a small one?

8.3 Simple Electric Circuits and Electrical Safety

There are two principal forms of electric current. In a battery circuit, such as in Fig. 8.5, the electron flow is always in one direction, from the negative terminal to the positive terminal. This type of current is called **direct current,** or **dc.** We use direct current in battery-powered devices such as flashlights, portable radios, and automobiles.

The other common type of current is **alternating current,** or **ac,** which is produced by constantly changing the voltage from positive to negative to positive and so on. (Although the usage is redundant, we commonly say "ac current" and "ac voltage.") Alternating current is produced by electric companies and is used in the home.

The frequency of changing from positive to negative voltages is usually at the rate of 60 cycles per second (cps) or 60 Hz (see Fig. 8.6b). The average voltage varies from 110 V to 120 V, and household ac voltage is commonly listed as 110 V, 115 V, or 120 V. The equations for Ohm's law (Eq. 8.4) and power (Eqs. 8.5 and 8.6) apply to both dc and ac circuits containing only resistances.

Once the electricity enters the home or business, it is used in circuits to power (energize) various appliances and other items. Plugging appliances, lamps, and other electrical applications into a wall outlet places them in a circuit. There are two basic ways of connecting elements in a circuit: in *series* and in *parallel*.

An example of a series circuit is shown in ● Fig. 8.7. The lamps are conveniently represented as resistances in the circuit diagram. In a **series circuit,** the same current passes through all the resistances. This is analogous to a liquid circuit with a single line connecting several components. The total resistance is simply the sum of the individual resistances. As with different height potentials, the total voltage is the sum of the individual voltage drops, and

$$V = V_1 + V_2 + V_3 + \cdots$$

$$V = IR_1 + IR_2 + IR_3 + \cdots$$

or $\qquad V = I(R_1 + R_2 + R_3 + \cdots)$

where the equation is written for three or more resistances. And, if we write $V = IR_s$, where R_s *is the total equivalent series resistance,* then, by comparison,

$$R_s = R_1 + R_2 + R_3 + \cdots \qquad \textbf{8.7}$$
resistances in series

What this equation means is that all the resistances in series can be replaced with a single resistance R_s, and the same current would flow and the same power would be dissipated. For example, $P = I^2 R_s$ is the power used in the whole circuit. (The resistances of the connecting wires are considered negligible.)

The example of lamps or resistances in series in Fig. 8.7 could as easily have been a string of Christmas tree lights, which used to be connected in a simple series circuit. When a bulb burned out, the whole string of lights went out, because there was no longer a completed path for the current, and the circuit was "open." Having a bulb burn out was like opening a switch in the circuit to turn off the lights. However, in most strings of lights purchased today, one light can burn out and the others remain lit. Why? We will explain this shortly in our discussion of parallel circuits.

FIGURE 8.7 Series Circuit

The light bulbs are connected in series, and the current is the same through each bulb.

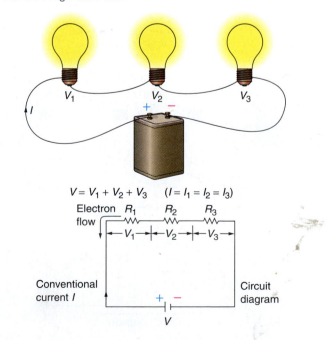

$V = V_1 + V_2 + V_3 \qquad (I = I_1 = I_2 = I_3)$

The other basic type of simple circuit is called a *parallel circuit,* as illustrated in ● Fig. 8.8. In a **parallel circuit,** the voltage across each resistance is the same, but the current through each resistance may vary. Notice that the current from the voltage source (battery) divides at the junction where all the resistances are connected together. This arrangement is analogous to liquid flow in a large pipe coming into a junction where it divides into several smaller pipes.

Because there is no buildup of charge at the junction, the charge leaving the junction must equal the charge entering the junction (law of conservation of charge), and we may write in terms of current,

$$I = I_1 + I_2 + I_3 + \cdots$$

Using Ohm's law (in the form $I = V/R$), we may write for the different resistances,

$$I = \frac{V}{R_1} + \frac{V}{R_2} + \frac{V}{R_3} + \cdots$$

or

$$I = V\left(\frac{1}{R_1} + \frac{1}{R_2} + \frac{1}{R_3} + \cdots\right)$$

The voltage V is the same across each resistance R because the voltage source is effectively connected "across" each resistance, and each gets the same voltage effect or drop.

Writing Ohm's law as $I = V/R_p$, where R_p *is the total equivalent parallel resistance,* then, by comparison,

$$\frac{1}{R_p} = \frac{1}{R_1} + \frac{1}{R_2} + \frac{1}{R_3} + \cdots \qquad \textbf{8.8}$$

resistances in parallel

For a circuit with only two resistances in parallel, this equation can be conveniently written as

$$R_p = \frac{R_1 R_2}{R_1 + R_2} \qquad \textbf{8.9}$$

(2 resistances in parallel)

As in the case of the series circuit, all the resistors could be replaced by a single resistance R_p without affecting the current from the battery and the power dissipation.

FIGURE 8.8 Parallel Circuit

The light bulbs are connected in parallel, and the current from the battery divides at the junction (where the three bulbs are connected together). The amount of current in each parallel branch is determined by the relative values of the resistances in the branches—the greatest current is in the path of least resistance.

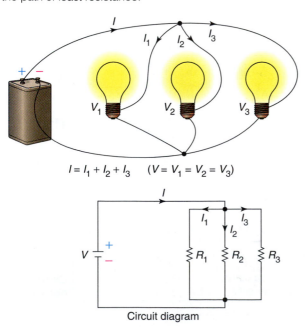

$$I = I_1 + I_2 + I_3 \qquad (V = V_1 = V_2 = V_3)$$

Circuit diagram

EXAMPLE 8.3

Resistances in Parallel

Three resistors have values of $R_1 = 6.0\ \Omega$, $R_2 = 6.0\ \Omega$, and $R_3 = 3.0\ \Omega$. What is their total resistance when connected in parallel, and how much current will be drawn from a 12-V battery if it is connected to the circuit?

SOLUTION

Let's first combine R_1 and R_2 into a single equivalent resistance, using Eq. 8.9:

$$R_{p1} = \frac{R_1 R_2}{R_1 + R_2} = \frac{(6.0\ \Omega)(6.0\ \Omega)}{6.0\ \Omega + 6.0\ \Omega} = 3.0\ \Omega$$

Hence an equivalent circuit is a resistance R_{p1} connected in parallel with R_3. Apply Eq. 8.9 to these parallel resistances to find the total resistance:

$$R_p = \frac{R_{p1} R_3}{R_{p1} + R_3} = \frac{(3.0\ \Omega)(3.0\ \Omega)}{3.0\ \Omega + 3.0\ \Omega} = 1.5\ \Omega$$

The same current will be drawn from the source for a 1.5-Ω resistor as for the three resistances in parallel.

The problem also can be solved by using Eq. 8.8. In this case, R is found by using the lowest common denominator for the fractions (zeros and units omitted for clarity):

$$\frac{1}{R_p} = \frac{1}{R_1} + \frac{1}{R_2} + \frac{1}{R_3} = \frac{1}{6} + \frac{1}{6} + \frac{1}{3}$$

$$= \frac{1}{6} + \frac{1}{6} + \frac{2}{6} = \frac{4}{6\,\Omega}$$

or

$$R_p = \frac{6\,\Omega}{4} = 1.5\,\Omega$$

The current drawn from the source is then given by Ohm's law, using R_p:

$$I = \frac{V}{R_p} = \frac{12\text{ V}}{1.5\,\Omega} = 8.0\text{ A}$$

CONFIDENCE EXERCISE 8.3

Suppose the resistances in Example 8.3 were wired in series and connected to the 12-V battery. Would the battery supply more or less current than it would for the parallel arrangement? What would be the current in the circuit in this case?

An interesting fact about resistances connected in parallel is that *the total resistance is always less than the smallest parallel resistance.* Such is the case in Example 8.3. Try to find a parallel circuit that proves otherwise. (Forget it, you'd be wasting your time.)

Home appliances are wired in parallel (● Fig. 8.9). There are two major advantages to the parallel circuit.

1. The same voltage (110–120 V) is available throughout the house. This makes it much easier to design appliances. (The 110–120-V voltage is obtained by connecting across the "hot," or high-voltage, side of the line to *ground,* or zero potential. This gives a voltage *difference* of 120 V, even if one of the "high" sides is at a potential of −120 V. The voltage for large appliances, such as central air conditioners and heaters, is 220 to 240 V, which is available by connecting across the two incoming potentials, as shown in Fig. 8.9. This potential is analogous to a height difference between two positions, one positive and one negative, for gravitational potential energy.)

2. If one appliance fails to operate, the others in the circuit are not affected because their circuits are still complete. In a series circuit, if one component fails, none of the others will operate because the circuit is incomplete, or "open."

FIGURE 8.9 Household Circuits

As illustrated here, household circuits are wired in parallel. For small appliances, the circuit voltage is 120 V. Because there are independent branches, any particular circuit element can operate when others in the same circuit do not. For large appliances, such as a central air conditioner or electric stove, the connection is between the +120 V and −120 V potential wires to give a voltage difference of 240 V.

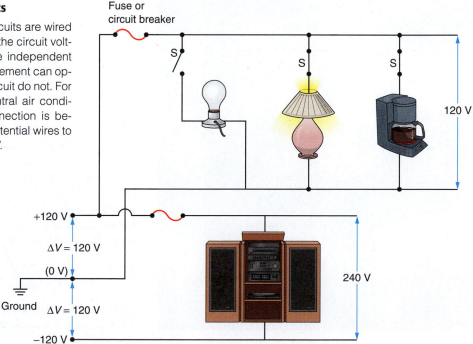

Question: Consider Christmas tree lights that remain on when one bulb burns out. How are they connected?

Answer: The bulbs could be wired in parallel. As shown in the household circuit in Fig. 8.9, if a bulb burns out, the other components in the parallel circuit continue to operate. However, the total resistance of a string of lights wired in parallel would be small, and an undesirably and dangerously large current would flow in the circuit. Also, parallel wiring would require a lot of additional wire, which is expensive. For cost-effectiveness, each bulb is wired in series with the other bulbs but in parallel with a "shunt" resistor, as illustrated in ● Fig. 8.10.

When the filament of a bulb burns out, the resistor provides a path for the current, shunting it around the defective bulb, and the other bulbs remain lit. Actually, the shunt resistor is insulated and not part of the circuit when its bulb is lit. How-

ever, when the bulb burns out, this places the voltage across the resistor, which causes sparking and burns off the shunt insulation material, so it makes contact and completes the circuit.

We also can have series-parallel combinations, which give intermediate equivalent total resistances, but we will not go into these.

Electrical Safety

Electrical safety for both people and property is an important consideration in using electricity. For example, in household circuits, as more and more appliances are turned on, there is more and more current and the wires get hotter and hotter. The fuse shown in the circuit diagram in Fig. 8.9 is a safety device that prevents the wires from getting too hot and possibly starting a fire. When there is a preset amount of current, the fuse filament gets so hot that it melts and opens the circuit.

FIGURE 8.10   Christmas Tree Lights and Shunt Resistors
Modern Christmas tree lights are wired with an insulated resistor in parallel with each bulb filament. When a filament burns out, the voltage across the resistor causes sparking that burns away the insulation. The resistor then becomes part of the circuit, *shunting* the burned-out bulb and allowing the other bulbs in the series to operate.

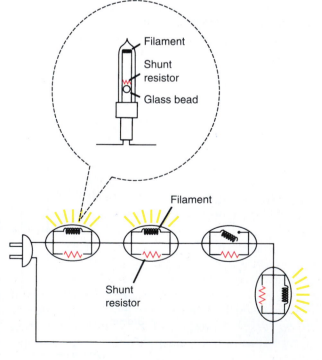

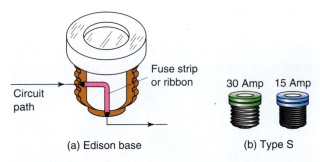

(a) Edison base   **(b) Type S**

FIGURE 8.11   Fuses

(a) An Edison-base fuse. If the current exceeds the fuse rating, joule heat causes the fuse strip or ribbon to burn out, and the circuit is opened. (b) Type-S fuses. Edison-base fuses have the same screw thread for different ratings, and so a 30-A fuse could be put into a 15-A circuit, which would be dangerous. (Why?) Type-S fuses have different threads for different fuse ratings and cannot be interchanged.

There are a couple of types of fuses commonly used in household circuits. The so-called *Edison-base fuse* has a base with threads similar to those on a light bulb (● Fig. 8.11a). As such, they are replaceable with any amp rating; that is, a 30-A fuse can be screwed into a socket that should have a 15-A fuse. Such a mixup could be dangerous, however, so *type-S fuses* are often used (Fig. 8.11b). With type-S fuses, a threaded adapter specific to a particular fuse is put into the socket. Different-rated fuses have different threads, and a 30-A fuse cannot be screwed into a 15-A socket.

These days the more popular *circuit breaker* is generally replacing fuses. It serves the same function as fuses. When the current in a circuit reaches a preset amperage, the circuit breaker triggers a switch that opens, or "breaks," the circuit so that there is no current. This is done by thermal expansion or magnetically. When the trouble is corrected, the circuit breaker can be reset to close the circuit.

Switches, fuses, and circuit breakers are always placed in the "hot," or high-voltage, side of the line. If placed in the ground side, there would be no current when a circuit was opened, but there would still be 120-V potential to the appliance, which could be dangerous if one came in contact with it.

However, even when wired properly, fuses and circuit breakers may not always give protection from electrical shock. A hot wire inside an appliance or power tool may break loose and come into contact with its housing or casing, putting it at a high voltage.

The fuse does not blow unless there is a large current. Should a person touch a casing that is conductive, as illustrated in ● Fig. 8.12a, a path is provided to ground, and the person receives a shock.

This condition is prevented by *grounding* the casing, as shown in Fig. 8.12b. Then, if a hot wire touches the casing, a current flows, and the fuse blows. This grounding process is the purpose of the three-prong plugs found on many electrical tools and appliances (● Fig. 8.13a).

A *polarized* plug is shown in Fig. 8.13b. You have probably noticed that some plugs have one blade or prong larger than the other and will only fit into a wall outlet one way. Polarized plugs are an older type of safety feature. Being polarized or directional, one side

FIGURE 8.12   Electrical Safety by Use of Dedicated Grounding

(a) Suppose an internal "hot" wire broke and came in contact with the metal casing of an appliance. Without a dedicated ground wire, the casing would be at a high potential without a fuse being blown or a circuit breaker being tripped. If someone touched the casing, a dangerous shock could result. (b) By grounding the case with a dedicated ground wire through a third prong on the plug, the circuit would be opened, and the casing would be at zero potential.

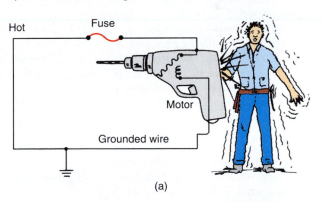

(a)

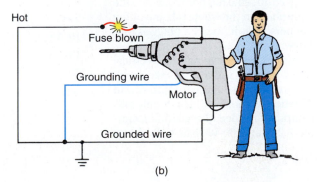

(b)

HIGHLIGHT: Electrical Safety

When working with electricity, common sense and a knowledge of fundamental electrical principles are important. Electric shocks can be very dangerous; they kill many people every year (Fig. 1).

The danger is proportional to the amount of electric current that goes through the body. The amount of current going through the body is given by Ohm's law as

$$I = \frac{V}{R_{\text{body}}} \qquad 1$$

where R_{body} is the resistance of the body.

A current of 0.001 A can be felt as a shock, and a current as low as 0.05 A can be fatal. A current of 0.10 A is nearly always fatal.

The amount of current, as indicated in Eq. 1, is very dependent on the body's resistance. The body's resistance varies considerably, mainly due to the dryness of the skin. Because our bodies are mostly water, skin resistance makes up most of the body's resistance.

A dry body can have a resistance as high as 500,000 Ω, and the current from a 110-V source will be 0.00022 A (or 0.22 mA). The danger occurs when the skin is moist or wet. Then the resistance of the body can go as low as 100 Ω, and the current will rise to 1.1 A. Injuries and death from shocks usually occur when the skin is wet. Therefore, appliances such as radios should not be used near a bathtub. Should a plugged-in radio happen to fall into the bathtub, then the whole tub, including the person in it, may be plugged into 110 V.

Although we have various electrical safety devices, it does not take much current to cause human injuries and fatalities. As shown in Table 1, only milliamps are needed. (Recall from Example 8.2 that a 60-W light bulb draws

FIGURE 1 Electrical Hazards
Electrical hazards, such as this frayed wire, can be dangerous and cause injury.

0.50 A or 500 mA in normal operation.) If you should come into contact with a hot wire and become part of a circuit, not only is your body resistance important (as discussed) but also how you are connected in the circuit. If the circuit is completed through your hand (say, finger to thumb), a shock and a burn can result. However, if the circuit is completed through the body from hand to hand or hand to foot, then the resulting effects may be more serious depending on the amount of current.

Notice that 15 to 25 mA can cause muscular freeze, and a person may not be able to let go of a hot wire. Muscles are controlled by nerves, which in turn are controlled by electrical impulses. Slightly larger currents can cause breathing difficulties, and just slightly larger currents can cause ventricular fibrillation or uncontrolled contractions of the heart. Greater than 100 mA (or 0.10 A) can result in death. Keep in mind for your personal electrical safety that a little current goes a long way.

TABLE 1 Effects of Electric Currents on Humans

Current (mA)	Effect*
1	Barely perceived
5–10	Mild shock
10–15	Difficulty in releasing
15–25	Muscular freeze, cannot release or let go
50–100	May stop breathing, ventricular fibrillation
>100	Death

*Effects vary with individuals.

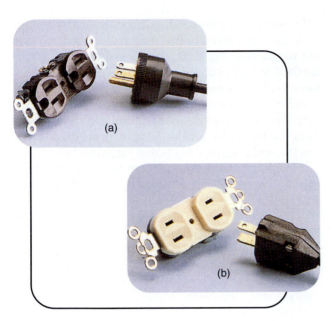

FIGURE 8.13 Electric Plugs

(a) A three-prong plug and socket. The third, rounded prong is connected to a dedicated grounding wire used for electrical safety. (b) A two-prong polarized plug and socket. Note that one blade, or prong, is larger than the other. The ground, or neutral (zero potential), side of the line is wired to the large-prong side of the socket. This distinction, or polarization, permits paths to ground for safety purposes.

of the plug is always connected to the ground side of the line. The casing of an appliance can be connected to ground in this way, with a similar effect as the three-wire system.

However, a dedicated grounding wire is better because the polarized system depends on the circuit and the appliance being wired properly, and there is a chance of error. Also, even though wired to the ground side of the line, this is still a current-carrying wire, whereas the dedicated ground is not.

An electric shock can be very dangerous, and *touching exposed electric wires should always be avoided.* Many injuries and deaths occur from receiving electric shocks. The effects are discussed in the chapter's second Highlight.

RELEVANCE QUESTION: *When you turn on more lights and appliances in your home, the current in the circuits is greater. Why is this, since you are adding more resistance with each component?*

8.4 Magnetism

LEARNING GOALS

▼ State the law of poles, and describe the magnetic field.

▼ Identify the cause of magnetism, and tell why some materials can be magnetized and some cannot.

▼ Analyze some aspects of Earth's magnetic field.

One of the first things one notices in examining a bar magnet is that it has two regions of magnetic strength or concentration—one at each end of the magnet—which we call *poles*. We designate one as the north pole, N, and the other as the south pole, S. This is so because the N pole of a magnet, when used as a compass, is the north-seeking pole (that is, it points north), and the S pole is the south-seeking pole.

When examining two magnets, we find that there are attractive and repulsive forces between them that are specific to the poles. The forces are described by the **law of poles:**

Like poles repel, and unlike poles attract.

In other words, N and S poles (N-S) attract, and N-N and S-S poles repel each other (● Fig. 8.14a). The strength of the attraction or repulsion depends on the strength of the magnetic poles. Also, in a manner similar to Coulomb's law, the strength of the magnetic force is inversely proportional to the square of the distance between the poles. Figure 8.14b shows some toy magnets that seem to defy gravity due to their magnetic repulsion.

All magnets have two poles; that is, they are *di*poles. Unlike electric charge, which occurs in single charges, magnets are always dipoles. A *magnetic monopole* would consist of a single N or S pole without the other. There is no known physical reason for magnetic monopoles not existing, but thus far their existence has not been confirmed experimentally. The discovery of a magnetic monopole would be an important fundamental development.

Every magnet produces a force on every other magnet. In order to discuss these effects, we introduce the concept of a magnetic field. A **magnetic field** (*B*) is a set of imaginary lines that indicates the direction in which a small compass needle would point if it were placed near a magnet. Hence the field lines are

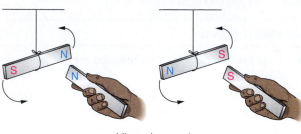

Like poles repel

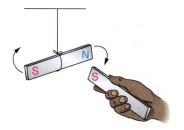

Unlike poles attract

(a)

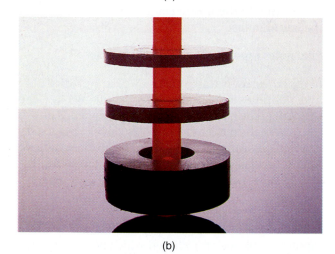

(b)

FIGURE 8.14   Laws of Poles

(a) Like poles repel, and unlike poles attract. (b) The adjacent poles of the circular magnets must be like poles. Why?

act as small compass needles. The outline of the magnetic field produced in this manner is shown for two bar magnets in Fig. 8.15b. The field concept also can be used for an *electric field* around charges, but this force field is not so easily visualized. The electric field is the electric force per unit charge. The electric and magnetic fields are vector quantities, and electromagnetic waves, as discussed in Chapter 6, are made up of electric and magnetic fields that vary with time.

Electricity and magnetism are discussed together in this chapter because they are linked. In fact, *the*

FIGURE 8.15   Magnetic Field

(a) Magnetic field lines may be plotted by using a small compass. The N pole of the compass needle points in the direction of the field at any point. (b) Iron filings become induced magnets and conveniently outline the pattern of the magnetic field.

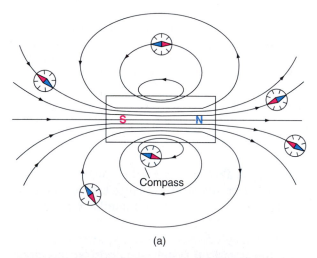

(a)

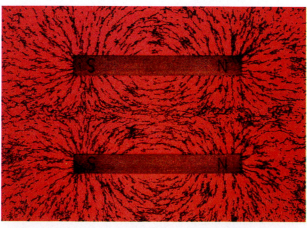

(b)

indications of the magnetic force—a force field, so to speak. ● Figure 8.15a shows the magnetic field lines around a simple bar magnet. The arrows in the field lines indicate the direction in which the north pole of a compass would point. The closer together the field lines, the stronger is the magnetic force.

Magnetic field patterns can be "seen" by using small iron filings. The iron filings are magnetized and

source of magnetism is moving and "spinning" electrons. Hans Oersted, a Danish physicist, first discovered in 1820 that a compass needle is deflected by an electric current-carrying wire. When a compass is placed near a wire in a simple battery circuit and the circuit is closed, there is current in the wire, and the compass needle is deflected from its north-seeking direction. When the circuit is opened, the compass needle goes back to pointing north again.

Also, it is found that the strength of the magnetic field is directly proportional to the magnitude of the current—the greater the current, the greater is the strength of the magnetic field. Hence a current produces a magnetic field that can be turned off and on at will. We can investigate such fields by using iron filings.

Different configurations of current-carrying wires give different magnetic field configurations. Some of them are shown in ● Fig. 8.16. A straight wire produces a field in a circular pattern around the wire. A single loop of wire gives a field not unlike that of a small bar magnet, and the field of a coil of wire with several loops is very similar to that of a bar magnet.

But what produces the magnetic field of a permanent magnet such as a bar magnet? In our simplistic model of the atom, electrons are pictured as going around the nucleus. This is electric charge in motion, or a current loop, so to speak, and it might be expected that this would be a source of a magnetic field. However, it is found that the magnetic field produced by orbiting atomic electrons is very small. Also, the atoms of a material are distributed such that the magnetic fields would be in various directions and generally cancel each other, giving a zero net effect.

Modern theory predicts the magnetic field to be associated with electron "spin." This effect is pictured as an electron spinning on its axis, in the same manner as Earth rotates on its axis. As such, we have charge in motion. A material has many atoms and electrons, and the magnetic spin effects of all these electrons usually cancel each other out. So most materials are not magnetic (do not become magnetized) or are only slightly magnetic. In some instances, however, the magnetic effect can be quite strong.

Materials that are highly magnetic are called **ferromagnetic.** Ferromagnetic materials include the elements iron, nickel, and cobalt, as well as certain alloys of these and a few other elements. In ferromagnetic materials, the magnetic fields of many atoms combine to give rise to **magnetic domains,** or local

FIGURE 8.16   Magnetic Field Patterns
Iron filing patterns near current-carrying wires outline the magnetic fields for (from top) a long straight wire, a single loop of wire, and a coil of wire (a solenoid).

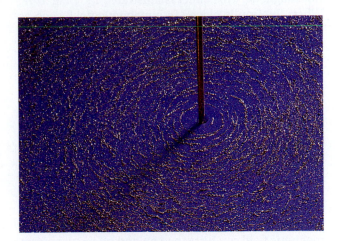

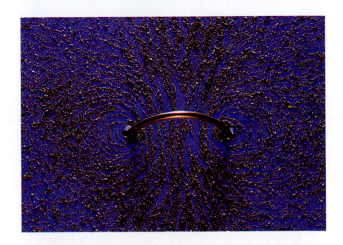

regions of alignment. A single magnetic domain acts like a tiny bar magnet.

In iron, the domains can be aligned or nonaligned. A piece of iron with the domains randomly oriented is not magnetic. This effect is illustrated in ● Fig. 8.17. When the iron is placed in a magnetic field, such as that produced by a current-carrying loop of wire, the domains line up, or those parallel to the field grow at the expense of other domains, and the iron is magnetized.

When the magnetic field is removed, the domains tend to return to a mostly random arrangement due to heat effects that cause disordering. The amount of domain alignment remaining after the field is removed depends on the strength of the applied magnetic field.

An application of this effect is an *electromagnet*, which basically consists of a coil of insulated wire wrapped around a piece of iron (● Fig. 8.18). Since a magnetic field can be turned on and off by turning an

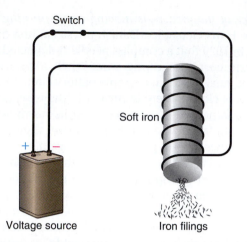

FIGURE 8.18   Electromagnet

A simple electromagnet consists of an insulated coil of wire wrapped around a piece of iron. When the switch is closed, there is a current in the wire, which gives rise to a magnetic field that magnetizes the iron, thus creating a magnet. When the switch is open, there is no current in the coil, and the iron is not magnetized.

FIGURE 8.17   Magnetization

(a) In a ferromagnetic material, the magnetic domains are generally unaligned, so there is no magnetic field. For simplicity, the domains are represented by small bar magnets. (b) In a magnetic field produced by a current-carrying loop of wire, the domains become aligned with the field (and the aligned domains may grow at the expense of others), and the material becomes magnetized.

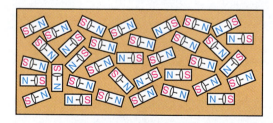

(a) Unmagnetized material

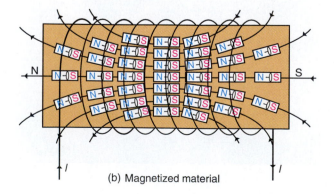

(b) Magnetized material

electric current on and off, we can control whether or not the iron will be a magnet. When the current is on, the magnetic field of the coil magnetizes the iron. The aligned domains add to the field, making it about 2000 times stronger.

Electromagnets have many applications. Large ones are used routinely to pick up and transfer scrap iron, and small electromagnets are used in magnetic relays and solenoids, which act as magnetic switches. Solenoids are used in automobiles to engage the starting motor.

One type of circuit breaker uses an electromagnetic switch. The strength of an electromagnet is directly proportional to the current in its coils. When there is a certain amount of current in the breaker circuit, an electromagnet becomes strong enough to attract and "trip" a metallic conductor, thus opening the circuit.

The iron used in electromagnets is called "soft" iron. This does not mean that it is physically soft, but rather that this type of iron can be magnetized but quickly becomes demagnetized. Certain types of iron, along with nickel, cobalt, and a few other elements, are known as "hard" magnetic materials. Once magnetized, they retain their magnetic properties for a long time.

Thus hard iron is used for permanent magnets. When permanent magnets are heated or struck, the domains are shaken from their alignment, and the magnet becomes weaker. In fact, above a certain temperature, called the **Curie temperature,** a material ceases to be ferromagnetic. The Curie temperature of iron is 770°C.*

A permanent magnet is made by "permanently" aligning the domains inside the material. One way of doing this is to heat a piece of hard ferromagnetic material above its Curie temperature and then apply a strong magnetic field. The domains line up with the field, and as the material cools, the domain alignment is frozen in, so to speak, producing a permanent magnet.

Earth's Magnetic Field

At the beginning of the seventeenth century, William Gilbert, an English scientist, suggested that Earth acted as a huge magnet. Today, we know that such a magnetic effect does exist for our planet. It is Earth's magnetic field that makes compasses point north.

Experiments have shown that a magnetic field exists within Earth and extends many hundreds of miles out into space. The aurora borealis and aurora australis (northern lights and southern lights), common sights in higher latitudes near the poles, are associated with Earth's magnetic field. This effect will be discussed in Chapter 19.

The origin of Earth's magnetic field is not known, but the most acceptable theory is that it is caused by Earth's rotation, which produces internal currents of electrically charged particles deep within Earth. It is not due to some huge mass of magnetized iron compound within Earth. Earth's interior is quite hot and above the Curie temperature, so materials are not ferromagnetic. Also, the magnetic poles slowly change their positions, which suggests changing currents.

Earth's magnetic field does approximate that of a current loop or a huge imaginary bar magnet, as illustrated in ● Fig. 8.19. Notice that the actual magnetic South (S) Pole is near the geographic North Pole. It is for this reason that the north pole of a compass needle, which is a little magnet, points north (law of poles). Because the "north-seeking" pole of the compass needle points toward the region where there is a concentration of magnetic field lines, we refer to this direction as *magnetic north.* Also, we say that mag-

netic north is in the direction of a magnetic "north" pole, which is near the geographic North Pole. The direction of the geographic pole is called *true north* or in the direction of Earth's north spin axis.

The magnetic and geographic poles do not coincide. Presently, the magnetic North Pole is some 13°, or about 1500 km (930 mi), from the geographic North Pole. In the Southern Hemisphere, the magnetic South Pole is displaced even more from its respective geographic pole.

Hence the compass does not point toward true north but toward magnetic north. The variation between the two directions is expressed in terms of **magnetic declination,** which is the angle between geographic (true) north and magnetic north (● Fig. 8.20). The declination may vary east or west of a geographic meridian (an imaginary line running from pole to pole).

It is important in navigation to know the magnetic declination at a particular location so that the magnetic compass direction can be corrected for true

FIGURE 8.19   Earth's Magnetic Field

Earth's magnetic field is thought to be caused by internal currents in the liquid outer core, in association with the planet's rotation. The magnetic field is similar to that of a giant bar magnet within Earth (but such a bar does not really exist). Note that magnetic north (toward which the compass points) and the geographic North Pole (Earth's axis of rotation) do not coincide.

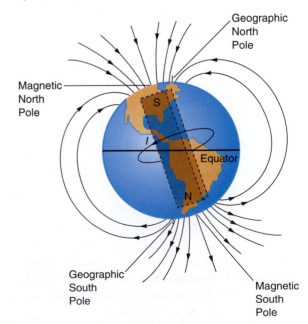

*The Curie temperature is named after Pierre Curie (1859–1906), the French scientist who discovered the effect. Pierre Curie was the husband of Marie (Madame) Curie. They both did pioneering work in radioactivity (Chapter 10).

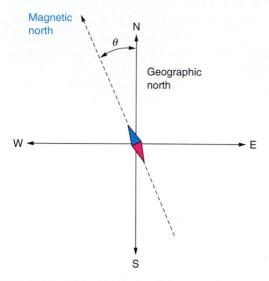

FIGURE 8.20   Magnetic Declination
The angle θ of magnetic declination is the angle between geographic (true) north and magnetic north (as indicated by a compass). The declination is measured in degrees east and west of geographic north.

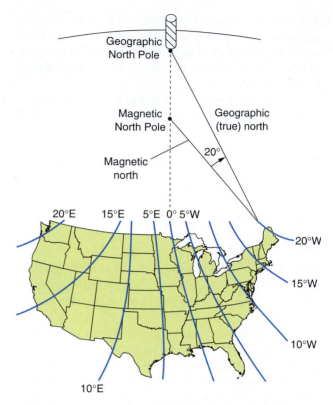

FIGURE 8.21   United States Magnetic Declination
The map shows isogonic (same magnetic declination) lines for the conterminous United States. For locations on the 0° line, magnetic north is in the same direction as geographic (true) north. On either side of this line, a compass has an easterly or westerly variation. For example, on the 20°W line, a compass has a westerly declination of 20°—that is, magnetic north is 20° west of true north.

north. This is provided on navigational maps that show lines of declination expressed in degrees east and west (● Fig. 8.21).

The magnetic field of Earth is relatively weak compared with that of magnets used in the laboratory. However, the field is strong enough to be used by certain animals (including ourselves) for orientation. For instance, it is believed that migratory birds and homing pigeons use Earth's magnetic field to aid them in their homeward flights. Iron compounds have been found in their brains.

8.5 Electromagnetism

LEARNING GOALS

▼ Identify some electromagnetic interactions and applications.

▼ Distinguish between motors and generators.

▼ Explain the principle and use of transformers.

The interaction of electrical and magnetic effects is known as **electromagnetism.** Electromagnetism is one of the most important aspects of physical science, and most of our current technology is directly related to this crucial interaction. Two basic principles of this interaction are as follows:

1. Moving electric charges (current) give rise to magnetic fields.
2. A magnetic field may deflect a moving electric charge.

The first principle forms the basis of an *electromagnet,* which was considered previously. Electromagnets are found in a variety of applications, such as doorbells, telephones, and devices used to move magnetic materials (see Fig. 8.18). Let's look at the mechanism in a telephone receiver as another common practical example. A simplified diagram of a telephone circuit is shown in ● Fig. 8.22.

FIGURE 8.22   Simple (One-Way) Telephone

A simplified diagram of a (one-directional) telephone circuit. Sound waves are converted into varying electrical impulses in a transmitter (microphone). The pulses travel along the telephone lines to the receiver, where they are converted back to sound waves by the actions the pulses have on an electromagnet that drives a diaphragm in a receiver (speaker).

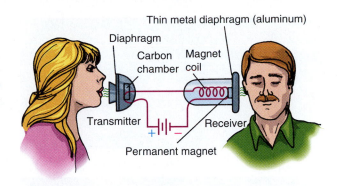

When a telephone number is dialed, a circuit is completed with another telephone's bell. By lifting the receiver of the ringing phone, the circuit between the speakers and receivers of the two telephones is completed. A telephone conversation involves converting the sound waves to varying electric current. This varying current travels along the wire to the other telephone's receiver, where it is converted back into sound.

Here's how this is done. The transmitter of a phone consists of a diaphragm that vibrates in response to the spoken sound waves. The diaphragm vibrates against a chamber that contains carbon granules. As the diaphragm vibrates, the pressure on the carbon granules varies, causing more or less electrical resistance in the circuit. The resistance is low when the granules are pressed together, and it increases as they spread apart. By Ohm's law, this varying resistance gives rise to a varying electric current in the telephone circuit.

At the receiver end, there is an electromagnet or magnetic coil and a permanent magnet, which is attached to a disk. The varying electric current gives the electromagnet varying magnetic strengths. The activated electromagnet attracts the permanent magnet with varying force, and the force variation causes the attached disk to vibrate. The vibrations set up sound waves in the air that closely resemble the original sound waves, and the voice is heard at the other end of the telephone.

Magnetic Force on Moving Electric Charge

The second of the previously mentioned electromagnetic principles may be stated in a qualitative way: A magnetic field can be used to deflect moving electric charges. A stationary electric charge in a magnetic field experiences no force, but when a moving charge

enters a magnetic field as shown in ● Fig. 8.23, it experiences a force. It is found that the magnetic force (F_{mag}) is perpendicular to the plane formed by the velocity vector (v) and the magnetic field (B).

In the figure, the force initially would be out of the page, and with an extended field, the negatively charged particle would follow a circular arc path. If the moving charge were positive, it would be deflected in the opposite direction, or into the page. Also, if a charge, positive or negative, is moving parallel to a magnetic field, there is no force on the charge.

This effect can be demonstrated experimentally as shown in ● Fig. 8.24. A beam of electrons is traveling in the tube from left to right and is made visible by a piece of fluorescent paper in the tube. In the upper photo, the beam is undeflected in the absence of a magnetic

FIGURE 8.23   Magnetic Deflection

Electrons entering a vertical magnetic field as shown experience a force F_{mag} that deflects them out of the page. (See text for description.)

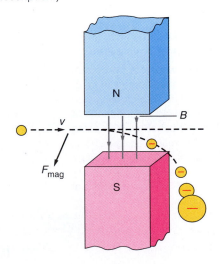

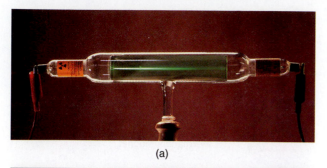

(a)

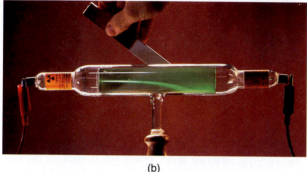

(b)

FIGURE 8.24   Magnetic Force on a Moving Charge

(a) The presence of a beam of electrons is made evident by a fluorescent strip in the tube that allows the beam to be seen. (b) The magnetic field of a bar magnet gives rise to a force on the electrons, and the beam is deflected.

field. In the lower photo, the magnetic field of a bar magnet causes the beam to be deflected downward. This means that the magnetic field is generally directed into the strip of paper. (If you rotated Fig. 8.24 toward you 90°, you would have the same effect.)

Motors and Generators

The electrons in a conducting wire also experience force effects caused by magnetic fields. In ● Fig. 8.25a, a nonconducting wire is shown in a magnetic field. Because there is no current or no net motion of the electrons, there is no force on the wire. However, with a current (moving here to the right), the wire experiences a force out of the page (Fig. 8.25b). This situation is similar to the one in Fig. 8.23, except that here the force causes the whole wire to be forced out of the page toward the viewer.

Hence a current-carrying wire in a magnetic field can experience a force. With a force available, it might quickly come to mind that we could use it to do work, and this is what is done in electric motors. Basically, a **motor** *is a device that converts electrical energy into mechanical energy.* We plug motors in, and we use the mechanical rotations of their shafts to do work. To help understand the electromagnetic-mechanical interaction of motors (of which there are many types), the diagram of a simple *dc motor* is shown in ● Fig. 8.26. Real motors have many loops or windings, but only one is shown here for simplicity. The battery supplies current to the loop, which is free to rotate in the magnetic field between the pole faces. The force on the current-carrying loop produces a torque, causing it to rotate.

Continuous rotation requires a split-ring commutator that reverses the polarity and the current in the loop each half cycle so that the loop has the appropriate force to rotate continuously (Fig. 8.26b). The inertia of the loop carries it through the positions of unstable conditions.

FIGURE 8.25   Magnetic Field and Force on a Current-Carrying Wire

(a) A stationary wire with no current does not experience a force in a magnetic field. (b) A current-carrying wire in a magnetic field experiences a force. With the electron current going to the right, the magnetic force F_{mag} would be out of the page, as illustrated here. Such a force on a current-carrying wire is the basic principle of the electric motor.

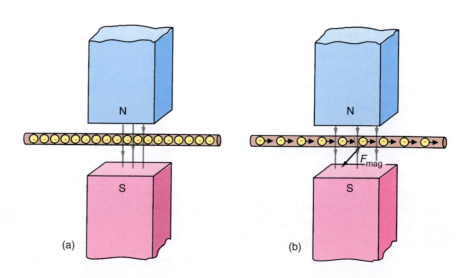

(a)   (b)

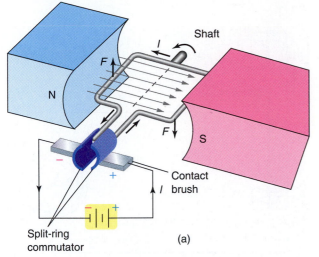

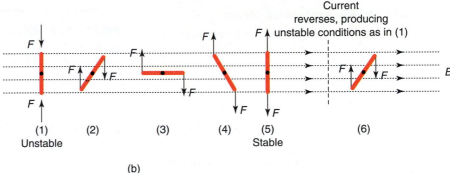

FIGURE 8.26   A dc Motor

(a) An illustration of a loop of a coil in a dc motor. When carrying a current in a magnetic field, the coil experiences a torque and rotates the attached shaft. The split-ring commutator effectively reverses the loop current each half-cycle so that the coil will rotate continuously. (b) The forces on the coil show why the current reversal is necessary.

When a rotating armature has many windings (loops), the effect is enhanced. The rotating loops cause a connected shaft to rotate, which is used to do mechanical work. The conversion of electrical energy to mechanical energy is enhanced by many loops of wire and stronger magnetic fields.

One might ask if the reverse is possible—that is, is the conversion of mechanical energy into electrical energy possible? Indeed it is, and this principle is the basis of electrical generation. Have you ever wondered how electricity is generated?

As illustrated in ● Fig. 8.27, suppose that an applied (mechanical) force sets a wire in a magnetic field in motion. The electrons in the wire are charges moving in a magnetic field and hence experience a magnetic force (F_{mag}) to the right. The electrons move in the conductor, and an electric current is set up, as is a voltage, since Ohm's law applies to the wire. Note that the current is generated without batteries, plugs, or other external voltage sources. This illustrates the basic principle of electrical generation.

A **generator** *is a device that converts mechanical work or energy into electrical energy.* A generator oper-

ates on what is called *electromagnetic induction.* This principle was discovered in 1831 by Michael Faraday, an English scientist. An illustration of his experiment is shown in ● Fig. 8.28. When a magnet is moved toward a loop of wire (or a coil for enhancement), it is observed that a current is induced in the wire, as indicated on the meter. Investigation shows that this is caused by a time-varying magnetic field through the loop.

The same effect is obtained by using a stationary magnetic field and rotating the loop in the field. The magnetic field through the loop varies with time, and a current is induced. A simple *ac generator* is illustrated in ● Fig. 8.29. When the loop is mechanically rotated, a voltage and current are induced in the loop that vary in magnitude and alternate back and forth, changing direction each half-cycle. Hence we have alternating current (ac). There are also dc generators, which are essentially dc motors operated in reverse. However, most electricity is generated as ac and converted, or *rectified,* to dc.

Generators are used in power plants to convert other forms of energy to electrical energy. For the most

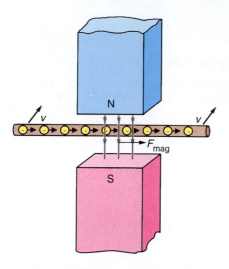

FIGURE 8.27 Motion and Induced Current

If a wire is moved perpendicularly to a magnetic field as illustrated here, the magnetic force causes the electrons in the wire to move, setting up a current in the wire. This is the basic principle of an electric generator.

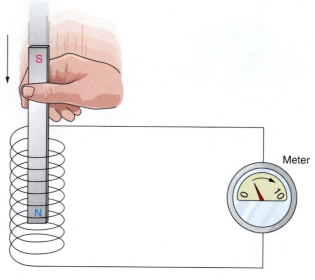

FIGURE 8.28 Electromagnetic Induction

An illustration of Faraday's experiment showing electromagnetic induction. The reading on the meter indicates a current in the circuit as the magnet is moved toward and into the coil.

part, fossil fuels and nuclear energy are used to heat water to generate steam that is used to turn turbines that supply mechanical energy in the generation process. The electricity is carried to homes and businesses, where it is either converted back to mechanical energy to do work or converted to heat energy.

But how is electrical energy or power transmitted? We have all seen the high-voltage (or high-tension) transmission lines running across the land. The voltage for transmission is stepped up by means of **transformers,** simple devices based on electromagnetic in-

duction (● Fig. 8.30). Basically, a transformer consists of two insulated coils of wire wrapped around an iron core, which concentrates the magnetic field when there is current in the input coil. With an ac current in the input or primary coil, there is a time-changing magnetic field as a result of the current going back and forth. The magnetic field goes through the secondary coil and induces a voltage and current.

Because the secondary coil has more windings than the primary coil, the induced ac voltage is greater than the input voltage, and we call this type of

FIGURE 8.29 An ac Generator

(a) An illustration of a coil loop in an ac generator. If the loop is mechanically turned, a current is induced, as indicated by the ammeter. (b) The current varies in direction each half-cycle and hence is called ac or alternating current.

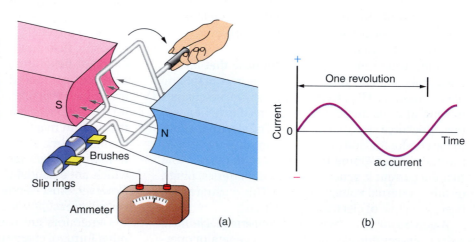

(a)   (b)

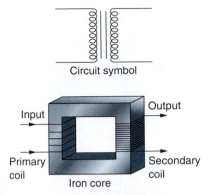

FIGURE 8.30 The Transformer
The circuit symbol for and the basic features of a transformer are shown. A transformer consists of two coils of insulated wire wrapped around a piece of iron. Alternating current in the primary coil creates a time-varying magnetic field, which is concentrated by the iron core, and the field passes through the secondary coil. The varying magnetic field in the secondary coil produces an alternating current output.

transformer a *step-up transformer*. However, when the voltage is stepped up, the secondary current is stepped down (by the conservation of energy, since $P = IV$). The factor of voltage step-up depends on the ratio of the number of windings on the coils, which can be easily controlled.

So why step up the voltage? Actually, it is the step down in current that is really of interest. Transmission lines have resistance and therefore I^2R losses. So, by stepping down the current, we reduce these losses and save energy that would be lost as joule heat. If you step up the voltage by a factor of 2, the current is reduced by a factor of 2, and with half the current, you would have only one-quarter the I^2R losses. (Why?) Thus, for power transmission, the voltage is stepped up to a very high voltage to get the corresponding current step-down and thereby avoid joule heat losses.

Of course, we cannot use such high voltages in our homes, which, in general, have 220–240-V service entries. Therefore, how do we get the voltage back down (and the current up)? In this case, we use a *step-down transformer*, which steps down the voltage and steps up the current. This is done by simply reversing the input and output coils. If the primary coil has more windings than the secondary coil, the voltage is stepped down.

The voltage change for a transformer is given by

$$V_2 = \left(\frac{N_2}{N_1}\right) V_1 \qquad \textbf{8.10}$$

(transformer voltage change)

where V_2 = secondary voltage
V_1 = primary voltage
N_1 = number of turns in primary coil
N_2 = number of turns in secondary coil

EXAMPLE 8.4

Finding Voltage Output for a Transformer

A transformer has 500 windings in its primary coil and 25 in its secondary coil. If the primary voltage is 4400 V, find the secondary voltage.

SOLUTION

Using Eq. 8.10 with $N_1 = 500$, $N_2 = 25$, and $V_1 = 4400$ V, we have

$$V_2 = \left(\frac{N_2}{N_1}\right) V_1 = \left(\frac{25}{500}\right)(4400 \text{ V}) = 220 \text{ V}$$

This result is typical of a step-down transformer on a utility pole near residences. The voltage is stepped down by a factor of 25/500 = 1/20, so the current is stepped up by an inverse factor of 20 (that is, 500/25 = 20).

CONFIDENCE EXERCISE 8.4

Suppose the transformer in Example 8.4 is used as a step-up transformer with a voltage input of 100 V. What would the voltage output be?

An illustration of stepping up and stepping down the voltage in electrical transmission is shown in ● Fig. 8.31. The voltage step-up and the corresponding current step-down, which reduces joule heat losses, are major reasons why we use ac electricity in power transmission. This can't be done with dc; that is, a transformer will not work on dc. (Why?) The first commercial electric company in this country was started by Thomas Edison in New York City and did

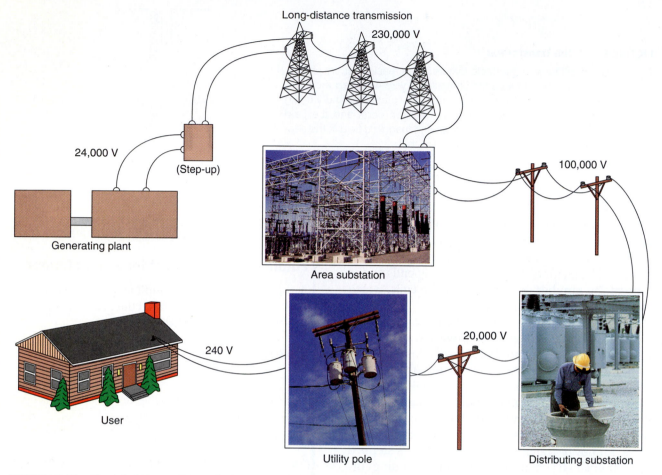

FIGURE 8.31 Electrical Power Transmission System

At the generating plant, the voltage is stepped up with a corresponding current step-down so as to reduce the I^2R losses in the lines for long-distance transmission. The high voltage is then stepped down in substations and finally to 240 V by the common utility-pole transformer (sometimes on the ground for buried lines) for household usage.

use direct current, but this would not be practical with today's long-distance transmissions.

Indeed, we are now in an electrical age of high technology and *electronics.* The latter is the branch of physics and engineering that deals with the emission and control of electrons. When you watch television, a beam of electrons is accelerated by a voltage difference and made to scan the screen by deflecting electric or magnetic fields to produce a picture. Electromagnets are often used. The TV signal input current to the electromagnets (at least four, for up-down and right-left motions) determines the strength of the magnetic fields and regulates the beam deflection. The beam scan produces 30 pictures each second on the phosphorescent screen to give the illusion of movement. Most color TVs have three electron beams that activate a triad of red, green, and blue dots to give a colored picture.

Our electronic instruments have become smaller and smaller, primarily because of the development of solid-state *diodes* and *transistors* (● Fig. 8.32). These devices control the direction of electron flow, and the transistor allows for the amplification of an input signal (as in a transistor radio). Solid-state diodes and transistors offer great advantages over the older vacuum tubes and have all but replaced them. Major advantages are smaller size, lower power consumption, and material economy.

Even further miniaturization has come about through *integrated circuits* (ICs). An integrated circuit

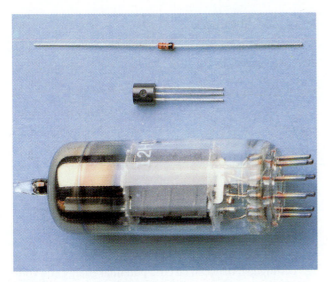

FIGURE 8.32   Solid-State Diode, Transistor, and Vacuum Tube

A diode (two leads) and transistor (three leads) are shown with a vacuum tube. Solid-state diodes and transistors have all but replaced vacuum tubes in most applications.

FIGURE 8.33   Integrated Circuits

A microprocessor integrated circuit, or chip, is held in the eye of a threaded needle.

may contain many diodes, transistors, and other components on a silicon *chip,* usually with dimensions of only a few millimeters (*microchip,* ● Fig. 8.33). Such chips can have many logic circuits and are the "brains" of computers and calculators. Microchips have many applications in processors that perform tasks almost instantaneously. For example, they are in our automobiles, computing the mileage range for the available gasoline, applying antilock brakes, and signaling the inflation of air bags when needed. What will the future bring? Think about it.

RELEVANCE QUESTION: *In what ways (name two), other than those mentioned, do microchips affect your life?*

Important Terms

electric charge (8.1)
electrons
protons
coulomb
current
ampere
law of charges
Coulomb's law

electric potential
   energy (8.2)
voltage
volt
resistance
ohm
Ohm's law
electric power

direct current (dc) (8.3)
alternating current (ac)
series circuit
parallel circuit
law of poles (8.4)
magnetic field
ferromagnetic
magnetic domains

Curie temperature
magnetic declination
electromagnetism (8.5)
motor
generator
transformer

Important Equations

Current: $I = \dfrac{q}{t}$

Coulomb's Law: $F = \dfrac{kq_1q_2}{r^2}$

$(k = 9.0 \times 10^9 \text{ N-m}^2/\text{C}^2)$

Voltage: $V = \dfrac{W}{q}$

Ohm's Law: $V = IR$

Electric Power: $P = IV = I^2R$

Resistance in Series: $R_1 = R_1 + R_2 + R_3 + \cdots$

Resistances in Parallel: $\dfrac{1}{R_p} = \dfrac{1}{R_1} + \dfrac{1}{R_2} + \dfrac{1}{R_3} + \cdots$

Two Resistances in Parallel: $R_p = \dfrac{R_1 R_2}{R_1 + R_2}$

Transformer (Voltages and Turns): $V_2 = \left(\dfrac{N_2}{N_1}\right)V_1$

Review Questions

8.1 Electric Charge and Current

1. What is the unit of electric charge?
   (a) newton   (c) ampere
   (b) coulomb   (d) volt

2. What is the unit of electric current?
   (a) newton   (c) ampere
   (b) coulomb   (d) volt

3. Two equal negative charges are placed equidistant on either side of a positive charge. What would the positive charge experience?
   (a) a net force to the right
   (b) a net force to the left
   (c) a zero net force

4. Name three particles that make up atoms. Which has the smallest mass? Which has no electric charge?

5. Which two particles that make up atoms have about the same mass? Which two have the same magnitude of electric charge?

6. Why is the mutual gravitational attraction among subatomic particles not a consideration in the study of electricity?

7. What is an electric current, and what are its units? (Give two equivalent units.)

8. Explain how a charged rubber comb attracts bits of paper and how a charged balloon sticks to a wall or ceiling.

9. Why are some materials good conductors and others are not?

8.2 Voltage and Electrical Power

10. What is the unit of voltage?
    (a) joule
    (b) joule/coulomb
    (c) amp-coulomb
    (d) amp/coulomb

11. In electrical terms, power has what units?
    (a) joule/coulomb
    (b) amp/ohm
    (c) amp-ohm
    (d) amp-volt

12. Distinguish between electric potential energy and voltage.

13. State Ohm's law, and give the unit of each term in this law.

14. How does electrical power depend on
    (a) current and voltage and
    (b) current and resistance?

15. Explain what is meant by I^2R losses and joule heat.

8.3 Simple Electric Circuits and Electrical Safety

16. Given three resistances, the greatest current flow in a battery circuit would be if the resistance were connected in
    (a) series.
    (b) parallel.
    (c) series-parallel.

17. Distinguish between alternating and direct currents.

18. Why are home appliances connected in parallel rather than in series?

19. Discuss the safety features of (a) fuses (b) circuit breakers (c) three-prong plugs and (d) polarized plugs.

20. What are the causes of electrical shocks, what are some of the consequences, and what precautions should be taken to avoid them?

8.4 Magnetism

21. Which of the following is true for a magnetic field?
    (a) determined by the law of poles
    (b) in the direction indicated by the north pole of a compass
    (c) initiated above the Curie temperature
    (d) a source of magnetism

22. What is the variation in the location of Earth's magnetic north pole from true north given by?
    (a) the law of poles
    (b) magnetic field
    (c) magnetic domains
    (d) magnetic declination

23. Compare the law of charges and the law of poles.

24. (a) What is a ferromagnetic material, and what are some examples?
    (b) Why does a permanent magnet attract pieces of ferromagnetic materials? What would happen if these pieces were above their Curie temperature?

25. (a) What does Earth's magnetic field resemble, and where are its poles?
    (b) What is magnetic declination, and why is it important in navigation?

8.5 Electromagnetism

26. What type of energy conversion does a motor perform?
    (a) chemical energy into mechanical energy
    (b) mechanical energy into electrical energy
    (c) electrical energy into mechanical energy
    (d) mechanical energy into chemical energy

27. What type of energy conversion does a generator perform?
    (a) chemical energy into mechanical energy
    (b) mechanical energy into electrical energy
    (c) electrical energy into mechanical energy
    (d) mechanical energy into chemical energy

28. How do telephone transmitters and receivers work?

29. Describe the basic principle of a dc electric motor.

30. Describe the basic principle of an ac generator.

31. What is the principle of a transformer, and how are transformers used?

32. What is the major reason for using ac electricity in our homes and businesses?

33. What would happen if electric power were transmitted from the generating plant to our home at 120 V?

Applying Your Knowledge _____

1. Are automobile headlights wired in series or parallel? How do you know?

2. An old saying about electrical safety states that you should keep one hand in your pocket when working with electricity. What does this mean?

3. What happens if you cut a bar magnet in half? Could it be continually cut in half to finally get two magnetic monopoles? Explain.

4. Why will transformers not operate on direct current?

Exercises _____

8.1 Electric Charge and Current

1. How many electrons make up one coulomb of charge?
   Answer: 6.25×10^{18}

2. An object has one million more electrons than protons. What is the net charge of the object?

3. There is a net passage of 4.8×10^{18} electrons by a point in a wire conductor in 0.25 s. What is the current in the wire?
   Answer: 3.1 A

4. A current of 1.50 A flows in a conductor for 6.0 s. How much charge passes a given point in the conductor during this time?

5. What are the forces on two charges of $+0.50$ C and $+2.0$ C, respectively, if they are separated by a distance of 3.0 m?
   Answer: 1.0×10^9 N, mutually repulsive

6. Find the force of electrical attraction between a proton and an electron that are 5.0×10^{-11} m apart (the arrangement in the hydrogen atom). Compare this to the gravitational force between these particles (see Section 3.4).

8.2 Voltage and Electrical Power

7. To separate a 0.25-C charge from another charge, 30 J of work is done. What is the electric potential energy of the charge?
   Answer: 30 J

8. What is the voltage of the 0.25-C charge in Exercise 7?

9. If an electrical component with a resistance of 40 Ω is connected to a 120-V source, how much current would flow through the component?
   Answer: 3.0 A

10. What battery voltage is necessary to supply 0.50 A of current to a circuit with a resistance of 24 Ω?

11. A car radio draws 0.25 A of current in the auto's 12-V electrical system.
    (a) How much electric power does the radio use?
    (b) What is the effective resistance of the radio?
    Answer: (a) 3.0 W (b) 48 Ω

12. A flashlight uses batteries that add up to 3.0 V and has a power output of 0.50 W.
    (a) How much current is drawn from the batteries?
    (b) What is the effective resistance of the flashlight?

13. How much does it cost to run a 1500-W hair dryer 30 minutes each day for one month (30 days) at a cost of 8¢ per kWh?
    Answer: $1.80

14. A refrigerator using 1000 W runs one-eighth of the time. How much does the electricity cost to run the refrigerator each month at 10¢ per kWh?

15. The heating element of an iron operates at 110 V with a current of 10 A.
    (a) What is the resistance of the iron?
    (b) What is the power dissipated by the iron?
    Answer: (a) 11 Ω (b) 1100 W

16. A 100-W light bulb is turned on. It has an operating voltage of 120 V.
    (a) How much current flows through the bulb?
    (b) What is the resistance of the bulb?
    (c) How much energy is used each second?

17. Two resistors with values of 20 Ω and 30 Ω, respectively, are connected in series and hooked to a 12-V battery.
    (a) How much current is in the circuit?
    (b) How much power is expended in the circuit?
    Answer: (a) 0.24 A (b) 2.9 W

18. Suppose the two resistors in Exercise 17 were connected in parallel. What would be (a) the current and (b) the power in this case?

19. A student in the laboratory connects a 10-Ω resistor, a 15-Ω resistor, and a 20-Ω resistor in series and hooks the arrangement to a 50-V dc source.
    (a) How much current is in the circuit?
    (b) How much power is expended in the circuit?
    Answer: (a) 1.1 A (b) 56 W

20. The student in Exercise 19 redoes the experiment but with the resistors connected in parallel.
    (a) What is the current?
    (b) What is the power?

21. An electrical component has two resistances of 10 Ω and 15 Ω connected in parallel.
    (a) What is the total resistance of the component?
    (b) If eight such components were connected in series and hooked to a 120-V source, how much current would be drawn from the source?
    Answer: (a) 6.0 Ω (b) 2.5 A

22. Three resistances, $R_1 = 60$ Ω, $R_2 = 30$ Ω, and $R_3 = 20$ Ω, are connected in series. (a) What is the total equivalent resistance of the arrangement? (b) If this series arrangement is connected in parallel to a 50-Ω resistance and the total combination is connected to a 12-V source, how much current is drawn from the source?

8.5 Electromagnetism

23. A transformer has 300 turns on its secondary and 50 turns on its primary. The primary is connected to a 12-V source.
    (a) What is the voltage output of the secondary?
    (b) If 3.0 A flows in the primary coil, how much current is there in the secondary coil?
    Answer: (a) 72 V (b) 0.50 A

24. A transformer has 250 turns on its primary and 50 turns on its secondary.
    (a) Is this a step-up or a step-down transformer?
    (b) If a voltage of 100 V is applied to the primary and a current of 0.25 A flows in these windings, what is the voltage output of and the current in the secondary?

25. A transformer with 1000 turns in the primary coil has to decrease the voltage from 4400 V to 220 V for home use. How many turns should there be on the secondary coil?
    Answer: 50

26. A power company transmits current through a 240,000-V transmission line. This voltage is stepped down at an area substation to 40,000 V by a transformer that has 900 turns on the primary coil. How many turns are on the secondary of the transformer?

Solutions to Confidence Exercises

8.1 $q = ne = (5.0 \times 10^{19} \text{ electrons})(1.6 \times 10^{-19} \text{ C/electron})$
$\qquad = 8.0 \text{ C}$

$\qquad I = \dfrac{q}{t} = \dfrac{8.0 \text{ C}}{40 \text{ s}} = 0.20 \text{ A}$

8.2 $P = IV = (10 \text{ A})(120 \text{ V}) = 1200 \text{ W or } 1200 \text{ J/s}$

8.3 Greater resistance, less current.

$\qquad I = \dfrac{V}{R_s} = \dfrac{12 \text{ V}}{6.0 \text{ } \Omega + 6.0 \text{ } \Omega + 3.0 \text{ } \Omega} = 0.80 \text{ A}$

8.4 $\qquad V_2 = \left(\dfrac{N_2}{N_1}\right)V_1 = \left(\dfrac{500}{25}\right)100 \text{ V} = 2000 \text{ V}$

Answers to Multiple-Choice Review Questions

1. b   3. c   11. d   21. b   26. c
2. c   10. b   16. b   22. d   27. b

ATOMIC PHYSICS

9

In all science, error precedes the truth; and it is better it should go first than last.

Horace Walpole (1717–1797)

The development of physics prior to about 1900 is now termed *classical physics,* or *Newtonian physics.* It was generally concerned with the *macrocosm*—the description and explanation of large-scale, observable phenomena such as the movement of cannonballs and planets.

As the year 1900 approached, scientists thought that the field of physics was in fairly good order. The principles of mechanics, wave motion, sound, and optics were reasonably well understood. Electricity and magnetism had been combined into electromagnetism, and light had been shown to be electromagnetic waves. Certainly, some rough edges remained, but it seemed that only a few refinements were needed.

However, as scientists probed deeper into the submicroscopic world of the atom (the *microcosm*), they observed strange things—strange in the sense that they could not be explained by the classical principles of physics. These discoveries were unsettling, because physicists became aware that radical new approaches were needed to describe and explain submicroscopic phenomena.

The development of physics since about 1900 is commonly termed *modern physics.* Chapter 9 looks at *atomic physics,* the part of modern physics that deals mainly with phenomena involving the electrons in atoms. Chapter 10 examines *nuclear physics,* which deals with the central core, or nucleus, of the atom.

Photo: The aurora borealis, or northern lights, is explained using concepts of atomic physics.

An appropriate way in which to start our examination of atomic physics is with a brief history of the concept of the atom. How that concept has changed provides an excellent example of how scientific theories adapt as new experimental evidence is gathered.

■

9.1 Early Concepts of the Atom

LEARNING GOAL

▼ Describe the atom models of Dalton, Thomson, and Rutherford.

About 400 B.C. the Greek philosophers debated whether matter was continuous or discrete (particulate). For instance, if one could repeatedly cut in half a sample of gold, would an ultimate particle of gold theoretically be reached that could not be divided further? If an ultimate particle is reached, matter is discrete; if not, matter is continuous.

Most of the philosophers, including the renowned Aristotle, decided that matter was continuous and could be divided again and again, indefinitely. A few, notably Leucippus and Democritus, thought that an ultimate, *indivisible* (Greek: *atomos*) particle would indeed be reached. The question was a purely philosophical one, and neither side could, or apparently

wanted to, present what we today would call scientific evidence to support one viewpoint or the other. Modern science differs from the approach of the ancient Greeks by its reliance not only on logic but also on the systematic gathering of facts by observation and experimentation, and by the rigorous testing of hypotheses.

The "continuous" model of matter of Aristotle and his followers prevailed for about 2200 years, until John Dalton (photo on page 227) in 1807 presented evidence that matter was discrete and must exist as particles. We will wait until Section 12.3 to examine Dalton's evidence for the atomic theory. For now, it is sufficient to point out that Dalton's major hypothesis was that each chemical element is composed of small indivisible particles called **atoms,** which are identical for that element but different (particularly in their masses and chemical properties) from atoms of other elements. Dalton's concept of the atom has been called the "billiard ball model," because he thought of atoms as essentially featureless spheres of uniform density (● Fig. 9.1a).

Dalton's billiard ball model had to be refined about 90 years later when the electron was discovered by J. J. Thomson (photo on page 227) at Cambridge University in 1897. Thomson studied electrical discharges in tubes of low-pressure gas called *gas-discharge tubes*, or *cathode-ray tubes* (● Fig. 9.2). He discovered that when high voltage was applied to the tube, a "ray" was

FIGURE 9.1   Dalton's, Thomson's, and Rutherford's Models of the Atom

(a) John Dalton's 1807 "billiard ball model" pictured the atom as an indivisible, uniformly dense, solid sphere. (b) J. J. Thomson's 1903 "plum pudding model" conceived of the atom as a sphere of positive charge in which were embedded negatively charged electrons. (c) Ernest Rutherford's 1911 "nuclear model" depicted the atom as having a dense center of positive charge, called the *nucleus*, around which electrons orbited. (The size of the nucleus relative to the atom is *greatly* exaggerated in this sketch.)

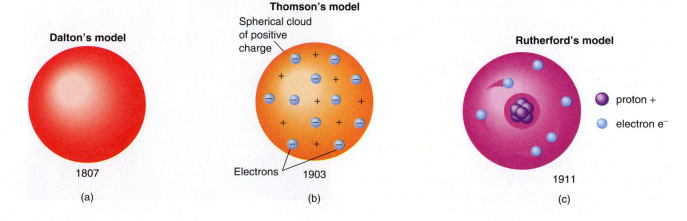

FIGURE 9.2   A Cathode-Ray Tube in Operation

Fast-moving electrons excite the gas in the tube, causing a glow between the electrodes. The green line is due to the electron beam hitting a screen coated with a fluorescent chemical.

produced at the negative electrode (the cathode) and sped to the positive electrode (the anode). Unlike electromagnetic radiation, the ray was deflected by electric and magnetic fields. Thomson concluded that the ray consisted of a stream of negatively charged particles, now called **electrons.**

Further experiments by Thomson and others showed that the electrons had a mass of 9.11×10^{-31} kg, a charge of -1.60×10^{-19} C, and that the electrons were being produced by the voltage "tearing" them from atoms of gas in the tube. Because identical electrons were being produced no matter what gas was in the tube, it became apparent that *atoms of all types contain electrons.*

An atom, as a whole, is electrically neutral; therefore, some other part of the atom must be positively charged. Indeed, further experiments detected these positively charged particles (now called *positive ions*) as they sped to and passed through holes in the negative electrode of a cathode-ray tube. In 1903, Thomson concluded that an atom was much like a sphere of plum pudding: The electrons were the raisins stuck randomly in an otherwise homogeneous mass of positively charged "pudding"—the rest of the atom (Fig. 9.1b).

Thomson's "plum pudding model" of the atom was modified in 1911, only eight years later. Ernest Rutherford (photo on page 227) discovered that 99.97% of the mass of an atom was concentrated in a tiny core called the *nucleus.* (Rutherford's classic experiment is discussed in Section 10.2.) Rutherford's nuclear model of the atom pictured the electrons as circulating in some way in the otherwise empty space around this tiny, positively charged core (Fig. 9.1c).

9.2 **The Dual Nature of Light**

LEARNING GOALS

▼ State Planck's hypothesis, and apply the equation for it.

▼ Describe and explain the photoelectric effect.

▼ Explain the meaning of the *dual nature of light.*

Before we can discuss the next improvement in the model of the atom, we must consider a radical development about the nature of light. Even before the turn of the twentieth century, scientists knew that visible light of all frequencies was emitted by the atoms of an incandescent (glowing hot) solid, such as the filament of a light bulb. For example, ● Fig. 9.3 shows that the radiation emitted by a hot bar of steel as it comes from the furnace has its maximum intensity in the red region of the visible spectrum.

FIGURE 9.3   Red-Hot Steel

The radiation component of maximum intensity of a hot solid determines its color, as shown here by red-hot steel coming out of a furnace.

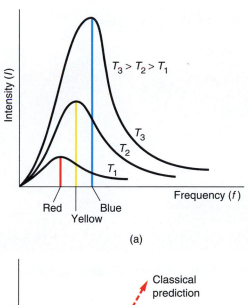

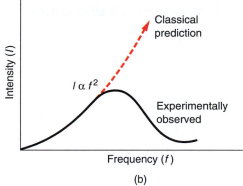

FIGURE 9.4   Thermal Radiation

(a) As the temperature of an incandescent solid increases, the component of maximum intensity shifts to a higher frequency. (b) Classically, it is predicted that intensity should be proportional to f^2, which indicates a much greater energy than is actually observed.

As the temperature is increased, more radiation is emitted, and the component of maximum intensity is shifted to a higher frequency. Consequently, a hot solid heated to higher and higher temperatures appears to go from a dull red to a bluish white (● Fig. 9.4a). This outcome is expected, because the hotter the solid, the greater are the electron vibrations in the atoms and the higher is the frequency of the emitted radiation.

However, according to classical wave theory, the intensity (I) of energy of the emitted radiation should be proportional to the second power of the frequency, $I \propto f^2$, which means that the intensity should increase quite rapidly as the frequency increases. But this is

not what is actually observed (Fig. 9.4b). This discrepancy was termed the *ultraviolet catastrophe*—"ultraviolet" because the difficulty occurred at high frequencies beyond the violet end of the visible spectrum and "catastrophe" because the energy intensity actually observed was very much less than theory predicted.

The dilemma was resolved in 1900 by Max Planck (pronounced "plonck"), a German physicist (● Fig. 9.5). He introduced a radical idea that explained the observed thermal radiation intensity distribution. In doing so, Planck took the first step toward a new theory called *quantum physics*. Classically, an electron oscillator may vibrate at any frequency or have any energy up to some maximum value. But *Planck's hypothesis* states that the energy is *quantized;* that is, an oscillator can have only discrete, or specific, amounts of energy. Moreover, Planck said that the energy (E) of an oscillator depends on its frequency (f) in accordance with the following equation:

FIGURE 9.5   Max Planck (1858–1947)

While a professor of physics at the University of Berlin in 1900, Planck proposed that the energy of thermal oscillators exists in only discrete amounts, or quanta. The important small constant h is called *Planck's constant.* Planck was awarded the Nobel Prize in physics in 1918 for his contributions to quantum physics.

energy = Planck's constant × frequency

$$E = hf$$ **9.1**

where h is a constant, called *Planck's constant*, with the very small value of 6.63×10^{-34} J-s.

Planck's hypothesis correctly accounted for the actually observed radiation curve shown in Fig. 9.4b. Thus Planck introduced the idea of a **quantum**—a discrete amount of energy (sort of a "packet" of energy).

Quantized energy is similar to the potential energy of a person on a staircase, who can have only the potential energy values determined by the height of each particular step. Continuous energy, on the other hand, is like the potential energy of a person on a ramp, who could stand at any height and thus have any value of potential energy (● Fig. 9.6).

In the latter part of the nineteenth century, scientists observed that electrons are emitted when certain

FIGURE 9.7 The Photoelectric Effect in Action
Radiant (solar) energy is converted directly into electrical energy in a solar-powered car racing through Australia's outback.

FIGURE 9.6 The Concept of Quantized Energy
(a) The woman on the staircase can stop only on one of the steps, so she can have four nonzero potential energy values. (b) When on the ramp, she could be at any height and have a continuous range of potential energy values.

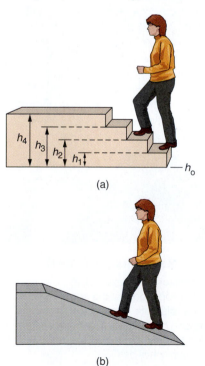

(a)

(b)

metals are exposed to light; the phenomenon was named the **photoelectric effect.** This direct conversion of light (radiant energy) into electrical energy now forms the basis of photocells used in the automatic door openers at your supermarket, in photographic light meters, and in solar energy applications (● Fig. 9.7).

As in the case of the ultraviolet catastrophe, certain aspects of the photoelectric effect could not be explained by classical theory. For example, the amount of energy necessary to free an electron from a photomaterial could be calculated. But according to classical theory, in which light is considered a wave with a continuous flow of energy, it would take an appreciable time for electromagnetic waves to supply the energy needed for an electron to be emitted. However, electrons flow from photocells almost immediately on being exposed to light. Also, it was observed that only light above a certain frequency would cause electrons to be emitted. According to classical theory, light of any frequency should be able to provide the needed energy.

In 1905, Albert Einstein (see the chapter's first Highlight) attacked the photoelectric effect problem. Applying Planck's hypothesis, he decided that light (and, in fact, all electromagnetic radiation) is quantized and consists of "particles," or "packets," of energy, rather than waves. Einstein coined the term **photon** to refer to such a quantum of electromagnetic radiation. He used Planck's relationship ($E = hf$) and stated that a quantum, or photon, of light contains a discrete amount of energy (E) equal to Planck's constant (h) times the frequency (f) of the light. The higher the frequency of the light, the greater is the energy of its photons. For example, because blue light has a higher frequency than red light, photons of blue light have more energy than do photons of red light. Let's see how photon energy is determined.

EXAMPLE 9.1

Determining Photon Energy from the Frequency

Find the energy in joules of the photons of red light of frequency 5.00×10^{14} Hz. (Recall that a hertz is a reciprocal second, $1/s$.)

SOLUTION

We are given the frequency, so we can find the energy directly by using Eq. 9.1.

$$E = hf = (6.63 \times 10^{-34} \text{ J-s})(5.00 \times 10^{14} \text{ 1/s})$$
$$= 33.2 \times 10^{-20} \text{ J}$$

CONFIDENCE EXERCISE 9.1

Find the energy in joules of the photons of blue light of frequency 7.50×10^{14} Hz.

By considering light to be composed of photons, Einstein was able to explain the photoelectric effect. The classical time delay necessary to get enough energy to free (knock loose) an electron is not a problem if the concept of photons of energy is used. A photon with the proper amount of energy could deliver the release energy instantaneously in a "packet." (An analogy of the delivery of wave and quantum energy is shown in ● Fig. 9.8.) Also, we can see why light with greater than a certain minimum frequency is required for emission of an electron. Because $E = hf$, a photon of light with a frequency smaller than this minimum value would not have enough energy to free an electron.

But how can light be composed of photons (discrete packets of energy) when it shows wave phenomena such as polarization, diffraction, and interference (Sections 7.3 and 7.4)—behavior that is explained

Wave nature   Quantum nature

FIGURE 9.8   Wave and Quantum Analogy
A wave supplies a continuous flow of energy, somewhat analogous to the stream of water from the garden hose. A quantum supplies its energy all at once in a "packet," or "bundle," somewhat like each bucketful of water thrown on the fire.

very well by assuming the wave nature of light but which *cannot* be explained by the photon (particle) concept? On the other hand, the photoelectric effect (and other newly found phenomena in which light interacts with matter) cannot be explained by invoking the wave nature of light; the photon concept is necessary. We have a confusing situation! Is light a wave, or is it a particle? The answer to this is couched in the term **dual nature of light,** which simply means that, to explain various phenomena, light sometimes must be described as a wave and sometimes as a particle (● Fig. 9.9).

Our macroscopic idea that something must be *either* a wave *or* a particle breaks down here. Light is

FIGURE 9.9   The Wave-Particle Duality of Light
Electromagnetic radiation (a beam of light) can be pictured in two ways: as a wave or as a stream of individual packets of energy called *photons.*

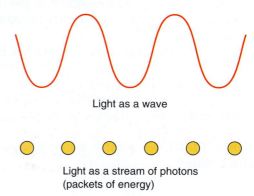

Light as a wave

Light as a stream of photons
(packets of energy)

HIGHLIGHT: Albert Einstein

Isaac Newton's only rival for the accolade of "greatest scientist of all time" is Albert Einstein, who was born in Ulm, Germany, in 1879 (Fig. 1). In high school, Einstein did poorly in Latin and Greek and was interested only in mathematics. His teacher told him, "You will never amount to anything, Einstein."

Albert Einstein attended college in Switzerland, graduated in 1901, and accepted a job as a junior official at the patent office in Berne, Switzerland. He spent his spare time working in theoretical physics. In 1905, five of his papers were published in the *German Yearbook of Physics,* and in that same year he earned his Ph.D. at the age of 26. One paper explained the photoelectric effect (Section 9.2), and he was awarded the 1921 Nobel Prize in physics for that contribution.

Another of Einstein's 1905 papers put forth his ideas on what came to be called the *special theory of relativity.* This paper dealt with what would happen to an object as it approached a speed that was a significant fraction of

FIGURE 1   Albert Einstein (1879–1955)
Einstein is shown here during a visit to Caltech in the 1930s.

the speed of light. Einstein asserted that to an outside observer the object would get shorter in the direction of

motion, it would become more massive, and a clock would run slower in the object's system. All this is against "common sense," but common sense is based on limited experience with objects of ordinary size moving at ordinary speeds. All three of Einstein's predictions have now been verified experimentally.

One other result came from the special theory of relativity: Energy and matter are related by what has come to be the most famous equation in scientific history: $E = mc^2$. We will explore this relationship and use this equation in Chapter 10, where we will also give some information about Einstein's later life.

In 1915, Einstein published his *general theory of relativity,* which deals mainly with the effect of a gravitational field on the behavior of light and has profound implications with regard to the structure of the universe. Once again, Einstein's predictions were verified experimentally, and the general theory of relativity is a cornerstone of modern physics and astronomy.

---

not really a wave, nor is it really a particle; it has characteristics of both, and we simply have no good, single, macroscopic analogy that fits the combination.

Therefore, in a specific type of experiment, scientists (renowned for being pragmatic) use whichever model of light works. In some experiments the wave model does the job; in other experiments, the particle model is necessary.

The concept of the dual nature of light has been around for almost a century now. It actually does an excellent job of explaining known and newly discovered phenomena and of making valid predictions that have served as the basis of new technologies.

RELEVANCE QUESTION: *What is the basic "quantum unit" of our money?*

9.3 Bohr Theory of the Hydrogen Atom

LEARNING GOALS

▼ Describe Bohr's model of the hydrogen atom.

▼ Explain the formation of line spectra.

Now that you have an idea of what is meant by quantum theory and photons, we can discuss the next advance in our understanding of the atom. Recall from Section 7.2 that when light from incandescent sources such as light bulb filaments is analyzed with a *spectrometer,* a *continuous spectrum* (light of all colors) is observed (● Fig. 9.10a).

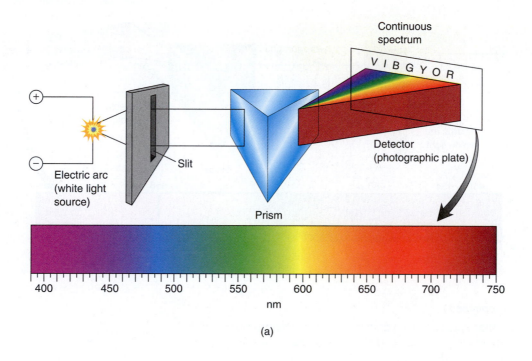

(a)

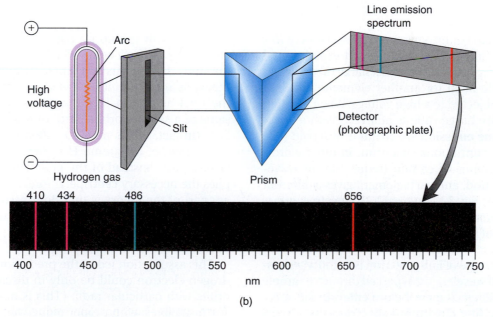

(b)

FIGURE 9.10 Three Types of Spectra

(a) A continuous spectrum containing all wavelengths of visible light (indicated by the initial letters of the colors of the rainbow).

(b) The line emission spectrum for hydrogen consists of four discrete wavelengths of visible radiation.

Continued at top of following page.

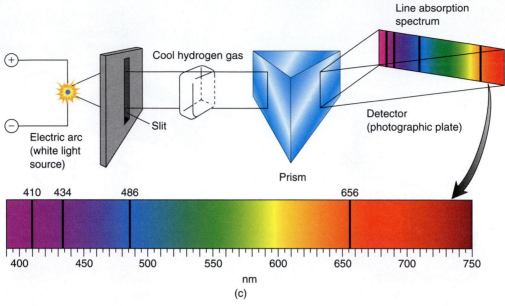

FIGURE 9.10 *(Continued)*

(c) The line absorption spectrum for hydrogen consists of four discrete "missing" wavelengths that appear as dark lines against a rainbow-colored background.

In addition to continuous spectra, two types of *line spectra* exist. In the late 1800s, much experimental work was being done with gas-discharge tubes, which contain a little neon (or another element) and emit light when subjected to a high voltage. When the light from a gas-discharge tube is analyzed with a spectrometer, a **line emission spectrum** is observed (Fig. 9.10b), not a continuous spectrum. In other words, only spectral lines of certain frequencies or wavelengths are found, and each element gives a different set of lines (see Fig. 7.14 on page 150). Spectroscopists did not understand why only discrete, characteristic wavelengths of light were emitted by atoms in various excited gases.

The second type of line spectrum is found when visible light of all wavelengths is passed through a sample of a cool gaseous element before entering the spectrometer. The **line absorption spectrum** that results has dark lines of missing colors (Fig. 9.10c). The dark lines are at exactly the same wavelengths as the bright lines of the *line emission spectrum* for that particular element. Compare, for instance, the emission and absorption spectra of hydrogen in Figs. 9.10b and c.

An explanation of the spectral lines observed for hydrogen was put forth in 1913 by the Danish physi-

cist Niels Bohr (photo on page 227). The hydrogen atom is the simplest atom, which is why Bohr chose it for study. Its nucleus is a single proton, and Bohr's theory assumes that its one electron revolves around the nuclear proton in a circular orbit—much the same as a satellite orbits Earth or a planet orbits the Sun. (In fact, Bohr's model is often called the *planetary model* of the atom.) However, in the atom, the electric force instead of the gravitational force supplies the necessary centripetal force. The revolutionary part of the theory is that Bohr assumed that the angular momentum of the electron is quantized. He correctly reasoned that a discrete line spectrum must be the result of a quantum effect.

This assumption led to the prediction that the hydrogen electron could be only in discrete (specific) orbits with particular radii. (This is not the case for Earth satellites; with proper maneuvering, a satellite can have any orbital radius.)

Bohr's possible electron orbits are characterized by whole-number values, $n = 1, 2, 3, \ldots$, where n is called the **principal quantum number** (● Fig. 9.11). The lowest-value orbit ($n = 1$) has the smallest radius, and the radii increase as the principal quantum number increases. (Notice in Fig. 9.11 that the orbits are

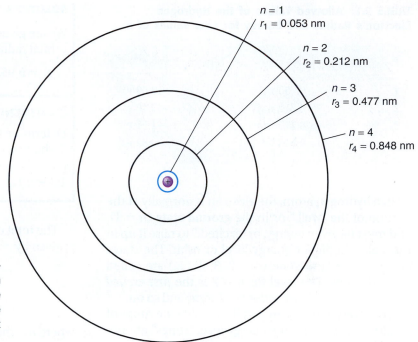

FIGURE 9.11 **Bohr Electron Orbits**

The Bohr hypothesis predicts only certain discrete orbits for the hydrogen electron. Each orbit is indicated by a quantum number *n*. The orbit shown in blue is the ground state. The orbits move outward and get farther apart with increasing *n*.

not evenly spaced; as *n* increases, not only does the distance from the nucleus increase but also the distance *between* orbits increases.)

However, a problem remained with Bohr's hypothesis. According to classical theory, an accelerating electron radiates electromagnetic energy. An electron in circular orbit has centripetal acceleration (Section 2.4) and hence should radiate energy continuously as it circles the nucleus. Such a loss of energy would cause the electron to spiral into the nucleus, similar to the death spiral of an Earth satellite in a decaying orbit because of energy losses due to atmospheric friction.

However, atoms do not continuously radiate energy, nor do they collapse. Therefore, Bohr made another nonclassical assumption: He hypothesized that the hydrogen electron does not radiate energy when in an allowed, discrete orbit but does so only when it makes a *quantum jump,* or *transition,* from one allowed orbit to another.

The allowed orbits of the hydrogen electron are commonly expressed in terms of *energy states,* or *energy levels,* with each state corresponding to a specific orbit (● Fig. 9.12). We characterize the energy levels as states in a *potential well* (Section 4.2). Similar to lifting a bucket in a water well, energy is required to lift

the electron to a higher level. If the top of the well is the zero reference level, the energy levels in the well will have negative values.

FIGURE 9.12 **The Energy-Level Diagram for the Hydrogen Atom**

Each Bohr orbit has a particular energy value, or energy level. The lowest level, which has the quantum number *n* = 1, is called the *ground state.* The higher energy levels, which have *n* values greater than 1, are called *excited states.*

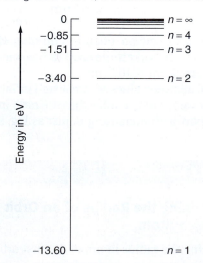

TABLE 9.1 Allowed Values of the Hydrogen Electron's Radius and Energy for Low Values of n

n	r_n	E_n
1	0.053 nm	−13.60 eV
2	0.212 nm	−3.40 eV
3	0.477 nm	−1.51 eV
4	0.848 nm	−0.85 eV

In a hydrogen atom, the electron is normally at the bottom of the "well," or in the **ground state** ($n = 1$), and must be given energy, or "excited," to raise it up in the well to a higher energy level or orbit. The states above the ground state ($n = 2, 3, 4, \ldots$) are called **excited states.** The level for $n = 2$ is the *first excited state;* for $n = 3$, the *second excited state;* and so on.

The levels in the energy well resemble the rungs of a ladder, except that the energy level "rungs" are not evenly spaced. Just as a person going up and down a ladder must do so in discrete steps on the ladder rungs, so a hydrogen electron must be excited (or de-excited) from one energy level to another by discrete amounts. If enough energy is applied to excite the electron to the top of the well, the electron is no longer bound to the nucleus, and the atom is *ionized.*

A mathematical development of Bohr's theory is beyond the scope of this textbook. However, the important results are the prediction of the radii and the energies of the allowed orbits. The radius of a particular orbit is given by

$$r_n = 0.053 \, n^2 \, \text{nm} \qquad \textbf{9.2}$$

where n is the principal quantum number of an orbit ($n = 1, 2, 3, \ldots$), and r is the orbit radius, measured in nanometers (1 nm = 10^{-9} m).

Several allowed values of r are listed in Table 9.1 for some low values of n. Notice that the radii indeed get farther apart with increasing n, just as Fig. 9.11 indicates.

EXAMPLE 9.2

Determining the Radius of an Orbit in a Hydrogen Atom

Determine the radius in nm of the first orbit ($n = 1$, the ground state) in a hydrogen atom.

SOLUTION

We are given the principal quantum number, so the orbital radius can be found directly by use of Eq. 9.2.

$$r_1 = 0.053 \, r^2 \, \text{nm} = 0.053 \, (1)^2 \, \text{nm} = 0.053 \, \text{nm}$$

CONFIDENCE EXERCISE 9.2

Determine the radius in nm of the second orbit ($n = 2$, the first excited state) in a hydrogen atom. Compare the value you compute with that shown in Table 9.1.

The total energy of an electron in an allowed orbit is given by

$$E_n = \frac{-13.60}{n^2} \, \text{eV} \qquad \textbf{9.3}$$

where n is the orbit's principal quantum number, and E is energy measured in electron volts (eV).[*]

EXAMPLE 9.3

Determining the Energy of an Orbit in the Hydrogen Atom

Determine the energy of an electron in the first orbit ($n = 1$, the ground state) in a hydrogen atom.

SOLUTION

The energy of a particular orbit, or energy level, in a hydrogen atom is calculated by using Eq. 9.3 and the n value for that orbit. So, for $n = 1$,

$$E_1 = \frac{-13.60}{(1)^2} \, \text{eV} = -13.60 \, \text{eV}$$

CONFIDENCE EXERCISE 9.3

Determine the energy of an electron in the second orbit ($n = 2$, the first excited state) in a hydrogen atom. Compare your answer with the value given in Table 9.1.

---

[*]An *electron volt* is the amount of energy an electron acquires when it is accelerated through an electric potential of one volt. The eV is a common unit of energy in atomic and nuclear physics, and 1 eV = 1.60×10^{-19} J.

Table 9.1 shows the energies for the hydrogen electron for low values of n (orbits nearest the nucleus). These values correspond to the energy levels shown in Fig. 9.12. Note that unlike the distances between orbits, the energy levels get *closer* together as n increases. Recall that the minus signs, indicating negative energy values, show that the electron is in a potential energy well. Because the energy value is -13.60 eV for the ground state, it would require this much energy input to ionize a hydrogen atom. We thus say that the hydrogen electron's *binding energy* is 13.60 eV.

But how did the hypothesis stand up to experimental verification? Recall that Bohr was trying to explain discrete line spectra. According to his hypothesis, an electron can make transitions only between two allowed orbits, or energy levels. In these transitions, the total energy must be conserved. If the electron is initially in an excited state, it will lose energy when it "jumps down" to a less excited (lower n) state. In this case, the electron's energy loss will be carried away by a photon—a quantum of light.

The total initial energy (E_{n_i}) must equal the total final energy ($E_{n_f} + E_{photon}$); thus we have

$$E_{n_i} = E_{n_f} + E_{photon}$$

(energy before = energy after)

or, rearranging, the energy of the emitted photon (E_{photon}) is the difference in the energies of the initial and final states:

$$E_{photon} = E_{n_i} - E_{n_f} \qquad \textbf{9.4}$$

A schematic diagram of the process of *photon emission* is shown in ● Fig. 9.13a. As can be seen, hydrogen's line emission spectrum results from the relatively few allowed energy transitions as the electron de-excites. Figure 9.13b illustrates the reverse process of *photon absorption* to excite the electron. Hydrogen's line absorption spectrum results from exposing hydrogen atoms in the ground state to visible light of all wavelengths (or frequencies). The hydrogen electrons absorb only those wavelengths that can cause electron transitions "up." These wavelengths are taken out of the incoming light, whereas the other, unusable wavelengths pass through to produce a spectrum of color containing dark lines. Of course, in a hydrogen atom a photon of the same energy emitted in a "down" transition will have to be absorbed in an "up" transition between the same two levels. So the dark lines in the hydrogen absorption spectrum ex-

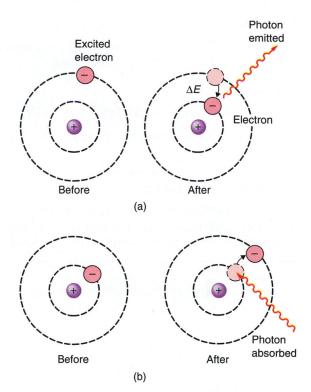

FIGURE 9.13   Photon Emission and Absorption

(a) When an electron in an excited hydrogen atom makes a transition to a lower energy level, or orbit, the atom loses energy by emitting a photon. (b) When a hydrogen atom absorbs a photon, the electron is excited into a higher energy level, or orbit.

actly match up with the bright lines in the hydrogen emission spectrum.

The transitions for photon emissions in the hydrogen atom are shown on an energy-level diagram in ● Fig. 9.14. The electron may "jump down" one or more energy levels in becoming de-excited; that is, the electron may go down the energy "ladder" using adjacent "rungs," or it may skip "rungs." (To show photon absorption and jumps up to higher energy levels, the arrows in Fig. 9.14 would point up.)

EXAMPLE 9.4

Determining the Energy of a Transition in a Hydrogen Atom

Use Table 9.1 and Eq. 9.4 to determine the energy of the photon emitted as an electron in the hydrogen atom jumps down from the $n = 2$ level to the $n = 1$ level.

SOLUTION

Table 9.1 shows values of -3.40 eV for the $n = 2$ level and -13.60 eV for $n = 1$. Thus, using Eq. 9.4,

$$E_{\text{photon}} = E_2 - E_1$$

$$= -3.40 \text{ eV} - (-13.60 \text{ eV}) = 10.20 \text{ eV}$$

The positive value indicates that the 10.20-eV photon is *emitted*.

CONFIDENCE EXERCISE 9.4

Use Table 9.1 and Eq. 9.4 to determine the energy of the photon absorbed as an electron in the hydrogen atom jumps up from the $n = 1$ level to the $n = 3$ level.

Hence, the Bohr hypothesis predicts that an excited hydrogen atom will emit light with discrete fre-

FIGURE 9.14 Spectral Lines for Hydrogen

The transitions among discrete energy levels by the electron in the hydrogen atom give rise to discrete spectral lines. For example, transitions down to $n = 2$ from $n = 3, 4, 5,$ and 6 give the four spectral lines in the visible region that form the Balmer series. Bohr correctly predicted the existence of both the Lyman and Paschen series.

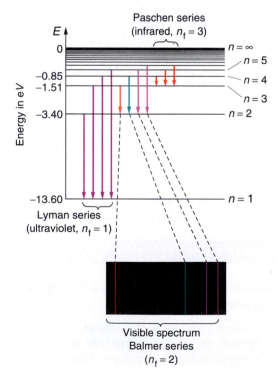

quencies (and hence wavelengths) corresponding to discrete transitions "down," whereas a hydrogen atom in the ground state will absorb light with discrete frequencies corresponding to discrete transitions "up."

The theoretical frequencies for the four allowed transitions that give lines in the visible region were computed using the energy values for the various initial and final energy levels and compared with the frequencies of the lines of the hydrogen emission and absorption spectra. The theoretical and the experimental values were identical—a triumph for the Bohr theory (Fig. 9.14).

Transitions to a particular lower level form a *transition series*. These series were named in honor of early spectroscopists who discovered or experimented with the hydrogen atom's spectral lines that belonged to particular regions of the electromagnetic spectrum. For example, the series of lines in the visible spectrum of hydrogen, which corresponds to transitions from $n = 3, 4, 5,$ and 6 back to $n = 2$, is called the *Balmer series* (Fig. 9.14).

Bohr's calculations predicted the existence of a series of lines of specific energies in the ultraviolet (UV) region of the hydrogen spectrum and another series of lines in the infrared (IR) region. Lyman and Paschen, two spectroscopists, discovered these series, whose lines were exactly at the wavelengths Bohr had predicted, and these series bear their names (Fig. 9.14).

Thus quantum theory and the quantum nature of light scored another success. As you might imagine, the energy-level arrangements for atoms other than hydrogen—those with more than one electron—are more complex, and we will consider them in Chapter 11. Even so, the line spectra for atoms of various elements are indicative of their energy-level spacings and provide characteristic line emission and line absorption "fingerprints" by which atoms may be identified by spectroscopy.

The composition of far distant stars can be determined from analysis of the dark absorption lines in their spectra. In fact, the element helium was discovered in the Sun before it was found on Earth! Pierre Janssen in 1868 detected a dark line in the solar spectrum that did not match the emission line of any known element, and he reasoned that the line must be that of a new element. The element was named *helium* after the Greek word for Sun, *helios*. Twenty-seven years later, William Ramsey found helium on Earth trapped in a uranium mineral.

Another interesting quantum phenomenon is auroras, called the *northern* and *southern lights,* which are caused by charged particles from the Sun entering Earth's atmosphere close to the magnetic north and south poles (see the chapter-opening photograph and Section 19.1). These particles interact with molecules in the air and excite some of their electrons to higher energy levels. When the electrons fall back down, some of the absorbed energy is emitted as visible radiation. Two other useful quantum phenomena, fluorescence and phosphorescence, are discussed in the chapter's second Highlight, on page 220.

RELEVANCE QUESTION: *Why, at night, under the mercury or sodium vapor lights in a mall parking lot, do cars seem to be peculiar colors?*

9.4 Microwave Ovens, X-Rays, and Lasers

LEARNING GOALS

▼ Describe and explain the operation of a microwave oven.

▼ Tell how X-rays are produced, and explain their spectra.

▼ Explain how laser light is produced.

In this section we will take a look at three important technological applications that involve quantum theory: microwave ovens, X-rays, and lasers.

Microwave Ovens

Much of modern physics and chemistry is based on the study of energy levels of various atomic and molecular systems. When light is emitted or absorbed, scientists study the emission or absorption spectrum to learn about the energy levels of the system, as in Fig. 9.14 for the hydrogen atom. Some scientists do research in *molecular spectroscopy*—the study of the spectra and energy levels of molecules. As you would expect, molecules of one substance produce a spectrum different from that produced by molecules of another substance. Molecules can have quantized energy levels because of molecular vibrations or rotations or because they contain excited atoms.

The water molecule has some rotational energy levels spaced very closely together. The energy differ-

ences are such that microwaves (Section 6.2), which have relatively low frequencies and energies, are absorbed by the water molecules. This principle forms the basis of the *microwave oven.* Because all foods contain moisture, their water molecules absorb microwave radiation, thereby gaining energy and rotating more rapidly, and thus heating and cooking the food. Molecules of fats and oils in a food also are excited by microwave radiation, so they, too, contribute to the cooking. The interior metal sides of the oven reflect the radiation and remain cool.

Because it is the water content of foods that is crucial in microwave heating, objects such as paper plates and ceramic or glass dishes do not get hot immediately in a microwave oven. However, they often become warm or hot after being in contact with hot food (heat transfer by conduction—Section 5.4).

Some people think that the microwaves penetrate the food and heat it throughout, but this is not the case. Microwaves penetrate only a few centimeters before being completely absorbed, so the interior of a large mass of food must be heated by conduction as in a regular oven. For this reason, microwave oven users are advised to let foods sit for a short time after microwaving. Otherwise, the outside of the food may be quite hot, while the center is disagreeably cool.

X-Rays

X-rays, which are used widely in industrial and medical fields (● Fig. 9.15), are another example of the technological use of quantum phenomena. **X-rays** are high-frequency, high-energy electromagnetic radiation (Section 6.2). They were discovered accidentally in 1895 by German physicist Wilhelm Roentgen ("RUNT-gin"). While working with a gas-discharge tube, Roentgen noticed that a piece of fluorescent paper across the lab was glowing, apparently from being exposed to some unknown radiation being emitted from the tube. He called it *X-radiation*—the X standing for "unknown."

In a modern X-ray tube, electrons are accelerated through a large electrical voltage toward a metal target (● Fig. 9.16a). When the electrons strike the target, they interact with the electrons in the target material, and the electrical repulsion decelerates the incident electrons. The result is an emission of high-frequency X-ray photons (quanta). In keeping with their mode of production, X-rays are called *Bremsstrahlung* ("braking rays") in German.

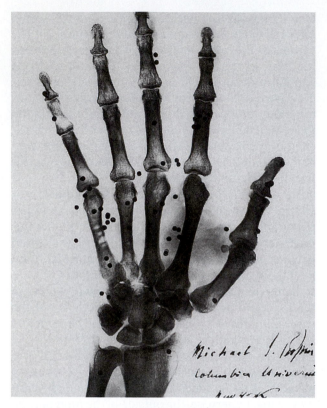

FIGURE 9.15 X-Rays Quickly Found Practical Use

Roentgen discovered X-rays in December 1895. By February 1896, they were being put to practical use. X-rays can penetrate flesh relatively easily and leave a skeletal image on film. The black spots in the "X-ray" are bird shot embedded in the subject's hand from a hunting accident.

An X-ray spectrum is illustrated in Fig. 9.16b. Notice that there is a cutoff wavelength λ_o below which no X-rays are emitted for a given tube voltage. Also, for very large tube voltages, intense "spikes," or spectral lines, are found. These spikes are characteristic of the target material and are called *characteristic X-rays*. Both features of the X-ray spectrum can be explained by quantum theory.

The low cutoff wavelength corresponds to the quantum of highest frequency or energy (both frequency and energy are inversely proportional to the wavelength; $E = hf = hc/\lambda$). Such quanta result when an incident electron is stopped completely and gives up all its energy. There can be no quanta of greater frequency (lower wavelength), because the maximum energy has been given up. This explanation is supported by the fact that the cutoff wavelength can be made smaller by increasing the tube voltage so as to give the incident electrons even more energy.

What about the explanation for the characteristic spectral lines? An atom of the target material contains many electrons. The electrons in the lower energy levels near the nucleus are shielded from the incident electrons by the electrons in the higher levels, so the incident electrons generally do not interact with the inner electrons. At large tube voltages, however, the incident electrons have sufficient energy to occasionally eject an electron from an inner orbit. This ejected electron leaves a vacancy that can be filled by an electron from a nearby outer orbit, and this, in turn, leaves a vacancy in that orbit. The transitions as electrons jump down to fill these vacancies give rise to the characteristic spectral lines, similar to

FIGURE 9.16 X-Ray Production and Spectrum

(a) X-rays are produced in a tube in which electrons from the cathode are accelerated toward the anode. On interacting with the atoms of the anode material, the electrons are slowed down, and the atoms emit energy in the form of X-rays. (b) An X-ray spectrum. (See text for description.)

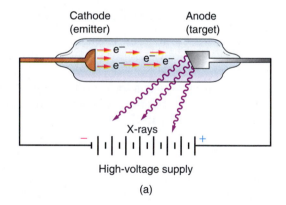

(a)

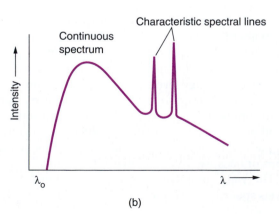

(b)

the emission lines produced by transitions from higher to lower levels in the hydrogen atom.

Lasers

Another device based on energy levels is the laser, the development of which was a great success for modern science. Scientific discoveries, such as X-rays, often have been made accidentally. X-rays were put to practical use before anyone understood the how or why of the phenomenon. Similarly, early investigators often applied a trial-and-error approach until they found something that worked; Edison's invention of the incandescent light is a good example of this approach. In contrast to X-rays and the light bulb, the idea of the laser was first developed "on paper" from theory around 1965 and then built with the full expectation that it would work as predicted.

The word **laser** is an acronym for **l**ight **a**mplification by **s**timulated **e**mission of **r**adiation. The amplification of light provides an intense beam. Ordinarily, when an electron in an atom is excited by a photon, it then emits a photon and returns to its ground state immediately. In this process, called *spontaneous emission,* one photon goes in and one photon comes out (● Fig. 9.17a and b).

However, some substances, such as ruby crystals and carbon dioxide gas, or combinations of substances, such as a mixture of the gases helium and neon, have *metastable* excited states—that is, some of their electrons can jump up into these excited energy levels and remain there briefly.

When many of the atoms or molecules of a substance have been excited into a metastable state by the input of the appropriate energy, we say a *population inversion* has occurred (● Fig. 9.18a). In such a condition, an excited atom can be stimulated to emit a photon (Figs. 9.17c and 9.18b). In a **stimulated emission** process—the key process of a laser—an excited atom is struck by a photon of the same energy as the allowed transition, and two photons are emitted (one in, two out—amplification). Of course, in this process we do not get something for nothing; energy was needed to excite the atom initially.

The light intensity is amplified due to the emitted photon being in phase with the stimulating photon and thus interfering constructively to produce maximum intensity (Section 7.4). Each of the two photons can then stimulate the emission of yet another identical photon. The result of many stimulated emissions

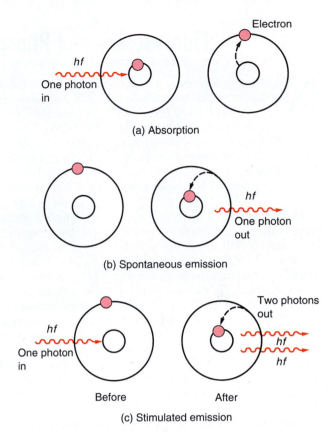

FIGURE 9.17 Spontaneous and Stimulated Emissions
(a) An atom absorbs a photon and becomes excited. (b) The excited atom may return spontaneously to its ground state almost instantaneously with the emission of a photon, or (c) if the excited atom is struck by a photon with the same energy as the initially absorbed photon, the atom is stimulated to emit a photon, and both photons leave the atom.

and reflections in a laser tube is a narrow, intense beam of laser light, due to the beam consisting of photons having the same energy and wavelength (it's *monochromatic*) and traveling in phase in the same direction (it's *coherent;* Fig. 9.18c).

Light from sources such as an incandescent bulb consists of all wavelengths and is *incoherent* because the excitation occurs randomly, the atoms emit randomly at different wavelengths, and the waves have no particular directional or phase relationships to one another.

Because a laser beam is so directional, it spreads very little as it travels. This feature has permitted us to reflect a laser beam back to Earth from a mirror placed on the Moon by astronauts, enabling accurate measurements of the distance to the Moon so that

HIGHLIGHT: Fluorescence and Phosphorescence

Now that energy levels, electron transitions, and the photon nature of light have been discussed, you can understand what the terms *fluorescence* and *phosphorescence* mean. Fluorescence is easily associated with those long, white lamp tubes that probably light your classroom.

Fluorescent lights are much more efficient than incandescent light bulbs, which radiate primarily in the nonvisible, infrared (IR) region and give off a lot of heat. The primary radiation emitted by electrically excited mercury atoms in a fluorescent tube is in the nonvisible, ultraviolet (UV) region. The UV radiation is absorbed by the white fluorescent material that coats the inside of the tube. This material reradiates at frequencies in the visible region, giving visible light for reading and other purposes.

Why does it reradiate in the visible region? During **fluorescence,** some of the electrons that have absorbed photons and been excited return to the original state by two or more steps (transitions), like a ball bouncing down a flight of stairs. Because each downward transition is smaller than the initial upward one, the emitted photons have lower energies and frequencies than the exciting photon, so some downward transitions emit photons in the visible region.

Other uses of fluorescence include the "black lights" used in discos and light displays. These tubes emit radiations in the violet (visible region) and ultraviolet (nonvisible region). Fluorescent materials in paints and dyes cause walls and clothes to glow and stand out in the visible region.

In fluorescent lamps, the conversion of the ultraviolet radiation to visible light is not complete, and the remnant of UV radiation can be used to commercial advantage. In grocery stores, which are generally illuminated with economical fluorescent lights, products such as laundry detergents appear to have extra-bright-colored boxes. Manufacturers hoping to influence purchase selection use fluorescent inks in the box coloring and printing so as to make the packaging appear brighter, or "stand out," under the fluorescent lights.

Some butterflies and other organisms manufacture fluorescent pigments that emit visible radiations when excited by the UV light in sunlight. A few minerals have fluorescent properties that are a help in their location and identification (Fig.1).

Now, what about **phosphorescence**? Phosphorescent materials are used in luminous TV remotes, key rings, toys, and so on—things that "glow in the dark". When exposed to light, the electrons of a phosphorescent material are excited to higher energy levels, and some go to metastable (temporarily stable) states. Electrons in some of

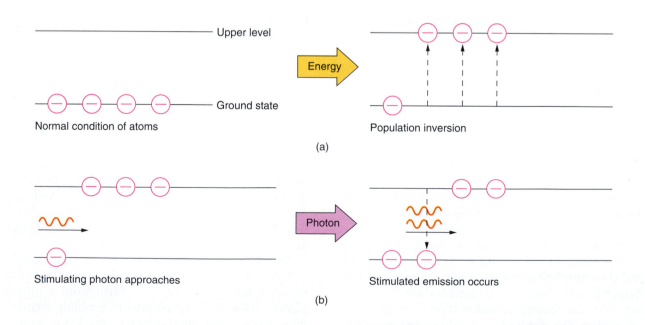

Upper level

Energy

Ground state

Normal condition of atoms

Population inversion

(a)

Photon

Stimulating photon approaches

Stimulated emission occurs

(b)

these metastable states can stay there several seconds, minutes, and even hours. As a result, photon-emitting transitions occur over relatively long periods of time, and the object phosphoresces (glows) long after the source of the photons needed for excitation—perhaps a light bulb—has been removed.

(a)

FIGURE 1   Mineral Fluorescence

(a) A mineral specimen containing willemite (Zn_2SiO_4) and calcite ($CaCO_3$), which are black and white, respectively, under visible light. (b) The same specimen in ultraviolet light. The specimen is from Franklin, New Jersey, which proclaims itself to be "the fluorescent mineral capital of the world."

(b)

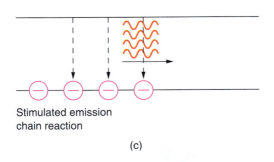

Stimulated emission
chain reaction

(c)

FIGURE 9.18   Steps in the Action of a Laser

(a) Atoms absorb energy and move to a higher energy level (a population inversion). (b) A photon approaches, and stimulated emission occurs. (c) The photons emitted cause other stimulated emissions, in a chain reaction. The photons are of the same wavelength, in phase, and moving in the same direction.

small fluctuations in the Moon's orbit can be studied. Similar measurements of light reflected from Moon mirrors help in determining the rate of plate tectonic movement on Earth (Section 22.2).

Laser light is used in an increasing number of applications. For instance, long-distance communications use laser beams in space and in optical fibers for telephone conversations. Lasers are used in medicine as diagnostic and surgical tools (● Fig. 9.19). In industry, the intense heat produced by focused laser light incident on a small area can drill tiny holes in metals and can weld machine parts. Laser "scissors" cut cloth in the garment industry. Laser printers produce computer printouts. Other applications occur in surveying, weapons systems, chemical processing, photography, and holography (the process of making three-dimensional images). Another common application is found at supermarket checkout counters,

FIGURE 9.19   Eye Surgery by Laser
A laser beam can be used to "weld" a detached retina into its proper place. In other surgical operations, a laser beam can serve as a scalpel, and the immediate cauterization prevents excessive bleeding.

where you have noticed a reddish glow produced by a helium-neon laser in an optical scanner used for reading the product codes on items. In research, laser light is increasingly being used to determine the details of chemical reactions.

You may own a laser yourself—a compact disc (CD) player. A laser "needle" is used to read the information (sound) stored on the disc in small dot patterns. The dots produce reflection patterns that are read by photocells and converted to electronic signals, which are changed to sound waves by the speaker system.

RELEVANCE QUESTION: *What instrument or appliance based on quantum phenomena has been of use to you recently?*

9.5 Heisenberg's Uncertainty Principle

LEARNING GOAL

▼ Explain the significance of Heisenberg's uncertainty principle.

Let's examine another important aspect of quantum mechanics. According to classical mechanics, there is no limit to the accuracy of a measurement. Theoretically, accuracy can be continually improved by refine-

ment of the measurement instrument or procedure to the point at which the measurement contains no uncertainty. This notion resulted in a deterministic view of nature. For example, if you know or measure at a particular time the position and velocity of a particle exactly, you can determine where it will be in the future and where it was in the past (assuming no future or past unknown forces).

However, quantum theory predicts otherwise and sets limits on measurement accuracy. This idea, developed in 1927 by the German physicist Werner Heisenberg, is called **Heisenberg's uncertainty principle:**

It is impossible to know a particle's exact position and velocity simultaneously.

This concept is often illustrated with a simple example. Suppose you want to measure the position and velocity of an electron, as illustrated in ● Fig. 9.20. If you are to see the electron and determine its location, at least one photon must bounce off the electron and come to your eye. In the collision process, some of the photon's energy and momentum are transferred to the electron. (This situation is analogous to a classical collision of billiard balls, which involves a transfer of momentum and energy.)

At the moment of collision, the electron recoils. The very act of measuring the electron's position has greatly altered its velocity. Hence the process of trying

FIGURE 9.20   Uncertainty
Imagine trying to determine accurately the location of an electron with a single photon, which must strike the electron and come to the detector. The electron recoils at the moment of collision, which introduces a great degree of uncertainty in knowing the electron's velocity or momentum.

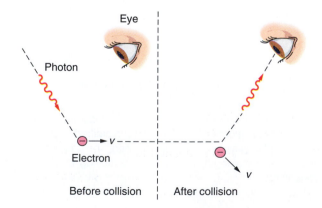

to locate the position accurately causes an uncertainty in knowing the electron's velocity.

Further investigation led to the conclusion that when the mass (m) of the particle, the minimum uncertainty in velocity (Δv), and the minimum uncertainty in position (Δx) are multiplied, a value on the order of Planck's constant ($h = 6.63 \times 10^{-34}$ J-s) is obtained; that is,

$$m(\Delta v)(\Delta x) \approx h$$

The bottom line of Heisenberg's uncertainty principle is that there *is* a limit on measurement accuracy that is philosophically significant, but it is of practical importance only when dealing with particles of atomic and subatomic size. As long as the mass is relatively large, Δv and Δx will be very small.

RELEVANCE QUESTION: *Would Heisenberg's uncertainty principle cause police officers a problem when using radar to determine a car's speed?*

9.6 Matter Waves

LEARNING GOAL

▼ Explain the term *dual nature of matter.*

As the concept of the dual nature of light developed, what was thought to be a wave was sometimes found to act as a particle. Can the reverse be true? In other words, can particles have a wave nature? This question was considered by the French physicist Louis de Broglie ("dee-BROY"), who in 1925 postulated that matter, as well as light, has properties of both waves and particles.

According to de Broglie's hypothesis, any moving particle has a wave associated with it whose wavelength is given by

$$\lambda = \frac{h}{mv} \qquad \text{9.5}$$

where λ = wavelength of the moving particle
    m = mass of the moving particle
    v = speed of the moving particle
    h = Planck's constant (6.63×10^{-34} J-s)

The waves associated with moving particles are called **matter waves,** or **de Broglie waves.**

Notice in Eq. 9.5 that the wavelength (λ) of a matter wave is inversely proportional to the mass of the particle or object—that is, the smaller the mass, the larger (longer) the wavelength. Thus the longest wavelengths are generally for particles with little mass. (Speed is also a factor, but particle masses have more effect because they can vary over a much larger range than speeds can.) However, Planck's constant is such a small number (6.63×10^{-34} J-s) that any wavelengths of matter waves are quite small. Let's use an example to see how small.

EXAMPLE 9.5

Finding the de Broglie Wavelength

Find the de Broglie wavelength for an electron ($m = 9.11 \times 10^{-31}$ kg) moving at 7.30×10^5 m/s.

SOLUTION

We are given the mass and speed, so the wavelength can be found by using Eq. 9.5.

$$
\begin{aligned}
\lambda &= \frac{h}{mv} \\[6pt]
&= \frac{6.63 \times 10^{-34} \text{ J-s}}{(9.11 \times 10^{-31}\text{ kg})(7.30 \times 10^5 \text{ m/s})} \\[6pt]
&= \frac{6.63 \times 10^{-34} \text{ kg-m}^2\text{-s/s}^2}{(9.11 \times 10^{-31}\text{ kg})(7.30 \times 10^5 \text{ m/s})} \\[6pt]
&= 1.0 \times 10^{-9} \text{ m} = 1.0 \text{ nm}
\end{aligned}
$$

This wavelength is several times larger than the diameter of the average atom, so although small, it is certainly significant as far as the size of an electron goes.

CONFIDENCE EXERCISE 9.5

Find the de Broglie wavelength for a 1000-kg car traveling at 25 m/s (about 56 mi/h).

Your answer to Confidence Exercise 9.5 should confirm that a moving, relatively massive object has a short wavelength. The 1000-kg car traveling at 56 mi/h has a wavelength of only about 10^{-38} m, which is certainly not significant relative to the size of the car. Because a wave is generally detected by its interaction with an object of about the same size as the wavelength, we can see why matter waves of common moving objects are not noticed.

De Broglie's hypothesis was met with skepticism at first, but it was verified experimentally in 1927 by G. Davisson and L. H. Germer in the United States. They showed that a beam of electrons exhibits a diffraction pattern. Because diffraction is a wave phenomenon, a beam of electrons must have wavelike properties.

For appreciable diffraction to occur, a wave must pass through a slit with a width approximately the size of the wavelength (Section 7.4). Visible light has wavelengths from about 400 nm to 700 nm, and slits with widths of these sizes can be made quite easily. However, as Example 9.5 showed, a fast-moving electron has a wavelength of about 1 nm. Slits of this width cannot be manufactured.

Fortunately, nature has provided us with suitably small slits in the form of crystal lattices. The atoms in these crystals are arranged in rows (or some other orderly arrangement), and the rows make natural "slits." Davisson and Germer bombarded nickel (Ni) crystals with electrons and obtained a diffraction pattern on a photographic plate. A diffraction pattern made by X-rays (electromagnetic radiation) and one made by an electron beam incident on a thin aluminum (Al) foil are shown in ● Fig. 9.21. The similarity in the diffraction patterns from the electromagnetic *waves* and from the electron *particles* is evident.

Electron diffraction demonstrates the **dual nature of matter**—namely, that moving matter has not only particle characteristics but also wave characteristics. But remember, the wave nature of ordinary-sized moving particles is too small to be measurable. The wave nature of matter becomes of practical importance only with small particles such as electrons and atoms.

An *electron microscope* is based on the theory of matter waves. This instrument uses a beam of electrons, rather than a beam of light, to view an object. The amount of fuzziness of an image is directly proportional to the wavelength that is used to view it. A typical beam wavelength in an electron microscope is 1 nm. This length is quite short compared with the wavelength of visible light (400 to 700 nm). Hence finer detail, as well as greater magnification, can be achieved using an electron microscope than can be obtained with an ordinary microscope. Results obtained with the type of electron microscope called a *scanning tunneling microscope* (STM) are quite remarkable. Some electron micrographs are shown in ● Fig. 9.22.

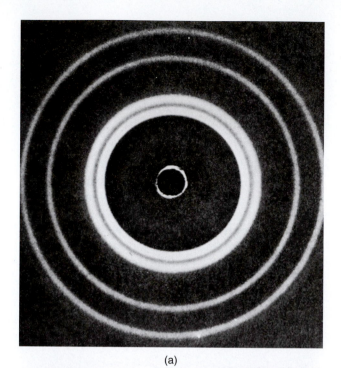

(a)

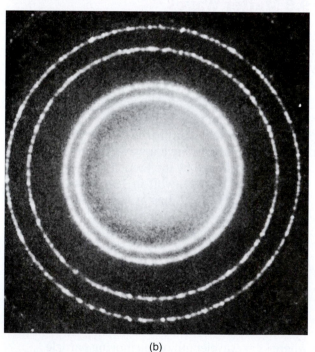

(b)

FIGURE 9.21 Diffraction Patterns

(a) The diffraction pattern produced by X-rays can be explained using a wave model of the X-rays. (b) The appearance of the diffraction pattern of electrons shows that electrons have a wave nature, which can be explained using de Broglie's concept of matter waves.

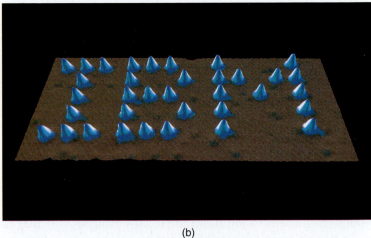

(a)   (b)

FIGURE 9.22 Electron Micrographs

(a) A dust mite is shown magnified by a factor of 150 in this scanning electron micrograph. Dust mites eat fragments of old human skin and are one source of allergens in the home. (b) A scanning tunneling microscope not only can image individual xenon atoms (the "cones") but also can move them around on a nickel surface. This work was done over 22 hours by scientists at IBM, at a temperature close to absolute zero so that the atoms would "stay put" after being placed in position.

9.7 The Quantum Mechanical Model of the Atom

LEARNING GOAL

▼ Describe the quantum mechanical model of the atom.

Bohr chose to analyze the hydrogen atom for an obvious reason—it's the simplest atom. It is increasingly difficult to analyze atoms with two or more electrons (*multielectron atoms*) and to determine their electron energy levels. The difficulty arises because, in multielectron atoms, more electrical interactions exist than in the hydrogen atom. Forces exist among the various electrons, and in large atoms, electrons in outer orbits are partially shielded from the attractive force of the nucleus by electrons in inner orbits. Bohr's theory, so successful for the hydrogen atom, did not give the correct results when applied to multielectron atoms.

Also, recall from the Section 9.3 discussion that Bohr deduced from the small number of lines in line emission spectra that electron energy levels are quantized, but he was unable to say *why* they are quantized. Bohr also stated that although classical physics says that the electron should radiate energy as it travels in its orbit, it just does not do that, but here also he

could offer no explanation. So a better model of the atom was needed.

As a result of the discovery of the dual natures of waves and particles, a new kind of physics—**quantum mechanics** or *wave mechanics*—based on the synthesis of wave and quantum ideas was born in the 1920s and 1930s. In accordance with Heisenberg's uncertainty principle, its concept of *probability* replaced the classical mechanics view that everything moves according to *exact* laws of nature.

De Broglie's hypothesis showed that waves are associated with moving particles and somehow govern or describe the particle behavior. In 1926, Erwin Schrödinger (photo on page 227), an Austrian physicist, presented a widely applicable mathematical equation that gave new meaning to de Broglie's matter waves. Schrödinger's equation is basically a formulation of the conservation of energy. The detailed form of the equation is quite complex, but it is written in its simple form as

$$(E_k + E_p)\Psi = E\Psi$$

where E_k, E_p, and E are the kinetic energy, potential energy, and total energy, respectively, and Ψ is the wave function.

Schrödinger's *quantum mechanical model* (or *electron cloud model*) of the atom focuses on the wave nature of the electron and treats it as a spread-out

FIGURE 9.23 The Electron as a Standing Wave

The hydrogen electron can be treated as a standing wave in a circular orbit around the nucleus. In order for the wave to be stable, however, the circumference must accommodate a *whole number* of wavelengths, as shown in (a) and (b). In (c), the wave would destructively interfere with itself, so this orbit is forbidden. This restriction on the orbits, or energies, explains why the atom is quantized. (From Zumdahl, Steven S., *Chemistry*, Fourth Edition. Copyright © 1999 by Houghton Mifflin Company. Used with permission.)

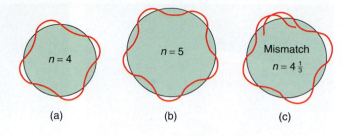

(a)   (b)   (c)

wave, with its energy levels being a consequence of the wave having to have a *whole number* of wavelengths to form standing waves (Section 6.5) in orbits around the nucleus (● Fig. 9.23a and b). Any orbit between two adjacent permissible orbits would require a fractional number of wavelengths and would not produce a standing wave (Fig. 9.23c). This requirement of a whole number of wavelengths explains the quantization that Bohr had to just assume. Furthermore, standing waves do not move from one place to another, so an electron in a standing wave is not accelerating and would not have to radiate light, which explains why Bohr's second assumption was correct.

In Schrödinger's equation, the symbol Ψ (Greek letter psi) is called the *wave function* and mathematically represents the wave associated with a particle. At first, scientists were not sure how Ψ should be inter-

preted. For the hydrogen atom, they concluded that Ψ^2 (the wave function squared), when multiplied by the square of the radius r, represents the *probability* that the hydrogen electron will be at a certain distance r from the nucleus. (In Bohr's theory the electron can only be in circular orbits with discrete radii given by $r = 0.053n^2$ nm.)

A plot of $r^2\Psi^2$ versus r for the hydrogen electron (● Fig. 9.24a) shows that the most probable radius for the hydrogen electron is one with $r = 0.053$ nm, which is the same value Bohr calculated in 1913 for the ground state orbit of the hydrogen atom. (In fact, all the energy levels for the hydrogen atom were found to be exactly the same as those Bohr had calculated.) The electron might be found at other radii, but with less likelihood—that is, with lower probability. This idea gave rise to the model of an *electron cloud*

FIGURE 9.24 $r^2\Psi^2$ Probability

(a) The square of the wave function (Ψ^2) multiplied by the square of the radius (r^2) gives the probability of finding the electron at that particular radius. As shown here, the radius of a hydrogen atom with the greatest probability of containing the electron is 0.053 nm, which corresponds to the first Bohr radius. (b) The probability of finding the electron at other radii gives rise to the concept of an "electron cloud," or probability distribution.

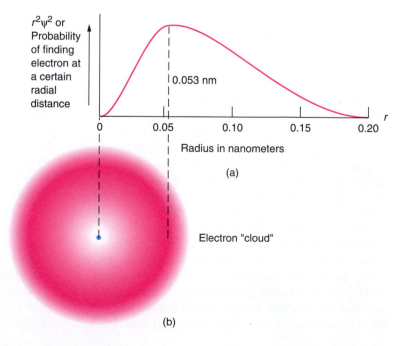

A SPOTLIGHT ON: The Changing Model of the Atom

Dalton, 1807 **(billiard ball model)**

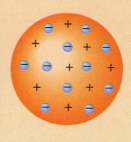

Thompson, 1903 **(plum pudding model)**

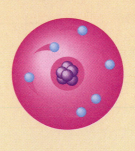

Rutherford, 1911 **(nuclear model)**

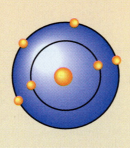

Bohr, 1913 **(planetary model)**

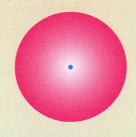

Schrödinger, 1926 **(electron cloud model)**

around the nucleus, with the cloud density reflecting the probability that the electron could be in that region (Fig. 9.24b).

Thus Bohr's simple planetary model has been replaced, for many purposes, by a more sophisticated, highly mathematical model that treats the electron as a wave and can explain more data and predict more accurately. The quantum mechanical model of the atom is harder to visualize than the Bohr model. The location of a specific electron becomes more vague than in the Bohr model and can only be given in terms of probability. However, the important point is that the quantum mechanical model allows the energy of the electrons in multielectron atoms to be determined accurately. For scientists, knowing the electron's energy is much more important than knowing its precise location.

The Spotlight feature on the previous page summarizes how our model of the atom has changed over the past 200 years.

Important Terms

atoms (9.1)	line emission	excited states	Heisenberg's uncertainty
electrons	spectrum (9.3)	fluorescence	principle (9.5)
quantum (9.2)	line absorption spectrum	phosphorescence	matter (de Broglie)
photoelectric effect	principal quantum	X-rays (9.4)	waves (9.6)
photon	number	laser	dual nature of matter
dual nature of light	ground state	stimulated emission	quantum mechanics (9.7)

Important Equations

Photon Energy: $E = hf$ $(h = 6.63 \times 10^{-34}$ J-s)

Hydrogen Electron Orbit Radii:

$r_n = 0.053 \, n^2$ nm $(n = 1, 2, 3, \ldots)$

Hydrogen Electron Energy:

$E_n = \dfrac{-13.60}{n^2}$ eV $(n = 1, 2, 3, \ldots)$

Photon Energy for Transition: $E_{\text{photon}} = E_{n_i} - E_{n_f}$

de Broglie Wavelength: $\lambda = \dfrac{h}{mv}$

Review Questions

9.1 Early Concepts of the Atom

1. About 400 B.C., which of the following persons championed the idea of the atom?
   (a) Aristotle (c) Democritus
   (b) Plato (d) Archimedes

2. Which of the following persons is associated with the "plum pudding model" of the atom?
   (a) Thomson (c) Bohr
   (b) Rutherford (d) Dalton

3. What main feature of Dalton's model of the atom was abandoned after Thomson's discovery?

4. How did Thomson know that electrons are negatively charged?

5. What main feature of Thomson's model of the atom was abandoned after Rutherford's discovery?

9.2 The Dual Nature of Light (and Highlight)

6. Planck developed his quantum hypothesis to explain which of these phenomena?
   (a) the ultraviolet catastrophe (c) line spectra
   (b) the photoelectric effect (d) uncertainty

7. Light of which of the following colors has the greatest photon energy?
   (a) red (b) orange (c) yellow (d) violet

8. What proof can be offered that light has a wave nature? A particle nature?

9. What do we mean when we say that something is *quantized*?

10. Distinguish between a proton and a photon.

11. How are the frequency and wavelength of an electromagnetic wave related?

12. Explain the difference between a photon of yellow light and one of green light in terms of energy, frequency, and wavelength.

13. Light shining on the surface of a photomaterial causes the ejection of electrons if the frequency of the light is above a certain minimum value. Why is there a certain minimum value?

14. What scientist won the Nobel Prize for explaining the photoelectric effect? Name another theory for which that scientist is famous.

9.3 Bohr Theory of the Hydrogen Atom (and Highlight)

15. The Bohr theory was developed to explain which of these phenomena?
    (a) energy levels (c) line spectra
    (b) the photoelectric effect (d) quantum numbers

16. In which of the following states does a hydrogen electron have the greatest energy?
    (a) $n = 1$ (b) $n = 3$ (c) $n = 5$ (d) $n = 7.5$

17. According to the Bohr theory, what is the approximate *diameter* in nm of the hydrogen atom in the ground state?

18. How did Bohr address the problem that, according to the classical approach, an orbiting electron should emit radiation?

19. Distinguish between a *ground state* for an electron and an *excited state*.

20. How many visible lines make up the emission spectrum of hydrogen? What are their colors?

21. How does the Bohr theory explain the discrete lines in the *emission* spectrum of hydrogen?

22. How does the Bohr theory explain the discrete lines in the *absorption* spectrum of hydrogen?

23. In the Bohr model for hydrogen, as n increases, the energy of a principal energy level _____ (increases/decreases) and the distance of the electron from the nucleus _____ (increases/decreases).

24. What is the difference between fluorescence and phosphorescence—in terms of both what is observed and why?

9.4 Microwave Ovens, X-Rays, and Lasers

25. Bombardment of a metal anode with high-energy electrons produces which of the following?
    (a) laser light (c) microwaves
    (b) X-rays (d) neutrons

26. The amplification of light in a laser depends on which of the following?
    (a) the photoelectric effect (c) spontaneous emission
    (b) microwave absorption (d) stimulated emission

27. Why does a microwave oven heat a potato but not a ceramic plate?

28. What does the acronym *laser* stand for?

29. What is unique about light from a laser source?

30. Why should you never look directly into a laser beam or its reflection?

31. Why are X-rays called "braking rays"?

32. Why does increasing the tube voltage shift the cutoff wavelength of an X-ray spectrum to a shorter wavelength?

9.5 Heisenberg's Uncertainty Principle

33. Limitations are put on measurements by which of the following?
    (a) Schrödinger's equation
    (b) de Broglie's hypothesis
    (c) Heisenberg's principle
    (d) special theory of relativity

34. According to the uncertainty principle, it is impossible to determine exactly and simultaneously a particle's
    (a) charge and mass.
    (b) position and velocity.
    (c) charge and position.
    (d) velocity and momentum.

35. How does the Heisenberg uncertainty principle change the classical deterministic view of the universe?

36. Why are the consequences of the uncertainty principle significant only for particles of atomic and subatomic size?

9.6 Matter Waves

37. Matter waves were first hypothesized by which of the following persons?
    (a) Schrödinger (c) Heisenberg
    (b) de Broglie (d) Einstein

38. What is a matter wave, and when is the associated wavelength significant?

39. How was a beam of electrons shown to have wavelike properties?

40. What useful instrument makes use of the wavelike properties of electrons?

9.7 The Quantum Mechanical Model of the Atom

41. Why did the Bohr model need improvement?
    (a) It only worked for the hydrogen atom.
    (b) It did not explain why the atom is quantized.
    (c) It did not explain why an electron does not emit radiation as it orbits.
    (d) All these answers are correct.

42. In the quantum mechanical model of the atom, the electron is treated as a _____.

43. What scientist is primarily associated with the quantum mechanical model of the atom?

44. The quantum mechanical model of the atom is also called the _____ model.

45. In the quantum mechanical model of the atom, the electron's "location" is stated only in terms of _____.

46. It is more important to know an electron's _____ than to know its location.

Applying Your Knowledge

1. Why are microwave ovens constructed so they will *not* operate when the door is open?

2. Photographers usually work in darkrooms lit by dim red lights. Why red lights and not, say, blue lights?

3. Color television tubes are shielded to protect viewers from X-rays. How could a TV set produce X-rays?

4. Leading up to an elevated loading dock at a warehouse are a set of steps and a ramp. You see a black cat resting on the ramp, while an orange cat sleeps on the steps. Immediately, the analogy between the classical contin-uous view of energy values and the modern view comes to your mind. Explain.

5. A friend tells you of seeing beautiful colors brought out in a picture when "black light" was shined on it. Which region of the spectrum is sometimes called "black light"? How would you explain to your friend what was happening at the atomic level to make the picture glow?

6. Your friend says that atoms do not exist because no one has ever seen one. What would be your reply?

Exercises

9.2 The Dual Nature of Light

1. The human eye is most sensitive to yellow-green light having a frequency of about 5.45×10^{14} Hz (a wavelength of about 550 nm). What is the energy in joules of the photons associated with this light?
   Answer: 3.61×10^{-19} J

2. Light having a frequency of about 5.00×10^{14} Hz (a wavelength of about 600 nm) appears orange to our eyes. What is the energy in joules of the photons associated with this light?

3. Photons of a certain ultraviolet light have an energy of 6.63×10^{-19} J. (a) What is the frequency of this UV light? (b) Use $\lambda = c/f$ to calculate its wavelength in nm.
   Answer: (a) 1.00×10^{15} Hz   (b) 300 nm

4. Photons of a certain infrared light have an energy of 1.66×10^{-19} J. (a) What is the frequency of this IR light? (b) Use $\lambda = c/f$ to calculate its wavelength in nm.

9.3 Bohr Theory of the Hydrogen Atom

5. What is the radius in nm of the electron orbit of a hydrogen atom for $n = 3$?
   Answer: 0.48 nm

6. What is the radius in nm of the electron orbit of a hydrogen atom for $n = 4$?

7. What is the energy in eV of the electron of a hydrogen atom for the orbit designated $n = 3$?   Answer: -1.51 eV

8. What is the energy in eV of the electron of a hydrogen atom for the orbit designated $n = 4$?

9. Use Table 9.1 to determine the energy in eV of the photon emitted when an electron jumps down from the $n = 4$ orbit to the $n = 2$ orbit of a hydrogen atom.
   Answer: 2.55 eV

10. Use Table 9.1 to determine the energy in eV of the photon absorbed when an electron jumps up from the $n = 1$ orbit to the $n = 4$ orbit of a hydrogen atom.

9.6 Matter Waves

11. Calculate the de Broglie wavelength of a 0.50-kg ball moving with a constant velocity of 26 m/s (about 60 mi/h).
    Answer: 5.1×10^{-35} m

12. What is the de Broglie wavelength for Earth ($m = 6.0 \times 10^{24}$ kg) as it moves in its orbit with a speed of 3.0×10^4 m/s?

Solutions to Confidence Exercises

9.1 $E = hf = (6.63 \times 10^{-34}\ \text{J-s})(7.50 \times 10^{14}\ 1/\text{s})$
$= 49.7 \times 10^{-20}\ \text{J}$

9.2 $r_n = 0.053\ n^2$ nm, or $r_2 = 0.053\ (2)^2$ nm $= 0.21$ nm

9.3 $E_2 = \dfrac{-13.60}{2^2}\ \text{eV} = \dfrac{-13.60}{4}\ \text{eV} = -3.40\ \text{eV}$

9.4 Table 9.1 shows values of -13.60 eV for the $n = 1$ level and -1.51 eV for $n = 3$. Thus
$E_{\text{photon}} = E_1 - E_3 = -13.60\ \text{eV} - (-1.51\ \text{eV})$
$= -12.09\ \text{eV}$ (the negative value indicating the photon is absorbed).

9.5 $\lambda = \dfrac{h}{mv} = \dfrac{6.63 \times 10^{-34}\ \text{J-s}}{(10^3\ \text{kg})(25\ \text{m/s})} = 2.7 \times 10^{-38}\ \text{m}$

Answers to Multiple-Choice Review Questions

1. c   6. a   15. c   25. b   33. c   37. b
2. a   7. d   16. c   26. d   34. b   41. d

NUCLEAR PHYSICS

10

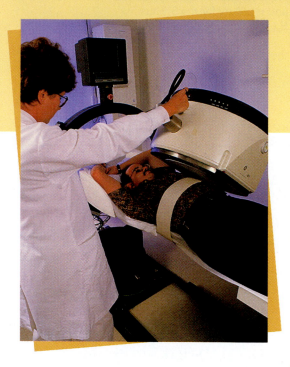

It is no good to try to stop knowledge from going forward. Ignorance is never better than knowledge.

Enrico Fermi (1901–1954)

CHAPTER OUTLINE

The atomic nucleus and its properties have an important impact on our society. The nucleus is involved with dating archaeological objects, treating and diagnosing cancer and other diseases, chemical analysis, radiation damage and nuclear bombs, the generation of electricity by nuclear energy and the subsequent disposal of nuclear waste, the formation of new elements, the shining of the Sun and other stars, and even the operation of the common household smoke detector. This chapter discusses these topics and includes Highlights on how radioactivity was discovered and how the atomic bomb was built.

An appropriate way in which to start our examination of nuclear physics is with a brief history of how the concept of *element* arose and how elements are symbolized. ■

Photo: Radioactive thallium-201 is used to diagnose heart disease.

10.1 **Symbols of the Elements**

LEARNING GOALS

▼ Name the elements of the ancient Greek philosophers.

▼ Be familiar with the names and symbols for 45 elements.

The Greek philosophers who lived during the period from about 600 to 200 B.C. were apparently the first people to speculate about what basic substance or substances make up matter. In the fourth century B.C., the Greek philosopher Aristotle developed the idea that all matter on Earth is composed of four "elements": earth, air, fire, and water. He was wrong on all four, and in Chapter 11 we will discuss the discovery and properties of true elements. In this chapter we will often refer to these true elements, so we must take a moment to discuss how elements are symbolized.

To designate the different elements, we use a symbol notation that was first conceived in the early 1800s by the Swedish chemist Jöns Jakob Berzelius ("bur-ZEE-lee-us"). He used one or two letters of the Latin name for each element. Thus sodium is designated Na for *natrium,* silver is Ag for *argentum,* and so forth (Table 10.1).

Since Berzelius' time, most elements have been symbolized by the first one or two letters of the English name. Examples are C for carbon, O for oxygen, and Ca for calcium. The first letter of a chemical symbol is always capitalized and the second is lowercase. Inside the front cover you will find a periodic table showing the positions, names, and symbols of the 112 known elements.

Although it is not expected that you learn the names and symbols of all 112 elements, you should become familiar with the names and symbols of the 45 elements listed in Table 10.2. (Making and using flashcards is an efficient and effective way to learn them.)

10.2 **The Atomic Nucleus**

LEARNING GOALS

▼ Describe the structure and composition of atoms.

▼ Calculate atomic masses of elements.

All matter encountered in day-to-day living is made up of atoms. An atom is composed of negatively charged particles, called **electrons,** that surround a positively charged nucleus. The **nucleus** is the central core of an atom. It consists of **protons,** which are positively charged, and **neutrons,** which are electrically neutral. An electron and a proton have the same magnitude of electrical charge, but the charge on the electron is designated to be negative, and that on the proton is said to be positive.

Protons and neutrons have almost the same mass and are about 2000 times more massive than an electron. Nuclear protons and neutrons are collectively called **nucleons.** Table 10.3 summarizes the basic properties of electrons, protons, and neutrons, all of which were discovered in England—the electron by J. J. Thomson in 1897 (Section 9.1; photo on page 227), the proton in 1918 by Ernest Rutherford (photo on page 227), and the neutron by James Chadwick in 1932.

Subsequent investigations have revealed that the electron seems to be a truly fundamental particle of matter. However, in the late 1960s, physicists using high-energy proton beams to probe nucleons found that they had three separate centers of charge within them. Additional theoretical and experimental work led to the conclusion that neutrons and protons are made up of combinations of two types of still smaller particles called *quarks* (● Fig. 10.1). Theoretically, six

TABLE 10.1   Chemical Symbols from Latin Names

Modern Name	Symbol	Latin Name
Antimony	Sb	*Stibium*
Copper	Cu	*Cuprum*
Gold	Au	*Aurum*
Iron	Fe	*Ferrum*
Lead	Pb	*Plumbum*
Mercury	Hg	*Hydrargyrum*
Potassium	K	*Kalium*
Silver	Ag	*Argentum*
Sodium	Na	*Natrium*
Tin	Sn	*Stannum*
Tungsten	W	*Wolfram*

TABLE 10.2   Names and Symbols of Common Elements

Name	Symbol	Name	Symbol	Name	Symbol
aluminum	Al	gold	Au	phosphorus	P
argon	Ar	helium	He	platinum	Pt
arsenic	As	hydrogen	H	plutonium	Pu
barium	Ba	iodine	I	potassium	K
beryllium	Be	iron	Fe	radium	Ra
bismuth	Bi	krypton	Kr	radon	Rn
boron	B	lead	Pb	silicon	Si
bromine	Br	lithium	Li	silver	Ag
calcium	Ca	magnesium	Mg	sodium	Na
carbon	C	manganese	Mn	strontium	Sr
cesium	Cs	mercury	Hg	sulfur	S
chlorine	Cl	neon	Ne	tin	Sn
chromium	Cr	nickel	Ni	uranium	U
copper	Cu	nitrogen	N	xenon	Xe
fluorine	F	oxygen	O	zinc	Zn

TABLE 10.3   Major Constituents of an Atom

Particle (symbol)	Charge (C)	Electronic Charge	Mass (kg)	Mass (u)	Location
Electron (e)	-1.60×10^{-19}	-1	9.109×10^{-31}	0.00055	Outside nucleus
Proton (p)	$+1.60 \times 10^{-19}$	$+1$	1.673×10^{-27}	1.00728	Nucleus
Neutron (n)	0	0	1.675×10^{-27}	1.00867	Nucleus

FIGURE 10.1   Quarks

(a) A proton is made of two "up" quarks and one "down" quark. The electrical charges of the three quarks add to +1. (b) A neutron is composed of one "up" quark and two "down" quarks. The charges of the three quarks add to zero.

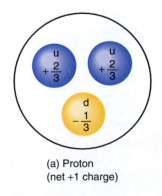

(a) Proton
(net +1 charge)

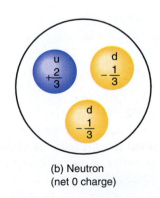

(b) Neutron
(net 0 charge)

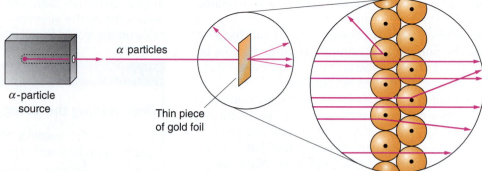

FIGURE 10.2   Rutherford's Apha-Scattering Experiment

Almost all the alpha particles striking the gold foil go right on through, but a few are deflected, and a very few bounce back.   These results led to the discovery of the nucleus.

types of quarks should exist, and five of them were found by 1984. The existence of the elusive sixth quark was verified experimentally in 1994.

That the atom consists of a nucleus surrounded by orbiting electrons was discovered in 1911, also by Rutherford. He was curious about what would happen when energetic alpha particles (helium nuclei) were allowed to bombard a very thin sheet of gold. J. J. Thomson's plum pudding model (Fig. 9.1b on page 205) predicted that the alpha particles would pass through the evenly distributed positive charges in the gold atoms with little or no deflection from their original paths.

Rutherford's experiment was conducted in a vacuum in an apparatus such as the one illustrated in ● Fig. 10.2. The behavior of the alpha particles could be determined by using a movable screen coated with zinc sulfide. When an alpha particle hit the screen, a small flash of light was emitted that could be observed with a low-power microscope. (A similar phenomenon causes TV screens to glow when hit by moving electrons.)

Rutherford found that by far the majority of the alpha particles went through the gold foil as if it were not even there. Relatively few of the positively charged alpha particles were deflected off course, however, and about 1 out of 20,000 actually bounced back. Rutherford could explain this behavior only by assuming that each gold atom had its positive charge concentrated in a small core of the atom, which he named the *nucleus.* He also assumed that electrons move around the nucleus like bees around a hive.

Rutherford's alpha-scattering experiment showed that a nucleus has a diameter of about 10^{-14} m (● Fig.

10.3). In contrast, the atom's outer electrons have orbits with diameters of about 10^{-10} m (see Section 9.3). Thus the diameter of an atom is approximately 10,000 times the diameter of its nucleus. Most of an atom's volume consists of empty space. The electrical repulsion between an atom's electrons and those of adjacent atoms keeps matter from collapsing. Electron orbits determine the size (volume) of atoms, but the nucleus contributes over 99.97% of the mass. If nuclei could be packed together into a sphere the size of a Ping-Pong ball, the ball would have a mass of about 2.5 *billion* metric tons. Such a large density can occur in neutron stars (Section 18.3), but no material on Earth is anywhere near so dense.

We designate the particles in an atom by certain numbers. The **atomic number,** symbolized by the letter Z, is the number of protons in the nucleus of each atom of that element. In fact, an **element** is defined as

FIGURE 10.3   A Representation of the Nucleus

The nucleus of an aluminum-27 atom consists of 13 protons (blue) and 14 neutrons (yellow), for a total of 27 nucleons. The diameter of this nucleus is 7.2×10^{-15} m, close to the 10^{-14} m diameter of an average nucleus.

$$\mapsto 7.2 \times 10^{-15}\,\text{m} \mapsto$$

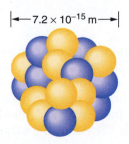

a substance in which all the atoms have the same number of protons (same atomic number). For an atom to be electrically neutral (have a total charge of zero), the number of electrons and protons must be the same. Therefore, the atomic number also represents the number of electrons in a neutral atom.

Electrons may be gained or lost by an atom, and the resulting particle (called an *ion*) will be electrically charged. However, because the number of protons has not changed, the particle is an ion of that same element. For example, if a *sodium atom* (Na) loses an electron, it becomes a *sodium ion* (Na$^+$), not an atom or ion of some other element.

The **neutron number** (N) is, of course, the number of neutrons in a nucleus. The **mass number** (A) is the number of protons plus neutrons in the nucleus; in other words, it's the total number of nucleons. Atoms of the same element can be different because of different numbers of neutrons in their nuclei. Forms of atoms that have the same number of protons (same Z, same element) but differ in their number of neutrons (different N, different A) are known as the **isotopes** of that element. *Isotope* literally means "same place," thus designating atoms that occupy the same place in the periodic table. Even though only 112 elements are known, the total of their isotopes is about 2000.

As shown below, the general designation for a specific nucleus places the mass number (A) to the upper left of the chemical symbol (the X, for generality). The atomic number (Z) goes at the lower left, as shown here.

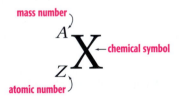

The number of neutrons (N) in a nucleus is easily determined by subtracting the atomic number (Z) from the mass number (A).

$$N = A - Z \qquad \textbf{10.1}$$

Thus, for example, it is common to refer to the uranium isotope $^{238}_{92}$U. Because it is a simple matter to obtain an element's atomic number from a periodic table, a specific isotope of an element also can be represented by just the chemical symbol and mass number (for example, ^{238}U) or by the name of the element followed by a hyphen and the mass number (for ex-

ample, uranium-238). The chemical symbols and names for all the elements are given in the periodic table on the inside front cover of this textbook.

EXAMPLE 10.1

Determining the Composition of an Atom

Determine the number of protons, electrons, and neutrons in the fluorine atom $^{19}_{9}$F.

SOLUTION

The atomic number Z is 9, so the number of protons is 9, as is the number of electrons. The mass number A is 19, so the number of neutrons $N = A - Z = 19 - 9 = 10$. The answer is 9 protons, 9 electrons, and 10 neutrons.

CONFIDENCE EXERCISE 10.1

Determine the number of protons, electrons, and neutrons in the carbon atom $^{14}_{6}$C, a radionuclide important to archaeologists and geologists (Chapter 24).

The isotopes of an element have the same chemical properties because they have the same number of electrons, but they differ somewhat in physical properties because they have different masses. ● Figure 10.4 shows the atomic composition of the three isotopes of hydrogen. They even have their own names: $^{1}_{1}$H is *protium* (or just *hydrogen*), $^{2}_{1}$H is *deuterium* (D), and $^{3}_{1}$H is *tritium* (T). The atomic nucleus in each case is referred to as a *proton, deuteron,* and *triton,* respectively; that is, a proton is the nucleus of a protium atom, and so on.

In a given sample of naturally occurring hydrogen, about one atom in 6000 is deuterium and about one atom in 10,000,000 is tritium. Protium and deuterium are stable atoms, whereas tritium is unstable (radioactive; Section 10.3). *Heavy water* (D_2O) consists of molecules made up of two atoms of deuterium and one atom of oxygen.

The Atomic Mass

Generally, each element occurs naturally as a combination of its isotopes (Table 10.4), and the mass and abundance of each isotope can be determined using

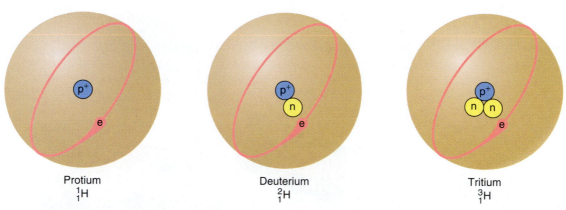

FIGURE 10.4   The Three Isotopes of Hydrogen

Each atom has one proton and one electron, but they differ in the number of neutrons in the nucleus. (*Note:* Not drawn to scale; the nucleus is shown much too large relative to the size of the whole atom.)

an instrument called a *mass spectrometer* (● Fig. 10.5). The weighted average mass (in *unified atomic mass units,* abbreviated u) of an atom of the element in a naturally occurring sample is called the **atomic mass** of the element and is given under its symbol in the periodic table.

All atomic masses are based on the ^{12}C atom, which is assigned a relative atomic mass of *exactly* 12 u. Naturally occurring carbon has an atomic mass slightly greater than 12.0000 u because it contains not only ^{12}C but also a little ^{13}C and a trace of ^{14}C. An isotope's mass number closely approximates its atomic mass (its actual mass in u), as you can see from Table 10.4 and Example 10.2.

TABLE 10.4   Common Isotopes of Some of the Lighter Elements

Isotope	Mass (u)	Percent Natural Abundance
^{1}H	1.0078	99.985
^{2}H	2.0140	0.015
^{6}Li	6.015	7.42
^{7}Li	7.016	92.58
^{9}Be	9.010	100.
^{10}B	10.013	19.7
^{11}B	11.009	80.3
^{12}C	12.000	98.89
^{13}C	13.003	1.11
^{16}O	15.995	99.76
^{17}O	16.999	0.04
^{18}O	17.999	0.20
^{20}Ne	19.992	90.51
^{21}Ne	20.994	0.27
^{22}Ne	21.991	9.22

EXAMPLE 10.2

Calculating an Element's Atomic Mass

Naturally occurring chlorine is a mixture consisting of 75.77% chlorine-35 (atomic mass = 34.97 u) and 24.23% chlorine-37 (atomic mass = 36.97 u). Calculate the atomic mass of the element chlorine.

SOLUTION

Calculate the contribution each chlorine isotope makes to the atomic mass by multiplying the *fractional abundance* of each (the percentage abundance divided by 100) by its atomic mass, then add the two answers to get the atomic mass of chlorine (35.46 u).

$$0.7577 \times 34.97\ u = 26.50\ u \quad (^{35}Cl)$$
$$0.2423 \times 36.97\ u = \underline{\ 8.96\ u} \quad (^{37}Cl)$$
$$\text{for a total of} \qquad 35.46\ u$$

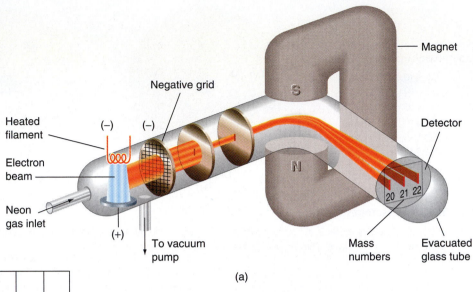

(a)

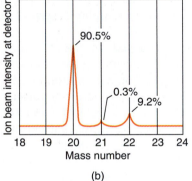

(b)

FIGURE 10.5   Schematic Drawing of a Mass Spectrometer
(a) A mass spectrometer, first invented in the 1920s, separates and measures the masses of isotopes of an element by ionizing the atoms with a beam of electrons and sending them through a magnetic field. Because the ions have the same charge but different masses, they interact differently with the field and form separate beams. (From Ebbing, Darrell D., *General Chemistry*, Sixth Edition. Copyright © 1999 by Houghton Mifflin Company. Used with permission.) (b) A mass spectrogram shows the three isotopes of neon and their relative abundances.

CONFIDENCE EXERCISE 10.2

Find the atomic mass of a hypothetical element X if it consists of 60.00% of ^{20}X (atomic mass = 20.00 u) and 40.00% of ^{22}X (atomic mass = 22.00 u).

The Strong Nuclear Force

In previous chapters, we studied two fundamental forces of nature—electromagnetic and gravitational. The electromagnetic force between a proton and an electron in an atom is about 10^{39} times greater than the corresponding gravitational force. The electromagnetic force is the only important force on the electrons in an atom and is responsible for the structure of atoms, molecules, and hence matter in general.

In a nucleus, the positively charged protons are packed closely together. According to Coulomb's law (see Section 8.1), like charges repel each other, so the repulsive electric forces in a nucleus are huge. The nucleus should fly apart.

Obviously, the nucleus generally does remain intact, and thus a third fundamental force must exist. This **strong nuclear force** (or just *strong force* or *nuclear force*) acts between nucleons—that is, between two protons, between two neutrons, and between a proton and a neutron. It holds the nucleus together. The exact equation describing the nucleon–nucleon interaction is unknown. However, for the very short nuclear distances of less than about 10^{-14} m, the interaction is strongly attractive; in fact, it is the strongest fundamental force known. Yet, at distances greater than about 10^{-14} m, the nuclear force is zero!

A typical large nucleus is illustrated in ● Fig. 10.6. A proton on the surface of the nucleus is attracted only by the six or seven nearest nucleons. Because the strong nuclear force is a short-range force, only the nearby nucleons contribute to the attractive force.

FIGURE 10.6   A Multinucleon Nucleus
The protons on the surface of the nucleus, such as those shown in the red semicircle, are attracted by the strong nuclear force of only the six or seven closest nucleons, but they are electrically repelled by all the other protons. When the number of protons exceeds 83, the electrical repulsion overcomes the nucleon attraction, and the nucleus is unstable.

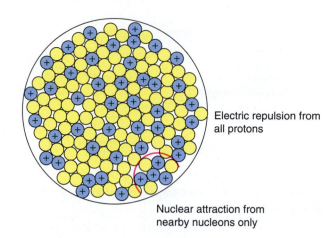

Electric repulsion from all protons

Nuclear attraction from nearby nucleons only

On the other hand, the repulsive electric force is long range and acts between any two protons, no matter how far apart they are in the nucleus. As more and more protons are added to the nucleus, the electric repulsive forces increase, yet the attractive nuclear forces remain constant because they are determined by nearest neighbors only. When the nucleus has more than 83 protons, the electric forces of repulsion overcome the nuclear attractive forces, and the nucleus is subject to spontaneous disintegration, or *decay*. That is, particles are emitted so as to adjust the neutron–proton imbalance.

10.3 Radioactivity and Half-Life

LEARNING GOALS

▼ Complete equations for radioactive decay.

▼ Identify which nuclides are radioactive.

▼ Apply the concept of half-life.

A specific type of nucleus, such as ^{238}U or ^{14}C, is referred to as a **nuclide.** Nuclides whose nuclei undergo spontaneous decay (disintegration) are called **radionuclides** (or *radioactive isotopes* or *radioisotopes*). The spontaneous process of nuclei undergoing a change by emitting particles or rays is called *radioactive decay,* or **radioactivity.** Substances that give off such radiation are said to be *radioactive.* (The chapter's first Highlight discusses the discovery of radioactivity by Becquerel and of two new radioactive elements by the Curies.)

Radioactive nuclei can disintegrate in three common ways: alpha decay, beta decay, and gamma decay

(Fig. 10.7). (Fission, another important decay process, will be discussed in Section 10.5.) In all decay processes, energy is given off, usually in the form of energetic particles that can produce heat. Equations for radioactive decay take the form

$$A \rightarrow B + b$$

In radioactive decay, the original nucleus (*A*) is sometimes called the *parent* nucleus, and the resulting nucleus (*B*) is referred to as the *daughter* nucleus. The *b* in the equation represents the emitted particle or ray.

Alpha decay is the disintegration of a nucleus into a nucleus of another element, with the emission of an *alpha particle,* which is a helium nucleus (^4_2He). Alpha decay is common for elements with atomic numbers greater than 83. An example of alpha decay is

$$^{232}_{90}Th \rightarrow ^{228}_{88}Ra + ^4_2He$$

Note that in this decay equation the sum of the mass numbers on each side of the arrow is the same; that is, $232 = 228 + 4$. Also, the sum of the atomic numbers on each side is the same, that is, $90 = 88 + 2$.

	Left Side		Right Side
A:	232	=	228 + 4
Z:	90	=	88 + 2

This principle holds for all nuclear decays and involves the conservation of nucleons and the conservation of charge, respectively.

In a nuclear decay equation, the sums of the mass numbers will be the same on each side of the arrow, as will be the sums of the atomic numbers.

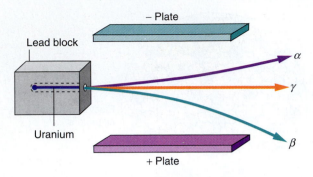

FIGURE 10.7 The Three Components of Radiation from Heavy Radionuclides

An electric field separates the rays from a sample of a heavy radionuclide, such as uranium, into alpha (α) particles (positively charged helium nuclei), beta (β) particles (negatively charged electrons), and neutral gamma (γ) rays (high-energy electromagnetic radiation).

EXAMPLE 10.3

Finding the Products of Alpha Decay

$^{238}_{92}$U undergoes alpha decay. Write the equation for the process.

SOLUTION

Step 1

First, we write the symbol for the parent nucleus, followed by an arrow.

$$^{238}_{92}\text{U} \rightarrow$$

Step 2

Because alpha decay involves the emission of $^4_2$He, write this symbol after the arrow, preceded by a plus sign and room for the symbol for the daughter nucleus.

$$^{238}_{92}\text{U} \rightarrow \_\_ + {}^4_2\text{He}$$

Step 3

Determine the mass number, atomic number, and chemical symbol for the daughter nucleus. The sum of the mass numbers on the left is 238. The sum on the right also must be 238, and so far only the 4 for the alpha particle shows. Therefore, the daughter must have a mass number of $238 - 4 = 234$. By sim-

ilar reasoning, the atomic number of the daughter must be $92 - 2 = 90$. From the periodic table, we find that the element with atomic number 90 is Th (thorium). The equation for the decay is

$$^{238}_{92}\text{U} \rightarrow {}^{234}_{90}\text{Th} + {}^4_2\text{He}$$

CONFIDENCE EXERCISE 10.3

Write the equation for the alpha decay of the radium isotope $^{226}_{88}$Ra.

Beta decay is the disintegration of a nucleus into a nucleus of another element, with the emission of a *beta particle,* which is an electron ($_{-1}^{0}$e). An example of beta decay is

$$^{14}_{6}\text{C} \rightarrow {}^{14}_{7}\text{N} + {}^{0}_{-1}\text{e}$$

Note that a beta particle, or electron, is assigned a mass number of 0 (because it contains no nucleons) and an atomic number of -1 (because its electric charge is opposite that of a proton's $+1$ charge). The mass numbers and atomic numbers on both sides of the arrow are equal in our example of beta decay because $14 = 14 + 0$ and $6 = 7 - 1$. In beta decay, a neutron ($^1_0$n) is transformed into a proton and an electron ($^1_0$n $\rightarrow$ $^1_1$p $+$ $^0_{-1}$e). The proton remains in the nucleus, and the electron is emitted.

Gamma decay occurs when a nucleus emits a *gamma ray* (γ) and becomes a less energetic form of the same nucleus. A gamma ray, being a photon of high-energy electromagnetic radiation, has no mass number or atomic number. Gamma radiation is similar to X-rays but is more energetic. An example of gamma decay is

$$^{204}_{82}\text{Pb}^* \rightarrow {}^{204}_{82}\text{Pb} + \gamma$$

The * following the symbol means that the nucleus is in an excited state, analogous to an atom being in an excited state with an electron in a higher energy level (Section 9.3). When the nucleus de-excites, one or more gamma rays are emitted, and the nucleus is left in a state of lower excitation and, ultimately, in a "ground (or stable) state" of the same nuclide. Note the absence of the asterisk in the symbol for the daughter in the preceding equation. Gamma decay generally occurs any time a nucleus is formed in an excited state, for example, as a product of alpha or beta decay.

In addition to alpha, beta, and gamma radiation, certain nuclear processes (generally involving synthetic radionuclides) emit *positrons* $\left(_{+1}^{0}e\right)$. For example,

$$_{9}^{17}F \rightarrow \, _{8}^{17}O + \, _{+1}^{0}e$$

Positrons are sometimes called *beta-plus* particles, because they are the so-called *antiparticle* of the electron and have the same mass but an electrical charge of +1. Table 10.5 lists five common forms of nuclear radiation.

With the exceptions of technetium (atomic number 43) and promethium (atomic number 61), at least one stable isotope exists for every element up to atomic number 83. A nucleus with atomic number greater than 83 is always radioactive and commonly undergoes a series of alpha, beta, and gamma decays until a stable nucleus is produced.

For example, the series of decays beginning with uranium-238 and ending with stable $_{82}^{206}Pb$ is illustrated in ● Fig. 10.8. (Other similar decay series end with either $_{82}^{207}Pb$, $_{82}^{208}Pb$, or $_{83}^{209}Bi$). Notice how the alpha (α) and beta (β) transitions are indicated in Fig.

TABLE 10.5 Nuclear Radiations

Name	Symbol	Charge	Mass Number
Alpha	$_{2}^{4}He$	2+	4
Beta	$_{-1}^{0}e$	1−	0
Gamma	γ	0	0
Positron	$_{+1}^{0}e$	1+	0
Neutron	$_{0}^{1}n$	0	1

10.8. The gamma decays that accompany the alpha and beta decays in the series are not apparent on the diagram because the neutron and proton numbers do not change in gamma decay.

Identifying Radionuclides

Which nuclides are unstable (radioactive) and which are stable? When we plot the number of protons (Z) versus the number of neutrons (N) for each stable

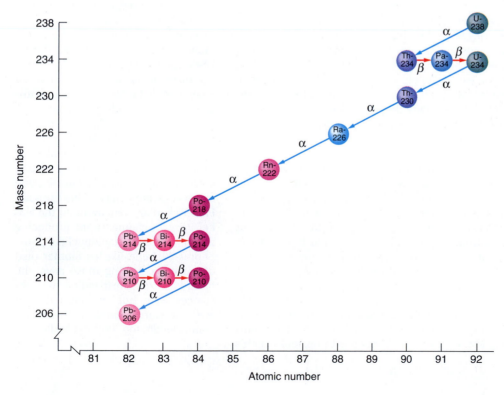

FIGURE 10.8 The Decay of Uranium-238 to Lead-206

Each radioactive nucleus in the series undergoes either alpha decay or beta decay. Finally, stable lead-206 is formed as the end product. (The gamma decays, which change only the energy of nuclei, are not shown.) (From Stoker, H. Stephen, *General Organic, and Biological Chemistry*. Copyright © 1998 by Houghton Mifflin Company. Used with permission.)

HIGHLIGHT: The Discovery of Radioactivity

In 1896 Henri Becquerel ("beh-KREL") in Paris heard of Wilhelm Roentgen's ("RUNT-gin's") recent discovery of X-rays (Section 9.4) and wondered if any of the fluorescent materials he was investigating might emit X-rays. He wrapped a photographic plate in black paper and put it in sunlight with a crystal of a fluorescent mineral on it. He knew that the ultraviolet light in sunlight could not penetrate the black paper but that X-rays could. The film was indeed fogged, and Becquerel decided (incorrectly) that the crystal was emitting X-rays as it fluoresced due to the sunlight hitting it.

His experiments were interrupted by a series of cloudy days. In a drawer he left a fresh, wrapped plate with an unexposed crystal resting on it. Becoming impatient, he decided to develop the plate and found, to his surprise, that it was heavily fogged. Whatever radiation the crystal was giving off had nothing to do with fluorescence. Becquerel (Fig. 1) traced the radiation to the uranium in the mineral.

In 1897, Marie Curie (Fig. 2), born Marie Sklodowska in Poland, began a search for naturally radioactive elements. (Marie was a top student at the Sorbonne in Paris when in 1895 she married the university's most famous physicist, Pierre Curie.) She noticed that some of the uranium ores she studied were much more radioactive than could be accounted for by their uranium content.

At this point Pierre abandoned his own research and became his wife's assistant. In July 1898 they isolated from the uranium ore a minute amount of a new element hundreds of times more radioactive than uranium. They called it *polonium* after Marie's native country. In December they found an even more radioactive element, which they named *radium*. That same year, Marie

FIGURE 1   Henri Becquerel (1852–1908)

FIGURE 2   Marie Curie (1867–1934)

Curie coined the term *radioactivity* for Becquerel's uranium radiation phenomenon.

To get enough radium to investigate thoroughly, the Curies used their life savings to buy tons of uranium ore. They obtained permission to work in an old wooden shed with a leaky roof, no floor, and inadequate heat. For four years they processed the tons of ore into smaller and smaller samples of more and more intensely radioactive material, while also taking care of their baby, Irene. Eventually, eight tons of ore gave them about 10 mg of radium and a smaller amount of polonium. Despite the obvious chance of wealth, the Curies refused to patent their process. They were too ill in 1903 to journey to Stockholm to receive the Nobel Prize in physics, which they shared with Becquerel.

In 1906 Pierre was killed by a horse-drawn wagon as he stepped from a carriage into a Paris street. Marie, now known as Madame Curie, took over his

professorship at the Sorbonne, becoming the first woman to teach there. In 1911 she received an unprecedented second Nobel Prize, this time in chemistry for her work on the properties of radium. Despite her fame, during World War I she and her daughter Irene organized medical units with X-ray equipment to locate shrapnel and broken bones in wounded soldiers.

Madame Curie's last years were spent supervising the Paris Institute of Radium. She died in 1934 of leukemia, which probably was caused by overexposure to radioactivity. Her death came one year before Irene and her husband Frederick Joliot were awarded the Nobel Prize in chemistry for producing the first synthetic radionuclide, phosphorus-30. Irene, like her mother, died of leukemia. During those days, the hazards of exposure to radioactive substances were not yet realized, and Madame Curie's early notebooks remain, literally, too "hot" to handle.

nuclide, the points (red dots) form a narrow band called the *band of stability* (● Fig. 10.9). For comparison, the straight red line in the figure represents equal numbers of protons and neutrons.

The increasing divergence of the band from the $N = Z$ line shows that as more protons are packed into the nucleus, the neutron-to-proton ratio must increase in order for stability to result. (Compare the

TABLE 10.6   The Pairing Effect in Stabilizing Nuclei

Proton Number (Z)	Neutron Number (N)	Number of Stable Nuclides
Even	Even	160
Even	Odd	52
Odd	Even	52
Odd	Odd	4

FIGURE 10.9   A Plot of Number of Neutrons (N) Versus Number of Protons (Z) for the Nuclides

The red dots representing stable nuclides trace out a *band of stability* that begins on a line where *neutrons* and *protons* are equal in number and gradually diverges from the line as the *number of protons* gets greater. (Note how the neutron/proton ratio increases from lithium-6 to cadmium-110 to mercury-202.) Because all nuclides with more than 83 protons are radioactive, the band ends at this number of protons. The blue dots represent known radionuclides. (From Zumdahl, Steven S., *Chemistry,* Fourth Edition. Copyright © 1999 by Houghton Mifflin Company. Used with permission.)

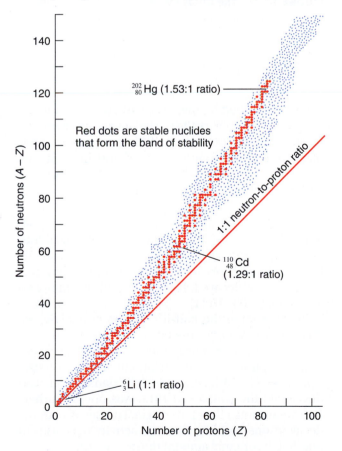

N/Z ratio for lithium-6, cadmium-110, and mercury-202 in Fig. 10.9.)

Each blue dot in Fig. 10.9 represents a known radionuclide. Note that the radionuclides cluster on each side of the band of stability and sometimes are found within it. No stable nuclides (red dots) are found past $Z = 83$, but numerous radionuclides with greater than 83 protons are known.

An inventory of the number of protons and number of neutrons in stable nuclides reveals an interesting pattern (Table 10.6). Most of the stable nuclides have both an even number of protons (p) and an even number of neutrons (n) in their nuclei. We refer to these as *even-even* nuclides. Practically all the other stable nuclides are either *even-odd* or *odd-even*. Nature seems to dislike *odd-odd* nuclides (only four stable ones exist), due apparently to the existence of energy levels in the nucleus that favor the pairing of two protons or two neutrons. Note that the descriptions such as *odd-odd* refer to the number of protons and neutrons, respectively, not to the atomic number and mass number. For example, because $N = A - Z$, an *odd* atomic number (say, 9) coupled with an *even* mass number (say, 20) means an *odd* number of protons (9) but also an *odd* number of neutrons (11).

A nuclide will be radioactive if it meets any of the following criteria.*

1. Its atomic number is greater than 83.

2. It has fewer n than p in the nucleus (except for $_1^1$H and $_2^3$He).

3. It is an *odd-odd* nuclide (except for $_1^2$H, $_3^6$Li, $_5^{10}$B, and $_7^{14}$N).

*A fourth criterion, which we will not use because it is difficult to apply, is that unless the mass number of the nuclide is relatively close to the element's atomic mass, the nuclide will be radioactive.

Identifying Radionuclides

Identify the radionuclide in each pair, and state your reasoning.

(a) $^{208}_{82}Pb$ and $^{222}_{86}Rn$

(b) $^{19}_{10}Ne$ and $^{20}_{10}Ne$

(c) $^{63}_{29}Cu$ and $^{64}_{29}Cu$

SOLUTION

(a) $^{222}_{86}Rn$ (Z above 83)

(b) $^{19}_{10}Ne$ (fewer n than p)

(c) $^{64}_{29}Cu$ (*odd-odd*)

CONFIDENCE EXERCISE 10.4

Predict which two of the following nuclides are radioactive.

$$^{232}_{90}Th \quad ^{24}_{12}Mg \quad ^{40}_{19}K \quad ^{31}_{15}P$$

Half-Life

Some samples of radionuclides take a long time to decay; others decay very rapidly (Table 10.7). In a sample of a given radionuclide, the decay of an individual nucleus is a random event. It is impossible to predict which nucleus will be the next to undergo a nuclear change. However, given a large number of nuclei, it is possible to predict how many will decay in a certain length of time. The rate of decay of a given radionu-

TABLE 10.7 The Half-Lives of Some Radionuclides

Radionuclide	Half-Life
Beryllium-8	6.7×10^{-15} s
Oxygen-19	26.9 s
Technetium-104	18.3 min
Radon-222	3.82 d
Strontium-90	29 y
Carbon-14	5730 y
Uranium-238	4.46×10^9 y
Indium-115	4.4×10^{14} y

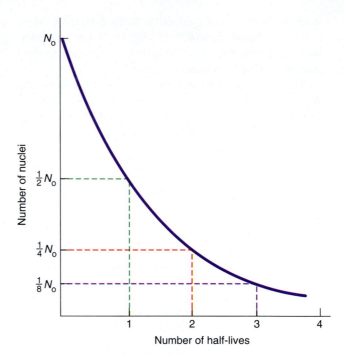

FIGURE 10.10 The Decay Curve for Any Radionuclide

Starting with the number of nuclei N_o of a radionuclide, after one half-life has elapsed, only one-half of the nuclei will remain undecayed. After two half-lives have gone by, only one-quarter of the original nuclei will remain, and so on.

clide is described by the term **half-life,** which is the time it takes for half of the nuclei of a sample to decay. In other words, after one half-life has gone by, one-half of the original amount of radionuclide remains undecayed; after two half-lives, $\frac{1}{2} \times \frac{1}{2} = \frac{1}{4}$ of the original amount is undecayed; and so on (● Fig. 10.10).

For example, thorium-234 has a half-life of 24 days (d). So, if we start with a 100-g sample of thorium-234, after 24 d (one half-life), 50 g will have changed into other atoms and 50 g of thorium-234 will remain. After another 24 d, one-half of the remaining thorium-234 will have decayed, leaving 25 g of thorium-234, and so on (● Fig. 10.11).

To determine the half-life of a radionuclide, we monitor the *activity*—the rate of emission of the decay particles, commonly measured in counts per minute (cpm)—with an instrument such as a *Geiger counter* (● Fig. 10.12, on page 246). (Hans Geiger, who developed the counter in 1913, was one of Rutherford's assistants.) If half the original nuclei of a sample decay in one half-life, then the activity decreases to one-half its original amount during that time.

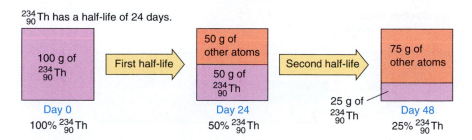

FIGURE 10.11   The Decay of Thorium-234 over Two Half-Lives

Thorium-234 has a half-life of 24 days. If 100 g of ^{234}Th is initially present, only 50 g of ^{234}Th will be undecayed 24 days (1 half-life) later. The other 50 g has been changed into another nuclide (and a little energy). After 48 days (2 half-lives) have gone by, only 25 g of ^{234}Th remains.

Let's take a look at some calculations involving half-life. The *original amount* of radionuclide sample is designated N_o, and the *final amount* of radionuclide sample at some later time is N. For both N_o and N we will use the units of either mass (grams) or activity (cpm) or use rational fractions in which N is stated as $\frac{1}{2}N_o$, $\frac{1}{4}N_o$, $\frac{1}{8}N_o$, and so on. The *elapsed time* is the time between the measurement of N_o and N. If told a radionuclide's half-life is, say, 12 y, keep the units straight by putting it into a calculation as 12 y/half-life (12 years per half-life).

For simplicity, we will consider only exercises in which the number of half-lives is a small whole number. In a given exercise, the quantity we are solving for will be one of the following: the number of half-lives, the final amount, or the elapsed time.

$$N_o \;\boxed{\text{first half-life}}\; \frac{N_o}{2} \;\boxed{\text{second half-life}}\; \frac{N_o}{4} \;\boxed{\text{third half-life}}\; \frac{N_o}{8}$$

Thus three half-lives have passed, and with $N_o/8$ remaining, the final amount of iodine-131 is $\frac{1}{8}$ of 40 mg, or 5 mg.

CONFIDENCE EXERCISE 10.5

Strontium-90 (half-life = 29 y) is one of the worst components of fallout from atmospheric testing of nuclear bombs because it concentrates in the bones. The last such bomb was tested in 1963. In the year 2021, how many half-lives will have gone by for the strontium-90 produced in the blast? What fraction of the strontium-90 will remain in that year?

EXAMPLE 10.5

Finding the Number of Half-Lives and the Final Amount

What fraction and mass of a 40-mg sample of iodine-131 (half-life = 8 d) will remain after 24 d?

SOLUTION

Step 1

Find the number of half-lives that have passed in 24 d.

$$\frac{24 \text{ d}}{8 \text{ d/half-life}} = 3 \text{ half-lives}$$

Step 2

Starting with the defined original amount N_o, halve it three times (because three half-lives have passed).

EXAMPLE 10.6

Finding the Elapsed Time

How long would it take a sample of ^{14}C to decay to one-fourth its original activity? The half-life of ^{14}C is 5730 y.

SOLUTION

The ^{14}C has been decaying for a time period equal to two half-lives, as shown by the number of arrows in the sequence below.

$$N_o \rightarrow \frac{N_o}{2} \rightarrow \frac{N_o}{4}$$

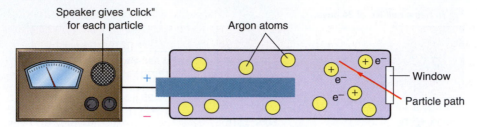

FIGURE 10.12 A Schematic Representation of a Geiger Counter

A Geiger counter detects ions formed as a result of a high-energy particle from a radioactive source entering the window and ionizing argon atoms along its path. The ions and electrons formed produce a pulse of current, which is amplified and counted and can produce the familiar series of "clicks" associated with the sound of the device.

To find the elapsed time, multiply the number of half-lives by the half-life.

$$(2 \text{ half-lives})(5730 \text{ y/half-life}) = 11,460 \text{ y}$$

CONFIDENCE EXERCISE 10.6

Technetium-99 is often used as a radioactive tracer to assess heart damage. Its half-life is 6.0 h. How long would it take a sample of technetium-99 to decay to one-sixteenth its original amount?

10.4 Nuclear Reactions

LEARNING GOALS

▼ Complete equations for nuclear reactions.

▼ State some uses of radionuclides.

Radioactive nuclei, through the emission of alpha and beta particles, spontaneously change (undergo *transmutation*) into nuclei of other elements. Scientists wondered whether the reverse process was possible; that is, could a particle be added to a nucleus to change it into that of another element? The answer was yes.

Ernest Rutherford produced the first such *nuclear reaction* in 1919 by bombarding nitrogen (^{14}N) gas with alpha particles from a radioactive source. Particles coming from the gas were identified as protons. Rutherford reasoned that an alpha particle colliding with a nitrogen nucleus occasionally can knock out a proton. The result is an *artificial transmutation* of a nitrogen isotope into an oxygen isotope. The equation for the reaction is

$$^{4}_{2}\text{He} + ^{14}_{7}\text{N} \rightarrow ^{17}_{8}\text{O} + ^{1}_{1}\text{H}$$

As this equation indicates, the conservation of mass number and the conservation of atomic number hold in nuclear reactions, just as they do in nuclear decay.

The general form of a nuclear reaction is

$$a + A \rightarrow B + b$$

where a is the particle that bombards nucleus A to form nucleus B and an emitted particle b. In addition to the particles listed in Table 10.5, common particles encountered in nuclear reactions are protons ($^{1}_{1}\text{H}$), deuterons ($^{2}_{1}\text{H}$), and tritons ($^{3}_{1}\text{H}$).

EXAMPLE 10.7

Completing an Equation for a Nuclear Reaction

Complete the equation for the proton bombardment of lithium-7.

$$^{1}_{1}\text{H} + ^{7}_{3}\text{Li} \rightarrow \underline{\quad} + ^{1}_{0}\text{n}$$

SOLUTION

The sum of the mass numbers on the left is 8. So far, only a mass number of 1 shows on the right, so the missing particle must have a mass number of $8 - 1 = 7$.

The sum of the atomic numbers on the left is 4. So far, the total showing on the right is 0. Thus the missing particle must have an atomic number of $4 - 0 = 4$. The atom with mass number 7 and atomic number 4 is an isotope of Be (beryllium). The completed equation is

$$^{1}_{1}\text{H} + ^{7}_{3}\text{Li} \rightarrow ^{7}_{4}\text{Be} + ^{1}_{0}\text{n}$$

Complete the equation for the deuteron bombardment of aluminum-27.

$$\,^2_1\text{H} + \,^{27}_{13}\text{Al} \rightarrow \,\_\!\_ + \,^4_2\text{He}$$

The reaction in Rutherford's experiment was discovered almost by accident, because it took place so infrequently. One proton is produced for about every one million alpha particles that shoot through the nitrogen gas. But think of the implications of its discovery: One element had been changed into another! This was the age-old dream of the alchemists—the original researchers into transmutation—although their main concern was to change common metals, such as lead, into gold.

Such artificial transmutations are now common. Large machines called *particle accelerators* use electric fields to accelerate charged particles to very high energies (● Fig. 10.13). The energetic particles are used to bombard nuclei and initiate nuclear reactions. Different reactions require different particles and different bombarding energies. One nuclear reaction that occurs when a proton strikes a nucleus of mercury-200 is

$$\,^1_1\text{H} + \,^{200}_{80}\text{Hg} \rightarrow \,^{197}_{79}\text{Au} + \,^4_2\text{He}$$

Thus gold (Au) has indeed been made from another element. Unfortunately, making gold by this process costs about one million dollars an ounce—much more than the gold is worth.

Neutrons produced in nuclear reactions can be used to induce other nuclear reactions. Because they have no electric charge, neutrons do not experience repulsive electrical interactions with nuclear protons, as would alpha-particle and proton projectiles. As a result, they are especially effective at penetrating the nucleus and inducing a reaction. For example,

$$\,^1_0\text{n} + \,^{45}_{21}\text{Sc} \rightarrow \,^{42}_{19}\text{K} + \,^4_2\text{He}$$

The *transuranium elements*—those with atomic number greater than 92—are all synthetic (as are Tc, Pm, At, and Fr). Elements 93 (neptunium) to 101 (mendelevium) can be made by bombarding a lighter nucleus with alpha particles or neutrons. For example,

$$\,^1_0\text{n} + \,^{238}_{92}\text{U} \rightarrow \,^{239}_{93}\text{Np} + \,^{\,\,0}_{-1}\text{e}$$

Beyond mendelevium, heavier bombarding particles are required. For example, element 109, meitnerium, is made by bombarding bismuth-209 with iron-58 nuclei.

$$\,^{58}_{26}\text{Fe} + \,^{209}_{83}\text{Bi} \rightarrow \,^{266}_{109}\text{Mt} + \,^1_0\text{n}$$

Atoms of hydrogen, helium, and lithium are thought to have been formed in the Big Bang (Section 18.7), whereas atoms of beryllium up through iron are made in the cores of stars by fusion (Section 18.3). Atoms of elements heavier than iron are believed to be formed during supernova explosions of stars, when neutrons are in abundance and can enter into nuclear reactions with medium-sized atoms to form larger ones.

Some Uses of Radionuclides

Americium-241, a synthetic transuranium radionuclide (half-life = 432 y) is used in the most common type of home smoke detector. As the americium-241

FIGURE 10.13 The Fermi National Accelerator Laboratory at Batavia, Illinois

The large circle is the main accelerator with a 6.4-km (4.0-mi) circumference. Magnets bend and focus a proton beam that travels around the circle 50,000 times a second. With each roundtrip, more energy is added by a radio-frequency system, producing highly energetic particles for use in nuclear research. In 1994, the elusive "top" quark—the last of the six quarks that theory said should exist—was discovered at this facility.

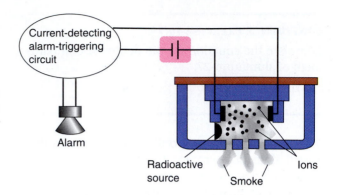

FIGURE 10.14   A Smoke Detector
In most smoke detectors, a weak radioactive source ionizes the air and sets up a small current. If smoke particles enter the detector, the current is reduced, causing an alarm to sound.

decays, the alpha particles emitted ionize the air inside part of the detector (● Fig. 10.14). The ions carry a small current and allow the 9-V battery to power a closed circuit. If smoke enters the detector, the ions become attached to the smoke particles and slow down, causing the current to decrease and an alarm to sound.

Radionuclides have many uses in medicine, chemistry, biology, agriculture, and industry. As an example, a radioactive isotope of iodine, ^{123}I, is used in a diagnostic measurement connected with the thyroid gland. The patient is administered a prescribed amount of the ^{123}I, and like regular iodine in the diet, it is absorbed by the thyroid gland. This allows doctors to monitor the iodine intake of the thyroid because the radioactive iodine can be traced as it is released into the bloodstream in the form of protein-bound iodine. Radiation is used in many other types of medical diagnoses; for example, in heart scans (see the chapter-opening photo).

Nuclear radiation also can be used to treat diseased cells, which generally can be destroyed by radiation more easily than healthy cells. By focusing an intense beam of radiation from cobalt-60 on a cancerous tumor, its cells can be destroyed and its growth impaired or stopped. In the case of thyroid cancer, ^{131}I is taken for a prescribed length of time so that its radiation can destroy the cancer cells.

In chemistry and biology, radioactive "tracers" such as ^{14}C (radiocarbon) and ^{3}H (tritium) are used to tag an atom in a certain part of a molecule so that it can be followed through a series of reactions. In this way, the reaction pathways of hormones, drugs, and other substances can be determined.

Neutron activation analysis is one of the most sensitive analytical methods in science. A beam of neutrons irradiates the sample, and each constituent element forms a specific radionuclide that can be identified by the characteristic energies of the gamma rays it emits. One advantage is that the sample is not destroyed by this analytical technique. An interesting use is in the analysis of human hair. As little as 10^{-9} g of arsenic can be detected in hair, so arsenic poisoners now run a high risk of detection.

In agriculture, less than lethal doses of radioactivity were used to cause sterility in male Mediterranean fruit flies in California and Florida, where they were destroying crops. When released, the sterilized males mated with females, but the eggs would not hatch, thus drastically reducing the number of fruit flies.

In industry, the penetrating ability of radiation has been used to gauge and adjust the thickness of metal sheets and plastic films. Plutonium-238 powers a tiny battery used in heart pacemakers. Tracer radionuclides help manufacturers test the durability of mechanical components and identify structural weaknesses in equipment. In environmental studies, radionuclides help detect groundwater movement through soil and trace the paths of industrial air and water pollutants.

In Chapter 24 we will discuss how radioactivity is used in dating rocks and ancient organic remains. In the next two sections, we discuss the controlled and uncontrolled release of nuclear energy. Other uses of radioactivity, some very ingenious, are too numerous to mention.

RELEVANCE QUESTION: One-third of all hospital patients in the United States get treatments or tests that involve nuclear technology. Has the use of radioactivity in medicine directly affected you or any of your family or friends?

10.5 Nuclear Fission

LEARNING GOALS

▼ Explain how nuclear fission occurs.

▼ Discuss the operation of nuclear fission reactors.

Fission is the process in which a large nucleus is "split" into two intermediate-sized nuclei, with the emission of neutrons and the conversion of mass into energy. As an example, consider the fission decay of ^{236}U. If ^{235}U is bombarded with low-energy neutrons, ^{236}U is formed momentarily:

$$^1_0n + ^{235}_{92}U \rightarrow ^{236}_{92}U$$

The ^{236}U immediately fissions into two smaller nuclei, emits several neutrons, and releases energy. ● Figure 10.15a illustrates the following typical fission of ^{236}U:

$$^{236}_{92}U \rightarrow ^{140}_{54}Xe + ^{94}_{38}Sr + 2\,^1_0n$$

This is just one of many possible fission decays of ^{236}U. Another is

$$^{236}_{92}U \rightarrow ^{132}_{50}Sn + ^{101}_{42}Mo + 3\,^1_0n$$

EXAMPLE 10.8

Completing an Equation for Fission

Complete the following equation for fission.

$$^{236}_{92}U \rightarrow ^{88}_{36}Kr + ^{144}_{56}Ba + \underline{\quad}$$

SOLUTION

The atomic numbers are balanced ($92 = 36 + 56$), so the other particle must have an atomic number of 0. The mass number on the left is 236, and the sum of the mass numbers on the right is $88 + 144 = 232$. Hence, if the mass numbers are to balance, there must be four additional units of mass on the right side of the equation. Because no particle with atomic number of 0 and mass number of 4 exists, the missing "particle" is actually four neutrons. The reaction is then

$$^{236}_{92}U \rightarrow ^{88}_{36}Kr + ^{144}_{56}Ba + 4\,^1_0n$$

Both the atomic numbers and the mass numbers are now balanced.

FIGURE 10.15   Fission and Chain Reaction

(a) In a fission reaction, such as that shown for uranium-235, a neutron is absorbed, and the nucleus splits into two lighter nuclei with the emission of energy and two or more neutrons. (b) If the emitted neutrons cause increasing numbers of fission reactions, an expanding *chain reaction* occurs.

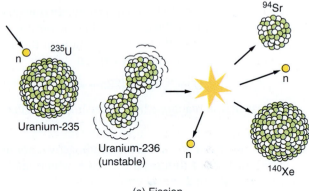

(a) Fission

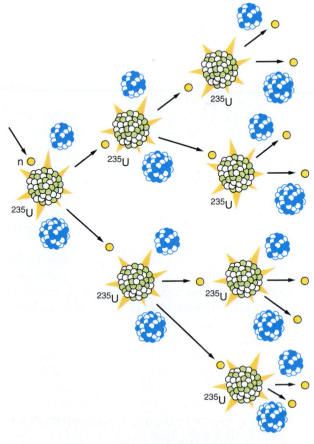

(b) Chain reaction

CONFIDENCE EXERCISE 10.8

Complete the following equation for fission.

$$^{236}_{92}\text{U} \rightarrow {}^{90}_{38}\text{Sr} + \underline{\quad} + 2\,^{1}_{0}\text{n}$$

The fast-fissioning ^{236}U is an intermediate nucleus and is often left out of the equation for the neutron-induced fission of ^{235}U; that is, the equation for the reaction in Example 10.8 is usually written

$$^{1}_{0}\text{n} + {}^{235}_{92}\text{U} \rightarrow {}^{88}_{36}\text{Kr} + {}^{144}_{56}\text{Ba} + 4\,^{1}_{0}\text{n}$$

Nuclear fission reactions have three important features:

1. The fission products are always radioactive. Some have half-lives of thousands of years, which leads to nuclear waste disposal problems.
2. Relatively large amounts of energy are produced (see Section 10.6).
3. Neutrons are released.

In an *expanding* **chain reaction,** one initial reaction triggers a growing number of subsequent reactions. In the case of fission, one neutron hits a nucleus of ^{235}U and forms ^{236}U, which can fission and emit two (or more) neutrons. These two neutrons can then hit two other ^{235}U nuclei, causing them to fission and release energy and four neutrons. These four neutrons can cause four fissions, releasing energy and eight neutrons, and so on (Fig. 10.15b). Each time a nucleus fissions, energy is released; as the chain expands, the energy output increases.

For a *self-sustaining* chain reaction each fission event needs to cause only one more fission event. This leads to a steady release of energy, not a growing release.

Of course, the process of energy production by fission is not as simple as just described. For a self-sustaining chain reaction to proceed, a sufficient amount and concentration of fissionable material (^{235}U) must be present. Otherwise, too many neutrons would escape from the sample before reaction with a nucleus (● Fig. 10.16a). The chain would be broken, so to speak. The minimum amount of fissionable material necessary to sustain a chain reaction is called the **critical mass.** The critical mass for pure ^{235}U is about 4 kg, which is approximately the size of a baseball. With a *subcritical mass,* no chain reaction

occurs. With a *supercritical mass,* the chain reaction grows and, under certain conditions, the mass can explode (Fig. 10.16b).

Natural uranium is composed of 99.3% ^{238}U and only 0.7% of the fissionable ^{235}U isotope. So that more fissionable ^{235}U nuclei are present in a sample, the ^{235}U is concentrated, or "enriched." Enriched uranium used in U.S. nuclear reactors for the production of electricity is about 3% ^{235}U. Weapons-grade uranium is enriched to 90% or more; this percentage provides many fissionable nuclei for a large and sudden release of energy. (The leftover ^{238}U is known as *depleted uranium* and can be used in armor-piercing shells.)

In a fission, or "atomic," bomb, a supercritical mass of highly enriched fissionable material must be formed and held together for a short time to get an explosive release of energy. Subcritical segments of the fissionable material in a fission bomb are kept separated before detonation so that a critical mass does not exist for the chain reaction. A chemical explosive is used to bring the segments together in an interlocking, supercritical configuration that holds them long enough for a large fraction of the material to undergo fission. The result is an explosive release of energy. The chapter's second Highlight discusses the building of the first atomic bombs.

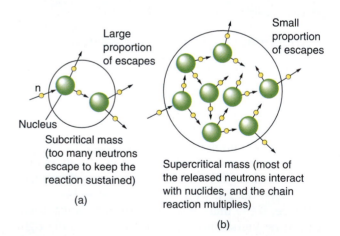

FIGURE 10.16 Subcritical and Supercritical Masses

(a) The mass of fissionable material is too small; thus most of the neutrons produced escape before causing another fission event. The process dies. (b) The mass of fissionable material is so large that each fission causes, on average, more than one additional fission. The process builds. (From Zumdahl, Steven S., *Chemistry,* Fourth Edition. Copyright © 1999 by Houghton Mifflin Company. Used with permission.)

Nuclear Reactors

A bomb is an example of *uncontrolled* fission. A nuclear reactor is an example of *controlled* fission, in which we control the growth of the chain reaction and the release of energy. The first commercial reactor for generating electricity went into operation in 1957 at Shippingsport, Pennsylvania. The basic design of a fission nuclear reactor is shown in ● Fig. 10.17.

Enriched uranium oxide fuel pellets are placed in metal tubes to form long *fuel rods,* which are placed in the reactor core, where fission takes place. Also in the core are *control rods* made of neutron-absorbing materials such as boron (B) and cadmium (Cd). The control rods are adjusted (inserted or withdrawn) so that only a certain number of neutrons are absorbed, ensuring that the chain reaction releases energy at the rate desired. For a steady rate of energy release, one neutron from each fission event should initiate only one additional fission event. If more energy is needed, the rods are withdrawn farther. When fully inserted into the core, the control rods absorb enough neutrons to stop the chain reaction, and the reactor shuts down.

A reactor's core is basically a heat source, and the heat energy is removed by a coolant flowing through the core. In U.S. reactors the coolant is most commonly water. The coolant flowing through the hot fuel assemblies transfers heat, which is used to produce steam to drive a turbogenerator, which produces electricity.

In addition, the coolant acts as a *moderator.* The ^{235}U nuclei react best with "slow" neutrons. The neutrons emitted from the fission reactions are relatively

FIGURE 10.17   A Nuclear Reactor

(a) A schematic diagram of the core of a nuclear reactor. The position of the control rods determines the level of energy production by regulating the number of neutrons available to cause additional fission. The fuel rods contain uranium oxide, enriched to about 3% $^{235}UO_2$. (b) A schematic diagram of the elements of an electrical generating plant. Note the core in place at the left. The heat from the fission of the ^{235}U atoms in the fuel rods is used to form steam, which drives a turbine and generates electricity.

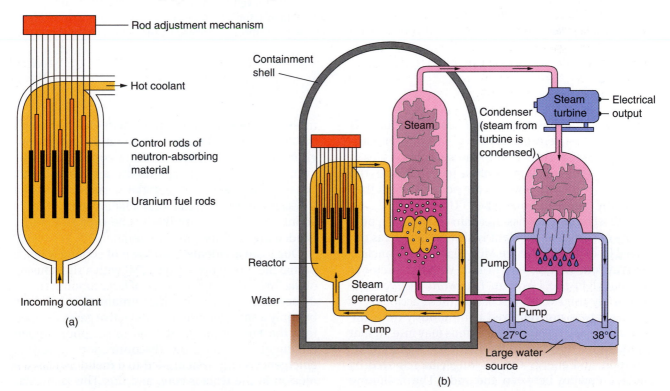

HIGHLIGHT: The Origins of the Bomb

In 1934, in Rome, Enrico Fermi and Emilio Segre bombarded uranium with neutrons and succeeded in making element 93, neptunium. In 1938, in Berlin, Otto Hahn and Fritz Strassman repeated the experiment and were surprised to find the element barium among the reaction products. Hahn described his findings in a letter to Lise Meitner, a former colleague who was an Austrian Jew but was living in Sweden. (Hahn had helped Meitner escape when the Nazis annexed Austria in 1938.)

Meitner surmised that Hahn had discovered a nuclear process in which the uranium atom was splitting. She informed her nephew, Otto Frisch, of her hypothesis, which she termed *nuclear fission*. Frisch passed the information on to Niels Bohr (photo on page 227), his colleague in Denmark, who was about to leave for a scientific conference at Princeton University.

When Bohr arrived at Princeton, he communicated the news of fission to Fermi, who had fled fascist Italy because his wife, Laura, was Jewish. (Fermi had persuaded Italian dictator Mussolini to let him take his family to Stockholm to see him receive the 1938 Nobel Prize in physics. After the ceremony, he and his family hastened off to the United States.) Bohr returned to Denmark but fled to the United States when Nazi armies overran Denmark in 1940. He almost died when he passed into a coma due to lack of oxygen while in a small plane flying him to England.

In 1939, at Columbia University, Leo Szilard (a Hungarian refugee) found that each fission produced more than one neutron, and thus a chain reaction could conceivably occur. Realizing that a chain reaction had tremendous explosive potential, in 1939 Szilard and Fermi composed a letter to President Roosevelt, informing him of their fears that Germany might be working on such a bomb. Afraid that Roosevelt would pay no attention to their letter, they took it to Albert Einstein, the most famous scientist in the world, who also had fled Nazi Germany and settled at Princeton University. Einstein signed the letter, and Roosevelt took it seriously.

Late in 1941, the *Manhattan Project* began—in top secrecy. On December 2, 1942, a group under Fermi's direction achieved the first self-sustaining fission reaction underneath the abandoned football stands at the University of Chicago (Fig. 1).

The major hurdle in the building of the bomb was the production of the fissionable material needed. It was the ^{235}U and not the more abundant ^{238}U that was undergoing fission. The natural uranium had to be enriched from its normal 0.7% ^{235}U to about 90% ^{235}U. This was accomplished at a sprawling, secret installation called Oak Ridge, in the hills of East Tennessee.

Plutonium-239, also fissionable and thus suitable for bomb building, became available when it was found to be formed by the beta decay of the neptu-

"fast," with energies that are not best suited for ^{235}U fission. The fast neutrons are slowed down, or moderated, by transferring energy to the water molecules in collision processes. After only a few collisions, the neutrons are slowed down to the point at which they efficiently induce fission in the ^{235}U nuclei.

With a continuous-fission chain reaction, the possibility of a nuclear accident is always present, as occurred in 1979 at the Three Mile Island (TMI) nuclear plant in Pennsylvania and in 1986 with the reactor at Chernobyl in the Ukraine. The word *meltdown* is commonly used when discussing these accidents. If heat energy is not removed continuously from the core of a fission reactor, the fuel rods may fuse. In that event, the reaction cannot be controlled with the control rods, and energy can no longer be removed by coolant flowing between the rods. The fissioning mass becomes extremely hot, melts down through the floor of the containment vessel, and enters the environment. Even under the worst conditions, however, a reactor cannot explode like a nuclear bomb, because the fissionable material present is far from sufficient purity. (Recall that reactor-grade fuel is about 3% ^{235}U, compared with the over 90% purity needed for weapons-grade material.)

Due to an accidental shutdown of cooling water, a partial meltdown occurred at TMI with a slight fusing of the fuel rods and the release of large amounts of radioactive material inside the containment building, but only a small amount of radioactive gases escaped into the environment. At Chernobyl, poor human judgment, including the disconnection of several emergency safety systems, led to a meltdown, an explosion in the reactor core, and fire. This particular

FIGURE 1   Nuclear Fission

The first self-sustaining nuclear fission reaction occurred in Chicago on December 2, 1942, as depicted in this painting. Fermi is the partially bald gentleman at the center close to the rail.

porized the 30-m steel tower on which it was placed and melted the sand around the site.

Some scientists were so awed by the blast that they argued against the bomb's use. However, fear of millions of casualties on both sides in an imminent invasion of Japan persuaded President Truman to order the dropping of two bombs. "Little Boy," a ^{235}U bomb, was dropped on Hiroshima on August 6, 1945. The energy released was equivalent to that of 20,000 *tons* of TNT (thus it is called a 20-kiloton bomb). The casualties numbered 100,000. Three days later, a second bomb ("Fat Man," a ^{239}Pu bomb) was dropped on Nagasaki. Five days later, Japan surrendered.

nium produced by bombardment of ^{238}U with neutrons. A series of large reactors was built at Hanford, Washington, to produce ^{239}Pu.

The first atomic bomb used ^{239}Pu as the fissionable material and was developed and tested in New Mexico in July 1945. The heat from the explosion va-

type of reactor used graphite (carbon) blocks for a moderator. Gas explosions caused the carbon to catch fire, radioactive material escaped with the smoke, and weather conditions caused radioactive fallout to spread over many countries. Several hundred deaths occurred in the immediate region, and it is estimated that as many as 50,000 additional cancer deaths will occur from the long-term effects of the radioactive fallout.

In addition to ^{235}U, the other fissionable nuclide of importance is ^{239}Pu (half-life = 2.4×10^4 y). This plutonium isotope is produced by bombardment of ^{238}U with fast neutrons. This means that ^{239}Pu is produced as nuclear reactors operate, because all the neutrons are not moderated or slowed down. Because it is fissionable, ^{239}Pu extends the time before the refueling of a reactor.

In a *breeder reactor* this process is promoted, and fissionable ^{239}Pu is produced from ^{238}U, which is otherwise useless for energy production. It is said that if the depleted uranium present at Oak Ridge, Tennessee, were converted to ^{239}Pu, the fission of the ^{239}Pu could equal the energy output of all the oil in Saudi Arabia. The ^{239}Pu can be chemically separated from the fission by-products and used as the fuel in an ordinary nuclear reactor or in weapons. Its use in weapons is a major concern, because it can be obtained from regular reactors.

It takes about 20 breeder reactors running for 1 year to produce enough plutonium to fuel an additional reactor for a year. Breeders run at higher temperatures than conventional reactors and generally use liquid sodium as the coolant. The withholding of federal funds killed the Clinch River Breeder Project in Tennessee in 1983, but breeder research continues at Argonne National Laboratory in Illinois, with some notable advances being made. Breeder reactors are currently being used in France and Germany, and they may someday be used in the United States.

In addition to concerns about the safety of operating nuclear fission power plants, the other major problem is what to do with the radioactive waste generated. Disposal of the waste is controversial. Several factors must be considered in order to evaluate the risk from radioactive waste.

1. *What is the half-life of the radionuclide and any daughter products?* Nuclides that decay quickly essentially disappear in a few days or weeks. Some, such as plutonium-239, have half-lives of thousands of years. A stable daughter product causes no more radiation problems. But if the daughter is radioactive, it will emit additional radiation.

2. *What type of radiation is emitted?* Because alpha, beta, and gamma radiation have different energies and penetrating abilities (see Section 10.7), the risk of biological damage depends on the type of radiation emitted. Although it is easy to shield alpha and beta radiation, several inches of lead or even more of concrete are required to shield against gamma rays.

3. *Is the radionuclide easily incorporated into the food chain?* Nuclides that decay slowly and are essential elements for growth are more likely to enter the food chain.

High-level radioactive waste—that composed of the long-half-life nuclides—will have to be stored safely for hundreds of thousands of years. It cannot be dumped in the ocean because the canisters would corrode and release their contents. Most of the waste is being held in temporary storage facilities at present. The favorite proposed solution to the growing problem of disposal is to encapsulate the waste in glass or ceramic and bury it deep underground in geologically stable rock formations. Congress has approved funds for the development of a permanent disposal site at Yucca Mountain in Nevada. However, many questions remain about this disposal method, and safety concerns about the transport of the waste to the burial site must be adequately addressed.

RELEVANCE QUESTION: *Where is the nuclear reactor closest to you?*

10.6 Nuclear Fusion

....

LEARNING GOALS

....

▼ Explain how nuclear fusion occurs.

▼ Tell the significance of *mass defect*.

▼ Calculate mass and energy changes in nuclear reactions.

....

Fusion is the process in which smaller nuclei combine to form larger ones, with the release of energy. Fusion is the source of energy of the Sun and other stars. In the Sun, the fusion process produces a helium nucleus from four protons (hydrogen nuclei). Also produced are two positrons. The thermonuclear process takes place in several steps, with the net result being

$$4\,^1_1\text{H} \rightarrow\,^4_2\text{He} + 2\,^0_{+1}\text{e} + \text{energy}$$

In the Sun, about 600 million tons of hydrogen are converted to 596 million tons of helium *every second.* The other 4 million tons of matter are converted to energy. Fortunately, the Sun has enough hydrogen to produce energy at its present rate for several billion more years.

Two other examples of fusion reactions are

$$^2_1\text{H} +\,^2_1\text{H} \rightarrow\,^3_1\text{H} +\,^1_1\text{H}$$

and

$$^2_1\text{H} +\,^3_1\text{H} \rightarrow\,^4_2\text{He} +\,^1_0\text{n}$$

In the first reaction, two deuterons fuse to form a triton and a proton. This is termed a D-D (deuteron-deuteron) reaction. In the second example (a D-T reaction), a deuteron and a triton form an alpha particle and a neutron (● Fig. 10.18).

Fusion involves no critical mass or size because there is no chain reaction to maintain. However, the repulsive force between two positively charged nuclei opposes fusing. This force is smallest for hydrogen fusion because the nuclei contain only one proton. To overcome the repulsive forces and initiate fusion, the kinetic energies of the particles must be increased by raising the temperature to about 100 million kelvins. At such high temperatures the hydrogen atoms are stripped of their electrons, and a **plasma** (a gas of electrons and protons or other nuclei) results. To achieve fusion, not only is a high temperature necessary but also the plasma must be confined at a high

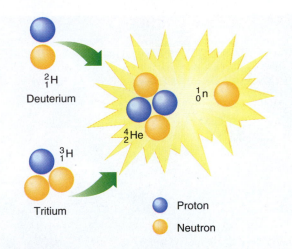

FIGURE 10.18   A D-T Fusion Reaction

The combination of a deuteron (D) and a triton (T) to produce an alpha particle and a neutron is one example of fusion. Nuclei of other elements of low atomic mass also can undergo fusion to produce heavier, more stable nuclei and release energy in the process.

enough density for the protons (or other nuclei) to collide frequently.

Large amounts of fusion energy have been released in an uncontrolled manner in a hydrogen bomb (H-bomb), where a fission bomb is used to supply the energy needed to initiate the fusion reaction ($\bullet$ Fig. 10.19). Unfortunately, controlled fusion for commercial use remains elusive. Controlled fusion might be accomplished by steadily adding fuel in small amounts to a fusion reactor. The D-T reaction requires the lowest temperature (about 100 million K) of any fusion reaction. For this reason, it is likely to be the first fusion reaction developed as an energy source.

Major problems arise in reaching such temperatures and confining the high-temperature plasma. If the plasma touches the reactor walls, it will cool rapidly. However, the walls will not melt, because even though the plasma is nearly 100 million K above the melting point of any material, the total quantity of heat that could be transferred from the plasma is very small because its concentration is extremely low.

One approach to controlled fusion is *inertial confinement,* a technique in which simultaneous high-energy laser pulses from all sides cause a fuel pellet containing deuterium and tritium to implode, resulting in compression and high temperatures. If the pel-

let stays intact for a sufficient time, fusion is initiated. Research on fusion initiated by lasers is being carried out at Los Alamos National Laboratory in New Mexico and at the Lawrence Livermore National Laboratory in California ($\bullet$ Fig. 10.20).

FIGURE 10.19   H-Bomb

The diagram shows the basic elements of a hydrogen bomb. To detonate an H-bomb, the TNT is exploded, forcing the ^{235}U together to get a supercritical mass and a fission explosion (a small atomic bomb, so to speak). The fusionable material is deuterium in the lithium deuteride (LiD). Heated to a plasma, D-D fusion reactions occur. Neutrons from the fission explosion react with the lithium to give tritium (^3H), and D-T fusion reactions also take place. The bomb is surrounded with ^{238}U, which tops off the explosion with a fission reaction. The result is shown in the photo.

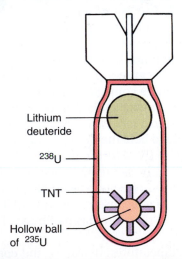

FIGURE 10.20   Laser Fusion
Using this gigantic apparatus at the Lawrence Livermore National Laboratory, 30 trillion watts of power are focused onto a tiny pellet of deuterium. The heat and compression give temperatures and densities near those found in the Sun's core, causing fusion for a brief instant.

Another approach to controlled fusion is *magnetic confinement*. Because a plasma is a gas of charged particles, it can be controlled and manipulated with electric and magnetic fields. A nuclear fusion reactor of the type called a *tokamak* uses a doughnut-shaped magnetic field to hold the plasma away from any material (● Fig. 10.21). Electric fields produce currents that raise the temperature of the plasma. The leading facilities for fusion research using magnetic confinement are the Massachusetts Institute of Technology (MIT) and the Princeton Plasma Physics Laboratory in New Jersey.

Plasma temperatures, densities, and confinement times have been problems with magnetic confinement, and commercial energy production is not expected until well into the twenty-first century. Even so, fusion is a promising energy source because of its advantages over fission:

1. *The low cost and abundance of deuterium,* which can be extracted inexpensively from water. Scientists estimate that the deuterium in the top two inches of water in Lake Erie could provide fusion energy equal to the combustion energy in all the world's oil reserves. On the other hand, uranium for fission is scarce, expensive, and hazardous to mine.
2. *Dramatically reduced nuclear waste disposal problems.* Some fusion by-products are radioactive because of nuclear reactions involving the neutrons that are formed in the D-T reactions, but they have relatively short half-lives compared with those of fission wastes.
3. *Fusion reactors could not get out of control.* In the event of a system failure in a fusion plant, the reaction chamber would immediately cool down, and energy production would halt.

The disadvantages of fusion compared with fission are that fission reactors are presently operational (commercial fusion reactors are at least decades away) and fusion plants will be more costly to build and operate than fission plants.

Nuclear Reactions and Energy

In 1905 Albert Einstein published his special theory of relativity, which deals with the changes in mass, length, and time as an object's speed approaches the speed of light (*c*). The theory also predicted that mass (*m*) and energy (*E*) are not separate quantities but are related by the equation

$$E = mc^2 \qquad \textbf{10.2}$$

The predictions proved correct. Scientists have indeed changed mass to energy and, on a very small scale, converted energy to mass.

FIGURE 10.21 Magnetic Confinement

An apparatus called a *tokamak* uses strong magnetic fields to confine the plasma for fusion to a doughnut-shaped region inside a vacuum vessel. (From Ebbing, Darrell D., *General Chemistry*, Sixth Edition. Copyright © 1999 by Houghton Mifflin Company. Used with permission.)

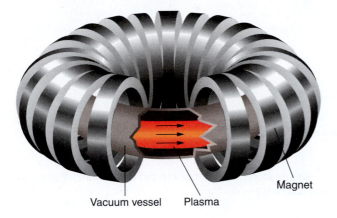

Vacuum vessel   Plasma   Magnet

For example, a mass of 1.0 g (0.0010 kg) has an equivalent energy of

$$E = mc^2 = (0.0010 \text{ kg})(3.00 \times 10^8 \text{ m/s})^2 = 90 \times 10^{12} \text{ J}$$

This 90 *trillion* joules is the same amount of energy that is released by the explosion of about 20,000 *tons* of TNT! Such calculations convinced scientists that nuclear reactions in which just a small amount of mass was "lost" were a potential source of vast amounts of energy.

The units of mass and energy commonly used in nuclear physics are different from those discussed in preceding chapters. Mass is usually given in atomic mass units, u, and energy is usually given in million electron volts, MeV (1 MeV = 1.60×10^{-13} J). With these units, Einstein's equation reveals that 1 u of mass has the energy equivalent of 931 MeV, so we can say that there are 931 MeV/u.

To determine the change in mass and hence the energy released or absorbed in any nuclear process, just add up the masses of all reactant particles and subtract from that sum the total mass of all product particles. If an increase in mass has taken place, the reaction is *endoergic* (absorbs energy) by that number of u times 931 MeV/u. If, as is more common, a decrease in mass has resulted, the reaction is *exoergic* (releases energy) by that number of u times 931 MeV/u. The decrease in mass in a nuclear reaction is often referred to as the **mass defect.**

EXAMPLE 10.9

Calculating Mass and Energy Changes in Nuclear Reactions

Calculate the mass defect and the corresponding energy released during this typical fission reaction:[*]

$$^{236}_{92}\text{U} \rightarrow {}^{88}_{36}\text{Kr} + {}^{144}_{56}\text{Ba} + 4\,{}^{1}_{0}\text{n}$$

(236.04556 u) (87.91445 u) (143.92284 u) (4 × 1.00867 u)

SOLUTION

The total mass on the left of the arrow is 236.04556 u. Adding the masses of the particles on the right gives 235.87197 u. The difference (0.17359 u) is the mass defect, which has been converted to

$$(0.17359 \text{ u})(931 \text{ MeV/u}) = 162 \text{ MeV of energy}$$

Thus, during the reaction, 0.17359 u of mass is converted to 162 MeV of energy.

CONFIDENCE EXERCISE 10.9

Calculate the mass defect and the corresponding energy released during a D-T fusion reaction:

$$^{2}_{1}\text{H} + {}^{3}_{1}\text{H} \rightarrow {}^{4}_{2}\text{He} + {}^{1}_{0}\text{n}$$

(2.0140 u)   (3.0161 u)   (4.0026 u)   (1.0087 u)

Your answer to Confidence Exercise 10.9 shows that 17.5 MeV is released in the fusion reaction, which is less than the 162 MeV released in the fission reaction of Example 10.9. However, we started with only 5.03 u in the fusion reaction but 236.05 u in the fission reaction. From a percentage standpoint, your calculations indicate that 0.0188/5.03 (× 100%) = 0.374% of the initial mass is converted to energy in the fusion reaction, whereas only 0.1736/236.05 (× 100%) = 0.07354% of the initial mass is converted to energy in the fission reaction.

[*]Because we are dealing with differences in mass, either the masses of the atoms or the masses of just their nuclei can be used. The number of electrons is the same on each side of the equation and thus does not affect the mass difference. We will use the masses of the atoms because they are more easily found in handbooks.

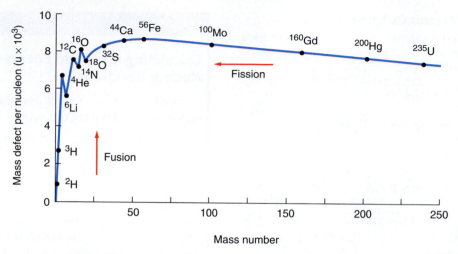

FIGURE 10.22 **The Relative Stability of Nuclei**

(See text for discussion. The *mass defect per nucleon* is simply the mass lost in the formation of the nucleus from its component neutrons and protons divided by the number of nucleons involved.)

The bottom line is that *kilogram for kilogram,* we
can get more energy from fusion than from fission.
(But recall that it takes a lot of energy to operate a fusion reactor.) Comparing fission and fusion with energy production by ordinary chemical reactions, the
fission of 1 kg of uranium-235 provides energy equal
to burning 2 million kg of coal, whereas the fusion of
1 kg of deuterium releases the same amount of energy
as the burning of 40 million kg of coal.

● Figure 10.22, which is a plot of *mass defect per
nucleon* for various nuclei as a function of the *mass
number* (the number of nucleons), shows that energy
can be released in both nuclear fission *and* nuclear
fusion. Note that fission of heavy nuclei at the far right
of the curve to intermediate-sized nuclei in the middle leads upward on the curve. Also, fusion of small
nuclei on the left to larger nuclei farther to the right
also leads upward on the curve. Any reaction that
leads upward on the curve in Fig. 10.22 releases energy, because such a reaction is accompanied by a
mass defect. Basically, each nucleon in the reactant
nucleus (or nuclei) loses a little mass in the process.

Any nuclear reaction in which the products are
lower on the curve than are the reactants can proceed
only with a net increase in mass and a corresponding
net absorption of energy. One type of nucleus cannot
give a net release of nuclear energy either by fission or
fusion. Of course, it is the one at the top of the curve,
^{56}Fe (iron-56). You can't go higher than the top, so no

net energy will be released either by splitting ^{56}Fe into
smaller nuclei or by fusing several ^{56}Fe into a larger
nucleus.

10.7 Biological Effects of Radiation

LEARNING GOALS

▼ Discuss how radiation affects organisms.

▼ Describe sources of radiation exposure.

Radiation that is energetic enough to knock electrons
out of atoms or molecules and form ions is classified
as *ionizing radiation.* Alpha particles, beta particles,
neutrons, gamma rays, and X-rays all fall into this category. Such radiation can damage or even kill living
cells, and it is particularly harmful when it affects
protein and DNA molecules involved in cell repro-
duction. Ionizing radiation is especially dangerous
because you cannot see, smell, taste, or feel it. Occupational exposures are often measured by the degree
of exposure of film badges.

The effects of radiation on living organisms can be
classified as follows.

1. **Somatic effects** are short-term and long-term effects on the recipient of the radiation.

2. **Genetic effects** are defects in the recipient's subsequent offspring.

When discussing the somatic effects of radiation, the SI unit used for equivalent absorbed dose is the sievert (Sv).[*] This unit takes into consideration the relative ionizing power of each type of radiation and its ability to affect humans. For the dose levels encountered in natural background radiation and diagnostic procedures such as X-rays, the millisievert (mSv) is an appropriate unit.

● Figure 10.23 shows the average radiation exposure (3.6 mSv) received annually in the United States from various sources, with about 82% coming from natural sources and 18% from human-made sources. However, individual exposures vary widely, depending on location, occupation, and personal habits.

The average U.S. citizen is exposed to natural and human-made background radiation of about 3 mSv per year. Sources of natural radiation include cosmic rays from outer space, so frequent flying in jetliners and living in high-altitude cities such as Denver, Colorado, provide more exposure to this part of the background radiation. Cosmic rays also form radionuclides such as carbon-14 and potassium-40. Carbon and potassium are essential elements in living organisms, and so these radionuclides become a part of all living organisms and thus another source of natural background radiation.

Other sources of natural background radiation include radionuclides in the rocks and minerals in our environment. One of the decay products of ^{238}U is radon-222. Radon gas and its radioactive daughters can be breathed into the lungs, where additional decays emit radiation. About 10,000 of the 130,000 annual lung cancer deaths in the United States are thought to be caused by indoor radon pollution. Exposure to radon varies greatly with location, because some soils and bedrocks are relatively rich in uranium, whereas others contain very little.

Human-made sources of radiation include X-rays and radionuclides used in medical procedures, fallout from nuclear testing, TVs, tobacco smoke, nuclear wastes, and emissions from power plants. Ironically, because fossil fuels contain traces of uranium and thorium and their daughters, more radioactivity is released into the atmosphere from power plants burning coal and oil than from nuclear power plants.

[*]An older and still widely used unit is the rem (*r*oentgen *e*quivalent, *m*an), and 1 rem = 1000 mrem = 0.01 Sv = 10 mSv.

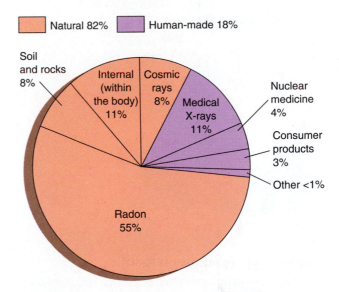

FIGURE 10.23 Sources of Exposure to Radiation
On average, each person in the United States receives a yearly radiation exposure of 3.6 mSv, of which 82% is from the natural sources shown in the chart. The other 18% is from human-made sources. (From Ebbing, Darrell D., and R. A. D. Wentworth, *Introductory Chemistry*, Second Edition. Copyright © 1998 by Houghton Mifflin Company. Used with permission.)

Table 10.8 lists the typical short-term somatic effects for an individual exposed to a single dose of radiation to the whole body. One-time exposure to radiation of up to 250 mSv gives no noticeable short-term somatic effects, but the cumulative effects of such exposures are not fully understood. The most common long-term somatic effect is an increased likelihood of developing cancer, particularly cancer of the blood or bone. Many early workers with radionuclides developed cancer from small doses of radiation

TABLE 10.8 Short-Term Somatic Effects of a Single Dose of Whole-Body Radiation

Dose (mSv)	Probable Effect
0–250	No detectable effects
250–1000	Temporary decrease in white blood cells
1000–2000	Vomiting, loss of hair
2000–6000	Vomiting, diarrhea, hemorrhaging, possible death
6000+	Death

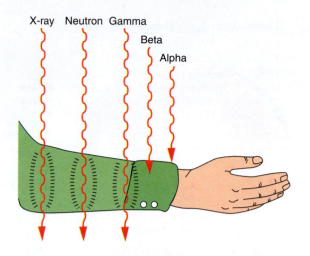

FIGURE 10.24   Penetration of Radiation

Alpha particles cannot penetrate clothing or skin. Beta particles can barely penetrate them. However, gamma rays, X-rays, and neutrons can penetrate an arm easily.

repeated over long periods of time. Some developed cancer up to 40 years after initial exposure.

A disturbing aspect of the situation is that there seems to be no lower limit, or *threshold,* below which the genetic effects of radiation are negligible. Thus any increase in radiation level is taken seriously. The federal government maintains strict safety standards for occupational exposure to radiation, but just what levels of exposure are "safe" remains controversial.

Shielding is used to decrease radiation exposure. Alpha and beta particles are electrically charged and can be stopped by materials such as paper and wood because of interactions with charged particles within these materials. Gamma rays, X-rays, and neutrons are not charged particles and are more difficult to stop (● Fig. 10.24). Shielding by thick lead or concrete is required for protection from these kinds of radiation.

RELEVANCE QUESTION: *What steps can you take to lower your exposure to ionizing radiation?*

Important Terms

electrons (10.2)	neutron number	radioactivity	critical mass
nucleus	mass number	alpha decay	fusion (10.6)
protons	isotopes	beta decay	plasma
neutrons	atomic mass	gamma decay	mass defect
nucleons	strong nuclear force	half-life	somatic effects (10.7)
atomic number	nuclide (10.3)	fission (10.5)	genetic effects
element	radionuclides	chain reaction	

Important Equations

Neutron Number = Mass Number − Atomic Number: $N = A - Z$

Mass-Energy: $E = mc^2$

Review Questions

10.1 Symbols of the Elements

1. What scientist devised the symbol notation we now use for elements?
   (a) Newton   (c) Dalton
   (b) Berzelius   (d) Einstein

2. Why do some symbols for elements seem to bear no relationship to their names?

10.2 The Atomic Nucleus

3. How many neutrons are in the nucleus of the atom $^{35}_{17}Cl$?
   (a) 35   (b) 17   (c) 18   (d) 52

4. The diameter of an average nucleus is about how many meters?
   (a) 10^{-5} m   (c) 10^{-10} m
   (b) 10^{-8} m   (d) 10^{-14} m

5. Name the three particles that comprise an atom. How do they compare in mass and charge?

6. What is the collective name of the two particles that comprise a nucleus?

7. Name the scientist who discovered the nucleus. In what year was the discovery made?

8. How does the diameter of an atom compare with that of its nucleus?

9. About what percentage of the mass of an atom is contained in the nucleus?

10. The chemical identity of an atom is determined by its number of _____.

11. What do the letters Z, A, and N stand for? State the equation that relates the three.

12. State the special names by which ^{1}H, ^{2}H, and ^{3}H are known.

13. Name the instrument used to detect and separate isotopes. On what basic principle does it operate?

14. On which atom are all atomic masses based? What mass is assigned to that atom?

15. Name the force that holds a nucleus together. At what distance does the force drop to zero?

10.3 Radioactivity and Half-Life (and Highlight)

16. What is the missing particle in the nuclear decay $^{179}_{79}$Au → $^{175}_{77}$Ir + __ ?
    (a) deuteron   (c) beta particle
    (b) neutron   (d) alpha particle

17. The majority of stable nuclides belong to which category?
    (a) odd-odd   (c) even-odd
    (b) even-even   (d) odd-even

18. Which of the following scientists discovered radioactivity?
    (a) Rutherford   (c) Becquerel
    (b) Libby   (d) Curie

19. How many half-lives would it take for a sample of a radionuclide to decrease its activity to $\frac{1}{32}$ of the original amount?
    (a) 32   (b) 16   (c) 6   (d) 5

20. Name the wife-and-husband team that discovered radium and polonium.

21. Use the letters A, B, and b to show the general form of a nuclear decay.

22. Give the symbol for an alpha particle, a beta particle, and a gamma ray.

23. List which of the three types of radiation—alpha, beta, or gamma—each of the following statements describes.
    (a) is not deflected by a magnet
    (b) has a negative charge
    (c) consists of ions
    (d) is similar to X-rays
    (e) has a positive charge

24. In a nuclear decay, the sums of the _____ will be the same on each side of the arrow, as will be the sums of the _____ .

25. No stable nuclides exist having Z greater than what number?

26. After two half-lives have gone by, what fraction of a sample of a radionuclide remains?

27. If uranium is radioactive, why does it occur naturally on Earth?

10.4 Nuclear Reactions

28. Which of the following completes the reaction $^2_1$H + $^{98}_{42}$Mo → __ + $^1_0$n?
    (a) $^{97}_{42}$Mo   (b) $^{99}_{43}$Md   (c) $^{99}_{43}$Tc   (d) $^{98}_{41}$Nb

29. Use the letters a, A, B, and b to show the general form of a nuclear reaction.

30. To what common use is americium-241 put?

31. Name the analytical procedure that bombards a sample with neutrons and measures the frequencies of gamma rays emitted.

10.5 Nuclear Fission (and Highlight)

32. What would be the appropriate procedure to decrease the heat output of a fission reactor core?
    (a) insert the control rods farther
    (b) remove a few fuel rods
    (c) increase the level of coolant
    (d) decrease the amount of moderator

33. Several of the subatomic particles named _____ are released when a nucleus fissions.

34. In terms of a chain reaction, explain what is meant by critical mass, subcritical mass, and supercritical mass.

35. What percentage of natural uranium is fissionable ^{235}U? To what percentage must that be enriched for use in a U.S. nuclear reactor? For use in nuclear weapons?

36. Name the scientist who led the successful effort to produce a self-sustaining chain reaction. Where, and in what year, did this occur?

37. What was the Manhattan Project?

38. Both control rods and moderators are involved with neutrons in a nuclear reactor. What is the role of each?

39. Why can't nuclear reactors accidentally undergo a nuclear explosion?

40. During operation, breeder reactors make fissionable fuel from nonfissionable material. What is the identity of the fuel made and the nonfissionable material used?

10.6 Nuclear Fusion

41. What is a very hot gas of nuclei and electrons called?
    (a) tokamak   (c) plasma
    (b) laser   (d) ideal gas

42. Name the nuclear process that allows stars to emit such enormous amounts of energy.

43. In discussions of nuclear fusion, for what do the letters D and T stand?

44. What is a plasma? Briefly describe two main approaches to forming and confining plasmas for controlled fusion.

45. Briefly tell three advantages and two disadvantages of energy production by fusion versus fission.

46. The equation $E = mc^2$ was developed by what scientist? Identify what each of the letters stands for.

47. What term is applied to a nuclear reaction if the products have less mass than the reactants? What term is applied to the difference in mass in such a case?

48. Show by means of a sketch how nuclear energy can be released by both fission and fusion.

10.7 Biological Effects of Radiation

49. Which of the following units is most closely associated with the biological effects of radiation?
    (a) the curie   (c) the becquerel
    (b) the sievert   (d) the cpm

50. What number of mSv received all at once virtually assures death to a human?

51. Distinguish between the somatic effects and the genetic effects of radiation exposure.

52. Name two natural sources of radiation exposure and two human-made sources.

Applying Your Knowledge

1. Suppose you pick up the morning newspaper and see a story reporting the discovery of a new element. The story states that two isotopes of the element have been formed: $^{274}_{114}X$ and $^{276}_{115}X$. What mistake has been made?

2. At dinnertime one evening you get a phone call urging you to invest in a new chemical procedure that produces the precious metal platinum from iron. Would you invest?

3. Even though fusion-based power plants are not yet a reality, our daily lives depend on fusion. Explain why.

4. What are the pros and cons of putting high-level radioactive waste on unmanned spaceships and shooting them into the Sun?

5. Suppose you have to make a choice between living downwind of a nuclear power plant or a coal-burning power plant. Which would you choose?

6. How would an instructor calculate the final grade for a student who has a total exam grade of 75 and a total lab grade of 85 if the exams count 70% of the grade and the labs count 30%? (How does the question fit into this chapter?)

Exercises

10.2 The Atomic Nucleus

1. For each of the following atoms, tell the number of nucleons, the number of protons, the number of electrons, the number of neutrons, and the chemical identity (symbol of element).
    (a) ^7_3X (b) $^{239}_{93}X$ (c) $^{31}_{15}X$ (d) $^{34}_{16}X$
       Answers: (a) 7, 3, 3, 4, Li   (b) 239, 93, 93, 146, Np
       (c) 31, 15, 15, 16, P   (d) 34, 16, 16, 18, S

2. For each of the following atoms, tell the number of nucleons, the number of protons, the number of electrons, the number of neutrons, and the chemical identity (symbol of element).
    (a) $^{90}_{38}X$ (b) $^{235}_{92}X$ (c) $^{11}_5X$ (d) $^{240}_{94}X$

3. On Earth, bromine occurs as 50.69% of ^{79}Br (atomic mass = 78.918 u) and 49.31% of ^{81}Br (atomic mass = 80.916 u). Calculate the atomic mass of bromine.
       Answer: 79.90 u

4. On Earth, potassium occurs as 93.26% of ^{39}K (atomic mass = 38.964 u), 0.012% of ^{40}K (atomic mass, 39.964 u), and 6.73% of ^{41}K (atomic mass = 40.962 u). Calculate the atomic mass of potassium.

10.3 Radioactivity and Half-Life

5. Complete the following equations for nuclear decay, and state whether the process is alpha decay, beta decay, or gamma decay.

   (a) $^{46}_{21}\text{Sc*} \rightarrow {}^{46}_{21}\text{Sc} + \underline{}$

   (b) $^{232}_{90}\text{Th} \rightarrow \underline{} + {}^{4}_{2}\text{He}$

   (c) $^{47}_{21}\text{Sc} \rightarrow {}^{47}_{22}\text{Ti} + \underline{}$

   Answers: (a) γ, gamma (b) $^{228}_{88}\text{Ra}$, alpha (c) $_{-1}^{0}\text{e}$, beta

6. Complete the following equations for nuclear decay, and state whether the process is alpha decay, beta decay, or gamma decay.

   (a) $^{237}_{93}\text{Np} \rightarrow \underline{} + {}_{-1}^{0}\text{e}$

   (b) $^{210}_{84}\text{Po} \rightarrow {}^{206}_{82}\text{Pb} + \underline{}$

   (c) $^{207}_{84}\text{Po*} \rightarrow {}^{207}_{84}\text{Po} + \underline{}$

7. Write the equation for the

   (a) alpha decay of $^{226}_{88}\text{Ra}$.

   (b) beta decay of $^{60}_{27}\text{Co}$.

   Answer: (a) $^{226}_{88}\text{Ra} \rightarrow {}^{222}_{86}\text{Rn} + {}^{4}_{2}\text{He}$ (b) $^{60}_{27}\text{Co} \rightarrow {}^{60}_{28}\text{Ni} + {}_{-1}^{0}\text{e}$

8. Thorium-229 ($^{229}_{90}\text{Th}$) undergoes alpha decay.
   (a) Write the equation.
   (b) The daughter formed in (a) undergoes beta decay. Write the equation.

9. Pick the radionuclide in each set. Explain your choice.

   (a) $^{249}_{98}\text{Cf}$ $^{12}_{6}\text{C}$

   (b) $^{79}_{35}\text{Br}$ $^{76}_{33}\text{As}$

   (c) $^{15}_{8}\text{O}$ $^{17}_{8}\text{O}$

   Answers: (a) $^{249}_{98}\text{Cf}$ ($Z > 83$) (b) $^{76}_{33}\text{As}$ (*odd-odd*)
   (c) $^{15}_{8}\text{O}$ (fewer n than p)

10. Pick the radionuclide in each set. Explain your choice.

   (a) $^{17}_{9}\text{F}$ $^{32}_{16}\text{S}$

   (b) $^{209}_{83}\text{Bi}$ $^{226}_{88}\text{Ra}$

   (c) $^{24}_{11}\text{Na}$ $^{14}_{7}\text{N}$

11. Technetium-99 (half-life = 6.0 h) is used in medical imaging. How many half-lives would go by in 36 h?

   Answer: six half-lives

12. How many half-lives would have to elapse for a sample of a radionuclide to decrease from an activity of 160 cpm to one of 5 cpm?

13. A thyroid cancer patient is given a dosage of ^{131}I (half-life = 8.1 d). What fraction of the dosage of ^{131}I will still be in his thyroid after 24.3 days?   Answer: $\frac{1}{8}$

14. A clinical technician finds that the activity of a sodium-24 sample is 480 cpm. What will be the activity of the sample 75 h later if the half-life of sodium-24 is 15 h?

15. Tritium (half-life = 12.3 y) is used to verify the age of expensive brandies. If an old brandy contains only $\frac{1}{16}$ of the tritium present in new brandy, how long ago was it produced?   Answer: 49 y

16. What is the half-life of thallium-206 if the activity of a sample drops from 4000 cpm to 250 cpm in 21.0 min?

17. Use the graph to find the half-life of radionuclide A. All you need is a sound understanding of the definition of half-life.

   Answer: For A, half-life = 22 d, because half of 40 g is 20 g, and 20 g is reached after 22 d.

18. Use the graph in Exercise 17 to find the half-life of radionuclide B.

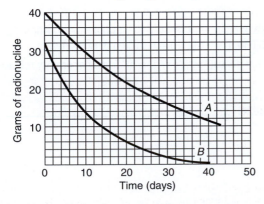

10.4 Nuclear Reactions

19. Complete the following nuclear reaction equations.

    (a) $^{4}_{2}\text{He} + ^{14}_{7}\text{N} \rightarrow ^{17}_{8}\text{O} + $ ___

    (b) $^{4}_{2}\text{He} + ^{27}_{13}\text{Al} \rightarrow ^{30}_{15}\text{P} + $ ___

    (c) ___ $+ ^{66}_{29}\text{Cu} \rightarrow ^{67}_{30}\text{Zn} + ^{1}_{0}\text{n}$

    (d) $^{1}_{0}\text{n} + ^{235}_{92}\text{U} \rightarrow ^{138}_{54}\text{Xe} + $ ___ $+ 5 ^{1}_{0}\text{n}$

    Answers: (a) $^{1}_{1}\text{H}$ (b) $^{1}_{0}\text{n}$ (c) $^{2}_{1}\text{H}$ (d) $^{93}_{38}\text{Sr}$

20. Complete the following nuclear reaction equations.

    (a) $^{16}_{8}\text{O} + ^{20}_{10}\text{Ne} \rightarrow $ ___ $+ ^{12}_{6}\text{C}$

    (b) $^{1}_{0}\text{n} + ^{28}_{14}\text{Si} \rightarrow $ ___ $+ ^{1}_{1}\text{H}$

    (c) ___ $+ ^{230}_{90}\text{Th} \rightarrow ^{223}_{87}\text{Fr} + 2 ^{4}_{2}\text{He}$

    (d) $^{4}_{2}\text{He} + ^{65}_{30}\text{Zn} \rightarrow $ ___ $+ 2 ^{1}_{0}\text{n}$

10.5 Nuclear Fission

21. Complete the following equation for fission.

    $$^{240}_{94}\text{Pu} \rightarrow ^{97}_{38}\text{Sr} + ^{140}_{56}\text{Ba} + \text{___}$$

    Answer: $3 ^{1}_{0}\text{n}$

22. Complete the following equation for fission.

    $$^{252}_{98}\text{Cf} \rightarrow \text{___} + ^{142}_{55}\text{Cs} + 4 ^{1}_{0}\text{n}$$

10.6 Nuclear Fusion

23. One of the fusion reactions that takes place as a star ages is called the triple alpha process.

    $$3 ^{4}_{2}\text{He} \rightarrow ^{12}_{6}\text{C}$$

    Calculate the mass defect (in u) and the energy produced (in MeV) each time the reaction takes place. (Atomic mass of $^{4}\text{He} = 4.00260$ u; atomic mass of $^{12}\text{C} = 12.00000$ u.)

    Answer: 0.00780 u, 7.26 MeV

24. Calculate the mass defect (in u) and the energy produced (in MeV) in the D-D reaction shown.

    $$^{2}_{1}\text{H} + ^{2}_{1}\text{H} \rightarrow ^{3}_{1}\text{H} + ^{1}_{1}\text{H}$$

    (2.0140 u) (2.0140 u) (3.0161 u) (1.0078 u)

Solutions to Confidence Exercises

10.1 6 protons, 6 electrons, 8 neutrons $(14 - 6)$

10.2 $(0.6000 \times 20.00 \text{ u}) + (0.4000 \times 22.00 \text{ u}) = 12.00 \text{ u} + 8.800 \text{ u} = 20.80 \text{ u}$

10.3 $^{226}_{88}\text{Ra} \rightarrow ^{222}_{86}\text{Rn} + ^{4}_{2}\text{He}$

10.4 $^{232}_{90}\text{Th}$ (above $Z = 83$) and $^{40}_{19}\text{K}$ (*odd-odd* nuclide)

10.5 $\dfrac{58 \text{ y}}{29 \text{ y/half-life}} = 2$ half-lives; $N_o \rightarrow \dfrac{N_o}{2} \rightarrow \dfrac{N_o}{4}$, or one-fourth will remain

10.6 First, see how many half-lives are needed to get to one-sixteenth of the original activity.

$$N_o \rightarrow \frac{N_o}{2} \rightarrow \frac{N_o}{4} \rightarrow \frac{N_o}{8} \rightarrow \frac{N_o}{16}$$

Four half-lives (count the arrows) are needed, so (4 half-lives)(6.0 h/half-life) = 24 h

10.7 $^{2}_{1}\text{H} + ^{27}_{13}\text{Al} \rightarrow ^{25}_{12}\text{Mg} + ^{4}_{2}\text{He}$

10.8 $^{236}_{92}\text{U} \rightarrow ^{90}_{38}\text{Sr} + ^{144}_{54}\text{Xe} + 2 ^{1}_{0}\text{n}$

10.9 $(2.0140 \text{ u} + 3.0161 \text{ u}) - (4.0026 \text{ u} + 1.0087 \text{ u}) = 5.030 \text{ u} - 5.0113 \text{ u} = 0.0188 \text{ u}$ of mass defect

$(0.0188 \text{ u})(931 \text{ MeV/u}) = 17.5$ MeV of energy released

Answers to Multiple-Choice Review Questions

1. b 4. d 17. b 19. d 32. a 49. b

3. c 16. d 18. c 28. c 41. c

THE CHEMICAL ELEMENTS

11

I approve of reasoning if it takes observed facts as its point of departure and methodically draws its conclusions from the phenomena.

Hippocrates (ca. 460–377 B.C.)

The division of physical science called **chemistry** deals with the composition and structure of matter (anything that has mass) and the reactions by which substances are changed into other substances. Chemistry had its beginnings early in history. When humans tamed fire more than 100,000 years ago, in a sense they began using chemistry. Egyptian hieroglyphs from 3400 B.C. show wine making, which requires a chemical fermentation process. By about 2000 B.C. the Egyptians and Mesopotamians produced and worked metals. The ancient Egyptians and Chinese prepared dyes, glass, pottery, and embalming fluids.

From about A.D. 500 to 1600, *alchemy* flourished. Its main objectives (never attained) were to change common metals into gold and to find an "elixir of life" to prevent aging. Modern chemistry began in 1774, with the work of Lavoisier (see the first Highlight in Chapter 12), and differs from alchemy by having reasonable objectives and avoiding mysticism, superstition, and secrecy.

Chemistry has five major divisions. *Physical chemistry,* the most fundamental, applies the theories of physics (especially thermodynamics) to the study of chemical systems in general. *Analytical chemistry* identifies what

Photo: Elements in watchglasses, clockwise from top, are Na, Cr, S, I, and Fe. Under the watchglasses are Zn and Cu strips. At top is Mg ribbon.

substances are present in a material and determines how much of each substance is present. *Organic chemistry* studies compounds that contain carbon. The study of all other chemical compounds is called *inorganic chemistry. Biochemistry,* where chemistry and biology meet, deals with the chemical reactions that occur in living organisms. These major divisions overlap, and smaller divisions exist, such as polymer chemistry and nuclear chemistry.

The 88 naturally occurring elements in our environment, either singly or in chemical combination, are the components of virtually all matter. The physical and chemical properties of the various elements affect us constantly. Our bodies and everything we wear, eat, breathe, and use are made of elements. In this chapter we will examine how matter is classified by chemists, discuss elements, develop an understanding of the periodic table, and learn how compounds are named. ■

11.1 Classification of Matter

LEARNING GOALS

▼ Explain how chemists classify matter.

▼ Distinguish among types of solutions and tell how they form.

In Chapter 5 we saw that matter can be classified by its physical phase: solid, liquid, gas, or plasma. However, the classification scheme summarized in ● Fig. 11.1 is useful in chemistry. It first divides matter into pure substances and mixtures.

A **pure substance** is a type of matter in which all samples have fixed composition and identical properties. Pure substances are divided into elements and compounds. An **element** is a substance in which all the atoms have the same number of protons—that is, the same atomic number (Section 10.2). Therefore, all samples of a given element have a fixed composition and identical properties. A **compound** is a substance composed of two or more elements chemically combined in a definite, fixed ratio by mass. All samples of a given compound have identical properties that are usually very different from the properties of the elements of which the compound is composed. Table 11.1 compares the properties of the compound zinc sulfide (ZnS) with those of zinc (Zn) and sulfur (S), the two elements into which it decomposes in a fixed ratio by mass of 2.04 parts zinc to 1 part sulfur.

A compound can be broken into its component elements only by chemical processes, such as the passage of electricity through melted zinc sulfide. Conversely, the formation of a compound from elements is also a chemical process.

A **mixture** is a type of matter composed of varying proportions of two or more substances that are just

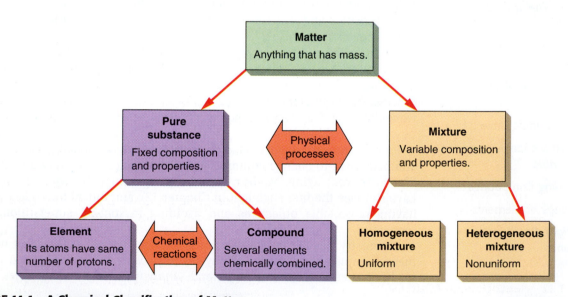

FIGURE 11.1 A Chemical Classification of Matter
(See text for description.)

TABLE 11.1 The Properties of a Compound Compared with Those of Its Component Elements

Property	Zinc Sulfide	Zinc	Sulfur
Appearance	White powder	Silvery metal	Yellow powder or crystals
Density (g/cm^3)	3.98	7.14	2.07
Melting point (°C)	1700	281	113
Conducts electricity as a solid	No	Yes	No
Conducts electricity as a liquid	Yes	Yes	No
Soluble in carbon disulfide	No	No	Yes

physically mixed, *not* chemically combined. Different samples of a particular mixture can have variable composition and properties. For example, a mixture of zinc and sulfur could consist of any mass ratio of zinc to sulfur; it would not be restricted to the 2.04-to-1 mass ratio found in the compound zinc sulfide. Mixtures are formed and broken down by physical processes such as dissolving and evaporation. For example, the mixture of zinc and sulfur could be separated simply by adding carbon disulfide (CS_2) to dissolve the sulfur, filtering off the zinc, and then evaporating the carbon disulfide to leave a deposit of sulfur crystals.

A *heterogeneous mixture* is one that is nonuniform; that is, at least two components can be observed. Examples are a pizza, a bottle of Italian salad dressing (● Fig. 11.2), and a pile of zinc and sulfur.

FIGURE 11.2 Heterogeneous Mixtures
Both the salad and the Italian dressing are heterogeneous mixtures.

A mixture that is uniform throughout is called a *homogeneous mixture,* or a **solution;** that is, it looks like it might be just one substance.* Most solutions we encounter consist of one or more substances (such as coffee crystals or salt) dissolved in water. However, a solution need not be a liquid; it also may be a solid or a gas. A metal *alloy* such as brass, a mixture of copper and zinc, is an example of a solid solution. Air, a mixture composed mainly of nitrogen and oxygen, is an example of a gaseous solution. Each appears uniform, but different samples of coffee, salt water, brass, and air can have different compositions. In a solution containing two substances, the liquid, or the substance present in the larger amount, is called the *solvent,* and the other substance is termed the *solute* (● Fig. 11.3).

Aqueous Solutions

A solution in which water is the solvent is called an *aqueous solution* (abbreviated *aq*). When a solute dissolves in water and is stirred thoroughly, the distribution of its particles (molecules or ions) is the same throughout the solution. If more solute can be dissolved in the solution at the same temperature, the solution is called an **unsaturated solution.** As more solute is added, the solution becomes more and more concentrated. Finally, when the maximum amount of solute is dissolved in the solvent, we have a **saturated solution.** Usually, some undissolved solute remains on the bottom of the container, and a dynamic

*Technically, to be a true solution the components must be mixed on the atomic or molecular level, so that even a magnifying glass or microscope would not reveal that the sample was several substances mixed.

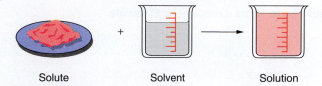

Solute   Solvent   Solution

FIGURE 11.3   A Solute and a Solvent Mix to Form a Solution

Solutions are homogeneous mixtures in which one substance (the solute) is dissolved in another (the solvent). The molecules or ions of the solute are dispersed evenly throughout the solvent.

(active) equilibrium is set up between solute dissolving and solute crystallizing (Fig. 11.4).

The **solubility** of a particular solute is the amount of solute that will dissolve in a specified volume or mass of solvent (at a given temperature) to produce a saturated solution. Solubility depends on the temperature of the solution. If the temperature is raised, the solubilities of practically all solids increase (Fig. 11.5). That is, hot water dissolves more solute than cold water.

When unsaturated solutions are prepared at high temperatures and then cooled, solubility decreases and may reach the saturation point, where excess solute normally begins crystallizing from the solution. However, if no crystals of the solid (or other nucleation sites) are already present in the solution, crystallization may not take place if the solution is cooled carefully. The solution will then contain a larger amount of solute than the solubility of the

FIGURE 11.4   A Saturated Solution

In a saturated solution containing excess solute, equilibrium exists between solute dissolving and solute crystallizing.

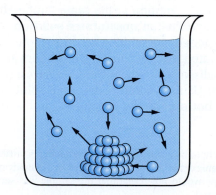

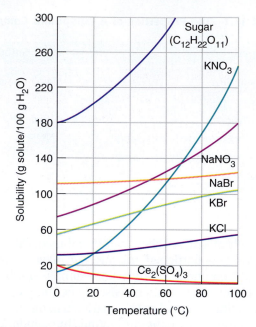

FIGURE 11.5   The Effect of Temperature on Solubilities of Salts in Water

As the graph indicates, the solubilities of the majority of salts increase as the solution's temperature rises.

solute dictates. A solution that contains more than the normal maximum amount of dissolved solute at its temperature is a **supersaturated solution.** Such solutions are unstable, and the introduction of a "seed" crystal will cause the excess solute to crystallize immediately (Fig. 11.6).

This is the basic principle behind seeding clouds to cause rain. If the air is supersaturated with water vapor, the introduction of certain types of crystals into the clouds greatly increases the probability that the water vapor will form raindrops (see Chapter 20).

Now let's consider the solubility of gases such as carbon dioxide (CO_2) in water. The solubility of gases increases with increasing pressure. This principle is used in the manufacture of soft drinks. CO_2 is forced into the beverage under high pressure. Then the beverage is bottled and capped tightly to maintain pressure on the CO_2. Once the bottle is opened, the pressure inside the bottle is reduced to normal atmospheric pressure, and the CO_2 starts escaping from the liquid, as evidenced by rising bubbles. If the bottle is open for some time, most of the CO_2 escapes, and the drink tastes flat.

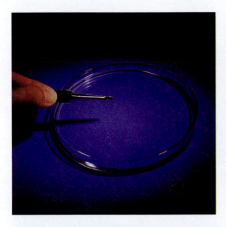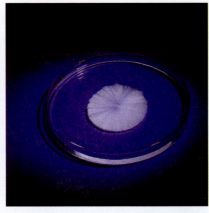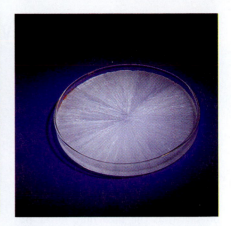

FIGURE 11.6 Supersaturated Solution

When a seed crystal is added to a supersaturated solution of sodium acetate, $NaC_2H_3O_2$ (left), the excess solid quickly crystallizes (middle), forming a saturated solution (right).

The solubility of gases in water decreases with increasing temperature. If an unopened soft drink is allowed to warm, the solubility of CO_2 decreases. When the bottle is opened, CO_2 may escape so fast that the beverage shoots out of the bottle. (See the chapter's first Highlight for another example.)

RELEVANCE QUESTION: *Name one compound and one homogeneous mixture you have encountered today.*

11.2 Discovery of the Elements

LEARNING GOALS

▼ Tell how the concept of *element* developed.

▼ Trace the history of the discovery of elements.

Section 10.1 mentioned that Aristotle's erroneous idea of four elements was dominant for almost 2000 years. Then, in 1661, Robert Boyle (● Fig. 11.7), an Irish-born chemist, developed a definition of element that took the concept from the realm of speculation and made it subject to laboratory testing. (This was the same era in which Isaac Newton was developing the laws that made physics a modern science.) In his book *The Skeptical Chemist,* Boyle proposed that the designation *element* be applied only to those substances that could not be separated into components by any method.

FIGURE 11.7 Robert Boyle (1627–1691)

Boyle wrote *The Skeptical Chemist* in 1661 and discovered Boyle's law of gases: The volume and pressure of a gas are inversely proportional.

HIGHLIGHT: The Tragedy of Lake Nyos

Nearly 2000 people were killed on August 21, 1986, when a cloud of gas suddenly erupted from Lake Nyos in Cameroon, Africa (Fig. 1). At first it was thought that the gas was hydrogen sulfide (H_2S), but it now seems clear that the huge, suffocating cloud was composed of the dense gas carbon dioxide (CO_2).

Why the cloud of CO_2 suddenly boiled from the lake may never be known for certain, but scientists think that the lake suddenly "turned over," bringing to the surface huge quantities of dissolved CO_2. Lake Nyos is deep, and layers of warm, less dense water near the surface float on the colder, denser water nearer the bottom of the lake. Normally, the lake remains this way, with little mixing among the layers.

Over hundreds or thousands of years, CO_2 must have seeped into the cold water at the bottom of the lake and dissolved in large amounts because of the great pressure present. On that fateful August day, a landslide,

FIGURE 1   Lake Nyos in Cameroon, Africa

wind, or cooling of the lake's surface due to monsoon clouds must have caused the water that was supersaturated with CO_2 to reach the surface and release tremendous quantities of gaseous CO_2. Being dense, colorless, and odorless, the gas spread out along the ground and replaced the oxygen-containing air. The people and animals nearby never knew what hit them.

Boyle championed the need for experimentation in science. It was he who initiated the practice of carefully and completely describing experiments so that anyone might repeat and confirm them. This procedure became universal in science. Without it, progress probably would have continued at a snail's pace. In addition, it was Boyle who first performed truly quantitative physical experiments, finding the inverse relationship between the pressure and volume of a gas (Section 5.6).

In the earliest of civilizations, 12 substances that later proved to be elements were isolated: gold, silver, lead, copper, tin, iron, carbon, sulfur, antimony, arsenic, bismuth, and mercury. Phosphorus is the first element whose date of discovery is known. It was isolated from urine by Hennig Brand, a German, in 1669,

eight years after Boyle's definition of element. By 1746, platinum, cobalt, and zinc had been discovered. The rest of the 1700s saw the discovery of ten more metals, the nonmetal tellurium, and the gaseous elements hydrogen, oxygen, nitrogen, and chlorine. ● Figure 11.8 shows how many elements were known at certain times in history.

About 1808, Humphry Davy, an English chemist, used electricity from a battery (a recent invention at the time) to break down compounds, thereby discovering six elements (Na, K, Mg, Ca, Ba, Sr), the record for one person. This is a good example of how advances in technology can result in advances in basic science.

By 1895, a total of 73 elements were known. From 1895 to 1898, the noble gases helium, neon, krypton,

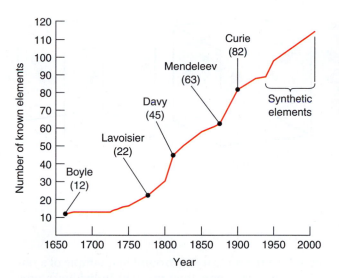

FIGURE 11.8   The Discovery of the Elements

The graph shows how many elements were known at various points in history. Note the number of elements known at the time of greatest contribution of well-known scientists.

and xenon ("ZEE-non") were discovered. The Curies (see the first Highlight in Chapter 10) discovered radium and polonium in 1898. The first synthetic element, technetium, was produced in 1937 by nuclear bombardment of the element molybdenum (Mo) with deuterons. About 20 more synthetic elements have followed, many made at Berkeley, California, by a team under the direction of Glenn Seaborg, giving a present total of 112 known elements. (Element 112 was produced in 1996 by German scientists who bombarded lead atoms with a beam of zinc ions.) More are expected to be created by nuclear bombardment using particle accelerators (Section 10.4).*

In Section 10.1 we stated that Berzelius is responsible for our present method of designating elements by using one or two letters of their names. Table 10.1 on page 233 listed the 45 elements whose names and symbols it is helpful to know. During the time lapse between a new element's confirmed discovery and its official naming, it is symbolized by the initial letters for the Greek prefixes that designate the element's

*In January 1999, Russian scientists reported the creation of a single atom of element 114 by bombardment of plutonium-244 with calcium-48 ions. However, the result has not been verified as of this writing.

atomic number. For example, element 110 is temporarily symbolized Uun (from the initial letters of *una* for one, *una* for one, and *nil* for zero).

11.3 **Occurrence of the Elements**

LEARNING GOALS

▼ State what elements are most common in our environment.

▼ Define the terms *molecule* and *allotrope*.

● Figure 11.9 shows that about 74% of the mass of Earth's crust is composed of only two elements—oxygen (47%) and silicon (27%). Earth's core is thought to be about 85% iron and 15% nickel. Earth's atmosphere close to the surface consists of 78% nitrogen, 21% oxygen, and almost 1% argon, together with traces of other elements and compounds (see Fig. 19.1 on page 501). The human body consists primarily of oxygen (65%) and carbon (18%).

FIGURE 11.9   Relative Abundance (by Mass) of Elements in Earth's Crust

Two elements, oxygen and silicon, account for about 75% of the mass of elements in Earth's crust.

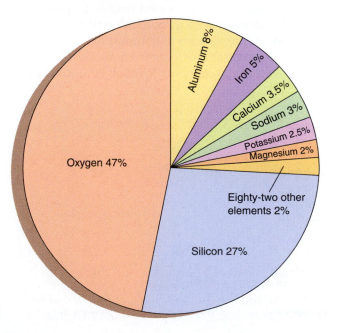

FIGURE 11.10   Atoms in Iron, Xenon, and Hydrogen

(a) Iron and other metallic elements are represented by just the symbol (such as Fe), because the individual atoms pack together in a repeating pattern. (b) Xenon and the other noble gases exist as single atoms, so they also are represented by just the symbol (such as Xe). (c) Hydrogen gas is composed of diatomic molecules, so the element is written as H_2 in chemical equations—not just H.

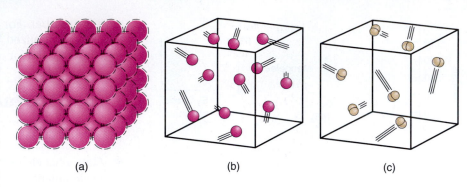

(a)   (b)   (c)

Analysis of electromagnetic radiation from stars, galaxies, and nebulas (interstellar clouds of gas and dust) indicates that hydrogen, the simplest element of all, accounts for about 75% of the mass of elements in the universe, with about 24% being helium, the next simplest element. All the other elements account for only 1%. Section 10.4 briefly discussed how elements are formed in stars, a process called *nucleosynthesis*.

Molecules

The individual units in a piece of metal are atoms. For example, the atoms in a sample of iron are packed together as shown in ● Fig. 11.10a, with each atom bonding equally to all its nearest neighbors. Thus, for a metallic element such as iron, just writing its symbol adequately represents its composition.

Also, we represent the composition of a noble gas just by writing the element's symbol, because if you could see the units flying around in a sample of a noble gas such as xenon, they would be individual atoms (Fig. 11.10b). (It is now possible to image and even manipulate individual xenon atoms, as shown in Fig. 9.22b on page 225.)

However, if you could observe the individual units in a sample of hydrogen gas, you would see that they consist of particles made up of two atoms of hydrogen combined chemically; they are *diatomic molecules* (Fig. 11.10c). A **molecule** is an electrically neutral particle composed of two or more atoms chemically combined (● Fig. 11.11). If the atoms are all of the same element, it is a molecule of an element (for example, H_2 or N_2). If the atoms are of different elements, it is a molecule of a compound (for example, H_2O or NH_3).

● Figure 11.12 illustrates the seven common elements that exist as diatomic molecules. Their atoms are too reactive to exist as independent individuals.

FIGURE 11.11   Representations of Molecules

Ball-and-stick models (top) and space-filling models (bottom) of the element hydrogen and three common compounds—water, ammonia, and methane.

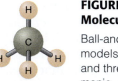

Hydrogen (H_2)   Water (H_2O)   Ammonia (NH_3)   Methane (CH_4)

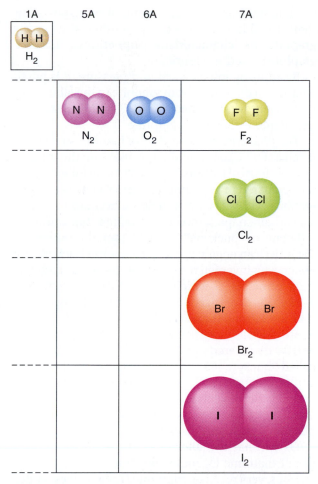

FIGURE 11.12 Elements That Exist as Diatomic Molecules

Hydrogen, nitrogen, oxygen, fluorine, chlorine, bromine, and iodine exist in nature as diatomic (two-atom) molecules.

FIGURE 11.13 Graphite and Diamond

Graphite (top of photo) is one component of pencil "lead." The diamond is synthetic. Carbon black (soot), charcoal, and coke are amorphous (seemingly noncrystalline) forms of graphite.

Note that six of the seven are close together and form a "7" shape at the right of the periodic table, making them easy to recall. When writing chemical equations, as we will do in Chapter 13, the formulas of these seven elements are written in the diatomic form (such as $H_2 + Cl_2 \rightarrow 2\,HCl$, and *not* $H + Cl \rightarrow HCl$).

Allotropes

Two or more forms of the same element that have different bonding structures in the same physical phase are called **allotropes.** Two of the three allotropes of

carbon are shown in ● Fig. 11.13. Diamonds are composed of pure carbon. ● Figure 11.14a shows that each carbon atom in diamond bonds to four of its neighbors. Picture the carbon atom in the middle of a geometric structure called a *regular tetrahedron,* a figure having four faces that are identical equilateral triangles. The four bonds of the carbon atom point toward the corners of the tetrahedron and form an angle of 109.5° to one another (Fig. 14.4 on page 354). This leads to a three-dimensional network that helps make diamond the hardest substance known.

Graphite, the black slippery solid that is a major component of pencil "lead," is also pure carbon. (Pencil "lead" contains no lead, Pb.) Figure 11.14b shows that the carbon atoms in graphite are bonded in a network of flat hexagons, giving an entirely different structure from that of diamond. Each carbon atom is bonded to three other carbon atoms that lie in the same plane, and thus each forms part of three hexagons. Each carbon atom is left with one loosely held outer electron, which enables graphite to conduct an electric current, whereas diamond does not. The

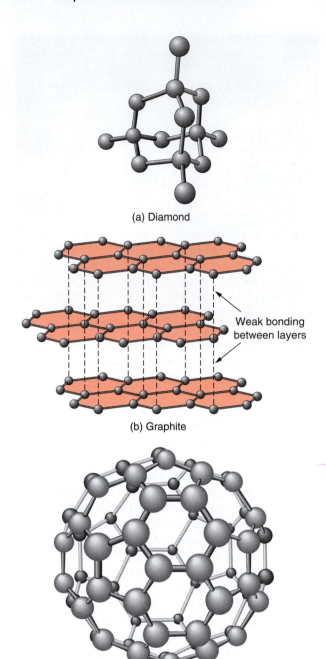

(a) Diamond

Weak bonding between layers

(b) Graphite

(c) Buckyball

FIGURE 11.14 Models of Three Allotropes of Carbon

(a) Diamond consists of a network of carbon atoms in which each atom bonds to four others in a tetrahedral fashion. (b) Graphite consists of a network of carbon atoms in which each atom bonds to three others and forms sheets of flat, interlocking hexagons. The sheets are held to each other only by weak electrical forces. (c) Buckminsterfullerene ("buckyball"), C_{60}, is a soccerball-like arrangement of interlocking hexagons and pentagons formed by carbon atoms.

weak forces between each plane of hexagons allow them to slide easily over one another, thus giving graphite its characteristic slipperiness. In fact, graphite is used as a lubricant.

By applying intense heat and pressure, it is possible to form diamonds from graphite in the laboratory. Conversely, when heated to 1000°C in the absence of air, diamond changes to graphite.

The *fullerenes,* a whole class of ball-like substances (such as C_{32}, C_{60}, C_{70}, and C_{240}), make up the third allotropic form of carbon. The most stable fullerene was first prepared in the laboratory in 1985 by vaporizing graphite with a laser. It is a 60-carbon-atom, hollow, soccerball-shaped molecule named buckminsterfullerene, or "buckyball" (Fig. 11.14c). It is named for the American engineer and philosopher Buckminster Fuller, who invented the geodesic dome, the architectural principle of which underlies the structure of buckyball. Scientists succeeded in isolating C_{60} in bulk in 1990 and found that it forms naturally in sooty flames such as those of candles. In 1991, Japanese scientists discovered structural relatives of buckyball called "buckytubes" because of their tubular shape and internal cavity. Various compounds of buckyball and its relatives act as insulators, conductors, semiconductors, and superconductors. Technological applications are highly probable.

Oxygen has two allotropes. It usually occurs as a gas of diatomic O_2 molecules but also can exist as gaseous, very reactive, triatomic O_3 molecules named *ozone* (● Fig. 11.15). Phosphorus, sulfur, tin, and several other elements also exist in allotropic forms.

RELEVANCE QUESTION: *Look around and try to identify one item that is not composed of one or more chemicals.*

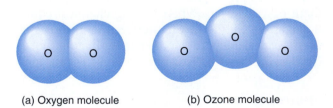

(a) Oxygen molecule   (b) Ozone molecule

FIGURE 11.15 Oxygen and Ozone

(a) The element oxygen normally exists as gaseous diatomic molecules (O_2). (b) Oxygen also can exist as gaseous, highly reactive, triatomic molecules (O_3) called *ozone.*

11.4 The Periodic Table

▼ Describe the major features and divisions of the periodic table.

▼ Use the periodic table to correlate and predict the properties of elements.

▼ State the electron configurations for hydrogen through argon, and tell the number of valence electrons for any representative element.

By 1869, a total of 63 elements had been discovered. However, a system for classifying the elements had not been established. As discussed in the chapter's second Highlight, it remained for the Russian chemist Dmitri Mendeleev ("men-duh-LAY-eff") to formulate a satisfactory classification scheme—the periodic table.

The *periodic table* puts the elements, in order of increasing atomic number, into seven horizontal rows, called **periods.** As a result, the elements' properties show regular trends, and similar properties occur *periodically*—that is, at definite intervals. Later in this section we will discuss some examples of these regular trends. The modern statement of the **periodic law** is

The properties of elements are periodic functions of their atomic numbers.

The vertical columns in the periodic table are called **groups.** At present, some debate exists about the designation of the groups. In 1986 the International Union of Pure and Applied Chemistry (IUPAC) decreed that the groups be labeled 1 through 18 from left to right, as shown in the periodic table inside the front cover. However, for many years in the United States, the groups have been divided into A and B subgroups, as is also shown in your periodic table and in ● Fig. 11.16. Which designation will ultimately prevail is still in question. We will use the A and B designation in our discussion.

The elements in the periodic table are classified in several ways, one of which is into representative, transition, and inner transition elements.

1. **Representative elements** are those in Groups 1A through 8A, shown in green in Fig. 11.16 and in the periodic table inside the front cover. Four of the representative element groups have commonly

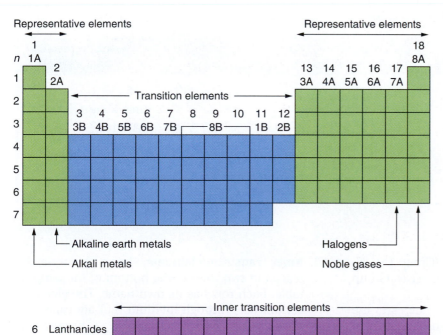

FIGURE 11.16   Names of Specific Portions of the Periodic Table

The representative elements are shown in green, the transition elements in blue, and the inner transition elements in purple. Note the location of the alkali metals (Group 1A, except H), the alkaline earth metals (Group 2A), the halogens (Group 7A), the noble gases (Group 8A), and the lanthanides and actinides.

HIGHLIGHT: Mendeleev and the Periodic Table

The first detailed and useful periodic table placed the elements in order of increasing atomic mass. It was published in 1869 by Dmitri Mendeleev, a Russian chemist (Fig. 1). The modern periodic table resembles the horizontal and vertical columns in his original table.

Mendeleev's table was useful because of the importance he placed on the occurrence of similar physical and chemical properties at regular intervals (that is, *periodically*) and because of his prediction of the properties of unknown elements. He even left vacancies in his table for them. When these elements, such as gallium, scandium, and germanium (which Mendeleev called eka-silicon, meaning "similar to silicon"), were discovered later, they had the properties and filled the vacancies Mendeleev had predicted. In Table 1, compare the properties Mendeleev predicted for eka-silicon with the properties of the element germanium (Ge)

Figure 1   Dmitri Mendeleev (1834-1907)
He developed the first useful periodic table of the chemical elements.

discovered in 1886, while Mendeleev was still living.

Mendeleev was unable to explain why similar chemical properties oc-

curred at regular intervals with increasing atomic mass. He realized that in several positions in his table the increasing atomic mass and the properties of the elements did not coincide. For example, on the basis of atomic mass, iodine should come *before* tellurium, but iodine's properties indicate that it should *follow* tellurium. Mendeleev thought that perhaps in these cases the atomic masses were incorrect. However, we now know that the periodic law is a function of the atomic number (and the corresponding electron configuration) of an element, not its atomic mass. Fortunately for Mendeleev, the atomic numbers generally increase as the atomic masses increase.

Mendeleev's periodic table, in modern form, is used by the entire scientific community. Element 101, mendelevium, is named for him.

TABLE 1   Predicted and Observed Properties of Germanium

Property	Mendeleev's Predictions for Eka-silicon (X)	Observed Properties of Germanium (Ge)
Atomic mass	72	72.6
Color	Dirty gray	Gray-white
Density	5.5 g/cm^3	5.35 g/cm^3
Oxide formula and density	XO_2; 4.7 g/cm^3	GeO_2; 4.23 g/cm^3
Chloride formula and density	XCl_4; 1.9 g/mL	$GeCl_4$; 1.84 g/mL
Boiling point of chloride	under 100°C	84°C

used special names: *alkali metals* (Group 1A), *alkaline earth metals* (Group 2A), *halogens* (Group 7A), and *noble gases* (Group 8A).

2. **Transition elements,** shown in blue, are the groups designated by a numeral and the letter B. Such familiar elements as iron, copper, and gold are transition elements.

3. **Inner transition elements,** shown in purple, are placed in two rows at the bottom of the periodic table. Each row has its own name. The elements cerium (Ce) through lutetium (Lu) are called the *lanthanides,* whereas thorium (Th) through lawrencium (Lr) make up the *actinides.* Except for uranium, few of the inner transition elements are

TABLE 11.2   Some General Properties of Metals and Nonmetals

Metals	Nonmetals
Good conductors of heat and electricity.	Poor conductors of heat and electricity.
Malleable—can be beaten into thin sheets.	Brittle—if a solid.
Ductile—can be stretched into wire.	Nonductile.
Possess metallic luster.	Do not possess metallic luster.
Solids at room temperature (Exception: Hg).	Solids, liquids, or gases at room temperature.
Usually have 1 to 3 valence electrons.	Usually have 4 to 8 valence electrons.
Lose electrons, forming positive ions.	Gain electrons to form negative ions or share electrons.

well known. However, the phosphors in color TV and computer screens are primarily compounds of lanthanides.

Metals and Nonmetals

Let's now look at another method for classifying elements—namely, into *metals* and *nonmetals*. This early but still useful classification was done originally on the basis of certain distinctive properties (Table 11.2). Our modern definition is that a **metal** is an element whose atoms tend to lose electrons during chemical reactions. A **nonmetal** is an element whose atoms tend to gain (or share) electrons.

As shown in ● Fig. 11.17, the metallic character of the elements increases as you go down a group and decreases across a period (from now on, *across a period* means moving left to right). As you would expect, the nonmetallic character of the elements shows the opposite trend—decreasing down a group and increasing across a period. Cesium is the most metallic element, and fluorine is the most nonmetallic. (Francium, Fr, which technically might be the most metallic element, is a synthetic element and unavailable in amounts greater than a few atoms.)

Most elements are metals. The actual dividing line between metals and nonmetals cuts through the periodic table like a staircase (Fig. 11.17). The elements boron, silicon, germanium, arsenic, antimony, and tellurium are called *semimetals,* or *metalloids.* They are located next to the staircase line and display properties of both metals and nonmetals. Several

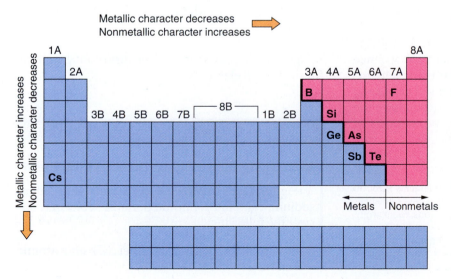

FIGURE 11.17   Metals and Nonmetals in the Periodic Table

The elements shown in blue, at the left of the staircase line, are metals. The nonmetals are shown in pink. Fluorine is the most reactive nonmetal, and cesium is the most reactive metal. Next to the staircase line are symbols identifying the six elements that are actually *semimetals* (those having some metallic and some nonmetallic properties). Note that hydrogen is considered a nonmetal.

semimetals are of crucial importance as semiconductors in the electronics industry.

The elements also can be classified according to whether they are solids, liquids, or gases at room temperature and atmospheric pressure. Only two, bromine and mercury, occur as liquids. Eleven occur as gases: hydrogen, nitrogen, oxygen, fluorine, chlorine, and the six noble gases. All the rest are solids, with the vast majority being metallic solids. The chapter-opening photograph on page 265 shows several common elements.

Electron Configuration and Valence Electrons

Section 9.3 stated that the main electron energy levels in atoms are designated by the principal quantum number n, which can have the values 1, 2, 3, and so on. We often refer to these energy levels as *shells;* for example, the first shell is the energy level where $n = 1$.

The chemical reactivity of the elements depends on the order of the electrons in the energy levels in their atoms, which is called the **electron configuration.** Chemistry students must know electron configurations in more detail than we need to. They must look into subdivisions of the shells, called *subshells* and *orbitals,* and even into the individual "spins" of electrons.* We will be content with learning how the electrons are arranged in the various shells—that is, just learning the *shell electron configurations.* To further keep things simple yet still show the basic ideas, we will deal only with the first 18 elements. However, it is necessary to be able to determine how many electrons are in the outermost shell of an atom of any representative element.

The outer shell of an atom is known as the **valence shell,** and the electrons in it are called the **valence electrons.** The valence electrons are extremely important because they are the ones involved in forming chemical bonds. Elements in a given group have the same number of valence electrons and, because of this, have similar chemical properties.

*The structure of the periodic table reflects the elements' electron configurations on the subshell level, where *s, p, d,* and *f* subshells can contain a maximum of 2, 6, 10, and 14 electrons, respectively. In Fig. 11.16, the block composed of Groups 1A and 2A has two columns, and the block composed of Groups 3A through 8A has six columns. The transition elements comprise a block having 10 columns, and the inner transition elements have 14 columns.

Our Chapter 12 discussion of chemical bonding will be limited to the representative (A group) elements. *The number of valence electrons for every element in a given A group is the same as the group number.* The only exception is helium, which has just two electrons in the outer shell, although it is in Group 8A. For the representative elements, the number of valence electrons increases by one as you proceed across a given period.

Here are the rules for writing shell electron configurations:

1. The number of electrons in an atom of a given element is the same as the element's atomic number. For example, the periodic table inside the front cover shows that lithium (Li) has atomic number 3, so we have three electrons to "place" in the proper shells.

2. For atoms of a given element, the number of shells that contain electrons will be the same as the period number. For example, lithium is in Period 2, so the three electrons are in two shells.

3. For the representative (A group) elements, the number of valence electrons (the electrons in the outer shell) is the same as the group number. For example, lithium is in Group 1A, so it has one valence electron. Clearly, if lithium's three electrons are to be distributed among two shells and the outer shell has one electron, the first shell must contain the other two. Thus the shell electron configuration of lithium is written 2,1 (that is, two electrons in the first shell and one in the second shell). In fact, the first shell of all atoms holds a maximum of two electrons, whereas the second shell holds a maximum of eight.

● Figure 11.18 shows the shell electron configurations for the first 18 elements.

EXAMPLE 11.1

Identifying Some Properties of an Element

● Figure 11.19 shows the very reactive element sodium, Na. Find Na in the periodic table, and answer the following:

(a) What are its atomic number (Z) and atomic mass?

FIGURE 11.18   Shell Distribution of Electrons for Periods 1, 2, and 3

Each atom's nucleus is shown in light blue. The number of shells containing electrons is the same as the period number. Each shell of electrons is shown as a partial circle, with a numeral showing the number of electrons in the shell. The number of valence electrons, those in the outer shell, is the same for all the atoms in the group and is equal to the group number (exception: He).

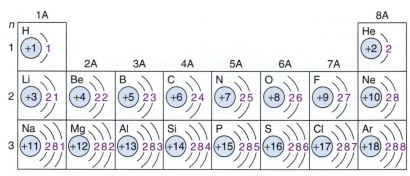

(b) Is it a representative, transition, or inner transition element? Metal or nonmetal?

(c) What period is it in, and what is its group number?

(d) How many total electrons are in an Na atom? How many protons?

(e) How many valence electrons are in it?

(f) How many shells of an Na atom contain electrons?

(g) What is sodium's shell electron configuration?

FIGURE 11.19   Sodium

This metallic element is so soft that it can be sliced with a knife. It is so reactive that an explosion or chemical burn can occur, so it must be handled with caution.

SOLUTION

(a) $Z = 11$, and atomic mass is 23.0 u.

(b) Representative (its group number has "A" in it); metal (at *left* of staircase line).

(c) Period 3 (it is in the third horizontal row down) and Group 1A.

(d) Eleven electrons and 11 protons (same as Z).

(e) One valence electron (same as the group number, for representative elements).

(f) Three shells contain electrons (same as the period number).

(g) Two electrons are always in the first shell, and the third shell of sodium must contain one valence electron, so that leaves 8 of the 11 for the second shell (which we already know can hold a maximum of 8). Thus sodium's shell electron configuration is 2,8,1.

CONFIDENCE EXERCISE 11.1

Find the element phosphorus (P) in the periodic table, and answer the following:

(a) What are its atomic number and atomic mass?

(b) It is a representative, transition, or inner transition element? Metal or nonmetal?

(c) What period is it in, and what is its group number?

(d) How many total electrons are in a P atom? How many protons?

(e) How many valence electrons are in a P atom?

(f) How many shells of a P atom contain electrons?

(g) What is phosphorus's shell electron configuration?

Atomic Size—Another Periodic Characteristic

We have discussed several periodic characteristics, such as metallic and nonmetallic character, electron configuration, and number of valence electrons. Let's now examine the periodic nature of atomic size, which ranges from a diameter of about 0.074 nm for hydrogen to about 0.47 nm for cesium, Cs.

● Figure 11.20 illustrates that atomic size increases down a group. (This is logical because each successive element of the group has an additional shell containing electrons.) Figure 11.20 also shows that atomic size decreases across a period. Notice that each Group 1A atom is large with respect to atoms of the other elements of that period. This is so because its one outer electron is loosely bound to the nucleus. As the charge on the nucleus increases (more pro-

tons) without adding an additional shell of electrons, the outer electrons are more tightly bound, thus decreasing the atomic size from left to right across the period.

Ionization Energy—Yet Another Periodic Characteristic

When an atom gains or loses electrons, it acquires a net electric charge and becomes an *ion*. The amount of energy that it takes to remove an electron from an atom is called its **ionization energy.** In general, the ionization energy increases across a period (● Fig. 11.21). The Group 1A elements have the lowest ionization energies. Their one valence electron is in an outer shell, so it does not take much energy to remove the electron. Elements located to the right of the Group 1A element in a particular period have additional pro-

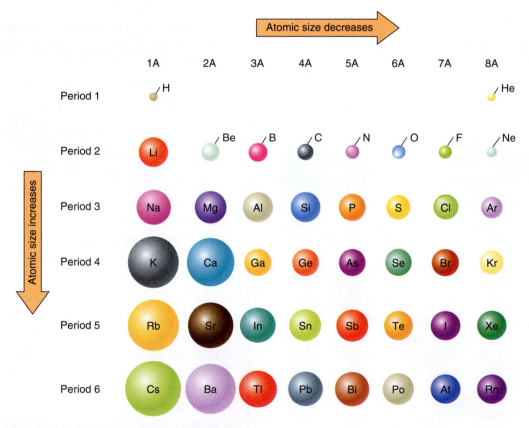

FIGURE 11.20 The Relative Sizes of Atoms of the Representative Elements

In general, atomic size decreases across a period and increases down a group. The diameter of the smallest atom, H, is about 0.074 nm, whereas that of the largest atom shown, Cs, is about 0.47 nm. (From Ebbing, Darrell, and R. A. D. Wentworth, *Introductory Chemistry*, Second Edition. Copyright © 1998 by Houghton Mifflin Company. Used with permission.)

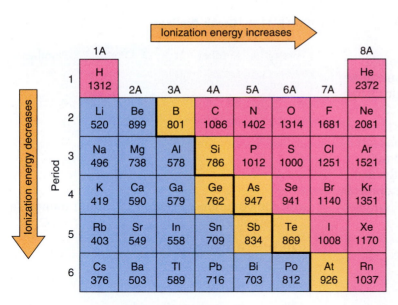

Ionization energy increases

Ionization energy decreases

Period

	1A	2A	3A	4A	5A	6A	7A	8A
1	H 1312							He 2372
2	Li 520	Be 899	B 801	C 1086	N 1402	O 1314	F 1681	Ne 2081
3	Na 496	Mg 738	Al 578	Si 786	P 1012	S 1000	Cl 1251	Ar 1521
4	K 419	Ca 590	Ga 579	Ge 762	As 947	Se 941	Br 1140	Kr 1351
5	Rb 403	Sr 549	In 558	Sn 709	Sb 834	Te 869	I 1008	Xe 1170
6	Cs 376	Ba 503	Tl 589	Pb 716	Bi 703	Po 812	At 926	Rn 1037

FIGURE 11.21   Ionization Energy—A Periodic Trend

The ionization energy (in kilojoules per mole) increases in a generally regular fashion across a period and decreases down a group. (From Ebbing, Darrell, and R. A. D. Wentworth, *Introductory Chemistry*, Second Edition. Copyright © 1998 by Houghton Mifflin Company. Used with permission.)

tons and electrons, but the electrons are added to the same shell. The added protons bind the electrons more and more strongly until the shell is completely filled.

Ionization energy decreases down a group because the electron to be removed is shielded from the attractive force of the nucleus by each additional shell of electrons. We will encounter another periodic characteristic—that of electronegativity—in Chapter 12.

11.5 Naming Compounds

LEARNING GOAL

▼ Name simple inorganic compounds.

Our discussion from this point on requires a basic understanding of compound *nomenclature* (from the Latin for "name" and "to call"). Elements combine chemically to form compounds. Each compound is represented by a *chemical formula,* which is written by putting the elements' symbols adjacent to each other—normally with the more metallic element first.

A *subscript* (subscripted numeral) following each symbol designates the number of atoms of the element in the formula. If an element has only one atom in the formula, the subscript 1 is not actually written, just understood. For instance, water is written H_2O, which means that a single molecule of water consists

of two atoms of hydrogen and one atom of oxygen. We often use chemical formulas, such as H_2O, to identify specific chemical compounds, but we also need *names* that unambiguously identify them.

Compounds with Special Names

Some compounds have such well-established special names that no systematic nomenclature can compete. Their names are just learned individually. Common examples to know are given in Table 11.3.

TABLE 11.3   Eleven Compounds with Special Names

Name	Formula
Water	H_2O
Ammonia	NH_3
Methane	CH_4
Nitrous oxide	N_2O
Nitric oxide	NO
Hydrochloric acid	$HCl(aq)$
Nitric acid	$HNO_3(aq)$
Acetic acid	$HC_2H_3O_2(aq)$
Sulfuric acid	$H_2SO_4(aq)$
Carbonic acid	$H_2CO_3(aq)$
Phosphoric acid	$H_3PO_4(aq)$

Naming a Compound of a Metal and a Nonmetal

For now, we will deal with naming compounds of metals that form only one ion, which are mainly those of Groups 1A (ionic charge 1+) and 2A (ionic charge 2+), plus Al, Zn, and Ag (ionic charges of 3+, 2+, and 1+, respectively). Section 12.4 will tell how to name compounds of metals that form more than one ion.

To name a binary (two-element) compound of a metal combined with a nonmetal, first give the name of the metal and then give the name of the nonmetal with its ending changed to -*ide* (Table 11.4). Examples are

NaCl	sodium chloride
Al_2O_3	aluminum oxide
Ca_3N_2	calcium nitride

Naming Compounds of Two Nonmetals

In a compound of two nonmetals, the less nonmetallic element (the one farther left or farther down in the periodic table) is usually written first in the formula and named first. The second element is named using its -*ide* ending. Generally, two nonmetallic elements form several binary compounds, which are distinguished by using Greek prefixes to designate the number of atoms of the element that occur in the molecule (Table 11.5). The prefix *mono*- is always omitted from the name of the first element in the compound, and is usually omitted from the second

TABLE 11.5 Greek Prefixes

Prefix	Number	Prefix	Number
mono-	1	penta-	5
di-	2	hexa-	6
tri-	3	hepta-	7
tetra-	4	octa-	8

(with the common exception of carbon monoxide, CO). Examples are

HCl	hydrogen chloride
CS_2	carbon disulfide
PBr_3	phosphorus tribromide
IF_7	iodine heptafluoride

Naming Compounds Containing Polyatomic Ions

An **ion** is an atom, or chemical combination of atoms, having a net electric charge because of a gain or loss of electrons. An ion formed from a single atom is a *monatomic ion,* whereas an electrically charged combination of atoms is a **polyatomic ion.** The names and formulas of the eight common polyatomic ions you should know are given in Table 11.6.

For a compound of a metal combined with a polyatomic ion, simply name the metal and then name

TABLE 11.4 The -*ide* Nomenclature for Common Nonmetals

Element Name	-*ide* Name
Bromine	Bromide
Chlorine	Chloride
Fluorine	Fluoride
Hydrogen	Hydride
Iodine	Iodide
Nitrogen	Nitride
Oxygen	Oxide
Phosphorus	Phosphide
Sulfur	Sulfide

TABLE 11.6 Some Common Polyatomic Ions

Name	Formula
Acetate	$C_2H_3O_2^-$
Hydrogen carbonate	HCO_3^-
Hydroxide	OH^-
Nitrate	NO_3^-
Carbonate	CO_3^{2-}
Sulfate	SO_4^{2-}
Phosphate	PO_4^{3-}
Ammonium	NH_4^+

Note: The names of positive polyatomic ions end in -*ium*. Many negative polyatomic ions have names ending in -*ate*. Another name for the hydrogen carbonate ion is *bicarbonate*.

the polyatomic ion. Examples are

$ZnSO_4$	zinc sulfate
$NaC_2H_3O_2$	sodium acetate
$Mg(NO_3)_2$	magnesium nitrate
K_3PO_4	potassium phosphate

When hydrogen is combined with a polyatomic ion, the compound is generally named as an acid, as is shown in Table 11.3.

The only common positive polyatomic ion present in compounds is the ammonium ion, NH_4^+. If the ammonium ion is combined with a nonmetal, change the ending of the nonmetal to *-ide*. If it is combined with a negative polyatomic ion, simply name each ion. Examples are

$(NH_4)_3P$	ammonium phosphide
$(NH_4)_3PO_4$	ammonium phosphate

The Spotlight feature on page 285 summarizes the basic nomenclature rules. Note the first question to ask yourself: Is it one of the 11 compounds whose special names I know? If the answer is yes, just name it. If no, then ask the second question: Is it a binary compound?

The naming of the complex compounds found in organic chemistry follows its own set of rules, as we will see in Chapter 14. And by the way, don't be upset if you run into a few compounds that seem to be named in violation of the rules given above. Space limitations prohibit complete coverage of the rules, and even scientists are not always consistent!

EXAMPLE 11.2

Naming Compounds

Name each of these six compounds in the preferred manner: (a) $H_2SO_4(aq)$, (b) $ZnCO_3$, (c) Na_2S, (d) SiO_2, (e) NH_3, (f) NH_4NO_3.

SOLUTION

(a) $H_2SO_4(aq)$ Sulfuric acid (a compound with a special name).

(b) $ZnCO_3$ Zinc carbonate (a compound of a metal with the carbonate polyatomic ion, so name the metal and then name the polyatomic ion).

(c) Na_2S Sodium sulfide (a binary compound of a metal with a nonmetal, so the metal is named first, then the *-ide* name of the nonmetal is given).

(d) SiO_2 Silicon dioxide (a binary compound of two nonmetals, so the Greek prefix system is preferred).

(e) NH_3 Ammonia (this common compound is always called by its special name.)

(f) NH_4NO_3 Ammonium nitrate (a compound that contains an ammonium ion with another polyatomic ion, so just name each one).

CONFIDENCE EXERCISE 11.2

Name the compounds AsF_5 and $CaCl_2$ in the preferred manner.

11.6 Groups of Elements

LEARNING GOALS

▼ List the basic properties of four groups of elements.

▼ Identify and describe the uses of some compounds.

Recall that a row of the periodic table is called a *period*, and a column of the table is termed a *group*. As we saw in Fig. 11.18, all the elements in a group have the same number of valence electrons. Therefore, if one element in a group reacts with a given substance, the other elements in the group usually react in a similar manner with that substance. The formulas of the compounds produced also will be similar in form. In this section we discuss four of these groups of elements: the noble gases (8A), the alkali metals (1A), the halogens (7A), and the alkaline earth metals (2A).

Noble Gases

The elements of Group 8A are known as the **noble gases.** They are monatomic; that is, they exist as single atoms (see Fig. 11.10b). The noble gases have the astounding chemical property of almost never forming compounds with other elements or even bonding to themselves. This is why the noble gas argon is used inside light bulbs, where the intense heat would cause the tungsten filament to react with most elements and quickly deteriorate (● Fig. 11.22).

We can conclude that noble gas electron configurations with eight electrons in the outer shell (or two if the element, like helium, uses only the first shell) are quite stable. This conclusion is of crucial importance when considering the bonding characteristics of atoms of other elements, as we will see in Chapter 12.

Helium is obtained from natural gas, and radon is a radioactive by-product of radium's decay, but the other noble gases are found only in the air. Recall that argon makes up almost 1% of air. A common use of the noble gases is in "neon" signs, which contain minute amounts of various noble gases or other gases in a sealed glass vacuum tube (● Fig. 11.23). When an electric current is passed through the tube, the gases glow—giving a color that is characteristic of the specific gas present.

The most striking physical property of the noble gases is their low melting and boiling points. For example, helium melts at $-272°C$ and boils at $-269°C$, just one degree and four degrees, respectively, above absolute zero. For this reason, liquid helium is often used in low-temperature research.

FIGURE 11.22   Elements and the Light Bulb

Argon gas surrounds the tungsten (W) filament in a light bulb. Argon is used because it is so inactive. Tungsten is used because it has the highest melting point of any metal. The gas carries heat away from the wire, which would otherwise overheat and boil away.

The Alkali Metals

The elements in Group 1A, except for hydrogen, are called the **alkali metals.** (Hydrogen, a gaseous *nonmetal*, is discussed at the end of this section.) Each

FIGURE 11.23   "Neon" Lights

When the atoms of a small amount of gas are subjected to an electric current, the gas glows, or fluoresces. The color of a particular light depends on the identity of the gas whose atoms are being excited. Neon gas glows a beautiful orange-red, whereas argon emits a blue light.

A SPOTLIGHT ON: Naming Compounds

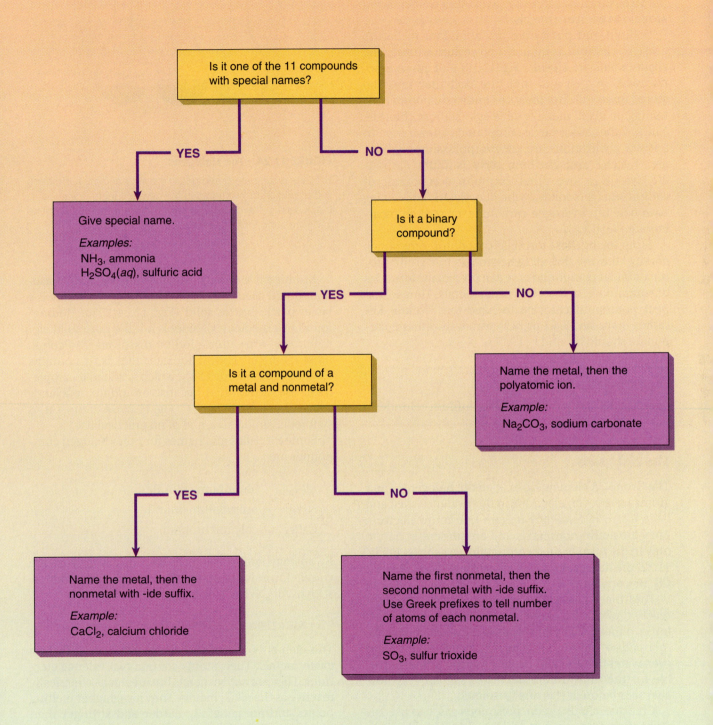

Is it one of the 11 compounds with special names?

YES

Give special name.

Examples:
NH_3, ammonia
$H_2SO_4(aq)$, sulfuric acid

NO

Is it a binary compound?

YES

Is it a compound of a metal and nonmetal?

NO

Name the metal, then the polyatomic ion.

Example:
Na_2CO_3, sodium carbonate

YES

Name the metal, then the nonmetal with -ide suffix.

Example:
$CaCl_2$, calcium chloride

NO

Name the first nonmetal, then the second nonmetal with -ide suffix. Use Greek prefixes to tell number of atoms of each nonmetal.

Example:
SO_3, sulfur trioxide

alkali metal atom has only one valence electron. The atom tends to lose this outer electron quite easily, so the alkali metals react readily with other elements and are said to be *active* metals.

Sodium and potassium are abundant in Earth's crust, but lithium, rubidium, and cesium are rare. The alkali metals are all soft (Fig. 11.19) and are so reactive with oxygen and moisture that they must be stored under oil. The most common compound containing an alkali metal is table salt (sodium chloride, NaCl). Other alkali metal compounds known even to ancient civilizations are *potash* (potassium carbonate, K_2CO_3) and *washing soda* (sodium carbonate, Na_2CO_3). Two other common compounds of sodium are *lye* (sodium hydroxide, NaOH) and *baking soda* (sodium hydrogen carbonate, or sodium bicarbonate, $NaHCO_3$).

From seeing these formulas of sodium compounds and by knowing that all the elements in a group produce similar compounds, we can predict the formulas of compounds of the other alkali metals. Thus we expect potassium chloride to have the formula KCl rather than, say, K_2Cl or KCl_2. Similarly, lithium carbonate should be, and is, Li_2CO_3.

CONFIDENCE EXERCISE 11.3

Predict the chemical formula for lithium chloride.

The Halogens

The Group 7A elements are called the **halogens.** Their atoms have seven electrons in their valence shell and have a strong tendency to gain one more electron. They are active nonmetals and are present in nature only in the form of their compounds. As shown in Fig. 11.12, the halogens consist of diatomic molecules (F_2, Cl_2, Br_2, and I_2).

Fluorine, a pale-yellow, poisonous gas, is the most reactive of all the elements. It corrodes even platinum, a metal that withstands most other chemicals. A stream of fluorine gas causes wood, rubber, and even water to burst into flame. Fluorine was responsible for the unfortunate deaths of several very able chemists before it was finally isolated.

Chlorine, a pale-green, poisonous gas, was used as a chemical weapon in World War I (● Fig. 11.24). In small amounts, it is used as a purifying agent in swimming pools and in public water supplies.

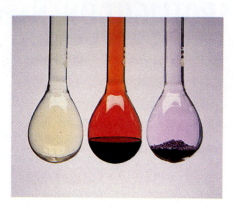

FIGURE 11.24   The Halogens
Chlorine is the pale-green gas in the left-hand flask. Bromine is the reddish-brown liquid in the center. Iodine is the violet-black solid in the right-hand flask. Note how readily bromine vaporizes and iodine sublimes.

Bromine is a reddish-brown, foul-smelling, poisonous liquid used as a disinfectant, whereas iodine is a violet-black, brittle solid (Fig. 11.24). The "iodine" found in a medicine cabinet is not the pure solid element; it is a *tincture*—iodine dissolved in alcohol. Most table salt is now iodized; that is, it contains 0.02% sodium iodide (NaI), which is added to supplement the human diet because an iodine deficiency causes thyroid problems (● Fig. 11.25). Astatine (At) is synthetic, radioactive, and of no practical use.

Some formulas and names for other halogen compounds are

$AlCl_3$	aluminum chloride
NH_4F	ammonium fluoride
$CaBr_2$	calcium bromide

By analogy with these formulas, we can predict the correct formulas of many other halogen compounds, such as AlF_3, NH_4Cl, and CaI_2.

The Alkaline Earth Metals

The elements in Group 2A are called the **alkaline earth metals.** Their atoms contain two valence electrons. They are active metals but are not as chemically reactive as the alkali metals. They have higher melting points and are generally harder and stronger than their Group 1A neighbors.

Beryllium occurs in the mineral beryl, some varieties of which make beautiful gemstones such as

Strontium compounds produce the red colors in fireworks, and barium compounds produce the green. When radioactive strontium-90 (from nuclear atmospheric testing fallout) is ingested, it goes into the bone marrow because of its similarity to calcium. If enough strontium-90 is ingested, it can destroy the bone marrow or cause cancer. Fortunately, radioactive fallout is minimal now. One barium compound encountered all too frequently is barium sulfate ($BaSO_4$), the white material in the thick solutions patients ingest prior to having X-rays taken of their gastrointestinal tracts (● Fig. 11.26). The barium sulfate improves the clarity of the X-rays.

Radium is an intensely radioactive element that glows in the dark. Its radioactivity is a deterrent to practical uses, although a few decades ago watch dials were painted with radium chloride ($RaCl_2$) so that they could be read in darkness.

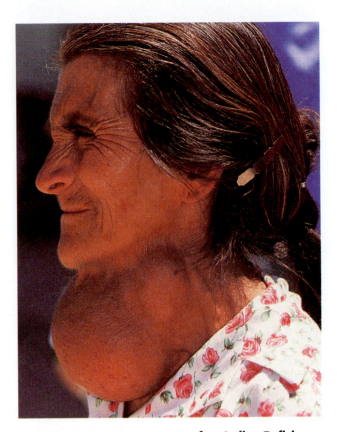

FIGURE 11.25   Consequences of an Iodine Deficiency
A lack of iodine in the diet can lead to an enlarged thyroid gland, a medical condition known as *goiter.*

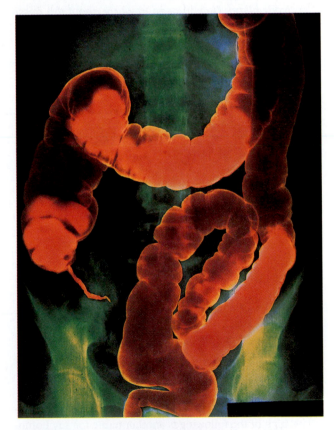

FIGURE 11.26   A Common Use of Barium Sulfate
The contrast in this radiograph of a patient's large intestine is produced by the introduction of $BaSO_4$, which absorbs X-rays much better than do body tissue and bone.

aquamarines and emeralds. Magnesium is used in safety flares because of its ability to burn with an intensely bright flame (see Fig. 13.1 on page 322). Both magnesium and beryllium are used to make light-weight metal alloys. The common over-the-counter medicine called *milk of magnesia* is magnesium hydroxide, $Mg(OH)_2$.

Calcium is used by vertebrates in the formation of bones and teeth, which are mainly calcium phosphate, $Ca_3(PO_4)_2$. Calcium carbonate ($CaCO_3$) is a mineral present in nature in many forms. The shells of marine creatures, such as clams, are made of calcium carbonate, as are coral reefs. Limestone, a sedimentary rock often formed from seashells and thus from calcium carbonate, is used to build roads and to neutralize acid soils. Tums tablets are mainly calcium carbonate. Also composed of calcium carbonate are the impressive formations of stalactites and stalagmites in underground caverns (see Fig. 21.24 on page 578).

Hydrogen

Although hydrogen is commonly listed in Group 1A, it is not considered an alkali metal. Hydrogen is a nonmetal that usually reacts like an alkali metal, forming HCl, H_2S, and so forth (similar to NaCl, Na_2S, and so on). Yet sometimes hydrogen reacts like a halogen, forming such compounds as NaH (sodium hydride) and CaH_2 (compare with NaCl and $CaCl_2$). It is the least dense of the elements and so was used in early dirigibles (● Fig. 11.27). At room temperature, hydrogen is a colorless, odorless gas and consists of diatomic molecules (H_2).

Although hydrogen burns in air, it is actually a safer fuel than gasoline. It probably will be the main fuel used in transportation at some point in the future. Because it burns to form only water, its use would greatly help solve our air pollution problems.

RELEVANCE QUESTION: You are invited for a ride in the Goodyear blimp. Do you need to worry about the possibility of the helium exploding?

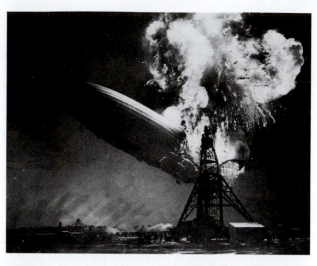

FIGURE 11.27 The *Hindenburg* Disaster, Lakehurst, New Jersey, May 6, 1937

The 800-ft-long airship burst into flame on landing after a trip from Germany. The disaster killed 35 of the 96 passengers and is responsible for the "*Hindenburg* syndrome"—reluctance to use hydrogen as a fuel. For size comparison, the Goodyear blimps are about 200 ft long. Airships nowadays use nonflammable helium to give them buoyancy.

Important Terms

chemistry
pure substance (11.1)
element
compound
mixture
solution
unsaturated solution
saturated solution

solubility
supersaturated solution
molecule (11.3)
allotropes
periods (11.4)
periodic law
groups
representative elements

transition elements
inner transition elements
metal
nonmetal
electron configuration
valence shell
valence electrons
ionization energy

ion (11.5)
polyatomic ion
noble gases (11.6)
alkali metals
halogens
alkaline earth metals

Review Questions

1. What does the field of chemistry study?
2. Name the five major divisions of chemistry and tell the focus of each.

11.1 Classification of Matter (and Highlight)

3. Which of the following is another name for any homogeneous mixture?
   (a) compound   (c) solution
   (b) colloid   (d) alloy

4. Refer to Fig. 11.5. Which substance has a *decreasing* solubility as temperature rises?
   (a) KNO_3 (c) NaBr
   (b) $Ce_2(SO_4)_3$ (d) sugar

5. A solute crystal dissolves when added to a solution. What type of solution was it?
   (a) saturated   (c) unsaturated
   (b) supersaturated   (d) presaturated

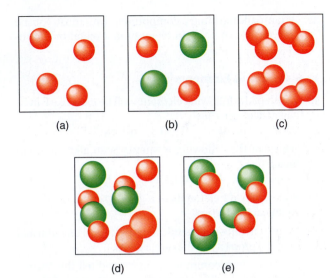

(a)   (b)   (c)

(d)   (e)

FIGURE 11.28

See Review Question 6.

6. Which illustrations in ● Fig. 11.28 represent mixtures? A compound? Only one element? Only diatomic molecules?

7. What characteristics distinguish pure substances from mixtures? Elements from compounds?

8. What type of process is involved in going from mixtures to pure substances, and vice versa? From elements to compounds, and vice versa?

9. Give one example each of a solid, liquid, and gaseous solution.

10. When sugar is dissolved in iced tea, what is the solvent and what is the solute? When would you become aware that the solution was saturated?

11. What is the chemical identity of the gas bubbles in a soft drink or beer? Why does an opened soft drink or beer go flat after a period of time?

11.2 Discovery of the Elements

12. Which of the following scientists in 1661 defined *element* in a manner that made it subject to laboratory testing?
    (a) Boyle   (c) Berzelius
    (b) Davy   (d) Mendeleev

13. Which of the following is the symbol for arsenic?
    (a) Ar   (b) AS   (c) As   (d) aR

14. Which of the following was the first synthetic element?
    (a) Np   (b) Tc   (c) Pu   (d) At

15. When Robert Boyle defined *element* in 1661, how many substances that turned out to be elements were already known? Name half of them.

16. For what does Humphry Davy hold a record, and what invention did he use?

17. How many elements are known at present? About how many occur naturally in our environment?

11.3 Occurrence of the Elements

18. Which one of these elements normally exists as a gas of diatomic molecules?
    (a) iodine   (c) sulfur
    (b) argon   (d) chlorine

19. Graphite is a(n) _____ of carbon.
    (a) isotope   (c) isomer
    (b) allotrope (d) compound

20. Name the two elements that predominate in (a) Earth's crust, (b) Earth's core, (c) the human body, (d) the air, and (e) the universe as a whole. In each case, list the more predominant element first.

21. What is the chemical formula for the ozone that protects us from UV rays?

11.4 The Periodic Table (and Highlight)

22. Consider the element calcium. Which statement is *false?*
    (a) It is in Period 4.
    (b) A Ca atom has electrons in 4 shells.
    (c) It is in Group 2A.
    (d) A Ca atom has 20 valence electrons.

23. Which one of these elements has the greatest atomic size?
    (a) Li   (b) F   (c) K   (d) Br

24. Which one of these elements has the greatest ionization energy?
    (a) Li   (b) F   (c) K   (d) Br

25. Name the scientist who receives the major credit for the development of the periodic table. In what year was his work published?

26. At first, the periodic table placed the elements in order of increasing atomic mass. What property is now used instead of atomic mass, and why?

27. State the periodic law and briefly explain what it means.

28. What formal term is applied to (a) the horizontal rows of the periodic table, and (b) the vertical columns?

29. Why are chemists so interested in the number of valence electrons in atoms?

30. How do metals differ from nonmetals with regard to (a) number of valence electrons, (b) conductivity of heat and electricity, and (c) phase?

31. How does metallic character change (a) across a period and (b) down a group? Name the most metallic element and the most nonmetallic element.

32. About six elements are intermediate in metallic and nonmetallic properties. What are these called and where, in general, are they located in the periodic table?

33. List the two elements that are liquids and the five (other than the six noble gases) that are gases at room temperature and atmospheric pressure.

34. How does the size of atoms change (a) across a period and (b) down a group?

11.5 Naming Compounds

35. Which of the following is the preferred name for Na_2SO_4?
    (a) sodium sulfide   (c) disodium sulfate
    (b) disodium sulfide   (d) sodium sulfate

36. Which of the following is the preferred name for Na_2S?
    (a) sodium sulfide   (c) disodium sulfate
    (b) disodium sulfide   (d) sodium sulfate

37. What is the name of the NH_4^+ ion?

38. Distinguish between (a) an atom and a molecule, (b) an atom and an ion, and (c) a molecule and a polyatomic ion.

39. When naming binary compounds, why must you be able to tell whether an element is a metal or a nonmetal?

11.6 Groups of Elements

40. Which of the following elements will be most like F in its chemical properties?
    (a) Cl   (b) Ne   (c) O   (d) H

41. Which of the following elements is an alkaline earth metal?
    (a) Fe   (b) Pb   (c) Ca   (d) K

42. Briefly, why do elements in a given group have similar chemical properties?

43. Name one unusual chemical property and one unusual physical property of the noble gases.

44. Name the four common halogens and tell the normal phase of each. Which one is the most reactive of all elements?

45. Explain why sodium iodide is added to table salt.

46. What is the formula for the common household compound known as *baking soda*?

47. Name the major compound (a) in bones and (b) in the shells of shellfish.

48. What is the *Hindenburg* syndrome?

Applying Your Knowledge

1. Attempts to make element 113 are in progress, and success may be announced at any time in your newspaper. Which known element should it most resemble?

2. Homogenized milk is composed of microscopic globules of fat suspended in a watery medium. Is homogenized milk a true solution (homogeneous mixture)? Explain.

3. Why might a locksmith put graphite in a car's door lock?

4. What two properties of the alkali metals probably preclude their use to form, say, the shafts of golf clubs?

5. Figure 11.24 on page 286 shows a photograph of chlorine, bromine, and iodine. Why should neither you nor the photographer attempt to take a picture of fluorine?

6. If you lived close to Lake Nyos and weren't well educated, to what might you ascribe an event that suddenly and invisibly kills a large number of people and animals in your vicinity?

Exercises

11.1 Classification of Matter (and Highlight)

1. Classify each of the following materials as an element, compound, heterogeneous mixture, or homogeneous mixture: (a) air, (b) water, (c) diamond, (d) soil.
   Answer: (a) homogeneous mixture   (b) compound (c) element   (d) heterogeneous mixture

2. Classify each of the following materials as an element, compound, heterogeneous mixture, or homogeneous mixture: (a) a fried egg, (b) ozone, (c) brass, (d) carbon dioxide.

3. Refer to Fig. 11.5 on page 268. Find the approximate solubility (g/100 g water) of KBr at 20°C.
   Answer: about 65 g/100 g H_2O

4. Refer to Fig. 11.5. Find the approximate solubility (g/100 g water) of KNO_3 at 80°C.

5. Refer to Fig. 11.5. Would the resulting solution be saturated or unsaturated if 100 g of sugar was stirred into 100 g of water at 40°C?
    Answer: unsaturated, about 240 g of sugar is soluble per 100 g H$_2$O at 40°C.

6. Refer to Fig. 11.5. Would the resulting solution be saturated or unsaturated if 150 g of NaNO$_3$ was stirred into 100 g of water at 60°C?

11.2 Discovery of the Elements

7. Give the symbol for each element: (a) sulfur, (b) sodium, (c) aluminum.   Answer: (a) S   (b) Na   (c) Al

8. Give the symbol for each element: (a) iron, (b) radon, (c) barium.

9. Give the name of each element: (a) N, (b) K, (c) Zn.
    Answer: (a) nitrogen   (b) potassium   (c) zinc

10. Give the name of each element: (a) Be, (b) Au, (c) Ar.

11.4 The Periodic Table (and Highlight)

11. Refer to the periodic table and give the period and group number of each of these elements: (a) magnesium, (b) zinc, (c) tin.
    Answer: (a) 3, 2A   (b) 4, 2B   (c) 5, 4A

12. Refer to the periodic table and give the period and group number for each of these elements: (a) cesium, (b) silver, (c) helium.

13. Classify each of these elements as representative, transition, or inner transition, state whether each is a metal or a nonmetal, and give its normal phase: (a) krypton, (b) iron, (c) uranium.
    Answer: (a) representative, nonmetal, gas   (b) transition, metal, solid   (c) inner transition, metal, solid

14. Classify each of these elements as representative, transition, or inner transition, state whether each is a metal or a nonmetal, and give its normal phase: (a) plutonium, (b) hydrogen, (c) mercury.

15. Give the total number of electrons, the number of valence electrons, and the number of shells containing electrons in (a) a silicon atom and (b) an arsenic atom.
    Answer: (a) 14, 4, 3   (b) 33, 5, 4

16. Give the total number of electrons, the number of valence electrons, and the number of shells containing electrons in (a) a calcium atom and (b) an oxygen atom.

17. Use the periodic table to find the atomic mass, atomic number, number of protons, and number of electrons for an atom of (a) lithium and (b) gold.
    Answer: (a) 6.94 u, 3, 3, 3   (b) 197.0 u, 79, 79, 79

18. Use the periodic table to find the atomic mass, atomic number, number of protons, and number of electrons for an atom of (a) neon and (b) lead.

19. Give the shell electron configuration for (a) helium and (b) aluminum.   Answer: (a) 2   (b) 2, 8, 3

20. Give the shell electron configuration for (a) carbon and (b) chlorine.

21. Arrange in order of increasing *nonmetallic* character (a) the Period 4 elements Se, Ca, and Mn, and (b) the Group 6A elements Se, Po, and O.
    Answer: (a) Ca, Mn, Se   (b) Po, Se, O

22. Arrange in order of increasing *nonmetallic* character (a) the Period 3 elements P, Cl, and Na, and (b) the Group 7A elements F, Br, and Cl.

23. Arrange in order of increasing ionization energy (a) the Period 5 elements Sn, Sr, and Xe, and (b) the Group 8A elements Ar, He, and Ne.
    Answer: (a) Sr, Sn, Xe   (b) Ar, Ne, He

24. Arrange in order of increasing ionization energy (a) the Group 1A elements Na, Cs, and K, and (b) the Period 4 elements As, Ca, and Br.

25. Arrange in order of increasing atomic size (a) the Period 4 elements Ca, Kr, and Br, and (b) the Group 1A elements Cs, Li, and Rb.
    Answer: (a) Kr, Br, Ca   (b) Li, Rb, Cs

26. Arrange in order of increasing atomic size (a) the Period 2 elements C, Ne, and Be, and (b) the Group 4A elements Si, Pb, and Ge.

11.5 Naming Compounds

27. Name each of these common acids: (a) H$_2$SO$_4$(*aq*), (b) HNO$_3$(*aq*), (c) HCl(*aq*).
    Answer: (a) sulfuric acid   (b) nitric acid   (c) hydrochloric acid

28. Name each of these common acids: (a) H$_3$PO$_4$(*aq*), (b) HC$_2$H$_3$O$_2$(*aq*), (c) H$_2$CO$_3$(*aq*).

29. Give the preferred names for (a) $CaBr_2$, (b) N_2S_5, (c) $ZnSO_4$, (d) KOH, (e) $AgNO_3$, (f) IF_7, (g) $(NH_4)_3PO_4$, and (h) Na_3P.

     Answer: (a) calcium bromide   (b) dinitrogen pentasulfide
               (c) zinc sulfate   (d) potassium hydroxide
               (e) silver nitrate   (f) iodine heptafluoride
               (g) ammonium phosphate   (h) sodium phosphide

30. Give the preferred names for (a) $Al_2(CO_3)_3$, (b) $(NH_4)_2SO_4$, (c) Li_2S, (d) SO_3, (e) Ba_3N_2, (f) $Ba(NO_3)_2$, (g) SiF_4, and (h) S_2Cl_2.

11.6 Groups of Elements

31. Sodium forms the compounds Na_2S, Na_3N, and $NaHCO_3$. What are the formulas for lithium sulfide, lithium nitride, and lithium hydrogen carbonate?

                    Answer: Li_2S, Li_3N, $LiHCO_3$

32. Magnesium forms the compounds $Mg(NO_3)_2$, $MgCl_2$, and $Mg_3(PO_4)_2$. What are the formulas for barium nitrate, barium chloride, and barium phosphate?

Solutions to Confidence Exercises

11.1 (a) $Z = 15$; atomic mass is 31.0 u.
     (b) representative; nonmetal
     (c) Period 3; Group 5A
     (d) 15 electrons; 15 protons
     (e) 5 valence electrons (same as group number)
     (f) 3 shells (same as period number)
     (g) 2,8,5

11.2 (a) AsF_5 is a binary compound of two nonmetals, so the -*ide* suffix will be used for F, and because fluorine has more than one atom showing in the formula, Greek prefixes are needed for it. The answer is *arsenic pentafluoride*. (b) $CaCl_2$ is a binary compound of a metal and a nonmetal. Name the metal, then use the -*ide* ending for the nonmetal. The answer is *calcium chloride*.

11.3 Potassium, sodium, and lithium are all in Group 1A. Because sodium chloride is NaCl and potassium chloride is KCl, lithium chloride's formula will be LiCl.

Answers to Multiple-Choice Review Questions

3. c   5. c   13. c   18. d   22. d   24. b   36. a   41. c
4. b   12. a   14. b   19. b   23. c   35. d   40. a

CHEMICAL BONDING

12

See plastic nature working to this end
The single atoms each to other tend
Attract, attracted to, the next in place
Form'd and impell'd its neighbor
to embrace.

Alexander Pope (1688–1744)

I n this chapter we focus on chemical bonding and its role in compound formation. As the chapter-opening quotation implies, virtually everything in nature depends on chemical bonds. The proteins, carbohydrates, fats, and nucleic acids that make up living matter are complex molecules held together by chemical bonds. Simpler molecules such as the ozone, carbon dioxide, and water in the air absorb radiant energy, thus keeping Earth at a livable temperature.

Various molecules and ions bond together to form the compounds that make up the minerals and rocks of Earth. Potassium ions in our heart cells help maintain the proper contractions, and in the extracellular fluids they help control nerve transmissions to muscles. Were it not for the hydrogen bonding that causes one water molecule to attract four others, water would not be a liquid and life would not exist on Earth.

Chemical bonding results from the electromagnetic forces among the various electrons and nuclei of the atoms involved, and chemists often must use quantum mechanics (Section 9.7) to solve the complicated problems that arise. However, as you will see, relatively simple concepts can correlate and explain much of the information encountered in chemical bonding.

Photo: The breaking and formation of chemical bonds are responsible for the vivid colors in this fireworks display over Capitol Hill.

We begin our study of chemical bonding with a discussion of two basic laws that describe mass relationships in compounds and helped lead John Dalton to the atomic theory. ■

12.1 Law of Conservation of Mass

▼ State and use the law of conservation of mass.

If the total mass involved in a chemical reaction is precisely measured before and after the reaction takes place, the most sensitive balances cannot detect any change (● Fig. 12.1). This generalization is known as the **law of conservation of mass:**

No detectable change in the total mass occurs during a chemical reaction.

As discussed in the chapter's first Highlight, this law was discovered in 1774 by Antoine Lavoisier ("lah-vwah-ZHAY"). Now, two centuries later, Lavoisier's law of conservation of mass is still valid and useful.

EXAMPLE 12.1

Using the Law of Conservation of Mass

The complete burning in oxygen of 4.09 g of carbon produces 15.00 g of carbon dioxide as the only product.

Carbon + oxygen ⟶ carbon dioxide

4.09 g + ? ⟶ 15.00 g

How many grams of oxygen gas must have reacted?

SOLUTION

The total mass before and after reaction must be equal. Because carbon dioxide was the only product and weighed 15.00 g, the total mass of carbon and oxygen that combined to produce it must have been 15.00 g. Because carbon's mass was 4.09 g, the oxygen that reacted with it must have contributed 15.00 g minus 4.09 g, or 10.91 g.

(a)

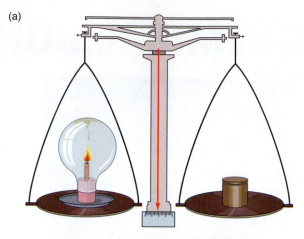

(b)

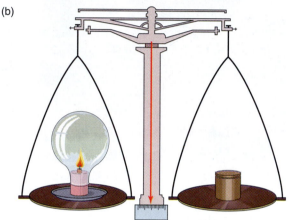

FIGURE 12.1 The Law of Conservation of Mass

When a candle is burned in an airtight container of oxygen, there is no detectable change in the candle's mass, as illustrated by the balance's pointer being in the same place (a) before the reaction and (b) after the reaction.

CONFIDENCE EXERCISE 12.1

If 111.1 g of calcium chloride is formed when 40.1 g of calcium is reacted with chlorine, what mass of chlorine combined with the calcium?

12.2 Law of Definite Proportions

▼ Calculate the formula masses of compounds.

▼ State and use the law of definite proportions.

Formula Mass

Recall from Section 10.2 that the *atomic mass* (abbreviated AM) of an element is the *average* mass assigned to each atom in its naturally occurring mixture of isotopes. The masses of elements are based on a scale that assigns the ^{12}C atom the value of exactly 12 u. The atomic masses for most elements have been determined to several decimal places, as shown in the periodic table on the inside front cover of this textbook, but for convenience we will round off these values to the nearest 0.1 u. For example, we will consider the atomic masses of hydrogen, oxygen, and calcium to be 1.0 u, 16.0 u, and 40.1 u, respectively.

The **formula mass** (abbreviated FM) of a compound or element is the sum of the atomic masses given in the formula of the substance. For example, the formula mass of O_2 is 16.0 u + 16.0 u = 32.0 u, and that of methane (swamp gas, CH_4) is 12.0 u + (4 × 1.0 u) = 16.0 u. If the formula of an element under consideration is given by just the element's *symbol* (for example, Fe or Xe), then the formula mass is just the atomic mass.*

EXAMPLE 12.2

Calculating Formula Masses

Find the formula mass of lead chromate, $PbCrO_4$, the bright yellow compound used in paint for the yellow lines on streets.

SOLUTION

Find the atomic masses of Pb, Cr, and O in the periodic table. The formula shows one atom of Pb (207.2 u), one atom of Cr (52.0 u), and four atoms of O (16.0 u), so FM = 207.2 u + 52.0 u + (4 × 16.0 u) = 323.2 u.

CONFIDENCE EXERCISE 12.2

Find the formula mass of hydrogen sulfide, H_2S, the gas that gives rotten eggs their offensive odor.

---

*Chemists often use the term *atomic weight* in place of *atomic mass, molecular weight* or *formula weight* instead of *formula mass,* and *amu* rather than *u.*

Law of Definite Proportions

In 1799, the French chemist Joseph Proust ("proost") discovered the **law of definite proportions:**

Different samples of a pure compound always contain the same elements in the same proportion by mass.

For example,

9 g H_2O is composed of 8 g oxygen and 1 g hydrogen.

18 g H_2O is composed of 16 g oxygen and 2 g hydrogen.

36 g H_2O is composed of 32 g oxygen and 4 g hydrogen.

In each case the ratio by mass of oxygen to hydrogen is 8 to 1.

Because the elements in a compound are present in a specific proportion or ratio by mass, they are also present in a specific percentage by mass. The general equation for calculating the percentage of any component in a total is

$$\% \text{ component} = \frac{\text{amount of component}}{\text{total amount}} \times 100\% \qquad \textbf{12.1}$$

For example, if there are 6 sophomores in a class of 30 students, the percentage of sophomores would be 6/30 × 100% = 20%. Of course, if one of the other two quantities is the unknown, the equation can be rearranged (Appendix III) and solved for that unknown.

Equation 12.1 can be used to find the percentage by mass of an element if the total mass of the compound and the mass contribution of the element are known. For example, if the mass of the compound is 44.0 g and it consists of 11.0 g of element X and 33.0 g of element Y, the percentage by mass of element X is (11.0 g/44.0 g) × 100%, or 25.0%.

The percentage by mass of an element X in a compound can be calculated from the compound's formula by using Eq. 12.2.

$$\% X \text{ by mass} = \frac{(\text{atoms of } X \text{ in formula}) \times (\text{AM}_X)}{\text{FM}_{\text{cpd}}} \times 100\% \qquad \textbf{12.2}$$

HIGHLIGHT: Lavoisier, The Father of Chemistry

In the seventeenth and eighteenth centuries, the phenomenon of combustion was studied intensely. Georg Stahl (1660–1734), a German, explained combustion by hypothesizing that a substance called *phlogiston* was emitted into the air when a material burned. If the material was in a closed container, it presumably stopped burning because the air around it became saturated with phlogiston.

A gas discovered in 1774 by Joseph Priestley, an Englishman, vigorously supported combustion, and Priestly called the gas "dephlogisticated air." The phlogiston hypothesis was discarded thanks to the quantitative experiments of the wealthy French nobleman Antoine Lavoisier in that same year, 1774, when he made chemistry a modern science, just as Galileo and Newton had done for physics more than a century earlier.

Lavoisier performed the first quantitative chemical experiments to explain combustion and to settle the question of whether mass was gained, lost, or unchanged during a chemical reaction. Unlike others who had tried to answer the same question, Lavoisier understood the nature of gases and took care to do his experiments in closed containers so that no substances could enter or leave. In addition, he used the most accurate balance that had ever been built.

FIGURE 1   The Lavoisiers

Antoine Lavoisier (1743–1794), known as the "father of chemistry," and his wife Marie-Anne, who helped him with experiments and produced the illustrations in his textbook.

After many experiments, he was able to reason inductively from his specific findings on individual reactions and formulate the following law: *No detectable change in the total mass occurs during a chemical reaction* (the law of conservation of mass). During the course of these investigations, Lavoisier established conclusively that when things burn, they are not losing phlogiston to the air but gaining *oxygen* (his name for "dephlogisticated air") from it. Sometimes the burned materials gained mass by forming solid oxides (this would have required phlogiston to have a negative mass), and sometimes they lost mass by forming gaseous oxides. The successes of Lavoisier caused the importance of measurement to be widely recognized by chemists.

In addition to introducing quantitative methods into chemistry, discovering the role of oxygen in combustion, and finding the law of conservation of mass, Lavoisier established the principles for naming chemicals. In 1789, he wrote the first modern chemistry textbook. Lavoisier is justly referred to as the "father of chemistry."

Lavoisier invested in a private tax-collection firm and married the daughter of one of its executives (Fig. 1). Because of this, and because he had made an enemy of a would-be scientist named Marat, who became a leader of the French Revolution, Lavoisier was guillotined in 1794 during the last months of the Revolution. When he objected upon arrest that he was a scientist, not a "tax-farmer," the arresting officer replied that "the republic has no need of scientists." Within two years of Lavoisier's death, the regretful French were unveiling busts of him.

---

where AM_X is the atomic mass of element X, and FM_{cpd} is the formula mass of the compound.

EXAMPLE 12.3

Finding the Percentage by Mass from a Compound's Formula

"Dry ice" is solid carbon dioxide (● Fig. 12.2). Find the percentage by mass of carbon and oxygen in CO_2.

SOLUTION

The periodic table shows that the atomic masses of C and O are 12.0 u and 16.0 u, respectively. Therefore, the formula mass of CO_2 is 12.0 u + (2 × 16.0 u) = 44.0 u. Using Eq. 12.2 from the preceding page,

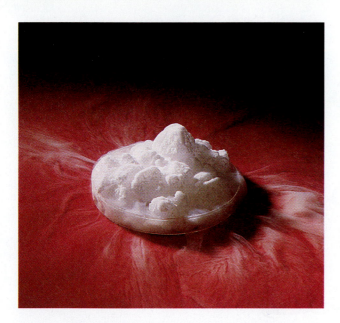

FIGURE 12.2   Dry Ice
Solid carbon dioxide has a temperature of about −78°C, so cold that it should never touch the bare skin. Like mothballs, it sublimes (passes directly from the solid phase into the gaseous phase). The white fog is water vapor condensing from the air in contact with the cold CO_2.

$$\% \text{ by mass of C} = \frac{1 \times \text{AM}_C}{\text{FM}_{CO_2}} \times 100\%$$

$$= \frac{1 \times 12.0 \text{ u}}{44.0 \text{ u}} \times 100\% = 27.3\%$$

Because oxygen is the only other component, the percentage of O and the percentage of C must add up to 100.0%. Thus the % O by mass must be 100.0% minus 27.3%, or 72.7%. (*Note:* When finding these percentages on a calculator, do *not* hit the % key after punching in 100.)

CONFIDENCE EXERCISE 12.3

Find the percentage by mass of aluminum and oxygen in aluminum oxide, Al_2O_3, the major compound in rubies.

Therefore, when a compound is broken down, its elements are found in a definite proportion by mass. Conversely, when the same compound is made from its elements, the elements will combine in that same proportion by mass. If the elements that are combined to form a compound are not mixed in the correct proportion, then one of the elements, called the **limiting reactant,** will be used up completely. The other, the **excess reactant,** will be only partially used up; some of it will remain unreacted.

In ● Fig. 12.3a, 10.00 g of Cu wire reacts completely with 5.06 g S to form 15.06 g of CuS. None of either reactant is left over. In Fig. 12.3b, 10.00 g of Cu reacts with 7.06 g S, but this is not the proper ratio. In accordance with the law of definite proportions, only 5.06 g of S can combine with 10.00 g of Cu. Therefore, 15.06 g of CuS is again formed, and the other 2.00 g of sulfur is in excess. (Note that the law of conservation of mass is satisfied, since the total mass before and after reaction is 17.06 g.) In Fig. 12.3c, it is the copper that is the excess reactant, and both the law of conservation of mass and the law of definite proportions again hold.

12.3   Dalton's Atomic Theory

LEARNING GOAL

▼   State Dalton's atomic theory.

In 1803, John Dalton (photo on page 227) proposed the following hypotheses to explain the laws of conservation of mass and definite proportions.

1. Each element is composed of small indivisible particles called *atoms,* which are identical for that element but are different (particularly in their masses and chemical properties) from atoms of other elements.
2. Chemical combination is the bonding of a definite, small number of atoms of each of the combining elements to make one molecule of the formed compound.
3. No atoms are gained, lost, or changed in identity during a chemical reaction; they are just rearranged.

If a chemical reaction is just a rearrangement of atoms, it is easy to see that the law of conservation of mass is explained.

These hypotheses also explain the law of definite proportions. If different samples of a particular compound are made up only of various numbers of the

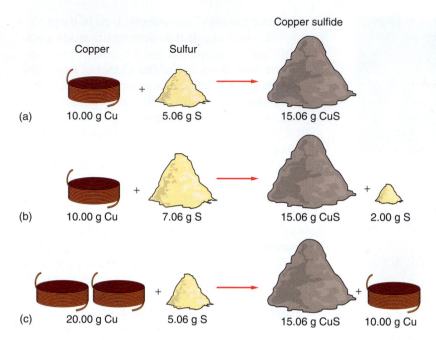

Copper sulfide

Copper   Sulfur

(a)   10.00 g Cu   5.06 g S   15.06 g CuS

(b)   10.00 g Cu   7.06 g S   15.06 g CuS   2.00 g S

(c)   20.00 g Cu   5.06 g S   15.06 g CuS   10.00 g Cu

FIGURE 12.3   The Law of Definite Proportions

When Cu and S react to form a specific compound, the law of definite proportions states that they must always react in the same ratio. If the ratio in which they are mixed is different, part of one reactant will be left over.

same basic molecule, it is easily seen that the mass ratio of the elements will have to be the same in every sample of that compound. For example, each molecule of H_2O is composed of one atom of oxygen (AM 16.0 u) and two atoms of hydrogen (AM 1.0 u, total mass of hydrogen = 2×1.0 u = 2.0 u). Thus each molecule of H_2O is composed of 16.0 parts oxygen by mass and 2.0 parts hydrogen by mass, or a ratio of 8.0 to 1.0. Because every pure sample of a molecular compound is simply a very large collection of identical molecules, the proportion by mass of any element in each sample will be the proportion by mass that it has in an individual molecule of that compound.

Because Dalton's hypotheses explained these two laws, it is no wonder that he believed that atoms exist and behave as he stated. In addition, Dalton saw a way to test his hypotheses—a necessary step in the scientific method. Suppose atoms exist and form molecules of two compounds in which the number of atoms of one element are the same (as for carbon in CO and CO_2) and the numbers of atoms of the other element are different (as for oxygen in CO and CO_2). Then the *mass ratio* of the second element in the two compounds would have to be a small whole number ratio (in this case, say, 16 g of oxygen to 32 g of oxygen, or a 1 to 2 mass ratio), thus reflecting the small whole number *atom ratio* of the second element (in this case, 1 oxygen atom to 2 oxygen atoms) in the two molecules.

Experiments by other scientists verified Dalton's prediction (now called the *law of multiple proportions*). More and more supporting evidence for Dalton's concept of the atom accumulated. Some modification of his original ideas occurred as new evidence arose, but modification is expected with all scientific ideas. Dalton's basic ideas have worked so well and for so long that we now call them the *atomic theory*. Whenever an explanation of a chemical occurrence is sought, our thoughts immediately turn to the concept of atoms. Dalton's atomic theory is the cornerstone of chemistry.

RELEVANCE QUESTION: *What was the last hypothesis you tested in your everyday life?*

12.4 Ionic Bonding

LEARNING GOALS

▼ Write formulas of ionic compounds.

▼ Describe the characteristics of ionic compounds.

▼ Use the Stock system to name compounds.

We saw in Chapter 11 that elements in the same group have the same number of valence (outer) electrons and form compounds with similar formulas—for example, the chlorides of Group 1A are LiCl, NaCl, KCl, RbCl, and CsCl. Because of this behavior, we conclude that the valence electrons are the ones involved in compound formation.

Recall from Section 11.6 that the noble gases (Group 8A) are unique in that, except for several compounds of Xe and of Kr, they do not bond chemically with atoms of other elements. Also, all noble gases are monatomic; that is, their atoms do not bond to one another to form molecules. We conclude that noble gas electron configurations (eight electrons in the outer shell, but just two for He) are uniquely stable.

The formation of the vast majority of compounds is explained by combining these two conclusions into the **octet rule:**

In forming compounds, atoms tend to gain, lose, or share valence electrons to achieve electron configurations of the noble gases; that is, they tend to get eight electrons (an octet) in the outer shell. Hydrogen is the exception; it tends to get two electrons in the outer shell, like the configuration of the noble gas helium.

Individual atoms can achieve a noble gas electron configuration in two ways: by transferring electrons or by sharing electrons. Bonding by transfer of electrons is discussed in this section, and bonding by sharing of electrons is treated in Section 12.5.

In the transfer of electrons, one or more atoms lose their valence electrons, and another one or more atoms gain these same electrons to achieve noble gas electron configurations. Compounds formed by this electron transfer process are called **ionic compounds.** This name is used because the loss or gain of electrons destroys the electrical neutrality of the atom and produces the net positive or negative electric charge that characterizes an ion.

In Section 11.4 we saw that atoms of metals generally have low ionization energies and thus tend to lose electrons and form positive ions. On the other hand, atoms of nonmetals tend to gain electrons to form negative ions. A nonmetal atom that needs only one or two electrons to fill its outer shell can easily acquire the electrons from atoms that have low ionization energy—that is, from atoms with only one or two electrons in the outer shell, such as those of Groups 1A and 2A. Atoms of some elements gain or lose as many as three electrons.

To illustrate, consider how common table salt, NaCl, is formed. When the nonmetal chlorine, a pale-green gas composed of Cl_2 molecules, is united with the metal sodium, an energetic chemical reaction forms the ionic compound sodium chloride (● Fig. 12.4). ● Figure 12.5a illustrates the loss of the valence electron from a neutral sodium atom to form a sodium ion with a 1+ charge. The neutral atom has the same number of protons (11) as electrons (11). Thus the net electric charge on the atom is zero. After the loss of an electron, the number of protons (11) is one more than the number of electrons (10), leaving a net charge on the sodium ion of 11 minus 10, or 1+. Similarly, Fig. 12.5b illustrates the gain of an electron by a neutral chlorine atom (17 p, 17 e) to form a chloride ion (17 p, 18 e) with a charge of 17 minus 18, or 1−. *The net electric charge on an ion is the number of protons minus the number of electrons.*

As Na^+ and Cl^- ions are formed, they are bonded by the attractive electric forces among the positive and negative ions. ● Figure 12.6a shows a model of sodium chloride that illustrates how the sodium ions and chloride ions arrange themselves in an orderly lattice to form a cubic crystal. (A *crystal* is a solid whose external symmetry reflects an orderly, geometric, internal arrangement of atoms, molecules, or ions.) Note that no neutral units of fixed size are present; it is impossible to associate any one Na^+ with one specific Cl^-. Thus it is somewhat inappropriate to refer to a "molecule" of sodium chloride or of any other ionic compound. Instead, we generally refer to one sodium ion and one chloride ion as being a *formula unit* of sodium chloride, the smallest combination of ions that gives the formula of the compound (see Fig. 12.6).

Similarly, the formula CaF_2 means that the compound calcium fluoride has one calcium ion (Ca^{2+}) for every two fluoride ions (2 F^-). The three ions are a formula unit of calcium fluoride.

The basic ideas of compound formation that we have been discussing are due mainly to the work of Gilbert Newton Lewis, who in 1916 developed *electron dot symbols* to help explain chemical bonding. In a **Lewis symbol,** the nucleus and inner electrons of an atom or ion are represented by the element's symbol, and the valence electrons are shown as dots arranged in four groups of one or two dots around the symbol

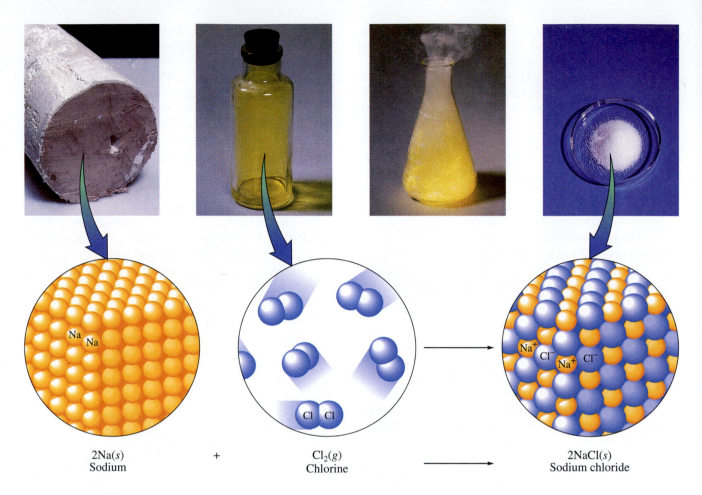

FIGURE 12.4   The Formation of Sodium Chloride

Metallic sodium (far left photo) and chlorine gas (middle left photo) react (middle right photo) to form the ionic compound named *sodium chloride* (far right photo).

(From Zumdahl, Steven S., *Chemistry*, Fourth Edition. Copyright © 1999 by Houghton Mifflin Company. Used with permission.)

(Table 12.1). It makes no difference on which side of the symbol various electron dots are put, but they are left unpaired to the extent possible.

Lewis structures use Lewis symbols to show valence electrons in molecules and ions of compounds. In a Lewis structure, a shared electron pair is indicated by two dots halfway between the atoms or, more often, by a dash connecting the atoms. Unshared pairs of valence electrons (called *lone pairs,* those not used in bonding) are shown as belonging to the individual atom or ion. Lewis structures are two-dimensional representations and therefore they do not show the actual three-dimensional nature of molecules.

The sodium and chlorine atoms of Fig. 12.5 are represented in Lewis symbols as

$$\text{Na·} \qquad \text{·}\ddot{\text{C}}\text{l:}$$

The sodium and chloride ions of Fig. 12.5 are represented in Lewis symbols as

$$\text{Na}^+ \qquad \text{:}\ddot{\text{C}}\text{l:}^-$$

where the + and − indicate the same charge as that on a proton and an electron, respectively. If the charge of the ion is greater than 1+ or 1−, it is represented by a numeral followed by the appropriate sign, as shown in the Lewis symbols for the magnesium ion and the oxide ion:

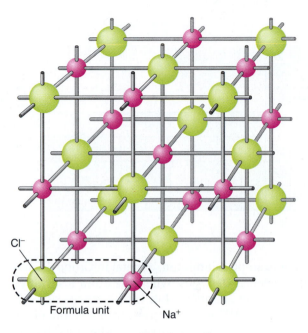

FIGURE 12.6   Sodium Chloride (NaCl)

Schematic diagram of the sodium chloride crystal. The oval of dashes indicates what is meant by a *formula unit* of NaCl.

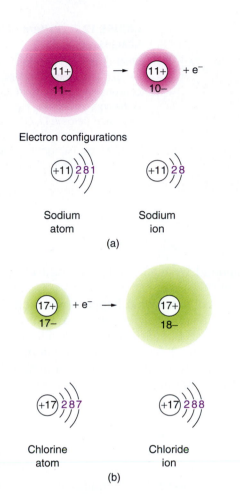

FIGURE 12.5   The Formation of (a) A Sodium Ion and (b) A Chloride Ion

After the transfer of one electron from the Na atom to the Cl atom, both ions have the electron configurations of a noble gas (eight electrons in the outer shell).

$$Mg^{2+} \qquad :\ddot{O}:^{2-}$$

The charge on most ions of the representative elements can be determined easily. First, metals form positive ions, or **cations** ("CAT-eye-ons"); nonmetals form negative ions, or **anions** ("AN-eye-ons"). Second, the positive charge on the metal's ion will be equal to the atom's number of valence electrons (its group number), whereas the negative charge on the nonmetal's ion will be the atom's number of valence electrons (its group number) minus 8 (● Fig. 12.7).

Group 8A elements already have eight valence electrons and thus form no ions in chemical reac-

tions. For Group 4A, carbon and silicon do not form ions during chemical reactions because it is so hard to either lose or gain four electrons. (We will work only with the ions shown in Fig. 12.7 and in Table 12.3 on page 305.)

The pattern of charges in Fig. 12.7 arises because valence electrons are being lost by metals and gained by nonmetals, generally to the extent necessary to get eight electrons in the outer shell—that is, to acquire an electron configuration isoelectronic with a noble gas. (Atoms or ions that have the same electron configuration are termed *isoelectronic.*)

TABLE 12.1   Lewis Symbols for the First Three Periods of Representative Elements

1A							8A
H·	2A	3A	4A	5A	6A	7A	He:
Li·	·Be·	·Ḃ·	·Ċ·	·N̈·	:Ö·	:F̈·	:N̈e:
Na·	·Mg·	·Äl·	·S̈i·	·P̈·	:S̈·	:C̈l·	:Är:

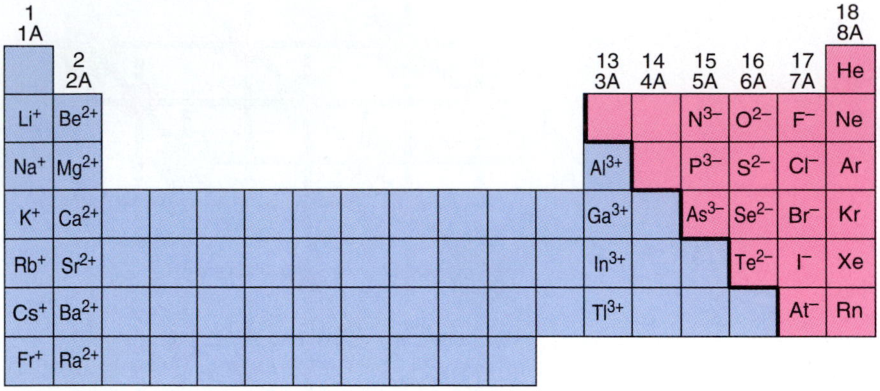

1 1A	2 2A												13 3A	14 4A	15 5A	16 6A	17 7A	18 8A
																		He
Li^+	Be^{2+}														N^{3-}	O^{2-}	F^-	Ne
Na^+	Mg^{2+}												Al^{3+}		P^{3-}	S^{2-}	Cl^-	Ar
K^+	Ca^{2+}												Ga^{3+}		As^{3-}	Se^{2-}	Br^-	Kr
Rb^+	Sr^{2+}												In^{3+}			Te^{2-}	I^-	Xe
Cs^+	Ba^{2+}												Tl^{3+}				At^-	Rn
Fr^+	Ra^{2+}																	

FIGURE 12.7 Pattern of Ionic Charges

Most of the representative elements exhibit a regular pattern of ionic charges when they form ions. (From Robinson, William R., Henry F. Holtzclaw, and Jerome D. Odom, *General Chemistry*, Tenth Edition. Copyright © 1997 by Houghton Mifflin Company. Used with permission.)

For example, aluminum, a metal in Group 3A, has three valence electrons. When the atom loses three electrons, an aluminum ion with a charge of 3+ is formed. The Al^{3+} is isoelectronic with an atom of the noble gas neon (Ne), because both have a 2,8 electron configuration (Fig. 11.18 on page 279). Sulfur, a nonmetal in Group 6A, has six valence electrons. When forming an ion, a sulfur atom gains two additional electrons. The sulfide ion (S^{2-}) thus has a negative charge of 6 − 8, or 2−, and is isoelectronic with the noble gas argon (Ar) because both have a 2,8,8 electron configuration.

In the formation of simple ionic compounds, one element loses its electrons and the other element gains them, resulting in ions. The ions are then held together in the crystal lattice by the electrical attractions among them—the **ionic bonds.** Using Lewis symbols and structures, the formation of NaCl can be represented as

$$Na\cdot + \cdot\ddot{C}l\colon \longrightarrow Na^+ \quad \colon\ddot{C}l\colon^-$$

In this example, the sodium atom lost one electron and the chlorine atom gained one. *In every ionic compound, the total charge in the formula adds up to zero and the compound exhibits electrical neutrality.* Thus, in the case of NaCl, the ratio of Na^+ to Cl^- must be 1 to 1 so that the compound will exhibit net electrical neutrality.

Another example of an ionic compound is calcium oxide, CaO. Its formation is represented as

$$\cdot Ca\cdot + \cdot\ddot{O}\colon \longrightarrow Ca^{2+} \quad \colon\ddot{O}\colon^{2-}$$

The calcium atom loses two electrons and the oxygen atom gains two. Thus, to get a net electrically neutral

compound, we again need a 1 to 1 ratio of ions of each element.

Now consider what happens when calcium and chlorine react. A calcium atom has two valence electrons to lose, but each atom of chlorine can gain only one. We must have two atoms of chlorine to accept the two electrons of the calcium atom and give a net electrically neutral compound. We have

$$\cdot Ca\cdot + \cdot\ddot{C}l\colon + \cdot\ddot{C}l\colon \longrightarrow Ca^{2+} \quad \colon\ddot{C}l\colon^- \quad \colon\ddot{C}l\colon^-$$

Thus the formula for calcium chloride is $CaCl_2$. All the ions in the compound have noble gas configurations, and the total charge is zero.

We can now see how formulas of ionic compounds arise. The numbers of atoms of the various elements involved in the compound are determined by the requirements that the total electrical charge be zero and that all the atoms have noble gas electron configurations.

In order for ionic compounds to be electrically neutral, each formula unit must have an equal number of positive and negative charges. Thus, to write correct formulas for ionic compounds, the anions and cations must be shown in the smallest whole number ratio that will equal zero charge. This principle, which works for both monatomic and polyatomic ions, can be mastered by studying Table 12.2, Example 12.4, and Confidence Exercise 12.4.

The only other information required to become adept at writing the formulas of ionic compounds is knowledge of the charges on common polyatomic ions (given in Table 11.6 on page 282) and on ions of the representative elements (follow the pattern in Fig. 12.7).

TABLE 12.2 Formulas of Ionic Compounds

General Cation Symbol	General Anion Symbol	Cation to Anion Ratio for Neutrality	General Compound Formula	Specific Example of Compound
M^+	X^-	1 to 1	MX	NaF
M^{2+}	X^{2-}	1 to 1	MX	MgO
M^{3+}	X^{3-}	1 to 1	MX	AlN
M^+	X^{2-}	2 to 1	M_2X	Na_2O
M^{2+}	X^-	1 to 2	MX_2	MgF_2
M^+	X^{3-}	3 to 1	M_3X	Na_3N
M^{3+}	X^-	1 to 3	MX_3	AlF_3
M^{2+}	X^{3-}	3 to 2	M_3X_2	Ca_3N_2
M^{3+}	X^{2-}	2 to 3	M_2X_3	Al_2O_3

EXAMPLE 12.4

Writing Formulas for Ionic Compounds

Write the formula for calcium phosphate, the major component of bones.

SOLUTION

Calcium is in Group 2A and so has an ionic charge of 2+. The phosphate ion is PO_4^{3-} (Table 11.6). Neutrality can be achieved with three Ca^{2+} and two PO_4^{3-}; the correct formula is $Ca_3(PO_4)_2$. (Note the parentheses around the phosphate ion show that the subscript 2 applies to the entire polyatomic ion.)

CONFIDENCE EXERCISE 12.4

In the nine blank spaces in the matrix below, write the formulas for the ionic compounds formed by combining each metal ion (M) with each nonmetal ion (X). This matrix covers the general formulas for all possible compounds formed from ions having charges of magnitude one through three.

	X^-	X^{2-}	X^{3-}
M^+			
M^{2+}			
M^{3+}			

Very strong forces of attraction exist among oppositely charged ions, so ionic compounds are always crystalline solids with high melting and boiling points. Another important property of ionic compounds is their behavior when an electric current is passed through them.

If a light bulb is connected to a battery by two wires, the bulb glows. Electrons flow from the negative terminal of the battery, through the light bulb, and back to the positive terminal. If one wire is cut, the electrons cannot flow, and the light bulb will not glow. If the ends of the cut wire are inserted into a solid ionic compound, the bulb does not light, because the ions are held in place and cannot move (● Fig. 12.8a). However, if the cut wires are inserted into a *melted* ionic compound, such as molten (liquid) NaCl, the bulb lights (Fig. 12.8b).

Ionic compounds in the liquid phase conduct an electric current because the ions are now free to move and carry charge from one wire to the other. *This is the crucial test of whether or not a compound is ionic: When melted, does it conduct an electric current? If so, it is ionic.* ● Figure 12.9 illustrates how the conduction takes place.

Many ionic compounds dissolve in water, thus forming aqueous solutions in which the ions are free to move. Like molten salts, such solutions conduct an electric current. (Table 12.5 on page 310 summarizes the properties of ionic compounds and compares them to those of covalent compounds.)

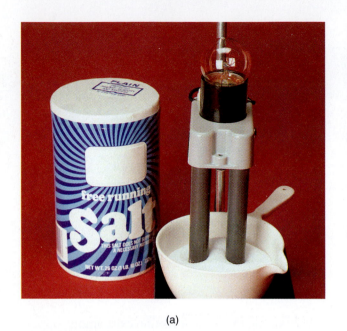

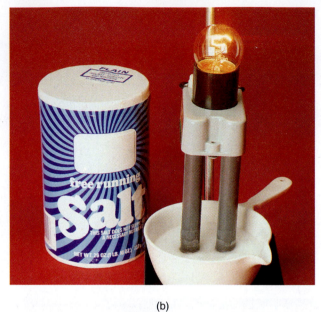

(a)

(b)

FIGURE 12.8 Ionic Compounds Conduct Electricity When Melted (But Not When Solid)

(a) When dry salt, an ionic compound, is used to connect the electrodes, the bulb in the conductivity tester does not light. (b) When the salt is melted, the ions are free to move, the circuit is completed, and the bulb lights. (Salt dissolved in water also will conduct electricity.)

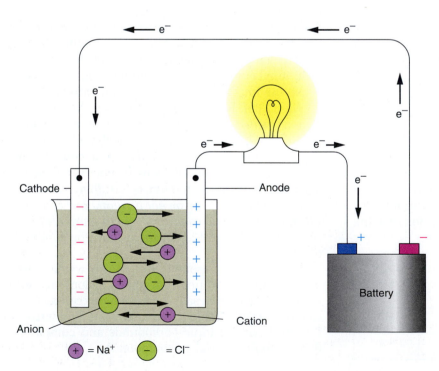

FIGURE 12.9 How Melted Salt (NaCl) Conducts Electricity

The negative ions (anions) move toward the anode, and the positive ions (cations) move toward the cathode. The movement of the ions closes the electric circuit and allows electrons to flow in the wire, as indicated by the lighted bulb.

TABLE 12.3   Some Metals That Form Two Ions

Copper	Cu^+	Cu^{2+}
Gold	Au^+	Au^{3+}
Iron	Fe^{2+}	Fe^{3+}
Nickel	Ni^{2+}	Ni^{3+}
Chromium	Cr^{2+}	Cr^{3+}

The Stock System

The basic rules for naming compounds were discussed in Section 11.5, where we stated that the rules must be expanded for metals that form two (or more) types of ions. Such metals form more than one compound with a given nonmetal or polyatomic ion. For simplicity, we will consider only the five metals listed in Table 12.3. To distinguish the compounds, use the **Stock system;** that is, *place in parentheses directly after the metal's name a Roman numeral giving the value of the metal's ionic charge.* For example,

$CrCl_2$ chromium(II) chloride

$CrCl_3$ chromium(III) chloride

Because the chloride ions have a 1− charge and the compound must be electrically neutral, the chromium ion in $CrCl_2$ must have a charge of 2+. Similar reasoning indicates that the chromium ion in $CrCl_3$ must have a charge of 3+. The two compounds, $CrCl_2$ and $CrCl_3$, have entirely different properties (Fig. 12.10).

EXAMPLE 12.5

Naming Compounds of Metals That Form Several Ions

A certain compound of gold and sulfur has the formula Au_2S. As shown in Table 12.3, gold is a metal that forms several ions, so it would be preferable to name Au_2S using the Stock system. What is its Stock system name?

SOLUTION

Sulfur has a 2− ionic charge (Fig. 12.7). Thus the two Au atoms must contribute a total of 2+, or 1+ each.

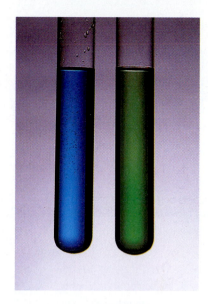

FIGURE 12.10   The Ions of Chromium
Solutions containing Cr^{2+} are usually blue; those of Cr^{3+} are generally green.

Therefore, gold's ionic charge in this compound is 1+, and the compound's Stock system name is gold(I) sulfide.

CONFIDENCE EXERCISE 12.5

Under an older system of nomenclature, CuF and CuF_2 are named cuprous fluoride and cupric fluoride, respectively. What would be their Stock system names?

12.5 Covalent Bonding

LEARNING GOALS

▼ Write formulas and Lewis structures for covalent compounds.

▼ Describe the characteristics of covalent compounds.

In the formation of ionic compounds, electrons are transferred, which produces ions. In the formation of **covalent compounds,** pairs of electrons are shared,

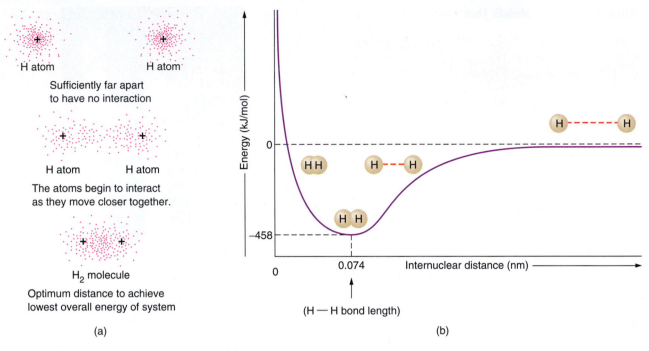

FIGURE 12.11   Bonding in the H₂ Molecule

(a) The interaction of two hydrogen atoms (electron cloud model). (b) Energy of the system is plotted as a function of distance between the nuclei of the two hydrogen atoms. As the atoms approach one another, the energy decreases until the distance reaches 0.074 nm. This is the distance at which the system is most stable. At closer distances, the energy begins to increase due to increased repulsion between the two nuclei.

which produces molecules. When a pair of electrons is shared by two atoms, we say that a **covalent bond** exists between the atoms. If the covalent bond is between atoms of the same element, the molecule formed is that of an element. As examples, let's examine the H_2 and Cl_2 molecules.

Hydrogen gas, H_2, is the simplest example of a molecule with a single covalent bond. When two hydrogen atoms are brought together, the result is attraction between opposite electrons and protons and repulsion between the two electrons and between the two protons. Once the nuclei are 0.074 nm apart, attraction is at a maximum (● Fig. 12.11). The two electrons no longer orbit individual nuclei. They are shared equally by both nuclei, and this holds the atoms together.

In Lewis symbols, the separated hydrogen atoms are written as

$$H \cdot \qquad \cdot H$$

The Lewis structure of the H_2 molecule is written as

$$H:H$$

The two dots between the hydrogen atoms indicate that these electrons are being shared, giving each atom a share in the electron configuration of the noble gas helium. (Shared electrons are counted as belonging to each atom.) The single covalent bond (one shared pair of electrons) also can be represented by a dash.

$$H—H$$

Not all atoms will share electrons. For instance, a helium atom, with two electrons already filling its valence shell, does not form a stable molecule with another helium atom or with any other atom. The molecule H_2 is stable, but He_2 is not. Stable covalent molecules are formed when the atoms share electrons in such a way as to give all atoms a share in a noble gas configuration, in other words, whenever the octet rule is followed.

Now consider the bonding in Cl_2. In Lewis symbols, two chlorine atoms are shown as

$$:\ddot{\text{C}}\text{l}\cdot \qquad \cdot\ddot{\text{C}}\text{l}:$$

Each chlorine atom requires one electron to complete its octet. If each shares its unpaired electron with the other, a Cl_2 molecule is formed.

$$:\ddot{\text{C}}\text{l}:\ddot{\text{C}}\text{l}: \quad \text{or} \quad :\ddot{\text{C}}\text{l}-\ddot{\text{C}}\text{l}:$$

In the Cl_2 molecule, each chlorine atom has six electrons (three lone pairs) plus two shared electrons, giving each an octet of electrons. (Remember, the shared electrons are counted as belonging to each atom.)

Covalent bonds between atoms of different elements form molecules of compounds. Consider the gas HCl, hydrogen chloride. The H and Cl atoms have the Lewis symbols

$$\text{H}\cdot \qquad \cdot\ddot{\text{C}}\text{l}:$$

If each shares its one unpaired electron with the other, they both acquire noble gas configurations, and the Lewis structure for hydrogen chloride is

$$\text{H}:\ddot{\text{C}}\text{l}: \quad \text{or} \quad \text{H}-\ddot{\text{C}}\text{l}:$$

Up to this point, we have seen that the noble gases tend to form no covalent bonds, that H forms one bond, and that Cl and the other elements of Group 7A form one bond. Let's take a look at the covalent bonding tendencies of oxygen, nitrogen, and carbon—representatives of Groups 6A, 5A, and 4A, respectively.

One of the most common molecules containing an oxygen atom is water, H_2O. The Lewis symbols of the individual atoms are

$$\text{H}\cdot \qquad \cdot\ddot{\text{O}}:$$
$$\qquad\qquad \dot{\text{H}}$$

When they combine, two single bonds are formed to the oxygen atom as each hydrogen atom shares its electron with the oxygen atom, which shares its two unpaired valence electrons. The Lewis structure of water is

$$\text{H}:\ddot{\text{O}}: \quad \text{or} \quad \text{H}-\ddot{\text{O}}:$$
$$\ \ \dot{\text{H}} \qquad\qquad\qquad |$$
$$\qquad\qquad\qquad\qquad\quad \text{H}$$

Oxygen and the other Group 6A elements tend to form two covalent bonds.

One of the most common compounds containing a nitrogen atom is ammonia, NH_3, an important industrial compound. (See the chapter's second Highlight.) The nitrogen atom has five valence electrons

$(\cdot\dot{\text{N}}\cdot)$ and shares a different one of its three unpaired electrons with three different H atoms (H·). The Lewis structure of ammonia is

$$\text{H}:\ddot{\text{N}}:\text{H} \quad \text{or} \quad \text{H}-\dot{\text{N}}-\text{H}$$
$$\quad\ \dot{\text{H}} \qquad\qquad\qquad\quad |$$
$$\qquad\qquad\qquad\qquad\qquad \text{H}$$

Nitrogen and the other Group 5A elements tend to form three covalent bonds.

A common and simple compound of carbon is methane, CH_4. By now, you can predict that a carbon atom with its four valence electrons $(\cdot\dot{\text{C}}\cdot)$ will share a different valence electron with four different hydrogen atoms (H·). The Lewis structure of methane is

$$\qquad\qquad\qquad\qquad \text{H}$$
$$\quad\ \ \ddot{\text{H}} \qquad\qquad\qquad |$$
$$\text{H}:\ddot{\text{C}}:\text{H} \quad \text{or} \quad \text{H}-\text{C}-\text{H}$$
$$\quad\ \ \ddot{\text{H}} \qquad\qquad\qquad |$$
$$\qquad\qquad\qquad\qquad \text{H}$$

Carbon and the other Group 4A elements tend to form four covalent bonds.

Covalent bonding is encountered mainly, but not exclusively, in compounds of nonmetals with other nonmetals. Writing Lewis structures for covalent compounds requires an understanding of the number of covalent bonds normally formed by common nonmetals. Table 12.4 summarizes our discussion of the number of covalent bonds to be expected from the elements of Groups 4A through 8A. (Exceptions are uncommon in Periods 1 and 2 but occur with more frequency starting with Period 3.)

When an element has 2, 3, or 4 *unpaired* valence electrons, its atoms will sometimes share more than one of them with another atom. Thus double bonds and triple bonds between two atoms are possible

TABLE 12.4 Number of Covalent Bonds Formed by Common Nonmetals

4A	5A	6A	7A	8A
$\cdot\dot{X}$	$\cdot\dot{X}$	$\cdot\ddot{X}$	$\cdot\ddot{X}$	$:\ddot{X}:$
4 bonds	3 bonds	2 bonds	1 bond	0 bonds
$-\overset{\mid}{\underset{\mid}{\text{C}}}-$	$-\overset{}{\underset{\mid}{\text{N}}}-$	$\ddot{\text{O}}-$	$:\ddot{\text{F}}-$	$:\ddot{\text{Ne}}:$

HIGHLIGHT: The Varied Uses of Ammonia

Ammonia, NH_3, is the gas that gives the sharp aroma to smelling salts and some household cleaners. It can be used to make compounds such as ammonium nitrate, NH_4NO_3, which is a prime fertilizer because it is inexpensive to produce and provides the abundant nitrogen needed for plant growth.

Many explosives, such as nitroglycerine and TNT, are comprised of compounds whose molecules contain several nitrogen atoms. When such compounds break down, they provide a lot of energy due in part to the formation of stable N_2 molecules held together by strong triple bonds. Moreover, N_2 and other decomposition products are gases, and confined hot gases lead to explosions!

During World War I, the development of a process for making ammonia directly from nitrogen and hydrogen gas allowed Germany to produce explosives and extend the war for a full 2 years. In 1947, a ship laden with ammonium nitrate exploded in the harbor of Texas City, Texas. The accident killed almost 600 people and injured another 3500.

The terrorist attacks on the World Trade Center in New York in 1993 and on the Alfred P. Murrah Federal Office Building in Oklahoma City in 1995

FIGURE 1   Oklahoma City Bombing
The Alfred P. Murrah Federal Office Building after the bomb attack in 1995.

were both carried out using hundreds of pounds of ammonium nitrate mixed with fuel oil (Fig. 1).

Of course, explosives have many constructive uses, for example, in mining coal and quarrying rock, digging tunnels and blasting roadways, and demolishing old buildings to make way for new ones.

Ammonium nitrate can be put to constructive use by a farmer growing food or to destructive use by a terrorist taking innocent lives. As in the utilization of guns, drugs, cars, and so on, the outcome depends on the goodwill or ill will of people.

(but, for geometric reasons, quadruple bonds are not). Consider the bonding in carbon dioxide, CO_2. The Lewis symbols for two oxygen atoms and one carbon atom are

$$:\ddot{O}\cdot \quad \cdot\dot{C}\cdot \quad \cdot\ddot{O}:$$

To get a stable molecule, eight electrons are needed around each atom. Therefore, we must have four electrons shared between the carbon atom and each oxygen atom. The Lewis structure of the CO_2 molecule is

$$:\ddot{O}::C::\ddot{O}: \quad \text{or} \quad :\ddot{O}=C=\ddot{O}:$$

Note that each atom in the CO_2 molecule has an octet of electrons around it. Perhaps you can see this more clearly if each atom and its octet of electrons is enclosed in a circle.

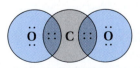

A sharing of two pairs of electrons between two atoms produces a *double bond*, represented by a double dash. A sharing of three pairs of electrons between two atoms produces a *triple bond,* represented by a triple dash. Nitrogen gas, N_2, is an example of a molecule with a triple bond. Two nitrogen atoms may be represented as

$$:\dot{N}\cdot \qquad \cdot\dot{N}:$$

To satisfy the octet rule, each nitrogen atom must share its three unpaired valence electrons with the other, forming a triple bond.

The Lewis structure for the N_2 molecule is

$$:N\!:\!:\!:\!N: \quad \text{or} \quad :N\!\equiv\!N:$$

In general, when drawing Lewis structures for covalent compounds, write the Lewis symbol for each atom in the formula. Realize that atoms forming only one bond can never connect two other atoms (just as a one-armed person cannot grab two people and serve as a "connection" between them). Try connecting the atoms using single bonds, remembering the number of covalent bonds each atom normally forms. If each atom gets a noble gas configuration, your job is done. If not, see if a double or triple bond will work.

Drawing Lewis Structures for Simple Covalent Compounds

Draw the Lewis structure for chloroform, $CHCl_3$, a covalent compound once used as an anesthetic but now replaced by less toxic compounds.

SOLUTION

A carbon atom (Group 4A) forms four bonds, H forms one bond, and Cl (Group 7A) forms one bond. Only C can be the central atom; thus

$$H\cdot \quad \cdot\dot{C}\cdot \quad \cdot\ddot{C}l: \quad \cdot\ddot{C}l: \quad \cdot\ddot{C}l:$$

gives

$$\begin{array}{c} H \\ :\ddot{C}l\!:\!\dot{C}\!:\!\ddot{C}l: \\ :\ddot{C}l: \end{array} \quad \text{or} \quad \begin{array}{c} H \\ | \\ :\ddot{C}l\!-\!C\!-\!\ddot{C}l: \\ | \\ :\ddot{C}l: \end{array}$$

Because of the nature of the bonding involved, covalent compounds have quite different properties from those of ionic compounds. Unlike ionic compounds, covalent compounds are composed of individual molecules with a specific molecular formula. For example, carbon tetrachloride consists of individual CCl_4 molecules, each composed of one carbon atom and four chlorine atoms. Although each covalent bond is strong *within* a molecule, the various molecules in a sample of the compound only weakly attract *one another.* Therefore, the melting points and boiling points of covalent compounds are generally low compared with those of ionic compounds. For instance, CCl_4 melts at $-23°C$ and is a liquid at room temperature, whereas the ionic compound NaCl melts at $801°C$. Many covalent compounds occur as liquids or gases at room temperature. Covalent compounds do not conduct electricity well, no matter what phase they are in. The general properties of ionic and covalent compounds are summarized in Table 12.5.

Some compounds contain both ionic and covalent bonds. Sodium hydroxide (lye, NaOH) is an example. In Chapter 11 we saw that it is not uncommon for several atoms to form a polyatomic ion such as the hydroxide ion, OH^-. The atoms *within* the polyatomic ion are covalently bonded, but the whole aggregation behaves like an ion in forming compounds. Because strong covalent bonds are present between atoms within polyatomic ions, it is difficult to break them up. Therefore, in chemical reactions they frequently act as a single unit, as we will see in Chapter 13. The Lewis structure of sodium hydroxide is

$$Na^+[:\ddot{O}:H]^-$$

A covalent bond exists between the O and the H in each hydroxide ion, but the hydroxide ions and the sodium ions are bound together in a crystal lattice by ionic bonds, and thus NaOH is an ionic compound.

Follow these rules to predict whether a particular compound is ionic or covalent.

TABLE 12.5   Comparison of Properties of Ionic and Covalent Compounds

Ionic Compounds	Covalent Compounds
Crystalline solids (made of ions)	Gases, liquids, or solids (made of molecules)
High melting and boiling points	Low melting and boiling points
Conduct electricity when melted	Poor electrical conductors in all phases
Many soluble in water but not in nonpolar liquids	Many soluble in nonpolar liquids but not in water

1. Compounds formed of only nonmetals are covalent (except ammonium compounds).
2. Compounds of metals and nonmetals are generally ionic (especially for Group 1A or 2A metals).
3. Compounds of metals with polyatomic ions are ionic.
4. Compounds that are gases, liquids, or low-melting-point solids are covalent.
5. Compounds that conduct an electric current when melted are ionic.

EXAMPLE 12.7

Predicting Bonding Type

Predict which compounds are ionic and which are covalent: (a) KF, (b) SiH_4, (c) $Ca(NO_3)_2$, (d) compound X, a gas at room temperature, (e) compound Y, which melts at 900°C and then conducts an electric current.

SOLUTION

(a) KF is ionic (Group 1A metal and a nonmetal).
(b) SiH_4 is covalent (only nonmetals).
(c) $Ca(NO_3)_2$ is ionic (metal and polyatomic ion).
(d) X is covalent (as are all substances that are gases or liquids at room temperature).
(e) Y is ionic (the compound has a high melting point, and the melt conducts electricity).

CONFIDENCE EXERCISE 12.7

Is PCl_3 ionic or covalent in bonding? What about MgF_2?

Polar Covalent Bonding

In covalent bonding, the electrons involved in the bond between two atoms are shared. However, unless the atoms are of the same element, the bonding electrons will spend more time around the more nonmetallic element; that is, the sharing is unequal. Such a bond is called a **polar covalent bond,** indicating that it has a slightly positive end and a slightly negative end.

Electronegativity (abbreviated EN) is a measure of the ability of an atom in a molecule to draw bonding electrons to itself. ● Figure 12.12 shows the numerical

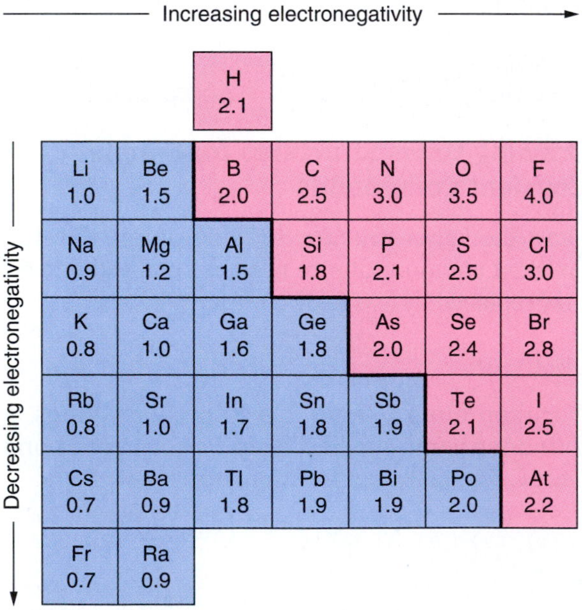

FIGURE 12.12   Electronegativity Values

In general, electronegativity increases across a period and decreases down a group. Fluorine is the most electronegative element.

values calculated for the electronegativities of the representative elements. Electronegativity shows a definite periodic trend; it increases across (from left to right) a period and decreases down a group, just as does nonmetallic character (Section 11.4).

Consider the covalent bond in HCl. The chlorine atom is more electronegative (EN = 3.0) than the hydrogen atom (EN = 2.1). Although the two bonding electrons are shared between the two atoms, they tend to spend more time at the chlorine end than at the hydrogen end of the molecule. We get a polar bond in which the polarity could be represented by an arrow:

$$H \xrightarrow{\;\;+\!\!\!\rightarrow\;\;} \ddot{C}l\!:$$

The head of the arrow points to the more electronegative atom and denotes the negative end of the bond. The "feathers" of the arrow make a plus-sign (at left above) that indicates the positive end of the bond. ● Figure 12.13 summarizes our discussion of ionic, covalent, and polar covalent bonding.

EXAMPLE 12.8

Showing the Polarity of Bonds

Use arrows to show the polarity of the covalent bonds in (a) water, H_2O, and (b) carbon tetrachloride, CCl_4.

SOLUTION

(a) Oxygen (EN = 3.5) is more electronegative than hydrogen (EN = 2.1), so the arrows denoting the polarity of the bonds would point as shown here.

$$H \xrightarrow{\;\;+\!\!\!\rightarrow\;\;} \ddot{O}\!: \\ \big\updownarrow \\ H$$

(b) Chlorine (EN = 3.0) is more electronegative than carbon (EN = 2.5), so the arrows denoting the polarity of the bonds would point as shown.

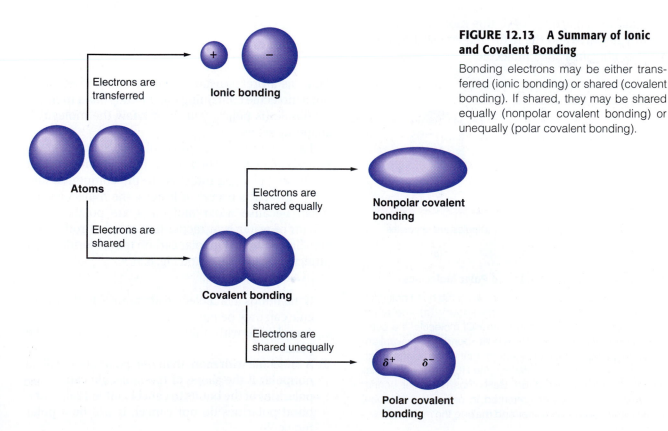

FIGURE 12.13 A Summary of Ionic and Covalent Bonding

Bonding electrons may be either transferred (ionic bonding) or shared (covalent bonding). If shared, they may be shared equally (nonpolar covalent bonding) or unequally (polar covalent bonding).

$$:\ddot{C}l:$$

(Lewis structure of CCl₄ with resonance-style arrows)

$$:\ddot{C}l \rightleftarrows C \leftrightharpoons \ddot{C}l:$$

$$:\ddot{C}l:$$

CONFIDENCE EXERCISE 12.8

Use arrows to show the polarity of the covalent bonds in ammonia, NH_3.

Polar Bonds and Polar Molecules

Molecules, as well as bonds, can have polarity. A molecule, as a whole, is polar if electrons are more attracted to one end of the molecule than to the other end. Such a molecule has a slightly negative end and a slightly positive end, and we say it has a *dipole,* or is a **polar molecule.** The slightly negative end of the polar molecule is denoted by a $\delta-$ (delta minus), and the slightly positive end by a $\delta+$ (delta plus). Consider the HCl molecule with its one polar bond (● Fig. 12.14a). With only one polar bond present, it should be obvious that the chlorine end of this molecule must be slightly negative and the hydrogen end slightly positive, resulting in a polar molecule.

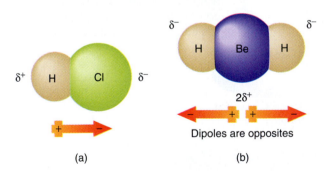

(a)   (b)

FIGURE 12.14   Polar Bonds and Polar Molecules
To predict the polarity of a molecule, both the bond polarities and the molecular shape must be known. (a) The arrow shown here for the one bond in an HCl molecule is a common symbol for the dipole of a polar bond. By convention, the arrow points to the more electronegative end of the bond. With only one polar bond present, the HCl molecule is polar overall. (b) Beryllium hydride, BeH₂, has two polar bonds. However, the dipoles are oriented in opposite directions, thereby canceling each other and making the molecule nonpolar overall.

Now consider the linear molecule beryllium hydride, BeH_2 (Fig. 12.14b). It has two polar bonds; however, the dipoles are oriented in opposite directions and thus cancel one another, making the molecule nonpolar overall.

As yet another example, consider the water molecule. If it were linear (it's not!), it would be nonpolar. The dipoles of the two bonds would cancel, because the central point of the positive charges of the bonds would be in exactly the same place as the center of the negative charges—right in the middle of the molecule. No charge separation means no molecular dipole.

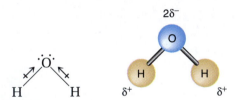

But the water molecule is actually angular (105° between the two bonds), so the bond polarities reinforce instead of canceling. The center of positive charge is midway between the two hydrogen atoms. The center of negative charge is at the oxygen atom. So, the charges are separated, and water molecules are polar.

(diagrams of water molecule, angular, with $2\delta^-$ on O and δ^+ on each H)

Thus, as a prerequisite for determining whether or not a molecule containing polar bonds has a molecular dipole (is polar), you must know the molecule's shape, or geometry.

The polarity of the molecules of a liquid compound can be tested by a simple experiment. Bring an electrically charged rod close to the compound and see if attraction occurs. If it does, the molecules are polar. Because water molecules are polar, a thin stream of water is attracted to a charged rod, but a similar stream of nonpolar carbon tetrachloride is not attracted (● Fig. 12.15 and Fig. 8.3).

In summary,

1. If the bonds in a molecule are nonpolar, the molecule can only be *nonpolar.*
2. A molecule with only one polar bond has to be polar.
3. A molecule with more than one polar bond will be nonpolar if the shape of the molecule causes the polarities of the bonds to cancel (Table 12.6). If the bond polarities do not cancel, it will be a polar molecule.

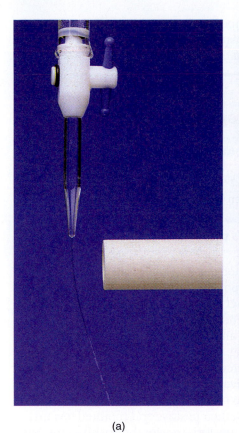

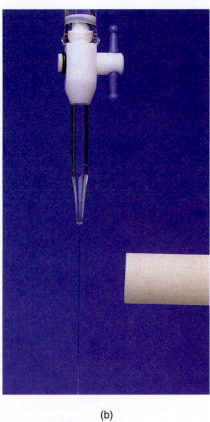

(a)   (b)

FIGURE 12.15   Polar and Nonpolar Liquids

(a) A stream of polar water molecules is deflected toward an electrically charged rod. (b) A stream of nonpolar CCl_4 molecules is undeflected.

TABLE 12.6   Types of Molecules with Polar Bonds but No Resulting Dipole

Type	Cancellation of Polar Bonds		Example	Ball-and-Stick Model
Linear molecules with two identical bonds	B—A—B	⟵—+ +—⟶	CO_2	
Planar molecules with three identical bonds 120 degrees apart	B above A, B—A—B 120°	(cancellation diagram)	SO_3	
Tetrahedral molecules with four identical bonds 109.5 degrees apart	tetrahedral B—A—B with B,B	(cancellation diagram)	CCl_4	

(From Zumdahl, Steven S., *Chemistry*, Fourth Edition. Copyright © 1999 by Houghton Mifflin Company. Used with permission.)

Predicting the Polarity of Molecules

The carbon tetrachloride molecule is tetrahedral, as shown in the accompanying sketch. All the bond angles are 109.5°, and the bonds are polar. Is the molecule polar?

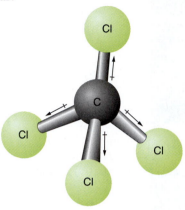

SOLUTION

Each bond has the same degree of polarity, the center of positive charge is at the C atom, and the center of negative charge is also there. The polar bonds cancel, and thus the molecule is nonpolar.

CONFIDENCE EXERCISE 12.9

Boron trifluoride, BF_3, is an exception to the octet rule. A single bond extends from each fluorine atom to the boron atom, which has a share in only six electrons. The three bonds are polar, yet the molecule itself is nonpolar. What must be the geometry of the molecule and the angle between the bonds?

12.6 Hydrogen Bonding

LEARNING GOALS

▼ Explain the statement, *Like dissolves like.*

▼ Predict when hydrogen bonding will occur.

▼ Describe how hydrogen bonding affects a compound's properties.

Why does water dissolve table salt but doesn't dissolve oil? The polar nature of water molecules causes them to interact with an ionic substance such as salt. The positive ends of the water molecules attract the negative ions, and the negative ends attract the positive ions (● Fig. 12.16). If the attraction of the water molecules overcomes the attractions among the ions, the salt dissolves. As Fig. 12.16 shows, the negative ions move into solution surrounded by several water molecules with their positive ends pointed toward the ion. Just the opposite is true for the positive ions. Such attractions are called *ion-dipole interactions*.

As you would expect, the molecules of two polar substances have a *dipole-dipole interaction* and tend to dissolve in one another. Similarly, it would not be surprising to find that two nonpolar substances mix well, such as oil and gasoline. The nonpolar molecules in oil have no more affinity for one another than

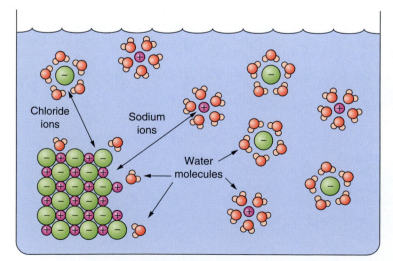

FIGURE 12.16 Sodium Chloride Dissolving in Water

The negative ends of the polar water molecules attract and surround the positive sodium ions (purple spheres). The positive ends of the water molecules attract and surround the negative chloride ions (green spheres).

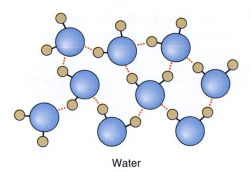

FIGURE 12.17 Hydrogen Bonding in Water

Weak forces of attraction (symbolized here by red dots) between the hydrogen atom of one water molecule and the oxygen atom of another molecule cause water to be a liquid. At room temperature, about 80% of the molecules are hydrogen-bonded at any one time.

they do for the nonpolar molecules in gasoline. These are examples of the well-known solubility principle, *Like dissolves like.* In general, however, polar substances and nonpolar substances (*unlike* substances) do not dissolve in one another, because the polar molecules tend to gather together and exclude the nonpolar molecules.

A **hydrogen bond** is a special kind of dipole-dipole interaction that can occur whenever hydrogen atoms are covalently bonded to small, highly electronegative atoms (only O, F, and N fit these criteria). Because the bond is so polar and a hydrogen atom is so small, the partial positive charge on hydrogen is highly concentrated. Thus a hydrogen atom has an electrical attraction for nearby O, F, or N atoms that may be in the same or neighboring molecules, because these atoms have highly concentrated partial negative charges.

For small molecules, such as H_2O, HF, and NH_3, hydrogen bonding is a weak force of attraction between a hydrogen atom in *one* molecule and an O, F, or N atom in *another* molecule (● Fig. 12.17). Such hydrogen bonds are *inter*molecular forces, not *intra*molecular forces.*

Hydrogen bonds are strong enough (about 5% to 10% of the strength of a covalent bond) to have a pronounced effect on the properties of the substance. In a given group, the boiling points of similar compounds generally increase with increasing formula mass. But note in ● Fig. 12.18 that the hydrogen

*For large molecules such as DNA and proteins, hydrogen bonds exist between some of the H atoms and O or N atoms in other parts of the same molecule that have twisted around into close proximity. The famous "unzipping" of DNA molecules is actually a breaking of intramolecular hydrogen bonds.

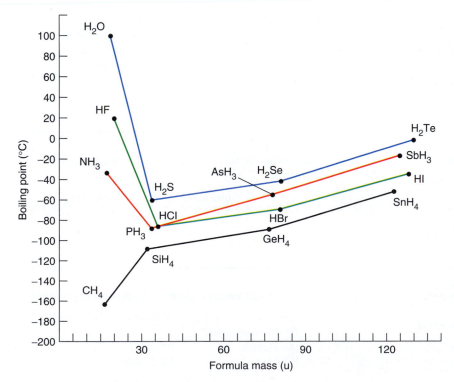

FIGURE 12.18 Hydrogen Bonding at Work

The boiling points of the hydrogen compounds of elements in Groups 4A, 5A, 6A, and 7A show that extensive hydrogen bonding exists and causes inordinately high boiling points in samples of H_2O, HF, and NH_3, but not in CH_4. Without hydrogen bonding, the boiling points of compounds in each group should increase with increasing formula mass.

FIGURE 12.19   Hydrogen Bonding and Density
Hydrogen bonding and the shape of the water molecule cause ice to have an open structure, making it less dense than liquid water. Thus ice floats, as shown on the left. For virtually every other substance, such as the benzene (C_6H_6) shown on the right, the solid is more dense than the liquid and thus does not float.

bonding in H_2O, HF, and NH_3 causes those compounds to have much higher boiling points relative to the other hydrogen compounds in their respective groups. However, because hydrogen bonding does not occur in CH_4, its boiling point shows a normal pattern relative to those of the other hydrogen compounds of Group 4A elements. Most substances with a formula mass as low as water's 18 u are gases. Without hydrogen bonding, water at room temperature would be a gas, not a liquid, because there would be so little attraction among the water molecules.

Almost all solids sink to the bottom in a sample of their liquid; why does ice float in water (● Fig. 12.19)? As the temperature of a sample of water drops, the hydrogen bonds align more and more in a hexagonal manner. This alignment gives ice an open structure, increases its volume, and makes it less dense than liquid water. (The Chapter 5 Highlight on page 98 gives more details.)

Important Terms

law of conservation of
   mass (12.1)
formula mass (12.2)
law of definite proportions
limiting reactant

excess reactant
octet rule (12.4)
ionic compounds
Lewis symbol
Lewis structures

cations
anions
ionic bonds
Stock system
covalent compounds (12.5)

covalent bond
polar covalent bond
electronegativity
polar molecule
hydrogen bond (12.6)

Important Equations

Percentage of Component in a Sample:

$$\% \text{ component} = \frac{\text{amount of component}}{\text{total amount}} \times 100\%$$

Percentage by Mass of Element X from Compound's Formula:

$$\% \; X \text{ by mass} = \frac{(\text{atoms of } X \text{ in formula}) \times \text{AM}_X}{\text{FM}_{\text{cpd}}} \times 100\%$$

Review Questions

1. Which one of these three forces is responsible for chemical bonding?
   (a) gravitational   (c) strong nuclear
   (b) electromagnetic

12.1   Law of Conservation of Mass (and Highlight)

2. The law of conservation of mass was discovered in 1774 by which scientist?
   (a) Dalton   (c) Lewis
   (b) Lavoisier   (d) Proust

3. State the law of conservation of mass, and give an example.

4. Give four reasons why Lavoisier is generally designated the "father of chemistry."

5. Lavoisier named a new element "oxygen" and tried to take credit for its discovery. Who actually discovered the element, and what did that scientist call it?

12.2 Law of Definite Proportions

6. A sample of compound AB decomposes to 48 g of A and 12 g of B. Another sample of the same compound AB decomposes to 24 g of A. The predicted number of grams of B obtained is
   (a) 12   (b) 8   (c) 6   (d) 3

7. State the law of definite proportions and give an example. Who discovered this law?

8. What is the chemical formula for *dry ice*? What are its two most striking physical properties?

9. When a sample of A reacts with one of B, a new substance AB is formed and no A but a little of B is left. Which is the limiting reactant?

12.3 Dalton's Atomic Theory

10. Dalton proposed that during a chemical reaction
    (a) no atoms are formed.
    (b) no atoms are destroyed.
    (c) no atoms are changed in identity.
    (d) all of the preceding answers

11. State the three parts of Dalton's atomic theory.

12. Dalton thought all atoms of a given element were alike. We now know that most elements consist of a mixture of isotopes (Table 10.4 on page 237). Explain why this has no effect on the explanation of the law of definite proportions.

12.4 Ionic Bonding

13. What is the normal charge on an ion of sulfur?
    (a) 6+   (b) 6−   (c) 2+   (d) 2−

14. An ionic compound formed between a Group 2A element M and a Group 7A element X would have which general formula?
    (a) M_2X (b) M_7X_2 (c) MX_2 (d) M_2X_7

15. Which is the formula for iron(III) bromide?
    (a) $FeBr$ (b) Fe_3Br (c) Fe_2Br_3 (d) $FeBr_3$

16. State the octet rule. What common element is an exception to the rule, and why?

17. Which electrons of an atom generally take part in the formation of compounds?

18. How is the number of valence electrons in an atom related to its tendency to gain or lose electrons during compound formation?

19. Ionic compounds are formed when individual atoms achieve a noble gas configuration by what general process?

20. A formula unit of sodium oxide (Na_2O) would consist of what particles?

21. What does the element's symbol in a Lewis symbol stand for? The number of dots used is the same as the atom's number of _____ (two words).

22. What is the difference between a Lewis symbol and a Lewis structure?

23. What is the basic difference in ions formed by metals and ions formed by nonmetals? What are the meanings of the terms *cation* and *anion*?

24. What is meant when we say F^-, Ne, and Na^+ are *isoelectronic*?

25. State the two principles used in writing the formulas of ionic compounds.

26. What is the crucial experimental test of whether or not a compound is ionic?

27. Why does an aqueous solution of table salt conduct electricity, whereas an aqueous solution of table sugar does not?

28. When is the use of the Stock system of nomenclature preferred?

12.5 Covalent Bonding (and Highlight)

29. How many covalent bonds are normally formed by a nitrogen atom?
    (a) 3   (b) 4   (c) 2   (d) 5

30. Which one is definitely a covalent compound?
    (a) NF_3 (b) $TiCl_2$ (c) Na_2O (d) $CaSO_4$

31. Briefly, how are covalent compounds formed, and what principle is used to predict if a formula represents a stable molecule?

32. What is the basic difference distinguishing single, double, and triple bonds?

33. How does electronegativity vary, in general, across (left to right) a period and down a group? What element has the highest electronegativity?

34. A covalent bond in which the electron pair is unequally shared is called by what name?

35. Could a molecule composed of two atoms joined by a polar covalent bond ever be nonpolar? Explain.

36. To predict whether or not a molecule consisting of two or more polar bonds is polar, you must know the _____ of the molecule.

37. Compare ionic and covalent compounds with regard to (a) phase and (b) electrical conductivity.

38. A certain compound has a boiling point of $-10°C$. Is it ionic or covalent?

12.6 Hydrogen Bonding

39. In which one of these compounds does hydrogen bonding occur?
    (a) H_2S (b) HF   (c) PH_3 (d) CH_4

40. State the short general principle of solubility, and explain what it means.

41. What is a *hydrogen bond*? About how strong are hydrogen bonds relative to covalent bonds?

Applying Your Knowledge

1. You decide to have hot dogs for dinner. In the grocery store, you find that buns come only in packages of 12, whereas the wieners come in packages of 8. How can your purchases lead to having no buns or wieners left over? (How does this problem fit in with this chapter?)

2. Why can't we destroy bothersome pollutants by just burning them or dissolving them in the ocean?

3. You are the supervisor in a car assembly plant where each car produced requires one body and four wheels. You discover that the plant has 200 bodies and 900 wheels. How many cars can be made from this inventory? (How does this problem fit in with this chapter?)

4. Every time you use your bottle of vinegar-and-oil salad dressing, you have to shake it up. Why?

5. Why might you be hesitant to store a ton of ammonium nitrate in your basement at the request of an acquaintance?

Exercises

12.1 Law of Conservation of Mass

1. Volcanoes emit much hydrogen sulfide gas (H_2S), which reacts with the oxygen in the air to form water and sulfur dioxide (SO_2). Every 68 tons of H_2S reacts with 96 tons of oxygen and forms 36 tons of water. How many tons of SO_2 are formed?
   Answer: 128 tons

2. Silver utensils are not attacked by oxygen unless H_2S or sulfur-containing foods such as eggs or mustard are present. The reaction that causes the tarnishing of silver in the presence of oxygen and H_2S leads to the formation of silver sulfide (the tarnish) and water. If 432 g of silver react with 68 g of H_2S and 32 g of oxygen, 36 g of water are formed. How many grams of silver sulfide are formed?

3. Calculate the formula mass (to the nearest 0.1 u) of these compounds.
   (a) carbon dioxide, CO_2
   (b) methane, CH_4
   (c) sodium phosphate, Na_3PO_4
   Answer: (a) 44.0 u   (b) 16.0 u   (c) 164.0 u

4. Calculate the formula mass (to the nearest 0.1 u) of these compounds.
   (a) potassium chloride, KCl
   (b) sodium hydrogen carbonate, $NaHCO_3$
   (c) calcium acetate, $Ca(C_2H_3O_2)_2$

12.2 Law of Definite Proportions

5. Find the % by mass of Cl in $MgCl_2$ if it is 25.5% Mg by mass.
   Answer: 74.5%

6. Find the % by mass of Cl in $AlCl_3$ if it is 20.2% Al by mass.

7. Determine the percentage by mass of each element in (a) salt, NaCl, and (b) sugar, $C_{12}H_{22}O_{11}$.
   Answer: (a) 39.3% Na, 60.7% Cl   (b) 42.1% C, 6.4% H, 51.5% O

8. Determine the percentage by mass of each element in (a) baking soda, $NaHCO_3$, and (b) iron(III) oxide, Fe_2O_3.

9. In a lab experiment, 6.1 g of Mg reacts with sulfur to form 14.1 g of magnesium sulfide. (a) Use Eq. 12.1 to calculate the percentage by mass of Mg in magnesium sulfide. (b) How many grams of sulfur reacted, and how do you know?

    Answer: (a) 43% (b) 8.0 g, law of conservation of mass

11. Refer to Exercise 9. How much magnesium sulfide would be formed if 6.1 g of Mg were reacted with 10.0 g of S, and how do you know?

    Answer: Still 14.1 g. The law of definite proportions indicates that the proper ratio is 6.1 g Mg to 8.0 g S, so 10.0 minus 8.0 = 2.0 g of S is in excess.

10. In a lab experiment, 7.75 g of phosphorus reacts with bromine to form 67.68 g of phosphorus tribromide. (a) Calculate the percentage by mass of P in phosphorus tribromide. (b) How many grams of bromine reacted, and how do you know?

12. Refer to Exercise 10. How much phosphorus tribromide would be formed if 10.00 g of phosphorus reacted with 59.93 g of bromine, and how do you know?

12.4 Ionic Bonding

13. Referring only to a periodic table, give the ionic charge expected for each of these representative elements: (a) S, (b) K, (c) Br, (d) N, (e) Mg, (f) Ne, (g) C, and (h) Al.

    Answer: (a) 2− (b) 1+ (c) 1− (d) 3−
    (e) 2+ (f) 0 (g) 0 (h) 3+

15. Write the Lewis symbols and structures showing how Na_2O forms from sodium and oxygen atoms.

    Answer: Na^+ Na^+ $:\ddot{O}:^{2-}$

17. Predict the formula for each ionic compound. (Recall the polyatomic ions in Table 11.6.)

  (a) cesium iodide   (d) lithium sulfide
  (b) barium fluoride   (e) beryllium oxide
  (c) aluminum nitrate   (f) ammonium sulfate

    Answer: (a) CsI (b) BaF_2 (c) $Al(NO_3)_3$
    (d) Li_2S (e) BeO (f) $(NH_4)_2SO_4$

19. Give the Stock system name for (a) $Ni(OH)_2$, (b) $CuCl_2$, and (c) AuI_3.

    Answer: (a) nickel(II) hydroxide (b) copper(II) chloride
    (c) gold(III) iodide

14. Referring only to a periodic table, give the ionic charge expected for each of these representative elements: (a) F, (b) O, (c) Ga, (d) He, (e) P, (f) Ca, (g) Rb, and (h) Si.

16. Write the Lewis symbols and structures showing how $BaBr_2$ forms from barium and bromine atoms.

18. Predict the formula for each ionic compound. (Recall the polyatomic ions in Table 11.6.)

  (a) ammonium phosphate   (d) aluminum phosphide
                               (e) sodium carbonate
  (b) potassium nitride   (f) magnesium bromide
  (c) strontium acetate

20. Give the Stock system name for (a) Fe_2S_3, (b) $CrBr_2$, and (c) $AuNO_3$.

12.5 Covalent Bonding

21. Referring only to a periodic table, give the number of covalent bonds expected for each of these representative elements: (a) S, (b) Ne, (c) Br, (d) N, and (e) C.

    Answer: (a) 2 (b) 0 (c) 1 (d) 3 (e) 4

23. Draw the Lewis structure for the rocket fuel hydrazine, N_2H_4. Show a structure with all dots, and then show one with both dots and dashes.

    Answer: H:N̈:N̈:H and H—N̈—N̈—H
           H H | |
                                  H H

25. Use your knowledge of the appropriate number of covalent bonds to predict the formula for a simple compound formed between chlorine and (a) hydrogen, (b) nitrogen, (c) sulfur, and (d) carbon.

    Answer: (a) HCl (b) NCl_3 (c) SCl_2 (d) CCl_4

22. Referring only to a periodic table, give the number of covalent bonds expected for each of these representative elements: (a) H, (b) Ar, (c) F, (d) Si, and (e) P.

24. Draw the Lewis structure for formaldehyde, H_2CO, a compound whose odor is known to most biology students due to its use as a preservative. Show a structure with all dots, and then show one with both dots and dashes.

26. Use your knowledge of the appropriate number of covalent bonds to predict the formula for a simple compound formed between sulfur and (a) hydrogen, (b) nitrogen, (c) bromine, and (d) carbon.

27. Predict which of these compounds are ionic and which are covalent. State your reasoning.
    (a) N_2H_4 (d) CBr_4
    (b) NaF   (e) $C_{12}H_{22}O_{11}$
    (c) $Ca(NO_3)_2$ (f) $(NH_4)_3PO_4$
    Answer: (a) covalent, two nonmetals (b) ionic, Group 1A metal and nonmetal (c) ionic, metal and polyatomic ion (d) covalent, two nonmetals (e) covalent, all nonmetals (f) ionic, two polyatomic ions

29. Use arrows to show the polarity of each bond in (a) SCl_2, and (b) CO_2. (*Hint:* Use the general periodic trends in electronegativity.)
    Answer: (a) S and Cl are in the same period, but Cl is farther right than S, so it is the more electronegative and the arrows would point to the two Cl atoms. (b) C and O are in the same period, but O is farther right and so it is the more electronegative. Thus the arrows would point to the two oxygen atoms.

28. Predict which of these compounds are ionic and which are covalent. State your reasoning.
    (a) CaO   (d) K_3N
    (b) PBr_3 (e) H_2S
    (c) $C_6H_4Cl_2$ (f) $NaC_2H_3O_2$

30. Use arrows to show the polarity of each bond in (a) BrCl and (b) NI_3. (*Hint:* Use the general periodic trends in electronegativity.)

Solutions to Confidence Exercises

12.1 The answer is 71.0 g of chlorine. Because the compound is composed of only calcium and chlorine, the difference in mass between 111.1 g of compound and 40.1 g of calcium must be the mass of the chlorine.

12.2 The answer is 34.1 u. FM = (2 × 1.0 u for H) + 32.1 u for S = 34.1 u.

12.3 The answer is 52.9% Al and 47.1% O.

$$\% \ Al = \frac{2 \times 27.0 \ u}{102.0 \ u} \times 100\% = 52.9\%;$$

$$\% \ O = 100.0\% \ minus \ 52.9\% = 47.1\%$$

12.4

	X^-	X^{2-}	X^{3-}
M^+	MX	M_2X	M_3X
M^{2+}	MX_2	MX	M_3X_2
M^{3+}	MX_3	M_2X_3	MX

12.5 The Stock system names are copper(I) fluoride and copper(II) fluoride. Cu is the symbol for the metal copper, and F is the symbol for the nonmetal fluorine. The name of the nonmetal changes to its -*ide* form, so both compounds are copper fluorides. Because the ionic charge of F is always 1−, copper's ionic charge must be 1+ in the first compound and 2+ in the second.

12.6 O (Group 6A) forms two bonds, and H forms one bond. Only the O atoms can connect to two atoms. Thus the structure of H_2O_2 must be

H:Ö: or   H—Ö:
:Ö:H   :Ö—H

12.7 PCl_3 is *covalent* (two nonmetals). MgF_2 is *ionic* (Group 2A metal and a nonmetal).

12.8 Nitrogen (EN = 3.0) is more electronegative than hydrogen (EN = 2.1), so the arrows denoting the polarity of the bonds would point as shown here.

H⟷Ṅ⟷H
|↕
H

12.9 In order for the three bond dipoles to exactly cancel, the BCl_3 molecule must be flat, with bond angles at 120°. Only in this way can the center of positive charge and the center of negative charge be at the same place. (See the SO_3 case in Table 12.6.)

Answers to Multiple-Choice Review Questions

1. b   6. c   13. d   15. d   30. a
2. b   10. d   14. c   29. a   39. b

CHEMICAL REACTIONS

13

All changes we produce consist in separating particles that are in a state of cohesion or combination, and joining those that were previously at a distance.

John Dalton (1766–1844)

Now that elements, compounds, and chemical bonding have been discussed, the groundwork is laid for an examination of chemical reactions. Our environment is composed of atoms and molecules that undergo chemical changes to produce the many substances we need and use. In producing new products, energy is released or absorbed.

For example, green plants absorb carbon dioxide from the air and, with energy from the Sun and chlorophyll as a catalyst, react with water from the soil to form glucose (a carbohydrate) and oxygen (see the chapter-opening photo). This complex chemical reaction is called *photosynthesis,* and the Sun's energy is stored in chemical bonds as *chemical energy.* Animal life on our planet inhales the oxygen and ingests the plants' carbohydrates to obtain energy. Chemical changes in the animals' cells then return water and carbon dioxide to the environment.

In this chapter, we will discuss general types of chemical reactions and the major principles that underlie chemical change. The Highlights discuss how heat packs work and the chemical reactions involved in tooth decay. ■

Photo: Green plants use chemical reactions to store the Sun's energy.

13.1 Balancing Chemical Equations

LEARNING GOALS

▼ Distinguish between chemical and physical changes.

▼ Balance chemical equations.

▼ Identify combination, decomposition, and hydrocarbon combustion reactions.

The characteristics of a substance are called its *properties. Physical properties* are those that do not describe the chemical reactivity of the substance. Among them are density, hardness, phase, color, melting point (m.p.), electrical conductivity, and specific heat. Physical properties can be measured without causing new substances to form.

Chemical properties tell that a substance can be transformed into another—that is, they describe the substance's chemical reactivity. A chemical property of wood is that it burns when heated in the presence of air, a chemical property of iron is that it rusts, and a chemical property of water is that it can be decomposed by electricity into hydrogen and oxygen.

Changes that do not alter the chemical composition are classified as *physical changes.* Examples are the freezing of water (it is still H_2O), the dissolving of table salt in water (it is still sodium ions and chloride ions), the heating of an iron bar (still Fe), and the evaporation of rubbing alcohol (still C_3H_8O). A change that alters the chemical composition of a substance and hence forms one or more new substances is called a *chemical change* or, more often, a **chemical reaction.** The decomposition of water, the rusting of iron, and the burning of magnesium in oxygen to form magnesium oxide are examples of chemical changes (● Fig. 13.1).

A chemical reaction is simply a rearrangement of atoms in which some of the original chemical bonds are broken and new bonds are formed to give different chemical structures (● Fig. 13.2). Generally, only an atom's valence electrons are involved directly in a chemical reaction. The nucleus, and hence the atom's identity as a particular element, is unchanged.

Consider the reaction shown by the generalized chemical equation

$$A + B \longrightarrow C + D$$

Reactants   **Products**

The arrow indicates the direction of the reaction and has the meaning of "react to form" or "yield." (See

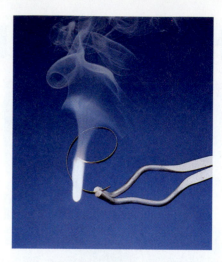

FIGURE 13.1 The Reaction of Magnesium with Oxygen

Magnesium metal (Mg) and oxygen (O_2) in the air combine to form magnesium oxide (MgO, the "smoke"). Magnesium is used in fireworks because of the bright light produced by the reaction.

Table 13.1 for the meanings of other symbols commonly seen in chemical equations.) Thus the equation is read, "Substances A and B react to form substances C and D." The original substances A and B are called the **reactants,** and the new substances C and D are called the **products.** *In any chemical reaction, three things take place:*

1. The reactants disappear or are diminished.
2. New substances appear as products that have different chemical and physical properties from the original reactants.

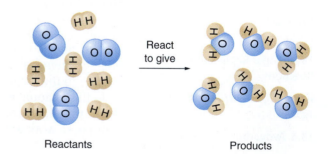

Reactants   Products

FIGURE 13.2 A Chemical Reaction Is a Rearrangement of Atoms

When hydrogen and oxygen form water, bonds are broken in the reactants and new bonds are formed to give the products. No atoms can be lost, gained, or changed in identity.

TABLE 13.1   Common Symbols in Chemical Equations

Symbol	Meaning
$+$	Plus, or and
$\longrightarrow$	React to form, or yields
(g)	Gas
(l)	Liquid
(s)	Solid
(aq)	Aqueous (water) solution
$\xrightarrow{MnO_2}$	Catalyst (MnO_2, in this case)
$\rightleftharpoons$	Equilibrium

3. Energy (heat, light, electricity, sound) is either released or absorbed, although sometimes the energy change is too small to be detected easily.

A chemical equation can be written for each chemical reaction. The correct chemical formulas for the reactants and products must be used *and cannot be changed.* For example, the decomposition of hydrogen iodide is initially written $HI \longrightarrow H_2 + I_2$. However, until the equation is *balanced,* it does not express the actual *ratio* in which the substances react and form. Most chemical reactions can be balanced by trial and error, using three simple principles.

1. The same number of atoms of each element must be represented on each side of the reaction arrow, because no atoms can be gained, lost, or changed in identity during a chemical reaction.

The equation $HI \longrightarrow H_2 + I_2$ is unbalanced because two atoms of both H and I are represented on the right side, but only one of each is shown on the left side.

2. You may manipulate only the *coefficients*—the numbers in front of the formulas, which designate the relative amounts of the substances—and not the *subscripts,* which denote the correct formulas of the substances.

Thus you *cannot* balance the preceding equation by changing the formula of HI to H_2I_2. However, you can place a coefficient of 2 before the formula of HI. The 2 HI represents two molecules of hydrogen iodide,

each made up of one hydrogen atom and one iodine atom. This gives $2\,HI \longrightarrow H_2 + I_2$ and balances the equation. (Just as with subscripts, a coefficient of 1 is not written, just understood.)

3. The final set of coefficients should be whole numbers (not fractions) and should be the smallest whole numbers that will do the job.

For example, $2\,HI \longrightarrow H_2 + I_2$ is appropriate, but *not* $HI \longrightarrow \frac{1}{2}\,H_2 + \frac{1}{2}\,I_2$ *or* $4\,HI \longrightarrow 2\,H_2 + 2\,I_2$.

The following tips will help.

1. *You must be able to count atoms.* Consider $4\,Al_2(SO_4)_3$. The subscript 2 multiplies the Al; the subscript 4 multiplies the O; the subscript 3 multiplies everything in parentheses; the coefficient 4 multiplies the whole formula. Therefore, a total of 8 Al atoms, 12 S atoms, and 48 O atoms are on hand. (If you were counting sulfate ions, there are 12.)

2. *Start with an element that is present in only one place on each side of the arrow.* For example, when balancing $C + SO_2 \longrightarrow CS_2 + CO$, start with S or O, not C.

3. Find the *lowest common denominator* (the smallest whole number into which each goes a whole number of times) for each element that is present in only one place on each side. Insert coefficients in such a way as to get the same number of atoms of that element on each side.

   For example, in $C + SO_2 \longrightarrow CS_2 + CO$, two atoms of sulfur show on the product side and only one shows on the reactant side. The lowest common denominator is 2, so put a coefficient 2 in front of the SO_2, giving $C + 2\,SO_2 \longrightarrow CS_2 + CO$. Next, take care of the oxygen. Four oxygen atoms show on the reactant side (in $2\,SO_2$) and only one on the product side (in CO). Putting a 4 before the CO gives $C + 2\,SO_2 \longrightarrow CS_2 + 4\,CO$. Finally, balance the carbons. Five atoms of carbon now show on the product side and only one appears on the reactant side, so place a 5 before the C on the reactant side to get

$$5\,C + 2\,SO_2 \longrightarrow CS_2 + 4\,CO$$

   A quick recheck shows that all is in balance (two S, four O, five C) and that it is not possible to get a set of smaller coefficients by dividing them by, say, 2 or 3.

4. When *polyatomic ions* remain intact during the reaction, balance them as a unit.

For example, in $Al + H_2SO_4 \longrightarrow Al_2(SO_4)_3 + H_2$, you would balance Al atoms, H atoms, and SO_4^{2-} (sulfate ions). (What coefficient would you put before the H_2SO_4?)

5. If you come to a point where everything would be balanced if it weren't for a *fractional* coefficient that has to be used in one place, multiply all the coefficients by whatever number is in the denominator of the fraction.

For example, in $C_2H_2 + O_2 \longrightarrow CO_2 + H_2O$, putting a 2 in front of the CO_2 and leaving an understood 1 in front of both H_2O and C_2H_2 would require a ½ in front of the O_2 (the oxygen is left for last because it is present in two places on the product side). This would give $C_2H_2 + \frac{1}{2}O_2 \longrightarrow 2\,CO_2 + H_2O$. But we generally don't want fractional coefficients, so multiply the whole equation by 2 (the number in the denominator), thus getting

$$2\,C_2H_2 + 5\,O_2 \longrightarrow 4\,CO_2 + 2\,H_2O$$

For practice in this fundamental chemical skill, let's go through the process of balancing two more

FIGURE 13.3 A Combination Reaction and a Decomposition Reaction

(a) Combination reactions have the format $A + B \longrightarrow AB$, as shown in this schematic of iron and sulfur combining to give iron(II) sulfide. (The FeS figure represents an ion pair, or formula unit, not a molecule.)

(b) Decomposition reactions have the format $AB \longrightarrow A + B$, as shown in this schematic of mercury(II) oxide decomposing to give mercury metal and oxygen gas. (The HgO figure represents an ion pair, or formula unit, not a molecule.)

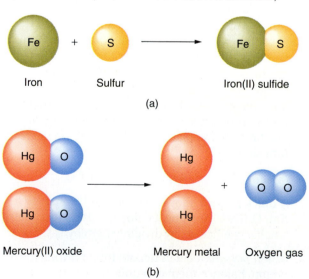

Iron + Sulfur ⟶ Iron(II) sulfide

(a)

Mercury(II) oxide ⟶ Mercury metal + Oxygen gas

(b)

equations in Example 13.1. Each example reaction also illustrates a basic type of reaction that we want you to understand.

1. A **combination reaction** (● Fig. 13.3a) occurs when at least two reactants combine to form just one product: $A + B \longrightarrow AB$.
2. A **decomposition reaction** (● Fig. 13.3b) occurs when only one reactant is present and breaks into two (or more) products: $AB \longrightarrow A + B$.

EXAMPLE 13.1

Balancing Equations

(a) An example of a *combination reaction* is the reaction of magnesium and oxygen to form magnesium oxide (see Fig. 13.1). Balance the equation. (Recall that elemental oxygen exists as diatomic molecules.)

$$Mg + O_2 \longrightarrow MgO$$

(b) Air bags in cars and trucks are inflated by the nitrogen gas produced by electrical ignition of sodium azide, NaN_3. Balance the equation for this *decomposition reaction*.

$$NaN_3(s) \longrightarrow Na(s) + N_2(g)$$

SOLUTION

(a) The magnesium atoms are balanced (one on the left and one on the right), but two oxygens show on the left and only one on the right. Because the lowest common denominator of 2 and 1 is 2, to balance the oxygen we place a 2 in front of the MgO.

$$Mg + O_2 \longrightarrow 2\,MgO$$

This step balances the oxygen, but now the magnesium is unbalanced (two on the right and one on the left). To rebalance the magnesium, we place a 2 in front of the Mg. The balanced equation is

$$2\,Mg + O_2 \longrightarrow 2\,MgO$$

(b) Each element is present in only one place on each side, so it does not matter which one we start with. We see that the sodium atoms are balanced already, but the nitrogen is not balanced. With three atoms of N on the left and two on the right, the lowest common denominator is 6. Therefore, we put a 2 in front of the NaN_3 and a 3 in front of the N_2, giving

$$2\,NaN_3(s) \longrightarrow Na(s) + 3\,N_2$$

The nitrogen is balanced at six atoms on each side, but the sodium is now unbalanced. However, we can balance it by putting a 2 in front of the Na. The balanced equation is

$$2\,NaN_3(s) \longrightarrow 2\,Na(s) + 3\,N_2$$

CONFIDENCE EXERCISE 13.1

The passsage of an electric current through water (the *electrolysis* of water) forms hydrogen gas and oxygen gas at the electrodes, as shown in ● Fig. 13.4. Is this a combination reaction, or is it a decomposition reaction? Balance the equation.

$$H_2O \longrightarrow H_2 + O_2$$

RELEVANCE QUESTION: *In what way is balancing a checkbook similar to balancing a chemical equation?*

FIGURE 13.4 The Electrolysis of Water

Bubbles of oxygen gas (the tube on the left) and hydrogen gas (the tube on the right) form as the battery provides an electric current that decomposes water into its component elements. This is an example of *electrolysis*, the use of an electric current to cause a chemical reaction.

13.2 Energy and Rate of Reaction

LEARNING GOALS

▼ Describe the role of energy in chemical reactions.

▼ State the factors that affect the rate of a reaction.

All chemical reactions involve a change in energy, as can be seen in Fig. 13.1. The energy change is related to the bonding energies between the atoms that form the molecules. During a chemical reaction, some chemical bonds are broken and others are formed. Energy must be absorbed to break bonds, and energy is released when bonds are formed. The energy is released or absorbed in the form of heat, light, electrical energy, or sound. *If a net release of energy to the surroundings occurs in a chemical reaction, it is an* **exothermic reaction.** An example of a common exothermic reaction is the burning of natural gas, which is composed primarily of methane, CH_4 (● Fig. 13.5).

$$CH_4 + 2\,O_2 \longrightarrow CO_2 + 2\,H_2O + energy$$

When methane burns in air, the chemical energy of the bonds in the products is less than the chemical energy of the bonds in the reactants; that is, more

FIGURE 13.5 A Common Exothermic Reaction

Natural gas consists mostly of methane (CH_4) with smaller amounts of other hydrocarbons. When hydrocarbons burn in abundant oxygen, the products are water and carbon dioxide. Energy is given off in the form of heat and light.

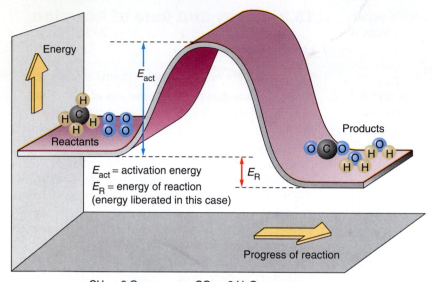

FIGURE 13.6   Exothermic Reaction

The reaction between methane and oxygen results in a net release of energy (E_R) because the bonds in the products (CO_2 and H_2O) have less total energy than the bonds in the reactants (CH_4 and O_2).

E_{act} = activation energy
E_R = energy of reaction
(energy liberated in this case)

$$CH_4 + 2\,O_2 \longrightarrow CO_2 + 2\,H_2O$$

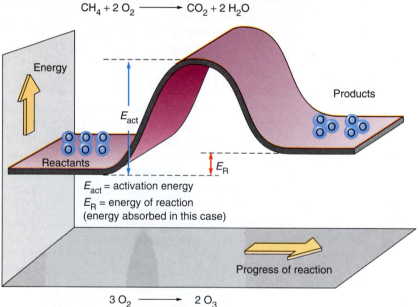

FIGURE 13.7   Endothermic Reaction

In the oxygen-ozone reaction, a net absorption of energy (E_R) occurs because the bonds in two ozone molecules have a higher total energy than the bonds in three oxygen molecules.

E_{act} = activation energy
E_R = energy of reaction
(energy absorbed in this case)

$$3\,O_2 \longrightarrow 2\,O_3$$

energy is given off when the new bonds form than is absorbed in breaking the old bonds (● Fig. 13.6).

If a net absorption of energy from the surroundings occurs during a chemical reaction, it is an **endothermic reaction.** An example of an endothermic reaction is the production of ozone, the triatomic molecule of oxygen.

$$3\,O_2 + energy \longrightarrow 2\,O_3$$

Energy is released when the bonds are formed in the ozone molecules, but the amount is less than that absorbed in breaking the oxygen molecules' bonds

(● Fig. 13.7). Thus there is a net absorption of energy from the surroundings, and the reaction is endothermic.

Ozone formation occurs in the upper atmosphere, where the energy is provided by ultraviolet radiation from the Sun. It also occurs near electric discharges, and ozone's pungent odor can be detected when we are near electrical sparking. Ozone is a worrisome pollutant at Earth's surface, but its presence in the stratosphere provides us with vital protection from ultraviolet radiation, as discussed in Section 20.5.

We are all aware of the necessity of striking a match to get it to ignite (● Fig. 13.8). One must contribute some energy—through friction—to initiate the chemical reaction. When burning methane gas in, say, a gas stove, a flame or spark is necessary to ignite the methane because the C—H and O—O bonds must be broken initially. Once the gas is ignited, the net energy released breaks the bonds of still more CH_4 and O_2 molecules, and the reaction proceeds continuously, giving off energy in the form of heat and light. The energy necessary to start a chemical reaction is called the **activation energy.**

The activation energy is a measure of the minimum kinetic energy colliding molecules must possess in order to react chemically. However, once the activation energy is supplied, an exothermic reaction can release more energy than was supplied. Think of a boulder resting next to a low barrier wall at the edge of a cliff. Without the initial input of energy to raise the boulder to the top of the barrier, it cannot fall over the cliff and release a much larger amount of energy than was initially supplied.

A similar situation exists for an exothermic chemical reaction. Once the activation energy is supplied, the formation of new chemical bonds can release more energy than was absorbed in breaking the original bonds. This is usually perceived on a large scale as the surroundings becoming warmer (● Fig. 13.9a). In

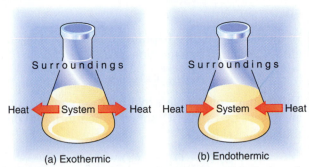

FIGURE 13.9 Heat Flow in Exothermic and Endothermic Reactions

(a) During exothermic reactions, the reaction vessel heats up, and the heat flows to the surroundings, which also get warmer. (b) During endothermic reactions, the reaction vessel gets cold, and heat flows in from the surroundings, which also cool down.

endothermic reactions, more energy is absorbed to break chemical bonds than is generated when new bonds are formed. In this case there is a net intake of energy of reaction (E_R) from the surroundings, which is usually perceived on a large scale as a cooling of the surroundings (Fig. 13.9b). The activation energy (E_{act}) is not fully recovered. Thus the "humps" (called *energy barriers*) in Figs. 13.6 and 13.7 indicate the activation energy that must be supplied for the reaction to proceed.

An explosion occurs when an exothermic chemical reaction liberates its energy almost instantaneously, simultaneously producing large volumes of gaseous products (see Chapter 12 Highlight on page 308). A **combustion reaction,** in which a substance reacts with oxygen to burst into flame and form an oxide, proceeds more slowly than an explosion and yet is still quite rapid. Common examples of combustion are the burning of natural gas, coal, paper, and wood. All carbon-hydrogen compounds (hydrocarbons) and carbon-hydrogen-oxygen compounds produce energy and give carbon dioxide and water when burned completely. We will focus on hydrocarbon combustion reactions.

FIGURE 13.8 Activation Energy
Rubbing the head of a match against a rough surface provides the activation energy necessary for the match to ignite.

EXAMPLE 13.2

Complete Hydrocarbon Combustion

One of the components of gasoline is the hydrocarbon named *heptane*, C_7H_{16}. Write the balanced equation for its complete combustion.

SOLUTION

Write the formula for heptane plus that of oxygen gas, O_2, followed by a reaction arrow.

$$C_7H_{16} + O_2 \longrightarrow$$

The products of complete hydrocarbon combustion are always CO_2 and H_2O, so write their formulas on the product side.

$$C_7H_{16} + O_2 \longrightarrow CO_2 + H_2O$$

Now balance the equation, starting with either C or H, and leaving O until last. The answer is

$$C_7H_{16} + 11\,O_2 \longrightarrow 7\,CO_2 + 8\,H_2O$$

CONFIDENCE EXERCISE 13.2

Write and balance the equation for the complete combustion of the hydrocarbon named *propane*, C_3H_8, a common fuel gas.

When the heptane (C_7H_{16}) in gasoline is burned completely, the combustion reaction is

$$C_7H_{16} + 11\,O_2 \longrightarrow 7\,CO_2 + 8\,H_2O$$

Given insufficient time or oxygen for complete combustion, sooty black carbon (C) and the poisonous gas carbon monoxide (CO) also will be products. The automobile engine has insufficient oxygen to always burn the hydrocarbon completely, so the reaction is sometimes

$$C_7H_{16} + 9\,O_2 \longrightarrow 4\,CO_2 + 2\,CO + C + 8\,H_2O$$

The black color of some exhaust gases indicates the presence of large amounts of carbon and shows that oxidation is incomplete. Because of the signifi-cant amounts of CO formed, running an automobile engine in a closed garage can be fatal. When tobacco burns, CO is one of the health-damaging gases that are released.

Rate of Reaction

The rate of a reaction depends on (1) temperature, (2) concentration, (3) surface area, and (4) the possible presence of a catalyst (a substance that can speed up a reaction). Let's examine the first factor, *temperature.*

To react, molecules (or atoms, or ions) must collide in the proper orientation and with enough kinetic energy to break bonds—the activation energy. The kinetic energy involved in any given collision may or may not be great enough for reaction. Molecules in reactions that have low activation energies react readily because of the greater number of the *effective collisions*—those in which the molecules collide with at least the minimum energy and proper orientation to break bonds and form new substances. Recall from Chapter 5 that if heat is added to a substance and raises its temperature, the average speed and kinetic energy of the molecules will be increased. For chemical reactions, this increased temperature causes more collisions and harder collisions, and the reaction rate increases dramatically. (Snakes, lizards, and other cold-blooded creatures warm themselves in sunlight in order to speed up their metabolism and become more active.)

Now let's take a look at the role that the second factor, *concentration*, plays in the rates of chemical reactions. Generally, the greater the concentration of the reactants, the greater is the rate of the reaction because the molecules are packed more closely. Thus more collisions will occur each second, and the reaction rate should increase (● Fig. 13.10). Astronauts

FIGURE 13.10 Concentration and Reaction Rate

The effect of concentration on reaction rate is apparent from the greater intensity of light coming from phosphorus burning in 100% oxygen (*left*) as compared with burning in air's 21% oxygen (*right*).

FIGURE 13.11   A Grain-Dust Explosion

A grain elevator exploded in 1982 in Council Bluffs, Iowa, killing 5 people. When finely divided grain dust is suspended in air in a confined space, the enormous surface area of the dust particles can cause such a fast combustion reaction that an explosion occurs.

Gus Grissom, Roger Chaffee, and Ed White lost their lives in 1967 when fire broke out in the Project Apollo spacecraft in which they were training on the ground. The environment of the capsule was 100% oxygen, and thus everything burned furiously. Since then, an environment with a much smaller oxygen concentration has been used in space capsules.

Surface area, the third factor, can play a surprisingly important role in the rate of a reaction. You would be startled to see a lump of coal or a pile of grain explode when a match was held to it. But get finely divided coal dust or grain dust in the air, and its enormous surface area will cause a combustion reaction that takes place with explosive speed (● Fig. 13.11). Knowledge of basic scientific principles can sometimes mean the difference between life and death. The chapter's first Highlight discusses another example of the importance of surface area.

The fourth and final factor in reaction rate is the possible presence of a **catalyst,** a substance that increases the rate of reaction but is not itself consumed in the reaction.* Some catalysts act by providing a surface on which the reactants are concentrated. The majority of catalysts work by providing a new reaction pathway with lower activation energy; in effect, they lower the energy barrier (● Fig. 13.12).

By definition, catalysts are not consumed in the reaction, but they are involved in it. They unite with a reactant to form an intermediate substance that takes part in the chemical process and then decomposes to release the catalyst in it original form. For example, in the manufacture of sulfuric acid, which is the chemical produced in the largest quantity in industry, sulfur

*A catalyst never slows down a reaction. A substance that reduces the rate of a chemical reaction is called an *inhibitor,* which generally acts by tying up a catalyst for the reaction or by interacting with a reactant to reduce its concentration. Unlike a catalyst, an inhibitor is used up in a reaction.

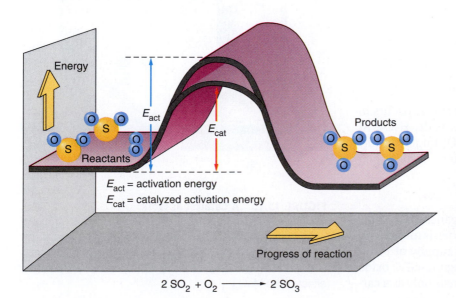

E_{act} = activation energy
E_{cat} = catalyzed activation energy

$2 SO_2 + O_2 \longrightarrow 2 SO_3$

FIGURE 13.12   How a Catalyst Works

A catalyst generally operates by providing a new reaction pathway with a lower activation energy requirement ($E_{cat} <$ E_{act}). Thus more collisions possess enough energy to break bonds, and the reaction rate increases.

HIGHLIGHT: The Chemistry of Heat Packs

A skier—trapped by a snowstorm—builds a snow cave for protection. Realizing her hands and feet are cold and in danger of frostbite, she removes four small heat packs from her pocket, strips the plastic cover from them to reveal four paper packets, and places one in each mitten and boot. Soon her hands and feet are warm again.

These "magic" packets of energy contain powdered iron (and other components) moistened by a little water (Fig. 1). When the plastic cover is removed, air can penetrate the paper packet, producing a common exothermic reaction—the rusting of iron. In simplified form, the reaction can be written:

$$4 \, Fe + 3 \, O_2 \longrightarrow 2 \, Fe_2O_3 + heat$$

Of course, an iron (steel) surface exposed to air and moisture inevitably rusts, but this process is much too slow to be useful in heat packs. However, recall the discussion in Section 13.2 about the effect of surface area on the rate of a chemical reaction. The iron in the heat packs is in the form of a fine powder! The large amount of surface area causes the reaction with oxygen to be fast enough to produce sufficient energy to keep our skier's hands and feet warm for up to 6 hours, giving her a good chance to be rescued.

FIGURE 1

A popular brand of portable heat pack.

---

dioxide must react with oxygen to form sulfur trioxide:

$$2 \, SO_2 + O_2 \longrightarrow 2 \, SO_3 \quad \text{(slow)}$$

Yet this reaction is very slow unless nitric oxide (NO) is added as a catalyst. The reaction then proceeds in two fast steps. The NO combines with oxygen to form nitrogen dioxide (NO_2). The NO_2 then reacts with SO_2 to form SO_3, releasing the NO catalyst, which can be used again. Adding together the two fast reactions shows that the net result is the same as the one slow reaction:

$$2 \, NO + O_2 \longrightarrow 2 \, NO_2 \quad \text{(fast)}$$
$$\underline{2 \, SO_2 + 2 \, NO_2 \longrightarrow 2 \, SO_3 + 2 \, NO \quad \text{(fast)}}$$
$$2 \, SO_2 + O_2 \longrightarrow 2 \, SO_3 \quad \text{net reaction (fast)}$$

Another example of the use of a catalyst occurs in the decomposition of hydrogen peroxide, H_2O_2. At room temperature, a solution of H_2O_2 imperceptibly decomposes, slowly producing water and releasing oxygen.

$$2 \, H_2O_2 \longrightarrow 2 \, H_2O + O_2 \quad \text{(slow)}$$

However, if a small amount of manganese(IV) oxide, MnO_2 (often called *manganese dioxide*), is mixed with the H_2O_2, the reaction takes place rapidly at room temperature (● Fig. 13.13). The manganese(IV) oxide is not consumed in the reaction but acts only as a cat-

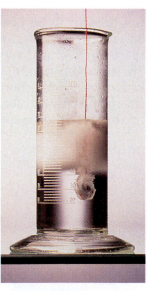

(a)   (b)

FIGURE 13.13   A Catalyst at Work

(a) Aqueous hydrogen peroxide (H_2O_2) decomposes to O_2 and H_2O at an imperceptible rate at room temperature.
(b) When a lump of manganese(IV) oxide is lowered into the solution, the reaction takes place rapidly, due to the catalytic effect of the MnO_2.

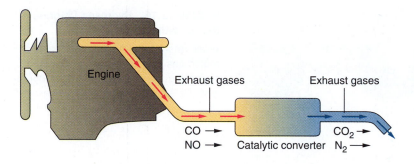

FIGURE 13.14 Catalytic Converters

Noxious gases, such as carbon monoxide (CO) and nitric oxide (NO), coming from a car's engine pass through a catalytic converter, where they are changed into less harmful carbon dioxide (CO_2) and harmless nitrogen (N_2) before emission into the air. Decreases in air pollution help all of us. (From Robinson, William F., Henry F. Holtzclaw, and Jerome D. Odom, *General Chemistry*, Tenth Edition. Copyright © 1997 by Houghton Mifflin Company. Used with permission.)

alyst. The presence of a catalyst is indicated by placing its formula over the reaction arrow.

$$2\ H_2O_2 \xrightarrow{MnO_2} 2\ H_2O + O_2 \quad \text{(fast)}$$

A common example of catalysis is the use of catalytic converters in cars. Beads of a platinum (Pt), rhodium (Rh), or palladium (Pd) catalyst are packed into a chamber through which the exhaust gases must pass before they leave the tailpipe. During the passage through the converter, noxious CO and NO are changed to CO_2 and N_2, which are normal components of the atmosphere (● Fig. 13.14). This results in a great decrease in air pollution.

Catalysts are used extensively in manufacturing, and they also play a crucial role in biochemical processes. The human body has many thousands of biological catalysts called *enzymes* that act to control various physiologic reactions. The names of enzyme catalysts usually end in -*ase.* During digestion, lactose (milk sugar) is broken down in a reaction catalyzed by the enzyme lactase. Many infants and adults, particularly those of African and Asian descent, have a deficiency of lactase and thus are unable to digest the lactose in milk. The bombardier beetle also uses enzymes (● Fig. 13.15).

RELEVANCE QUESTION: *In terms of reaction rate, why are some foods normally stored in a refrigerator?*

13.3 Acids and Bases

LEARNING GOALS

▼ Describe the properties of acids and bases.

▼ Write chemical equations for three types of double-replacement reactions.

The classification of substances as acids or bases originated early in the history of chemistry. An acid, when dissolved in water, has the following properties:

1. Conducts electricity.
2. Changes the color of litmus dye from blue to red.
3. Tastes sour. (But never taste an acid or anything else in a lab!)
4. Reacts with a base to neutralize its properties.
5. Reacts with active metals to liberate hydrogen gas.

A base, when dissolved in water, has the following properties:

1. Conducts electricity.
2. Changes the color of litmus dye from red to blue.
3. Reacts with an acid to neutralize its properties.

One of the first theories formulated to explain acids and bases was put forth in 1887 by Svante Arrhenius ("ar-RAY-nee-us"), a Swedish chemist. He

FIGURE 13.15 Bombardier Beetle

An enzyme catalyzes the highly exothermic reaction that gives the hot spray of chemicals with which the beetle protects itself.

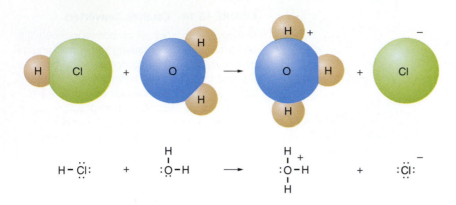

FIGURE 13.16 Hydrogen Chloride Reacts with Water

When gaseous hydrogen chloride (HCl) molecules are added to water (H_2O), hydronium ions (H_3O^+) and chloride ions (Cl^-) are formed.

proposed that the characteristic properties of aqueous solutions of acids and bases are due to the hydrogen ion (H^+) and the hydroxide ion (OH^-), respectively.

According to the *Arrhenius acid–base concept,* when a substance such as colorless, gaseous hydrogen chloride (HCl) is added to water, virtually all the HCl molecules ionize into H^+ ions and Cl^- ions.

$$HCl \longrightarrow H^+ + Cl^-$$

The acidic properties of HCl are due to the H^+ ions. Actually, when hydrogen chloride is placed in water, hydrogen ions are transferred from the HCl to the water molecules, as shown by the following equation:

$$HCl + H_2O \longrightarrow H_3O^+ + Cl^-$$

The H_3O^+, called the *hydronium ion,* and the Cl^- are formed in this reaction (● Fig. 13.16). An Arrhenius acid is a substance that gives hydrogen ions, H^+, (or hydronium ions, H_3O^+) in water.

Acids are classified as strong or weak. A *strong acid* is one that ionizes almost completely in solution (● Fig. 13.17a). For example, hydrochloric acid (HCl), nitric acid (HNO_3), and sulfuric acid (H_2SO_4) are common strong acids. In water, they ionize virtually completely.

Every reaction is to some extent reversible; that is, while the reactants are forming the products, some of the products are reacting to form the reactants. If in a given reaction the reverse reaction is significant, a *double arrow* ($\rightleftharpoons$) is placed between the reactants and products. When two competing reactions or

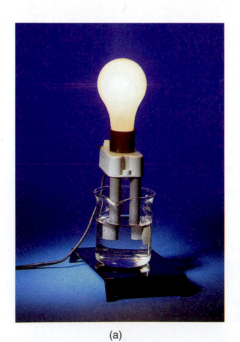

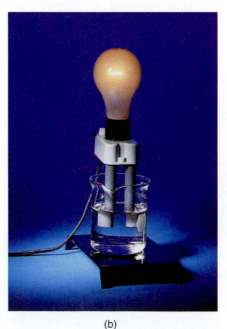

(a)   (b)

FIGURE 13.17 Strong and Weak Acids

(a) A strong acid such as HCl ionizes almost completely in water, so the bulb glows brightly. (b) A weak acid such as acetic acid ionizes only slightly, so the bulb glows dimly.

FIGURE 13.18   Dynamic Equilibrium

Each juggler throws clubs to the other at the same rate at which he receives clubs. Because clubs are thrown continuously in both directions (hence the term *dynamic*), the number of clubs moving in each direction is constant and the number of clubs each juggler has at a given time remains constant. *Note:* In a chemical reaction, the number of "clubs" (molecules) on each side does not have to be equal; they just have to be changing from one side to another at the same rate.

processes are occurring at the same rate, we say that the system is in dynamic **equilibrium** (● Fig. 13.18).

A *weak acid* is one that does not ionize to any great extent; that is, at equilibrium, only a small fraction of its molecules react with H_2O to form H_3O^+ ions (Fig. 13.17b). Acetic acid, $HC_2H_3O_2$, is the most common weak acid. In aqueous solution we have

$$HC_2H_3O_2 + H_2O \rightleftharpoons H_3O^+ + C_2H_3O_2^-$$

Note how the shorter, right-pointing equilibrium arrow indicates that ionization is not substantial for acetic acid in water.

Acids are useful compounds. For example, sulfuric acid is used in refining petroleum, processing steel, and manufacturing fertilizers and numerous other products. A dilute solution of hydrochloric acid is present in the human stomach to help digest food. Many weak acids are present in our foods—citric acid ($H_3C_6H_5O_7$) in citrus fruits, carbonic acid (H_2CO_3) and phosphoric acid (H_3PO_4) in soft drinks, and acetic acid ($HC_2H_3O_2$) in vinegar (● Fig. 13.19).

When pure sodium hydroxide (NaOH, commonly known as *lye*), a white solid, is added to water, it dis-

FIGURE 13.19   Some Common Acids

Vinegar, citrus fruits and juices, sauerkraut, and soft drinks are examples of common household acids.

FIGURE 13.20 Some Common Bases

Baking soda, bleach, soap, cleaning supplies, and milk of magnesia are examples of common household bases.

solves, releasing Na^+ and OH^- into the solution. The basic properties of NaOH solutions are due to the hydroxide ions, OH^-. An Arrhenius **base** is a substance that produces hydroxide ions, OH^-, in water. However, a substance need not *initially* contain hydroxide ions in order to have the properties of a base. For example, ammonia (NH_3) contains no OH^-, but in water solutions it is a weak base. Its molecules react to a slight extent with water to form OH^-.

$$NH_3 + H_2O \rightleftharpoons NH_4^+ + OH^-$$

Common household bases are Drano, which contains NaOH; Windex, which has NH_3 in it; and baking soda, $NaHCO_3$ (● Fig. 13.20).

Water ionizes, but only slightly, as shown in the following equation:

$$H_2O + H_2O \rightleftharpoons H_3O^+ + OH^-$$

Thus all aqueous solutions contain both H_3O^+ and OH^-. For pure water, the concentrations of H_3O^+ and OH^- are equal, and the liquid is *neutral*. An *acidic solution* contains a higher concentration of H_3O^+ than OH^-. If the concentration of OH^- is higher than that of H_3O^+, it is a *basic solution* (● Fig. 13.21).

It is common practice to designate the relative acidity or basicity of a solution by citing its **pH** (power of hydrogen), which is a measure (on a logarithmic scale) of the concentration of hydrogen ion (or hydronium ion) in the solution. A pH of 7 indicates a neutral solution. Values from 6 down to -1 indicate increasing acidity, with each drop of 1 in value meaning a *tenfold* increase in acidity. Similarly, pH values from 8 up to 15 indicate increasing basicity, with each increase of 1 in value meaning a tenfold increase in basicity. ● Figure 13.22 illustrates this concept and shows the pH values of some common solutions.

Most body fluids have a normal pH range, and a continued deviation from normal usually indicates some disorder in a body function. Thus the pH value can be used as a means of diagnosis. For example, if the pH of blood is not between 7.35 and 7.45, illness or death can result.

The pH of a solution is usually measured by a pH meter that can find the value to several decimal places (● Fig. 13.23). However, approximate values can be found by the use of an acid–base indicator, which is a chemical that changes color over a narrow pH range. For example, *litmus* is red below pH 5 and blue above pH 8.

An important property of an acid is the disappearance of its characteristic properties when brought into contact with a base, and vice versa. This is known as an **acid–base reaction:** The H^+ of an acid unites with the OH^- of a base to form water, while the cation of the base combines with the anion of the acid to form a salt. Thus we can generalize: An acid and a hydroxide base react to give water and a salt. (The chap-

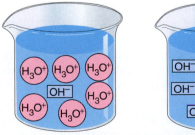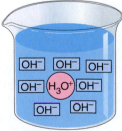

FIGURE 13.21 Acidic and Basic Solutions

An acidic solution (left) has a higher concentration of H_3O^+ than OH^-. In a basic solution (right), the situation is just the reverse.

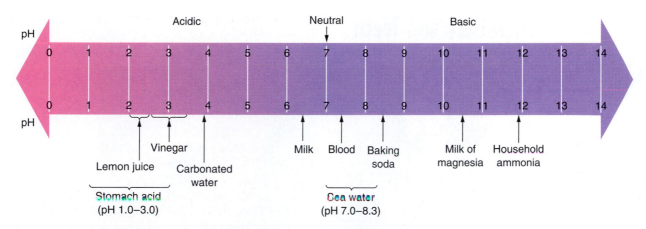

Vinegar
Lemon juice   Carbonated
                water
Milk Blood Baking
              soda
Milk of   Household
magnesia   ammonia

Stomach acid
(pH 1.0–3.0)

Sea water
(pH 7.0–8.3)

FIGURE 13.22   The pH Scale

A solution having a pH of 7 is neutral, a solution with pH less than 7 is acidic, and a solution with a pH greater than 7 is basic. (From Ebbing, Darrell D., and R. A. D. Wentworth,

Introductory Chemistry, Second Edition. Copyright © 1998 by Houghton Mifflin Company. Used with permission.)

ter's second Highlight discusses an example of the importance of equilibrium and acid–base reactions.)

Of course, you are aware that NaCl is called "salt," but actually there are many salts. *A* **salt** *is an ionic compound composed of any cation except H^+ and any anion except OH^-*. Examples are potassium chloride (KCl) and calcium phosphate, $Ca_3(PO_4)_2$. The salt may remain dissolved in water (KCl does), or it may form solid particles (precipitate) if it is insoluble like $Ca_3(PO_4)_2$. Some salts occur in our environment as *hydrates*—salts that contain molecules of water

bonded in their crystal lattices. A common example is the blue, crystalline hydrate named copper(II) sulfate pentahydrate, $CuSO_4 \cdot 5H_2O$ (● Fig. 13.24). If this hydrate is heated, *anhydrous* $CuSO_4$, a white powder, is formed as water is expelled from the crystals.

FIGURE 13.24   A Hydrate Salt

Copper sulfate pentahydrate ($CuSO_4 \cdot 5H_2O$) is the deep blue, crystalline hydrate salt on the left. When the hydrate is heated, the water is driven off, and white, powdery, anhydrous copper(II) sulfate is formed (right). As shown at right, when water is added to the white anhydrous salt, the blue hydrated salt forms again.

FIGURE 13.23   Two pH Meters

Each pH meter's digital display gives the pH of the solution in the beaker. Which solution is acidic?

HIGHLIGHT: Chemistry and Teeth

Several of the concepts discussed in this chapter—polyatomic ions, equilibrium, and acid–base reactions—are involved when discussing tooth enamel and tooth decay. Tooth enamel is composed of fibrous protein and the mineral hydroxyapatite, which has the formula $Ca_5(PO_4)_3OH$. You should recognize both the phosphate and hydroxide polyatomic ions in the formula. Research has shown that this compound constantly dissolves (demineralizes) and re-forms (mineralizes) in the saliva at the surface of a tooth. This is an example of a dynamic equilibrium.

$$Ca_5(PO_4)_3OH \rightleftharpoons$$
$$5\,Ca^{2+} + 3\,PO_4^{3-} + OH^-$$

Of course, if demineralization and mineralization occur at the same rate, there is no net loss of enamel. However, if demineralization gains the upper hand, tooth decay can result (Fig. 1). Demineralization is caused primarily by weak acids formed in the saliva as bacteria metabolize carbohydrates. From our discussion of acid–base reactions, you can see readily that an acidic solution would attack the OH^- ions in hydroxyapatite, and many toothpastes contain the basic substance baking soda ($NaHCO_3$) to neutralize the acid.

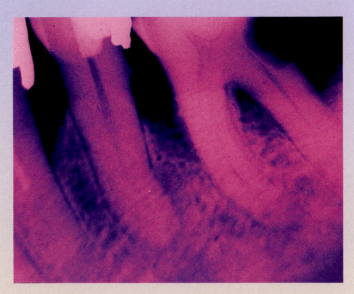

FIGURE 1

An X-ray showing decay (dark area) in the molar at right.

However, if the saliva contains very small amounts of fluoride ion, F^-, an even more important effect is observed. The F^- ions partially replace the OH^- ions in the remineralization process, and this forms fluorapatite, $Ca_5(PO_4)_3F$, which is harder than hydroxyapatite and more resistant to tooth decay. This is why many municipal water supplies are fluoridated and why many brands of toothpaste contain fluoride as one of their cavity-fighting ingredients.

Of course, to prevent tooth decay, it is important to remove the bacteria and food particles from teeth by brushing and flossing. As a sign in one dentist's office states, "You don't have to floss all your teeth—just the ones you want to keep."

Acid–Base Reactions

If you have an "acid stomach" (excess HCl), you might take a milk of magnesia tablet, $Mg(OH)_2$, to neutralize it. Write the balanced chemical equation for the reaction.

SOLUTION

You know that HCl is an acid and that $Mg(OH)_2$ is a base (note the hydroxide ion). Therefore, the products of this acid–base reaction must be water and a salt. Write the formulas for both reactants, put in the reaction arrow, and then write the formula for water.

$$HCl + Mg(OH)_2 \longrightarrow H_2O + \text{(a salt)}$$

Now determine the correct formula for the salt. (*Note:* If you ever have trouble balancing the equation for an acid–base reaction, you probably have an incorrect formula for the salt and may need to refer to Section 12.4.) Write the cation from the base first in the salt formula; then follow it with the anion of the acid: $Mg^{2+}Cl^-$, in this case. You can see that a 1 to 1 ratio of ions does not give electrical neutrality; two Cl^- are needed for each Mg^{2+}. So the correct formula for the salt is $MgCl_2$. Add it to the product side, and then balance the equation. The answer is

$$2\,HCl + Mg(OH)_2 \longrightarrow 2\,H_2O + MgCl_2$$

CONFIDENCE EXERCISE 13.3

Aluminum hydroxide, found in Di-Gel and Mylanta, is another popular antacid ingredient. Complete and balance the equation for the reaction of this base with stomach acid.

$$HCl + Al(OH)_3 \longrightarrow$$

A chemical common to many households is sodium hydrogen carbonate, $NaHCO_3$, commonly known as *baking soda,* an excellent odor absorber. Acids act on the hydrogen carbonate ion of baking soda to give off carbon dioxide, a gas.

$$NaHCO_3 + H^+ \text{(from an acid)} \longrightarrow Na^+ + H_2O + CO_2(g)$$

FIGURE 13.25   An Acid–Carbonate Reaction on a Large Scale

Nitric acid (20,000 gal) was spilled from a railroad tank car in Denver in 1983. Firefighters used an airport snowblower to throw sodium carbonate (Na_2CO_3) onto the HNO_3 and neutralize it.

This reaction is involved in the leavening process in baking. Baking powders contain baking soda plus an acidic substance such as $KHC_4H_4O_6$ (cream of tartar). When this combination is dry, no reaction occurs, but when water is added, CO_2 is given off. This is called an **acid–carbonate reaction:** An acid and a carbonate (or hydrogen carbonate) react to give carbon dioxide, water, and a salt.

● Figure 13.25 shows how this type of reaction can be used to neutralize acid spills.

EXAMPLE 13.4

Acid–Carbonate Reactions

Another way of relieving an overacid stomach is to take an antacid tablet (such as Tums) that contains calcium carbonate, $CaCO_3$ (● Fig. 13.26). Write the balanced equation for the reaction between stomach acid (HCl) and such an antacid ($CaCO_3$).

SOLUTION

This is an acid–carbonate reaction, so the products are carbon dioxide, water, and a salt. Start by writing the formulas for both reactants, the reaction arrow, and the formulas for carbon dioxide and water.

FIGURE 13.26   An Acid–Carbonate Reaction on a Small Scale

Calcium carbonate in a Tums antacid tablet reacts with an acidic solution of HCl to give CO_2 gas, H_2O, and dissolved $CaCl_2$.

Leave a space for the formula of the salt.

$$HCl + CaCO_3 \longrightarrow CO_2 + H_2O +$$ **(a salt)**

As in the preceding example, determine the correct formula for the salt. The Ca^{2+} and Cl^- will form $CaCl_2$. Add the formula of the salt, and then balance the equation.

$$2\,HCl + CaCO_3 \longrightarrow CO_2 + H_2O + CaCl_2$$

CONFIDENCE EXERCISE 13.4

Complete and balance the equation for the reaction of nitric acid with sodium carbonate, the reaction shown in Fig. 13.25.

$$HNO_3 + Na_2CO_3 \longrightarrow$$

Double-Replacement Reactions

Acid–base and acid–carbonate reactions are types of **double-replacement reactions**—those in which the positive and negative components of the two compounds "change partners" (● Fig. 13.27). The general format is

$$AB + CD \longrightarrow AD + CB$$

For example, note that in the acid–base reaction of HCl and NaOH, the positive part of the acid (H^+) attaches to the negative part of the base (OH^-) to form HOH (another way of writing H_2O). The positive part of the base (Na^+) attaches to the negative part of the acid (Cl^-).

$$HCl + NaOH \longrightarrow HOH + NaCl$$

As another example, in the acid–carbonate reaction of H_2SO_4 and K_2CO_3, the positive part of the acid

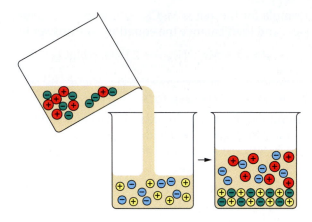

FIGURE 13.28 A Double-Replacement Precipitation Reaction

The positive ions of one soluble substance combine with the negative ions of another soluble substance to form a precipitate when the solutions are mixed. The other positive and negative ions generally remain dissolved.

(H^+) attaches to the negative part of the carbonate (CO_3^{2-}) to form H_2CO_3 (which immediately decomposes to CO_2 and H_2O). The positive part of the carbonate (K^+) attaches to the negative part of the acid (SO_4^{2-}).

$$H_2SO_4 + K_2CO_3 \longrightarrow CO_2 + H_2O + K_2SO_4$$

When aqueous solutions of two salts are mixed, the "changing of partners" often results in the formation of a **precipitate** (*s* or *ppt*), an insoluble solid that appears when two liquids (usually aqueous solutions) are mixed. ● Figure 13.28 illustrates how a double-replacement precipitation reaction might look if we could see the ions. The formation of a precipitate of insoluble lead(II) iodide and soluble potassium nitrate from aqueous solutions of potassium iodide and

FIGURE 13.27 The Format of a Double-Replacement Reaction

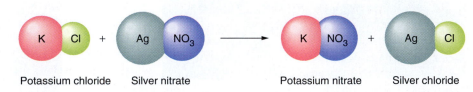

Potassium chloride   Silver nitrate   Potassium nitrate   Silver chloride

Double-replacement reactions have the format $AB + CD \longrightarrow AD + CB$, as shown in this schematic of aqueous solutions of potassium chloride and silver nitrate "changing partners" to form potassium nitrate and silver chloride. (The silver chloride is insoluble in water and thus precipitates.) Note that each compound sketch represents an ion pair, or formula unit, and not a molecule.

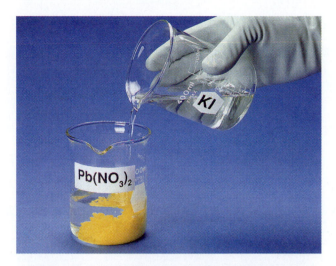

FIGURE 13.29 **The Yellow Precipitate Is Lead(II) Iodide**

When a clear, colorless solution of potassium iodide is poured into a clear, colorless solution of lead(II) nitrate, yellow lead(II) iodide precipitates.

lead(II) nitrate is shown in ● Fig. 13.29 and discussed in Example 13.5. Table 13.2 contains a useful, but not complete, list of water-soluble salts.

EXAMPLE 13.5

Double-Replacement Precipitation Reactions

Write the equation for the double-replacement reaction shown in Fig. 13.29.

SOLUTION

First, write the correct formulas for the reactants using your knowledge of nomenclature and ionic charges.

$$KI(aq) + Pb(NO_3)_2(aq) \longrightarrow$$

Now switch ion partners and put them together so that the positive ion is first and in such a ratio that each compound will be net electrically neutral. To show that the lead(II) iodide is insoluble, place (s), for solid, after its formula. To show that potassium nitrate is soluble (it contains nitrate ions), put (aq) after its formula. Finally, balance the equation. The answer is

$$2 KI(aq) + Pb(NO_3)_2(aq) \longrightarrow PbI_2(s) + 2 KNO_3(aq)$$

TABLE 13.2 Some Water-Soluble Salts

Salts Containing	Examples
Alkali metal ions $(Li^+, Na^+, K^+, Rb^+, Cs^+)$	$NaCl, K_2SO_4$
Ammonium ions (NH_4^+)	$NH_4Br, (NH_4)_2CO_3$
Nitrate ions (NO_3^-)	$AgNO_3, Mg(NO_3)_2$
Acetate ions $(C_2H_3O_2^-)$	$Al(C_2H_3O_2)_3$

CONFIDENCE EXERCISE 13.3

Complete and balance the equation for the double-replacement reaction between aqueous solutions of sodium sulfate and barium chloride.

$$Na_2SO_4(aq) + BaCl_2(aq) \longrightarrow$$

RELEVANCE QUESTION: *The ideal pH for a swimming pool is between 7.2 and 7.6, the pH of tears. Is this close to neutral? On which side of neutral is this range?*

13.4 **Single-Replacement Reactions**

LEARNING GOALS

▼ Define the terms *oxidation* and *reduction.*

▼ Write equations for single-replacement reactions.

We have seen that combustion reactions involve the addition of oxygen to a substance. *When oxygen combines with a substance (or when an atom or ion loses electrons), we call the process* **oxidation.** *Conversely, when oxygen is removed from a compound (or when an atom or ion gains electrons), the process is called* **reduction.** For example, an important step in the steel-making industry is the reduction of hematite ore (Fe_2O_3) with coke (C) in a blast furnace:

$$2 Fe_2O_3 + 3 C \longrightarrow 4 Fe + 3 CO_2$$

Oxygen is removed from the iron(III) oxide, so we say that the iron oxide has been reduced. At the same time, oxygen has reacted with the carbon to form carbon dioxide; that is, the carbon has been oxidized. We call such a chemical change an *oxidation-reduction reaction,* or, for short, a *redox reaction.*

Redox reactions do not always have to involve the gain and loss of oxygen. So that the term will apply to a larger number of reactions, redox reactions are also identified in terms of electrons lost (oxidation) or gained (reduction). Because all the electrons lost by atoms or ions must be gained by other atoms or ions, it follows that oxidation and reduction occur at the same time and at the same rate.

The relative *activity* of any metal is its tendency to lose electrons to ions of another metal or to hydrogen ions. It is determined by placing the metal in a solution containing the ions of another metal and observing whether the test metal replaces the one in solution. If it does, it is more active, because it has given electrons to the ions of the metal in solution. For example, ● Fig. 13.30 shows that when zinc metal (Zn) is placed in a solution containing Cu^{2+} ions, the Zn loses electrons to the Cu^{2+} ions, producing copper metal (Cu) and Zn^{2+} ions.

$$Zn + Cu^{2+} \longrightarrow Cu + Zn^{2+}$$

TABLE 13.3   Activity Series

Metals	Ion Found	
Lithium	Li^+	
Potassium	K^+	
Calcium	Ca^{2+}	
Sodium	Na^+	
Magnesium	Mg^{2+}	
Aluminum	Al^{3+}	
Zinc	Zn^{2+}	
Chromium	Cr^{3+}	
Iron	Fe^{2+}	Increasing activity
Nickel	Ni^{2+}	
Tin	Sn^{2+}	
Lead	Pb^{2+}	
HYDROGEN*	H^+	
Copper	Cu^{2+}	
Silver	Ag^+	
Platinum	Pt^{2+}	
Gold	Au^{3+}	

*Hydrogen is in capital letters because the activities of the metals are often determined in relation to the activity of hydrogen.

(a)   (b)

FIGURE 13.30   Zinc Replaces Copper Ions

(a) A strip of Zn is about to be placed into aqueous, blue $CuSO_4$. (b) A few minutes later, the single-replacement redox reaction has formed metallic copper and colorless $ZnSO_4$ solution.

Thus the Zn has lost electrons and has been oxidized; the Cu^{2+} has gained electrons and has been reduced.

On the other hand, copper metal placed in a solution of Zn^{2+} leads to no reaction. Thus zinc is more active than copper. If similar experiments are carried out for all metals (and hydrogen), an **activity series** can be obtained (Table 13.3).

If element *A* is listed *above* element *B* in the activity series, *A* is more active than *B* and will replace *B* in a compound *BC*. This is called a **single-replacement reaction** and has the general format

$$A + BC \longrightarrow B + AC$$

We will deal only with the most common type of single-replacement reaction, the type in which element *A* is a metal (● Fig. 13.31). In such cases, *A* will lose its valence electrons and thus be oxidized; *B* will gain these electrons and thus be reduced. All single-replacement reactions are also redox reactions.

Zinc metal   Copper(II) sulfate   Copper metal   Zinc sulfate

FIGURE 13.31 The Format of a Single-Replacement Reaction

Single-replacement reactions have the format $A + BC \longrightarrow B + AC$, as shown in this schematic of zinc reacting with an aqueous solution of copper(II) sulfate. The copper precipitates, but the other product, zinc sulfate, is soluble. The sketches of $CuSO_4$ and $ZnSO_4$ represent an ion pair, or formula unit, and not a molecule. See Fig. 13.30 for photos of the actual reaction.

EXAMPLE 13.6

Single-Replacement Reactions

Refer to the activity series (Table 13.3) and predict if placing copper metal in a solution of silver nitrate will lead to a reaction. If you predict that it will, complete and balance the equation.

SOLUTION

Copper is above silver in the activity series, and thus a reaction will occur, as shown in ● Fig. 13.32. The copper atoms will be oxidized to Cu^{2+} ions, and the Ag^+ ions will be reduced to Ag atoms. The nitrate ions, NO_3^-, stay intact and dissolved in solution, but we can write them as if they were joined to the Cu^{2+} ions to give $Cu(NO_3)_2(aq)$. The unbalanced equation for the reaction is

$$Cu + AgNO_3(aq) \longrightarrow Ag + Cu(NO_3)_2(aq)$$

Because the nitrate ions stay intact during the reaction, they can be balanced as a unit. Two nitrates are showing on the product side, so place a 2 before $AgNO_3$ on the left. Complete the balancing by adding a 2 before the Ag on the right.

$$Cu + 2\,AgNO_3(aq) \longrightarrow 2\,Ag + Cu(NO_3)_2(aq)$$

CONFIDENCE EXERCISE 13.6

Suppose you place a strip of aluminum metal in a solution of copper(II) sulfate, $CuSO_4(aq)$, and also put a strip of copper metal in a solution of aluminum sulfate, $Al_2(SO_4)_3(aq)$. Refer to Table 13.3 and predict in which case a single-replacement reaction will take place; then complete and balance the equation for the reaction.

(a)

(b)

FIGURE 13.32 A Single-Replacement Reaction Between Copper Wire and Silver Nitrate

(a) A copper wire is shown immediately after placement in a solution of silver nitrate. (b) Later, the reaction is complete. One product, metallic silver, is evident on the surface of the wire. Copper(II) nitrate, the other product, remains dissolved in solution (note the characteristic blue color of Cu^{2+}).

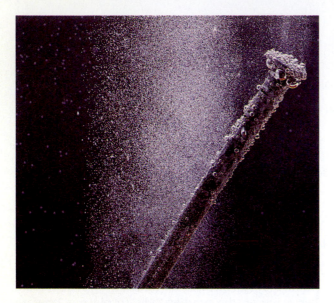

FIGURE 13.33   Metals and Acids

A single-replacement reaction occurs when an iron nail reacts with an aqueous solution of sulfuric acid to form H_2 gas and aqueous $FeSO_4$.

The metals above hydrogen in Table 13.3 will undergo a single-replacement reaction with acids to give hydrogen gas and a salt of the metal (● Fig. 13.33). For example,

$$Fe + H_2SO_4(aq) \longrightarrow H_2 + FeSO_4(aq)$$

In fact, metals above magnesium in Table 13.3 react vigorously with water and produce hydrogen gas and

FIGURE 13.34   Active Metals and Water

Metals, such as calcium, that are above magnesium in the activity series react with water to form hydrogen gas and the metal hydroxide. Note that no reaction is occurring between magnesium and water (left beaker), but hydrogen gas is being formed at a rapid rate as calcium reacts (right beaker).

the metal hydroxide (● Fig. 13.34). (The metal hydroxide is a base, or *alkali,* and that is why these two groups have "alkali" and "alkaline" in their names.)

This is an appropriate point to summarize the reaction types we have covered in this chapter. Please examine Table 13.4 and the Spotlight feature carefully.

TABLE 13.4   A Summary of Reaction Types

Reaction Type	Example
Combination	$2\,Mg + O_2 \longrightarrow 2\,MgO$
Decomposition	$2\,HgO \longrightarrow 2\,Hg + O_2$
Hydrocarbon combustion (complete)	$C_2H_4 + 3\,O_2 \longrightarrow 2\,CO_2 + 2\,H_2O$
Single-replacement	
(a) two metals	$Zn + CuSO_4 \longrightarrow Cu + ZnSO_4$
(b) metal and acid	$Fe + 2\,HCl \longrightarrow H_2 + FeCl_2$
Double-replacement	
(a) precipitation	$BaCl_2 + Na_2SO_4 \longrightarrow BaSO_4(s) + 2\,NaCl$
(b) acid–base	$2\,HCl + Ca(OH)_2 \longrightarrow 2\,H_2O + CaCl_2$
(c) acid–carbonate	$H_2SO_4 + Na_2CO_3 \longrightarrow H_2O + CO_2 + Na_2SO_4$ (The H_2CO_3 formed decomposes.)

A SPOTLIGHT ON: Reaction Types

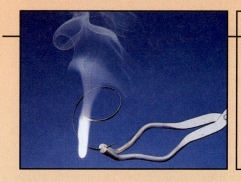

Combination Reaction

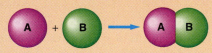

Example: $2 Mg + O_2 \longrightarrow 2 MgO$

Magnesium reacts with oxygen to form magnesium oxide.

Hydrocarbon Combustion

Example:
$CH_4 + O_2 \longrightarrow CO_2 + 2 H_2O$

A hydrocarbon reacts with oxygen to form carbon dioxide and water.

Decomposition Reaction

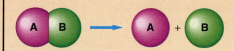

Example: $2 H_2O \longrightarrow 2 H_2 + O_2$

Water decomposes to form hydrogen and oxygen.

Single-Replacement Reaction

Example: $Cu + 2 AgNO_3(aq) \longrightarrow 2 Ag + Cu(NO_3)_2(aq)$

Copper reacts with silver nitrate to form silver and copper(II) nitrate.

Double-Replacement Reaction

Example: $2 KI(aq) + Pb(NO_3)_2(aq) \longrightarrow PbI_2(s) + 2 KNO_3(aq)$

Potassium iodide reacts with lead(II) nitrate to form lead(II) iodide and potassium nitrate.

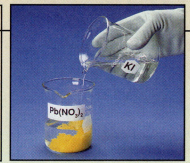

13.5 Avogadro's Number

▼ State the relationships among mole, mass, and Avogadro's number.

Just as bakers often talk of their products in terms of dozens, chemists often speak of moles of chemicals. The SI defines the **mole** (abbreviated *mol*) as the quantity of a substance that contains as many elementary units as there are atoms in exactly 12 g of carbon-12. This turns out to be 6.02×10^{23} carbon-12 atoms, and this huge number is referred to as **Avogadro's number.**

The concept of mole is like that of dozen. Just as when you hear *dozen* you think 12 units, when you hear *mole* you should think 6.02×10^{23} units. Also, a mole of any substance has a mass equal to the same number of grams as the formula mass (see Section 12.2) of the substance. For example, a mole of copper atoms (FM 63.5 u) is 6.02×10^{23} atoms and 63.5 g, and a mole of water (FM 18.0 u) is 6.02×10^{23} water molecules and 18.0 g (● Fig. 13.35).

The number 6.02×10^{23} is called *Avogadro's number* not because Italian physicist Amedeo Avogadro discovered it but in honor of him. Avogadro was the first person to use the term *molecule*. In 1811 he used the concept of molecules of elements to explain the newly discovered law of combining gas volumes, a law that threatened to torpedo Dalton's atomic theory. (Dalton at first proposed that all elements consisted of independent atoms and thus thought that the formula for hydrogen gas was H and not H_2 as Avogadro proved later.)

How can we comprehend such a large number as 6.02×10^{23}? Pour out 6.02×10^{23} BBs on the United States, and they would cover the entire country to a depth of about 4 miles!

How do we *know* how many particles are in one mole? The unit for measuring electric charge is the coulomb (C), and it takes 96,485 C to reduce 1 mole of singly charged ions to atoms. For example, by *electrolysis* (chemical change accomplished by use of an electric current), 96,485 C will produce 1 mole (23.0 g) of sodium metal from molten sodium chloride (see Fig. 12.9 on page 304). It takes one electron to reduce each sodium ion to a sodium atom:

$$Na^+ + e^- \longrightarrow Na$$

Thus 96,485 C must be the total charge on one mole of electrons. A single electron has a negative charge of 1.6022×10^{-19} C. Therefore, Avogadro's number must be

$$\frac{96,485 \text{ C/mol}}{1.6022 \times 10^{-19} \text{ C/electron}}$$

$$= 6.0220 \times 10^{23} \text{ electrons/mol}$$

This is the number of electrons in a mole of electrons, or the number of units in a mole of anything.

FIGURE 13.35 One Mole of Seven Substances

Each sample consists of 6.02×10^{23} formula units. *Clockwise from the top:* 63.5 g of Cu, 55.8 g of Fe, 253.8 g of I_2, 32.1 g of S, 27.0 g of Al, and (in the center) 200.6 g of Hg.

Important Terms

chemical
 properties (13.1)
chemical reaction
reactants
products
combination reaction
decomposition reaction
exothermic
 reaction (13.2)

endothermic reaction
activation energy
combustion reaction
catalyst
acid (13.3)
equilibrium
base
pH
acid–base reaction

salt
acid–carbonate reaction
double-replacement
 reactions
precipitate
oxidation (13.4)
reduction
activity series

single-replacement
 reaction
mole (13.5)
Avogadro's number

Review Questions

1. Name the reactants, products, and catalyst for photosynthesis. What is the source of the necessary energy?

13.1 Balancing Chemical Equations

2. The density of lead is 11.3 g/cm^3. This statement is an example of a _____ of lead.
   (a) physical property   (c) physical change
   (b) chemical property   (d) chemical change

3. When iron rusts in the presence of oxygen and water, a _____ is occurring.
   (a) physical property   (c) physical change
   (b) chemical property   (d) chemical change

4. When the equation $MnO_2 + CO \longrightarrow Mn_2O_3 + CO_2$ is balanced, what is the sum of all written and "understood" coefficients?
   (a) 9   (b) 4   (c) 5   (d) 8

5. Hydrogen is a (1) colorless, (2) odorless (3) gas that has a (4) very low density and (5) reacts with oxygen to form water and (6) with oils to form fats. Which of these six properties are physical and which are chemical?

6. What three things occur in every chemical reaction?

7. In a chemical reaction, what happens to chemical bonds and the identities of atoms?

8. When balancing chemical reactions, some numbers can be manipulated and some cannot. Distinguish these types of numbers by name, and give an example of each.

9. The following reaction occurs when a butane cigarette lighter is operated: $C_4H_{10} + O_2 \longrightarrow CO_2 + H_2O$. When balancing the equation, you should *not* start with which element?

10. How many lead, nitrogen, and oxygen atoms are indicated by $2 Pb(NO_3)_2$? How many nitrate ions?

11. What is inappropriate about each of the "balanced" equations shown?
    (a) $C_4H_{10} + \frac{13}{2} O_2 \longrightarrow 4 CO_2 + 5 H_2O$
    (b) $4 H_2O \longrightarrow 2 O_2 + 4 H_2$
    (c) $Na + H_2O \longrightarrow NaOH + H$

12. Use the letters A and B to describe the general format of (a) combination reactions and (b) decomposition reactions.

13.2 Energy and Rate of Reaction

13. What would a substance be called that increases the rate of a reaction, but is not consumed?
    (a) a reactant   (c) an allotrope
    (b) a catalyst   (d) a redox agent

14. Distinguish between exothermic and endothermic reactions.

15. What is absorbed during bond breaking but liberated during bond formation?

16. A collision between two molecules that have the potential to react may or may not result in a reaction. Explain.

17. List the four major factors that influence the rate of a chemical reaction.

18. Explain why chemical reactions proceed faster (a) as the temperature is increased, and (b) as the concentrations of the reactants are increased.

19. Name the two products of the complete combustion of hydrocarbons. What is the other reactant, and how do the products change when this reactant is in short supply?

20. Heating a mixture of lumps of sulfur and zinc does not lead to reaction nearly as fast as heating a mixture of powdered sulfur and zinc. Explain.

21. What is the role of a catalyst in a chemical reaction? Describe, in brief, how it accomplishes its role.

22. What do sucrase and cholinesterase have in common?

23. Tell what is indicated by each of these six symbols sometimes seen in chemical reactions: (*aq*), (*s*), (*l*), (*g*), $\longrightarrow$, and $\rightleftharpoons$, and by a chemical formula written above a reaction arrow.

13.3 Acids and Bases

24. When aqueous solutions of these four acids are tested for electrical conductivity, which solution gives results different from the other three?
    (a) nitric acid
    (c) hydrochloric acid
    (b) acetic acid
    (d) sulfuric acid

25. Which scientist first postulated the existence of ions?
    (a) Lavoisier
    (c) Arrhenius
    (b) Le Châtelier
    (d) Lewis

26. What would be the pH of a solution ten times as acidic as one of pH 4?
    (a) 3   (b) 14   (c) 5   (d) −6

27. What is the pH of a neutral aqueous solution? How many times as acidic is a solution of pH 3 than one of pH 6?

28. What color will litmus be in a solution of pH 10? Of pH 4?

29. What is the name of the electronic instrument used to measure pH? Which solution shown in Fig. 13.23 is basic?

30. List five general properties of acids and three of bases.

31. In the Arrhenius theory, how are acids and bases defined? Distinguish among hydrogen ions, hydronium ions, and hydroxide ions.

32. What is the chemical identity of (a) stomach acid, (b) milk of magnesia, (c) lye, (d) baking soda, and (e) vinegar?

33. The reaction of an acid with a hydroxide base gives what two products? When writing an equation for such a reaction, what is the most common mistake made?

34. What do we call salts that contain molecules of water bonded in their crystal lattices?

35. The reaction of an acid with a carbonate or hydrogen carbonate always gives what three products?

36. Use the letters $A, B, C,$ and D to illustrate the general format of a double-replacement reaction.

37. Describe what is seen in a precipitation reaction. What is happening on the atomic level to cause what is observed?

13.4 Single-Replacement Reactions

38. *Oxidation* can be defined as which of the following?
    (a) a gain of electrons
    (c) a loss of electrons
    (b) a loss of oxygen
    (d) both (a) and (b)

39. The reaction

    $$3\,Zn + 2\,Au(NO_3)_3(aq) \longrightarrow 2\,Au + 2\,Zn(NO_3)_2(aq)$$

    will occur if Zn is ———— Au in the activity series.
    (a) above
    (c) below
    (b) to the right of
    (d) to the left of

40. Describe oxidation and reduction from the standpoint of gain and loss of (a) oxygen and (b) electrons.

41. Use the letters $A, B,$ and C to illustrate the general format of a single-replacement reaction.

42. Metals above hydrogen in the activity series react with acids to give what two products? What general type of reaction does this exemplify?

43. Why are the precious metals gold and silver found as elements in nature, whereas the metals sodium and magnesium are found in nature only in compounds?

13.5 Avogadro's Number

44. One mole of hydrogen peroxide, H_2O_2, would consist of how many molecules?
    (a) 6.02×10^{23}
    (c) $34.0 \times 6.02 \times 10^{23}$
    (b) 1
    (d) 34.0

45. One mole of hydrogen peroxide, H_2O_2, would consist of how many grams?
    (a) 6.02×10^{23}
    (c) $34.0 \times 6.02 \times 10^{23}$
    (b) 17.0
    (d) 34.0

Applying Your Knowledge

1. How might tooth decay be increased by drinking soft drinks, and what could be done to prevent the decay?

2. Explain why a bag of charcoal briquettes contains the warning: Do not use for indoor heating or cooking unless ventilation is provided for exhausting fumes to the outside.

3. An Alka-Seltzer tablet contains solid citric acid and sodium hydrogen carbonate. What happens, and why, when the tablet is dropped into water?

4. The human body converts sugar into carbon dioxide and water at body temperature 98.6°F, or 37.0°C. Why are much higher temperatures required for the same conversion in the laboratory?

5. Why does blowing on or fanning hot charcoal briquettes cause them to glow more brightly?

6. Why is an open box of baking soda often placed in the refrigerator?

Exercises

13.1 Balancing Chemical Equations

1. Identify each of the following as a physical or chemical change.
   - (a) melting ice
   - (b) burning a match
   - (c) fermenting wine
        Answer: (a) physical (b) chemical (c) chemical

2. Identify each of the following as a physical or chemical change.
   - (a) dissolving sugar in water
   - (b) rusting steel
   - (c) magnetizing a sewing needle

3. Balance these chemical equations. (Each answer shows the correct coefficients in order.)
   - (a) $CuCl_2 + H_2 \longrightarrow Cu + HCl$
   - (b) $Fe + O_2 \longrightarrow Fe_2O_3$
   - (c) $Al + H_2SO_4 \longrightarrow Al_2(SO_4)_3 + H_2$
   - (d) $CaC_2 + H_2O \longrightarrow Ca(OH)_2 + C_2H_2$
   - (e) $KNO_3 \longrightarrow KNO_2 + O_2$
   - (f) $C_6H_6 + O_2 \longrightarrow CO_2 + H_2O$
        Answer: (a) 1, 1, 1, 2 (b) 4, 3, 2 (c) 2, 3, 1, 3 (d) 1, 2, 1, 1
                 (e) 2, 2, 1 (f) 2, 15, 12, 6

4. Balance these chemical equations.
   - (a) $SO_2 + O_2 \longrightarrow SO_3$
   - (b) $NH_3 + O_2 \longrightarrow N_2 + H_2O$
   - (c) $C_4H_{10} + O_2 \longrightarrow CO_2 + H_2O$
   - (d) $Pb(NO_3)_2 \longrightarrow PbO + NO_2 + O_2$
   - (e) $Al + Fe_3O_4 \longrightarrow Al_2O_3 + Fe$
   - (f) $Ba(OH)_2 + H_2SO_4 \longrightarrow BaSO_4 + H_2O$

5. Identify the following reactions from Exercise 3 as combination, decomposition, or hydrocarbon combustion: 3(b), 3(e), and 3(f).
        Answer: 3(b) combination, 3(e) decomposition,
                 3(f) hydrocarbon combustion

6. Identify the following reactions from Exercise 4 as combination, decomposition, or hydrocarbon combustion: 4(a), 4(c), and 4(d).

7. (a) Nitrogen and hydrogen react to give ammonia in a combination reaction. Write and balance the equation.
   (b) Electrolysis can decompose KCl into its elements. Write and balance the equation.
        Answer: (a) $N_2 + 3 H_2 \longrightarrow 2 NH_3$ (b) $2 KCl \longrightarrow 2 K + Cl_2$

8. (a) Aluminum and bromine react to give aluminum bromide in a combination reaction. Write and balance the equation.
   (b) Heating decomposes $MgCO_3$ to magnesium oxide and carbon dioxide. Write and balance the equation.

13.2 Energy and Rate of Reaction

9. Write and balance the reaction for the complete combustion of propane, C_3H_8.
        Answer: $C_3H_8 + 5 O_2 \longrightarrow 4 H_2O + 3 CO_2$

10. Write and balance the reaction for the complete combustion of ethane, C_2H_6.

13.3 Acids and Bases

11. Complete and balance the following acid–base and acid–carbonate reactions:
    - (a) $HNO_3 + KOH \longrightarrow$
    - (b) $HC_2H_3O_2 + K_2CO_3 \longrightarrow$
    - (c) $H_3PO_4 + NaOH \longrightarrow$
    - (d) $H_2SO_4 + CaCO_3 \longrightarrow$
        Answer: (a) $HNO_3 + KOH \longrightarrow H_2O + KNO_3$
                (b) $2 HC_2H_3O_2 + K_2CO_3 \longrightarrow H_2O + CO_2$
                    $+ 2 KC_2H_3O_2$
                (c) $H_3PO_4 + 3 NaOH \longrightarrow 3 H_2O + Na_3PO_4$
                (d) $H_2SO_4 + CaCO_3 \longrightarrow H_2O + CO_2 + CaSO_4$

12. Complete and balance the following acid–base and acid–carbonate reactions:
    - (a) $HCl + Ba(OH)_2 \longrightarrow$
    - (b) $HCl + Al_2(CO_3)_3 \longrightarrow$
    - (c) $H_3PO_4 + LiHCO_3 \longrightarrow$
    - (d) $Al(OH)_3 + H_2SO_4 \longrightarrow$

13. Complete and balance the following double-replacement reactions. (*Hint:* Information in Table 13.2 will help you identify the precipitate.)
    - (a) $AgNO_3(aq) + HCl(aq) \longrightarrow$
    - (b) $BaCl_2(aq) + K_2CO_3(aq) \longrightarrow$
        Answer: (a) $AgNO_3(aq) + HCl(aq) \longrightarrow AgCl(s) + HNO_3(aq)$
                 (b) $BaCl_2(aq) + K_2CO_3(aq) \longrightarrow BaCO_3(s)$
                     $+ 2 KCl(aq)$

14. Complete and balance the following double-replacement reactions. (*Hint:* Information in Table 13.2 will help you identify the precipitate.)
    - (a) $Na_2CO_3(aq) + Pb(NO_3)_2(aq) \longrightarrow$
    - (b) $K_3PO_4(aq) + CuSO_4(aq) \longrightarrow$

13.4 Single-Replacement Reactions

15. Refer to the activity series (Table 13.3) and predict in each case whether the single-replacement reaction shown will, or will not, actually occur.
    (a) $Na + KCl(aq) \longrightarrow K + NaCl(aq)$
    (b) $Ni + CuBr_2(aq) \longrightarrow Cu + NiBr_2(aq)$
    (c) $2\,Al + 6\,HCl(aq) \longrightarrow 3\,H_2 + 2\,AlCl_3(aq)$
    Answer: (a) will not   (b) will   (c) will

16. Refer to the activity series (Table 13.3) and predict in each case whether the single-replacement reaction shown will, or will not, actually occur.
    (a) $Zn + Fe(NO_3)_2(aq) \longrightarrow Fe + Zn(NO_3)_2(aq)$
    (b) $Pb + FeCl_2(aq) \longrightarrow Fe + PbCl_2(aq)$
    (c) $2\,Ag + 2\,HNO_3(aq) \longrightarrow H_2 + 2\,AgNO_3(aq)$

17. Refer to the activity series (Table 13.3). Complete and balance the equation for the following single-replacement reactions.
    (a) $Ni + Pt(NO_3)_2(aq) \longrightarrow$
    (b) $Zn + H_2SO_4(aq) \longrightarrow$
    Answer: (a) $Ni + Pt(NO_3)_2(aq) \longrightarrow Pt + Ni(NO_3)_2(aq)$
    (b) $Zn + H_2SO_4(aq) \longrightarrow H_2 + ZnSO_4(aq)$

18. Refer to the activity series (Table 13.3). Complete and balance the equation for the following single-replacement reactions.
    (a) $Mg + HCl(aq) \longrightarrow$
    (b) $Al + FeSO_4(aq) \longrightarrow$

13.5 Avogadro's Number

19. Two moles of hydrogen sulfide, H_2S, would consist of how many molecules?
    Answer: 12.04×10^{23} molecules

20. Three moles of ammonia, NH_3, would consist of how many molecules?

21. Fill in the blanks in the following table for the element sodium, Na.

Moles	Atoms	Mass
1.00		
	18.06×10^{23}	
		46.0 g

Answer 6.02×10^{23} atoms, 23.0 g; 3.00 mol, 69.0 g; 2.00 mol, 12.04×10^{23} atoms

22. Fill in the blanks in the following table for the compound carbon dioxide, CO_2.

Moles	Molecules	Mass
2.00		
		44.0 g
	3.01×10^{23}	

Solutions to Confidence Exercises

13.1 $2\,H_2O \longrightarrow 2\,H_2 + O_2$; a decomposition reaction

13.2 $C_3H_8 + 5\,O_2 \longrightarrow 3\,CO_2 + 4\,H_2O$

13.3 $3\,HCl + Al(OH)_3 \longrightarrow 3\,H_2O + AlCl_3$

13.4 $2\,HNO_3 + Na_2CO_3 \longrightarrow CO_2 + H_2O + 2\,NaNO_3$

13.5 $Na_2SO_4(aq) + BaCl_2(aq) \longrightarrow BaSO_4(s) + 2\,NaCl(aq)$

13.6 Examination of the activity series (Table 13.3) shows that Al is higher than Cu, so a reaction takes place only in the first beaker. It may be written

$$2\,Al + 3\,CuSO_4(aq) \longrightarrow 3\,Cu + Al_2(SO_4)_3(aq)$$

Answers to Multiple-Choice Review Questions

2. a   4. c   24. b   26. a   39. a   45. d
3. d   13. b   25. c   38. c   44. a

ORGANIC CHEMISTRY

14

There's hardly a thing that man can name
Of use or beauty in life's small game
But you can extract from alembic or jar
From the physical basis of black coal tar
Oil and ointment, wax and wine
And the lovely colors called aniline
You can make anything from salve to a star
If you only know how, from black coal tar.

Punch Magazine, *1884*

The vast majority of chemicals are organic compounds. Scientists originally defined these as compounds of plant or animal origin—hence the name. However, it proved possible to make many of them in the laboratory from minerals (inorganic compounds). Therefore, the definition of *organic compounds* changed to just *compounds that contain carbon,* and their study is known as **organic chemistry.** The study of the chemical compounds and reactions that occur in living cells is now called *biochemistry.* It is a close relative of organic chemistry because carbon is the basic constituent of the complex molecules of the carbohydrates, fats, proteins, and nucleic acids.

Millions of organic compounds have been identified, and others are being added continually to the list as they are isolated from natural sources or synthesized in the laboratory. Many of them are found in things we use daily—food, fuels, drugs, detergents, perfumes, synthetic fibers, and so on.

Organic chemistry is a gigantic field. In this chapter, we introduce enough of the fundamental concepts, compounds, and reactions to impart a basic appreciation and comprehension of this fascinating area. ■

Photo: Nylon, an organic compound, is used to make parachutes and many other items.

14.1 Bonding in Organic Compounds

LEARNING GOAL

▼ Use bonding rules to identify valid and incorrect structural formulas.

In addition to carbon, the most common elements in organic compounds are hydrogen, oxygen, nitrogen, sulfur, and the halogens. Because these are all non-metals, organic compounds are covalent in bonding (Section 12.5). Examination of the Lewis symbols and application of the octet rule show that these elements should bond as summarized in Table 14.1.

Any structural formula that follows the bonding rules probably represents a known or possible compound. Any structure drawn that breaks one of these rules is unlikely to represent a real compound.

EXAMPLE 14.1

Spotting Incorrect Structural Formulas

Two structural formulas are shown. Which one does not represent a real compound? Why?

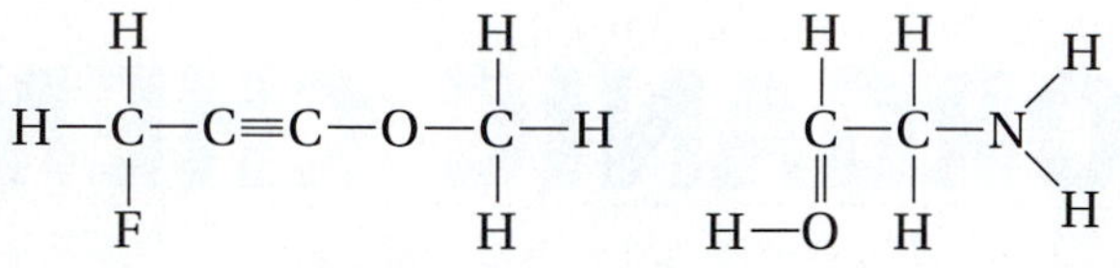

(a)   (b)

SOLUTION

In structure (a), each hydrogen and halogen has one bond, each carbon has four, and the oxygen has two. This is a valid structure. Structure (b) cannot be correct. Although the nitrogen has three bonds, the hydrogens one each, and the carbons four each, the oxygen has three bonds when it should have only two.

CONFIDENCE EXERCISE 14.1

The following structural formula for caffeine, a stimulant found in coffeee beans and tea leaves, appears in a recent chemistry book. Check the number of bonds to each atom and determine if any bonding rules are violated.

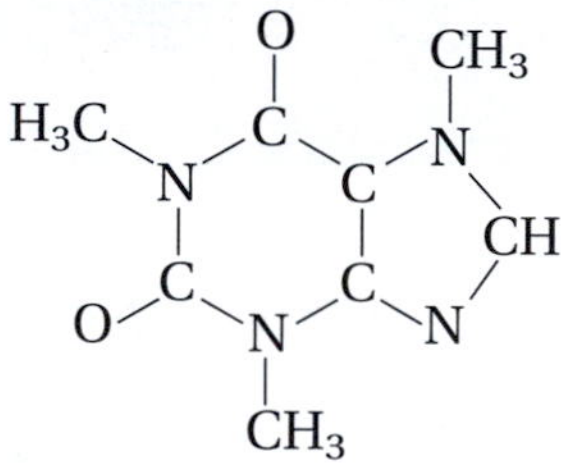

TABLE 14.1   Numbers and Types of Bonds for Common Elements in Organic Compounds

Element	Total Number of Bonds	Distribution of Total Number of Bonds and Examples		
C	4	4 singles —C̶—	2 singles, 1 double —C=	1 single, 1 triple —C≡
N	3	3 singles —N—	1 single, 1 double —N=	1 triple N≡
O (or S)	2	2 singles O—	1 double O=	
H or halogens	1	1 single H—, Cl—, etc.		

14.2 Aromatic Hydrocarbons

LEARNING GOALS

▼ Identify the structures of benzene and its relatives.

▼ State some uses and properties of aromatic hydro-carbons.

The simplest organic compounds are the **hydrocarbons,** which contain only carbon and hydrogen. For purposes of classification, all other organic compounds are considered to be *derivatives* of hydrocarbons. Thus the first class of compounds discussed in organic chemistry is the hydrocarbons, which are first divided into aromatic hydrocarbons or aliphatic ones (● Fig. 14.1).

Hydrocarbons that possess one or more benzene rings are called **aromatic hydrocarbons** (many have pungent aromas). The most important one is benzene (C_6H_6) itself, a clear, colorless liquid with a distinct odor. It is a **carcinogen** (a cancer-causing agent) and has these structural formulas and shorthand symbols:

Benzene
(Lewis structure)

Benzene
(Kekulé symbol)

Benzene
(modern symbol)

The Lewis structure and Kekulé symbol indicate that the ring has alternating single and double bonds between the carbon atoms. However, the properties of the benzene molecule and advanced bonding theory show the six electrons are shared by *all* the carbon atoms in the ring, forming an electron cloud that ex-

tends above and below the plane of the ring (● Fig. 14.2). This sharing of six "delocalized" electrons by all the ring atoms lends a special stability to benzene and its relatives. Thus the modern shorthand symbol for benzene is a hexagon with a circle inside, although the Kekulé symbol is still often used.

Some other aromatic hydrocarbons are toluene (or methylbenzene, used in model airplane glue), naphthalene (used in mothballs), and phenanthrene (used in the synthesis of dyes, explosives, and drugs). Polycyclic hydrocarbons such as phenanthrene are so stable that they have even been identified in interstellar space, dust shells around stars, and meteorites.

CH_3

Toluene
(methylbenzene)

Naphthalene

Phenanthrene

Benzene is obtained mainly from petroleum, but other aromatic hydrocarbons are commonly obtained from coal tar, a by-product of soft coal (see the chapter-opening quotation). When other atoms or groups of atoms are substituted for one or more of the hydrogen atoms in the benzene ring, a vast number of different compounds can be produced. These compounds include such things as perfumes, explosives, drugs, solvents, insecticides, and lacquers. An example is the explosive called TNT.

CH_3
O_2N NO_2
NO_2

TNT (2,4,6-trinitrotoluene)

FIGURE 14.1 Classification of Hydrocarbons

(See text for description.)

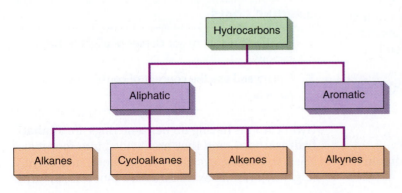

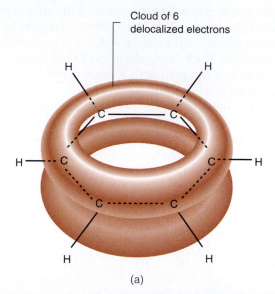

Cloud of 6 delocalized electrons

(a)

(b)

FIGURE 14.2   Benzene

(a) A representation of benzene shows that it is a flat molecule with six delocalized electrons forming an electron cloud above and below the plane of the ring. (b) A space-filling model of benzene.

To draw the structures for simple benzene derivatives, draw a benzene ring and attach the substituent to the ring in the position indicated by the number before the substituent's name. (If only one substituent is on the ring, the number 1 is omitted.)

EXAMPLE 14.2

Drawing Structures for Benzene Derivatives

Draw the structural formula for 1,3-dibromobenzene.

SOLUTION

First, draw a benzene ring.

Second, attach a bromine atom (*bromo* means "bromine atom") to the carbon atom at the ring position you choose to be number 1.

Third, attach a second bromine atom (the name says *di*, meaning "two") to ring position 3 (you may number either clockwise or counterclockwise from carbon 1), and you have the answer.

CONFIDENCE EXERCISE 14.2

Draw the structural formula for 1-chloro-2-fluorobenzene.

RELEVANCE QUESTION: Aromatic hydrocarbons are present in the black smoke emitted from diesel engines. What possible health effect might there be on someone breathing a lot of this smoke?

14.3 Aliphatic Hydrocarbons

LEARNING GOALS

▼ Describe the four major classes of aliphatic hydrocarbons.

▼ Explain and use the concept of constitutional isomers.

Hydrocarbons having no benzene rings are **aliphatic hydrocarbons,** and they are divided into four major classes: alkanes, cycloalkanes, alkenes, and alkynes (Fig. 14.1). We will examine each major class in turn.

Alkanes

The **alkanes** are hydrocarbons that contain only single bonds. They are said to be *saturated hydrocarbons* because their hydrogen content is at a maximum. Alkanes have a composition that satisfies the general formula

$$C_nH_{2n+2}$$

where n = the number of carbon atoms

$2n + 2$ = the number of hydrogen atoms

That is, the number of hydrogen atoms present in a particular alkane is twice the number of carbon atoms plus two. Note how the general formula applies to the alkanes listed in Table 14.2.

Methane (CH_4) is the first member ($n = 1$) of the alkane series. Ethane (C_2H_6) is the second member, propane (C_3H_8) the third, and butane (C_4H_{10}) the fourth. After butane, the number of carbon atoms is indicated by Greek prefixes such as *penta-, hexa-,* and so on. *The names of alkanes always end in -ane.* Methane through butane are gases, pentane through about $C_{17}H_{36}$ are liquids, and the rest are solids. The alkanes generally are colorless and, because they are nonpolar, do not dissolve in water.

A hydrocarbon's structure is easy to visualize when we write its **structural formula** (a graphic representa-

TABLE 14.2 **The First Eight Members of the Alkane Series**

Name	Molecular Formula	Condensed Structural Formula
Methane	CH_4	CH_4
Ethane	C_2H_6	CH_3CH_3
Propane	C_3H_8	$CH_3CH_2CH_3$
Butane	C_4H_{10}	$CH_3(CH_2)_2CH_3$
Pentane	C_5H_{12}	$CH_3(CH_2)_3CH_3$
Hexane	C_6H_{14}	$CH_3(CH_2)_4CH_3$
Heptane	C_7H_{16}	$CH_3(CH_2)_5CH_3$
Octane	C_8H_{18}	$CH_3(CH_2)_6CH_3$

tion of the way the atoms are connected to one another) instead of its *molecular formula* (which tells only the type and number of atoms). For example, the structural formulas for methane, ethane, and pentane are shown in the second row in ● Fig. 14.3. Each dash represents a covalent bond (two shared electrons). To save time and space, *condensed* structural formulas are often used, as in the third row of Fig. 14.3 (and the third column of Table 14.2).

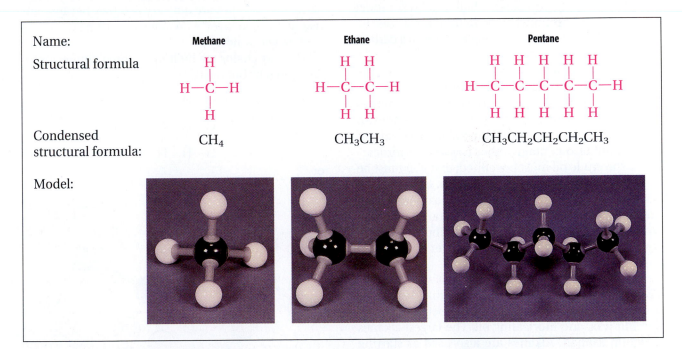

FIGURE 14.3 **Models of Three Alkanes**
Structural formulas, condensed structural formulas, and ball-and-stick models of methane, ethane, and pentane.

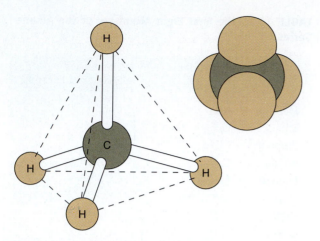

FIGURE 14.4   Methane

The tetrahedral geometry of four single bonds to a carbon atom is emphasized in the ball-and-stick and space-filling models for methane.

The fourth row of Fig. 14.3 shows ball-and-stick models of methane, ethane, and pentane. The four single bonds of each carbon point to the corners of a regular tetrahedron (a geometric figure with four identical equilateral triangles as faces). Carbon's four single bonds form angles of 109.5°, not 90°, as may appear from two-dimensional structural formulas. ● Figure 14.4, which shows a ball-and-stick model and a space-filling model of methane, emphasizes the tetrahedral geometry of four single bonds to a carbon atom.

The alkanes make up many well-known products. Methane is the principal component of natural gas (Fig. 13.5 on page 354), which is used in many homes for heating and cooking. Propane and butane are also used for that purpose. Petroleum is made up chiefly of alkanes but also contains other classes of hydrocarbons. The crude oil must be refined; that is, it must be separated into fractions by distillation. Each fraction is still a very complex mixture of hydrocarbons. Gasoline consists of the alkanes from pentane to decane ($n = 5$ to $n = 10$). At oil refineries, additional gasoline is made by catalytic "cracking" of larger alkanes into smaller ones (● Fig. 14.5).

Kerosene contains the alkanes with $n = 10$ to 16. The alkanes with higher values of n make up other products such as diesel fuel, fuel oil, petroleum jelly, paraffin wax, and lubricating oil. The largest alkanes make up asphalt. Alkanes are also used as starting materials for products such as paints, plastics, drugs, detergents, insecticides, and cosmetics.

A substituent that contains one less hydrogen atom than the corresponding alkane is called an **alkyl group,** and is given the general symbol R. The name of the alkyl group is obtained by dropping the *-ane* suffix and adding *-yl*. For example, *methane* becomes *methyl*, and *ethane* becomes *ethyl*. The open bonds in the methyl and ethyl groups indicate that these groups are bonded to another atom; they do not have an independent existence.

$$\begin{array}{ccc} & H & \\ & | & \\ H-\!\!&C&\!\!-H \\ & | & \\ & H & \end{array} \qquad \begin{array}{ccc} & H & \\ & | & \\ H-\!\!&C&\!\!- \quad \text{or} \quad CH_3- \\ & | & \\ & H & \end{array}$$

Methane   **Methyl group**

$$\begin{array}{cc} H & H \\ | & | \\ H-C-C-H \\ | & | \\ H & H \end{array} \qquad \begin{array}{cc} H & H \\ | & | \\ H-C-C- \quad \text{or} \quad CH_3CH_2- \\ | & | \\ H & H \end{array}$$

Ethane   **Ethyl group**

Alkanes are highly combustible. Like all hydrocarbons, when ignited, they react with the oxygen in air, forming carbon dioxide and water and releasing heat (Section 13.2). If the combustion is not complete, carbon monoxide and black, sooty carbon are formed. Otherwise, alkanes are not very reactive, because any reaction would involve breaking the strong C—H and C—C single bonds.

The way chains are built up in three dimensions is illustrated by butane (C_4H_{10}):

$$\begin{array}{cccc} H & H & H & H \\ | & | & | & | \\ H-C-C-C-C-H \\ | & | & | & | \\ H & H & H & H \end{array}$$

or

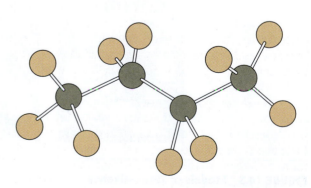

Butane

Each end carbon atom has three hydrogen atoms attached, and two hydrogens are bonded to each of the middle two carbon atoms.

However, another arrangement of the atoms of the butane molecule is possible. Isobutane (2-methyl-propane) also has the molecular formula C_4H_{10}, but its structural formula is

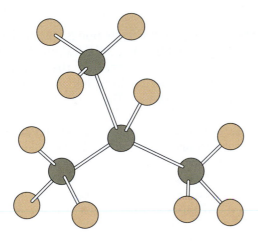

or

Isobutane (2-methylpropane)

FIGURE 14.5   Catalytic-Cracking Unit at a Petroleum Refinery

Catalytic cracking breaks larger hydrocarbons into smaller molecules to increase the yield of gasoline.

Examination of the two-dimensional representations of the butane and isobutane molecules and their three-dimensional ball-and-stick models shows that the structures are indeed different. Butane has a *continuous-chain* (or *straight-chain*) structure in which no carbon is bonded directly to more than two others, whereas isobutane has a *branched-chain* structure in which one carbon atom is bonded directly to three others. Although they have the same molecular formula, these compounds have different physical and chemical properties (for example, butane boils at $-0.5°C$, whereas isobutane boils at $-11.6°C$).

Isobutane and butane are **constitutional isomers,** compounds that have the same *molecular* formula but different *structural* formulas. In other words, they have the same number and type of each atom but differ in how these atoms are connected to one another. Constitutional isomers exist whenever two or more structural formulas can be built from the same number and type of each atom without violating the octet rule.

The phenomenon of constitutional isomerism is somewhat akin to using the same amounts of wood and brick to build houses that are entirely different in structure. Because of the ability of carbon atoms to bond to many other carbon atoms and atoms of other elements in so many different ways, the number of possible organic compounds is infinite, as Table 14.3 indicates.

Because of the number and complexity of organic compounds, a consistent method of nomenclature was developed so that communication would be effective. The IUPAC system of nomenclature for organic compounds begins with the rules for alkanes. Let's examine the basic rules and see how they allow the writing of a structure from the name, and vice versa. If these rules seem confusing the first time you read them, don't worry. Some examples will make them clear.

1. The longest continuous chain of carbon atoms (the "backbone") is found, and the compound is

TABLE 14.3 Number of Possible Isomers of Alkanes

Molecular Formula	Total Isomers
CH_4	1
C_2H_6	1
C_3H_8	1
C_4H_{10}	2
C_5H_{12}	3
C_6H_{14}	5
C_7H_{16}	9
C_8H_{18}	18
C_9H_{20}	35
$C_{10}H_{22}$	75
$C_{15}H_{32}$	4,347
$C_{20}H_{42}$	366,319
$C_{30}H_{62}$	4.11×10^9

TABLE 14.4 Substituents in Organic Compounds

Formula of Substituent	Name of Substituent
Br—	Bromo
Cl—	Chloro
F—	Fluoro
I—	Iodo
CH_3—	Methyl
CH_3CH_2—	Ethyl

named as a derivative of the alkane with this number of carbon atoms.

2. The positions and names of the substituents (single atoms or groups; see Table 14.4) that have replaced hydrogen atoms on the backbone chain are added. When more than one type of substituent is present, either on the same carbon atom or on different carbon atoms, the substituents are listed in alphabetical order. If more than one of the same type of substituent is present, the prefixes *di-, tri-, tetra-, penta-,* and so forth are used to indicate how many.

3. The carbon atoms on the backbone chain are numbered by counting from the end of the chain nearest the substituents. The position of attachment of each substituent is identified by giving the number of the carbon atom in the chain. Each substituent must have a number. Commas are used to separate numbers from other numbers, and hyphens are used to separate numbers from names.

For example, consider this structure:

$$\overset{5}{CH_3}-\overset{4}{CH_2}-\overset{3}{CH_2}-\overset{2}{CH}-\overset{1}{CH_3}$$
$$|$$
$$CH_3$$

2-Methylpentane

The longest continuous chain of carbon atoms is five, so the compound is named as a pentane derivative.

Attached to the pentane backbone is a methyl group (see Table 14.4). Numbering the backbone from right to left gives a smaller number for the methyl substituent than numbering from left to right and so would be the correct way. The compound's name is 2-methylpentane.

Rather than use these rules to assign names to the structural formulas for compounds, we will use them in reverse to draw structural formulas from the names. This is somewhat simpler and still gets the major points across about the relationship between structure and IUPAC name.

EXAMPLE 14.3

Drawing a Structure from a Name

Draw the structural formula for 2,3-dimethylhexane.

SOLUTION

The end of the name is *hexane,* so draw a continuous chain of six carbon atoms joined by five single bonds, and add enough bonds to each C atom so that all have four.

$$-\overset{|}{\underset{|}{C}}-\overset{|}{\underset{|}{C}}-\overset{|}{\underset{|}{C}}-\overset{|}{\underset{|}{C}}-\overset{|}{\underset{|}{C}}-\overset{|}{\underset{|}{C}}-$$

In your mind, number the carbons 1 through 6, starting from either end you wish, and attach a methyl group (CH_3—) to carbon 2 and another to carbon 3. Hydrogen atoms are added to the remaining bonds, giving

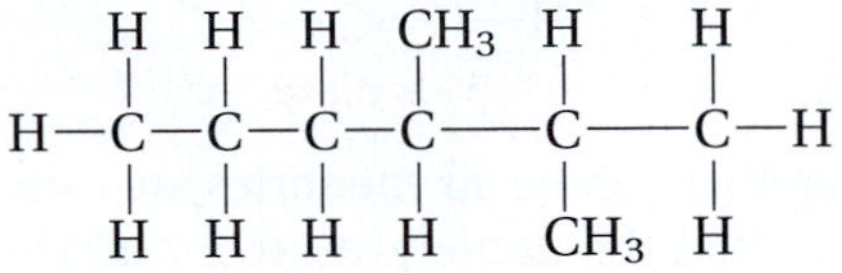

$$H-\underset{\underset{H}{|}}{\overset{\overset{H}{|}}{C}}-\underset{\underset{H}{|}}{\overset{\overset{H}{|}}{C}}-\underset{\underset{H}{|}}{\overset{\overset{H}{|}}{C}}-\underset{\underset{CH_3}{|}}{\overset{\overset{CH_3}{|}}{C}}-\underset{\underset{H}{|}}{\overset{\overset{H}{|}}{C}}-\underset{\underset{H}{|}}{\overset{\overset{H}{|}}{C}}-H$$

CONFIDENCE EXERCISE 14.3

The octane rating for gasoline assigns a value of 100 to the combustion of the "octane" whose IUPAC name is 2,2,4-trimethylpentane. Draw the structure of this important hydrocarbon.

Cycloalkanes

The **cycloalkanes,** members of a second series of saturated hydrocarbons, have the general molecular formula C_nH_{2n} and possess *rings* of carbon atoms, with each carbon atom bonded to a total of four carbon or hydrogen atoms. The smallest possible ring occurs with cyclopropane, C_3H_6; then come cyclobutane, cyclopentane, and so forth. Note the inclusion of the prefix *cyclo* when naming cycloalkanes. ● Figure 14.6 shows the names, molecular formulas, structural formulas, and condensed structural formulas for the first four cycloalkanes.

You will soon be able to recognize automatically that a carbon atom is at each corner of the condensed figure and that enough hydrogen atoms are assumed to be attached to each carbon to give a total of four single bonds. To draw the structure of a cycloalkane derivative, draw the geometric figure with the number of sides indicated by the compound's name. Then place each substituent on the ring in the numbered position indicated in the name. (As in the case of benzene derivatives, if only one substituent is on the cycloalkane ring, the number 1 is omitted.) For example, the structure of 1-chloro-2-ethylcyclopentane is

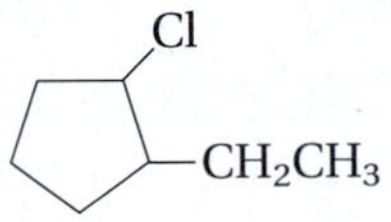

Alkenes

Hydrocarbons that have a double bond between two carbon atoms are called **alkenes.** Imagine that a hydrogen atom has been removed from each of two adjacent carbon atoms in an alkane, thereby allowing these C atoms to form an additional bond between them. So the general formula for the alkene series is C_nH_{2n} (the same as for cycloalkanes). The series begins with ethene, C_2H_4 (more commonly known as *ethylene*), shown by the structural formula

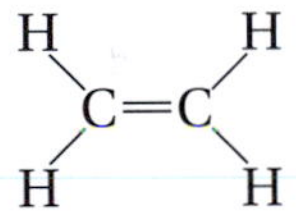

Ethene (ethylene)

Some of the simpler alkenes are listed in Table 14.5. Note that the -*ane* suffix for alkane names is changed to -*ene* for alkenes. A number preceding the name indicates the carbon atom on which the double bond

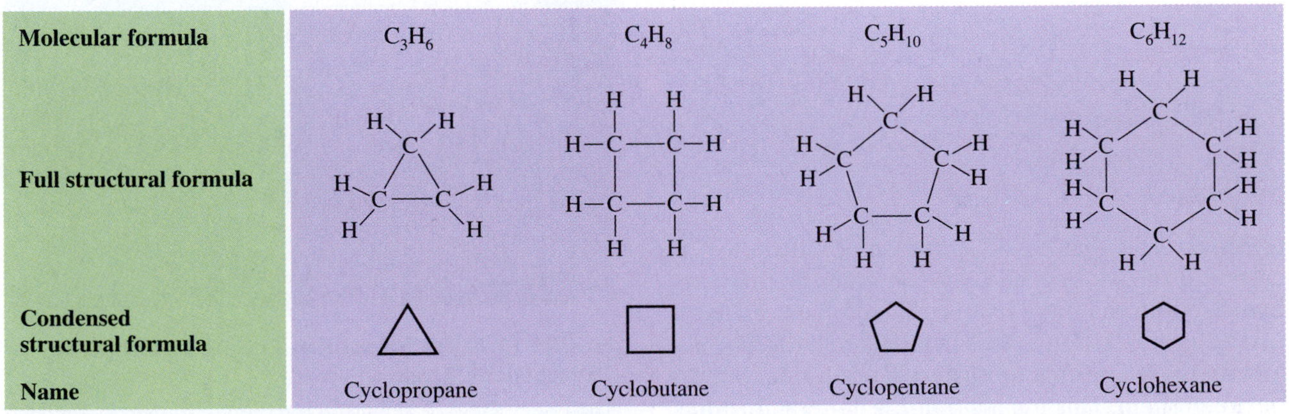

Molecular formula	C_3H_6	C_4H_8	C_5H_{10}	C_6H_{12}
Full structural formula				
Condensed structural formula				
Name	Cyclopropane	Cyclobutane	Cyclopentane	Cyclohexane

FIGURE 14.6 The First Four Cycloalkanes

TABLE 14.5 Some Members of the Alkene Series

Name	Molecular Formula	Condensed Structural Formula
Ethene (ethylene)	C_2H_4	$CH_2=CH_2$
Propene	C_3H_6	$CH_3CH=CH_2$
1-Butene	C_4H_8	$CH_3CH_2CH=CH_2$
2-Butene	C_4H_8	$CH_3CH=CHCH_3$
1-Pentene	C_5H_{10}	$CH_3(CH_2)_2CH=CH_2$

starts. The carbons in the chain are numbered starting at the end that gives the double bond the lower number. For example, 1-butene and 2-butene have the structural formulas

1-Butene   **2-Butene**

Alkenes are very reactive. They are termed *unsaturated hydrocarbons* because a characteristic reaction is the *addition of hydrogen* (using a platinum catalyst) to the double-bonded carbons to form the corresponding alkane. For example,

Ethene   **Ethane**

Addition of hydrogen

Alkynes

Hydrocarbons that have a triple bond between two carbon atoms are called **alkynes.** Imagine that two hydrogen atoms have been removed from each of two adjacent carbon atoms in an alkane, thereby allowing these C atoms to form two additional bonds between them. This means that the general formula for the alkyne series is C_nH_{2n-2}. The simplest

alkyne is ethyne (more commonly called *acetylene,* ● Fig. 14.7):

$$H—C≡C—H$$

Ethyne (acetylene)

Some members of the alkyne series are listed in Table 14.6. Note that the nomenclature for alkynes follows rules similar to those of the alkenes. (Cycloalkenes and cycloalkynes do exist, but we will not discuss them.)

Alkynes are unsaturated hydrocarbons and, like alkenes, add hydrogen. Alkynes can add *two* molecules of hydrogen across the triple bond. For example,

$$H—C≡C—H + 2\,H_2 \longrightarrow CH_3CH_3$$

Ethyne   **Ethane**

The left-hand side of the Spotlight feature on page 371 summarizes the names and characteristic structural features of the hydrocarbons.

FIGURE 14.7 An Ironworker Cutting Steel with an Oxyacetylene Torch

The combustion of acetylene (ethyne) produces the intense heat needed to fuse metals.

TABLE 14.6   Some Members of the Alkyne Series

Name	Molecular Formula	Condensed Structural Formula
Ethyne (acetylene)	C_2H_2	$HC \equiv CH$
Propyne	C_3H_4	$CH_3C \equiv CH$
1-Butyne	C_4H_6	$CH_3CH_2C \equiv CH$
2-Butyne	C_4H_6	$CH_3C \equiv CCH_3$
1-Pentyne	C_5H_8	$CH_3(CH_2)_2C \equiv CH$

14.4 Derivatives of Hydrocarbons

LEARNING GOALS

▼ Identify the structures of some derivatives of hydrocarbons.

▼ Write equations for ester and amide formation.

The characteristics of organic molecules depend on the number, type, and arrangement of their atoms. Any atom, group of atoms, or organization of bonds that determines specific properties of a molecule is called a **functional group.** Generally, the functional group is the reactive part of a molecule, and its presence signifies certain predictable properties.

The double bond in an alkene and the triple bond in an alkyne are functional groups. Other functional groups (for example, a chlorine atom or an —OH group) may be attached to a carbon atom in place of a hydrogen atom. (See the Spotlight feature on page 371.) Organic compounds that contain atoms other than C and H are called *derivatives* of hydrocarbons.

In the general structural formula for a derivative, the letter R usually represents an alkyl group to which the functional group is attached. Let's take a look at some interesting and useful derivatives of hydrocarbons.

Alkyl Halides

The general formula for an **alkyl halide** is R—X, where X is a halogen atom and R is an alkyl group. Recall that when naming organic compounds, a fluorine atom is

designated by the term *fluoro,* a chlorine atom by *chloro,* and so forth (Table 14.4).

The alkyl halides called **CFC**s are **c**hloro**fluoro**car**bons, such as dichlorodifluoromethane (Freon-12), and have been used commonly in air conditioners, refrigerators, heat pumps, and so forth. They are unreactive gases at Earth's surface.

$$F - \overset{\displaystyle F}{\underset{\displaystyle Cl}{C}} - Cl$$

Dichlorodifluoromethane (a CFC)

The major problem with the use of CFCs is that they travel upward into the stratosphere, where Earth's protective ozone (O_3) layer is formed by the action of sunlight on ordinary oxygen (O_2). The ozone layer absorbs much of the solar ultraviolet rays (UV) that can harm plant and animal life on Earth's surface (● Fig. 14.8). The manner in which CFCs react with and destroy ozone is discussed in Section 20.5 and Chapter 20's third Highlight.

Chloroform is another familiar alkyl halide. It was used in the past as a surgical anesthetic but has been found to be a carcinogen. A similar compound, carbon tetrachloride, was used in the past in fire extinguishers and fabric cleaners. However, it causes liver

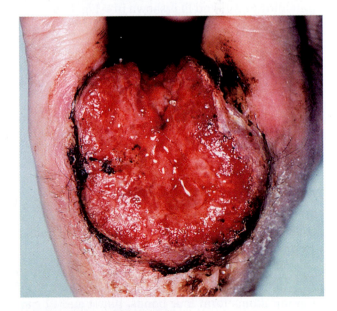

FIGURE 14.8   A Skin Cancer

Health risks are associated with overexposure to ultraviolet radiation from the Sun.

FIGURE 14.9 Liquid Fluorocarbons
Oxygen is so soluble in this liquid fluorocarbon that the submerged mouse can breathe by absorbing oxygen from it. When the mouse is taken out, the fluorocarbon in its lungs vaporizes, and it starts breathing air once more.

damage, so its use has been greatly reduced. An interesting use of some liquid fluorocarbons (compounds containing only C and F) is shown in ● Fig. 14.9.

**Chloroform
(trichloromethane)**

**Carbon tetrachloride
(tetrachloromethane)**

EXAMPLE 14.4

Drawing Constitutional Isomers

To better understand constitutional isomerism, let's draw the structural formulas for the two alkyl halide isomers having the molecular formula $C_2H_4Cl_2$.

SOLUTION

Carbon atoms form four bonds, but H and Cl can form only one each. Thus H and Cl can never connect two other atoms; they can only stick to the backbone. Therefore, first draw the two-carbon backbone using a single bond, and then add enough

bonds so that each carbon has four.

Count the open bonds. They total six, just right for the attachment of four H atoms and two Cl atoms. (If there were too many open bonds, an alkene, alkyne, or ring structure would be tried.)

Ignore the H atoms for the time being and see how many different ways you can put on the two Cl atoms. You will find two different ways. (*Note:* You must remember the tetrahedral geometry of the four bonds to C. *There is no difference in the bonds you are drawing "out," "up," or "down." Each bond is really at 109.5° from another, not 90° or 180.°*)

Add the four hydrogens to each structural formula, and the question is answered (● Fig. 14.10). Note how the names of the two isomers denote their structures.

1,2-Dichloroethane   **1,1-Dichloroethane**

CONFIDENCE EXERCISE 14.4

Two constitutional isomers of C_3H_7F exist. Draw the structure for each.

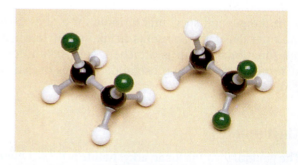

FIGURE 14.10 Constitutional Isomers
The two constitutional isomers of dichloroethane ($C_2H_4Cl_2$) are shown in these models.

Alcohols

Alcohols are organic compounds containing a **hydroxyl group,** —OH, attached to an alkyl group. The general formula for an alcohol is R—OH, and their IUPAC names end in *-ol*. Thousands of alcohols exist, some with only one hydroxyl group and others with two or more. The simplest alcohol is methanol, which is also called *methyl alcohol* or *wood alcohol*. It is poisonous, but it has its uses (● Fig. 14.11).

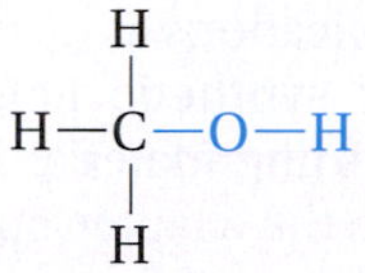

$$H-\underset{\underset{H}{|}}{\overset{\overset{H}{|}}{C}}-O-H \quad \text{or} \quad CH_3OH$$

Methanol (methyl alcohol)

Ethanol (CH_3CH_2OH) is also called *ethyl alcohol* or *grain alcohol*. It is a colorless liquid that mixes with water in all proportions and is the least toxic and most economically important of all alcohols. Ethanol is used in alcoholic beverages and in the production of many substances, including perfumes, dyes, and varnishes.

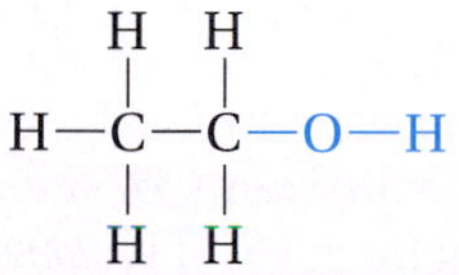

$$H-\underset{\underset{H}{|}}{\overset{\overset{H}{|}}{C}}-\underset{\underset{H}{|}}{\overset{\overset{H}{|}}{C}}-O-H \quad \text{or} \quad CH_3CH_2OH$$

Ethanol (ethyl alcohol)

Ethylene glycol, a compound widely used as an antifreeze and coolant, is an example of an alcohol with

FIGURE 14.11   Methanol (CH_3OH) Is the Simplest Alcohol

Methanol is used for fuel in some types of racing cars.

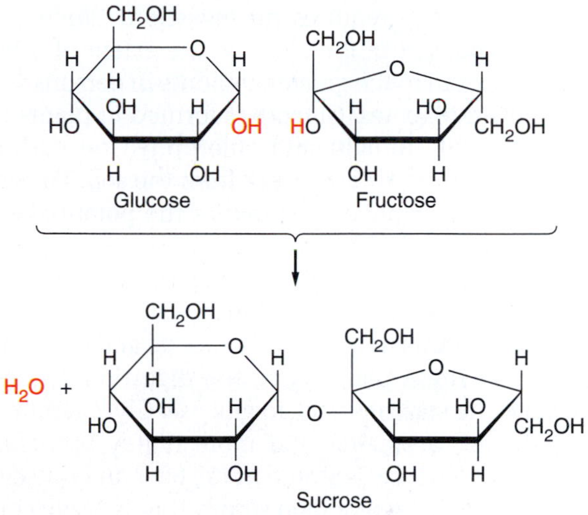

FIGURE 14.12   The Structures of Glucose, Fructose, and Sucrose

A molecule of glucose can react with a molecule of fructose to form sucrose (ordinary table sugar) and water.

two hydroxyl groups. It causes kidney failure if ingested, and its sweet taste can act as a lure to animals and children.

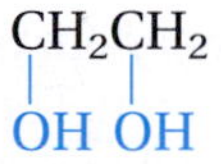

$$\underset{\text{OH OH}}{CH_2CH_2}$$

Ethylene glycol (1,2-ethanediol)

Carbohydrates, an important class of compounds in living matter, contain multiple hydroxyl groups in their molecular structures, and their names end in *-ose*. One of the primary uses of carbohydrates in cells is as an energy source. The most important carbohydrates are the sugars, the starches, and cellulose. Two important simple sugars are the isomers glucose ($C_6H_{12}O_6$) and fructose ($C_6H_{12}O_6$). When a glucose molecule is bonded to a fructose molecule by removal of the three colored atoms, as shown in ● Fig. 14.12, a molecule of sucrose (common cane sugar) is formed.

Fructose (fruit sugar) is the sweetest of all sugars and is present in fruits and honey. If sucrose is given an arbitrary sweetness value of 100, fructose is rated 173. The artificial sweetener named aspartame has a value of 15,000, and saccharin is rated 35,000. No wonder a little artificial sweetener goes a long way!

Glucose, also known as *dextrose,* is found in sweet fruits, such as grapes and figs, and in flowers and honey. Carbohydrates must be digested into glucose for circulation in the blood. Hospitalized patients are sometimes fed intravenously with glucose solutions because glucose requires no digestion. Glucose is normally present in the blood to the extent of 0.1%, but it occurs in much greater amounts in persons suffering from diabetes. Glucose is formed in plants by the action of sunlight and chlorophyll on carbon dioxide from the air and water from the soil. The energy from the sunlight is stored as the potential energy of the chemical bonds.

Starch is a polymer consisting of long chains of up to 3000 glucose units. It is a noncrystalline substance formed by plants in their seeds, tubers, and fruits. After we eat these plant parts, the digestion process converts the starches back to glucose. Glycogen (animal starch), a smaller and more highly branched polymer of glucose, is stored in the liver and muscles of animals as a reserve food supply that is easily converted to energy.

Cellulose, another polymer of glucose, has the same general formula, $(C_6H_{10}O_5)_n$, as starch. However, its structure is slightly different, so it has different properties. Cellulose is the main component of the cell walls of plants and received its name for this reason. It is the most abundant organic substance found in our environment. Cellulose cannot be digested by humans because our digestive systems do not contain the necessary enzymes to break the linkages in the molecular chain. The bacteria in the digestive tracts of termites and herbivores (such as cows or deer) do have the necessary enzymes to break the linkages between the glucose units and thus obtain nutrition from cellulose. Cellulose is contained in many commercial products, such as rayon, cellophane, explosives, and paper.

Amines

Amines ("ah-MEANS") are organic compounds that contain nitrogen and are basic (alkaline). The general formula for an **amine** is R—NH_2, but one or two additional alkyl groups could be attached to the nitrogen atom of the **amino group**, —NH_2, in place of one or more hydrogen atoms. (Recall that a nitrogen atom forms three bonds.) Examples are methylamine, dimethylamine, and trimethylamine. A condensed formula is shown for dimethylamine, and an even more condensed formula represents trimethylamine.

Methylamine **Dimethylamine** **Trimethylamine**

$(CH_3)_3N$

The simple amines have strong odors. The odor of raw fish comes from the amines it contains. Decaying flesh forms putresine (1,4-diaminobutane) and cadaverine (1,5-diaminopentane), two amines with especially foul odors.

Putresine (1,4-diaminobutane)

Amines have many applications as medicinals and as starting materials for synthetic fibers. Aniline (aminobenzene) is the starting material for a whole class of synthetic dyes (see the chapter-opening quotation), as well as many other useful compounds.

Aniline (aminobenzene)

Amines occur widely in nature as drugs, such as nicotine, the addictive substance in tobacco, and coniine, the poison that killed Socrates (● Fig. 14.13).

Amphetamines, such as Benzedrine, are synthetic amines that are powerful stimulants of the central nervous system. They raise the level of glucose in the blood, thereby fighting fatigue and reducing appetite. Although these drugs have legitimate medical usages, they are addictive and can lead to insomnia, excessive weight loss, and paranoia. Other drugs are discussed in the chapter Highlight.

Benzedrine

Carboxylic Acids

The **carboxylic acids** ("CAR-box-ILL-ic") contain a **carboxyl group** and have the general formula RCOOH, in which the bonding is

Carboxyl group **Carboxylic acid (general formula)**

The simplest carboxylic acid, formic acid (methanoic acid), is the cause of the painful discomfort from

FIGURE 14.13 *The Death of Socrates*

The hemlock that Socrates drank contained the deadly alkaloid named *coniine*. In 1787 the French artist Jacques David ("dah-VEED") painted *The Death of Socrates*. David also painted the picture of Lavoisier and his wife in Chapter 12's first Highlight.

insect bites or bee stings. Vinegar is a 5% solution of acetic acid (ethanoic acid) in water. The structural formulas for formic and acetic acids are

$$H-\overset{\overset{\displaystyle O}{\|}}{C}-O-H \qquad CH_3-\overset{\overset{\displaystyle O}{\|}}{C}-O-H$$

Formic acid
(methanoic acid)   **Acetic acid**
                            (ethanoic acid)

The chapter Highlight discusses some interesting aspects of odors of carboxylic acids.

Esters

An **ester** is a compound that has the general formula

$$R-\overset{\overset{\displaystyle O}{\|}}{C}-O-R'$$

where R and R' (read "R prime") are any alkyl groups. R and R' may be identical, but they are usually different.

Unlike amines, most esters possess pleasant odors. The fragrances of many flowers and the pleasing taste of ripe fruits are due to one or more esters (Table

14.7). Wintergreen mints and Pepto-Bismol get their fragrance from the ester named *methyl salicylate*, commonly called *oil of wintergreen*.

$$\overset{\overset{\displaystyle OH}{|}}{\underset{\overset{\|}{O}}{\underset{|}{C}}}-O-CH_3$$

Methyl salicylate (oil of wintergreen)

A carboxylic acid and an alcohol react to give an ester and water. This is referred to as *ester formation*. The reaction mixture must be heated, and sulfuric acid is used as a catalyst. In the general case,

$$R-\overset{\overset{\displaystyle O}{\|}}{C}-O-H + H-O-R' \xrightarrow{H_2SO_4}$$

Carboxylic acid   **Alcohol**

$$H_2O + R-\overset{\overset{\displaystyle O}{\|}}{C}-O-R'$$

Water   **Ester**

HIGHLIGHT: Drugs, Dogs, and Body Odor

A *drug* is a compound that can produce a physiologic change in human beings or animals. A group of drugs called *alkaloids* are nitrogen-containing, basic (alkaline) compounds found in plants. Alkaloids include morphine, as well as the substances nicotine and coniine, which have been mentioned in this chapter. Morphine is the addictive narcotic present in crude opium, which is extracted from the Oriental poppy (Fig. 1). It is used to relieve pain and induce sleep.

FIGURE 2 Drug Sniffing
A drug-detecting dog at work.

Heroin, produced synthetically from morphine by an ester formation reaction with acetic acid, is a narcotic that suppresses the central nervous system. Drug-sniffing dogs can find hidden heroin by detecting the vinegary odor of acetic acid, trace amounts of which remain in the heroin (Fig. 2).

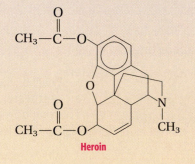

Heroin

An unsaturated carboxylic acid (3-methyl-2-hexenoic acid) produced by skin bacteria, particularly those found in armpits, is largely responsible for "body odor," but other acids also contribute. Bloodhounds are able to distinguish among the different proportions of carboxylic acids that cause each individual to emit a different odor.

$$CH_3{-}CH_2{-}CH_2{-}\underset{\underset{CH_3}{|}}{C}{=}CH{-}COOH$$

3-Methyl-2-hexenoic acid

Morphine

FIGURE 1 Morphine
An alkaloid, morphine, is in the resin that seeps from opium poppy pods.

TABLE 14.7   Odors of Esters

Name	Formula	Odor*
Ethyl formate	$HCOOCH_2CH_3$	Rum
Pentyl acetate	$CH_3COOCH_2CH_2CH_2CH_2CH_3$	Banana
Octyl acetate	$CH_3COOCH_2CH_2CH_2CH_2CH_2CH_2CH_2CH_3$	Orange
Methyl butyrate	$CH_3CH_2CH_2COOCH_3$	Apple
Ethyl butyrate	$CH_3CH_2CH_2COOCH_2CH_3$	Pineapple
Pentyl butyrate	$CH_3CH_2CH_2COOCH_2CH_2CH_2CH_3$	Apricot

*Natural flavors are generally complex mixtures of esters and other constituents.

(From Ebbing, Darrell D., *General Chemistry*, Sixth Edition. Copyright © 1999 by Houghton Mifflin Company. Used with permission.)

Note that the net result of the reaction (proved by tracing oxygen isotopes) is that *the —OH from the carboxyl group unites with the H from the hydroxyl group to form water. The remaining two fragments then bond together* by means of the bond (dash) that was attached to the H of the hydroxyl group of the alcohol.

EXAMPLE 14.5

Writing an Equation for Ester Formation

Complete the equation for the sulfuric acid–catalyzed reaction between acetic acid and ethanol.

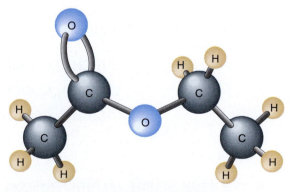

FIGURE 14.14   The Ethyl Acetate Molecule

Ethyl acetate, an ester, is used as a solvent in lacquers and other protective coatings. (From Ebbing, Darrell D., *General Chemistry*, Sixth Edition. Copyright © 1999 by Houghton Mifflin Company. Used with permission.)

SOLUTION

"Lasso" the —OH from the acid and the H from the hydroxyl group of the alcohol to form water. Attach the acid fragment to the alcohol fragment by means of the bond (dash) that was attached to the H of the hydroxyl group of the alcohol. This gives the structural formula of the ester called *ethyl acetate* (● Fig. 14.14). You would recognize its odor as that of fingernail polish.

CONFIDENCE EXERCISE 14.5

Complete the equation for this ester formation reaction that forms artificial banana flavoring.

Fats are esters composed of the trialcohol named *glycerol*, $CH_2(OH)CH(OH)CH_2(OH)$, and long-chain

FIGURE 14.15 Fat and Oils

The partial hydrogenation of the double bonds in the molecules of a liquid vegetable oil (left) produces a semisolid substance similar to an animal fat (right).

carboxylic acids known as *fatty acids.** A typical fatty acid is stearic acid, $C_{17}H_{35}COOH$, a component of beef fat. Stearic acid's structure is that of a long-chain hydrocarbon containing 17 carbon atoms (and their associated hydrogen atoms) attached to a carboxyl group. It is written below in condensed form to save space. When stearic acid is combined with glycerol, the triester named *glyceryl tristearate* (a fat) is obtained. The reaction is

$$3\ C_{17}H_{35} \overset{\displaystyle O}{\underset{}{-C}} -OH + \begin{matrix} H-O-CH_2 \\ H-O-CH \\ H-O-CH_2 \end{matrix} \longrightarrow$$

Stearic acid   **Glycerol**

$$3\ H_2O + \begin{matrix} C_{17}H_{35}-\overset{O}{\overset{\|}{C}}-O-CH_2 \\ C_{17}H_{35}-\overset{O}{\overset{\|}{C}}-O-CH \\ C_{17}H_{35}-\overset{O}{\overset{\|}{C}}-O-CH_2 \end{matrix}$$

Glyceryl tristearate (a fat)

Fats that come from animals are generally solids at room temperature, but those that come from plants and fish are usually liquid (● Fig. 14.15). These liquid fats are usually referred to as *oils*. Their molecules are composed of hydrocarbon chains with double bonds

between some of the carbon atoms; that is, they are unsaturated. These oils can be changed to solid (saturated) fats by a process called *hydrogenation*. In this process, hydrogen is added to the carbon atoms that have the double bonds (Section 14.3), and the hydrocarbon chains become saturated, or nearly so. Thus liquid fats (oils) are esters of glycerol and unsaturated acids, and solid fats are esters of glycerol and saturated acids. When cottonseed oil (a liquid) is hydrogenated, margarine (a solid) is obtained. The reaction is

$$\begin{matrix} CH_3(CH_2)_7CH\!=\!CH(CH_2)_7COOCH_2 \\ CH_3(CH_2)_7CH\!=\!CH(CH_2)_7COOCH + 3\ H_2 \longrightarrow \\ CH_3(CH_2)_7CH\!=\!CH(CH_2)_7COOCH_2 \end{matrix}$$

Cottonseed oil

$$\begin{matrix} CH_3(CH_2)_{16}COOCH_2 \\ CH_3(CH_2)_{16}COOCH \\ CH_3(CH_2)_{16}COOCH_2 \end{matrix}$$

Margarine

*The reaction of glycerol with nitric acid forms *nitroglycerin*, a thick, pale-yellow liquid that is a treacherous explosive. In 1866 a young Swedish inventor, Alfred Nobel, discovered that nitroglycerin was much safer to handle when soaked into porous silica. This material was made into sticks and called *dynamite*. Nobel became rich from his discovery and used some of his wealth to establish the Nobel Prizes, which were first awarded in 1901.

Fats and oils are used in the diets of humans and other organisms. In the digestive process, the fats are broken down into glycerol and acids, which are absorbed into the bloodstream and oxidized to produce energy that may be used immediately or stored for future use. Fats are also used by the body as insulation to prevent loss of heat and are important components of cell membranes. The metabolism of 1 gram of fat produces 9 kcal of energy, whereas 1 gram of protein or carbohydrate produces only 4 kcal. It is well established that a diet heavy in saturated fats is unhealthy because the fats lead to a buildup of cholesterol, a waxy substance that can clog the arteries.

When fats are treated with sodium hydroxide (lye, NaOH), the ester linkages break to give glycerol and sodium salts of fatty acids. *The sodium salts of fatty acids are* **soap** (● Fig. 14.16). A typical soap is sodium stearate, whose condensed structural formula is

$$CH_3(CH_2)_{16}\overset{\displaystyle O}{\overset{\displaystyle \|}{C}}-O^-\ Na^+$$

Sodium stearate

If you wish to dissolve stains made by a nonpolar compound such as grease, you must use either a nonpolar solvent or a soap or detergent. One end of the soap or detergent molecule is highly polar and dissolves in the water, whereas the other part of the molecule is a long, nonpolar hydrocarbon chain that dissolves in the grease (● Fig. 14.17). The grease is then emulsified and swept away by rinsing.

FIGURE 14.16 Making Soap

Reacting a fat with lye (NaOH) breaks the ester linkages and gives glycerol and soap (the sodium salts of fatty acids).

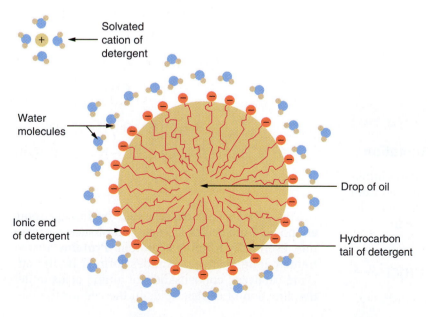

Solvated cation of detergent

Water molecules

Ionic end of detergent

Drop of oil

Hydrocarbon tail of detergent

FIGURE 14.17 Like Dissolves Like

The ionic ends of detergent molecules dissolve in the polar water, and the long nonpolar chains of the detergent molecules dissolve in the grease. The emulsified grease droplets can then be rinsed away.

Soaps have the disadvantage of forming precipitates when used in acidic solutions or with hard water, which contains ions of calcium, magnesium, and iron. The modern **synthetic detergents**, which are soap substitutes, contain a long hydrocarbon chain that is nonpolar (such as $C_{12}H_{25}$—) and a polar group such as sodium sulfate ($—OSO_3^- Na^+$). The detergent sodium lauryl sulfate has the condensed structural formula

$$CH_3(CH_2)_{11}—O—SO_3^-\ Na^+$$
Sodium lauryl sulfate

Synthetic detergents do not form precipitates with the calcium, magnesium, and iron ions. Therefore, they are effective cleansing agents in hard water.

Amides

Amides ("AM-eyeds"), another class of nitrogen-containing organic compounds, have the general formula

$$R-\overset{\overset{\displaystyle O}{\|}}{C}-\overset{\overset{\displaystyle H}{|}}{N}-R'$$

The reaction for *amide formation* is very similar to that of ester formation. A carboxylic acid and an amine react to give water and an amide. The general reaction is

$$R-\overset{\overset{\displaystyle O}{\|}}{C}\overset{}{-}O-H + H-\overset{\overset{\displaystyle H}{|}}{N}-R' \longrightarrow$$
Carboxylic acid   **Amine**

$$H_2O + R-\overset{\overset{\displaystyle O}{\|}}{C}-\overset{\overset{\displaystyle H}{|}}{N}-R'$$
Water   **Amide**

EXAMPLE 14.6

Writing an Equation for Amide Formation

Phenacetin, used as a pain reliever until it was implicated in kidney damage, is an amide formed from acetic acid and the amine shown. Complete the equation for the reaction to form Phenacetin.

$$CH_3-\overset{\overset{\displaystyle O}{\|}}{C}-O-H + H-\overset{\overset{\displaystyle H}{|}}{N}-\bigcirc-OCH_2CH_3 \longrightarrow$$
Acetic acid   **(An amine)**

SOLUTION

"Lasso" the —OH from the acid and an H from the amino group to form water. Attach the acid fragment to the amine fragment by means of the bond (dash) left on the amine fragment by removal of the H. This gives the structure of the amide called *Phenacetin*.

$$CH_3-\overset{\overset{\displaystyle O}{\|}}{C}\overset{}{-}O-H + H\overset{}{-}\overset{\overset{\displaystyle H}{|}}{N}-\bigcirc-OCH_2CH_3 \longrightarrow$$
Acetic acid   **(An amine)**

$$H_2O + CH_3-\overset{\overset{\displaystyle O}{\|}}{C}-\overset{\overset{\displaystyle H}{|}}{N}-\bigcirc-OCH_2CH_3$$
Water   **Phenacetin**

CONFIDENCE EXERCISE 14.6

Complete this equation for an amide formation reaction.

$$CH_3CH_2\overset{\overset{\displaystyle O}{\|}}{C}-O-H + H-\overset{\overset{\displaystyle H}{|}}{N}-CH_3 \longrightarrow$$

An **amino acid** is an organic compound that contains both an amino group and a carboxyl group. Over 20 natural amino acids exist, 8 of which are essential in the human diet. The simplest amino acids are glycine and alanine. When a molecule of glycine combines with a molecule of alanine, a molecule of water is eliminated and an amide linkage, —CONH—, is formed in the resulting molecule called a *dipeptide*.

$$H-\overset{\overset{\displaystyle H}{|}}{\underset{\underset{\displaystyle H}{|}}{N}}-\overset{\overset{\displaystyle H}{|}}{\underset{\underset{\displaystyle H}{|}}{C}}-\overset{\overset{\displaystyle O}{\|}}{C}-OH + H-\overset{\overset{\displaystyle H}{|}}{N}-\overset{\overset{\displaystyle H}{|}}{\underset{\underset{\displaystyle CH_3}{|}}{C}}-\overset{\overset{\displaystyle O}{\|}}{C}-OH \longrightarrow$$
Glycine   **Alanine**

$$H_2O + H-\overset{\overset{\displaystyle H}{|}}{\underset{\underset{\displaystyle H}{|}}{N}}-\overset{\overset{\displaystyle H}{|}}{\underset{\underset{\displaystyle H}{|}}{C}}-\overset{\overset{\displaystyle O}{\|}}{C}-\overset{\overset{\displaystyle H}{|}}{N}-\overset{\overset{\displaystyle H}{|}}{\underset{\underset{\displaystyle CH_3}{|}}{C}}-\overset{\overset{\displaystyle O}{\|}}{C}-OH$$
Water   **A dipeptide**

This process can be repeated by linking more amino acid molecules to each end of such a dipeptide, eventually forming a protein. **Proteins** are extremely long-chain polyamides formed by the enzyme-catalyzed condensation of amino acids under the direction of nucleic acids in the cell (or the biochemist in the lab). Their formula masses range from a few thousand (insulin, 6000 u) up to millions for the most complex (hemocyamine, 9 million u). Proteins function in living organisms both as structural components such as muscle fiber, hair, and feathers, and as enzymes (biological catalysts).

The Spotlight feature on page 371 summarizes the names and general formulas of the derivatives of the hydrocarbons we have studied.

RELEVANCE QUESTION: *While shopping in the grocery store, you pass a rack of fresh pineapples. As you savor the aroma, your thoughts turn immediately to which class of hydrocarbon derivatives?*

14.5 Synthetic Polymers

LEARNING GOALS

▼ Explain the importance of polymers.

▼ Show how addition and condensation polymers are formed.

Chemists have long tried to duplicate the compounds of nature. As the science of chemistry progressed and formulas and basic components became known, chemists were able to synthesize some of these natural compounds by reactions involving the appropriate elements or compounds. During this early trial-and-error period, there were probably as many serendipitous (accidental, but fortunate) as deliberate discoveries.

Attempts to synthesize natural compounds led to the discovery of **synthetics**—materials whose molecules have no duplicates in nature. The first synthetic was a polymer prepared by Leo Baekeland in 1907 and commercially known as Bakelite, a common electrical insulator.

Baekeland's discovery triggered serious efforts to prepare synthetic materials. Chemists became aware that substituting different atoms or groups in a molecule would change its properties. For example, substituting a chlorine atom for a hydrogen atom in ethane produces chloroethane, which has very different properties from ethane. By knowing the general properties of the substituted groups, a chemist often can tailor a molecule to satisfy a given requirement.

As a result of this scientific approach, multitudes of synthetic compounds have been constructed. Probably the best known of these is the group of synthetic polymers that can be molded and hardened—the *plastics*. Plastics have become an integral part of our modern life, being used in clothing, shoes, buildings, autos, sports equipment, art, electrical appliances, toothbrushes, toys, and myriad other things.

Some plastics are used as heat shields for space vehicle reentry, where the temperatures are in excess of 8000°C. These materials transmit virtually no heat, because slow, layer-by-layer decomposition of their molecules uses excess heat as the latent heat of vaporization (Section 5.3). Let's now examine how, in general, polymers are formed and discuss a few of the more common ones.

Molecules containing large numbers of atoms and having exceedingly high formula masses are often made up of repeating units of smaller molecules that have bonded together to form long, chain-like structures. The fundamental repeating unit is termed the **monomer,** and the long chain made up of the repeating units is called the **polymer.**

The two major types of polymers are the addition polymers and the condensation polymers. **Addition polymers are those formed when molecules of an alkene monomer add to one another.** Under proper reaction conditions, often including a catalyst, one bond of the monomer's double bond opens up. This allows the monomer to attach itself by single bonds to two other monomer molecules, and then each end attaches to another monomer, and so on and on. The polymerization of ethylene (ethene) to polyethylene is illustrated below, where the subscripted n means that the unit shown in the brackets is repeated thousands of times, as thousands of the monomer molecules join. (The atoms that eventually terminate the polymer are not usually shown.)

Ethylene (ethene)   **Polyethylene**

Polyethylene is the simplest synthetic polymer. Because of its chemical inertness, it is used for chemical storage containers and many other packaging applications (● Fig. 14.18). Milk jugs and some other containers are made of high-density polyethylene and are stamped with a 2 inside a triangle. This number indicates that the plastic container can and should be recycled. On the other hand, low-density polyethylene, which is used for such items as trash and laundry bags, bears the number 4 and is rarely recycled.

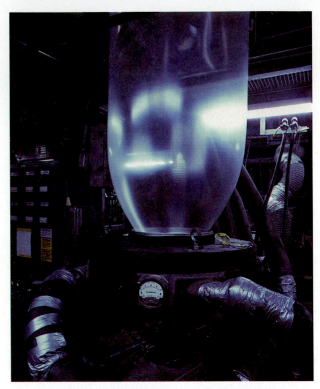

FIGURE 14.18   The Formation of Polyethylene
Polyethylene and similar polymers are formed from large continuous tubes blown from the hot liquid polymer.

Another common polymer that has found many uses, especially in coating cooking utensils, is Teflon, a hard, strong, chemically resistant fluorocarbon resin with a high melting point and low surface friction (● Fig. 14.19). Its monomer (tetrafluoroethene) and the polymer structure are

$$\underset{\textbf{Tetrafluoroethene}}{\overset{F}{\underset{F}{C}}=\overset{F}{\underset{F}{C}}} \xrightarrow{\text{Catalyst}} \underset{\textbf{Teflon}}{\left[\begin{array}{c} F\ \ F \\ -C-C- \\ F\ \ F \end{array}\right]_{n}}$$

Table 14.8 on page 372 shows the names, monomers, and uses of some common addition polymers.

EXAMPLE 14.7

Drawing the Structure of an Addition Polymer

An addition polymer can be prepared from vinylidene chloride, $CH_2=CCl_2$. Draw the structure of the polymer.

SOLUTION

In forming the polymer, one bond of the double bond opens and adds to another molecule of the monomer on each end. This happens repeatedly as more and more monomers add to each side of the growing chain, so the polymer's structure is shown as

$$\left[\begin{array}{c} H\ \ Cl \\ -C-C- \\ H\ \ Cl \end{array}\right]_{n}$$

CONFIDENCE EXERCISE 14.7

Draw the structure of the monomer from which this addition polymer was made.

$$\left[\begin{array}{c} H\ \ Cl \\ -C-C- \\ H\ \ H \end{array}\right]_{n}$$

Condensation polymers are constructed from molecules that have two or more reactive groups. Generally, one molecule attaches to another by an ester or amide linkage. Water is the other product, hence the name *condensation polymer*. Of course, if a

FIGURE 14.19   Teflon Has Many Uses
Teflon-coated frying pans are noted for their nonstick surfaces.

A SPOTLIGHT ON: Hydrocarbons and Their Derivatives

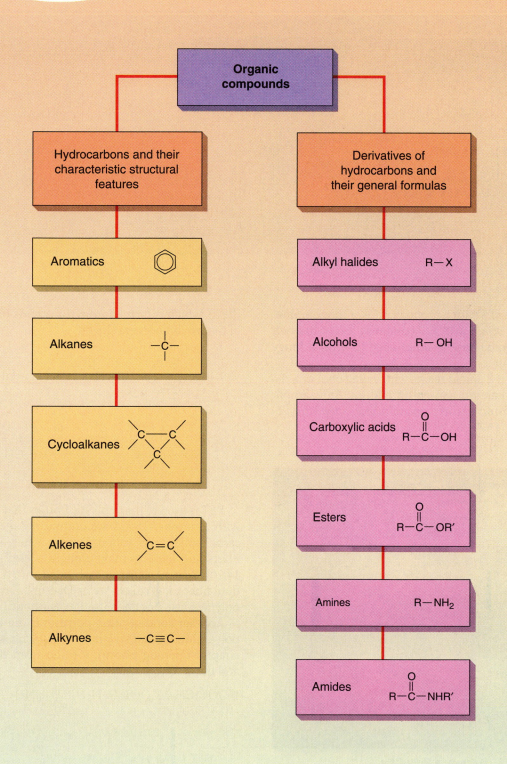

TABLE 14.8   Some Addition Polymers

Polymer	Monomer	Uses
Polyethylene	$CH_2{=}CH_2$	Bottles, plastic tubing
Polypropylene	$CH_2{=}CH$ with CH_3	Bottles, carpeting, textiles
Polytetrafluoroethylene (Teflon)	$CF_2{=}CF_2$	Nonstick surface for frying pans
Poly(vinyl chloride) (PVC)	$CH_2{=}CH$ with Cl	Plastic pipes, floor tile
Polyacrylonitrile (Orlon, Acrilan)	$CH_2{=}CH$ with CN	Carpets, textiles
Polystyrene	$CH_2{=}CH$ with C_6H_5	Plastic foam insulation, cups

monoacid reacts with a monoalcohol or monoamine, the reaction stops with the condensation of the two molecules, and there is no chance to form a long-chain polymer.

However, if a diacid reacts with a dialcohol or a diamine, the reaction can go on and on. An example is the polyester named *polyethylene terephthalate*

(PET), which is formed from thousands of molecules of the two monomers shown.

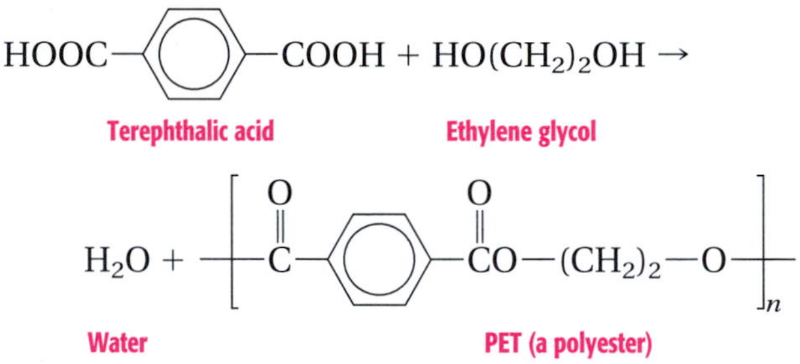

$$HOOC{-}\!\!\bigcirc\!\!{-}COOH + HO(CH_2)_2OH \rightarrow$$

Terephthalic acid   **Ethylene glycol**

$$H_2O + \left[\begin{matrix} O \\ \| \\ C \end{matrix} \!\!-\!\!\bigcirc\!\!-\!\! \begin{matrix} O \\ \| \\ CO \end{matrix} \!\!-\!\!(CH_2)_2\!\!-\!\!O \right]_n$$

Water   **PET (a polyester)**

PET is used to make plastic bottles that are marked with a number 1 and are commonly recycled. This same polymer is called *Dacron* when drawn into fibers and used to make polyester clothing (● Fig. 14.20). When Dacron is fashioned into a film rather than fibers, it is called *Mylar*. Dacron, a truly versatile polymer, also has been used in synthetic heart valves.

Another condensation polymer is the widely used polyamide named *nylon* (● Fig. 14.21; also see the chapter-opening photo), formed from the diamine and the diacid shown.

$$HOOC(CH_2)_4COOH + H_2N(CH_2)_6NH_2 \rightarrow$$

Adipic acid   **Hexamethylenediamine**

$$H_2O + \left[\begin{matrix} O \\ \| \\ C \end{matrix}(CH_2)_4\!\!\begin{matrix} O \\ \| \\ C \end{matrix}\!\!-\!\!\begin{matrix} H \\ | \\ N \end{matrix}(CH_2)_6\!\!\begin{matrix} H \\ | \\ N \end{matrix} \right]_n$$

Water   **Nylon (a polyamide)**

FIGURE 14.20   Dacron, a Polyester

Fibers of Dacron are shown being spun from the reaction vessels.

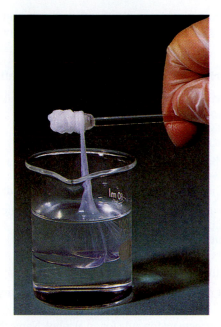

FIGURE 14.21   Nylon, a Polyamide, Was First Synthesized in 1935 at DuPont

Here a strand of nylon is being drawn from the interface (boundary) of the two reactants, where the reaction occurs.

Velcro is made of two nylon strips, one having thick loops that are slit open to form "hooks" and the other having thin, closed loops that entangle the slit fibers when the sides are pressed together (● Fig. 14.22).

The inventor of Velcro took his idea from noticing how cockleburrs clung to his clothing when he walked through a field.

RELEVANCE QUESTION: What addition or condensation polymer did you use today?

FIGURE 14.22   A Scanning Electron Micrograph of Velcro

The "hooks" of one nylon surface entangle with the loops of the other.

Important Terms

organic chemistry
hydrocarbons (14.2)
aromatic hydrocarbon
carcinogen
aliphatic
  hydrocarbon (14.3)
alkanes
structural formula
alkyl group

constitutional isomers
cycloalkanes
alkenes
alkynes
functional group (14.4)
alkyl halide
CFCs
alcohols
hydroxyl group

carbohydrates
amine
amino group
carboxylic acids
carboxyl group
ester
fats
soap
synthetic detergents

amides
amino acid
proteins
synthetics (14.5)
monomer
polymer
addition polymers
condensation polymers

Review Questions

1. Distinguish between organic chemistry and biochemistry.

14.1 Bonding in Organic Compounds

2. How many covalent bonds does a carbon atom form?
   (a) 1   (b) 2   (c) 3   (d) 4

3. Tell the number of covalent bonds formed by an atom of each of these common elements in organic compounds: C, H, O, S, N, a halogen.

14.2 Aromatic Hydrocarbons

4. Which of the following is the most common aromatic compound?
   (a) Benzedrine   (c) butane
   (b) ethylene   (d) benzene

5. A *carcinogen*
   (a) fights bacteria.   (c) ripens fruit.
   (b) depletes ozone.   (d) causes cancer.

6. Is benzene a solid, liquid, or gas? What is its source? Give its molecular formula and the preferred representation of its structural formula.

14.3 Aliphatic Hydrocarbons

7. C_6H_{14} is the molecular formula for which of the following hydrocarbons?
   (a) hexane   (c) hexyne
   (b) hexene   (d) benzene

8. Which of the following is an ethyl group?
   (a) CH_3—
   (b) CH_3CH_2—
   (c) CH_2CH_3—
   (d) none of the previous answers

9. What structural feature distinguishes an aromatic hydrocarbon from an aliphatic hydrocarbon?

10. Give the general molecular formulas for alkanes, cycloalkanes, alkenes, and alkynes. Write the structural formula for one example of each of these classes.

11. Name and give the molecular formulas for the first eight members of the alkane series. Which one is the principal component of natural gas?

12. Describe the geometry and bond angles when a carbon atom forms four single bonds.

13. What term is applied to two or more compounds having the same molecular formula but different structural formulas?

14. When drawing a structure from the name of an alkane or alkane derivative, what part of the name should you look at first?

15. Use both full and condensed structural formulas to show the difference between methane and a methyl group, and ethane and an ethyl group.

16. In a structural formula, what does R stand for?

17. Name the compound represented by a square, and give its molecular formula and full structural formula.

18. Both ethene and ethyne are often called by their more common names. Tell the common names, and draw the structural formula for each compound.

19. Distinguish between saturated and unsaturated hydrocarbons.

20. Describe an addition reaction that alkenes and alkynes undergo. Why can't alkanes undergo addition reactions?

14.4 Derivatives of Hydrocarbons (and Highlight)

21. Which of the following is the general formula for an alcohol?
    (a) RX   (c) ROH
    (b) ROR   (d) RCOOH

22. A carboxylic acid and an amine react to give water and which of the following?
    (a) ether   (b) ester   (c) alkyl group   (d) amide

23. Any compound that can produce a physiologic change in humans or animals is called
    (a) an antibiotic.   (c) a drug.
    (b) an alkaloid.   (d) an amphetamine.

24. Give the general formula for an alkyl halide.

25. What does CFC stand for? What is the primary use of CFCs? Why is their use a problem?

26. Give the general formula for an alcohol. Name the characteristic group it contains. What suffix is used in the IUPAC names of alcohols?

27. Give the general formula for an amine. Name the characteristic group it contains. What property makes the simple amines unpopular?

28. Give the general formula for a carboxylic acid. Name the characteristic group it contains. Name and give the structural formula for the carboxylic acid found in vinegar.

29. Give the general formula for an ester. Why are esters popular? Name the ester found in wintergreen mints.

30. What two classes of organic compounds react to form esters? What is the other product? Use general formulas to show how the reaction takes place.

31. What two classes of organic compounds react to form amides? What is the other product? Use general formulas to show how the reaction takes place.

32. Name the type of biochemical compound that (a) contains an abundance of hydroxyl groups, (b) is a triester of glycerol, (c) has multiple amide linkages.

33. What two simpler sugars combine to form sucrose? Which sugar is the monomer of both starch and cellulose? Why can herbivores digest cellulose but humans cannot?

34. Name the monomers of proteins. Name and write the structural formula for the simplest one of these monomers.

35. What are the basic differences and similarities in structure and physical properties of fats and oils? How is an oil converted to a fat?

36. What is the relationship between a fat and a soap, and between a soap and a synthetic detergent?

14.5 Synthetic Polymers

37. Which of the following is an addition polymer?
    (a) nylon   (c) Dacron
    (b) Teflon   (d) Velcro

38. Which of the following is a condensation polymer?
    (a) nylon   (c) Styrofoam
    (b) Teflon   (d) polyethylene

39. Describe from a structural standpoint the basic difference in the method of formation of addition and condensation polymers.

40. Name a well-known synthetic fiber that is a polyester and one that is a polyamide. Name two addition polymers.

Applying Your Knowledge

1. Although any life elsewhere in the universe is probably based on carbon, science fiction writers have speculated that it could be based on another element with similar bonding properties. What element do you think they chose?

2. You overhear someone comment that a lot of "cat cracking" takes place in Louisiana. Should you call the Society for Prevention of Cruelty to Animals?

3. Your friend tells you that he has read that something called *ethylene* is used to ripen fruit and vegetables. Impress him by drawing the structural formula for the compound.

4. Why is celery such a good snack food for someone on a diet? (*Hint:* "CEL-ery.")

5. Which class of compounds discussed in this chapter do you think the first *vitamins* belonged to?

6. If you get tears in your eyes when you peel onions, the water-soluble compound thiopropanal S-oxide (CH_3CH_2CHSO) is the culprit. What simple strategy would allow you to peel onions without irritating your eyes?

Exercises

14.1 Bonding in Organic Compounds

1. Which of these structural formulas is valid, and which is incorrect?

Answer: (a) is valid. (b) is incorrect because Cl should have one bond, not two.

2. Which of these structural formulas is valid, and which is incorrect?

14.2 Aromatic Hydrocarbons

3. Draw the structural formula for 1-bromo-2-methylbenzene.

Answer:

4. Draw the structural formula for 1,3,5-trifluorobenzene.

14.3 Aliphatic Hydrocarbons

5. Classify each of the following hydrocarbon structural formulas as an alkane, cycloalkane, alkene, alkyne, or aromatic.

   (a)

   (b)

   (c) $CH_3-C\equiv C-CH_3$

   (d) $CH_3CH_2CH_3$

   (e)

   Answer: (a) alkene (b) cycloalkane (c) alkyne (d) alkane (e) aromatic

7. Use each of the following names to classify each hydrocarbon as an alkane, cycloalkane, alkene, alkyne, or aromatic.
   (a) 2-methylbutane
   (b) 3-methyl-1-pentyne
   (c) 1,1-dimethylcyclobutane
   (d) 3-octene
   (e) 1,3-dimethylbenzene
   Answer: (a) alkane (b) alkyne (c) cycloalkane (d) alkene (e) aromatic

9. State whether the structural formulas shown in each case represent the *same compound*, or are *constitutional isomers*, or are *neither*.
   (a) $CH_3CH_2CH_3$ and

   (b) CH_3CH_2OH and $CH_3CH_2CH_2OH$
   (c) $CH_2=CHCH_3$ and

   Answer: (a) same compound (b) neither (c) isomers

11. Given the IUPAC name, draw the structural formula for each alkyl halide.
    (a) 1,1-dibromo-2-fluorobutane
    (b) 1-chloro-2-ethylcyclopentane

6. Classify each of the following hydrocarbon structural formulas as an alkane, cycloalkane, alkene, alkyne, or aromatic.

   (a) $CH_3CH_2CH_2CH_2CH_3$

   (b)

   (c)

   (d)

   (e) $H-C\equiv C-CH_2CH_3$

8. Use each of the following names to classify each hydrocarbon as an alkane, cycloalkane, alkene, alkyne, or aromatic.
   (a) 2-methyl-2-hexene
   (b) methylcyclohexane
   (c) 3,3,4-triethylhexane
   (d) ethylbenzene
   (e) 2-heptyne

10. State whether the structural formulas shown in each case represent the *same compound*, or are *constitutional isomers*, or are *neither*.
    (a) $CH_3CH_2NH_2$ and $CH_3NHCH_2CH_3$
    (b) $HOOCCH_2CH_3$ and CH_3CH_2COOH
    (c) CH_3CHCH_3 and $CH_3CH_2CH_2OH$

12. Given the IUPAC name, draw the structural formula for each alkyl halide.
    (a) 1-bromo-3-chlorobenzene
    (b) 1,1-dichlorocyclobutane

Answer: (a)

(b)

13. Two constitutional isomers of continuous-chain butenes exist: 1-butene and 2-butene. How many constitutional isomers of continuous-chain pentenes exist? Name each and draw its structural formula.

Answer: 1-pentene

$$\underset{H}{\overset{H}{}}\!C\!=\!C\underset{CH_2CH_2CH_3}{\overset{H}{}}$$

and 2-pentene

$$\underset{H}{\overset{CH_3}{}}\!C\!=\!C\underset{H}{\overset{CH_2CH_3}{}}$$

(3-Pentene would be the same as 2-pentene, and 4-pentene the same as 1-pentene.)

14. How many constitutional isomers of continuous-chain hexenes exist? Name each and draw its structural formula.

15. Complete the equation and name the product.

$$\underset{H}{\overset{H}{}}\!C\!=\!C\underset{H}{\overset{CH_3}{}}\; +\; H_2 \xrightarrow{\;Pt\;}$$

Answer:

$$H-\overset{\overset{\displaystyle H}{|}}{C}-\overset{\overset{\displaystyle H}{|}}{C}-\overset{\overset{\displaystyle H}{|}}{C}-H;\ propane$$

16. Complete the equation and name the product.

$$CH_3-C\equiv C-CH_3 + 2\,H_2 \xrightarrow{\;Pt\;}$$

14.4 Derivatives of Hydrocarbons

17. Identify each structural formula as belonging to an alkyl halide, alcohol, amine, carboxylic acid, ester, or amide.
    (a) $CH_3CH_2NH_2$

    (b) $CH_3CH_2\overset{\overset{\displaystyle O}{\|}}{C}-\overset{\overset{\displaystyle H}{|}}{N}-CH_3$

    (c) $CH_3CH_2\overset{\overset{\displaystyle O}{\|}}{C}-O-\triangleleft$

    (d) $CH_3CH_2\overset{\overset{\displaystyle O}{\|}}{C}-O-H$

    (e) $CH_3\underset{\underset{\displaystyle Cl}{|}}{C}HCH_3$

    (f) $CH_3\underset{\underset{\displaystyle OH}{|}}{C}HCH_2CH_3$

    Answer: (a) amine (b) amide (c) ester (d) carboxylic acid (e) alkyl halide (f) alcohol

18. Identify each structural formula as belonging to an alkyl halide, alcohol, amine, carboxylic acid, ester, or amide.
    (a) CH_3CH_2COOH

    (b) $CH_3\overset{\overset{\displaystyle O}{\|}}{C}-O-CH_2CH_3$

    (c) $CH_3CH_2CH_2OH$

    (d) CF_3CF_3

    (e) $CH_3CH_2NH_2$

    (f) $CH_3\overset{\overset{\displaystyle O}{\|}}{C}-\overset{\overset{\displaystyle H}{|}}{N}-CH_3$

19. Draw the constitutional isomers for each of the following: (a) C_2H_6O (two) and (b) C_3H_7Cl (two).

> Answer: (a) CH_3CH_2—OH and CH_3—O—CH_3
> (b) $CH_3CH_2CH_2Cl$ and $CH_3CH(Cl)CH_3$

20. Draw the constitutional isomers for each of the following: (a) C_2H_7N (two) and (b) C_5H_{12} (three).

21. Complete the equations.

(a) $CH_3CH_2\overset{\displaystyle O}{\overset{\|}{C}}$—O—H + H—$OCH_2CH_3$ $\xrightarrow{H_2SO_4}$

(b) (benzene ring)—$CH_2\overset{\displaystyle O}{\overset{\|}{C}}$—O—H + H—$\overset{\displaystyle H}{\overset{|}{N}}$—$CH_3$ $\longrightarrow$

> Answer: (a) $CH_3CH_2\overset{\displaystyle O}{\overset{\|}{C}}$—O—$CH_2CH_3$ + H_2O
>
> (b) (benzene ring)—$CH_2\overset{\displaystyle O}{\overset{\|}{C}}$—$\overset{\displaystyle H}{\overset{|}{N}}$—$CH_3$ + H_2O

22. Complete the equations.

(a) $CH_3\underset{\underset{\displaystyle CH_3}{|}}{CH}\overset{\displaystyle O}{\overset{\|}{C}}$—OH + H—$OCH_2$—(benzene ring) $\xrightarrow{H_2SO_4}$

(b) $CH_2\overset{\displaystyle O}{\overset{\|}{C}}$—OH + H—$\overset{\displaystyle H}{\overset{|}{N}}$—$CH_2CH_2$—(cyclopentane ring) $\longrightarrow$
    with F attached below CH_2

14.5 Synthetic Polymers

23. Polystyrene, or Styrofoam, is an addition polymer made from the monomer styrene. Show by means of an equation how styrene polymerizes to polystyrene.

> Answer:

24. Acrilan is an addition polymer made from the monomer acrylonitrile (cyanoethene). Show by means of an equation how acrylonitrile polymerizes to Acrilan.

25. Draw the structure of a portion of the chain of Kevlar, used in bulletproof vests, which is made by the condensation polymerization of the monomers shown.

H_2N—(benzene ring)—NH_2 and HO—$\overset{\displaystyle O}{\overset{\|}{C}}$—(benzene ring)—$\overset{\displaystyle O}{\overset{\|}{C}}$—OH

> Answer:
> $\left[\text{HN—(benzene ring)—NH—}\overset{\displaystyle O}{\overset{\|}{C}}\text{—(benzene ring)—}\overset{\displaystyle O}{\overset{\|}{C}}\right]_n$

26. The polyester formed from lactic acid (shown below) is used for tissue implants and surgical sutures that will dissolve in the body. Draw the structure of a portion of this polymer.

HO—$\underset{\underset{\displaystyle CH_3}{|}}{CH}$—$\overset{\displaystyle O}{\overset{\|}{C}}$—OH

Solutions to Confidence Exercises

14.1 Five of the carbons have only three bonds each and need four. The two oxygens have only one bond each and need two. One nitrogen has only two bonds and needs three. (Four double bonds were omitted.)

14.2 Attach a chlorine atom to the benzene ring. That carbon automatically becomes the 1 position. The carbon next to it must now be the 2 position, so attach a fluorine atom there. The final structure is

14.3 The compound 2,2,4-trimethylpentane has five carbons in a chain connected by single bonds. Two methyl groups are attached to carbon 2, and one methyl group is on carbon 4. Hydrogen atoms are at the end of all remaining bonds necessary to give each carbon atom four bonds. The final structure is

14.4 When the three carbon atoms are connected by single bonds, enough bonds remain to connect eight singly bonded atoms. Seven hydrogen atoms and one fluorine atom fill that requirement, so the only question is how can the fluorine atom be attached to give the two isomers? The result is 1-fluoropropane and 2-fluoropropane, as shown. (Putting the F on one end of the

chain rather than on the other would make no difference; that is, "3-fluoropropane" is really 1-fluoropropane.)

1-Fluoropropane   **2-Fluoropropane**

14.5 "Lasso" the —OH from the acid and the H from the alcohol to form water, then connect the remaining fragments by the bond from the alcohol's oxygen atom to obtain the products shown.

14.6 "Lasso" the —OH from the acid and the H from the amine to form water, then connect the remaining fragments by the bond from the amine's nitrogen atom to obtain the products shown.

14.7 The monomer must be the alkene that corresponds to the basic unit shown in the polymer. That is, the monomer must have the structure

Answers to Multiple-Choice Review Questions

2. d   5. d   8. b   22. d   37. b
4. d   7. a   21. c   23. c   38. a

THE SOLAR SYSTEM

15

In questions of science the authority of a thousand is not worth the humble reasoning of a single individual.

Galileo Galilei (1564–1642)

Astronomy is defined as the scientific study of the universe beyond Earth's atmosphere. The **universe** is everything—that is, all energy, matter, and space. Our awareness of the entire universe is limited, but we do know that our very existence is dependent on the most important object beyond Earth's atmosphere—the Sun. Our ancestors worshiped this light- and heat-giving object that rises in the east and sets in the west each and every day.

The Sun supplies the energy for the existence and maintenance of nearly all living organisms on planet Earth.* Solar energy influences Earth's weather, which, in turn, influences our minds and bodies as we adapt to the daily and seasonal changes. Thus the chemical and physical properties of the Sun are a major field of study for astronomers.

Primitive humans were certainly aware of the rising and setting of the Sun and the changing face of the Moon. As civilizations developed, solar observations showed that the *year* lasted about 365 days, and the 29.5-day cycle of the Moon became known as the *month*. These two facts are relevant in our daily

*Living organisms have been found more than two kilometers below Earth's surface. They survive in a hot environment where sunlight, oxygen, and water are absent.

Photo: A photomontage of the terrestrial planets, the Jovian planets, and Earth's Moon.

lives. We celebrate birthdays every year and pay our bills monthly.

Early civilizations used celestial observations to construct a calendar that provided the correct time for planting and harvesting agricultural crops and determined the dates for religious holidays. One of the best-known primitive structures, believed to have been used for astronomical observations, is Stonehenge, located on Salisbury Plain about 120 km west of London, England. Construction of Stonehenge started about 2800 B.C. and was completed some 700 years later. The structure provided a method for determining the time of sunrise and moonrise at certain times of the year.

The science of astronomy has a long history of human effort and experiences and has always played a major role in human development. Before the nineteenth century, the history of science was, for the most part, the history of astronomy. Today, as the twenty-first century begins, astronomy is advancing at a tremendous rate. Engineering technology has improved to the level where scientists and engineers are constructing huge, land-based telescopes (with mirrors 10 meters in diameter) with sophisticated electronic detectors interfaced with computers that store, analyze, and enhance the received images.

Most electromagnetic radiation coming from outer space is absorbed by Earth's atmosphere and fails to reach ground-based detectors. Only optical and radio waves, plus some radiation in the short wavelength region of the infrared spectrum, are able to pass through Earth's atmosphere. Other regions of the electromagnetic spectrum (the far infrared, ultraviolet, X-ray, and gamma ray) are detected by space-based instruments.

The new technology is providing astronomers with a wealth of data concerning the planets, stars, galaxies, and other celestial bodies detected in the vast volume of the observable universe. From these data, astronomers have concluded that planet Earth is but one of nine planets orbiting an average-size star located in a vast group of stars called the Milky Way Galaxy. The Milky Way is one of about 50 billion galaxies grouped in clusters and superclusters scattered throughout an enormous volume of space that presently has no observable end.

Chapters 15–18 present fundamental concepts relative to astronomy that will expand your point of view of the universe and aid you in thinking about your place and time in the world of which you are a part. ◼

15.1 The Solar System: An Overview

LEARNING GOALS

▼ Name some early astronomers, and outline some historical theories of the solar system.

▼ State and explain Kepler's laws of planetary motion.

▼ Describe the composition, structure, and motions of the planets.

The **solar system** is a complex system of moving masses held together by gravitational forces. At the center of this system is a star called the Sun, which is the dominant mass. Revolving around the Sun are nine rotating planets, over 60 satellites (moons), thousands of asteroids, vast numbers of comets, meteoroids, interplanetary dust particles, gases, and a solar wind composed of charged particles.

The rotating and revolving motions of planet Earth were concepts not readily accepted at first. In early times most people were convinced that Earth was motionless and that the Sun, Moon, planets, and stars revolved around Earth, which was considered the center of the universe. This concept, or model, of the solar system is called the Earth-centered, or **geocentric, model.** Its greatest proponent was Claudius Ptolemy, about A.D. 140.

Nicolaus Copernicus (1473–1543), a Polish astronomer (● Fig. 15.1), developed the theory of the

FIGURE 15.1 Nicolaus Copernicus

Copernicus was a Polish astronomer, gifted in mathematics, who developed the heliocentric (Sun-centered) theory of the solar system.

Sun-centered model, or **heliocentric model,** of the solar system. Although he did not prove that Earth revolves around the Sun, he did provide mathematical proofs that could be used to predict future positions of the planets.

After the death of Copernicus in 1543, the study of astronomy was continued and developed by several astronomers, three of whom made their appearance in the last half of the sixteenth century. Notable among these was the Danish astronomer Tycho Brahe ("BRAH-uh") (1546–1601), who built an observatory on the island of Hven near Copenhagen and spent most of his life observing and studying the stars and planets (● Fig. 15.2). Brahe is considered the greatest practical astronomer since the Greeks. His measurements of the planets and stars, all made with the unaided eye (the telescope had not been invented),

FIGURE 15.3   Johannes Kepler (1571–1630)
Kepler, a German mathematician and astronomer, formulated the basic quantitative laws that describe planetary motion.

proved to be more accurate than any made previously. Brahe's data, published in 1603, were edited by his colleague Johannes Kepler (1571–1630) (● Fig. 15.3), a German mathematician and astronomer who had joined Brahe during the last year of his life. After Brahe's death, his lifetime of observations were at Kepler's disposal and provided him with the data necessary for the formulation of three laws known today as Kepler's laws of planetary motion.

Kepler was interested in the irregular motion of the planet Mars. He spent considerable time and energy before coming to the conclusion that the uniform circular orbit proposed by Copernicus was not a true representation of the observed facts. Perhaps because he was a mathematician, he saw a simple type of geometric figure that would fit the observed motions of Mars and the other known planets.

Kepler's first law, known as the **law of elliptical paths,** states

All planets move in elliptical paths around the Sun with the Sun at one focus of the ellipse.

An ellipse is a figure that is symmetrical about two unequal axes (● Fig. 15.4). An ellipse can be drawn by

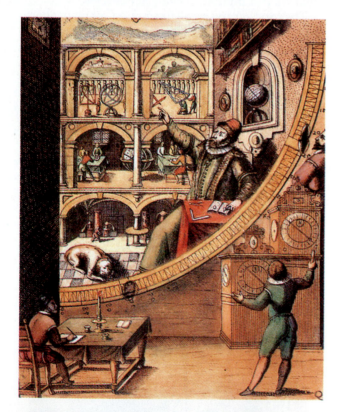

FIGURE 15.2   Tycho Brahe in His Observatory
The Danish astronomer is known for his very accurate observations, made with the unaided eye, of the positions of stars and planets. (The telescope had not been invented.) The instrument shown, a large quadrant, was used to make these measurements. Light passes through a small window (upper left in photo) and onto the quadrant.

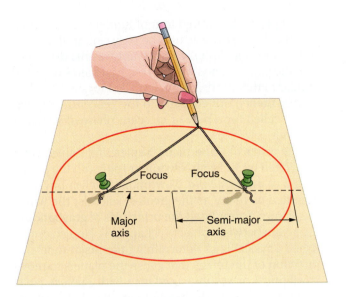

FIGURE 15.4   Drawing an Ellipse

An ellipse can be drawn using two thumbtacks, a loop of string, a pencil, and a sheet of paper.

using two thumbtacks, a closed piece of string, paper, and pencil. The points where the two tacks are positioned are called the *foci* of the ellipse. The longer axis, which passes through both foci, is called the *major axis.* Half the major axis is called the *semimajor axis.* When dealing with Earth's elliptical orbit, the semimajor axis is also the average distance between Earth and the Sun.

When speaking or writing about distances in the solar system, we use the **astronomical unit** (AU), the average distance between Earth and the Sun. One AU is 1.5×10^8 km. (The astronomical unit is illustrated in Fig. 15.11, on page 391.)

Kepler's first law gives the shape of the orbit but fails to predict when the planet will be at a particular position in the orbit. Kepler, aware of this, set about to find a solution from the mountain of data he had at his disposal. After a tremendous amount of work, he discovered what is now known as Kepler's second law, the **law of equal areas,** which states

An imaginary line (radial vector) joining a planet to the Sun sweeps out equal areas in equal periods of time.

As illustrated in ● Fig. 15.5, the speed of a revolving planet will be greatest when the planet is closest to

the Sun. This closest point in its orbit is called *perihelion* ("per-i-HEE-lee-on") and occurs for Earth about January 4. The speed of a planet is slowest when it is farthest from the Sun. This point in its orbit is called *aphelion* ("a-FEE-lee-on") and occurs about July 5 for Earth.

After the publication of his first two laws in 1609, Kepler began a search for a relationship among the motions of the different planets and an explanation to account for these motions. Ten years later he published *De Harmonice Mundi (Harmony of the Worlds),* in which he stated his third law of planetary motion, known as the **harmonic law:**

The square of the sidereal period of a planet is proportional to the cube of its semimajor axis (one-half the major axis).

This can be written as

$$(\text{period})^2 \propto (\text{semimajor axis})^3$$

or, in equation form,

$$T^2 = kR^3 \qquad \textbf{15.1}$$

where T = the sidereal period (time of one revolution with respect to a star)

R = the length of the semimajor axis

k = a constant of proportionality (the same value for all planets)

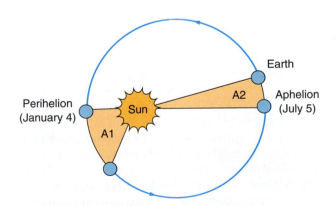

FIGURE 15.5   Kepler's Law of Equal Areas

An imaginary line joining a planet to the Sun sweeps out equal areas in equal periods of time. Area A1 equals area A2. Earth has a greater orbital speed in January than in July.

If the sidereal period of the planet is measured in years and the semimajor axis in astronomical units, k is equal to 1 y^2/AU^3.

EXAMPLE 15.1

Calculating the Period of a Planet

Calculate the period of a planet whose orbit has a semimajor axis of 1.52 AU.

SOLUTION

Step 1

Use Eq. 15.1 and substitute in the values for k and R.

$$T^2 = kR^3 = \frac{1\ y^2}{AU^3} \times \frac{(1.52\ AU)^3}{1}$$

Step 2

Cube 1.52 AU and cancel the AU^3s.

$$T^2 = 3.51\ y^2$$

Step 3

Take the square root of both sides.

$$T = 1.87\ y$$

CONFIDENCE EXERCISE 15.1

Calculate the period of a planet whose orbit has a semimajor axis of 30 AU.

Galileo Galilei (1564–1642), Italian astronomer, mathematician, and physicist who is usually called just Galileo, was one of the greatest scientists of all time. (See the Highlight on page 32.) The most important of his many contributions to science were in the field of mechanics. He originated the basic idea for the formulation of Newton's first law of motion (Chapter 3), and he founded the modern experimental approach to scientific knowledge (Chapter 1). The motion of objects (Section 2.3), especially the planets, was of prime interest to Galileo. His concepts of motion and the forces that produce motion opened up an entirely new approach to astronomy. In this field he is noted for his contributions to the heliocentric theory of the solar system.

In 1609 Galileo became the first person to observe the Moon and planets through a telescope. With the telescope, he discovered four of Jupiter's moons, thus proving that Earth was not the only center of motion in the universe. Equally important was his discovery that the planet Venus went through a change of phase similar to that of the Moon, as called for by the heliocentric theory, but contrary to the geocentric theory, which called for a new or crescent phase of Venus at all times.

The works of Copernicus, Kepler, and Galileo were integrated by Sir Isaac Newton in 1687 with the publication of the *Principia*. Newton, an English physicist regarded by many as the greatest scientist the world has known, formulated the principles of gravitational attraction between objects (Chapter 3). He also established physical laws determining the magnitude and direction of the forces that cause the planets to move in elliptical orbits in accordance with Kepler's laws. Newton invented calculus and used it to help explain Kepler's first law. He also used the law of conservation of angular momentum (Section 3.5) to explain Kepler's second law. Newton's explanations of Kepler's laws unified the heliocentric theory of the solar system and brought an end to the confusion.

Newton's correction to Kepler's third law contains as a factor the sum of the masses of the two revolving bodies and the quantity $4\pi^2/G$, which is a value for the proportionality constant in Eq. 15.1. In symbol notation, Newton's correction of Kepler's third law is

$$T^2 = \left[\frac{4\pi^2}{(m_1 + m_2)G}\right]R^3$$

or $\qquad (m_1 + m_2)\dfrac{T^2}{R^3} = \dfrac{4\pi^2}{G}$ **15.2**

where
$\quad T$ = orbital period of a planet

$\quad R$ = distance between the planet and the Sun

$\quad G$ = gravitational constant ($= 6.67 \times 10^{-11}$ N-m^2/kg^2)

$\quad m_1$ = mass of one body

$\quad m_2$ = mass of the second body

This law can be used if the two bodies revolve mutually about each other. In Eq. 15.2, if we let m_2 be the mass of the Sun and m_1 be the mass of Earth and then neglect the mass of Earth (since the Sun's mass is 333,000 times greater than Earth's), we can obtain a good approximation of the Sun's mass.

EXAMPLE 15.2

Calculating the Mass of the Sun

Calculate the mass of the Sun, in kilograms, using Eq. 15.2. Use Earth's orbital period (T) and average distance (R) from the Sun.

SOLUTION

Step 1

Leave out m_1, and rearrange Eq. 15.2 to solve for the mass of the Sun.

$$m_{Sun} = \frac{4\pi^2 R^3}{GT^2}$$

Step 2

Substitute in the numerical values for G, R, and T. Since the mass of the Sun is to be in kilograms, SI units must be used.

$$m_{Sun} = \frac{4 \times (3.14)^2 \times (1.5 \times 10^{11}\ \text{m})^3}{\left(6.67 \times 10^{-11}\ \frac{\text{N-m}^2}{\text{kg}^2}\right) \times (3.16 \times 10^7\ \text{s})^2}$$

Step 3

Solve for the mass of the Sun.

$$m_{Sun} = 2.0 \times 10^{30}\ \text{kg}$$

Notice that if we subtract 6.0×10^{24} kg (0.0000060×10^{30} kg), the mass of Earth, from 2.0×10^{30} kg, we still obtain 2.0×10^{30} kg. Thus the mass of Earth has a negligible effect on the answer obtained for the mass of the Sun.

CONFIDENCE EXERCISE 15.2

Calculate the mass of the Sun in kilograms. Use Mars' orbital period (T) and average distance (R) from the Sun (see page 397 for T).

The Sun is the dominant mass of the solar system, possessing 99.87% of the mass of the system. The distribution of the remaining 0.13% of the solar system's mass is shown in Table 15.1. Note that more than half is the mass of Jupiter.

Planets that have orbits smaller than Earth's are classified as "inferior" and those with orbits greater than Earth's as "superior." Another classification system calls Mercury, Venus, Earth, and Mars the inner planets, or **terrestrial planets,** because they resemble Earth. Jupiter, Saturn, Uranus, and Neptune are then classified as the outer planets, or **Jovian planets,** because they resemble Jupiter. (The Roman god Jupiter was also called Jove.) Since Pluto does not resemble Earth or Jupiter, some astronomers have suggested that it be classified as an asteroid.

TABLE 15.1   The Solar System

Name	Mean Distance from the Sun, Astronomical Units	Diameter, Earth = 1	Mass with Respect to Earth = 1
Sun			333,000.
Mercury	0.387	0.38	0.055
Venus	0.723	0.95	0.82
Earth	1.00	1.00	1.00
Mars	1.524	0.53	0.11
Asteroids	2.767		
Jupiter	5.203	11.2	318.
Saturn	9.555	9.41	94.3
Uranus	19.19	3.98	14.54
Neptune	30.110	3.81	17.2
Pluto	39.44	0.27	0.002

The relative distances of the planets from the Sun are shown in ● Fig. 15.6. The orbits are all elliptical, but nearly circular, except for that of Pluto. Notice that Pluto's orbit actually goes inside Neptune's orbit. The position of Pluto is shown for 2002. Look at Fig. 15.6 and notice how far Jupiter is from the Sun compared with the distance from the Sun to Mars. Also note that the distance from Saturn to Neptune is greater than that from the Sun to Saturn.

When viewed from above the solar system (that is, when viewed from above the North Pole of Earth), the planets all revolve counterclockwise around the Sun. This motion is west-to-east (eastward) revolution, or **prograde motion.** The planets also rotate with a counterclockwise, or prograde, motion when viewed from above the North Pole, with the exception of Venus and Uranus. These two planets have **retrograde motion**—that is, the motion is east to west (westward), or clockwise, as viewed from above the North Pole of Earth.

The relative sizes of the planets are shown in ● Fig. 15.7. Note how huge the Jovian planets are compared with the terrestrial planets. Compare the size of Pluto, the smallest planet, with the size of the other planets.

The inclinations of the orbits of the planets relative to Earth's orbit are shown in ● Fig. 15.8. Note that the solar system is disk-shaped rather than spherical. Also note the large inclination of Pluto's orbit.

A mnemonic for remembering the order of the planets from the Sun is the following: *My Very Excellent Mother Just Sent Us Nine Pizzas*—Mercury, Venus, Earth, Mars, Jupiter, Saturn, Uranus, Neptune, and Pluto.

The period of time required for a planet to travel one complete orbital path is referred to as either the sidereal or synodic period. The **sidereal period** is defined as the time interval between two successive **conjunctions** of the planet with a star (meaning the planet and star are together on the same meridian) *as observed from the Sun.* The **synodic period** is the time interval between two successive conjunctions (either inferior or superior) of the planet with the Sun (where the planet and the Sun are on the same meridian) *as observed from Earth.*

The relationship between the sidereal and synodic periods of a planet is illustrated in ● Fig. 15.9. At position P_1 Mercury and Earth are observed, from the Sun, on the meridian with the star S. Mercury, observed

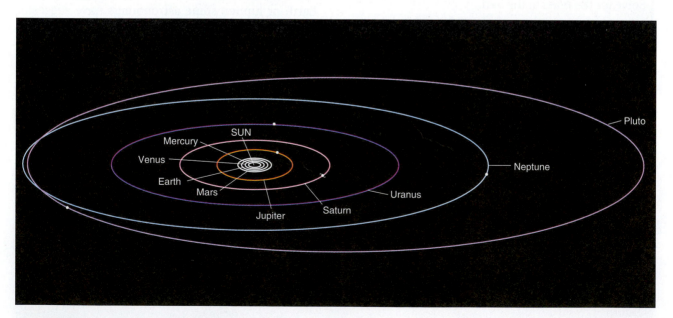

FIGURE 15.6 The Orbits of the Planets
All the planets revolve counterclockwise around the Sun, as observed from a position in space above the North Pole of Earth. The orbits of the planets are drawn to scale.

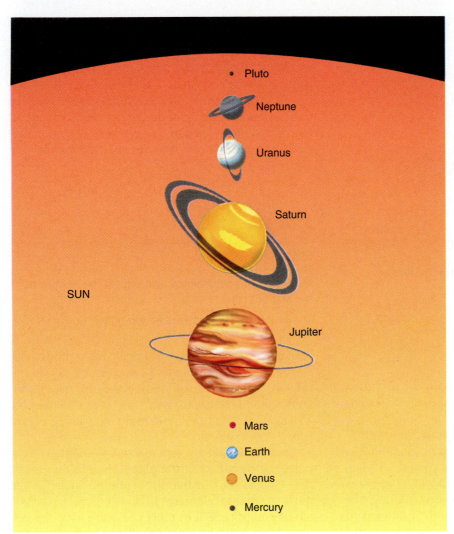

FIGURE 15.7   The Solar System
The nine planets and the Sun are drawn to scale, and their colors are similar to their surface colors.

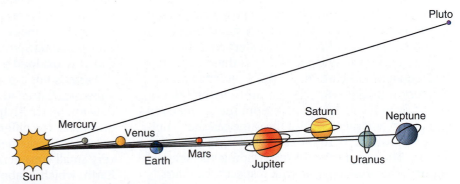

FIGURE 15.8   The Solar System
This diagram shows the inclination of the planets' orbits with the orbital plane of Earth.

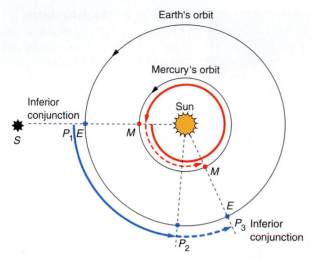

FIGURE 15.9 The Sidereal and Synodic Periods of Mercury

This diagram illustrates the difference between sidereal and synodic periods of the inferior planet Mercury. (See the text for an explanation.)

from Earth at this same instant, is on the meridian with the Sun. From position P_1 Mercury revolves eastward (counterclockwise) around the Sun through 360° back to position P_1. This motion and time period are represented by the solid-color pink circle in Fig. 15.9. This revolution is the sidereal period for Mercury—the time (88 Earth days) that Mercury requires to make one revolution around the Sun. During this time, Earth revolves approximately 87° eastward to position P_2. Mercury continues revolving eastward from position P_1 to position P_3. During this same period, Earth revolves from position P_2 to position P_3. This movement and time period are represented by the dashed blue lines in Fig. 15.9. At position P_3 an observer on Earth again sees Mercury on the meridian with the Sun. The total time for Mercury to revolve from P_1 back to P_1 and then to P_3 is the synodic period, equal to 116 Earth days. This is the time it takes Mercury to make one orbit around the Sun as observed from Earth, or the time from the *conjunction* (the planet and the Sun have the same celestial longitude) at position P_1 to the next conjunction at position P_3. The true period of revolution is the sidereal period—the orbital period with respect to the stars.

Opposition is the term used to describe the position of a superior planet when it is 180° from the Sun,

that is, when the planet is on the opposite side of Earth from the Sun.

RELEVANCE QUESTION: *Most of the information concerning the solar system, including the color photographs, was obtained over the past several years by manned and unmanned spacecraft missions throughout the solar system. Can you name one significant impact that these missions and the data collected had on your life?*

15.2 The Planet Earth

LEARNING GOALS

▼ Identify some chemical and physical properties of planet Earth.

▼ Define and explain Earth's two major motions.

Planet Earth is a solid, spherical, rocky body with oceans and an atmosphere. Of the nine planets in our solar system, Earth is unique. It is the only planet with large amounts of surface water, an oxygen-containing atmosphere, a temperate climate, and living organisms.

Because oxygen is a very reactive element, it dominates the chemistry of the planet. In addition to the 21% oxygen (O_2) in the atmosphere, oxygen is the most abundant element in Earth's crust. Oxygen atoms comprise 90% or more of Earth's rocks, by volume. Most common rocks are silicates, which are any one of numerous minerals that have the oxygen and silicon tetrahedron combined usually with one or more metals (Section 21.1). The crusts and mantles of all interior planets are composed of silicates. When oxygen combines with another substance, we call the process *oxidation* (Section 13.4). Consequently, we live in an oxidized environment.

Earth is not a perfect sphere but rather an oblate spheroid, flattened at the poles and bulging at the equator. Earth's shape is due primarily to its rotation on its axis. The difference between the diameter at the poles and the diameter at the equator (about 43 km) is very small considering the total average diameter of Earth, which is about 12,900 km. The ratio of 43 to 12,900 is 1/300, which is a rather small fraction. If Earth were represented by a basketball with a diameter of approximately $\frac{1}{4}$ m, the eye would not detect the difference of $\frac{1}{1200}$ m between the two diameters. Earth is a more nearly perfect sphere than the average basketball.

The fraction of incident sunlight reflected by an object is called its **albedo.** Earth's albedo is 0.33, and the Moon's albedo is only 0.07. This indicates that the Moon's surface reflects 7% of the incoming sunlight falling on its surface. Earth reflects more light (33%) because the clouds and water areas are much better reflecting surfaces than the dull, dark surface of the Moon. The planet Venus, the third brightest object in the sky (only the Sun and Moon are brighter), has an albedo of 0.76. Although the Moon's albedo is much smaller (0.07), the Moon is much closer to Earth than the bright, but distant, Venus.

Full Earth, as viewed from the Moon, appears about 16 times larger in diameter than the full moon viewed from Earth and reflects more than 70 times as much bright light.

Although we are unable to sense directly the motion of our home planet, Earth is undergoing several motions simultaneously. Two that have major influences on our daily lives will be explained in this section: (1) the daily rotation of Earth on its axis and (2) the annual revolution of Earth around the Sun. A third motion, precession, is discussed in Section 16.5.

When studying astronomy, it is important to know the difference between rotation and revolution. A mass is said to be in **rotation** when it spins on an internal axis. Examples are a spinning toy top and a Ferris wheel at an amusement park. **Revolution** is the movement of one mass around another. Earth revolves around the Sun, and the Moon revolves around Earth.

Earth revolves eastward around the Sun and sweeps out a plane called the *orbital,* or *ecliptic, plane.* This motion of Earth produces an apparent annual westward motion of the Sun on the *celestial sphere,* the apparent sphere of the sky. The apparent annual path of the Sun on the celestial sphere is called the **ecliptic.** The word is derived from eclipse, because eclipses of the Sun and Moon occur when the Moon is on or near the great circle forming the apparent annual path of the Sun.

Earth is rotating eastward around a central internal axis that is tilted 23.5° from a line that is perpendicular to its orbital plane. Earth's rotational period of 24 hours and the axis tilt (23.5°) provide excellent distri-

bution of the solar energy that is radiated onto its surface. Later discussion (Section 16.4) will show why the 23.5° tilt of the axis and the revolution of Earth around the Sun are the reasons for the four seasons we experience annually.

The fact that Earth rotates on its axis was not generally accepted until the nineteenth century. A few scientists had considered the possibility, but since no definite proof was available to support their beliefs, their ideas were not accepted.

In 1851 an experiment demonstrating the rotation of Earth was performed in Paris by Jean Foucault ("foo-KOH") (1819–1868), a French engineer, using a 61-m pendulum. Today, any pendulum used to demonstrate the rotation of Earth is called a **Foucault pendulum.** Even more noticeable results can be seen if the experiment is performed at Earth's North or South Pole.

Picture a large, one-room building with a ceiling over 61 m high located at the North Pole (● Fig. 15.10). Fastened to the ceiling precisely above the North Pole is a swivel support having very little friction, to which a 61-m fine steel wire is attached. Connected to the lower end of the wire is a massive iron ball with a short, sharp steel needle attached permanently to its underside. On the floor, under the pendulum, is a layer of fine sand that is slightly furrowed by the needle as the pendulum swings back and forth.

Someone starts the pendulum swinging by displacing it to one side with a strong fine thread, with one end attached to the side of the ball and the other end attached to one wall of the building where a 24-hour clock is mounted. To prevent any sideways motion, the iron ball is allowed to become motionless before it is released by burning the thread. Extreme care is taken to prevent any lateral external forces from being applied to the upper support point of the 61-m wire. As the pendulum swings freely back and forth, the needle point traces its path in the layer of sand.

After a few minutes, the plane of the swinging pendulum appears to be rotating clockwise, as shown by the markings in the sand. At the end of 1 hour, the plane has rotated 15° clockwise from its original position. When 6 hours have elapsed, the plane of the pendulum appears to have rotated 90° clockwise and is parallel to the wall that holds the 24-hour clock. With the passing of each hour, the plane appears to rotate another 15° clockwise. At the end of 24 hours, it has made an apparent rotation of 360°.

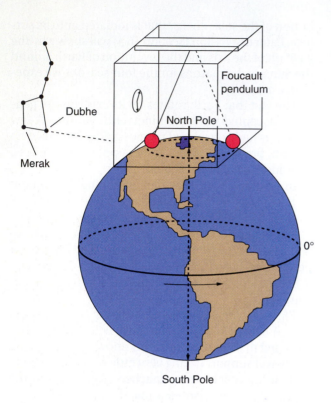

FIGURE 15.10   Foucault Pendulum

The drawing illustrates a Foucault pendulum positioned in a room at the North Pole of Earth. To an observer in the room, the pendulum will appear to change its plane of swing by 360° every 24 h. The photograph shows a Foucault pendulum at the Smithsonian Institution in Washington, D.C. As the pendulum swings back and forth, its plane of swing appears to change, as noted by the consecutive knocking over of the red markers positioned in a circle.

A person who believes in a motionless Earth would argue that the pendulum actually rotated 360°, because one rotation of the swinging pendulum has been observed by anyone stationed in the large room. We can understand the apparent rotation if we make the walls of the building out of a transparent material such as glass and perform the experiment sometime during the winter months for the Northern Hemisphere. The North Pole has 24 hours of darkness during these months, and the stars are always visible. When starting the pendulum this time, we take care to place the iron ball in direct line with the stars Dubhe and Merak, the pointers in the cup of the Big Dipper. As the minutes pass, we observe, as before, the apparent rotation of the plane of the swinging pendulum in a clockwise direction with reference to the large room and clock on the wall.

We also observe, through the transparent walls of the room, that the pendulum still swings in the same direct line with the stars Dubhe and Merak; that is, the pendulum, Dubhe, and Merak are all in the same plane. *The pendulum has not rotated with reference to fixed stars.* No forces have been acting on the pendulum to change its plane of swing. Only the force of gravity has been acting vertically downward, which

keeps the pendulum swinging. Therefore, the pendulum does not really rotate westward or clockwise, but the building and Earth rotate eastward, or turn counterclockwise, once during the 24-hour period, as viewed from above the North Pole.

The Foucault pendulum is an experimental proof of Earth's rotation on its axis. What experimental observation would prove that Earth revolves around the Sun?

As Earth orbits the Sun once a year, the apparent positions of nearby stars change with respect to more distant stars. This effect is called *parallax*. In general, **parallax** is the apparent motion, or shift, that occurs between two fixed objects when the observer changes position. To see parallax for yourself, hold your finger at a fixed position in front of you. Close one eye, move your head from side to side, and notice the apparent motion between your finger and some distant object. Note also that the apparent motion becomes less as you move your finger farther away. ● Figure 15.11 is an illustration of the parallax of a nearby star as measured from Earth relative to stars that are more distant.

The motion of Earth as it revolves around the Sun leads to an apparent shift in the positions of the

A second proof of Earth's orbital motion around the Sun is the telescopic observation of a systematic change in the position of all stars annually. The observed effect (called the *aberration of starlight*) is due to the finite speed of light and the motion of Earth around the Sun. The **aberration of starlight** is defined as the apparent displacement in the direction of light coming from a star owing to the orbital motion of Earth.

The great distances to stars are measured in a unit called the *parsec*. The name comes from the first three letters of the word *parallax* plus the first three letters of the measuring unit, the second, which is used to measure angle.

A circle contains 360°; a degree is divided into 60 equal parts, each of which is called a *minute*. The minute is further divided into 60 equal divisions, each of which is called a *second*. Thus one second is an angular measurement equal to $\frac{1}{3600}$ of a degree. One **parsec** is defined as the distance to a star when the star exhibits a parallax of 1 s. The parsec is explained in greater detail in Section 18.2.

The Greek mathematician and astronomer Eratosthenes ("er-uh-TOS-theh-neez") calculated the circumference of Earth about 250 B.C. Eratosthenes was living in Alexandria, Egypt, located 5000 stadia (the *stadium* was a Greek measure of length) almost due north of Syene (now Aswan), Egypt. He had received reports that deep wells at Syene were lighted all the way to the bottom on the first day of summer, which meant the Sun there was directly overhead (at the zenith). Eratosthenes discovered that, on the same day of the year at Alexandria, a vertical stick cast a shadow that positioned the Sun 7.2° south of his zenith, or $\frac{1}{50}$ of a circle. Thus he was able to calculate the circumference of Earth to be 250,000 stadia (5000 × 50).

After repeated measurements, Eratosthenes increased the circumference to 252,000 stadia. If we assume that the stadium used was $\frac{1}{6}$ km, the circumference calculated by Eratosthenes was 42,000 km, very close to the correct value of 40,000 km. ● Figure 15.12 illustrates the principles involved in Eratosthenes' calculations.

RELEVANCE QUESTION: *Earth rotates around an internal axis and revolves around the Sun. Identify two distinct events each motion contributes to your life.*

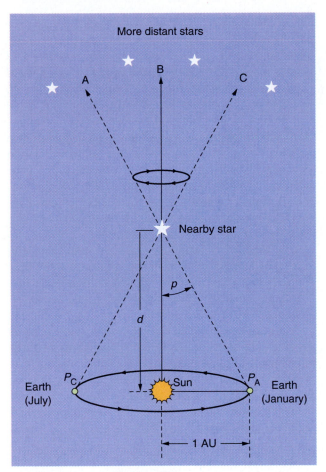

FIGURE 15.11   Stellar Parallax
The parallax of a star is the apparent displacement of a star that is located fairly close to Earth with respect to more distant stars. When the observer is at P_A, the star appears in the direction A. As Earth revolves counterclockwise, the star appears to be displaced and appears in the direction indicated for different positions of Earth. Positions P_A and P_C are six months apart. The angle of parallax p is also shown.

nearby stars with respect to more distant stars. Because the stars are at very great distances from Earth, the parallax angle is very small and cannot be seen with the unaided eye. It was first observed with a telescope in 1838 by Friedrich W. Bessel (1784–1846), a German astronomer and mathematician. The observation of parallax was indisputable proof that Earth really does go around the Sun. Today, the measurement of the parallax angle is the best method we have of determining the distances to nearby stars.

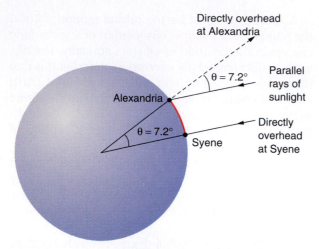

FIGURE 15.12 The Method Eratosthenes Used to Determine the Size of Earth

The 7.2° zenith angle measured at Alexandria by Eratosthenes was equal to the angle θ, because the rays of sunlight come in parallel to one another. The 7.2° angle is $\frac{1}{50}$ of a circle.

15.3 The Terrestrial Planets

LEARNING GOAL

▼ List and compare the physical characteristics of the terrestrial planets.

As you learned earlier, Mercury, Venus, Earth, and Mars are called the *terrestrial planets* because their physical and chemical characteristics resemble Earth in certain respects. ● Figure 15.13 shows these four planets and Earth's Moon along with the Jovian planets. All four terrestrial planets are relatively small in size and mass. They are composed of rocky material (silicates) and metals (having cores of mostly iron and nickel). All four are relatively dense (average density of about 5.0 g/cm^3) and have solid surfaces and weak magnetic fields.

Their orbits are, comparatively speaking, close together, and they are relatively close to the Sun. None has a ring system, and only Earth and Mars have moons. Although the terrestrial planets have some similarities, they are also very different from one another. Planet Earth alone has an abundance of surface water and an atmosphere that is 21% oxygen. The other terrestrials have no surface water and no free oxygen in their atmospheres.

Mercury

Mercury is the closest planet to the Sun and has the shortest period of revolution (88 days). The early Greeks named Mercury after the speedy messenger of the gods, and it is the fastest moving of all the planets because of its position closest to the Sun.

Mercury, at its greatest eastern or western elongation, can be seen only just after sunset or just before sunrise. The greatest elongation (the angular distance between Mercury and the Sun as viewed from Earth) is only 28°. When Mercury is near eastern elongation, it will appear above the western horizon just after sunset. At western elongation, Mercury will be on the eastern horizon shortly before sunrise.

The surface of Mercury appears similar to the Moon, as can be seen from the photo on page 396. Mercury has a high density, almost as high as Earth's. This high density indicates that it probably has an inner core of iron, as does Earth.

Mercury's rotation period is exactly two-thirds as great as its period of revolution. Thus it rotates three times while circling the Sun twice. This period probably results from tidal gravitational effects from the Sun. As it rotates, the side facing the Sun has temperatures of approximately 700 K (427°C), while the dark side is at about 100 K (−173°C).

Venus

Venus is our closest planetary neighbor, approaching Earth at a distance of 42 million kilometers at inferior conjunction. It is the third brightest object in the sky, exceeded only by the Sun and our Moon. Because of its brightness, Venus was named in honor of the Roman goddess of beauty.

Venus and Earth resemble one another in several ways. They have similar average density, mass, size, and surface gravity. But the similarities end there. Venus is covered with a dense atmosphere that is 96% carbon dioxide, less than 4% nitrogen, and has traces of argon, oxygen, and water vapor. At the surface of Venus the atmospheric pressure is a tremendous 90 atm, and the temperature is 750 K, or about 480°C. The high temperature is due mainly to the large amount of carbon dioxide in the atmosphere, which produces the "greenhouse effect" (see Section 19.2), so life as we know it cannot exist. Both temperature and pressure decrease with an increase in altitude.

The surface of Venus can never be seen by an observer on Earth because of dense, thick clouds that

FIGURE 15.13   The Terrestrial Planets, Jovian Planets, and Earth's Moon

This is a montage of photos taken by NASA spacecraft. At the top (from left to right) are Mercury, Venus, Earth, Mars, and Earth's Moon. At the bottom (from left to right) are Neptune, Uranus, Saturn, and Jupiter. The images are not reproduced to the same scale.

cover the planet. The clouds are composed mainly of sulfuric acid (H_2SO_4) droplets, along with some water droplets. The droplets do not fall out as rain because of the extremely high atmospheric pressure. The top layer of clouds contains large amounts of yellowish sulfur dust, giving Venus its yellowish or yellow-orange color when viewed from Earth. Orbital spacecraft observations of the outer cloud layer revealed that Venus' atmosphere makes one rotation every four Earth days in retrograde direction. This rotation is extremely fast compared with the −243 days for rotation of the solid planet. (The minus sign indicates retrograde motion.) Why Venus is rotating retrograde and very slowly is a mystery. One possibility is that Venus was struck by a large object during the formation of the solar system, which stopped the planet's rotation and produced a slow rotation in the opposite direction.

Although thick clouds conceal the surface of Venus from our eyes and light-sensitive instruments, surface features have been obtained with *radar* (*ra*dio *d*etecting *a*nd *r*anging) imaging. In 1990, NASA sent the radar imaging spacecraft *Magellan* to Venus and put the spacecraft in a near-polar orbit. Over a four-year period, thousands of radar images were taken of the planet's surface. The chapter Highlight (page 400) gives information on how spacecraft are launched to Venus and other planets in the solar system.

The *Magellan* radar images, which are two-dimensional black-and-white images, reveal Venus' surface as hot, black rock with relatively few large craters. About 1000 craters larger than a few kilometers in diameter were detected. Astronomers think that craters smaller than about six kilometers in diameter do not appear in the image because small incoming objects are consumed in Venus' thick atmosphere. Other surface features identified were fractures and fault lines with high walls and cliffs, mountain chains that extend hundreds of kilometers in length, and volcanic plains that cover more than 80% of the planet's surface. No active volcanoes appeared in *Magellan* radar images, but most surface rocks appear to be volcanic in origin, indicating that volcanism was the last geologic process to take place on the planet. No surface feature in the images appears older than about one billion years; most features seem to be approximately 400 million years old.

● Figure 15.14 shows Venus without clouds. The surface is dominated by vast plains and lowlands with some highlands. The planet's highest mountain, Maxwell Montes, is shown at the top of the figure. The mountain stands 11 kilometers above the mean

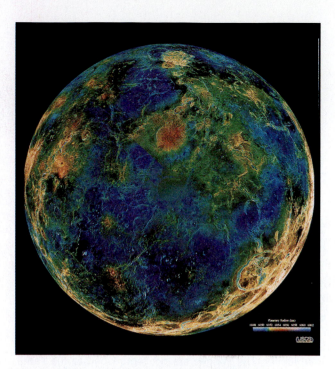

FIGURE 15.14 Hemispheric View of Venus

As revealed by more than a decade of radar investigations culminating in the 1990–1994 *Magellan* mission, this image of Venus is centered at the South Pole. The effective resolution of the image is about 3 km. This composite image, processed to improve contrast and emphasize small features, was color coded to represent elevation. Higher elevations above the average radius of the planet (approximately 6000 km) are shown in orange colors.

elevation. Another high region, Aphrodite Terra, can be seen extending along the right edge of the figure. This region is just north of the equator.

Mars

Viewed from Earth, Mars has a reddish color and was named after the bloody Roman god of war. Mars is about 1.5 times as far from the Sun as Earth. Its axis is tilted at an angle of 24°, which is very close to Earth's 23.5° angle of tilt. Mars rotates once every 24.5 h, which is very close to a single Earth day. It takes 687 days (about 23 Earth months) for Mars to revolve once around the Sun. The mass of Mars is about one-tenth the mass of Earth.

Mars has two small satellites, or moons, named Phobos ("fear") and Deimos ("panic") after the mythical horses that pulled the chariot of the god Mars. The moons are very small. Phobos is about 32 km in diameter and Deimos is about half that size. They are

irregularly shaped and extensively cratered. Both revolve eastward around Mars. Phobos circles Mars in only 7 h and 39 min, and Deimos revolves once in 30 h and 18 min. Like our Moon, they keep one side always facing the planet, because their periods of rotation and revolution are equal.

Two outstanding features of the surface of Mars are the polar caps (see page 397) and the 12 or more extinct volcanoes. In winter the polar caps are composed of frozen carbon dioxide (CO_2) and water ice. In summer the frozen CO_2 changes to vapor, leaving behind a residual polar cap of water ice. The depth of the water ice is unknown. Olympus Mons, "Mount Olympus," shown in ● Fig. 15.15, is the largest known volcano in the solar system. It rises 24 km above the plain and has a base with a diameter greater than 600 km. The volcano is crowned with an 80-km-wide crater. The largest volcano on Earth is Mauna Loa on the island of Hawaii. Mauna Loa's base rests on the ocean floor 5 km below the surface of the Pacific Ocean and extends upward another 4.2 km above the level of the ocean. Thus Mauna Loa is about one-third the height of Mount Olympus.

Another major feature of the surface is the large canyon called *Valles Marineris,* shown in ● Fig. 15.16. The canyon is about 4000 km long and 6 km deep. Its length is equivalent to the width of the United States. The canyon is almost four times as deep as the Grand Canyon. This tremendous gash in Mars' surface is thought to be a gigantic fracture formed by stress within the planet.

FIGURE 15.15 Olympus Mons, "Mount Olympus"

This huge Martian volcano (the largest in the solar system) is about 24 km high and 600 km wide at its base. The caldera is 80 km across at the summit.

FIGURE 15.16   Valles Marineris
This enhanced color mosaic shows the great canyon Valles Marineris. The canyon is 4000 km in length. Geologists believe that it is a fracture in the planet's crust caused by internal forces.

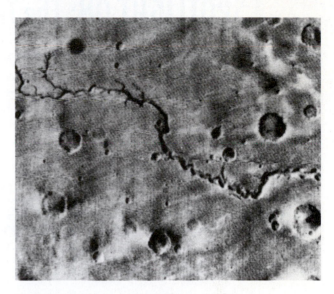

FIGURE 15.17   An Ancient Channel on Mars
Though not unique on the Martian surface, this meandering "river" is the most convincing piece of evidence that a fluid once flowed on Mars, draining a large area and eroding a deep channel. The feature is some 575 km long and 5 to 6 km wide.

Another interesting feature, an ancient channel, is shown in ● Fig. 15.17. It is believed to have been formed by a moving liquid—probably water. If water did flow to form the ancient channel, where did the water go? The polar caps, which are predominantly dry ice (CO_2), contain some water ice. Is there water somewhere else on the planet? One theory suggests that additional water is hidden below the polar caps in the form of permafrost similar to that found in the far northern regions of Earth. To answer the question, NASA launched the Mars Polar Lander (MPL) from Cape Canaveral on January 3, 1999. The MPL is scheduled to arrive at the planet's south-polar ice cap in December of 1999. The spacecraft will release a pair of two-kilogram Deep Space 2 penetrators that will pierce Mar's surface in the search for water.

Has life existed on Mars? Presently, the debate is over the evidence obtained from a 1.9-kg meteorite from Mars. Scientists do agree that meteorite ALH84001, which was found in an Antarctic ice field in 1984, came from Mars and landed on Earth some 13,000 years ago. Using lasar spectroscopy and electron microscopy, NASA scientists have discovered that the meteorite contains tiny globules of calcium carbonate, a mineral that crystallizes in the presence of water. This indicates that water, a necessary ingredient for life, existed on Mars. Scientists also discov-ered organic molecules, which are carbon-based compounds on which life on Earth is based. A powerful electron microscope exposed images of minerals inside the tiny globules that appear to be residue of biological activity. Moreover, the microscope revealed images of tiny structures that the scientists considered to be microfossils of Martian microbes.

The evidence is inspiring but not conclusive. The basic disagreement in the debate concerns the evidence found in the meteorite made of rock formed 3 to 4 billion years ago on Mars. Has the meteorite been contaminated from the Antarctic ice field? Obviously, the answer is yes. The meteorite has lain in the ice for 13,000 years, exposed to terrestrial contaminants. But are there components in the meteorite that are not found on Earth? The debate continues. Most likely, the truth will be obtained only when uncontaminated soil samples are brought back from Mars.

The terrestrial planets are illustrated and their physical properties summarized in the following Spotlight.

RELEVANCE QUESTION: *Name two common life-supporting substances used continuously by your body. Why is Earth the only planet that we know of in the solar system that possesses them?*

A SPOTLIGHT ON: The Terrestrial Planets

Diameter (equatorial)	12,102 km
Mass	4.870×10^{24} kg
Density (mean)	5250 kg/m³
Distance from Sun (mean)	1.082×10^{8} km
Rotation period (retrograde)	243.01 days
Sidereal period	224.70 days
Surface temperature (mean)	480°C
Magnetic field	No
Composition of atmosphere	96% CO_2, 3% N_2, 0.1?% H_2O

Venus The brightest planet

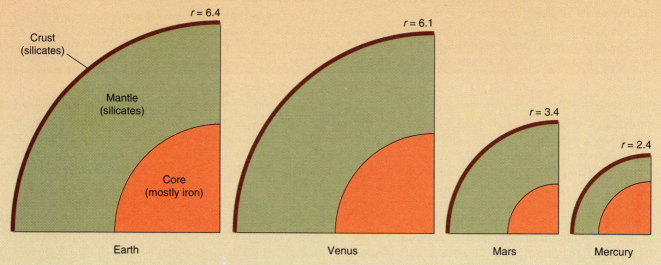

Crust (silicates)

Mantle (silicates)

Core (mostly iron)

$r = 6.4$

$r = 6.1$

$r = 3.4$

$r = 2.4$

Earth

Venus

Mars

Mercury

The internal structure of the terrestrial planets.
The approximate radius (r) of each planet is given in units of 10^3 km.

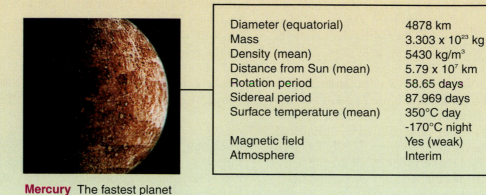

Diameter (equatorial)	4878 km
Mass	3.303×10^{23} kg
Density (mean)	5430 kg/m³
Distance from Sun (mean)	5.79×10^{7} km
Rotation period	58.65 days
Sidereal period	87.969 days
Surface temperature (mean)	350°C day
	-170°C night
Magnetic field	Yes (weak)
Atmosphere	Interim

Mercury The fastest planet

Earth The blue planet

Diameter (equatorial)	12,756 km
Mass	5.976×10^{24} kg
Density (mean)	5520 kg/m^3
Distance from Sun (mean)	1.496×10^8 km
Rotation period	23.9345 hours
Sidereal period	365.256 days
Surface temperature (mean)	20°C
Magnetic field	Yes
Number of satellites	1
Composition of atmosphere	78% N_2, 21% O_2, 1% Ar

Moon

Diameter	3476 km
Mass	7.349×10^{22} kg
Density (mean)	3340 kg/m^3
Distance from Earth (mean)	384,400 km
Orbital velocity (mean)	3680 km/h
Sidereal period	27.322 days
Synodic period	29.531 days
Surface temperature (mean)	130°C day
	-180°C night
Magnetic field	No
Atmosphere	No

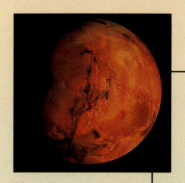

Mars The red planet

Diameter (equatorial)	6786 km
Mass	6.42×10^{23} kg
Density (mean)	3950 kg/m^3
Distance from Sun (mean)	2.279×10^8 km
Rotation period	1.026 days
Sidereal period	686.98 days
Surface temperature (mean)	20°C maximum
	-140°C minimum
Magnetic field	Very weak
Number of satellites	2
Composition of atmosphere	95% CO_2, 3% N_2, 1.6% Ar

Phobos

Deimos

15.4 The Jovian Planets and Pluto

LEARNING GOALS

▼ List and compare the physical properties of the Jovian planets.

▼ Identify the major differences between the terrestrial and the Jovian planets.

▼ List the physical properties of Pluto.

The four Jovian planets are large compared with the terrestrial planets, are gaseous, and have no solid surface. Figure 15.13 shows the four Jovian planets. They are composed mainly of hydrogen and helium, and all have a very low density (on average, 1.2 g/cm^3). The planets possess strong magnetic fields, have many moons and rings, and are far from the Sun with orbits far apart. All four planets are believed to have rocky cores with a layer of ices above the rocky core. Upper layers of molecular and metallic hydrogen apply high pressure to and create high temperature in the ice layers and rock core, producing ice and rock that are much different from the ice and rock on Earth. The rock is believed to be composed mainly of iron and silicates; the ices are believed to be methane, ammonia, and water. Thus the Jovian planets are very different from the terrestrial planets. Table 15.2 lists the significant differences between the terrestrial and the Jovian planets.

Pluto does not resemble Earth or Jupiter, and some astronomers have suggested that Pluto be classified as an asteroid. Other astronomers have suggested a new category called *ice dwarfs*.

When the planets first began to coalesce around 5 billion years ago, the predominant elements were the two least massive ones—hydrogen and helium. The heat from the Sun allowed these two elements to escape from the inner planets; that is, the velocities of the molecules of these elements were sufficient to allow them to escape the planets' gravitational pulls. Thus the inner planets were left with mostly rocky cores, giving them a high density. The four large outer planets were much colder, and they retained their hydrogen and helium, which now surround their ice layers and rocky cores. Thus the four large outer planets consist primarily of hydrogen and helium in various forms, and this composition gives them much lower densities. The Spotlight on pages 404 and 405 summarizes the Jovian planets.

Jupiter

Jupiter, named after the supreme Roman god of heaven because of its brightness and giant size, is the largest planet of the solar system, in both volume and mass. The motion about its axis is faster than that of any other planet, taking only 10 h to make one rotation.

Jupiter's diameter is 11 times as large as Earth's, and it has 318 times as much mass; however, its den-

TABLE 15.2 Significant Differences Between Terrestrial and Jovian Planets

Terrestrial Planets	Jovian Planets
Small diameter (approximately 5000–13,000 km)	Large diameter (approximately 50,000–143,000 km)
Rocky	Gaseous; mainly hydrogen and helium
Solid surface	No solid surface
Relatively high density (3.9–5.5 g/cm^3)	Relatively low density (less than 1.7 g/cm^3)
Relatively close to Sun	Great distance from Sun
Relatively high temperature environment	Cold temperature environment
Close proximity (greatest separation less than 1.2 AU)	Widely separated (greatest separation 25 AU)
Weak magnetic fields (if any)	Strong magnetic fields
Three moons total	Many moons (63 known)
No rings	All have rings
Slow rotation	Fast rotation

sity is only 1.3 g/cm³. Jupiter consists of a rocky core, a layer of ice, a layer of hydrogen in liquid metallic form (because it is at high pressure and temperature), and an outer layer of molecular hydrogen (see page 405). Above the molecular hydrogen is a thin layer of clouds composed of hydrogen, helium, methane, ammonia, and several other substances. The mean surface temperature at the top of the clouds is about 125 K.

In Fig. 15.13 the cloud features are easily seen. Jupiter's clouds have many different patterns—bands, ovals, and light and dark areas in white, yellow, orange, red, and brown. Convection currents are present, and we see the tops of updrafts (the lighter areas) and downdrafts (the darker areas). The Great Red Spot, which appears yellow in this figure, stands out. The spot has an erratic movement and changes color and shape. Sometimes it completely disappears. The most recent theory of the Great Red Spot states that it is a huge counterclockwise storm similar to a hurricane on Earth but lasting hundreds of years.

On July 16, 1994, the first impact of more than 20 fragments of Comet Shoemaker-Levy 9 bombarded Jupiter's cloud tops, creating Earth-sized, dark-brown dots. Most astronomers believe that Comet Shoemaker-Levy 9 was about 1.5 km in diameter before its breakup.

The cometary fragments smashed into the atmosphere at 60 kilometers per second (about 130,000 mph) and were vaporized in the impact, as was a large amount of Jupiter's atmosphere along the impact path. This set off explosions creating huge clouds that were easily visible from Earth. ● Figure 15.18 shows the impact on July 18 of fragment G, one of the largest.

Jupiter possesses a tremendous magnetic field; at the top of the atmosphere it is 25 times as strong as Earth's field. Its rotation axis is inclined by only a few degrees, so Jupiter does not experience seasonal effects as do Earth and Mars.

Jupiter has many moons—16 or more, depending on where the line is drawn between a large rock and a small moon. In addition, some 50,000 km above the planet's clouds is a very faint planetary ring of dust particles that is bright enough to be seen from Earth. The four largest moons, first discovered by Galileo in 1610, are called the *Galilean moons* of Jupiter. In order of increasing distances from Jupiter, they are Io, Europa, Ganymede (the largest moon in the solar system), and Callisto. They were photographed close up

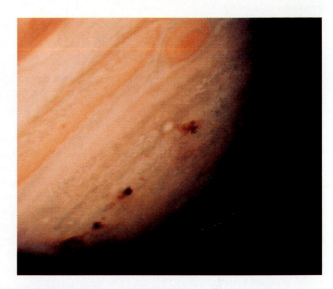

FIGURE 15.18   Comet Shoemaker-Levy Fragments Impact Jupiter's Atmosphere

This is Hubble's image of several of the impact sites of comet Shoemaker-Levy (SL9). The comet was broken into 21 major fragments. One of the impact sites was as large as the planet Earth.

by the *Voyager 1* and *Voyager 2* spacecraft in 1979. Photos of the moons are shown on page 404.

One of the most spectacular findings of the *Voyager* missions was that Io has many active volcanoes. The volcanoes occur because the gravitational attraction of other nearby moons—notably Europa—causes Io's orbit to vary so that it is closer to, then farther from, Jupiter. The resulting changes in Jupiter's gravitational force cause stresses in the interior rock of Io, and a great deal of frictional heat is generated, resulting in volcanoes.

Europa is the smallest of the Galilean satellites (diameter 3138 km). The moon has physical properties like no other in the solar system. It is covered with a smooth layer of ice, which is believed to be floating on an ocean of liquid water. The surface is crisscrossed with ridges and cracks, as shown in ● Fig. 15.19. There are signs of craters and iceberg-like debris. The various surface features are believed to be caused by gravitational tidal flexing among Europa, Jupiter, and the other Galilean satellites.

Water is a necessary ingredient for life. If the water on Europa has been there throughout its existence, then life has had time to evolve, especially with heat

HIGHLIGHT: Space Flights in the Solar System

Planet Earth orbiting the Sun has both kinetic and potential energy. Kinetic energy is due to its orbital motion, and potential energy is due to its position in the Sun's gravitational field. The greater the distance a planet is from the Sun, the greater is the total energy. Compared with Earth, the planets Mercury and Venus, having smaller orbits, have smaller total energies. Planets with orbits greater than Earth's have greater total energies.

Viewing the solar system from the North Celestial Pole, Earth orbits the Sun in a counterclockwise, or eastward, direction with an average orbital speed of 29 km/s (65,000 mi/h). If we want to send a space probe to Mercury, Venus, or the Sun, the probe must lose energy. Since our launching platform is moving with a speed of 29 km/s, we want to launch the probe in the direction opposite Earth's orbital motion. This will lower the probe's speed with respect to the Sun and decrease its energy, and it will fall toward the Sun. If we want to send a space probe outward from the Earth to Mars or any planet with an orbit greater than Earth's, the probe must gain energy. Thus we launch the probe in the direction of Earth's orbital motion to increase its speed and energy. This was done for the Mars *Pathfinder*, with the rover *Sojourner* aboard, which landed on Mars on July 4, 1997, and for NASA's Mars *Global Surveyor*, which arrived at Mars on September 11, 1997, and was placed in a circular orbit 400 km above the planet's surface. The *Global Surveyer* is presently mapping the planet and is producing close-up images with high resolution. Other Mars orbiters and landers are scheduled to be launched in 1999 and 2001.

To send a space probe to Mercury, we must give it a speed, relative to Earth, of 7.3 km/s (16,000 mi/h) in the direction opposite Earth's orbital motion. This speed, plus other parameters, will place the probe in an elliptical orbit that coincides with Mercury's orbit.

Before we can send a space probe off to Mercury, it has to be launched from Earth's surface with enough speed to escape Earth's gravity. The minimum vertical launch speed required for a space probe to escape Earth's gravity is 11.2 km/s (25,000 mi/h). Since this is greater than 7.3 km/s, the speed of the space probe must be reduced so that it can be sent on the proper course to Mercury after escaping Earth's gravity. The speed of the probe also must be adjusted to compensate for the gravitational force of Mercury.

Another parameter is launch time. Since Earth and Mercury are moving eastward around the Sun at different speeds, they must be in proper relative positions at the time of launch, or the space probe will miss its target. A term often used to describe this period is "launch window." This term refers to the interval of time when the space probe and the target planet are located in the proper positions for a successful mission.

Space probe flights have been made to all planets in the solar system except Pluto. All these flights have taken place in the ecliptic plane to take advantage of Earth's orbital motion to assist in projecting the probes on their journeys.

Scientists also want to examine all areas of the Sun. To accomplish this, the space probe must circle the Sun from pole to pole. This means that the probe has to be propelled out of the ecliptic plane. Launching a probe from Earth to enter a polar orbit around the Sun requires a velocity component of 29 km/s opposite Earth's orbital motion plus a perpendicular velocity to send the probe out of the ecliptic plane. Presently no launch vehicle can produce the thrust necessary to achieve these velocities. An alternative is to use the planet Jupiter to assist the probe to obtain the necessary velocities.

On October 6, 1991, the *Ulysses* space probe was released from the space shuttle *Discovery* and propelled into space on its way to the Sun by way of Jupiter. *Ulysses* arrived at Jupiter in February 1992 and looped around Jupiter in the opposite direction to the planet's orbital motion, thereby reducing its energy. At the same time, Jupiter's gravity flung *Ulysses* down out of the ecliptic plane toward the south polar region of the Sun, where it arrived in June 1994.

The probe's instruments, including solar wind plasma and ion detectors, magnetometers, energetic particle detectors, radio and plasma wave instruments, solar X-ray and gamma-ray burst detectors, and cosmic dust sensors, are examining every latitude of the Sun by circling it, pole to pole, from a distance of approximately 2 AU.

The space probe has detected shock waves millions of kilometers in diameter, which are generated by material ejected from the Sun's southern latitudes. Solar winds with speeds nearly double those ejected from the equatorial region have been detected flowing through large holes in the Sun's southern corona, and measurements indicate the Sun's magnetic field at the equator is not as strong as it is at the poles.

Ulysses has an orbital period of 6.2 years. The probe, after returning to its aphelion point at Jupiter in 1997–98, will travel back to the Sun for a second sweep over the Sun's poles in the years 2000–2001.

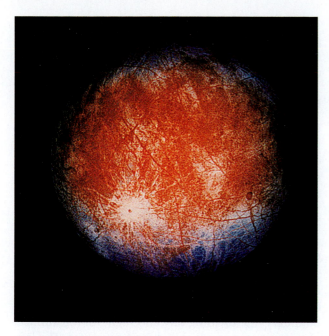

FIGURE 15.19   Europa

This image of Europa was taken by the *Galileo* spacecraft. White in this image indicates fine-grained, fragmented ice. The streaks are typically 20 to 40 km wide. (See the text for more details.)

FIGURE 15.20   Saturn's Rings in False Color

Possible variations in chemical composition from one part of Saturn's ring system to another are visible in this *Voyager 2* photograph as subtle color variations that can be recorded with special computer-processing techniques. This highly enhanced color view is assembled from colorless, orange, and ultraviolet frames obtained at a distance of 8.8 million km. The C ring and Cassini division appear blue in the photo.

from the satellite's hot core providing a warm temperature beneath the icy crust.

Saturn

The most distinctive feature of Saturn is its system of three prominent rings, which are less than 50 m thick and believed to be composed of particles of ice and ice-coated rocks ranging in size from a few micrometers to approximately 10 m in diameter. (See photo on page 404.) These rings have been viewed by Earth-based observers for centuries and are the most spectacular celestial sight that can be seen with a small telescope. The rings, inclined by 27° to Saturn's orbital plane, are identified by the letters A, B, and C.

Within a certain distance of any planet's center, called the *tidal stability limit,* or the **Roche limit,** the planet's tidal force* (due only to gravity), acting on a

large, solid, revolving object, will tear the object apart because the tidal forces are greater than the binding forces (gravitational only) holding the object together. The Roche limit is directly proportional to the radius of the planet and is a function of the density of the planet and the orbiting object. The Roche limit for Saturn is about 2.5. Thus a large, solid object within 2.5 times the planet's radius will be broken into fragments. Saturn's outer A ring is at 2.3 Saturn radii.

The *Voyager 1* and *Voyager 2* flights showed the rings to be very complicated systems of many individual ringlets (● Fig. 15.20). The photograph, taken from *Voyager 2* at a distance of 8.8 million km, shows the possible variations in the chemical composition from one part of Saturn's ring system to another.

The structure of Saturn itself is similar to that of Jupiter; that is, it has a small, solid core surrounded by a layer of ice, a layer of metallic hydrogen, and an outer layer of liquid hydrogen and helium. Saturn's density is only 0.7 g/cm^3, so it would float in water. Its temperature near the top of the clouds is approximately 120 K. Like Jupiter, Saturn radiates more heat than it gets from the Sun.

---

*Tidal force is a differential gravitational force that tends to deform or stretch a body. (See Section 17.6 for a specific explanation of tidal force.)

FIGURE 15.21   Saturn's Great White Spot

The Hubble Space Telescope's Wide-Field Planetary Camera used blue and infrared light to record this view of Saturn in November of 1990. This picture combines two colors to show the lower parts of the clouds in blue and the region with high clouds in red.

Saturn's mass is 95 times that of Earth, and its diameter is nine times larger. It rotates about once every 11 h. It has a magnetic field that is 1000 times stronger than Earth's.

Saturn's atmosphere is similar in chemical composition to that of Jupiter. The atmosphere has a banded structure, but it appears dull and not as bright as Jupiter's atmosphere. This is due to Saturn's surface temperature being lower than that of Jupiter.

● Figure 15.21 shows an image of Saturn taken by the Hubble Space Telescope. The photo shows clearly the banded structure and the Great White Spot, a rare and unusual cloud formation occurring over the equatorial region. The clouds are believed to be composed mainly of ammonia ice crystals.

Outside the main visible rings lie 20 or more moons of Saturn. Many of them are quite small, with diameters between about 30 and 100 km, and were discovered in 1980 by *Voyager 1*.

The most interesting moon of Saturn is Titan, its largest moon, with a diameter of 5150 km and a density of 1.9 g/cm^3. It is the only satellite in the solar system known to have a dense, hazy atmosphere, which is probably due to its low surface temperature of 94 K.

The main constituent of Titan's atmosphere is nitrogen (about 90%); also present are argon (less than 10%), methane (less than 2%), and traces of other hydrocarbons.

Uranus

Uranus was discovered in 1781 by William Herschel (1738–1822), an English astronomer. The name was chosen in keeping with the tradition of naming planets for the gods of mythology. Uranus was the father of the Titans and the grandfather of Jupiter.

● Figure 15.22 shows two pictures of Uranus taken by *Voyager 2*. The right photo has been processed to show Uranus as human eyes would see it from the spacecraft. The blue-green color results from the absorption of red light by methane gas in Uranus' atmosphere. The darker shading at the lower left of the disk corresponds to the day-night boundary. This boundary line dividing day and night on the surface of a planet or moon is called the **terminator.** The left photo uses false colors and contrast enhancement to show distinctive details in the polar region. The false-color picture reveals a dark polar hood surrounded by a series of progressively lighter concentric bands. One possible explanation is that a brownish haze is concentrated over the polar region.

The internal structures of Uranus and Neptune are similar, but differ from those of Jupiter and Saturn. Uranus and Neptune are much smaller and less mas-

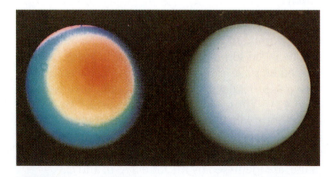

FIGURE 15.22 Uranus

These two pictures of Uranus were taken by the narrow-angle camera of *Voyager 2* when the spacecraft was 9.1 × 10^6 km from the planet. The photo on the right shows Uranus as the human eye would see it from the vantage point of the spacecraft. The photo on the left employs enhanced color. (See the text for details.)

FIGURE 15.23   Miranda, Uranus' Fifth Largest Satellite
Uranus' innermost large moon, Miranda, is roughly 470 km in diameter and exhibits a variety of geologic forms—some of the most bizarre forms in the solar system. Chevron-shaped regions and folded ridges in circular racetrack patterns are visible on the satellite's surface. There are large scarps, or cliffs, ranging up to 5 km in height; they are clearly visible in the lower right part of the photo. Next to them is a deep canyon approximately 50 km wide.

sive than Jupiter and Saturn. Also, their rocky cores are relatively large compared with their total size (see page 405).

Uranus has a ring system that is very thin. The rings are composed mainly of boulder-sized particles 1 m or larger in diameter, with very few dust-size particles present. Because of the lack of dust particles, the rings do not reflect light as well as the rings of Saturn, which are filled with tiny particles 1 cm and smaller. *Voyager 2* also recorded some very narrow sections of rings.

Uranus has five major satellites. They are, in order of increasing distance from the planet, Miranda, the smallest, having a diameter of 470 km and orbiting at a distance of 129,000 km; Ariel; Umbriel; Titania, the largest, with a diameter of 1587 km; and Oberon, the most distant, orbiting at 584,000 km. The satellites have densities of about 1.6 g/cm^3, which indicates a composition of a mix of water ice and rock.

The surface features of the satellites show that, with the exception of Umbriel, the moons have been tectonically active in the past. ● Figure 15.23 shows Miranda's surface, with large curvilinear regions of

grooves and ridges plus regions that appear chevron-shaped. The satellite's surface is pockmarked with craters.

Neptune

Neptune was discovered in 1846 by Johann G. Galle (1812–1910), a German astronomer at the Berlin Observatory. Partial credit is also shared by Englishman John Couch Adams and Frenchman U. J. J. Leverrier, two mathematicians. Using Newton's law of gravitation, Adams and Leverrier made calculations that produced information on where to look for a suspected planet that was disturbing the motion of Uranus. The name Neptune was proposed by D. F. Arago, the French physicist who had suggested that Leverrier begin the critical calculations.

In 1989 *Voyager 2* arrived at Neptune and sent back to Earth a photographic record of Neptune's clouds, storms, Great Dark Spot (see ● Fig. 15.24) similar to Jupiter's Great Red Spot, large wind systems, eight satellites, five rings, and thin layers of dust.

FIGURE 15.24   *Voyager's* Image of Neptune's Atmosphere
This image of Neptune was taken when the *Voyager 2* spacecraft was 6.1 million km from the planet. The dominant storm system in the atmosphere is the Great Dark Spot.

A SPOTLIGHT ON: The Jovian Planets

Diameter (equatorial)	142,980 km
Mass	1.90×10^{27} kg
Density (mean)	1330 kg/m³
Distance from Sun (mean)	7.783×10^{8} km
Rotation period (equatorial)	9 h 50 min 28 s
Sidereal period	11.86 years
Surface temperature (mean)	-110°C
Magnetic field	Yes (strong)
Number of satellites	16 known
Composition of atmosphere	86% H_2, 13.8% He

Jupiter The largest planet

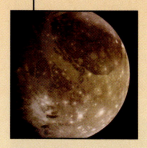

Ganymede

Callisto

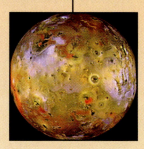

Io

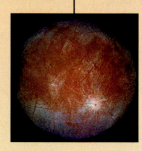

Europa

Diameter (equatorial)	120,540 km
Mass	5.69×10^{26} kg
Density (mean)	690 kg/m³
Distance from Sun (mean)	1.427×10^{9} km
Rotation period (equatorial)	10 h 39 min 25 s
Sidereal period	29.458 years
Surface temperature (mean)	-180°C
Magnetic field	Yes
Number of satellites	20 known
Composition of atmosphere	92.4% H_2, 7.4% He

Saturn The beautiful planet

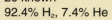

Enceladus

Dione

Rhea

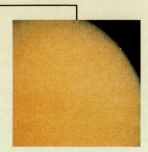

Titan

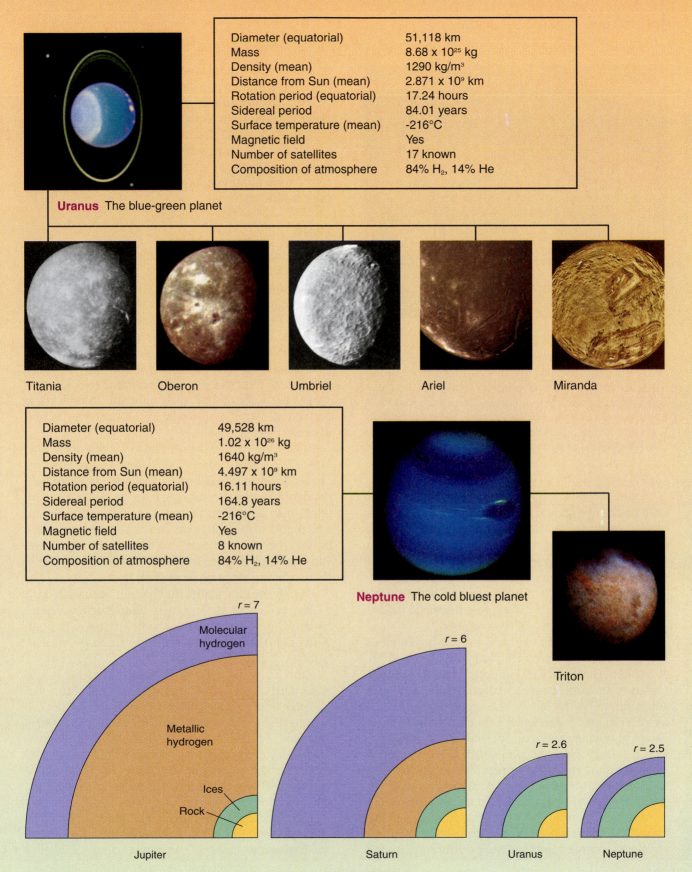

Diameter (equatorial)	51,118 km
Mass	8.68×10^{25} kg
Density (mean)	1290 kg/m³
Distance from Sun (mean)	2.871×10^9 km
Rotation period (equatorial)	17.24 hours
Sidereal period	84.01 years
Surface temperature (mean)	-216°C
Magnetic field	Yes
Number of satellites	17 known
Composition of atmosphere	84% H_2, 14% He

Uranus The blue-green planet

Titania   Oberon   Umbriel   Ariel   Miranda

Diameter (equatorial)	49,528 km
Mass	1.02×10^{26} kg
Density (mean)	1640 kg/m³
Distance from Sun (mean)	4.497×10^9 km
Rotation period (equatorial)	16.11 hours
Sidereal period	164.8 years
Surface temperature (mean)	-216°C
Magnetic field	Yes
Number of satellites	8 known
Composition of atmosphere	84% H_2, 14% He

Neptune The cold bluest planet

Triton

$r = 7$

Molecular hydrogen

Metallic hydrogen

Ices

Rock

$r = 6$

$r = 2.6$

$r = 2.5$

Jupiter   Saturn   Uranus   Neptune

Internal Structure: The appropriate radius (r) is given in units of 10^4 km.

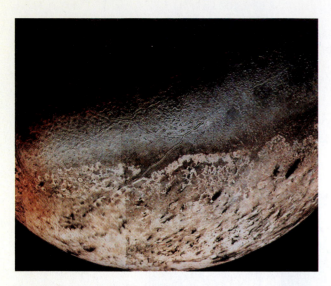

FIGURE 15.25 Triton's South Polar Cap
Neptune's largest satellite is primarily a white object with a pinkish cast in some areas. The pinkish color is probably due to frozen nitrogen. The land areas are strange and complex, and a scarcity of craters indicates the surface may have been melted or flooded by icy "slush." A number of high-resolution images were combined to produce this image of the south polar region.

Neptune can be regarded as a twin to Uranus. Not only are the two similar in size and in the composition of their atmospheres, but their internal structures are thought to be similar also. Each planet has a rocky core surrounded by a mantle of water, methane, and ammonia ice. The mantle is surrounded by a layer of gas composed mainly of hydrogen and helium. Data from *Voyager 2* revealed Neptune's magnetic field to be comparable to that of its twin planet.

Because of its thick methane-rich hydrogen cloud cover, Neptune's surface features could not be photographed. However, excellent photographs were obtained of its largest moon, Triton, which is slightly smaller than Earth's Moon. The satellite has a diameter of 2700 km, a density of 2.07 g/cm^3 and is composed of silicates and ices. Surface details are shown in ● Fig. 15.25. Note the white polar cap of frozen nitrogen and the strange, complex land forms. The surface temperature is about 37 K, which is below the freezing point of nitrogen. Triton has a thin, gaseous atmosphere composed of nitrogen and a small amount of methane. Because Triton orbits Neptune retrograde, some scientists believe the satellite is a

planetary body, similar to Pluto, that has been captured by Neptune's gravitational field.

Orbiting Neptune is a system of five rings and a sheet of dust in the equatorial region. The rings are not optically visible from Earth but were thought to exist from their occultation (blocking out) of starlight. But the observations were questionable. In 1989 *Voyager 2* took photographs of the rings and confirmed their reality.

Pluto

Pluto, named for the god of outer darkness, is the most distant planet from the Sun. It was discovered in 1930 at the Lowell Observatory in Arizona by C. W. Tombaugh, who did a thorough search near the position predicted by theoretical calculations. The planet had been predicted because discrepancies appeared in the orbital motions of Uranus and Neptune. Spectroscopic investigations indicate that the planet is covered with methane ice. The surface temperature ranges from about 50 K near aphelion to 60 K near perihelion. Pluto is presently in the warmest part of its orbit. (See Fig. 15.6.)

In June 1978 a satellite of Pluto was discovered by James W. Christy of the U.S. Naval Observatory. Named Charon ("KEHR-on"), this moon is about half the diameter of Pluto, making it the largest satellite in relation to its parent planet. Simultaneously, Pluto was found to be much smaller than previously believed. The image shown in ● Fig. 15.26 was taken with the Hubble Space Telescope and for the first time shows Charon separate from Pluto.

Pluto is the only planet that has not been visited by a space probe. A space probe to Pluto is scheduled to be launched December 2004 with an arrival date of 2013 or 2016. The arrival date depends on the type of launch vehicle.

Pluto has a diameter of 2300 km (Charon's is approximately 1200 km) and a mass of 1.29×10^{22} kg. The planet has a bright polar cap (probably frozen nitrogen) and a patchwork of bright and dark surface areas with complex markings. Charon's surface is covered with water ice. Pluto and its satellite have an average density of approximately 2 g/cm^3 and are composed of silicates and ices. The ices are methane, nitrogen, carbon monoxide, and water.

There are close similarities (diameters, densities, atmospheres, and surface features) between Pluto and Neptune's satellite Triton. Some scientists suggest

FIGURE 15.26   Pluto and Its Satellite Charon
Pluto was the first solar system object to be observed by the Hubble Space Telescope. This photo shows, for the first time, Pluto separate from its satellite. The circular halo around the planet is caused by a defect (spherical aberration) in the telescope's primary mirror that has since been corrected.

that both are planetary bodies (large asteroids) captured from interplanetary space.

Should Pluto be classified as a true planet? There is some disagreement about this question because of Pluto's highly tilted and eccentric orbit. Pluto went inside Neptune's orbit in late 1978 and exited in 1999. Thus for more than 20 years Neptune was the most distant planet from the Sun. Also, the physical characteristics of Pluto are not similar to those of the other eight major planets.

But just what *is* a planet? The term originated with ancient observers of the stars. These observers of the nighttime sky viewed, with the unaided eye, starlike objects (Mercury, Venus, Mars, Jupiter, and Saturn) that moved with respect to the stars. They called the starlike objects *planets*, from a Greek word meaning "wanderer." Earth was not included because the observers thought they were at the center of the universe. The other planets—Uranus, Neptune, and Pluto—were unknown to them. Presently, a planet is defined, in general terms, as any of the nine spherical bodies orbiting the Sun or similar-sized bodies orbiting other stars. Thus, by definition, Pluto is a planet.

On February 5, 1999, the International Astronomical Union (a sixty-one-nation organization) officially closed the debate and classified Pluto as a planet.

RELEVANCE QUESTION: *During the week of July 16–20, 1994, more than 20 fragments of Comet Shoemaker-Levy 9 smashed down on the cloud tops of planet Jupiter. What is the potential danger to both planet Earth and you from celestial impacts of asteroids or comets?*

15.5 Other Solar System Objects

LEARNING GOAL

▼ Describe asteroids, meteoroids, comets, and interplanetary dust.

Thus far this chapter has dealt with the planets, their satellites, and their ring systems. The Sun is the predominant mass of the solar system and holds the system together with its strong gravitational field whose influence may extend outward to 100,000 AU and beyond. The Sun supplies energy to all members of the solar system, thereby controlling their temperatures as well as their orbital motions. This section will consider asteroids, meteoroids, comets, and interplanetary dust. The Sun and the solar wind, the stream of charged particles expelled from the Sun and flowing outward through the solar system, will be discussed in Chapter 18.

Asteroids

In 1801, the first of the many planetary bodies between the orbits of Mars and Jupiter was discovered by Giuseppi Piazzi, an Italian astronomer. This small body is named Ceres after the protective goddess of Sicily. Ceres is 940 km in diameter, has an orbital period of 4.6 y, and is the largest of more than 2000 named and numbered objects that orbit the Sun between Mars and Jupiter. These objects are called **asteroids,** or *minor planets*. The three largest are Ceres (940 km), Pallas (580 km), and Vesta (540 km). Ceres and Pallas have very low albedos. Only Vesta, which has a relatively high albedo, can be seen with the unaided eye and then only when it is simultaneously at opposition and perihelion.

The first close-up image of an asteroid was taken in 1991 by the Jupiter-bound *Galileo* spacecraft. ● Figure 15.27 is a portrait of the irregularly shaped aster-

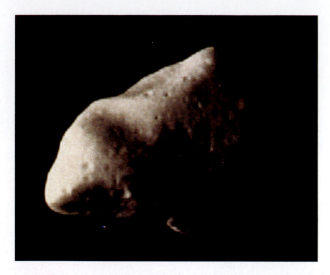

FIGURE 15.27 Asteroid Gaspra

This image of Gaspra was taken by the Jupiter-bound *Galileo* spacecraft in October 1991, from a distance of about 16,000 km. This is the first close-up view of an asteroid. Gaspra is a stony asteroid whose surface is covered with rocks moderately less gray than those on Earth's Moon. The asteroid is approximately 19 by 12 by 11 km and irregular in shape.

oid named 951 Gaspra, which is about 11 km wide and 19 km long. Gaspra is classified as a stony asteroid. Its surface is covered to a depth of about 0.9 m with a loose, rocky, gray material called *regolith.*

The diameters of the known asteroids range from that of Ceres (940 km) down to only a few kilometers, but most asteroids are probably less than a few kilometers in diameter. There are perhaps billions the size of boulders, marbles, and grains of sand. Only the largest asteroids are spherical in shape; the others are irregular in shape. Eros, which can approach Earth to within 1 AU, is about the size and shape of the island of Manhattan.

Like the planets, asteroids revolve counterclockwise around the Sun, as seen from above the ecliptic plane, with an average inclination to the ecliptic plane of 10°. About 100,000 asteroids exist that can be detected with Earth-based telescopes, but the total mass of all the asteroids orbiting between Mars and Jupiter is much less than the mass of Earth's Moon. Although most asteroids move in an orbit between Mars and Jupiter, some have orbits that range beyond Saturn or inside the orbit of Mercury.

Asteroids are believed to be early solar-system material that never collected into a single planet. One piece of evidence supporting this view is that there seem to be several different kinds of asteroids. Those at the inner edge of the belt appear to be stony, whereas the ones farther out are darker, indicating more carbon content. A third group may be composed mostly of iron and nickel.

Meteoroids

Meteoroids are interplanetary metallic and stony objects that range in size from a fraction of a millimeter to a few hundred meters. They are probably the remains of comets and fragments of shattered asteroids. They circle the Sun in elliptical orbits and strike the Earth from all directions at very high speeds. Their high speed, which is increased by Earth's gravitational force, produces great frictional heating when the meteoroids enter Earth's atmosphere.

A meteoroid is called a **meteor,** or "shooting star," when it enters Earth's atmosphere and becomes luminous because of the tremendous heat generated by friction with the air. Most meteoroids are vaporized in the atmosphere, but some larger ones may survive the flight through the atmosphere and strike Earth's surface, in which case they become known as **meteorites.** When a large meteorite strikes Earth's surface, a large crater is created. ● Figure 15.28 is a photo-

FIGURE 15.28 The Barringer Meteorite Crater Near Winslow, Arizona

The crater is 1300 m across and 180 m deep, and its rim is 45 m above the surrounding land.

FIGURE 15.29   The Tucson Ring Meteorite

This 623-kg meteorite was found by the first Spanish explorers near Tucson, Arizona, in 1851. Indians had known of it 300 years prior to this. The Smithsonian acquired the meteorite early in this century.

graph of a sizable meteorite crater near Winslow, Arizona, which scientists estimate to be more than 25,000 years old.

The largest known meteorite, with a mass of more than 55,000 kg, fell in southwest Africa. The largest known meteorite found in North America, with a mass of about 36,000 kg, was found near Cape York, Greenland, in 1895 and is on display at the Hayden Planetarium in New York City.

Meteorites vary in size and shape and are classified into three broad groups: irons, stones, and stony-irons. About 94% of all meteorites that fall on Earth are stones. Stones are composed of silicate minerals and other minerals and have the appearance of ordinary rocks found in Earth's crust. Thus they are difficult to identify. Irons are mostly iron with 5–20% nickel. Stony-irons are mixtures of iron and stony materials, as the name implies. An iron meteorite is shown in ● Fig. 15.29.

Comets

Comets are named from the Latin words *aster kometes,* which mean "long-haired stars." They are the solar system members that periodically appear in our sky for a few weeks or months and then disappear. A **comet** is a reasonably small object composed of dust and ice, and revolves about the Sun in a highly elliptical orbit. As it comes near the Sun, some of the surface vaporizes to form a gaseous head and usually a long tail.

A comet consists of four parts: (1) the *nucleus,* typically a few kilometers in diameter and composed of rocky or metallic material, as well as solid ices of water, ammonia, methane, and carbon dioxide; (2) the head, or *coma,* which surrounds the nucleus, can be as much as several hundred kilometers in diameter and is formed from the nucleus as it approaches within about five astronomical units of the Sun; (3) the long, voluminous, and magnificent *tails,* composed of ionized molecules, dust, or a combination of both, and which can be millions of kilometers in length; and (4) a spherical cloud of hydrogen surrounding the coma, believed to be formed from the dissociation of water molecules in the nucleus. The sphere of hydrogen in some comets may have a radius exceeding that of the Sun (● Fig. 15.30).

Comets are visible by reflected sunlight and the fluorescence of some of the molecules comprising the

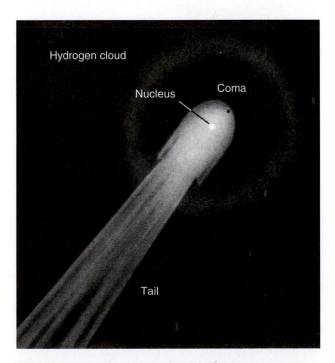

FIGURE 15.30   The Principal Parts of a Comet

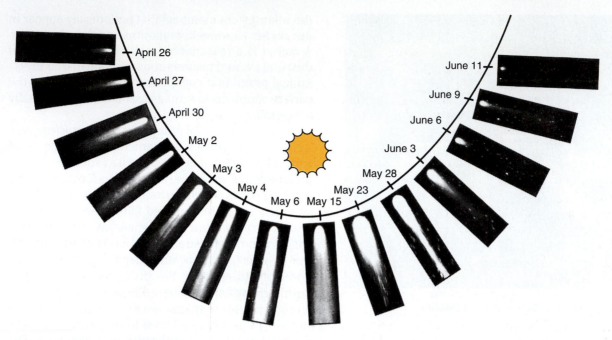

FIGURE 15.31 Halley's Comet

These 14 views of Halley's comet were taken between April 26 and June 11, 1910. Note the change in size of the coma and tail.

comet. As comets approach the Sun and move around it, the amount of material in the coma and tail gets larger (● Fig. 15.31). This increase in size is evidently caused by the Sun heating a thin outer shell of the comet's nucleus—perhaps by (1) the solar wind, which consists of streams of particles (electrons, protons, and the nuclei of light elements) given off by the Sun and driven outward at speeds of hundreds of kilometers per second, and (2) radiation pressure generated by the radiant energy given off by the Sun.

The years 1996 and 1997 were the best ever for observing comets. The brightest comet was Hale-Bopp, named for the codiscoverers. Hale-Bopp was discovered in July 1995 and reached maximum brightness (first magnitude) for a few weeks around its perihelion (closest approach to the Sun) date of April 1, 1997.

In January 1996, Japanese amateur astronomer Yuji Hyakutake discovered Comet Hyakutake, which reached maximum brightness (second magnitude) for a few weeks around its perihelion date of May 1,

1996. This comet and Hale-Bopp were observed by people around the world.

Halley's Comet, named after the British astronomer Edmond Halley (1656–1742), is one of the brightest and best-known comets. Halley was the first to suggest and predict the periodic appearance of the same comet. He observed the comet that bears his name in 1682 and, using Newton's law of motion, correctly predicted its return in 76 years. Halley's comet has appeared every 76 years, including 1910 and 1986.

Scientists believe that comets originate and evolve from dirty, icy objects that were part of the primordial debris thrown outward into interstellar space when the solar system was formed. These dirty, icy objects are believed to be more dirt than ice and are often described as "frozen mudballs." We observe these objects when they enter the vicinity of the Sun and develop a long plasma tail. The source region from which comets originate is called the **Oort Cloud,** in honor of Jan Hendrik Oort, a Dutch astronomer who proposed its existence in 1950. The highly eccentric orbits of typical comets indicate that their greatest distance from the Sun is about 50,000 AU. This source region is near the outer limits of the Sun's gravitational influence, and the objects within this region can be perturbed by passing stars.

What objects, if any, are located in the vast volume of space between Pluto's orbit (39 AU) and 50,000 AU? This vast volume of space is called the *Kuiper* (rhymes with "piper") *Belt* in honor of Gerard P. Kuiper (1905–1973), the Dutch-born American astronomer who suggested that this space is a region containing solar debris that serves as a reservoir of available cometary material. The debris cannot be detected presently by astronomers because of its great distance and small size. The objects in this volume of space are held more firmly by the Sun's gravity and are not influenced by nearby stars.

Where is the outer boundary of the solar system? As indicated above, the Oort Cloud is located in the vicinity of 50,000 AU. Does the Sun's gravitational force field have any influence beyond this limit? The answer is yes, according to some astrophysicists who have made some computer simulations that indicate that the outer limit for the influence of the Sun's gravitational field is between 80,000 and 100,000 AU. This limit is where the Sun's gravitational force field is balanced by that of the Milky Way. Beyond this boundary, the Sun's gravitational force field cannot hold any object.

Interplanetary Dust

In addition to the planets and other large bodies discussed thus far, the tremendous volume of the solar system's space is occupied by very small, solid particles known as *micrometeoroids,* or **interplanetary dust.** Two celestial phenomena, which can be observed with the unaided eye or photographed, show that the dust particles do exist.

Perhaps on a very clear, dark night, just after sunset in the western sky, you have observed the first of these phenomena—*zodiacal light*—a faint band of light along the zodiac (ecliptic). The band of light can also be seen just before sunrise. The faint glow is due to reflected sunlight from the dust particles.

The other phenomenon is called the *Gegenshein,* which means "counterglow," and is also due to reflected sunlight from dust particles. This faint glow is observed on the ecliptic exactly opposite the Sun. It is more difficult to observe than zodiacal light but appears as a diffuse, oval spot with an average angular size of nine degrees.

RELEVANCE QUESTION: *A physical science class is taking a geology field trip to identify surface rocks. Which of the following—meteoroid, meteor, or meteorite—would it be possible to find during the trip?*

15.6 The Origin of the Solar System

LEARNING GOAL

▼ Describe the theory for the origin of the solar system that is most widely accepted by astronomers.

Any theory that purports to explain the origin and development of the solar system must account for the system as it presently exists. The preceding sections in this chapter have given a general description of the system in its present state, which, according to our best measurements, has lasted for about 4.5 billion years.

If a theory for the origin is to have validity, the following questions concerning major properties of the solar system must have acceptable answers:

1. What was the origin of the material used to form the system?
2. What were the forces that acted to form the system?
3. Why are the planets in isolated, almost circular orbits that are located nearly in the same plane?
4. Why do the Sun, the planets, and the satellites of the planets all revolve in the same direction?
5. Why do the Sun, all planets except two, and nearly all satellites of the planets rotate in the same direction?
6. What determined the chemical and physical properties of the planets, and why are the terrestrial planets so much different from the Jovian planets?
7. What is the origin of the asteroids, which have properties that are different from the terrestrial and Jovian planets?
8. How do the comets and meteoroids fit in with the theory?

Presently, most astronomers believe that the formation of the solar system began with a large, swirling

FIGURE 15.32   The Formation of the Solar System

This drawing illustrates the formation of the solar system according to the condensation theory. The protostellar cloud of gas and dust (1) in gravitational collapse developed into a flattened, rotating disk (2) called the primordial nebula. Additional contraction produced a disk (solar nebula) whose masses condensed and accreted to cause the planets to form in the low-temperature regions while the Sun formed in the central part.

volume of cold gases and dust—a rotating **primordial nebula,*** positioned in space among the stars of the Milky Way, that contracted under the influence of its own gravity. Through the process of condensation and accretion, the nebula evolved into the system we observe today. This explanation is known as the **condensation theory,** and it is supported by the fact that today we observe such nebulae throughout the universe. What initiated the process of forming the nebula from interstellar matter is unknown.

The interstellar dust played a major role in the condensation process by allowing condensation to take place before the gas had a chance to disperse. The collection of particles was slow at first but became faster and faster as the central mass increased in size. As the particles moved inward, the rotation of the mass had to increase to conserve angular momentum. Because of the rapid turning, the cloud began to flatten and spread out in the equatorial plane (see ● Fig. 15.32). Kepler's third law states that the central part must move faster than the outer parts. This motion sets up shearing forces, which, coupled with variations in density, produced the formation of other masses that moved around the large central portion, the "protosun," sweeping up more material and forming the protoplanets.

The "protoearth" was perhaps 1000 times more massive than planet Earth. Temperature played a major role in the formation of protoplanets through condensation. The terrestrial protoplanets, receiving more heat than the Jovian protoplanets, were greatly affected by this heating and were formed out of rocky or metallic material. The Jovian protoplanets, being at great distances and receiving little radiation, were formed in a cold environment out of low-density, icy material. Even today they appear in a similar protoplanet stage. Planetary satellites are believed to have been formed from similar accretion of matter surrounding their protoplanets. (See the Highlight in Section 17.2 for a discussion of the origin of Earth's Moon.) Theory suggests that the formation of the solar system took place over a 100-million-year period beginning about five billion years ago.

During the early stages of development, the space between the protosun and the protoplanets was filled with large amounts of gas and dust, shielding the protoplanets from the protosun, which was beginning to fuse hydrogen into helium and radiate energy. With the passing of time, the space between the protosun and the protoplanets became transparent due to the accumulation of material by the planets and the ejection of dust and gas by radiation from the Sun and by the solar wind.

The preceding discussion of the origin of the solar system is a theoretical explanation that is constructed from observational evidence. Much of the knowledge concerning the formation process comes from complete chemical and physical examination of meteorites, which are the least-changed fragments of the initial solar system.

As mentioned previously, astronomers do not know what initiates the formation of a star from interstellar matter. Does gravitational attraction alone bring about condensation and accretion, or are there other factors involved, such as a shock wave moving through the nebula cloud? What is the role of electric and magnetic fields? These and other questions remain to be answered.

RELEVANCE QUESTION: Day after day you live in the only known solar system in the universe that harbors life. One of the following answers is probably true. Which one do you think is most probable? Justify your answer.

A solar system is formed through
(a) normal stellar evolution.
(b) a rare event.
(c) special conditions.

15.7 Other Planetary Systems

LEARNING GOAL

▼ Describe the latest information concerning other planetary systems.

Are there other planetary systems in the universe? As mentioned in the preceding section, our solar system is believed to have originated from a rotating solar nebula that was disk-shaped during the early stages of its formation. ● Figure 15.33 shows a star called Beta Pictoris that is about 50 light-years (ly) from Earth. This star, with its apparently dust-laden disk, is one clue we have that other planetary systems may exist.

---

*Some textbooks call this a solar nebula. The flattened, rotating disk of gas and dust around the protosun, from which the planets were formed, is called the *solar nebula*.

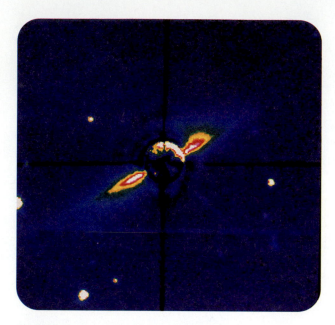

FIGURE 15.33   Beta Pictoris Disk

This optical photograph shows material around Beta Pictoris, a star about 50 ly from the solar system. Light from the star has been blocked out by a coronagraph, an instrument placed in front of the optical telescope. More detailed observation of the material may reveal a planetary system.

Detecting extrasolar planets is no easy task. One method used to detect a star with a companion planet is to observe a star's motion. A star with a large planet has a small wobble superimposed on its motion due to gravitational effects. The change in motion is very small and difficult to detect. It is best detected by the Doppler shifts that change the pattern of the star's spectrum. As the star approaches the observer, the wavelengths are compressed. As the star moves away from the observer, the wavelengths are lengthened (see Section 6.4). The astronomer must detect the changes generated by the Doppler shifts, which emerge from motions at speeds of only a few meters per second. To detect a star's wobble caused by a planet the size of Jupiter requires a measurement of about three meters per second. Detecting smaller planets than Jupiter is not possible using present technology. Since gravitational forces produce the wobble, the size of the wobble gives the planet's mass, and the wobble's cycle time is used to determine the orbital period. Using the period and Kepler's third law, the planet's average distance from the star can be determined.

Pulsars are very dense, rapidly rotating stars with precise periods. The rotation rate is observable because the pulsar emits pulses of radio waves that sweep past Earth like the beam of a rotating searchlight. When the beam shows a regular variation in pulse arrival time, the variations indicate gravitational disruption by a rotating object about the pulsar.

Using the Arecibo Observatory in Puerto Rico, astronomers at Pennsylvania State University reported in 1992 the discovery of two objects revolving about a millisecond pulsar. One object (a planet) orbits the pulsar at a distance of 0.4 AU every 66.6 Earth days, and the other at 0.5 AU every 98.2 days. The method of detection was an observable Doppler shift in the spectrum of the pulsar. The pulsar, named PSR 1257 +12 (see Section 18.2), is located some 1300 ly from Earth in the constellation Virgo.

The 1992 data were confirmed in 1994 with the detection of additional wavering of the pulsar's radio signals due to an extra wobble in the pulsar's motion caused by gravitational forces between the two planets. A third planet with a mass of about 0.02 Earth mass and orbiting 0.19 AU from the pulsar also has been detected. Based on confirmed evidence, the planets became the first detected beyond our solar system. The data also indicated that other objects may be orbiting the pulsar.

The second reported candidate for an extrasolar planetary system comes from astronomers at the Geneva Observatory in Switzerland. The astronomers are studying 51 Pegasi, a type G2-3 main-sequence star some 40 ly from Earth. The Sun is a type G2 star. Thus this is the first planet discovered beyond our solar system that is revolving around a star similar to our Sun. In October 1995 they found 51 Pegasi's line-of-sight velocity changing periodically by some 70 m/s every 4.23 days. Their data indicate that a huge planet about 90,000 kilometers in diameter with a surface gravity seven times Earth's is revolving around the star some seven million kilometers (0.051 AU) away. The planet is about half as massive as Jupiter.

In January 1996 astronomers at San Francisco State University announced the discovery of two separate planetary objects. One planet orbits a star known as 47 Ursae Majoris some 40 to 50 ly from Earth in the Big Dipper. The planet is about 3.5 times the mass of Jupiter with an orbital radius of about 2 AU. The other orbits a star called 70 Virginis some 55 to 70 ly away in the constellation Virgo. This planet is eight times as

TABLE 15.3   Extrasolar Planets

Star	Distance (light-years)	Star Mass (Sun = 1)	Minimum Planet Mass (Jupiter = 1)	Orbital Semimajor Axis (AU)	Orbital Period (days)
51 Pegasi	50	1.0	0.45	0.051	4.23
70 Virginis	59	0.95	6.8	0.47	117
47 Ursae Majoris	46	1.1	2.4	2.1	1098
Rho' Cancri	44	0.85	0.93	0.11	14.6
Rho Coronae Borealis	57	1.0	1.1	0.23	39.6
16 Cygni B	72	1.0	1.7	1.7	802
Upsilon Andromedae	57	1.25	0.65	0.056	4.6
Tau Boötes	49	1.25	3.7	0.045	3.3
HD 114762	90	1.15	11.6	0.36	84

massive as Jupiter with an orbital radius of 0.47 AU. Technically, it is known as a *brown dwarf* star.

Table 15.3 gives the present data on these two extrasolar planets and seven others that have been detected recently around stars similar to the Sun.

A different approach to the discovery of other planetary systems is the search for signals from extraterrestrial intelligence. Scientists have been scanning the skies for years hoping to detect a radio signal from another solar system. The early searches were in the electromagnetic spectrum close to 21 cm, the wavelength at which interstellar hydrogen emits radiation, and near 18 cm, the wavelength radiated by the hydroxyl (OH) radical. Together, these form the water molecule (HOH). When the background noise level of the Milky Way is plotted as a function of wavelength, there is a minimum noise level at these two wavelengths. This is the quietest part of the electromagnetic spectrum and is referred to as the *water hole*.

The *s*earch for *extra*terrestrial *i*ntelligence, called SETI, has become more sophisticated. NASA is using equipment that can scan eight million channels at the same time, over frequencies ranging from 1.2 to 10 GHz. The NASA project is scanning the entire sky, aiming at 773 sunlike stars within 100 ly of the Sun. Perhaps this comprehensive search will detect evidence of extraterrestrials in other planetary systems.

The private scientific community is also involved in searching for radio signals from beyond our atmosphere. Project META (*M*egachannel *Extra*terrestrial *A*rray) began in 1985, but as of this date (1999), nothing has been found. The META project is funded by The Planetary Society. There is also a META II project that scans the southern skies with an 8.4-million-channel system analyzer located near Buenos Aires.

The next advance in the program is the operation of the BETA (*B*illion-channel *Extra*terrestrial *A*rray). (Actually, there are only 240 million channels.) The project is a joint venture of The Planetary Society, NASA, and the Bosack/Kruger Foundation. The electronic equipment will cover the entire water hole. This includes all wavelengths between those of hydrogen and the hydroxyl radical.

Our receiving technology is becoming very sophisticated. Presently, antennae, detectors, and computers are merging into one system for collecting data over a broad spectrum of wavelengths, time, and space.

RELEVANCE QUESTION: *Life forms similar to ours here on Earth may exist somewhere in the Milky Way or in another, more distant galaxy. Can we ever be sure there is intelligent life elsewhere in the universe? Can we ever be sure there is not? Which possibility do you think is more likely? Which possibility would you prefer?*

Important Terms

astronomy	terrestrial planets	rotation	asteroids (15.5)
universe	Jovian planets	revolution	meteoroids
solar system (15.1)	prograde motion	ecliptic	meteor
geocentric model	retrograde motion	Foucault pendulum	meteorites
heliocentric model	sidereal period	aberration of starlight	comet
law of elliptical paths	conjunctions	parallax	Oort Cloud
astronomical unit	synodic period	parsec	interplanetary dust
law of equal areas	opposition	Roche limit (15.4)	primordial nebula (15.6)
harmonic law	albedo (15.2)	terminator	condensation theory

Important Equations

Kepler's Third Law: $T^2 = kR^3$

Newton's Version of Kepler's Third Law:

$$(m_1 + m_2)\frac{T^2}{R^3} = \frac{4\pi^2}{G}$$

Review Questions

15.1 The Solar System: An Overview

1. Which of the following is true of the solar system?
   (a) It is a heliocentric system.
   (b) It is held together by gravitational forces.
   (c) It contains planets classified as terrestrial and Jovian.
   (d) All of the above.

2. Opposition refers
   (a) only to inferior planets.
   (b) only to superior planets.
   (c) to the position of a planet when it is observed on the same celestial longitude as the Sun.
   (d) to an inferior planet when it is observed 180° from the Sun.

3. Which one of the following is *not* one of Kepler's laws?
   (a) law of elliptical paths
   (b) law of equal areas
   (c) law of radial velocities
   (d) harmonic law

4. Name the major components of the solar system.

5. What is the significant difference between the geocentric and the heliocentric models of the solar system?

6. What is the explanation of Kepler's second law? (*Hint:* Refer to a conservation law.)

7. State the difference between prograde and retrograde motions.

8. Why is Pluto *not* classified as either a terrestrial or Jovian planet?

9. How does the orbital speed of a planet vary with distance from the Sun?

10. Distinguish between sidereal and synodic period.

11. What is the time called when a planet and the Sun have the same celestial longitude?

15.2 The Planet Earth

12. Which of the following are abundant on Earth but not on the other eight planets?
    (a) oxygen   (c) life
    (b) water   (d) all of the above

13. The Foucault pendulum is an experimental proof of Earth's
    (a) revolution.   (c) precession.
    (b) rotation.   (d) retrograde motion.

14. Albedo refers to the fraction of incident sunlight _____ by a surface.
    (a) absorbed   (c) reflected
    (b) dispersed   (d) refracted

15. State and explain two major motions of planet Earth.

16. Before 1900, what proof was there that Earth (a) rotated and (b) revolved?

17. What is the ecliptic?

18. What does the parsec measure? State the definition of one parsec.

19. State why the parallax of a star *cannot* be seen with the unaided eye.

20. The observation of stellar parallax by astronomers provided proof of which one of Earth's motions?

21. Define one astronomical unit.

15.3 The Terrestrial Planets

22. Which of the following statements concerning the terrestrial planets is true?
    (a) All are relatively small and have relatively small mass.
    (b) All are dense and rocky.
    (c) All have physical and chemical properties similar to Earth.
    (d) All of the above.

23. Which of the following statements concerning the terrestrial planets is false?
    (a) All have permanent or interim atmospheres.
    (b) All have magnetic fields except Venus.
    (c) All rotate counterclockwise as viewed from above the North Pole.
    (d) They are relatively close to the Sun with orbits close together as compared to the outer planets.

24. Which of the following is *not* a physical characteristic of a terrestrial planet?
    (a) small diameter
    (b) solid surface
    (c) relatively low density
    (d) relatively high-temperature environment

25. Name the terrestrial planets. Why are they called the terrestrial planets?

26. Why does Venus have a high surface temperature?

27. What evidence is there that a fluid once flowed on Mars?

28. How do the atmospheric pressures on Venus, Earth, and Mars differ?

29. Why do we observe Mercury and Venus only around sunset and sunrise?

30. Venus is the third brightest object in the sky, exceeded only by the Sun and the Moon. Venus is not at its brightest either at full phase or when it is closest to Earth. Explain why.

15.4 The Jovian Planets and Pluto

31. Which of the following statements concerning the Jovian planets is false? All Jovian planets
    (a) are composed mainly of hydrogen and helium.
    (b) have strong magnetic fields.
    (c) rotate faster than the terrestrial planets.
    (d) have a relatively high-temperature environment.
    (e) have rings.

32. Which of the following is *not* a physical characteristic of a Jovian planet?
    (a) gaseous
    (b) weak magnetic field
    (c) no solid surface
    (d) fast rotation

33. The basic physical reason for the existence of the Roche limit is
    (a) Newton's second law.
    (b) tidal forces.
    (c) Kepler's second law.
    (d) all of the above.

34. Describe the internal structure of the Jovian planets.

35. Give another name for "tidal stability limit."

36. What are tidal forces?

37. What is unusual about the satellite Io?

38. What is unique about the satellite Titan?

39. Which planet has the greatest number of satellites? How many known satellites does the planet have?

40. Titan is the second largest moon in the solar system. The Jovian moon Ganymede is larger. Why does Titan have a thick atmosphere whereas Ganymede does not?

41. How do the interior structures of Uranus and Neptune differ from those of Jupiter and Saturn?

42. How do the magnetic fields of the Jovian planets compare with those of the terrestrial planets?

43. Give five major differences between the terrestrial and the Jovian planets.

44. Name the satellite that is believed to have an abundance of water.

45. Which planet has the most elliptical orbit?

46. Which planet is the largest? Which satellite of a planet is the largest?

47. Name the planet to which each of the following moons belongs: Phobos, Ganymede, Titan, Triton, and Miranda.

48. Which planet has the greatest average orbital speed? the least?

49. Name the planets that are visible to the unaided eye.

50. Which planet has the fewest daylight hours during one rotation? Give the approximate number of daylight hours that occur at the planet's equator.

51. Name the satellite that has a dense, hazy atmosphere.

52. Name the planets that are known to have rings.

15.5 Other Solar System Objects

53. Which of the following is true? Asteroids
    (a) are believed to be initial solar system material that never collected into a single planet.
    (b) are located mainly in orbits around the Sun between Mars and Jupiter.
    (c) range in size from hundreds of kilometers down to the size of sand grains.
    (d) are generally irregular in shape.
    (e) all of the above.

54. Which of the following is true? Comets
    (a) are composed of dust and ice.
    (b) revolve around the Sun in highly elliptical orbits.
    (c) are observable only when relatively close to the Sun.
    (d) usually have a long tail when they are close to the Sun.
    (e) all of the above.

55. Which of the following is true? Meteoroids are
    (a) small, solid, interplanetary metallic and stony objects.
    (b) usually smaller than a kilometer.
    (c) known as meteors when they enter Earth's atmosphere.
    (d) known as meteorites when they strike Earth's surface.
    (e) all of the above.

56. How do asteroids differ from meteoroids?

57. About how many asteroids have been named and numbered? Where are most of them located?

58. Describe the physical characteristics of a comet at (a) aphelion and (b) perihelion.

59. What is the origin of most observed comets?

60. What is the Oort Cloud? Where is it located?

61. What is the difference between meteoroids, meteors, and meteorites?

62. What is the evidence that interplanetary dust exists?

63. What is the estimated outer limit of the solar system?

15.6 The Origin of the Solar System

64. Which of the following is *not* a basic factor that must be taken into account in an acceptable theory for the origin of the solar system?
    (a) the forces needed to form the system
    (b) the origin of the material used to form the system
    (c) the present size and structure of the system
    (d) the presence of interstellar gases and dust
    (e) the origin of elements

65. The outer boundary of the solar system is thought to be as much as _____ astronomical units from the Sun.
    (a) 10,000   (c) 75,000
    (b) 40,000   (d) 100,000

66. The flattened, rotating disk of gas and dust around the protosun from which the planets were apparently formed is called the _____ nebula.
    (a) primordial
    (b) planetary
    (c) solar
    (d) interplanetary

67. Distinguish between accretion and condensation as referred to in the condensation theory of the formation of the solar system.

68. Distinguish between the primordial nebula and the solar nebula.

69. What role do dust particles play in the condensation theory for the formation of the solar system?

70. Where do scientists obtain the best evidence concerning the condensation and accretion processes in the formation of the solar system?

71. What causes interstellar matter to form a rotating nebula?

72. How old is the solar system?

15.7 Other Planetary Systems

73. Which of the following is *not* a useful method for detecting an extrasolar planetary system?
    (a) the observation of a star's motion
    (b) the observation of Doppler shifts in the spectrum of a star
    (c) the detection in visible light of a very large orbiting planet
    (d) the detection of alien electromagnetic signals

74. Which of the following physical properties of a planet is obtained from the Doppler shift produced by a star and its planet?
    (a) mass
    (b) period of revolution
    (c) average orbital distance
    (d) all of the above

75. In October 1995, the first planet revolving around a star similar to our Sun was discovered. The star's name is
    (a) 51 Pegasi.
    (b) 70 Virginis.
    (c) 47 Ursae Majoris.
    (d) Rho' Cancri.

76. What evidence do astronomers have concerning the actual existence of another planetary system?

77. What system of detection was used by astronomers to conclude that Pulsar 1257 +12 has planets revolving around it?

78. What is the BETA project?

Applying Your Knowledge

1. Give some reasons why our knowledge of the solar system has increased considerably in the past few years.

2. Describe your observation of one or more planets during this school year.

3. A Foucault pendulum suspended from the ceiling of a high tower located at Earth's equator is put in motion in a north-south plane. Explain what you will observe of any apparent deviation or the lack thereof of the pendulum from the north-south plane during a 24-hour period.

4. The planet Mars rotates from west to east the same as Earth. Its moon, Phobos, revolves from west to east the same as our Moon. Explain why you will see Phobos rise on the western horizon (not the eastern horizon as our Moon) when you are standing on Mars' equator.

Exercises

1. Calculate the period T of a planet whose orbit has a semimajor axis of 5.2 AU.   Answer: 11 y

2. Calculate the period T of a planet whose orbit has a semimajor axis of 19 AU.

3. Calculate the length R of the semimajor axis of a planet whose period is 225 days.   Answer: 0.387 AU

4. Calculate the length R of the semimajor axis of a planet whose period is 165 years.

5. Determine the period of revolution of Earth, if Earth's distance from the Sun were 2 AU rather than 1 AU. Assume that the mass of the Sun remains the same.
          Answer: 2.8 y

6. Determine the period of revolution of Earth, if Earth's distance from the Sun were 4 AU rather than 1 AU. Assume that the mass of the Sun remains the same.

7. Determine the period of revolution of Earth, if the Sun's mass were 0.5 times its present value. Assume that the distance to the Sun remains 1 AU.   Answer: 1.4 y

8. Determine the period of revolution of Earth, if the Sun's mass were 4 times its present value. Assume that the distance to the Sun remains 1 AU.

9. Use Kepler's third law to show that the closer a planet is to the Sun, the shorter its period.
          Answer: $\dfrac{T^2}{R^3} = k$ or $T^2 = kR^3$

10. Use Kepler's third law to show that the closer a planet is to the Sun, the faster its speed around the Sun.

11. List the planets in order of increasing distance from the Sun.
          Answer: Mercury, Venus, Earth, Mars, Jupiter, Saturn, Uranus, Neptune, Pluto.

12. List the following distances in order of increasing length.
    (a) Sun to Earth   (c) Jupiter to Saturn
    (b) Mars to Jupiter   (d) Saturn to Uranus

13. List the planets in order of decreasing size (largest first).
          Answer: Jupiter, Saturn, Uranus, Neptune, Earth, Venus, Mars, Mercury, Pluto.

14. List the planets in order of decreasing density (densest first).

Solutions to Confidence Exercises

15.1 $T^2 = \dfrac{1\,\text{y}^2}{\text{AU}^3} \times \dfrac{(30\,\text{AU})^3}{1}$

$T^2 = 27{,}000\,\text{y}^2$

$T = 164\,\text{y}$

15.2

$$m_{\text{Sun}} = \frac{4 \times (3.14)^2 \times (1.52 \times 1.5 \times 10^{11}\,\text{m})^3}{\left(6.67 \times 10^{-11}\dfrac{\text{N-m}^2}{\text{kg}^2}\right) \times \left(1.88\,\text{y} \times 3.16 \times 10^7\,\dfrac{\text{s}}{\text{y}}\right)^2}$$

$$m_{\text{Sun}} = \frac{4 \times 9.87 \times 11.85 \times 10^{33}\,\text{m}^3}{\left(6.67 \times 10^{-11}\,\text{kg-}\dfrac{\text{m}}{\text{s}^2} \times \dfrac{\text{m}^2}{\text{kg}^2}\right)\left(3.53\,\text{y}^2 \times 9.99 \times 10^{14}\,\dfrac{\text{s}^2}{\text{y}^2}\right)}$$

$$m_{\text{Sun}} = \frac{47.4 \times 10^{30}\,\text{kg}}{23.5} = 2.0 \times 10^{30}\,\text{kg}$$

Answers to Multiple-Choice Review Questions

1. d   3. c   13. b   22. d   24. c   32. b   53. e   55. e   65. d   73. c   75. a

2. b   12. d   14. c   23. c   31. d   33. b   54. e   64. e   66. c   74. d

PLACE AND TIME

16

What place have you?
What place have I?
What time have you?
What time have I?
Only here and now
Have you and I.

J T S (1968)

In physical science we observe and examine events that take place in our environment. These events occur at different places and at different times. Some occur nearby and are observed immediately, whereas others occur at great distances and are not observed until a later time. For example, you see a flash of lightning, but you do not hear the sound of thunder until later. A star explodes at some great distance from Earth, and years later the event is observed as the radiation reaches Earth. Thus the events taking place in our environment are separated in space and time.

Albert Einstein was the first to point out that space and time are related and exist as a single entity. Einstein associated gravitational fields with space and developed the concept of four-dimensional space, which has given us an entirely different concept of the reality of our environment.

In this chapter, we introduce the concepts used in two- and three-dimensional reference systems. These concepts are then expanded to describe the reference systems relating to the places (locations) of objects on Earth's surface and the places of planets, stars, galaxies, and other celestial objects beyond Earth's atmosphere.

Our five senses make it possible for us to know and understand objects and their places relative to one another in the physical world. The concept of time,

Photo: Falling sand illustrates the march of time.

on the other hand, is rather evasive. We relate to the daily cycle of the Sun and the yearly cycle of Earth revolving around the Sun, and we use these periodic changes to measure what we call *time*. Thus we think of time in reference to changes we observe in our environment. This chapter introduces and explains the time concepts that are used in our daily living. See the Spotlight at the end of the chapter. ■

16.1 Cartesian Coordinates

LEARNING GOAL

▼ Explain the Cartesian coordinate system.

The location of an object in our environment requires a reference system that has one or more dimensions. A one-dimensional system is depicted by the number line shown in ● Fig. 16.1, which illustrates two fundamental features of every coordinate system. A straight line is drawn, which may extend to plus infinity in one direction and minus infinity in the opposite direction. For the line to represent a coordinate system, an origin must be indicated, and unit length along the line must be expressed. Temperature scales, left-right, above ground–below ground, and profit-loss are all examples of one-dimensional coordinate systems.

A two-dimensional system is shown in ● Fig. 16.2, in which two number lines are drawn perpendicular to each other and the origin is assigned at the point of intersection. The two-dimensional system is called a **Cartesian coordinate system,** in honor of the French philosopher and mathematician René Descartes (1596–1650), the inventor of coordinate geometry. It is also referred to as a *rectangular coordinate system.* The horizontal line is normally designated the x-axis and the vertical line, the y-axis. Every position or point in the plane is assigned a pair of coordinates (x and y), which gives the distance from the two lines, or axes. The x number gives the distance from the y-axis, and the y number gives the distance from the x-axis. Many cities in the United States are laid out in the

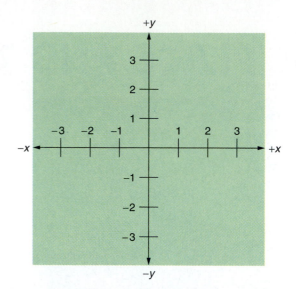

FIGURE 16.2   A Two-Dimensional Reference System

Cartesian coordinate system. For example, one street runs east and west, corresponding to the x-axis, and another street runs north and south, which corresponds to the y-axis.

Our interest in the Cartesian coordinate system results from our desire to determine the location of any position on the surface of spherical Earth and the location of any object on the celestial sphere, which is the apparent sphere of the sky, a sphere with a very large radius centered on the observer. A spherical surface is a curved surface on which all points are equidistant from a point called the center. The location of any position on a spherical surface can be found by using two reference circles analogous to the coordinate axes mentioned above.

RELEVANCE QUESTION: *The streets in your hometown are most likely laid out in a coordinate system that divides the town into four regions. Two streets are used. One may be called Main Street, which generally divides the town north and south or east and west. Name the two streets that divide your hometown into four regions. In which region (NE, NW, SE, SW) is your home located?*

16.2 Latitude and Longitude

LEARNING GOALS

▼ Define and explain *latitude* and *longitude*.

▼ Solve latitude and longitude exercises relative to Earth's surface.

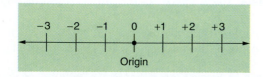

FIGURE 16.1   A One-Dimensional Reference System

The location of an object on the surface of Earth is established by means of a coordinate system known as *latitude and longitude.* Because Earth is turning about an axis, we can use as north-south reference points the *geographic poles,* which are defined as those imaginary points on the surface of Earth where the axis projects from the sphere.

The *equator,* defined with respect to the poles, is an imaginary line circling Earth at the surface, halfway between the North and South Geographic Poles. The equator is a great circle, that is, a circle on the surface of Earth located in a plane that passes through the center of Earth. Any such plane would divide Earth into two equal halves.

The **latitude** of a surface position is defined as the angular measurement in degrees north and south of the equator. The latitude angle is measured from the center of Earth relative to the equator (● Fig. 16.3). Lines of equal latitude are circles drawn around the surface of the sphere parallel to the equator. Any number of such circles can be drawn. They become smaller as the distance from the equator becomes greater. These circles are called **parallels,** and when we travel due east or west, we follow a parallel. Latitude has a minimum value of 0° at the equator and a maximum value of 90° north or 90° south at the poles.

Imaginary lines drawn along the surface of Earth running from the North Geographic Pole, perpendicular to the equator, to the South Geographic Pole are known as **meridians.** Meridians are half circles, which are portions of a great circle, because the half circle is located in the same plane as the center of Earth. An infinite number of lines can be drawn as meridians.

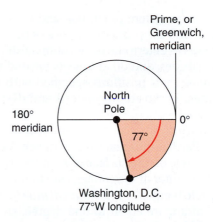

FIGURE 16.4   Diagram Showing the Longitude of Washington, D.C.

Longitude is defined as the angular measurement, in degrees, east or west of the reference meridian, which is called the prime, or Greenwich, meridian. Longitude has a minimum value of 0° at the prime meridian and a maximum value of 180° east and west (● Fig. 16.4).

The latitude and longitude of one point (Washington, D.C., 39°N, 77°W) are shown in Figs. 16.3 and 16.4. They are combined in ● Fig. 16.5 and shown in a cutaway of Earth. The **Greenwich (prime) meridian** was chosen as the zero meridian because a large optical telescope was located at Greenwich, England (a suburb of London). Also, England ruled the seas when

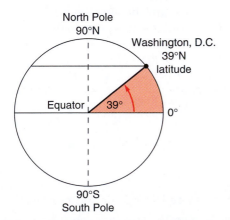

FIGURE 16.3   Diagram Showing the Latitude of Washington, D.C.

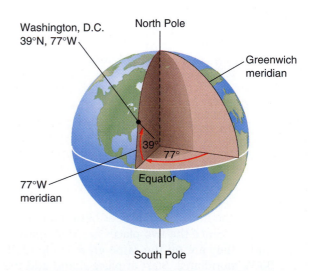

FIGURE 16.5   Diagram Showing the Latitude and Longitude of Washington, D.C.

the coordinate system of latitude and longitude was originated, and the primary purpose of the system at the time was to determine the location of ships at sea.

Stars, galaxies, and other objects beyond the solar system have their positions specified with celestial coordinates. These are introduced and discussed in Section 18.2.

The shortest surface distance between any two points on Earth is the great circle distance. A **great circle** is any circle on the surface of a sphere whose center is at the center of the sphere. One minute of arc of a great circle is equal to one *nautical mile* (n mi). Sixty minutes of arc are equal to one degree, so 60 n mi equals 1°. Thus the distance between two places on Earth's surface can be determined if the great circle angle between them is known.

EXAMPLE 16.1

Determining the Distance Between Two Places

Determine the number of nautical miles between place *A* (10°S, 90°W) and place *B* (70°N, 90°E).

SOLUTION

Step 1

Draw a diagram showing the latitude and longitude of place *A* and place *B*.

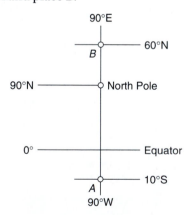

Step 2

Determine the number of degrees between place *A* and place *B*. Since the two places are 180° apart in longitude, they are on the same great circle (90°E and 90°W meridians). Start at place *A* and add the angle values.

$$10° + 90° + 30° = 130°$$

Step 3

Calculate the number of nautical miles (n mi) between place *A* and place *B*.

$$60 \text{ n mi} = 1°$$

Therefore,

$$130° \times \frac{60 \text{ n mi}}{1°} = 7800 \text{ n mi}$$

CONFIDENCE EXERCISE 16.1

Determine the number of nautical miles between place *A* (43°N, 84°W) and place *B* (34°N, 84°W).

EXAMPLE 16.2

Determining Your Location After Traveling Meridians and Parallels

You begin a journey at 40°N, 90°W and travel 300 n mi directly south, then 300 n mi directly west, then 300 n mi directly north, and then 300 n mi directly east. Where will you be with respect to your starting point?

SOLUTION

The key to solving this exercise is knowing that all meridians are one-half of a great circle, and they come together at the poles (see Fig. 16.11, on page 429). Also, parallels are circles. The equator is a great circle, but all other parallels are small circles. Parallels become smaller circles as the latitude increases.

Draw a diagram showing the meridians and parallels and record the data given in the exercise on the diagram.

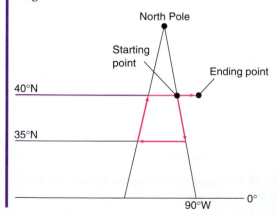

The completed diagram shows you the answer. When traveling directly north or south, you travel a great-circle route. The (300 n mi) traveled is the same when going north or south, and the angle (5°) passed through is the same. When traveling the parallels, your distance (300 n mi) is the same whether traveling west or east. But the angle passed through is greater traveling east because a smaller circle is traveled.

CONFIDENCE EXERCISE 16.2

A traveler starts at place *A* (10°N, 90°W) and travels 1200 n mi due south, then 1200 n mi due east, then 1200 n mi due north, and then 1200 n mi due west. Where will the traveler end up with respect to place *A*?

In 1589, Gerardus Mercator (1512–1594), Flemish cartographer and geographer, used the word *atlas* to describe a collection of maps. Today we use Mercator projection maps chiefly for navigation but also for many other purposes. Maps have become an indispensable reference for showing the location of a place on the surface of Earth. Atlases also show countries, cities, roads, rivers, mountains, lakes, and many other details. The places on maps are shown relative to one another, and the fundamental frame of reference is the lines of latitude and longitude.

Most maps are drawn with north at the top, south at the bottom, east at the right, and west at the left. With these references, direction can be determined, and the maps also can be used to obtain distance when a scale is provided.

RELEVANCE QUESTION: *Without referring to your textbook, give the approximate latitude and longitude of your present location. At the same time, give the latitude and longitude of the place on the other side of Earth opposite your position.*

16.3 Time

LEARNING GOALS

▼ Interpret the concept of time.

▼ Explain and solve exercises pertinent to local solar and standard time.

The continuous measurement of time requires the periodic movement of some object as a reference. See the discussion of time in the chapter Highlight. In 1964, the Twelfth General Conference on Weights and Measures, meeting in Paris, adopted an atomic definition of the *second* as the international unit of time. The definition is based on the frequency of radiation associated with a specific transition of the cesium-133 atom. When a transition from one energy state to another occurs, the atom emits or absorbs radiation whose frequency is proportional to the energy difference in the two states (Chapter 9). The cesium-133 atom provides a highly accurate and stable reference frequency of 9,192,631,770 cycles per second. The National Institute of Standards and Technology cesium-beam frequency generator is shown in ● Fig. 16.6.

For everyday purposes, we are interested in Earth as a time reference because our lives are influenced

FIGURE 16.6   Cesium-Beam Frequency Generator

The National Institute of Standards and Technology cesium-beam frequency generator for establishing the time standard. It is accurate to one part in 10^{13}.

HIGHLIGHT: The Concept of Time

In Chapter 1 time was defined as the continuous forward flowing of events. This definition implies that time relates to motion, has a forward direction, and is not quantized.

Things change from one position to another, and an interval of time is noted for the change to occur. This is something we observe in our physical world, and we call this changing of position *motion.* Everything appears to be in motion. The things we say are at rest are at rest only in reference to something else. A book on a table is at rest, but the table and book are on a moving Earth. Our thoughts concerning the concept of time are meaningless unless we include the concept of motion.

Do the events that take place in our physical world have a forward direction? That is, does time flow in a forward direction? If the answer is yes, how can we differentiate between forward and backward events? The answer can be found in the concept of entropy.

One of the most important laws of nature is the second law of thermodynamics, which can be stated in many ways. One way, which gives the direction of time, is with the entropy concept. Entropy is a measure of disorder (see Chapter 5), and the second law of thermodynamics states that the total

entropy of the universe increases in every natural process. Thus, as events occur (the flow of time), disorder becomes greater.

We do not observe and cannot measure extremely short (less than 10^{-13} s) intervals of time. Does this mean that things we cannot measure are meaningless? Not necessarily. For example, calculations concerning the Big Bang theory (see Section 18.7) for the origin of the universe specify that during the extremely brief interval 1.35×10^{-43} s (known as *Planck time*) following the Big Bang all forces were unified. Although we cannot measure this time interval, scientists use it in theories concerning the origin and expansion of the universe.

Is time absolute? The answer is a definite no. One of the outcomes from Albert Einstein's special theory of relativity was *time dilation.* This concept refers to the measurement of time by observers in different reference frames. For example, clocks run slower (time is stretched out) in a moving reference frame as observed by a stationary observer. Also, biological processes proceed at a slower rate on a moving reference frame than on a stationary one.

Scientists performed an experiment that measured the time difference be-

tween clocks, four moving and one stationary. Four cesium-beam atomic clocks were flown on jet flights around the world twice, once eastward and once westward. The times recorded by the clocks on the jet flights were compared with the corresponding clock at the U.S. Naval Observatory, and time differences were recorded that were in good agreement with relativity theory.

Einstein's special theory of relativity also binds three-dimensional space with time, ranking time as a basic reference coordinate and labeling it as a fourth dimension. Thus, to give the complete location of an event, we must state where it is in three spatial dimensions plus the time it is there. Space and time are relative concepts. They are dependent on the relative velocity of the reference frame in which they are measured. Space and time are related in a fundamental way in what we call the four-dimensional continuum, or space-time.

Although the concept of time is difficult to comprehend, the existence of time, as we experience it, is related to change. We, as individuals, experience the universe in our own time frame, which is different from everybody else's.

by the day and its subdivisions of hours, minutes, and seconds. The day has been defined in two basic ways. In the first definition, the solar day is the elapsed time between two successive crossings of the same meridian by the Sun. This period is also known as the **apparent solar day,** since this is what appears to happen. Because Earth travels in an elliptical orbit, its orbital speed is not constant; therefore, the apparent days are not the same duration. As a remedy for this, the mean solar day is computed from all the apparent days during 1 y.

The **sidereal day** is defined as the elapsed time between two successive crossings of the same meridian

by a star other than the Sun. ● Figure 16.7 shows the difference between the solar and sidereal days.

Because Earth rotates 365.25 times during one revolution, the magnitude of the angle through which it revolves in 1 day is

$$\frac{360°}{365.25 \text{ days}} = 0.985°/\text{day}$$

or slightly less than 1°/day. Earth must rotate through 360° plus an angle of this same magnitude to complete one rotation with respect to the Sun. Therefore, the solar day is longer than the sidereal day by ap-

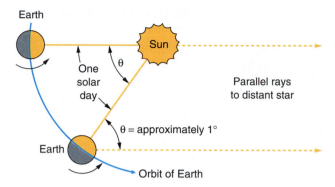

FIGURE 16.7 The Difference Between the Solar Day and the Sidereal Day

One rotation of Earth on its axis with respect to the Sun is known as one solar day. One rotation of Earth on its axis with respect to any other star is known as one sidereal day. Note that Earth turns through an angle of 360° for 1 sidereal day and approximately 361° for 1 solar day.

proximately 4 min because Earth rotates 360°/24 h, or 15°/h, or 1°/4 min.

The earliest measurement of solar time was accomplished with a simple device known as a *gnomon* (from the Greek meaning "a way of knowing"), which is a vertical rod erected on level ground that casts a shadow when the Sun is shining (● Fig. 16.8). The vertical pointer of the sundial is a gnomon. By the third century B.C., the water clock had been invented. In hot and dry regions of Earth, the flow of sand was used to count the hours. Burning candles also were used, because they burn at a fairly constant rate.

The 24-h day, as we know it, begins at midnight and ends 24 h later at midnight. By definition, when the Sun is on an observer's meridian, it is 12 noon *local solar time* at this meridian. The hours before noon are designated A.M. (**ante meridiem,** before midday) and those after noon, P.M. (**post meridiem,** after midday). The time of 12 o'clock should be stated as 12 noon or 12 midnight, with the dates. For example, we should write 12 midnight, December 3–4, to distinguish that time from, say, 12 noon, December 4.

Our modern civilization runs efficiently because of our ability to keep accurate time. Since the late nineteenth century, most countries of the world have adopted the system of **standard time zones.** This scheme theoretically divides the surface of Earth into 24 time zones, each containing 15° of longitude or 1 h, since the planet rotates 15°/h.

The first zone begins at the prime meridian, which runs through Greenwich, England, and extends 7.5°

each side of the prime meridian. The zones continue east and west from the Greenwich meridian, with the centers of the zones being multiples of 15°. The actual widths of the zones vary because of local conditions, but all places within a zone have the same time, which is the time of the central meridian of that zone. For example, Washington, D.C., is located at 77°W longitude, which is within 7.5° of the 75° meridian (● Fig. 16.9).

Time zones for the conterminous United States are shown in Fig. 16.9. The boundaries change in some areas when Daylight Saving Time is adopted from April through October each year. The major area (eastern part) of Alaska and the Hawaiian Islands are 2 h ahead of Pacific Standard Time. Alaska's western coast and the Aleutian Islands are 3 h ahead of Pacific Standard Time.

● Figure 16.10 shows the time and date on Earth for any Tuesday at 7 A.M. (PST), 8 A.M. (MST), 9 A.M. (CST), and 10 A.M. (EST). As Earth turns eastward, the Sun appears to move westward, taking 12 noon with it. Twelve midnight is 180° or 12 h east of the Sun, and as 12 noon moves westward, 12 midnight follows, bringing the new day.

When you travel west into a different time zone, the time kept by your watch will be 1 h fast or ahead of the standard time of the westward zone; therefore, you must move the hour hand back 1 h if the watch is to have the correct time. This process will be necessary as you continue west through additional time zones. A trip all the way around Earth in a westward direction will mean the loss of 24 h, or one complete day. When you travel east, the opposite is true; that is, your watch will be 1 h slow for each zone, and the hour hand is set ahead 1 h.

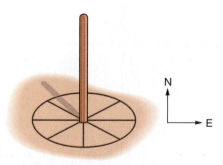

FIGURE 16.8 A Gnomon

This device is simply a vertical rod positioned so as to cast a shadow to indicate the time of day.

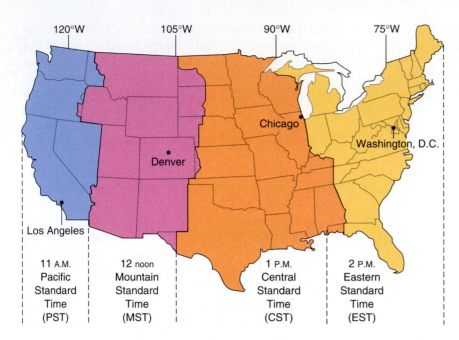

FIGURE 16.9 Time Zones of the Conterminous United States

A better understanding of why a day is lost in traveling around Earth in a westward direction can be obtained if we take a make-believe trip. Suppose we leave Dulles International Airport in Washington, D.C., by jet plane at exactly 12 noon local solar time on Tuesday, December 5, and fly west at a speed equal to the apparent westward speed of the Sun. Because Washington is located at 39°N latitude, the plane must travel the 39°N parallel west at about 1300 km/h (800 mi/h).

As we leave the airport, we observe the Sun out the left window of the plane, or toward the south. One

hour after leaving the airport, we notice the Sun can still be seen out the left window at the same altitude. Six hours later, with our watches indicating 7 P.M., the Sun still has the same apparent position as observed from the left window of the plane. Because the plane is flying at the same apparent speed as the Sun, the Sun will continue to be observed out the left window of the plane.

Twenty-four hours later we arrive back in Washington with the Sun in the same apparent position. During the 24-h trip, our time remained at 12 noon local solar time. If we had been observing the time with a

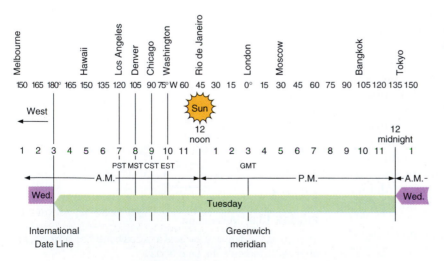

FIGURE 16.10 Diagram Showing the Times and Dates on Earth for Any Tuesday at 10 A.M. EST

As time passes, the Sun appears to move westward; thus, 12 noon moves westward and midnight follows (180° or 12 h) behind, bringing the new day, Wednesday, with it.

sundial, it would have remained at 12 noon. We are aware of the passing of the 24 h because we kept track of the time with our watches, but we did not see the Sun set or rise, and we did not pass through 12 midnight; therefore, the time to us is still 12 noon Tuesday. For friends meeting us at the airport, it is 12 noon Wednesday.

To remedy situations like this one, the **International Date Line** (IDL) was established at the 180° meridian. When one crosses the IDL traveling westward, the date is advanced into the next day; when one crosses the IDL traveling eastward, one day is subtracted from the present date.

If the local solar time is known at one longitude, the local time at another longitude can be determined by remembering that there are 15° for each hour of time, or 4 min for each degree. Should the calculation extend through midnight, or should the International Date Line be crossed, the date would change.

One problem of practical importance is to find the time and date in a distant city when you know the time and date in another time zone (Fig. 16.11).

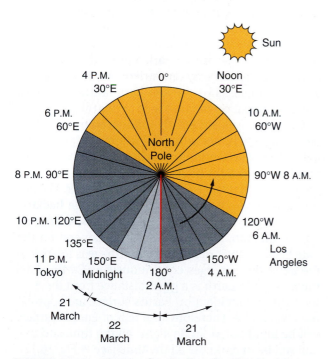

FIGURE 16.11 Diagram for Finding the Time and Date at a Specific Longitude
An example of finding the time and date in Tokyo, knowing that the time and date in Los Angeles are 6 A.M., March 21. The North Pole is at the center of the circle.

This problem may be encountered when you are trying to make a long-distance telephone call and don't want to awaken someone in the middle of the night.

The following example is explained using the views presented by the common highway map that gives a bird's-eye view of Earth's surface. On the highway map, the vertical lines represent meridians and north is toward the top of the map. East is to the right and west is to the left.

EXAMPLE 16.3

Finding Local Solar Time and the Date at a Given Longitude

What are the local solar time and the date at 72°W when the local solar time at 108°W is 12 noon, October 10?

SOLUTION

Step 1

Draw two vertical lines a small distance apart to represent meridians and record at the lines the given data plus what you are to find.

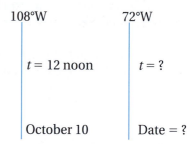

Step 2

Determine the difference in longitude between the two meridians.

$$108°W - 72°W = 36°$$

Step 3

Determine the amount of time corresponding to 36°. Since 15° is equivalent to 1 h and 1° is equivalent to 4 min, the time difference for 36° equals 30° + 6° is

$$30° \times \frac{1\,h}{15°} = 2\,h$$

$$6° \times \frac{4\,min}{1°} = 24\,min$$

$$\Delta t = 2\,h\ and\ 24\,min$$

Step 4

Determine the time and date at 72°W. The Sun is at 102°W. Since the Sun appears to travel westward, the time at 72°W is past the hour of 12 noon by 2 h and 24 min. Therefore, the time at 72°W is 2:24 P.M. Since midnight does not occur between the two given longitudes, the date at 72°W is October 10.

CONFIDENCE EXERCISE 16.3

Determine the local solar time and date at 114°W, when the local solar time at 82°W is 12 noon, October 20.

Because our watches and clocks keep standard time, the time shown on most watches and clocks around the world will display the same minutes but a different hour.* For example, if your watch shows twenty past 10:00 A.M., then most clocks worldwide will display twenty minutes past some hour. Many airports and hotels have six or more clocks on a wall showing the time at certain major cities around the world.

To calculate the standard time at a different longitude, use your standard time meridian and the standard time meridian of the distant longitude and proceed as in Example 16.3.

During World War I, the clocks of many countries were set ahead 1 h during the summer months to give more daylight hours in the evening, thus conserving fuel used to generate electricity for lighting. This practice has now become standard for all but a few of the 50 states of the United States. During the summer months in this country, time known as **Daylight Saving Time** (DST) begins at 2 A.M. on the first Sunday of April and ends at 2 A.M. on the last Sunday of October. The change to Daylight Saving Time helps conserve

*There are a few places in the world where time varies from the standard time zone by half an hour and other places by less than half an hour.

energy; it also reduces injuries and saves lives by preventing early-evening traffic accidents.

During a period of 1 y, the Sun appears to change its overhead position from 23.5°N southward across the equator to 23.5°S. From this latitude, the Sun appears to turn back northward, crossing the equator to 23.5°N, where it appears to turn again on a southward journey. The parallel 23.5°S is called the *Tropic* (turning place) *of Capricorn*. The constellation (star pattern) Capricorn is where the Sun appears to turn and travel northward. Similarly, the parallel 23.5°N is called the *Tropic of Cancer*.

As Earth revolves around the Sun, the Sun is directly overhead at different latitudes on different dates because of the tilt of Earth's axis. At the times of the vernal (spring) equinox and autumnal (fall) equinox, that is, at the times the Sun is overhead at the equator, all latitudes on Earth's surface are having approximately 12 h of daylight and 12 h of darkness.

When the Sun is overhead between latitudes 0° and 23.5°N, the northern latitudes experience more daylight hours than dark hours. Thus the amount of daily solar radiation striking Earth's surface at this time in the Northern Hemisphere is greater because (1) there are more daylight hours and (2) the Sun's rays are more vertical. In the meantime, the southern latitudes experience more dark hours than daylight hours, and the Sun's rays are striking Earth's surface at a greater angle. Thus the southern latitudes receive less solar radiation per unit area at this time.

Our daily lives are greatly influenced by the number of daylight hours we enjoy. We appreciate the daylight hours for working, playing, and driving. The number of daylight hours at any place on Earth depends on the latitude and the day of the year. The duration of daylight is a function of latitude because Earth's axis is tilted with reference to the incoming rays of the Sun. Also, daylight hours depend on the date because the orientation of Earth's axis relative to the Sun's rays changes continuously as Earth orbits the Sun. Since Earth is a great distance from the Sun, the light rays incident on Earth's surface are approximately parallel. Therefore, one half of Earth's surface will be illuminated (in daylight) all the time and one half will be in darkness all the time (see ● Fig. 16.12). Since Earth is rotating, most places on the planet's surface will receive sunlight each day. The amount of daylight or darkness each place experiences depends on its latitude and the day of the year.

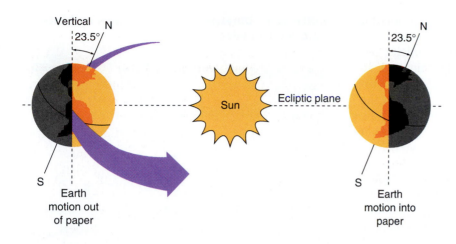

FIGURE 16.12   Earth's Positions, Relative to the Sun, for the Summer and Winter Solstices

As Earth revolves around the Sun, Earth's axis remains tilted 23.5° from the vertical. This inclination of the axis, in conjunction with Earth's orbital motion, causes a change in seasons on Earth. When Earth is at the position shown with its motion out of the page, the Northern Hemisphere has summer and the Southern Hemisphere has winter. When Earth is at the position shown with its motion into the paper, the Northern Hemisphere has winter and the Southern Hemisphere has summer.

Does height above sea level have any effect on the duration of daylight? The number of daylight hours does depend on the height of the observer above sea level. When observers rise vertically above sea level, they extend their horizon. For an observer at height h above the surface of the sea, the distance d to the horizon (where sunrise and sunset occur) is determined approximately by taking the square root of the product $2hR$, where R is the radius of the Earth. All quantities must be in the same units.

Note in ● Fig. 16.13 that on June 21 or 22 Earth's axis is tilted toward the Sun, and on December 22 or 23 Earth's axis is tilted away from the Sun, whereas on March 20 or 21 and September 22 or 23 Earth is broadside to the Sun. Thus, during the spring and summer months, the Northern Hemisphere will experience more daylight hours than dark hours. Note in Fig. 16.12 that the equator has approximately 12 hours of daylight and 12 hours of darkness every day of the year. As Earth rotates, the light and dark areas shown in the diagram are always equal for an observer at the equator. During the spring and summer months, the North Pole has 24 hours of daylight every day. Thus, during these months in the northern latitudes, there is a minimum number of daylight hours (12) at the equator and a maximum number (24) at the North Pole. Thus a person traveling continuously northward from the equator will experience, each day, more daylight hours.

FIGURE 16.13   Earth's Positions, Relative to the Sun, and the Four Seasons

As Earth revolves around the Sun, its north-south axis remains pointing in the same direction. On March 21 and September 22 the Sun is directly above the equator, and every place on Earth has 12 hours of daylight and 12 hours of darkness. On June 21 the Sun's declination is 23.5°N, and the Northern Hemisphere has more daylight hours than dark hours; it is then summer in the Northern Hemisphere and winter in the Southern Hemisphere. On December 23 the Sun's declination is 23.5°S, and the Southern Hemisphere has more daylight hours than dark hours; it is then winter in the Northern Hemisphere and summer in the Southern Hemisphere. The Sun is drawn slightly off center to indicate that Earth is slightly closer to the Sun during winter in the Northern Hemisphere than during summer.

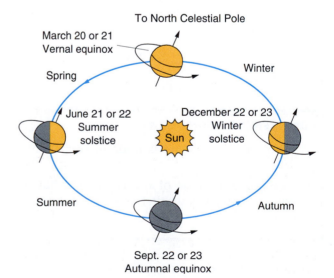

Table 16.1 gives the number of daylight hours for some northern latitudes on June 21 and December 22. The daylight hours are rounded off for simplicity. Values for other latitudes can be roughly estimated. To find the approximate time of sunrise and sunset,* divide the hours of daylight by two, then subtract the half value from 12 noon to obtain sunrise. Add the half value to 12 noon to obtain sunset. Remember that there are approximately the same number of daylight hours before noon as after noon. Also, there are approximately the same number of dark hours before midnight as after midnight. Table 16.1 also can be used for the Southern Hemisphere. Simply label the June column December and the December column June and change all N latitudes to S latitudes.

TABLE 16.1   Duration of Daylight Hours at Certain Northern Latitudes

	June 21	December 22
90°N	24 hours	0 hours
60°N	19 hours	6 hours
50°N	16 hours	8 hours
40°N	15 hours	9 hours
30°N	14 hours	10 hours
20°N	13 hours	11 hours
10°N	12.5 hours	11.5 hours
0°	12 hours	12 hours

The daylight hours are approximate values rounded off to whole numbers. Values for other latitudes can be roughly interpolated.

EXAMPLE 16.4

Determining the Number of Daylight Hours at a Given Latitude and Date

(a) Determine the approximate number of daylight hours at 30°N on June 21.

(b) Determine the approximate time of sunrise at 30°N on June 21.

SOLUTION

(a) Use Table 16.1 to find the number of daylight hours.   Answer: 14 hours

(b) The number of daylight hours before 12 noon is approximately the same as the number past 12 noon. Therefore,

$$\frac{14 \text{ hours}}{2} = 7 \text{ hours}$$

$$12 \text{ noon} - 7 \text{ hours} = 5 \text{ A.M.}$$

CONFIDENCE EXERCISE 16.4

(a) Give the approximate latitude of your hometown.

(b) Determine the time of sunrise in your hometown on the twenty-first day of the present month.

---

*Sunrise or sunset occurs when the top of the Sun, not its center, is on the horizon.

RELEVANCE QUESTION: We speak of time past, time present, and time future. In what way is the Hubble Space Telescope a time machine?

16.4 The Seasons

LEARNING GOALS

▼ Explain Earth's four seasons.

▼ Calculate the altitude of the Sun.

The spinning Earth is revolving around the Sun in an orbit that is elliptical yet nearly circular. When Earth makes one complete orbit around the Sun, the elapsed time is known as one *year.* We are concerned with two different definitions of the year in this textbook. The **tropical year,** or the year of the seasons, is the time interval from one vernal equinox to the next vernal equinox; that is, the tropical year is the elapsed time between one northward crossing of the Sun above the equator and the next northward crossing of the Sun above the equator. With respect to the rotation period of Earth, the tropical year is 365.2422 mean solar days.

The **sidereal year** is the time interval for Earth to make one complete revolution around the Sun with respect to any particular star other than the Sun. The sidereal year is equal to 365.2536 mean solar days. This is approximately 20 min longer than the tropical year. The reason for the difference will be explained in Section 16.5.

The axis of the spinning Earth is not perpendicular to the orbital plane of Earth as it revolves around the Sun; it is tilted 23.5° from the vertical, as illustrated in Fig. 16.12. This position of the axis with respect to the orbital plane produces a change in the Sun's overhead position throughout the year and causes our changing seasons. Figures 16.13 and ● 16.14 illustrate the apparent positions of the Sun over the course of 1 y.

During the summer months for the Northern Hemisphere, the Sun's rays strike Earth's northern latitudes from a more overhead position. Thus it is hotter in the summer when the Sun's rays are most direct, and it is colder in the winter when the Sun's rays are the least direct. When it is summer in the Northern Hemisphere, it is winter in the Southern Hemisphere, and vice versa.

The noon Sun's overhead position is never greater than 23.5° latitude, and the Sun always appears due south at 12 noon local solar time for an observer lo-

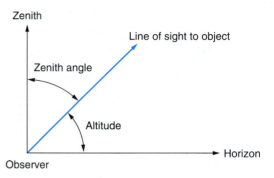

FIGURE 16.15 Zenith Angle and Altitude
Because the zenith is perpendicular to the horizon, the zenith angle plus the altitude equals 90°.

cated in the conterminous United States. When the Sun is at 23.5° north or south, it is at its farthest point from the equator. This farthest point of the Sun from the equator is known as the *solstice* (meaning "the Sun stands still"). The most northern point is called the **summer solstice,** and the most southern position is known as the **winter solstice.** This discussion applies to the Northern Hemisphere. In the Southern Hemisphere, dates for summer and winter solstices are reversed from those shown in Fig. 16.13.

As Earth orbits the Sun, the Sun's position overhead varies from 23.5° north to 23.5° south of the equator. When it is directly over the equator, the days and nights have about 12 h each around the world. These dates are called the *equinoxes*. The **vernal equinox** occurs on or about March 21, whereas the **autumnal equinox** occurs on or about September 22 each year. These dates are labeled in Fig. 16.13.

When the Sun is observed at 12 noon local solar time, it is on the observer's meridian and appears at its maximum altitude above the southern horizon on that day for all observers north of the Sun. The angle measured from the horizon to the line of sight to the Sun at noon is called its **altitude.** The angle from the **zenith** (position directly overhead) to the line of sight to the Sun at noon is called its **zenith angle** (● Fig. 16.15). The zenith is 90° from the **horizon** (the dividing line where Earth and sky appear to meet); therefore, the sum of the zenith angle and the altitude is 90°. The altitude of the Sun can be measured easily. If the Sun's position is known, the observer's latitude can be determined. The relationships among the preceding terms are illustrated in ● Fig. 16.16.

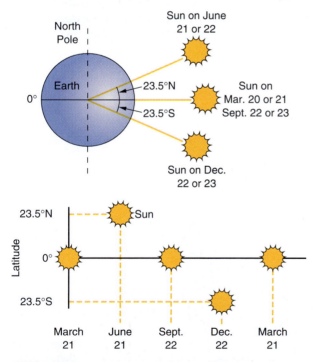

FIGURE 16.14 Diagrams of the Sun's Position (Degrees Latitude) at Four Different Times of the Year
The upper drawing shows a greatly magnified Earth with respect to the Sun, showing the Sun's position on the dates indicated. The lower graph plots the Sun's position (degrees latitude) versus time (months).

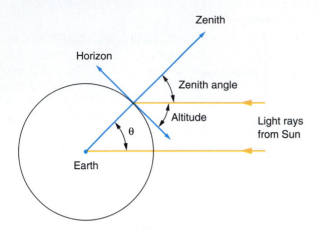

FIGURE 16.16 The Relationships Among Zenith Angle, Altitude, Horizon, and the Angle θ.

Because the incoming rays of light from the Sun are parallel, the angle θ and the zenith angle are equal in magnitude.

The maximum and minimum altitudes of the Sun for an observer in Washington, D.C. (39°N), can be determined by using data from Fig. 16.16. The solutions are as shown in ● Figs. 16.17 and 16.18. The relationship between the two solutions is illustrated in ● Fig. 16.19. The altitude of the Sun for all other days of the year, as observed from Washington, would be between these two values. A similar solution will give the Sun's altitude from any latitude.

The seasons have an effect on the lives of everyone. Our way of life is adjusted by the seasons' progressions. Many of our holidays were originally celebrated as commemorating a certain season of the year. The celebration that has evolved into our Easter was originally a celebration of the coming of spring and a renewal of life. Halloween originally commemorated the beginning of the winter season, and Thanksgiving commemorated the end of the harvest.

The ancient festival of the winter solstice (on December 21 or 22) and the beginning of the northward movement of the Sun evolved into our Christmas on December 25. The reasons for celebrating our various holidays have changed over the course of time, but the original dates were set by nature's annual timepiece—the movement of Earth around the Sun.

RELEVANCE QUESTION: The concept of time is related to change (see Section 1.3). Humans are aware of changes that are periodic in nature. The daily cycle of night and day is an example. Within you there are life processes that are periodic and life-dependent. Name two that have short periods and are necessary for life. Name two others that are life-dependent and periodic daily.

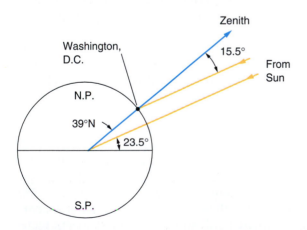

FIGURE 16.17 Finding the Approximate Altitude of the Sun as Observed from Washington, D.C., on June 21

Because the angle between the Sun and the observer is 39° − 23.5° = 15.5°, the altitude of the Sun is 90° − 15.5° = 74.5°.

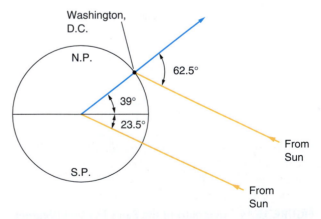

FIGURE 16.18 Finding the Approximate Altitude of the Sun as Observed from Washington, D.C., on December 21

Because the angle between the Sun and the observer is 39° + 23.5° = 62.5°, the altitude of the Sun is 90° − 62.5° = 27.5°.

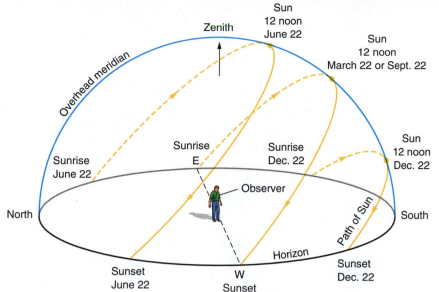

Zenith

Sun
12 noon
June 22

Overhead meridian

Sun
12 noon
March 22 or Sept. 22

Sunrise
E

Sunrise
Dec. 22

Sun
12 noon
Dec. 22

Sunrise
June 22

Observer

North

Path of Sun

South

Sunset
June 22

W
Sunset

Horizon

Sunset
Dec. 22

FIGURE 16.19   The Sun's Daily Path
The apparent path of the Sun across the sky is shown on June 22 (time of the summer solstice), December 22 (time of the winter solstice), and March or September 22 (times of the vernal and autumnal equinoxes) as observed from the Northern Hemisphere. Note that on the first day of spring and the first day of fall the Sun rises exactly east and sets exactly west.

16.5 Precession of Earth's Axis

LEARNING GOALS

▼ Define and explain *precession.*

▼ Describe the precession of Earth's axis.

Many of us are acquainted with the action of a toy top that has been placed in rapid motion and allowed to spin about its axis. After spinning a few seconds, the top begins to wobble or do what physicists call *precess* (● Fig. 16.20). The top, a symmetrical object, will continue to spin about an axis if the center of gravity remains above the point of support. When the center of gravity is not in a vertical line with the point of support, the axis slowly changes its direction. This slow rotation of the axis is called **precession.**

Because Earth is spinning rapidly, it bulges at the equator and cannot be considered a perfect sphere. The Moon and Sun apply a gravitational torque to Earth that tends to bring Earth's equatorial plane into its orbital plane. Because of this torque, the axis of Earth slowly rotates clockwise or westward about the vertical or the north ecliptic pole (● Fig. 16.21). The period of the precession is 25,800 years; that is, it takes 25,800 years for the axis to precess through 360°. As the axis changes its direction, the equinoxes move westward along the ecliptic. This movement is called the *precession of the equinoxes.*

Because the precession is clockwise and Earth is revolving counterclockwise around the Sun, the trop-

ical year is approximately 20 min shorter than the sidereal year. As the axis precesses, Polaris will no longer be the north star. The star Vega in the constellation Lyra will be the north star some 12,000 years from now. The Southern Cross, a constellation of stars located within 27° of the present South Celestial Pole, will then be visible from Washington, D.C.

Precession of Earth's axis does not have an influential effect on the seasons. The inclination of Earth's axis to the ecliptic remains the same throughout the precession period. Thus the Northern and Southern Hemispheres are alternately tilted toward and away from the Sun.

Precession does change the position of Earth in its orbit with respect to the stars. As Earth's axis precesses, the stars seen at various seasons will change.

Axis

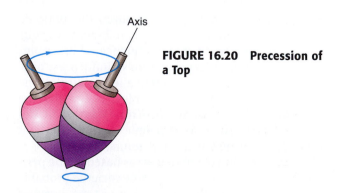

FIGURE 16.20   Precession of a Top

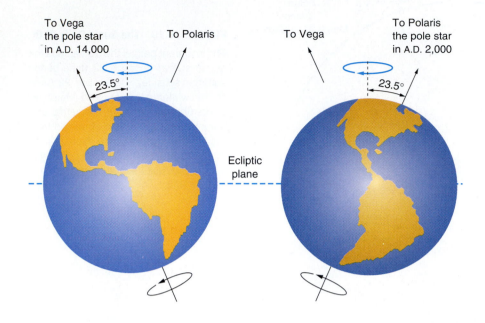

FIGURE 16.21 Precession of Earth's Axis Clockwise About a Line Perpendicular to Earth's Orbit

Earth's axis is presently pointing toward the star Polaris, which we call the pole, or north, star. In approximately 12,000 y the axis will be pointing toward the star Vega in the constellation Lyra.

This change is shown in ● Fig. 16.22 on page 438. The stars we see on summer nights will slowly change within a period of 25,800 y. In A.D. 14,900, stars seen on summer nights will be those we presently see on winter nights. Because of precession, the 12 constellations in the zodiac will slowly cycle through different months, with a 1-month change occurring every 2150 y (25,800/12). The stars seen overhead on June 21 some 2150 y ago are seen overhead on July 21 today and will be seen overhead on August 21 in another 2150 y.

16.6 The Calendar

LEARNING GOALS

▼ Give a brief history of the calendar.

▼ Identify the origin of the names for the days of the week.

The measurement of time requires the periodic movement of some object as a reference. Various lengths of time had a direct influence on life in ancient civilizations. Because of these influences, people had more than one reference for measuring the events of their lives. It is reasonable to believe that the first unit for the measurement of time was the day. Because of the need for food, people spent the daylight hours hunting; during the dark hours, they slept.

A longer period of reference was based on the periodic movement of the Moon. Some societies probably used the Moon as a basic division of time, because they were unable to count the days over a longer period. The first appearance of the crescent moon and the time of the full moon were times of worship for many primitive tribes.

Our month of today originated from the periodic phases of the Moon, which requires 29.5 solar days to orbit Earth. The plans for the first calendar seem to have originated before 3000 B.C. with the Sumerians, who ruled Mesopotamia. Their calendar was based on the motion of the Moon, which divided the year into 12 lunar months consisting of 30 days each. Because 30 × 12 = 360 days, and the year actually contains 365.25 days, corrections had to be made to keep the calendar adjusted to the seasons. The Babylonians, who followed the Sumerians, adjusted the length of the months and added an extra month when needed. This Babylonian calendar set the pattern for many of the calendars adopted by ancient civilizations.

The calendar we use today originated with the Romans. The early Roman calendar contained only 10 months, and the year began with the coming of spring. The months were named March, April, May, June, Quintilis, Sextilis, September, October, November, and December. The winter months of January and February did not exist. This was the period of waiting for spring to arrive. About 700 B.C., the month of January was added at the beginning of the year, before March, and February at the end of the year, after December. About 275 y later, in 425 B.C., the two months were changed to the present order.

A SPOTLIGHT ON: Place and Time

Cartesian Coordinate System

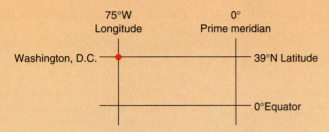

One solar day: one rotation of Earth relative to the Sun

One sidereal day: one rotation of Earth relative to a star

Seasons for the Northern Hemisphere

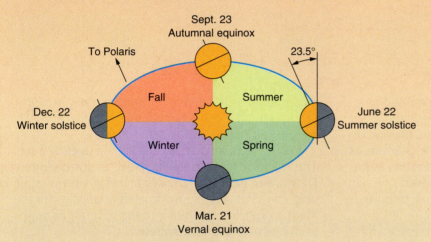

The Calendar

One year: one revolution of Earth around the Sun (365.25 days)

One month: periodic phase changes of the Moon (29.5 days)

One week: seven celestial bodies visible to the unaided eye—Sun, Moon, Mars, Mercury, Jupiter, Venus, Saturn

The Precession of Earth's Axis

(period = 25,800 years)

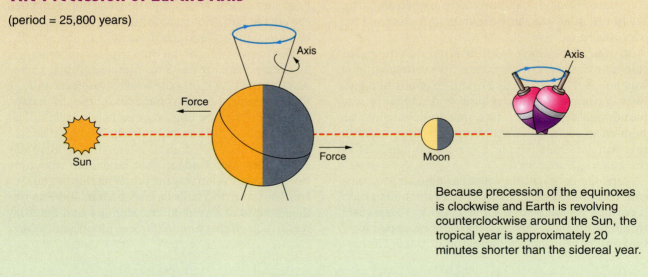

Because precession of the equinoxes is clockwise and Earth is revolving counterclockwise around the Sun, the tropical year is approximately 20 minutes shorter than the sidereal year.

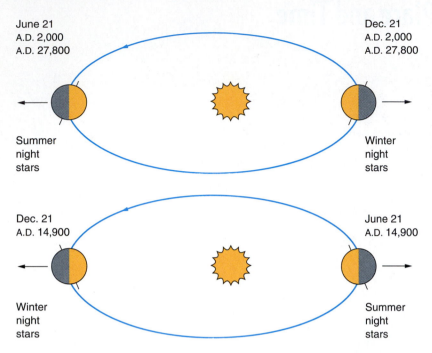

June 21
A.D. 2,000
A.D. 27,800

Summer
night
stars

Dec. 21
A.D. 2,000
A.D. 27,800

Winter
night
stars

Dec. 21
A.D. 14,900

Winter
night
stars

June 21
A.D. 14,900

Summer
night
stars

FIGURE 16.22   The Precession of Earth's Axis

The precession of Earth's axis will cause future generations to see different stars in the summer than are seen today. After 12,900 y, our winter night stars will be their summer night stars, and vice versa. After yet another 12,900 y, the constellations will be similar to what we see now.

The so-called *Julian calendar* was adopted in 45 B.C., during the reign of Julius Caesar. Augustus Caesar (adopted son of Julius Caesar), who ruled the Roman Empire after Julius, renamed the month Quintilis as July in honor of Julius, and the month Sextilis as August in honor of himself. He also removed one day from February, which he added to August to make it as long as July.

The Julian calendar had 365 days in a year. In every year divisible by 4, an extra day was added to make up for the fact that it takes approximately 365.25 days for Earth to orbit the Sun. Thus 1999, 2001, 2002, and 2003 have 365 days; 2000 and 2004 have 366 days. The Julian calendar was fairly accurate and was used for over 1600 y.

In 1582, Pope Gregory XIII realized that the Julian calendar was slightly inaccurate. The vernal equinox was not falling on March 21, and religious holidays were coming at the wrong time. A discrepancy was found, and the pope decreed that 10 days would be skipped to correct it. The discrepancy arose because there are 365.2422 days in a year and not 365.25, as the Julian calendar used.

Pope Gregory had the calendar adjusted to coincide with the seasons and proposed a method to keep it correct. He decreed that every 400 y, 3 leap years would be skipped. The leap years to be skipped were the century years not evenly divisible by 400. The corrections make the calendar accurate to 1 day in 3300 y. Our present-day calendar with these leap-year designations is called the **Gregorian calendar.**

In our calendar we have seven days in each week. The origin of the seven-day week is not definitely known, and not all cultures have had it. One possible origin is that it takes approximately seven days for the Moon to go from one phase to the next (for example, from new to first-quarter phase).

A more likely origin is the nighttime sky. As the ancients watched the sky night after night, they saw that exactly seven celestial bodies moved relative to the fixed stars. These seven objects visible to the unaided eye are the Sun, Moon, and five visible planets: Mars, Mercury, Jupiter, Venus, and Saturn.

Our present days of the week can still be connected by their names to the Sun, Moon, and five visible planets. Table 16.2 lists the heavenly objects and the English, French, and Saxon names for the seven days. Note that either the English or French word is similar to the name of the heavenly object.

Our Tuesday, Wednesday, Thursday, and Friday come from Tiw, Woden, Thor, and Fria, who were Nordic gods. Woden was the principal Nordic god, Tiw and Thor were the gods of law and war, and Fria was the goddess of love. Sunday, Monday, and Saturday retain their connection to the Sun, Moon, and Saturn.

TABLE 16.2 The Days of the Week

Celestial Object	English Name	French Name	Saxon Name
Sun	Sunday	Dimanche	Sun's day
Moon	Monday	Lundi	Moon's day
Mars	Tuesday	Mardi	Tiw's day
Mercury	Wednesday	Mercredi	Woden's day
Jupiter	Thursday	Jeudi	Thor's day
Venus	Friday	Vendredi	Fria's day
Saturn	Saturday	Samedi	Saturn's day

Important Terms

Cartesian coordinate
    system (16.1)
latitude (16.2)
parallels
meridians
longitude
Greenwich (prime)
    meridian

great circle
apparent solar day (16.3)
sidereal day
ante meridiem
post meridiem
standard time zones
International Date Line
Daylight Saving Time

tropical year (16.4)
sidereal year
summer solstice
winter solstice
vernal equinox
autumnal equinox
altitude
zenith

zenith angle
horizon
precession (16.5)
Gregorian calendar (16.6)

Review Questions

16.1 Cartesian Coordinates

1. A coordinate system must have
   (a) an indicated origin.
   (b) two dimensions.
   (c) a unit length expressed.
   (d) both (a) and (c).

2. Which of the following is true? A Cartesian coordinate system
   (a) is a two-dimensional system.
   (b) normally designates the horizontal line the x-axis.
   (c) normally designates the vertical line the y-axis.
   (d) all of the above.

3. Name two fundamental features of every coordinate system.

4. Give three examples of a one-dimensional reference system.

5. Draw a three-dimensional reference system.

6. What is the angle between each of the three axes in Question 5?

7. State a reference point, then use a three-dimensional reference system to give the position of your textbook in the room where you are located.

16.2 Latitude and Longitude

8. Which of the following is true? Latitude
   (a) is a linear measurement.
   (b) can have greater numerical values than longitude can.
   (c) is measured in an east-west direction.
   (d) can have negative values.
   (e) none of the above.

9. Which of the following is true? Longitude
   (a) can have a maximum value of 180°.
   (b) is an angular measurement.
   (c) has east or west values.
   (d) is measured along parallels.
   (e) all of the above.

10. Meridians
    (a) run north and south.
    (b) are infinite in number.
    (c) are half circles.
    (d) all of the above.

11. Which of the following is false?
    (a) Parallels become smaller as the distance from the equator becomes greater.
    (b) Parallels are all small circles.
    (c) Parallels are infinite in number.
    (d) All of the above are false.

12. What are the minimum and maximum values for latitude and longitude?

13. How is 0° defined for latitude and longitude?

14. What is the name of a line of equal longitude?

15. What is the longitude of the North Pole?

16. What are the approximate dates of sunrise and sunset at the North Pole?

17. Are meridians great circles? Explain.

18. Are any parallels great circles?

16.3 Time

19. The 24-h apparent solar day
    (a) begins at midnight.
    (b) begins first at the 180° meridian.
    (c) is not exactly 82,400 seconds every solar day.
    (d) all of the above.

20. The number of daylight hours at a specific place on Earth's surface is dependent on the
    (a) latitude.   (c) height above sea level.
    (b) date.   (d) all of the above.

21. The solar day is approximately
    (a) 1 minute longer than the sidereal day.
    (b) 2 minutes longer than the sidereal day.
    (c) 3 minutes longer than the sidereal day.
    (d) 4 minutes longer than the sidereal day.

22. During the month of January, the number of daylight hours at Washington, D.C., is _____ at Orlando, Florida.
    (a) more than   (b) less than   (c) the same as

23. How many time zones are there in the conterminous United States?

24. What do A.M. and P.M. mean?

25. Is it correct to state the time as 12 A.M.? Explain your answer.

26. What are some advantages of Daylight Saving Time?

27. What is the direction of rotation (clockwise or counterclockwise) of the shadow cast by a sundial located at (a) 30°N and (b) 30°S?

28. How is 12 noon defined?

29. Define one solar day.

30. Define one sidereal day.

31. How does the day of the year change when one crosses the IDL traveling westward?

32. What periodic motion is used as a reference for our year?

33. What two factors determine the number of daylight hours at any latitude on Earth's oceans?

16.4 The Seasons

34. Select the correct answer. The seasons
    (a) are a function of the inclination of Earth's axis.
    (b) are a function of Earth's revolving around the Sun.
    (c) would be more severe if the solstices were at 25°.
    (d) all of the above.

35. For an observer at 40°N, the Sun
    (a) is never directly overhead.
    (b) has an altitude of 50° on March 21.
    (c) has a zenith angle of 63.5° on December 22.
    (d) all of the above.

36. From September 23 to March 20, in what direction must an observer in the Northern Hemisphere look to see sunrise?
    (a) east   (b) northeast   (c) southeast

37. From March 23 to September 20, in what direction must an observer in the Northern Hemisphere look to see sunset?
    (a) west   (b) northwest   (c) southwest

38. Give two major reasons why there are four different seasons each year.

39. Give the times (dates) during the year when a place on the equator receives the (a) maximum daily amount of solar radiation and (b) minimum daily amount.

40. What is the position directly overhead of the observer called?

41. What is the altitude of Polaris (the north star) for an observer at the equator (0° latitude)? at the North Pole (90°N)?

42. What is the altitude of Polaris for an observer at Washington, D.C. (39°N)?

16.5 Precession of Earth's Axis

43. Precession of Earth's axis
    (a) is counterclockwise as viewed from above the North Pole.
    (b) changes the angle between the axis and the vertical.
    (c) has no important effect on Earth's seasons.
    (d) none of the above.

44. Precession of Earth's axis
    (a) is also called precession of the equinoxes.
    (b) is in a westward direction.
    (c) makes the tropical year approximately 20 minutes shorter than the sidereal year.
    (d) all of the above.

45. As Earth's axis changes direction, the equinoxes move _____ along the ecliptic.
    (a) westward
    (b) eastward
    (c) rapidly

46. What evidence is there to support precession of Earth's axis?

47. How long does it take for Earth to precess once?

48. Precession of Earth's axis *does* or *does not* have an influential effect on the seasons.

16.6 The Calendar

49. Which of the following is a natural unit of the calendar?
    (a) the day, which is based on the period of rotation of Earth
    (b) the month, which is based on the period of revolution of the Moon

(c) the year, which is based on the period of revolution of Earth around the Sun
(d) all of the above

50. The Gregorian calendar
    (a) is our present calendar.
    (b) is accurate to one day in 6000 years.
    (c) skips a leap year every century year.
    (d) all of the above.

51. What is the origin of the month?

52. What is the origin of the seven-day week?

53. How often is there a leap year in the Gregorian calendar?

54. What are the origins of the dates for Halloween (October 31) and Christmas (December 25)?

55. The calendar we use today originated with the _____.

Applying Your Knowledge

1. If Earth rotated in a clockwise direction as observed from above the North Pole, how would the solar day compare with the sidereal day?

2. Is there a place on Earth's surface where 1° of latitude and 1° of longitude have approximately equal numerical values in nautical miles? Explain.

3. October is the tenth month of the year, but *octo* means "eight." Explain.

4. Can you travel continuously eastward and circle Earth? Why or why not?

5. Can you travel continuously southward and circle Earth? Why or why not?

Exercises

16.2 Latitude and Longitude

1. What is the minimum angle between place *A* (50°N, 90°W) and place *B* (60°N, 90°E)?   Answer: 70°

2. What is the minimum angle between place *A* (20°N, 75°W) and place *B* (30°S, 75°W)?

3. What is the number of nautical miles between place *A* and place *B* in Exercise 1?   Answer: 4200 n mi

4. What is the number of nautical miles between place *A* and place *B* in Exercise 2?

5. Suppose you start at Washington, D.C. (39°N, 77°W), and travel 300 n mi due north, then 300 n mi due west, then 300 n mi due south, and then 300 n mi due east. Where will you arrive with respect to your starting point—at your starting point, or north, south, east, or west of it?   Answer: west of starting point

6. Suppose you start at Washington, D.C. (39°N, 77°W), and travel 600 n mi due south, then 600 n mi due west, then 600 n mi due north, and then 600 n mi due east. Where will you arrive with respect to your starting point?

7. What are the latitude and longitude of the point on Earth that is opposite Washington, D.C. (39°N, 77°W)?   Answer: 39°S, 103°E

8. What are the latitude and longitude of the point on Earth that is opposite Tokyo (36°N, 140°E)?

16.3 Time

9. What are the local solar time and date at (40°N, 118°W) when the local solar time at (35°N, 80°W) is 6 P.M. on October 8?   Answer: 3:28 P.M., October 8

10. What are the local solar time and date at (40°N, 110°W) when the local solar time at (30°N, 70°W) is 1 A.M. on October 16?

11. When it is 9 P.M. standard time on November 26 in Moscow (56°N, 38°E), what are the standard time and date in Tokyo (36°N, 140°E)?

Answer: 3 A.M., November 27

12. When it is 10 A.M. standard time on February 22 in Los Angeles (34°N, 118°W), what are the standard time and date in Tokyo (36°N, 140°E)?

13. A professional basketball game is to be played in Portland, Oregon. It is televised live in New York beginning at 9 P.M. EST. What time must the game begin in Portland?

Answer: 6 P.M. PST

14. If the polls close during a presidential election at 7 P.M. EST in New York, what is the standard time in California?

15. If an Olympic event begins at 10 A.M. PST on July 28 in Los Angeles (34°N, 118°W), what standard time and date will it be in Moscow (56°N, 38°E)?

Answer: 9 P.M., July 28

16. The time of an Olympic event begins at 1 P.M. EST in Atlanta, Georgia (34°N, 84°W). Determine the standard time in Milano, Italy (45.5°N, 9°E).

17. Determine the approximate time of sunset at 40°N on July 21.

Answer: 7 P.M.

18. Determine the approximate time of sunrise at 40°N on April 21.

19. What is the altitude angle of the Sun for someone at Atlanta, Georgia (34°N), on June 21?

Answer: 79.5°

20. What is the altitude angle of the Sun for someone at Atlanta, Georgia (34°N), on December 22?

21. What is the latitude of someone in the United States who sees the Sun at an altitude of 71.5° on June 21?

Answer: 42°N

22. What is the latitude of someone in the United States who sees the Sun at an altitude of 65.5° on December 22?

23. Determine the month and day when the Sun is at maximum altitude for an observer at Washington, D.C. (39°N). What is the altitude of the Sun at this time?

Answer: on or about June 21, 84.5°

24. Determine the month and day when the Sun is at minimum altitude for an observer at Washington, D.C. (39°N). What is the altitude of the Sun at this time?

Solutions to Confidence Exercises

16.1 $43°N - 34°N = 9°$ $\quad \dfrac{9° \times 60 \text{ n mi}}{1°} = 540 \text{ n mi}$

16.2 Since all meridians are great circle routes and the traveler goes east and later west on the same size parallel, the traveler will be at place A at the end of the journey.

16.3

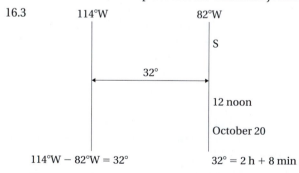

$114°W - 82°W = 32°$

$32° = 2\text{ h} + 8\text{ min}$

$12\text{ noon} - 2\text{ h} - 8\text{ min} = 9{:}52 \text{ A.M., October 20}$

16.4 The answers are a function of the student's hometown latitude.

Answers to Multiple-Choice Review Questions

THE MOON

17 ● ● ● ● ● ●

That's one small step for man,
one giant leap for mankind.

Neil Armstrong (1969)

The exact origin of the word *moon* seems to be unknown, but many writers believe it is related to the measurement of time. We do know that the length of our present month is based on the motion and the phases of the Moon, that primitive people worshipped the Moon, and that many societies today base their religious ceremonies on the new and full phases of the Moon. We also know that the human reproductive cycle seems to be synchronized to the lunar cycle, with the ovaries producing ova about every 28 days.

On July 20, 1969, humans first landed on the Moon, and Apollo 11 astronauts placed a retroreflector (an optical reflector designed to return the reflected ray of a laser beam exactly parallel to the incident ray) on the Moon's surface. The retroreflector is part of a lunar-ranging experiment that measures the distance to the Moon with a precision of ±15 cm.

Measurements with this high precision can be taken over long periods of time and will show the variation in the orbital distance of the Moon in great detail. Such measurements can be used to (1) determine the rate of continental drift on Earth (latitude and longitude of a place can be determined with great accuracy), (2) detect any change in the location of the North Pole, (3)

Photo: The Lunar Prospector spacecraft orbits the Moon at an altitude of 100 km.

determine the orbit of the Moon more exactly, and (4) determine whether the gravitational constant (*G*) is changing with time.

The Moon appears as the second-brightest object in the sky because it is very close to us. The Moon's average distance from Earth is about 384,000 km (240,000 mi). Because of its nearness and its influence on our lives, this chapter is devoted to the study of the Moon. ● Figure 17.1 shows an astronaut with the Lunar Rover collecting samples of the lunar surface. See the Spotlight feature at the end of the chapter. ■

17.1 General Features

LEARNING GOAL

▼ Describe the general physical properties of the Moon.

The Moon at its brightest is a wondrous sight as it reflects the Sun's light back to our eyes. The Moon appears quite large to Earth observers. In fact, our Moon is the largest of any inner planet's. Mercury and Venus have no moons, and those of Mars are quite small. Our Moon is the fifth largest in the solar system.

The Moon revolves around Earth in approximately 29.5 solar days, and it rotates at the same rate as it revolves. For this reason, we see only one side of the

FIGURE 17.1 Geologist-Astronaut Harrison Schmitt on the Moon

This surface view of the Moon was taken at the Taurus-Littrow landing site during the Apollo 17 mission. The Lunar Rover is at the right of the huge boulder.

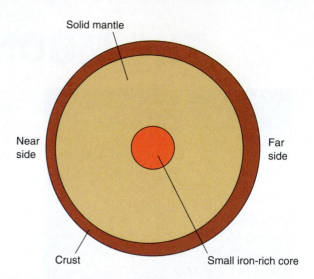

FIGURE 17.2 The Internal Structure of the Moon

Geologists theorize that the Moon consists of a crust that varies in depth from about 64 km on the near side to 125 km on the far side from Earth, a solid mantle rich in silicate materials, and a small (less than 720 km in diameter), perhaps solid, iron-rich core.

Moon. An observer on the side that faces Earth would always be able to see Earth, but the Sun would appear to rise and set and rise again once every 29.5 days. Thus all sides of the Moon are heated by the Sun's rays.

The Moon is nearly spherical, with a diameter of 3476 km (2160 mi), a distance slightly greater than one-fourth Earth's diameter. The slow rotation of the Moon, coupled with the tidal bulge caused by Earth's gravitational pull on the solid material, produces a very slight oblateness. The best measurements indicate a difference of less than 1.5 km between the polar and equatorial diameters.

The mass of the Moon is $\frac{1}{81}$ that of Earth, and its average density is 3.3 g/cm^3. Earth's average density is 5.5 g/cm^3. The surface gravity of the Moon is only one-sixth that of Earth. Therefore, your weight on the surface of the Moon would be about one-sixth your weight on Earth's surface (see Fig. 1.6). The Moon's interior is thought to be made up of a small, perhaps solid, iron-rich core, a solid mantle, and a crust that is about 64 km thick on the near side and 125 km thick on the far side (● Fig. 17.2).

The Moon does not possess a magnetic field; at least, none was detected by instruments carried by the Apollo astronauts. Surface rocks brought back by astronauts show some magnetism, indicating that the Moon had a slight magnetic field when the rocks solidified. The origin of this previous magnetic field is unknown.

Except for its phases, the Moon's most predominant feature is the appearance of its surface, which is marked with craters, basins, plains, rays, rills, mountain ranges, and faults. These features vary in size, shape, and structure. The most outstanding are the craters that are clearly visible to an Earth observer with low-power binoculars or a telescope.

Craters and Basins

Craters are the best-known characteristics of the Moon's surface. About 30,000 can be seen with an Earth-based telescope. They range in size from microscopic to hundreds of kilometers in diameter. Craters about 1 km in diameter tend to have smooth, bowl-shaped interiors. In larger craters, the floor flattens, and in still larger ones, the floor develops a central peak. ● Figure 17.3 shows details of a large lunar crater.

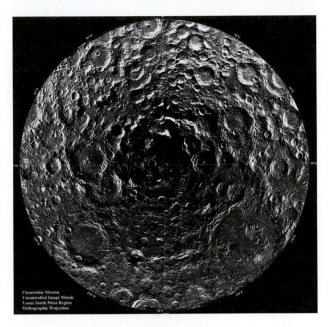

FIGURE 17.4   South Pole–Aitken Basin

FIGURE 17.3   A Lunar Crater

The large crater in the center of the picture has a diameter of about 81 km. The rugged terrain is typical of the far side of the Moon.

The word **crater** (Greek *krater*) means bowl-shaped, and lunar craters are small- and large-diameter depressions believed to be caused by the impact and explosion of small and large meteorites. The craters are rather shallow (their depths are small in comparison with their diameters), and their floors are located below the lunar surface. Measuring the volume of the material around a crater's rim and comparing it with the volume of the crater itself shows that they are approximately the same. This result supports the impact hypothesis.

The largest impact features are the multiring basins. ● Figure 17.4, which is compiled from images taken by the spacecraft *Clementine* during its 71-day lunar orbit in 1994, shows the South Pole–Aitken impact basin. The basin, unknown before the *Clementine* voyage, is the largest known in the solar system. The basin has a diameter of 2500 km (1550 mi) and a depth that averages about 12 km (7.4 mi). Its location at the south pole plus its great depth keeps the basin floor from sunlight and view. Another multiring basin is shown at the four o'clock position in Fig. 17.4. This one has a diameter of 320 km (198 mi) and is named Schrödinger after Erwin Schrödinger (1887–1961), German professor of physics and cofounder of quantum mechanics (Chapter 9). This is the second youngest impact basin on the Moon, the youngest being Orientale, which is about 1000 km across and was formed 3.8 billion years ago.

Plains

The lunar surface contains thousands of craters ranging in diameter from a few feet to the 240-km (150 mi) Clavius. The large, flat areas called *maria* (a Latin word meaning "seas"), named by Galileo, are believed to be craters formed by the impact of huge objects from space and later filled with lava. These areas, which are now called **plains,** appear to be very dark because the Moon's surface is a poor reflector. The plains, which are similar to black asphalt, appear darker than their surroundings because they have a relatively low albedo. The average albedo of the Moon is about 0.07, which means the surface reflects only 7% of the light received from the Sun. Fourteen major plains on the side facing Earth cover over 50% of the visible lunar surface. Most of the plains are located in the northern hemisphere and can be seen clearly with the unaided eye during the full phase of the Moon.

The surface of the Moon is a terrain of rolling, rounded knolls composed of a layer of loose debris or soil called *regolith,* which has a depth less than 10 m (33 ft) on the flat lunar plains. The lunar highlands, because they are older, have a thicker layer of regolith. The rock samples brought back by Apollo astronauts are similar to the volcanic rock found on Earth.

Rays

Some craters are surrounded by streaks, or **rays,** that extend outward over the surface. They are believed to be pulverized rock that was thrown out when the crater was formed. The rays appear much brighter than the crater, and we know that powdered rock reflects light better than regular-sized rock. The rays also become darker with age. Photographs show that in cases where rays from one crater overlap those of another, the ones on top appear to be brighter.

The ray system of a crater has an average diameter of about 12 times the diameter of the crater. Lunar photographs also show that the ray systems are marked with small craters called *secondary craters,* which are believed to have been formed by debris thrown out from the primary crater during the explosion caused by an impinging object from space.

Rills

Another feature of the lunar surface is long, narrow trenches, or valleys, called **rills.** They vary from a few meters to about 5 km in width and extend hundreds of kilometers in length with little or no variation of width. Some rills are rather straight, whereas others follow a circular path. They have very steep walls and fairly flat bottoms that are as much as 0.8 km below the lunar surface. Moonquakes are thought to cause the formation of rills. Similar separations of Earth's surface are produced by earthquakes.

Mountain Ranges and Faults

The **mountain ranges** on the lunar surface have peaks as high as 6100 m, and all formations seem to be components of circular patterns bordering the great plains. This pattern indicates that they were not formed and shaped by the same processes as mountain ranges on Earth, which were formed by internal forces.

The craters and other surface features of the Moon were formed a long time ago when the solar system was filled with large amounts of matter. They are much the same now as they were when formed because of the absence of erosion. Some changes have resulted from the impact of projectiles from space.

A **fault** is a break or fracture in the surface of the Moon along which movement has occurred. The motion along a fault can be vertical, horizontal, or parallel. Several faults are observed on the lunar surface. A very large cliff on the eastern side of Mare ("MAH-ray") Nubium, shown in ● Fig. 17.5, is the result of slippage of the Moon's crust along a fault. This cliff (called the Straight Wall) is about 113 km long and 244 m high, with its side inclined about 40° to the horizontal.

RELEVANCE QUESTION: *When you observe the full moon with the unaided eye, what two general surface features are most noticeable to you?*

17.2 Composition and History of the Moon

LEARNING GOALS

▼ Identify the composition of the Moon.

▼ Explain the latest theory for the origin of the Moon.

Before the Apollo program to land an astronaut on the Moon was begun in the early 1960s, very little was known about the composition and history of the

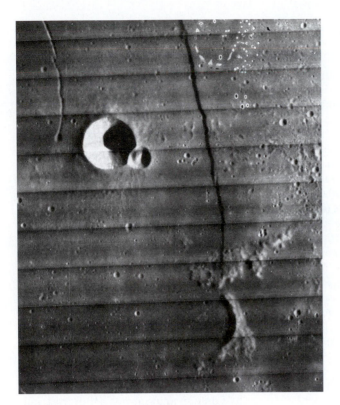

FIGURE 17.5   The Straight Wall

This photo shows a unique steep slope on the eastern side of Mare Nubium. The wall is about 113 km long and 244 m high. Rima Birt 1 (top left) is an irregular trough or rill. The photo was taken by *Lunar Orbiter V.*

lands were formed between 4.4 and 3.9 billion years ago, whereas those from the plains have ages between 3.9 and 3.2 billion years. No rocks older than 4.4 billion years or younger than 3.1 billion years have been found.

Almost all the craters on the Moon are now known to have resulted from the bombardment of meteorites of various sizes. Because the Moon has no atmosphere or water on its surface, very little erosion takes place. Therefore, once formed, a crater remains for billions of years or until a meteorite hits and forms a new crater on top of it. In contrast, Earth has only a few remaining meteorite craters. For the most part, the craters left by meteorites striking Earth have been eroded away.

The Moon's plains and a few of its craters (about 1%) were produced by volcanic eruptions. The plains are composed of black volcanic lava that covered many craters. Most of the plains are on the near side of the Moon. The fact that fewer volcanic eruptions occurred on the far side is probably correlated with the fact that the Moon's crust is thicker there (see Fig. 17.2).

The oldest rocks on the Moon were formed about 4.4 billion years ago when the Moon's crust became cool enough to solidify. Between 4.4 and 3.9 billion years ago, the Moon was bombarded intensely by many meteorites. This was the period when most of

Moon. Exploration by the Luna and Apollo programs has changed all that.

The first landing on the Moon was July 20, 1969, when the landing craft of Apollo 11 settled in Mare Tranquillitatis. After that, five other Apollo lunar landing missions (Apollo 12, 14, 15, 16, and 17)* were completed. The astronauts collected and brought back to Earth 379 kg of lunar material and erected scientific instruments that have collected data for many years.

The rock samples, such as the one shown in ● Fig. 17.6, have enabled us to have a much better understanding of the Moon's composition and history. (See the chapter Highlight.) Samples from the plains, or lowlands, have yielded ages considerably younger than those from the highlands. Rocks from the high-

*A 1995 movie told the story of the "missing number," Apollo 13.

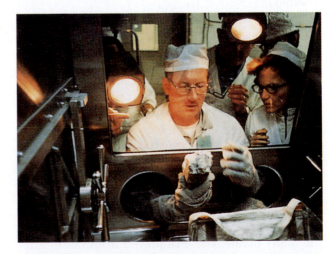

FIGURE 17.6   Scientists Examine a Lunar Sample

The samples are stored in an atmosphere of dry nitrogen, thus isolating them from oxygen and moisture (water) to prevent chemical reactions.

HIGHLIGHT: The Origin of the Moon

What is the origin of the Moon? Earth's Moon is an anomaly in the solar system because the Earth-Moon system is one of a kind. Earth is the only inner planet that has a large satellite, a satellite as large as one-fourth the diameter of the parent planet. Mercury and Venus do not have satellites, and Mars has two small ones that appear to be asteroids captured from the nearby asteroid belt. The outer planets are huge bodies with satellites that are small compared with the planets. Pluto and its relatively large satellite Charon are an oddity, and both are considered by some astronomers to be asteroids.

Several theories have been suggested for the origin of the Moon. Some of the facts that must be considered when formulating a theory are the following:

1. The lunar samples collected by the Apollo astronauts show that the Moon's chemical composition is similar to that of Earth's mantle (the interior region of Earth between the core and the crust). (See Section 22.3.)
2. The percentage abundances of the isotopes of oxygen (^{16}O, ^{17}O, ^{18}O) found in the lunar rock samples are similar to those of Earth, indicating that Earth and the Moon were formed at about the same distance from the Sun.
3. Water was not found in the lunar rock samples.
4. The lunar rocks show a small percentage of volatile elements (those which are driven off by extreme heating, such as sodium and potassium) and a high abundance of refractory elements (elements that are not easily vaporized).
5. Compared to Earth, the Moon samples show a lower abundance of iron.
6. The Moon has an average density of 3.3 g/cm^3. Earth's average density is 5.5 g/cm^3. Earth's mantle has a density of about 3.5 g/cm^3, the crust about 3.0 g/cm^3, and the core about 15 g/cm^3.
7. The oldest Earth and lunar rocks were formed at about the same time.

One theory at present is that Earth and the Moon were created at about the same time (the *sister theory*), and the Moon coalesced from particles revolving around Earth. This theory fails to explain why Earth has a great abundance of iron and water, whereas the Moon has very little iron and no water.

Another explanation for the origin of the Moon is the *great impact theory* (see the Spotlight feature on page 460). This theory proposes that a planet-sized object, about the size of Mars, struck Earth with a glancing blow 4.4 to 4.5 billion years ago. The impact ejected enough matter (most of it coming from Earth's mantle) into orbit to form the Moon. This would account for the similar densities of Earth's mantle and the Moon and for the Moon's low abundance of iron. Also, the impact of a large object generates tremendous heat, which would drive off water and other volatile substances.

The impact theory raises the possibility that other collisions or near collisions may have taken place in the solar system. For example, a collision or near collision could account for the large inclination of Uranus' axis and the planet's retrograde motion. Also, Venus' retrograde motion may have been caused by an outside disturbance. The planet Mercury, which has only an iron core and a thin crust, may have lost its mantle due to a collision.

The impact theory is gaining support within the scientific community, but lunar samples were collected from only nine surface areas by the Luna and Apollo programs. This is a very small sample of the Moon's surface. Perhaps, when additional information is obtained from the Moon, the impact theory or a better explanation can be established.

its craters were formed. The Moon's surface was virtually pulverized, leaving little evidence of the original crust.

During the period 3.9 to 3.1 billion years ago, the Moon's interior had heated up enough from radioactive effects to cause volcanic eruptions that formed the many plains. The lava flowed from these eruptions and covered much of the Moon's lowlands. During this period, meteorite bombardment became less intense, because fewer and fewer rock fragments were left near the Earth-Moon system.

By 3.1 billion years ago, the Moon's mantle had become so thick that it could no longer be penetrated by molten rock. The Moon has been geologically quiet since that time. Meteorites have continued to bombard the surface and have formed a layer of dust several meters thick on the surface.

RELEVANCE QUESTION: *What has been each working person's active role in obtaining modern data concerning the composition and origin of the Moon? (You may be unaware of it.)*

17.3 Lunar Motions

LEARNING GOALS

▼ Describe the Moon's orbit and motions.

▼ Explain sidereal and synodic months.

The Moon revolves eastward around Earth in an elliptical orbit in a little over 29.5 solar days, or almost 27.33 sidereal days. Its orbital plane does not coincide with that of Earth but is tilted at an angle of approximately 5° with respect to Earth's orbital plane (● Fig. 17.7). The 5° tilt allows the Moon to be overhead at any latitude between 28.5°N and 28.5°S. The Moon rotates eastward as it revolves, making one rotation during one revolution. Figure 17.7 is an illustration of the eastward motion of the Moon and the inclination of the orbital plane to the orbit of Earth.

There are two different lunar months. The period of the Moon with respect to a star other than the Sun is approximately 27.33 days; this is called the sidereal period, or **sidereal month.** It is the actual time taken for the Moon to revolve 360°. The period of the Moon with respect to the Sun is approximately 29.5 days. This period is called the **synodic month,** or the month of the phases. The Moon revolves more than 360° during the synodic month (● Fig. 17.8).

To an observer on Earth, the Moon appears to rise in the east and set in the west each day. This apparent motion of the Moon is due to Earth rotating eastward on its axis once each day. The times at which it rises and sets are discussed in the following section.

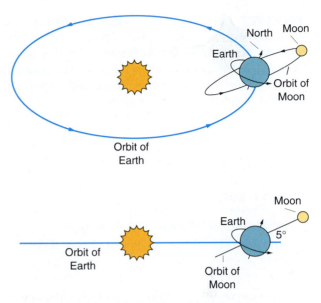

FIGURE 17.7   The Relative Motions of the Moon and Earth

The top diagram is a view from above Earth's orbital plane; the lower diagram is a view from within Earth's orbital plane.

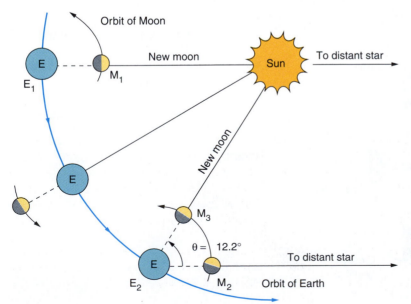

FIGURE 17.8   Diagram Illustrating the Difference Between the Sidereal and Synodic Months

With Earth at position E_1 and the Moon at position M_1, the Sun, the Moon, and Earth are all in the same plane. At this time the Moon is in its new phase. As Earth revolves eastward to position E_2, the Moon has revolved through 360° in approximately 27.3 days to position M_2, and 1 sidereal month—one revolution with respect to a distant star—has elapsed. The Moon must revolve through 360° plus the angle θ before arriving at position M_3. At this time the Sun, the Moon, and Earth will be in the same plane, the Moon will be in new phase, and 1 synodic month will have passed. The time for 1 synodic month is approximately 29.5 days.

RELEVANCE QUESTION: *The Moon is the second brightest object in the sky, outranked only by the Sun. At night, the Moon ranks first in brightness but is light-dependent on the Sun. As the Moon orbits Earth, what influences do the changes in appearance and brightness that occur night after night have on you?*

17.4 Phases of the Moon

LEARNING GOALS

▼ Define and explain the phases of the Moon.

▼ Solve exercises relating to different phases of the Moon.

The most outstanding feature presented by the Moon to an Earth observer is the periodic change in its appearance. One-half of the Moon's surface is always reflecting light from the Sun, but only once during the lunar month does the observer see all the illuminated half. Throughout most of the Moon's period of revolution, only a portion of its illuminated side is presented to us.

The starting point for the Moon's synodic month, or month of phases, is arbitrarily taken at the new-phase position. The new phase of the Moon occurs when Earth, Sun, and Moon are in the same plane, with the Moon positioned between the Sun and Earth. They are not necessarily in a straight line. At this position the dark side of the Moon is toward Earth, and the Moon cannot be seen from this planet. Because the Sun is on the observer's meridian with the Moon, the new moon occurs at 12 noon local solar time.

The **new moon** actually occurs just for an instant—the instant it is on the same meridian as the Sun. We often speak, however, of a phase of the Moon lasting for a full day of 24 h.

The Moon revolves eastward from the new-phase position, and for the next 7.375 solar days (one-fourth of 29.5 days) it is seen as a waxing crescent moon. The term **waxing phase** means that the illuminated portion of the Moon is getting larger; **waning phase** means that the illuminated portion is getting smaller as observed from Earth. A **crescent moon** has less than one-quarter of its surface illuminated. A **gibbous moon** occurs when more than one-quarter of the Moon's surface appears illuminated. ● Figures 17.9 and 17.10 illustrate how the phases occur and how they appear to an observer on Earth.

The Moon is in the *waxing crescent phase* and appears as a crescent moon to an Earthbound observer when it is less than 90° east of the Sun. The Moon is in **first-quarter phase** when it is 90° east of the Sun and appears as a quarter moon on the observer's meridian at 6 P.M. local solar time with the illuminated side

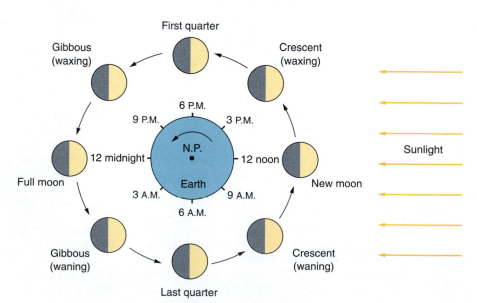

FIGURE 17.9 Phases of the Moon

Diagram shows the position of the Moon relative to Earth and the Sun during one lunar month as observed from a position in space above Earth's North Pole.

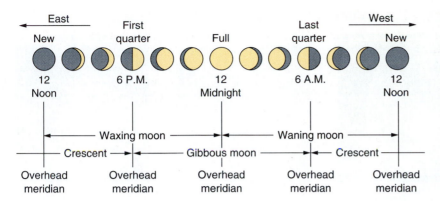

FIGURE 17.10 Phases of the Moon

Diagram illustrating the phases of the Moon as observed from any latitude north of 28.5°N. The observer is looking south; therefore, east is on the left. The Sun's position can be determined by noting the local solar time the Moon is on the overhead meridian. The time period represented in the drawing is 29.5 days. Compare this drawing with Fig. 17.11.

toward the west. The first-quarter phase has a duration of only an instant, because the Moon can only be 90° east of the Sun for an instant (Fig. 17.9).

From the first-quarter position, the Moon enters the *waxing gibbous phase* for 7.375 solar days. During this phase, it appears larger than a quarter moon but less than a full moon. When the Moon is 180° east of the Sun, it will be in full phase and will appear as a **full moon** to the Earthbound observer. The full moon appears on the observer's meridian at 12 midnight local solar time.

From the full-phase position, the Moon enters the *waning gibbous phase* and remains in that phase for 7.375 solar days. The appearance of the Moon during this phase is the same as in the waxing gibbous phase, except that the illuminated side is toward the east, and the Moon is seen in the sky at a different time. When the Moon is 270° east of the Sun, it will be in the **last-quarter phase.** The last-quarter phase appears on the observer's meridian at 6 A.M. local solar time, with the illuminated side of the Moon toward the east.

From the last-quarter position, the Moon enters the *waning crescent phase* and remains in this phase for 7.375 solar days. During this phase, the illuminated portion appears smaller than a quarter moon. Its appearance is the same as the waxing crescent moon, except that the illuminated side is toward the east, and the Moon appears in the sky at a different time.

Figure 17.10 illustrates the Moon's appearance and position above the southern horizon as observed from a northern latitude greater than 28.5° north. The Moon is shown in the first drawing on the left in the new-phase position. It is shown on the observer's meridian at 12 noon local solar time and shaded gray,

illustrating that it cannot be seen at this time because it is on the same meridian as the Sun. The next two positions illustrate the waxing crescent phase. Note that the illuminated area (colored yellow) appears larger for an Earth observer as the Moon approaches the first-quarter phase and that the illuminated side is toward the west where the Sun is located.

When the Moon is in the first-quarter phase, it is on the observer's meridian at 6 P.M. local solar time. The Sun will be at or near the western horizon at this time. The Moon revolves eastward, entering the waxing gibbous phase, as shown by the next two positions. Note that the illuminated area is larger than a quarter moon and still increasing in size of face. The illuminated side is still toward the west.

The next position shows the Moon at full phase and on the observer's meridian at 12 midnight. If the date is at the time of the vernal (spring) or autumnal (fall) equinox, the Moon will rise on the eastern horizon at 6 P.M., when the Sun is setting in the west, and the Moon will set at 6 A.M., as the Sun is seen rising on the eastern horizon.

The Moon continues revolving eastward, entering the waning gibbous phase. Note that the size of the illuminated area is decreasing for an Earth observer and that the illuminated side is toward the east—just the opposite from the waxing gibbous moon. When the Moon is 90° west of the Sun (same as 270° east of the Sun), it will be on the observer's meridian at 6 A.M. local solar time, as shown in the next position. The Moon appears as a quarter moon, but note that the illuminated side is toward the east. The Sun will be rising at or near this time.

After the last-quarter phase, the Moon enters the waning crescent phase, and the size of its face continues

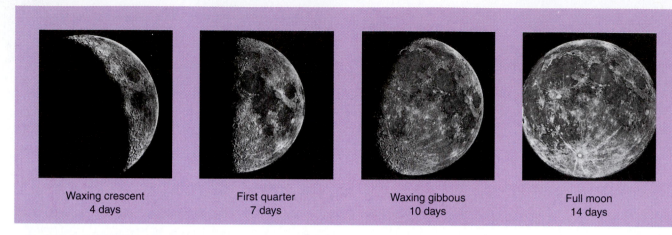

Waxing crescent
4 days

First quarter
7 days

Waxing gibbous
10 days

Full moon
14 days

FIGURE 17.11 Phases of the Moon
These eight photographs show the Moon at different times of the lunar month. They are arranged so that the view, from Earth, of the phases are the way they appear in a close-up view by the unaided eye. Compare these photos with Fig. 17.10.

to decrease. Note that the illuminated side of the waning crescent moon is toward the east. The last position shows the Moon back to the new-phase position.
● Figure 17.11 shows eight photographs of the Moon as it appears in a close-up view by the unaided eye. Compare these photographs with Fig. 17.10.

Table 17.1 summarizes the times for the various phases of the Moon to rise, be overhead, and set. An example of what an observer in the United States sees when looking at the first-quarter phase is shown in ● Fig. 17.12. Figures similar to 17.12 for the other phases can be drawn using the information in Table 17.1.

Because the Moon revolves around Earth every 29.5 solar days, it gains 360° on the Sun in that time, or

12.2°/day. Thus the Moon is on the observer's meridian about 50 min later each day, because Earth must rotate through 360° plus 12.2° before the Moon appears on the overhead meridian (● Fig. 17.13). The average time of moonrise is thus delayed about 50 min each day. The actual time depends on the latitude of the observer, with greater variation noted in the higher latitudes. The variation depends on the angle between the Moon's path and the horizon.

The approximate altitude of the full moon can be found by recognizing that the full moon will be on the opposite side of Earth from the Sun (● Fig. 17.14). Thus, when the Sun is low in the sky in the winter, the full moon will be high in the sky. In the summer, the Sun is high in the sky and the full moon is low in the sky.

TABLE 17.1 Times for the Various Phases of the Moon to Rise, Be Overhead, and Set When the Sun Is at the Vernal or Autumnal Equinox

Phase	Approximate Rising Time	Approximate Time Overhead	Approximate Setting Time
New moon	6 A.M.	Noon	6 P.M.
First-quarter moon	Noon	6 P.M.	Midnight
Full moon	6 P.M.	Midnight	6 A.M.
Last-quarter moon	Midnight	6 A.M.	Noon

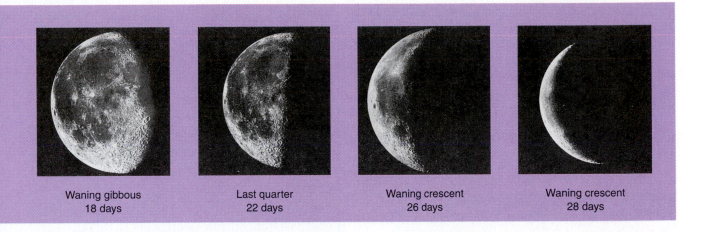

| Waning gibbous 18 days | Last quarter 22 days | Waning crescent 26 days | Waning crescent 28 days |

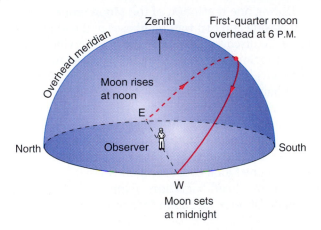

FIGURE 17.12 The Rising and Setting of the First-Quarter Moon

Diagram illustrates the first-quarter moon rising, on the overhead meridian, and setting during the time of the vernal or autumnal equinox for an observer in the United States. The side of the Moon facing west is illuminated by sunlight.

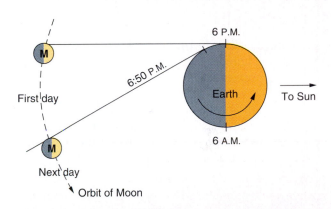

FIGURE 17.13 Moon's Rising Time

The Moon rises about 50 min later each day because, as Earth rotates, the Moon is revolving around Earth. For example, the full moon rises at about 6 P.M. on March 21 and at about 6:50 P.M. on March 22.

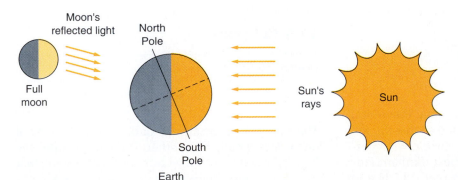

FIGURE 17.14 Relative Positions of the Full Moon, Earth, and the Sun

This side view of Earth, the full moon, and the Sun during the winter months in the Northern Hemisphere shows the Sun's rays striking the Southern Hemisphere most directly. The Moon's reflected light falls most directly on the Northern Hemisphere.

What is the maximum altitude of any phase of the Moon as observed from the United States? The answer, of course, depends on the latitude of the observer. The closer the observer's latitude is to 28.5°N, the greater is the maximum altitude of the Moon. If the Moon is overhead at 28.5°N and the observer's latitude is 28.5°N, then the zenith angle is zero and the Moon's altitude is 90°. The following example illustrates how to determine the maximum altitude of the Moon.

EXAMPLE 17.1

Calculating the Maximum Altitude of the Moon as Observed from a Given Latitude

Calculate the maximum altitude of the full moon as observed from Washington, D.C. (39°N, 77°W).

SOLUTION

Maximum altitude refers to the maximum angle above the horizon or the minimum zenith angle. To be at the minimum zenith angle, the full moon must be as close to 39°N (the observer's latitude) as possible. This is accomplished when the Sun is as far south as possible, since the full moon is 180° from the Sun's position. The Sun's most southern position is latitude 23.5°S on December 21 ± 2 days.

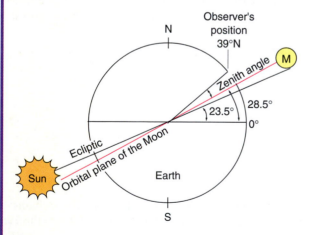

Step 1

Draw a diagram that illustrates the data. The orbital plane of the Moon must be oriented so that the full moon is at 28.5°N latitude. *Note:* Over one precession cycle of 18.6 y, the Moon's most northern latitude is 23.5°N ± 5° or varies between 18.5°N and 28.5°N.

Step 2

Calculate the zenith angle.

$$\text{zenith angle} = 39°N - 28.5°N = 10.5°$$

Step 3

Calculate the altitude of the Moon.

$$\text{altitude of Moon} = 90° - \text{zenith angle}$$
$$= 90° - 10.5° = 79.5°$$

CONFIDENCE EXERCISE 17.1

Calculate the maximum altitude of the full moon as observed from Atlanta, Georgia (34°N, 84°W).

During one revolution of the Moon around Earth, two waxing and two waning phases of the Moon take place. To discover the phase one is observing, determine the time the Moon is on the overhead meridian. Look at Fig. 17.9. If the Moon is on the overhead meridian at 3 P.M., the Moon must be in the waxing crescent phase. If the Moon is on the overhead meridian at 9 P.M., the Moon must be in the waxing gibbous phase. Likewise for the other waxing and waning phases. By studying Fig. 17.9, one can determine the phase of the Moon by knowing the time it is observed on the overhead meridian. Experience is a good teacher. Observe the Moon for a month.

RELEVANCE QUESTION: *Easter Sunday is an annual Christian festival. The time of the celebration is time-dependent on a phase of the Moon. The festival is held on the first Sunday after the first full moon after the Sun crosses the vernal equinox moving northward, which occurs on or about March 21. Name some other related influences the Moon has on our lives.*

17.5 Eclipses

LEARNING GOAL

▼ Describe and explain solar and lunar eclipses.

The word **eclipse** means the darkening of the light of one celestial body by another. The Sun provides the light by which we see objects in our solar system; that is, nonluminous objects in our solar system are observed by reflected light from the Sun. Because the

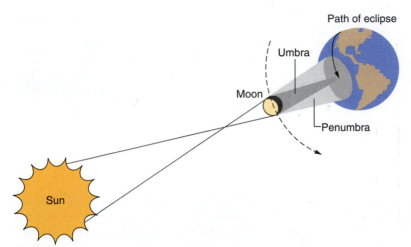

FIGURE 17.15 A Total Solar Eclipse

This diagram shows the positions of the Sun, the Moon, and Earth during a total solar eclipse. The umbra and penumbra are, respectively, the dark and semidark shadows cast by the Moon on the surface of Earth.

light from the Sun falls on objects in the solar system, objects cast shadows that extend away from the Sun. The size and shape of the shadow depend on the size and shape of the object and its distance from the Sun. Earth and the Moon, being spherical bodies, cast conical shadows, as viewed from space.

If we examine the shadow cast by Earth or the Moon, we discover two regions of different degrees of darkness. The darkest and smallest region is known as the **umbra** (● Fig. 17.15). An observer located within this region is completely blocked from the Sun during a solar eclipse. The semidark region is called the **penumbra.** An observer positioned in this region can see only a portion of the Sun during a solar eclipse.

A **solar eclipse** occurs when the Moon is at or near new phase and is in or near the ecliptic plane. When these two events occur together, the Sun, the Moon, and Earth are nearly in a straight line. The Moon's shadow will then fall on Earth, and the Sun's rays will be hidden from those observers in the shadow zone. A *total eclipse* occurs in the umbra region and a *partial eclipse* in the penumbra region (Figs. 17.15, ● 17.16, and ● 17.17).

The length of the Moon's shadow varies as the Moon's distance to the Sun varies. The average length of the Moon's umbra is 373,000 km, which is slightly less than the mean distance between Earth and the Moon. Because the umbra is shorter in length than

FIGURE 17.16 An Annular Eclipse of the Sun

When the umbra of the Moon's shadow does not reach all the way to the surface of Earth, we observe an annular eclipse of the Sun.

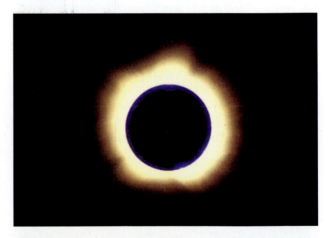

FIGURE 17.17 Total Solar Eclipse Showing Solar Corona

The solar corona, which can be photographed during a total solar eclipse, is composed of hot gases that extend millions of miles into space. This photo shows only the brightest inner part of the solar corona.

FIGURE 17.18   A Lunar Eclipse

This diagram shows the positions of the Sun, Earth, and the Moon during an eclipse of the Moon.

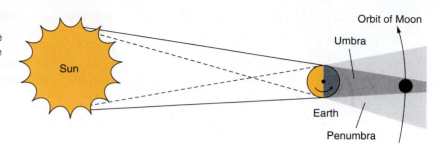

the mean distance from Earth to the Moon, an eclipse of the Sun can occur in which the umbra fails to reach Earth. An observer positioned on Earth's surface directly in line with the Moon and Sun sees the Moon's disk projected against the Sun, and a bright ring, or *annulus,* appears outside the dark Moon. This condition is called an **annular eclipse.** (See Fig. 17.16.)

Around the zone of the total (very dark) annular eclipse appears the larger, semidark region of the penumbra. The penumbra region may be as large as 10^4 km in diameter at the surface of Earth. The maximum diameter of the umbra at Earth's surface is about 270 km. This maximum value can exist only when the Sun is farthest from Earth, which is in early July, and the Moon is at *perigee,* or at its closest distance to Earth.

The motion of the Moon and Earth are such that the shadow of the Moon moves generally eastward during the time of the eclipse with a speed of about 1600 km/h. Thus the region of the total eclipse does not remain long at any one place. The greatest possible value is about 7.5 min, and the average is about 3 or 4 min.

A **lunar eclipse** occurs when the Moon is at or near full phase and is in or near the ecliptic plane (● Fig. 17.18). The Sun, Earth, and the Moon will be positioned in a nearly straight line, with Earth between the Sun and the Moon. Thus the shadow formed by Earth conceals the face of the Moon. The average

length of Earth's shadow is about 1.4×10^6 km, and the diameter of the shadow at the Moon's position is great enough to place the Moon in total eclipse for a time slightly greater than 1.5 h. A partial eclipse of the Moon can last as long as 3 h 40 min.

The orbital plane of the Moon is inclined to the ecliptic (the annual path of the Sun) at an angle slightly greater than 5°. Therefore, the path of the Moon crosses the ecliptic plane at two points as it makes its monthly journey around Earth. The points where the Moon's path crosses the ecliptic plane are known as *nodes.* The point of crossing going northward is called the **ascending node,** and the point of crossing going southward is called the **descending node** (● Fig. 17.19). A solar or lunar eclipse can occur only at or near the nodal points, because Earth, the Moon, and the Sun must be in a nearly straight line. This positioning occurs only at or near the points where the Moon crosses the ecliptic plane.

The orbital plane of the Moon is precessing westward, or clockwise, if viewed from above the orbital plane. The precession of the Moon's orbit causes the nodal points to move westward along the ecliptic, making one complete cycle in 18.6 y.

Predicting total solar eclipses is complicated because the line of nodes gradually moves westward and changes the nodes' direction in space. Data concerning the next five total solar eclipses are given in Table 17.2.

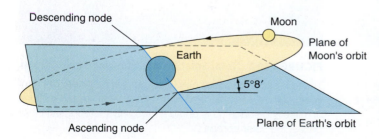

FIGURE 17.19   The 5°8' Angle Between the Orbital Planes of Earth and the Moon

This diagram illustrates the intersection of Earth's and the Moon's orbital planes. The angle between the planes is 5°8'. The Moon passes through the intersection twice during its monthly journey around Earth. The intersecting points are called *nodes.*

TABLE 17.2   Total Solar Eclipses from 1999 Through 2008

Date	Location Where Visible
August 11, 1999	Central Europe
June 21, 2001	Southern Africa
December 4, 2002	South Africa, Australia
November 23, 2003	Antarctica
March 29, 2006	West and Northern Africa
August 1, 2008	Siberia, North China

RELEVANCE QUESTION: *Since the Sun and Moon have about the same apparent or angular size, a total solar eclipse can be observed from Earth's surface. What is a safe method for observing a solar eclipse?*

17.6 Ocean Tides

LEARNING GOALS

▼ Define and describe tidal force.

▼ Explain ocean tides.

Anyone who has been to the seashore for a day's visit is aware of the rising and falling of the surface level of the ocean. The alternate rise and fall of the ocean's surface level is called the *tides.*

People related tides to the passage of the Moon in the first century A.D., but all efforts to explain the phenomenon failed until the seventeenth century, when Newton applied his law of universal gravitation to the problem. He related the alternate rise and fall of the ocean's surface level with the motions of Earth, the Moon, and the Sun. A few of the many factors contributing to the height that the ocean rises and falls at a particular location are

1. The force of attraction between the Moon and solid Earth.
2. The rotation of Earth on its axis.
3. The position of Earth, the Moon, and the Sun with respect to one another.
4. The varying distance between Earth and the Moon.
5. The inclination of the Moon's orbit.
6. The varying distance between Earth and the Sun.
7. The variation in the shape of coastlines and relief of ocean basins.

There are generally two high and two low tides daily because of the Moon's gravitational attraction and the motion of the Moon and Earth.

The reason for two daily tides can be understood by visualizing what shape Earth and its surface of water would take if there were no external gravitational forces and Earth had no motion (● Fig. 17.20a). If Earth did not rotate, there would be no centripetal

(a)

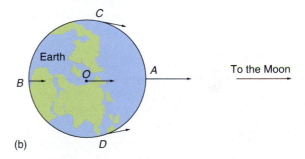

(b)

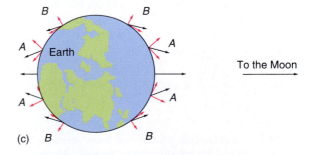

(c)

FIGURE 17.20   Tides on Earth

(a) The shape of stationary Earth with all water surfaces unaffected by any external gravitational forces. (b) Gravitational forces of the Moon (arrows) acting on Earth, which is considered as having a mass concentrated at a point. (c) The resultant tidal forces (arrows) act on Earth to produce an oblate (elongated at the center) contour. This view of Earth is from above the North Pole. (See text for details.)

force. On a nonrotating Earth, with no external gravitational forces, no forces would be exerted on Earth or the surface water and hence no tides would occur. How, then, do the gravitational force of the Moon and the motion of Earth produce two daily tides?

The answer can be found if the Moon's mass is considered to be concentrated at a point (Fig. 17.20b). The Sun is also a factor in causing tides, but it is left out of this explanation to simplify the results. The magnitude and direction of the Moon's gravitational forces acting on Earth at points A, B, C, D, and O are shown. The magnitude is indicated by the length of the arrow drawn to represent the gravitational force. Note that the forces at A, O, and B are all in the same direction toward the Moon, but the magnitudes are not the same. The force at A is the greatest because it is closest to the Moon.

Remember that Newton's law of gravitational attraction says the force between two masses is inversely proportional to the square of the distance between the two masses. Thus the force at O is less than the force at A, and the force at B is less than the force at O. Also, forces at C and D have approximately the same magnitude but are slightly less than the force at O. The direction of the forces at C and D are toward the Moon, as shown. There is a net differential force between points A and B that stretches solid Earth. This differential gravitational force is called the **tidal force.*** The gravitational forces of the Moon are not the same on all masses that make up Earth. The net effect of all gravitational forces causes Earth to round out (bulge) in the direction toward the Moon and also in the direction opposite the Moon. The effect of the tidal forces is much more noticeable on the oceans, since water flows easily over Earth's solid surface. Figure 17.20c shows the tidal forces with their vertical and horizontal components.

Although the horizontal components are very small, the forces acting over a period of hours will produce movement of the water that results in tidal bulges in the oceans. It is important to note that the tidal bulges are due to the horizontal components of the tidal forces. The high tides are not a result of the Moon's gravitational forces lifting the ocean's water away from Earth. The horizontal components cause the water to flow over Earth's surface toward areas nearest the Moon and toward areas opposite the Moon. Thus the oceans rise higher in these areas and will be correspondingly lower in areas from which the water is flowing.

When the Sun, Earth, and Moon are positioned in a nearly straight line, the gravitational force of the Moon and Sun combine to produce higher high tides and lower low tides than usual; that is, the variations between high and low tides are greatest at this time. These tides of greatest variation are called **spring tides,** and they occur at the new and full phases of the Moon. When the Moon is at first- or last-quarter phase, the Sun and Moon are 90° with respect to Earth. At these times the tidal forces of the Moon and Sun tend to cancel one another, and there is a minimum difference in the height of the surface of the ocean. In this case the tides are known as **neap tides.**

Note that two spring tides and two neap tides take place each lunar month, because the Moon passes through each of its phases once a month. A spring tide occurs at new moon, a neap tide at first quarter, another spring tide at full moon, and a second neap tide at last quarter (● Fig. 17.21).

The height of the tide also varies with latitude (● Fig. 17.22). The tide is highest at the Moon's overhead position and on the other side of Earth opposite the position of the Moon. The time of high tide does not correspond to the time of the meridian crossing of the Moon. The bulge is always a little ahead (eastward) of the Moon because of Earth's rotation. Because Earth rotates faster than the Moon revolves, Earth carries the tidal bulge forward in the direction it is rotating, which is eastward.

The action of the tides produces a retarding motion on Earth's rotation, slowing it and lengthening the solar day about 0.002 s per century. Because the conservation of angular momentum applies, the decrease of Earth's angular momentum must appear as an increase in the Moon's angular momentum. A measurement of the Moon's orbit shows that the semimajor axis is increasing about 1.3 cm/y. Thus 1 billion years ago the solar day was 5.6 h shorter, and the Moon was 13,000 km (8100 mi) closer to Earth.

---

*The tidal force—a differential gravitational force—occurs when two separate bodies interact gravitationally. The mass of body one (Earth) produces a difference in the gravitational force from position to position within body two (Moon). The force difference tends to stretch or change the shape of body two, producing body tides (bulges). For example, since the masses that constitute Earth are at varying distances from the masses that make up the Moon, Earth's gravitational forces vary throughout the Moon, producing body tides in the Moon. Likewise, the Moon's gravitational force varies throughout Earth, producing body and ocean tides.

(a) Spring tides

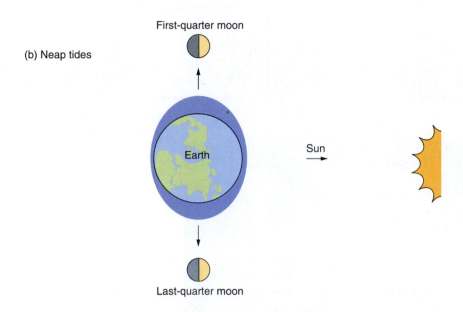

Full moon

Earth

New moon

Sun

First-quarter moon

(b) Neap tides

Earth

Sun

Last-quarter moon

FIGURE 17.21   Spring and Neap Tides
This diagram shows the relative position of Earth, the Moon, and the Sun at the times of spring and neap tides. The tidal bulge is highly exaggerated.

RELEVANCE QUESTION: *You live near the ocean and observe the twice daily rise and fall of the ocean tides. Suppose the Moon's orbit were to increase. (It is believed that the distance between the Moon and Earth was much* *less in the past.) Comment on what you would observe concerning the change in height of the ocean tides. Justify your answer.*

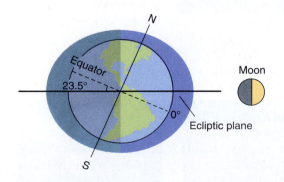

Equator

23.5°

N

S

0°

Ecliptic plane

Moon

Sun

FIGURE 17.22   Spring Tide at the Time of Summer Solstice
This diagram shows the position of Earth, the new moon, and the Sun at the time of summer solstice. The maximum height of the tidal bulge is at 23.5°N and 23.5°S, because the Sun is overhead at 23.5°N. If a spring tide were to occur three months later, when the Sun is on the equator, the maximum height of the bulge would be in the equatorial region. (The tidal bulge is exaggerated in the diagram.)

A SPOTLIGHT ON: The Moon

General Features

Craters
Basins
Plains
Rays
Rills
Mountain Ranges
Faults

Data

Mass	7.35×10^{22} kg
Diameter	3480 km
Density (mean)	3340 kg/m³
Surface temperature	
Day	130°C
Night	-180°C
Distance from Earth	384,000 km
Synodic period	29.5 days

Moon as seen from Earth.

Origin: The Great Impact Theory

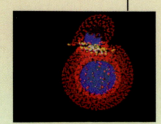
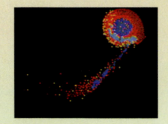
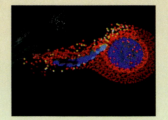
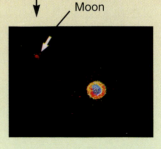

Moon

Motions

Rotation: Once every 29.5 solar days.
Revolution: Once every 29.5 days.

Phases

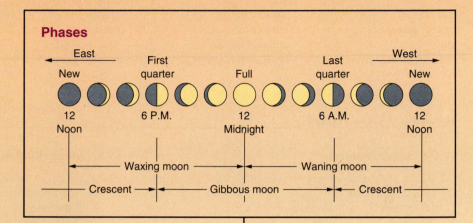

←— East

New   First quarter   Full   Last quarter   West —→ New

12 Noon   6 P.M.   12 Midnight   6 A.M.   12 Noon

←——— Waxing moon ———→ ←——— Waning moon ———→

←— Crescent —→ ←——— Gibbous moon ———→ ←— Crescent —→

Ocean Tides

Spring: Occur at new and full moon.
Neap: Occur at first and last quarter.

Ocean tides occur twice daily.

Eclipses

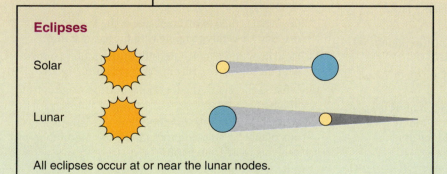

Solar

Lunar

All eclipses occur at or near the lunar nodes.

Important Terms

crater (17.1)	synodic month	full moon	lunar eclipse
plains	new moon (17.4)	last-quarter phase	ascending node
rays	waxing phase	eclipse (17.5)	descending node
rills	waning phase	umbra	tidal force (17.6)
mountain ranges	crescent moon	penumbra	spring tides
fault	gibbous moon	solar eclipse	neap tides
sidereal month (17.3)	first-quarter phase	annular eclipse	

Review Questions

17.1 General Features

1. Which of the following is *not* a general physical feature of the Moon's surface?
   (a) craters   (d) terminator
   (b) regolith   (e) rays
   (c) plains

2. The internal structure of the Moon
   (a) consists of a solid mantle that is rich in silicates.
   (b) consists of an inner and outer mantle.
   (c) includes a small, perhaps solid, iron-rich core.
   (d) both a and c.

3. Most craters on the surface of the Moon are believed to be caused by
   (a) faults.
   (b) meteorites.
   (c) volcanoes.

4. The largest impact features on the Moon are the
   (a) multiring basins.
   (b) plains.
   (c) rays.
   (d) maria.

5. How does the surface gravity of the Moon compare with the surface gravity of Earth?

6. What is the average distance between the Moon and Earth in kilometers?

7. Distinguish between the Moon-surface features rays and rills.

8. What causes each of the following on the Moon: (a) craters, (b) plains, (c) rays, and (d) rills?

9. Name the three major components of the Moon's internal structure.

10. An examination of the planets and their moons reveals something exceptional about the Earth-Moon system. What is exceptional about the Earth-Moon system?

17.2 Composition and History of the Moon

11. Which of the following statements is false?
    (a) The average density of the Moon is less than Earth's.
    (b) Experimental evidence indicates that the oldest Earth and lunar rocks were formed about the same time.
    (c) The Moon samples show a large abundance of iron.
    (d) The percentage of oxygen isotopes found in lunar samples is similar to that of Earth.

12. Which of the following must be considered when formulating a theory concerning the origin of the Moon?
    (a) similar chemical composition of Earth's mantle
    (b) no water found in lunar samples
    (c) lunar rocks show a small percentage of volatile elements
    (d) Earth and Moon rocks approximately the same age
    (e) all of the above

13. The most acceptable theory for the formation of the Moon is the _____ theory.
    (a) sister   (c) fission
    (b) great impact   (d) capture

14. What two facts does the sister theory for the origin of the Moon fail to explain?

15. How does the great impact theory explain the two facts referred to in Question 14?

16. Distinguish between volatile elements and refractory elements.

17. (a) What is the approximate age of the Moon?
    (b) How does this compare with the age of Earth?

18. How does the average density of the Moon compare with the average density of Earth's mantle?

19. What is the great impact theory for the origin of the Moon?

20. Why can scientists learn more about the early history of our planet by studying rocks from the Moon than by studying rocks from Earth?

17.3 Lunar Motions

21. Which of the following statements is false?
    (a) The Moon rotates and revolves eastward as observed from the Celestial North Pole.
    (b) The orbital plane of the Moon is tilted about 5° to Earth's orbital plane.
    (c) The difference between the sidereal and synodic months is about two days.
    (d) The Moon revolves in an elliptical orbit.
    (e) None of the above is false.

22. The rising of the Moon in the east and the setting in the west is due to
    (a) the orbital motion of the Moon.
    (b) the rotational motion of the Moon.
    (c) Earth's rotation.
    (d) none of the above

23. The sidereal period of rotation of the Moon is equal to
    (a) 29 days.
    (b) 30 days.
    (c) its sidereal period of revolution.
    (d) its synodic period of revolution.

24. What is the period of rotation of the Moon with respect to its period of revolution?

25. (a) What is the numerical value of the Moon's sidereal period?
    (b) How does the sidereal period compare with the synodic period?
    (c) Explain the difference.

26. What is the numerical value of the angle between the Moon's orbital plane and Earth's orbital plane?

27. Why does the Moon appear to rise in the east and set in the west every day?

28. How often would an observer on the Moon see (a) sunrise and (b) earthrise?

17.4 Phases of the Moon

29. During one month, the Moon passes through how many phases?
    (a) 4   (b) 6   (c) 8   (d) none of the preceding

30. The angular difference between the maximum and minimum altitude of the full moon as observed from 35°N is how many degrees?
    (a) 19   (b) 26.5   (c) 57   (d) 63.5

31. The full moon has its greatest northern declination and its greatest altitude at the time of the
    (a) summer solstice.   (c) vernal equinox.
    (b) winter solstice.   (d) autumnal equinox.

32. The Moon is in first-quarter phase when it is _____ of the Sun.
    (a) 90° west   (c) 90° east
    (b) 180° east   (d) none of the above

33. When do the following occur? (a) The new moon sets. (b) The full moon rises. (c) The last-quarter moon sets.

34. Why does the Moon rise 50 min later each day?

35. Which phase of the Moon (a) is overhead at 6 P.M., (b) sets at 6 A.M., and (c) rises at midnight?

36. What is the difference between a waxing and a waning moon?

17.5 Eclipses

37. An eclipse
    (a) occurs due to the darkening of the light of one celestial body by another.
    (b) requires the interplay of a minimum of three bodies.
    (c) can be total, partial, or annular.
    (d) all of the above

38. Which of the following is *not* a contributing factor in causing eclipses?
    (a) rotation of Earth about an axis
    (b) the inclination of the Moon's orbit
    (c) the varying distance between Earth and the Moon
    (d) the varying distance between Earth and the Sun
    (e) none of the above

39. A solar eclipse occurs when the Moon is at or near new phase and is in or near the
    (a) winter solstice.   (c) vernal equinox.
    (b) summer solstice.   (d) ecliptic plane.

40. During a total eclipse of the Sun by the Moon, the center of totality goes across the surface of Earth mainly from
    (a) east to west.   (c) west to east.
    (b) north to south.   (d) south to north.

41. Distinguish between umbra and penumbra.

42. What is the period of precession of the Moon's orbital plane? What is the direction of the precession?

43. What is an annular eclipse?

44. Why doesn't an eclipse occur every time there is a new or full moon?

45. Why do lunar eclipses last much longer than solar eclipses?

17.6 Ocean Tides

46. Tidal bulges in the ocean are due
    (a) to the Moon's gravitational force lifting the ocean water away from solid Earth.
    (b) to the horizontal components of tidal forces that cause water to flow over Earth's surface toward areas nearest the Moon and toward areas opposite the Moon.
    (c) mainly to gravitational forces between the Sun and Earth.
    (d) none of the above

47. Tidal forces
    (a) are differential gravitational forces.
    (b) cause ocean tides.
    (c) cause solid Earth to stretch.
    (d) all of the above

48. Spring tides on Earth take place
    (a) only during the spring season.
    (b) only during the times of new moons.
    (c) near the times of full and new moons.
    (d) near the times of first- and third-quarter moons.

49. What is the origin of tidal forces?

50. What are some factors that contribute to the height of ocean tides?

51. Why are there two high and two low tides each day?

Applying Your Knowledge

1. Suppose you are on the Moon facing Earth. Will you observe phases of Earth? If yes, describe them.

2. If you are at the south pole of the Moon and you see the right half of Earth bright with sunshine and the left half dark, what is the phase of the Moon for an Earth observer?

3. Which side of Earth's shadow does the Moon enter, east or west? Explain.

4. An observer in Washington, D.C., sees a waxing crescent moon slightly north of the vernal equinox. Name the observer's season, and explain your answer.

Exercises

17.1 General Features

1. If a person weighs 800 N on Earth, what is the person's weight on the Moon?
    Answer: 133 N

2. If a person weighs 160 lb on Earth, what is the person's weight (in lb) on the Moon?

17.3 Lunar Motions

3. How many days are in 12 lunar months (synodic months)?
    Answer: Approximately 354

4. How many days are in 12 lunar months (sidereal months)?

17.4 Phases of the Moon

5. (a) Determine the maximum altitude of the full moon as observed from Jackson, MS (32.5°N, 88°W).
   (b) What is the approximate date when this can occur?
    Answer: (a) 81°   (b) December 22

6. (a) Determine the maximum altitude of the full moon as observed from Columbia, SC (34°N, 81°W).
   (b) What is the approximate date when this can occur?

7. (a) Determine the maximum altitude of the first-quarter moon as observed from Atlanta, GA (34°N, 84°W).
   (b) What is the approximate date when this can occur?
    Answer: (a) 84.5°   (b) March 21

8. (a) Determine the maximum altitude of the last-quarter moon as observed from Orangeburg, SC (33.5°N, 81°W).
   (b) What is the approximate date when this can occur?

9. (a) Determine the longitude of an observer who sees the full moon on the overhead meridian with the local solar time at 105°W is 12 noon on September 22.
   (b) What are the longitude and local solar time of an observer who sees the full moon rising?
   (c) What are the longitude and local solar time of an observer who sees the full moon setting?
    Answer: (a) 105°E   (b) 15°W   (c) 165°E

10. (a) Determine the longitude of an observer who sees the new moon on the overhead meridian on March 21 when the local solar time at 120°W is 12 midnight.
    (b) What are the longitude and local solar time of an observer who sees the new moon rising?
    (c) What are the longitude and local solar time of an observer who sees the new moon setting?

11. Consider a person in the United States who sees the first-quarter phase.
    (a) Which side of the Moon is illuminated, east or west?
    (b) What phase does an observer in Australia see at the same time, and which side is bright?
        Answer: (a) west   (b) first-quarter phase; west (left) side

12. Consider a person in the United States who sees the last-quarter phase.
    (a) Which side of the Moon is illuminated?
    (b) What phase does an observer in Australia see at the same time, and which side is bright?

13. Determine the month and day when the full moon is at minimum altitude for an observer at Atlanta, GA (34°N). What is the altitude of the full moon at this time?
        Answer: June 21–22; 27.5°

14. Determine the month and day when the full moon is at maximum altitude for an observer at Atlanta, GA (34°N). What is the altitude of the full moon at this time?

17.5 Eclipses

15. Draw a diagram illustrating a total solar eclipse. Include the orbital paths of Earth and the Moon, and indicate the approximate time of day the eclipse is taking place.
        Answer:

16. Draw a diagram illustrating a total lunar eclipse. Include the orbital paths of Earth and the Moon, and indicate the approximate time of day that the eclipse is taking place.

17.6 Ocean Tides

17. A high tide is occurring at Charleston, SC (33°N, 84°W).
    (a) What other longitude is also experiencing a high tide?
    (b) What two longitudes are experiencing low tide?
        Answer: (a) 100°E   (b) 10°E, 170°W

18. A low tide is occurring at Galveston, TX (29°N, 95°W).
    (a) What other longitude is also experiencing low tide?
    (b) What two longitudes are experiencing high tide?

Solution to Confidence Exercise

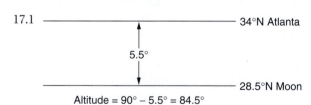

17.1

34°N Atlanta

5.5°

28.5°N Moon

Altitude = 90° − 5.5° = 84.5°

Answers to Multiple-Choice Review Questions

1. d   3. b   11. c   13. b   22. c   29. c   31. b   37. d   39. d   46. b   48. c
2. d   4. a   12. e   21. e   23. c   30. c   32. c   38. a   40. c   47. d

THE UNIVERSE

18

Merely to realize there are more things in heaven and earth than are dreamed of in one's philosophy is hardly an end in itself. The end should be to expand one's philosophy so as to include them.

Lord Rayleigh (1842–1919)

What is the universe? What are its composition, structure, size, and age? What are its primary building components, and what is the composition of these building components? What is the origin of the universe, and what is its future? Has the universe always existed, and will it exist forever, or does it go through the cycle of birth, life, death, and rebirth? What are galaxies? What are stars? These are some of the topics introduced and discussed in this chapter.

We live in a world that is life-dependent on astronomical objects, especially the stars. The Sun (a star) provided the energy for the generation of life on Earth and supplies us continuously with heat and light. Also, day after day the Sun keeps planet Earth in orbit. Hydrogen and most of the helium in the universe are primordial. The remaining elements were formed by nuclear fusion in the cores of stars and in stars when they exploded billions of years ago and up to present time. The elements formed by the stars make up most of the remaining mass of the universe, including our bodies.

The study of the stars is the oldest science. Thousands of years ago under the clear desert skies of the Near East, people watched the stars in awed wonder. The earliest scientists plotted the positions and brightnesses of the stars as Earth went through its calendar of seasons. The Sun and Moon were

Photo: A deep field (10–12 billion light years away) image by Hubble of several hundred never-before-seen galaxies.

worshiped as gods, and the days of the week were named after the Sun, Moon, and visible planets, as discussed in Chapter 16. Yet it is only recently that we have understood the most basic nature of stars.

Sixty years ago not even Einstein knew what made the stars shine. Today, we know that all stars go through a cycle of stages. Stars are born, they radiate energy, and then they expand, contract, possibly explode, and eventually die out. Our knowledge of how all this happens has been made possible through our study of the atomic nucleus and by applying the laws of science from many diverse fields.

The stars and galaxies of our universe give off all different kinds of electromagnetic radiation. This radiation was discussed in Chapter 6 and includes radio waves, microwaves, infrared, visible, ultraviolet, X-rays, and gamma rays. Up until 1931 we looked only at the visible light given off by the stars. The advent of radio telescopes led to the discovery of quasars and pulsars in 1960 and 1968, respectively. Most of the other regions of the electromagnetic spectrum are absorbed by our atmosphere. Satellites and balloons going above our atmosphere have enabled us to study other forms of radiation emitted by stars, and new developments are occurring frequently. ■

18.1 The Sun

LEARNING GOALS

▼ Describe the Sun's structure, and list its physical properties.

▼ Write the net reaction for the fusion of hydrogen to helium.

The **Sun** is a star—a self-luminous sphere of gas held together by its own gravity and energized by nuclear reactions in its interior. It moves through space with its family of planets at a speed of approximately 250 km/s in the direction of the constellation Lyra. Viewed from Earth, at its mean distance, the Sun's angular diameter is about 0.5°. The Sun's equator is inclined about 7° from the orbital plane of Earth. The glowing ball of gas rotates on its axis every 25 Earth days at its equator. The period of rotation is longer at higher latitudes.

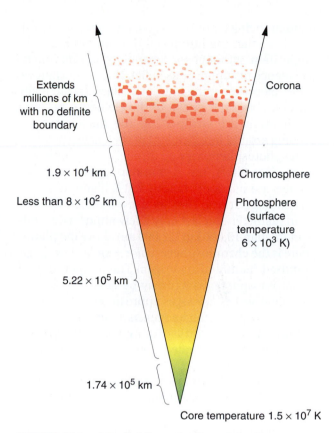

Extends millions of km with no definite boundary

1.9×10^4 km

Less than 8×10^2 km

5.22×10^5 km

1.74×10^5 km

Corona

Chromosphere

Photosphere (surface temperature 6×10^3 K)

Core temperature 1.5×10^7 K

FIGURE 18.1   A Radial Cross Section of the Sun
The boundary between layers is not sharply defined.

A cross-sectional view of the Sun is shown in ● Fig. 18.1. The Sun's temperature is believed to be about 15 million kelvins at its center and decreases radially outward to the visible surface of the Sun, which is called the **photosphere.** The temperature of the photosphere has been measured at about 6000 K.

The interior of the Sun is so hot that individual atoms do not exist, because high-speed collisions continually knock the electrons loose from the atomic nuclei. The interior is composed of high-speed nuclei and electrons moving about more or less independently, similar to a gas. A gas, you will recall, is composed of rapidly moving atoms or molecules. The Sun's high-speed, charged nuclei and electrons form a fourth phase of matter called a **plasma,** with an average density of 1.4 g/cm^3. Note that this density is 1.4 times as great as water!

The Sun's photosphere is about 94% hydrogen, 5.9% helium, and 0.1% heavier elements, the most abundant being carbon, oxygen, nitrogen, and neon. We believe that the interior of the Sun has a similar composition, although we have no good experimental evidence to support this belief. Note that when we discuss elements in the interior of stars, we are really speaking about the nuclei of the elements, because the electrons are stripped off the nuclei and are speeding around independent of the nuclei.

The photosphere, viewed through a telescope with appropriate filters, has a granular appearance. The granules are hot spots (about 100 K higher than the surrounding surface) that are a few hundred kilometers in diameter and last only a few minutes. Extending more than 19,000 km (12,000 mi) above the photosphere is the **chromosphere** (*color + sphere*), which is composed mainly of hydrogen. The temperature of the chromosphere, which may be heated by energy from shock waves, averages approximately 5×10^4 K. The chromosphere can be seen as a thin, red crescent for only the few seconds during which the photosphere has been concealed from view during a solar eclipse.

At the time of a total solar eclipse, the chromosphere and photosphere are hidden by the Moon, and the **corona** (outer solar atmosphere) can be seen as a white halo. (See Fig. 17.17.) The corona receives energy also by shock waves, and its temperature exceeds 1×10^6 K. This extreme temperature is sufficient to give protons, electrons, and ions enough energy to escape the Sun's atmosphere. The charged particles are projected into space, giving rise to a radial flow of radiation, which is controlled by the Sun's magnetic field. This flow of radiation is called the **solar wind,** and wind speeds exceed 400 km/s (893,000 mi/h) as the radiation passes Earth's orbit on its way through the solar system. Measurements made by the *Voyager* spacecraft have confirmed that the solar wind and the accompanying magnetic field extend outward to at least 50 AU from the Sun's surface. This volume of space over which the solar wind extends is called the **heliosphere.**

A very distinct feature of the Sun's surface is the periodic occurrence of sunspots. **Sunspots** are patches (thousands of kilometers in diameter) of cooler material on the surface of the Sun. Each has a central darker part, called the *umbra*, and a lighter border, called the *penumbra*. ● Figure 18.2 is a photograph of the whole solar disk. These large sunspots last for several weeks before disappearing from view.

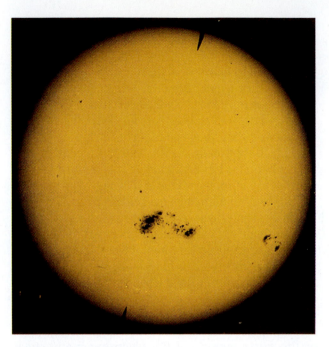

FIGURE 18.2 Sunspots
The very large sunspot group of March–April 1947.

The number of sunspots appearing on the Sun varies over a 22-y period. A period begins with the appearance of a few spots or groups near 30° latitude in both hemispheres of the Sun. The number of spots increases, with a maximum generally between 100 and 200 occurring about 4 y later near an average latitude of 15°. As time passes, the number of spots decreases until, in about 4 more years, only a few are observed near 8° latitude.

About this same time a few spots begin to appear at 30°, and the number begins to increase again, indicating an 11-y cycle. But there is a notable difference. The sunspots have an associated magnetic field that is different in appearance from the previous one. Studies indicate that if a sunspot has a north magnetic pole during the initial increase and decrease, the next 11-y cycle will show a south magnetic pole associated with the sunspot.

Another distinct feature of the Sun's surface is the appearance of **prominences** that seem to be connected with violent storms in the chromosphere. They are very evident to the astronomer during solar eclipses, at which time they appear as great eruptions at the edge of the Sun. They are red, have an associated magnetic field, and may appear as streamers,

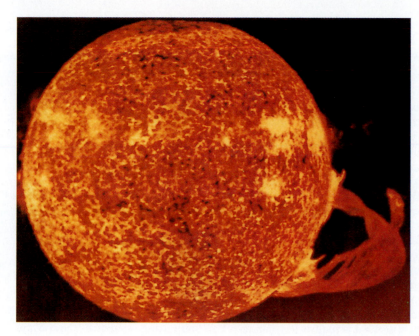

FIGURE 18.3   The Sun During a Major Solar Eruption

This ultraviolet photograph showing several flares and a large prominence was taken from the *Skylab* orbiting space station.

loops, spiral or twisted columns, fountains, curtains, or haystacks. They extend outward for thousands of miles from the surface, occasionally reaching a height of 1 million miles. An extraordinarily large prominence is shown in ● Fig. 18.3.

The chief property of the Sun, of course, is the fact that it radiates energy. However, it was not until 1938 that scientists came to understand the source of the radiation. We now know that the Sun radiates energy because of nuclear fusion reactions inside its core (see Chapter 10).

The Sun's core is made up mostly of hydrogen nuclei, or protons, or in nuclear notation, $_1^1H$. These protons are moving at very high speeds, and they occasionally fuse together, as shown in ● Fig. 18.4. The products of this nuclear reaction are a deuteron (proton and neutron together, or $_1^2H$), a positive electron (or positron, designated $_{+1}^0e$), and a neutrino (designated by the symbol ν, the Greek letter nu). A **neutrino** is an elementary particle that has no charge, has very little mass, travels at or near the speed of light, and hardly ever interacts with other particles such as electrons or protons. (See the chapter's first Highlight.) This first reaction is fairly rare, and for this reason, the Sun's hydrogen burns relatively slowly. In

fact, scientists believe that the Sun has been radiating energy for about 5 billion years.

Once the deuteron is formed, it quickly reacts with a proton to form a helium-3 nucleus ($_2^3He$) with the

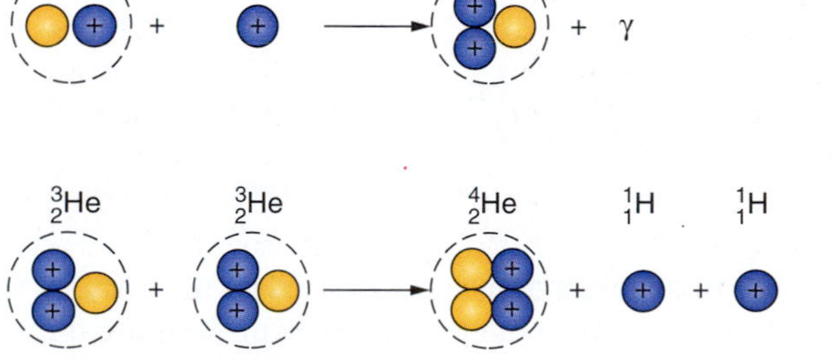

FIGURE 18.4   Nuclear Fusion

The reactions that make up the proton-proton chain. See text for description.

HIGHLIGHT: Neutrinos

The concept of the neutrino was formulated by Wolfgang Pauli, Austrian theoretical physicist, in 1930 to satisfy the requirements of quantum mechanics in the process of nuclear beta decay (Section 10.2). He suggested that an electrically neutral and massless unknown particle was produced to account for missing energy when neutrons decay. One year later Enrico Fermi, Italian-born American physicist, named the unknown particle *neutrino*, which means "little neutral one." Twenty-five years later, in 1956, the neutrino was discovered by direct observation at Los Alamos Scientific Laboratory of a reaction induced by a neutrino.

Three different types of neutrinos are known. Neutrinos from nuclear beta decay produce an electron when reacting with a proton and are called *electron neutrinos*. A second type of neutrino is associated with the muon and is called the *muon neutrino*. (Muons are short-lived subatomic particles produced by nuclear collisions. They are similar to the electron in most respects but are approximately 207 times as massive.) A third type of neutrino is the *tau neutrino*. (The tau particle is an extremely short-lived subatomic particle similar to the electron but 3500 times as massive.) The muon and tau particles are essentially more massive versions of the electron, and they both can decay into the less massive electron.

Neutrinos originate from the Sun and other stars and from the reactions of cosmic rays with Earth's atmosphere. Since they are extremely small and carry no electrical charge, they pass easily through matter. They penetrate Earth from all directions, and trillions pass through our bodies every second.

Particle physicists, over the past several years, have suggested that the "massless" neutrino actually may possess a tiny amount of mass—conceivably up to one ten-millionth the mass of an electron. To test this hypothesis, detectors have been constructed in deep underground mines in several countries and at the South Pole.

In June 1998, scientists in Japan reported that their Super-Kamiokande detector had provided indirect evidence that the neutrino has mass. The Super-Kamiokande, the largest neutrino detector ever constructed, is located more than 3000 feet underground in a zinc mine about 125 miles west of Tokyo. The detector is made up of a huge stainless steel tank (130 feet high and 120 feet in diameter) that is filled with 12.5 million gallons of pure water.

When all three neutrinos enter the Super-Kamiokande tank, a few collide with a proton or a neutron in a water molecule. An electron neutrino yields an electron, a muon neutrino yields a muon, and a tau neutrino yields a tau. The electron and muon in turn zoom through the water faster than the speed

of light in water, generating a streak of blue light that is detected by the 11,000 light amplifiers that line the inside of the tank. The experimenters identify the electron and muon by analyzing the energy and spatial geometry of the streaks of light. The Super-Kamiokande cannot detect tau neutrinos.

Theoretical values for the number of electron neutrinos and muon neutrinos do not agree with the experimental values detected by the Super-Kamiokande. A shortage of muon neutrinos is detected. This indicates that some of the muon neutrinos are changing to tau neutrinos. For this to occur, the neutrino must possess mass.

Supporting evidence may come soon from a neutrino detector that officially opened on April 28, 1998, in Sudbury, Ontario. This detector is using heavy water (deuterium oxide), which can interact with all three types of neutrinos.

The Super-Kamiokande experiment has provided strong evidence that neutrinos have a very small mass. Because of this strong evidence, and due to the fact there is an enormous number of neutrinos in the universe, cosmologists believe neutrinos may account for most or all of the "dark matter" they have been seeking (see Section 18.5). Also, a neutrino with mass may cause scientists to revise their model for the structure of matter and to modify their theory on how stars shine.

---

emission of gamma rays, designated by the symbol γ. Next, two helium-3 nuclei fuse to form the more common helium-4 nucleus and two protons.

In each of these three fusion reactions, energy is liberated by the conversion of mass. These three reactions are called the **proton-proton chain** and can be written as

$$\begin{aligned}
{}_{1}^{1}\text{H} + {}_{1}^{1}\text{H} &\rightarrow {}_{1}^{2}\text{H} + {}_{+1}^{0}\text{e} + \nu + \text{energy} &\text{(slow)} \\
{}_{1}^{2}\text{H} + {}_{1}^{1}\text{H} &\rightarrow {}_{2}^{3}\text{He} + \gamma + \text{energy} &\text{(fast)} \\
{}_{2}^{3}\text{He} + {}_{2}^{3}\text{He} &\rightarrow {}_{2}^{4}\text{He} + {}_{1}^{1}\text{H} + {}_{1}^{1}\text{H} + \text{energy} &\text{(fast)}
\end{aligned}$$

If we multiply the first two reactions by 2 and add both sides, we get the net reaction, which is

$$4\,^1_1\text{H} \rightarrow\,^4_2\text{He} + 2\,(_{+1}^{\ 0}e + \gamma + \nu) + \text{energy}$$

In the net reaction four protons form a helium nucleus, two positrons, two high-energy gamma rays, two neutrinos, and a great deal of energy. The energy factor simply means that the particles on the right possess more kinetic and radiant energy than the particles on the left. In our reaction we have converted mass into energy in conformity with Einstein's equation $E = mc^2$.

Every second in the Sun's interior about 6.0×10^{11} kg of hydrogen are converted into helium and energy. Even at this rate we expect the Sun to radiate energy from hydrogen fusion for about another 5 billion years.

EXAMPLE 18.1

Calculating the Time Required to Convert Part of the Sun's Hydrogen into Helium and Energy

The Sun has a mass of 2.0×10^{30} kg, about 90% of which is hydrogen. Calculate the time, in years, it would take to convert 10% of the Sun's hydrogen into helium and energy.

SOLUTION

Step 1

Given:

The Sun's mass of 2.0×10^{30} kg. Calculate 90% to obtain the amount of hydrogen.

$$2.0 \times 10^{30}\text{ kg} \times 0.90$$
$$= 1.8 \times 10^{30}\text{ kg (mass of hydrogen)}$$

Step 2

Calculate 10% of the Sun's hydrogen.

$$1.8 \times 10^{30}\text{ kg} \times 0.10 = 1.8 \times 10^{29}\text{ kg}$$

Step 3

Wanted:

The number of years required to convert 10% of the hydrogen into helium and energy. Since the rate of fusion is 6.0×10^{11} kg/s, use conversion factors to find the number of seconds in 1 y.

$$\frac{60\text{ s}}{1\text{ min}} \times \frac{60\text{ min}}{1\text{ h}} \times \frac{24\text{ h}}{1\text{ day}} \times \frac{365.24\text{ day}}{1\text{ y}}$$
$$= 3.16 \times 10^7\text{s/y}$$

Step 4

Determine the number of years required to convert 10% of the Sun's hydrogen into helium and energy. Take the amount converted each second and multiply by the number of seconds in 1 y. Multiply this answer by N (the number of years required). This is equal to the amount of hydrogen converted.

$$6.0 \times 10^{11}\frac{\text{kg}}{\text{s}} \times 3.16 \times 10^7\frac{\text{s}}{\text{y}} \times N$$
$$= 1.8 \times 10^{29}\text{ kg}$$

Solving for N:

$$N = \frac{1.8 \times 10^{29}\text{ kg}}{6.0 \times 10^{11}\frac{\text{kg}}{\text{s}} \times 3.16 \times 10^7\frac{\text{s}}{\text{y}}}$$
$$= 9.5 \times 10^9\text{ y}$$

CONFIDENCE EXERCISE 18.1

A certain star has a mass of 4.0×10^{30} kg. Calculate the time, in years, it would take to convert 15% of its hydrogen into helium and energy if 85% of its mass is hydrogen. Assume that 5.0×10^{12} kg of hydrogen is converted each second.

Most everything on planet Earth that we call life is water-dependent, and radiant energy from the Sun drives the natural sequence through which water passes into the atmosphere as water vapor, precipitates to Earth in liquid or solid form, and ultimately returns to the atmosphere through evaporation. Animals require oxygen, plants require carbon dioxide, and both require water. Thus animals and plants are life-dependent on energy from the Sun because energy from the Sun controls our atmosphere, food supply, water resources—our total environment.

The total energy that leaves the Sun in 1 second is about 4×10^{26} joules. The amount reaching Earth is about 1.4×10^3 watts per square meter. This value is known as the *solar constant*. A variation in the solar constant of as little as 0.5% would have catastrophic effects on our ability to stay alive.

RELEVANCE QUESTION: *Name some ways your daily life on planet Earth might be affected by a decrease of 0.5% in the solar constant.*

FIGURE 18.5 Star Trails

This time-exposure photograph shows star trails at the south celestial pole.

18.2 The Celestial Sphere

A view of the stars on a clear dark night makes a deep impression. The stars appear as bright points of light on a huge dome overhead. As the time of night passes, the dome seems to turn westward as part of a great sphere, with the observer at the center. The apparent westward motion of the stars is due to the eastward rotation of Earth. The unaided eye is unable to detect any relative motion among the stars on the apparent sphere or to perceive their relative distances from Earth. The stars all appear to be mounted on a very large sphere with Earth at the center.

This huge, apparently moving, imaginary sphere has been named the **celestial sphere,** and the way it appears depends on the observer's position on Earth. An observer positioned at 90°N (the North Pole) would see Polaris, the north star, directly overhead. From this latitude all stars on the celestial sphere appear to move in concentric circles about the north star, never going below the horizon (● Fig. 18.5); they never set. An observer located at 40°N latitude would observe the north star 40° above the northern horizon, and all stars within 40° of the north star would appear to move in concentric circles, never going below the horizon (● Fig. 18.6). Detailed observations reveal that the celestial sphere rotates (apparently) about an axis that is an extension of Earth's polar axis, with the celestial equator lying in the same plane as Earth's equator.

The position of a star or other object beyond our solar system is described by assigning three space coordinates. The first and second are *declination* and *right ascension,* which are angular coordinates representing the direction of the star with respect to the celestial equator and the vernal equinox, respectively.

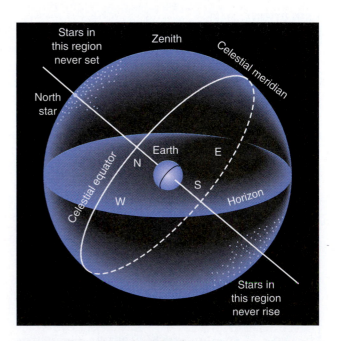

FIGURE 18.6 The Celestial Sphere

This drawing illustrates the celestial sphere as seen by an observer on Earth at a latitude of 40°N.

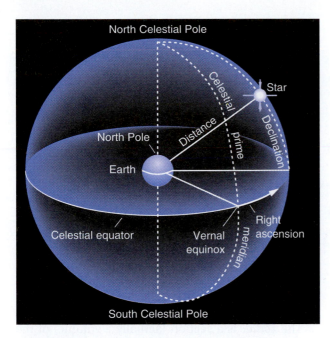

FIGURE 18.7   The Three Celestial Coordinates

This drawing illustrates the three celestial coordinates: declination (DEC), right ascension (RA), and distance.

The third coordinate is *distance,* which determines the star's linear distance from the Sun (● Fig. 18.7).

Declination (DEC) is the angular measure in degrees north or south of the celestial equator. It has a minimum value of zero at the celestial equator and increases to a maximum of 90° north and 90° south. All angles measured north of the equator have (+) values, and all angles measured south of the equator have (−) values. For example, the star in Fig. 18.7 has a declination of +37°.

Right ascension (RA) is the angular measure in hours, with the hours divided into minutes and seconds. Right ascension begins with 0 h at the celestial prime meridian and continues eastward to a maximum value of 24 h, which coincides with the starting point. Sirius, the brightest appearing star, has a right ascension of 6 h, 42.9 min. The **celestial prime meridian** is an imaginary half-circle running from the North Celestial Pole to the South Celestial Pole and crossing perpendicular to the celestial equator at the point of the vernal equinox.

The distance coordinate is usually measured in astronomical units, in parsecs, or in light-years. We have defined an *astronomical unit* (AU) as the mean distance of Earth from the Sun, which is 1.5×10^8 km

$(9.3 \times 10^7$ mi). A **light-year** (ly) is the distance traveled by light in one year. One light-year equals approximately 9.5×10^{12} km (6×10^{12} mi), calculated by multiplying the speed of light by the number of seconds in one year. One *parsec* (pc) is defined as the distance to a star when the star exhibits a parallax of one second of arc (● Fig. 18.8). One parsec equals 3.26 ly, or 206,265 AU:

$$1 \text{ pc} = 3.26 \text{ ly}$$
$$= 2.06 \times 10^5 \text{ AU}$$

The star in Fig. 18.8, observed from two positions, appears to move against the background of more distant stars. This apparent motion is, as you know, called *parallax.* The angle *p* measures the parallax in seconds of arc. The definition of a parsec provides an

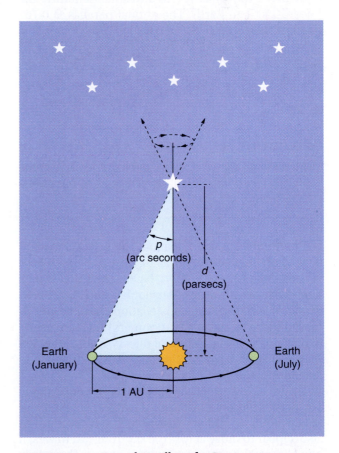

FIGURE 18.8   Annual Parallax of a Star

The shaded angle represents the annual parallax of the star, measured in arc seconds. By definition, when angle *p* is equal to 1 arc second, the distance *d* to the star is equal to 1 pc. The basic relationship can be written *d* = 1/*p*.

easy method for determining the distance to a celestial object, because merely taking the reciprocal of the angle p, measured in seconds, gives the distance in parsecs. That is,

$$d = \frac{1}{p} \qquad \text{18.1}$$

where d = distance in parsecs

   p = parallax angle in seconds of arc

EXAMPLE 18.2

Calculating the Distance (in Parsecs) to a Star

Calculate the distance to Proxima Centauri, the nearest star to Earth. The annual parallax is 0.762 second of arc.

SOLUTION

$$d = \frac{1}{p} = \frac{1}{0.762} = 1.31 \text{ pc (4.27 ly)}$$

CONFIDENCE EXERCISE 18.2

Calculate the distance (in parsecs) to Sirius, the brightest star in the night sky. The annual parallax is 0.376 second of arc.

Prominent groups of stars in the celestial sky appear to an Earth observer as distinct patterns. These groups, called *constellations,* have names that can be traced back to the early Babylonian and Greek civilizations. Although the constellations have no physical significance, today's astronomers find them useful in referring to certain areas of the sky. In 1927 they set specified boundaries for the 88 constellations so as to encompass the complete celestial sphere.

We are all aware of the apparent daily motion of the Sun across the sky, and when the Moon is visible, we are aware of its apparent motion. The constellations also appear to move across the sky from east to west if one observes the stars for an hour or two. Their daily motion is due to the eastward rotation of Earth on its axis. The constellations also have an annual motion resulting from Earth's revolution about the Sun. We

observe the constellations Pisces, Aquarius, and Capricornus in the autumn night sky. In the winter months, Orion (the Hunter) is seen, along with Gemini, Taurus, and Aries. Sagittarius (the Archer) is a summer constellation. Some other familiar constellations are Andromeda, Cassiopeia, Cygnus, Ursa Major, and Ursa Minor.

Some familiar groups of stars also are part of a constellation or part of different constellations. These groups are called *asterisms.* The Big Dipper, which is part of Ursa Major, is an example. (See ● Fig. 18.9.) In terms of brightness, six of the stars in the Big Dipper are second magnitude (see Section 18.3), and the other is third magnitude. Another example is the Summer Triangle, which is formed by the bright stars Altair, Deneb, and Vega (magnitudes 1, 1, and 0, respectively). These stars are in three different constellations: Altair is in Aquila (the Eagle), Deneb is in Cygnus (the Swan), and Vega is in Lyra (the Lyre).

The *zodiac* is a section (actually a volume) of the sky extending around the ecliptic 8° above and 8° below the ecliptic plane (● Fig. 18.10). The zodiac is divided into 12 equal sections, each 30° wide and 16° high. Each section has its apex at the Sun and extends outward to infinity. The boundaries of the zodiac were specified such that the Sun, Moon, visible planets, and most of the asteroids travel within its limits. Occasionally, however, the planets Pluto and Venus are outside the boundaries of the zodiac.

RELEVANCE QUESTION: *How is the concept of a celestial sphere meaningful to you, in spite of the fact that it does not really exist?*

18.3 Stars

LEARNING GOALS

▼ Identify and explain the different classes of stars.

▼ Explain the Hertzsprung-Russell (H-R) diagram.

Hipparchus of Nicaea, a Greek astronomer and mathematician, was antiquity's greatest known observer of the stars. He measured the celestial latitude and longitude of more than 800 stars and compiled the first star catalog, which was completed in 129 B.C. He assigned the stars, with respect to their brightness, to six magnitudes. The apparent brightness, or **magnitude,**

FIGURE 18.9   Ursa Major and Ursa Minor

The Big Dipper and the Little Dipper make up parts of the constellations Ursa Major and Ursa Minor, respectively. Therefore, they are considered asterisms rather than constellations. These constellations can be seen throughout the year from the United States. Can you find Leo the Lion at the bottom of the figure?

brighter first-magnitude star gave off about 100 times as much radiant energy. From this observation, a definition of the magnitude scale was made in which each magnitude difference is equal to the fifth root of 100. This definition can be written as

$$\text{magnitude difference} = \sqrt[5]{100} = 2.512$$

For example, a first-magnitude star is 2.512 times as bright as a second-magnitude star, and 2.512×2.512 times as bright as a third-, and so on. Note when giving the magnitude, the greater the negative number, the brighter is the star, and the greater the positive number, the dimmer is the star. On this scale the brightest object in the sky, our Sun, has a magnitude of -26.7; the full moon, a magnitude of -12.7; the planet Venus, a magnitude of -4.2; and the brightest star, Sirius, which is 8.7 ly distant, has an apparent magnitude of -1.43. Sirius is the closest star to Earth (except for the Sun) that can be seen by an observer in the United States.

Alpha Centauri A (magnitude -0.01) and its close companion Alpha Centauri B (magnitude 1.33) revolve around each other. To the unaided eye, they

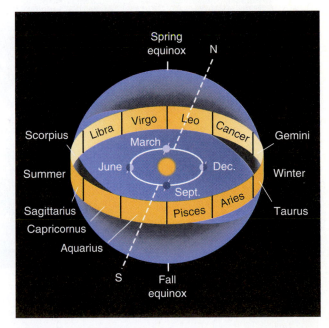

FIGURE 18.10   Signs of the Zodiac

The drawing illustrates the boundaries of the zodiacal constellations. Each of the 12 sections of the zodiac is 30° wide and 16° high, or 8° above and below the ecliptic plane.

refers to the brightness of a star as observed from Earth. The brightest ones were listed as stars of the first magnitude, those not quite as bright as second magnitude, the next less bright as third magnitude, and so on, down to the sixth magnitude, which are stars barely visible to the unaided eye. About 6000 stars are visible to the unaided eye.

A modified version of Hipparchus' scale is used today. When a comparison was made between a first-magnitude star and a sixth-magnitude star, the

appear as a single star. Close to Alpha Centauri A and B is a faint red dwarf star called Proxima Centauri, which is slightly closer to Earth. The three stars, which are 4.3 ly away, are the closest stars to Earth. They are located near 14 h right ascension and −61° declination, and so are seen only by observers in the Southern Hemisphere.

The energy output of a star is measured by its absolute magnitude. **Absolute magnitude** is defined as the apparent magnitude a star would have if it were placed 10 pc (32.6 ly) from Earth. The absolute magnitude of the Sun is +4.83. If the annual parallax (from which distance is calculated) and the apparent brightness of a star can be measured, the absolute magnitude can be calculated.

When the absolute magnitudes or brightnesses of stars are plotted against their surface temperatures or colors, we get an **H-R diagram,** named after Ejnar Hertzsprung, a Danish astronomer, and Henry Rus-

sell, an American astronomer. An H-R diagram for stars is shown in ● Fig. 18.11. Note that the temperature axis is reversed; that is, the temperature increases to the left instead of to the right.

Most stars on an H-R diagram get brighter as they get hotter. These stars form the **main sequence,** a narrow band going from upper left to lower right. Stars above the main sequence that are cool and yet very bright must be unusually large to be so bright. So they are called **red giants.** Stars below the main sequence that are hot yet very dim must be very small, so they are called **white dwarfs.**

The spectra (see Sections 7.2 and 9.3) of stars vary considerably, and this variation is mostly a function of the temperature in the stars' outer layers. The pattern of the absorption lines in the spectrum can be used to determine a star's temperature. Stars are placed in seven different spectral classes that range from type O to type M and indicate a temperature range of 50,000

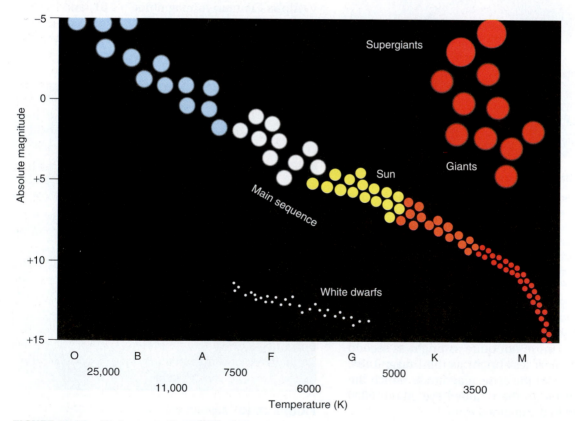

FIGURE 18.11 Hertzsprung-Russell Diagram

This diagram shows the absolute magnitude, spectral class, temperature, and color of various classes of stars. Note that the temperature increases from right to left. See the text for details.

TABLE 18.1   Spectral Sequence

Spectral Class	Color	Temperature (K)
O	Violet	> 28,000
B	Blue	10,000–28,000
A	Blue	7500–10,000
F	Blue to white	6000–7500
G	White to yellow	5000–6000
K	Orange to red	3500–5000
M	Red	< 3500

to 2000 K. The spectral classes are shown in Table 18.1. The Sun is a type G star. The majority of known stars are small, cool, red, type M stars called **red dwarfs.**

Stars are plasmas having a chemical composition of mostly hydrogen, with some helium and a very small percentage of other elements. Their surface temperatures range from about 2000 to 50,000 K. They vary in mass and size. Stars range in mass from 0.04 to 75 solar masses. Most have a range of between 0.1 and 5 solar masses. Normal stars range in size from white dwarfs, which are about 12,800 km (8000 mi) in diameter, to supergiants such as Antares, Betelgeuse, and Rigel, which are millions of miles in diameter.

Most stars are part of multiple systems. A **binary star** system consists of two stars orbiting each other. Systems with three or more stars also occur, but they are not nearly as common as binary stars.

Many stars in the sky are observed to vary in brightness over a period of time. The first one noted was Delta Cephei in the constellation Cepheus. Delta Cephei varies in brightness with a period of approximately 5.4 days. From its least bright magnitude of 5.2, it gradually becomes brighter for about 2 days and attains a magnitude of 4.1, after which it decreases in brightness until it reaches its minimum in another 5.4 days. The cycle begins again and goes through the same sequence. Presently, over 500 known stars vary in magnitude with a fixed period of between 1 and 50 days, and they are known as **cepheid variables,** after Delta Cephei.

The importance of the cepheid variables is the fact that a definite relationship exists between the period

and the average absolute brightness. Generally, the longer the period, the brighter is the cepheid variable. By measuring the period, astronomers can calculate absolute brightness. Then distance can be computed from absolute brightness. Thus astronomers can calculate the distance to any galactic system in which cepheid variables can be detected.

The period-luminosity relationship for cepheid-variable stars was discovered in 1912 by an American astronomer, Henrietta Swan Leavitt (1868–1921). This relationship provided Edwin P. Hubble (1889–1953), an American astronomer at the Mount Wilson Observatory in California, with the knowledge to show that some observed white patches of light (called *nebulae,* plural for *nebula,* the Latin word for "cloud") were actually galaxies beyond the Milky Way.

The distance to more remote galaxies is calculated by means of the redshift in the galaxy's spectrum. When a photograph is taken of an excited element in the gaseous phase by a spectrograph located in an Earth-based laboratory, a normal line emission spectrum is obtained. When a similar photograph is taken of a galaxy containing this same element, the spectrum may show a displacement of the normal lines toward the red end of the spectrum (longer wavelengths) or toward the blue end of the spectrum (shorter wavelengths). The displacement is due to the expansion of the universe, and the direction of the displacement depends on whether the galaxy is moving toward or away from the observer.

The displacement of two lines on the calcium spectrum labeled H and K for three different galaxies is shown in ● Fig. 18.12. A small shift of the lines means a low velocity of recession, a greater shift means a greater velocity, and so on. The interesting point concerning the redshift is a correlation of velocity and the apparent brightness of the galaxy. The fainter the galaxy, the greater is the velocity. Generally speaking, the farther away a galaxy is located, the less is its apparent brightness. Thus a velocity and magnitude correlation yields a velocity and distance correlation. This provides us with a means for measuring great distances that are far beyond the limits of cepheid-variable observation.

Many stars appear dim and insignificant but suddenly increase in brightness in a matter of hours by a factor of 100 to 1 million. A star undergoing such a drastic change in brightness is called a **nova,** or "new" star. A nova is the result of an explosion on the surface

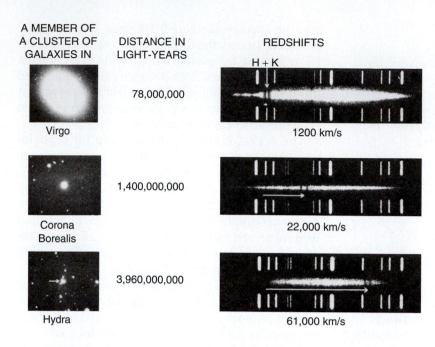

A MEMBER OF A CLUSTER OF GALAXIES IN	DISTANCE IN LIGHT-YEARS	REDSHIFTS
Virgo	78,000,000	H + K 1200 km/s
Corona Borealis	1,400,000,000	22,000 km/s
Hydra	3,960,000,000	61,000 km/s

FIGURE 18.12   Three Elliptical Galaxies and Their Respective Redshifts

On the left are three individual elliptical galaxies. From top to bottom they show the galaxies at increasing distance from the observer. On the right, the spectrum (the broad white band) of each galaxy is shown between an upper and lower comparison spectrum. The H and K lines of ionized calcium are the two dark vertical lines in each galaxy's spectrum. The arrows indicate the shift in the calcium H and K lines. The redshifts are expressed as velocities.

of a white dwarf star caused by matter falling onto its surface from the atmosphere of a larger binary companion. A nova is not a new star but a faint white dwarf that temporarily increases in brightness.

Occasionally, a star explodes and throws off large amounts of material that may be so great that the star is destroyed. Such a gigantic explosion is known as a **supernova.** Only three supernovae have been observed in our galaxy. The best known is the Crab Nebula in the constellation Taurus. This nebula is expanding at the rate of approximately 112 million km (70 million mi) per day. Because we know the average angular radius and the expansion rate, the original time of the explosion can be calculated. The result agrees closely with Chinese and Japanese records that report the appearance of a bright new star in the constellation Taurus in A.D. 1054.

In 1987, a supernova was observed in the Large Magellanic Cloud. It is the first supernova observed in this galaxy, which is the closest galaxy to our Milky Way. The Large Magellanic Cloud is 160,000 ly away. The supernova was designated 1987A because it was the first one to be observed in 1987.

The discovery of supernova 1987A is important to astronomers because they are observing a stellar explosion from the beginning, and they will be able to observe and examine the debris coming from the explosion and the remains of the star after the debris has cleared. Thus it will provide a history of a stellar explosion and supply valuable information concerning many branches of physics.

In addition to the novae and supernovae, there are the planetary nebulae,* which possess a large, slowly expanding, ringlike envelope, as shown in ● Fig. 18.13. These nebulae, as observed with a telescope, appear greenish.

The Orion nebula shown in ● Fig. 18.14, one of the brightest in the night sky, is known as an *emission nebula,* a glowing cloud of interstellar gas. This type of nebula is more irregular and turbulent than the planetary nebulae. The gas glows due to the presence of neighboring stars that ionize the gas. The Orion nebula is over 25 ly in diameter and contains hundreds of stars. It is in a dense core of material found within a molecular cloud such as the Orion nebula that the formation of a star begins.

Astronomers know that stars are born, radiate energy, expand, possibly explode, and then die. In general, this is their life cycle. However, the exact details depend on a star's initial composition (the percentage amounts of hydrogen, helium, and heavier elements) and on its mass. The greater the mass of a star, the faster it moves through its life cycle.

---

*The name is misleading. The expanding layers of the star have nothing to do with planets. Astronomers in the eighteenth century viewed these objects as round bodies—that is, like planets—not as points of light, such as stars.

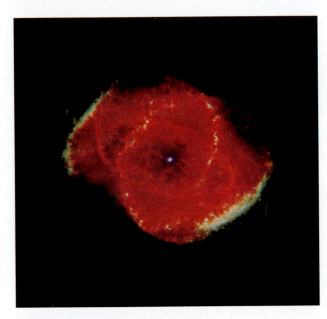

FIGURE 18.13   The Cat's Eye Nebula

The planetary nebula NGC 6543, located in the constellation Draco, is the best known surface-brightness nebula in the celestial sky. This image was taken by the Hubble Space Telescope.

FIGURE 18.14   Great Nebula in Orion

Considered the brightest of the emission nebulae, this one is located near the middle star in Orion's sword. It is a gaseous and dusty nebula some 25 ly across and 400 pc away.

The general evolution of a star with a mass typical of our Sun is shown in ● Fig. 18.15. The birth of a star, according to accepted theory, begins with the condensation of interstellar material (mostly hydrogen) in a nebula because of the gravitational attraction between the interstellar material, radiation pressure from nearby stars, and supernova shock waves. The size of the star formed depends on the total mass available, which, in turn, determines the rate of contraction. As the interstellar mass condenses and loses gravitational potential energy, the temperature rises and the material gains thermal energy. As the star continues to decrease in size, the temperature continues to increase until a thermonuclear reaction begins and hydrogen is converted to helium, as discussed in Section 18.1.

The star's position in the main sequence of the H-R diagram, its temperature, and the radiation given off

FIGURE 18.15   Hertzsprung-Russell Diagram

This diagram illustrates the evolution of a star with the same mass as our Sun. See the text for an explanation.

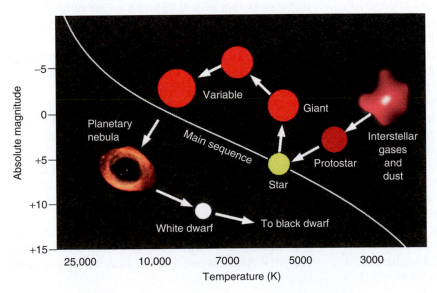

are determined by the mass of the star. On the main sequence of the H-R diagram, the star continues to convert hydrogen into helium. This process is called *hydrogen burning,* but it is a nuclear "burning," or fusion, not a chemical fire. The lifetimes of main-sequence stars range from 1 million to 200 billion years—about 10 billion years for a star like our Sun.

As the hydrogen in the core is converted into helium, the core begins to contract and heat up. This heats the surrounding shell of hydrogen and causes the fusion of the hydrogen in the shell to proceed at a more rapid rate. This rapid release of energy causes the star to expand and enter the first red-giant phase of its evolution.

Eventually, the core gets so hot that helium can fuse into carbon, and soon other nuclear reactions occur in which all elements up through iron are created. The creation of the nuclei of elements inside stars is called **nucleosynthesis.**

During the red-giant phase, the star varies in temperature and brightness, after which the star becomes very unstable and the outer layers expand, forming a beautiful planetary nebula. See Fig. 18.15 for the path taken during the expansion process. The expanding shell of matter eventually diffuses into interstellar space, and the star core becomes a *white dwarf,* which eventually cools to a black dwarf.

When a star becomes a white dwarf, it is very small. It has gravitationally collapsed as far as it can while still obeying the laws of physics. The star is about the size of Earth and is so dense that a single teaspoonful of matter weighs five tons. Because it is very small, the star is not very luminous.

More massive stars develop into supergiants that may become supernovae. When a massive star's nuclear fuel is depleted, the interior collapses catastrophically to form a small-diameter neutron star. During the process, the outer layers of the star bounce off the rigid inner core and then expand into space, destroying the star and giving rise to a supernova. (See ● Fig. 18.16.) The supernova phase does not occur for the vast majority of stars—those of relatively small mass.

Large-mass stars (between 10 and 25 times the Sun's mass) have more gravitational attraction, and they collapse to a size of approximately 20 km in diameter. The electrons and protons in this superdense star combine to form neutrons, and this **neutron star** is composed of about 99% neutrons. A

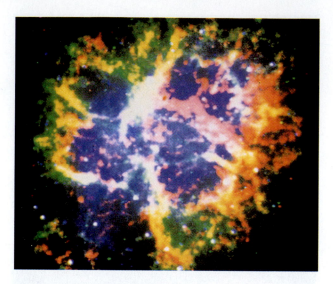

FIGURE 18.16   Crab Nebula
The Crab Nebula is the remnant of a supernova explosion observed by Chinese astronomers in A.D. 1054. False colors have been added to the photograph to bring out details.

teaspoonful of a neutron star would weigh one billion tons. Because the angular momentum of the star must be conserved, the small size of the neutron star dictates that it must be spinning rapidly. Rapidly rotating neutron stars give out radio waves in pulses and are called **pulsars.** The radio waves are detected and measured on Earth by large radio telescopes. (See ● Fig. 18.17.) Pulsars pulse with a constant period that may be between 0.03 and 4 s. One of the fastest-spinning pulsars discovered is located at the center of the Crab Nebula. (See Fig. 18.16.) It is identified with the remains of the A.D. 1054 supernova. The period of this pulsar is slowly increasing, indicating that the rotating neutron star is gradually slowing down.

If the core remaining after a supernova explosion is greater than about three times the mass of the Sun, the star is thought to end up as a gravitationally collapsed object even smaller and more dense than a neutron star. Such a star is so dense that light cannot escape from its surface because of its intense gravitational field. Thus it would appear black and is called a *black hole.* (See Section 18.4.)

After a star explodes in a supernova and the core goes into its end stage, what becomes of the ejected material? Astronomers believe that it is thrown into

FIGURE 18.17 The Very Large Array (VLA) Radio Telescope

This Y-shaped array is constructed with three arms, each containing nine parabolic reflectors 80 ft in diameter. Each railroad arm extends 13 mi into the desert near Socorro, New Mexico. The VLA is computer-linked as a single antenna and produces very sharp images of distant celestial objects.

space and eventually becomes seed material for future stars. In fact, because of its surface composition and age, astronomers believe our Sun is the product of a second generation of stellar evolution. In other words, one or more stars went through their life cycles and exploded as supernovae. The ejected material later coalesced into our Sun and solar system.

A recent discovery about stars is that the vast space between the galaxies is not as dark and empty as astronomers have thought. In the past few years, astronomers have detected in the Virgo cluster of galaxies planetary nebulae (the ejected shells of old, extremely hot, giant stars) and red giant stars that are not in any galaxy. These are extragalactic stars probably thrown out by the interaction of colliding galaxies or are stars pulled from galaxies passing one another at high speed.

How numerous are these extragalactic stars? The astronomers conducting the research in the Virgo cluster estimate they equal 20 to 70 percent of the number of stars within a galaxy. The total population of extragalactic stars may be equal to or greater than the number in galaxies because there is an extremely huge volume of space between galaxies.

RELEVANCE QUESTION: *On Earth during daylight hours the Sun dominates the sky. When the day ends, the stars take over, and the sky is brightened with thousands of twinkling lights. About 6000 stars can be seen over a period of one year with the unaided eye. Many of them are double stars, which are gravitationally bound and revolve around one another. Suppose our solar system had double stars (suns) and planet Earth was positioned to* *receive sunlight over the entire 24-hour day. As inhabitants of Earth's surface, would we ever be aware of the thousands of stars that are now seen in the night sky? Justify your answer.*

18.4 Gravitational Collapse and Black Holes

The possibility of a remarkable phenomenon has been proposed to explain some strange astronomical findings. The phenomenon is called **gravitational collapse** and is the collapse of a very massive body because of its attraction for itself. Recall that Newton's law of universal gravitation was given in Chapter 3 as

$$F = G \frac{m_1 m_2}{r^2}$$

where F = force of gravity between m_1 and m_2
m_1 = first mass
m_2 = second mass
G = universal gravitational constant
r = distance between the centers of m_1 and m_2

From this equation it is evident that if the distance r is made smaller, the force of gravity increases.

In a large mass such as Earth or the Sun, all parts are always attracting each other. Normally, in large masses, electromagnetic forces tend to keep the various particles such as atoms apart. What would happen if the body were so massive that the gravitational forces of attraction were stronger than the electromagnetic forces of repulsion? All parts would be drawn closer together. According to the equation, the forces increase as the distance r decreases. The closer the particles move together, the greater the gravitational force becomes. The whole mass would continue to contract until the original volume of mass became a fantastically massive point called a **singularity.** Gravitational collapse does not take place in

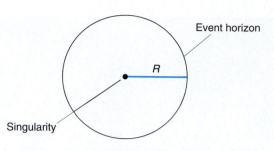

FIGURE 18.18   Configuration of a Nonrotating Black Hole

The black dot depicts a singularity surrounded by the event horizon at a distance R. Anything located within the event horizon cannot escape. Thus space inward from the event horizon is known as a *black hole*.

just any star. According to theory, for a star to collapse and form a *black hole*, the star's mass at the time of collapse must be greater than three solar masses.

The singularity is surrounded by an invisible spherical boundary known as the **event horizon.** Any matter or radiation within the event horizon cannot escape the influence of the singularity. See ● Fig. 18.18 for a sketch of a singularity and its event horizon for a nonrotating black hole. The value R (called the **Schwarzschild radius**), the radial distance the event horizon is located from the singularity, can be determined by equating the escape velocity equation ($v = \sqrt{2GM/R}$) to the speed of light. We obtain

$$R = \frac{2GM}{c^2} \qquad \text{18.2}$$

where G = universal gravitational constant
M = mass of the collapsed star
c = speed of light

EXAMPLE 18.3

Calculating the Radius of a Black Hole

Calculate the distance between the singularity and the event horizon for a nonrotating black hole having a mass of 7.0×10^{31} kg.

SOLUTION

Use Eq. 18.2.

$$R = \frac{2GM}{c^2}$$

$$R = \frac{2\left(6.67 \times 10^{-11} \, \cancel{kg} \times \frac{m}{s^2} \times \frac{m^2}{kg^2}\right) 7.0 \times 10^{31} \, \cancel{kg}}{(3.0 \times 10^8)^2 \frac{m^2}{s^2}}$$

$$R = \frac{2 \times 6.67 \times 10^{-11} \times 7.0 \times 10^{31} \times 10^{-16}}{9.0} \, m$$

$$= 1.0 \times 10^5 \, m$$

CONFIDENCE EXERCISE 18.3

Calculate the distance between the singularity and the event horizon for a nonrotating black hole having a mass of 8.0×10^{32} kg.

Anything located within the event horizon cannot escape. The event horizon is a one-way boundary because matter and radiation can enter but cannot leave. Thus space inward from the event horizon is a **black hole.**

Because most stars are rotating, a black hole will probably also be rotating. And because angular momentum is conserved, the rotation rate will increase as the star gets smaller, so the rotation of the black hole will be extremely rapid and the space near the hole enormously curved.

Because nothing can escape the influence of a black hole, how is one detected? One method is the detection of X-rays coming from the vicinity of a black hole. The detection of X-rays by Earth satellites from a double-star system in which one companion may be a black hole has provided astronomers with data indicating that black holes probably do exist.

For example, the supergiant star that is located some 8000 ly from Earth in the constellation Cygnus has a companion called Cygnus X-1 that cannot be detected directly, indicating that it may be a black hole. The detected X-rays coming from the double-star system are generated by gases captured by Cygnus X-1 from the supergiant star. The captured gases, accelerating to extremely high speeds, go into orbit around Cygnus X-1, forming a spiraling flat disk of matter called an *accretion disk*. These gases become extremely hot because of internal friction and emit high-energy X-rays (● Fig. 18.19a).

Calculations from experimental data indicate that Cygnus X-1 is smaller than Earth and more than seven times as massive as the Sun. This object is too massive to be a white dwarf or a neutron star, leaving the probability that Cygnus X-1 is a black hole.

Another candidate for a black hole is the dark companion of an orange dwarf star located in our Galaxy in the constellation Monoceros. It is estimated to be located 3200 ly from Earth. The orange dwarf star circles the dark companion (considered a black hole of 8 solar masses) every 7.75 hours.

One of the most recent candidates for a supermassive black hole is at the center of the spiral galaxy M77 (see ● Fig. 18.19b). This galaxy is some 50 million light years away and is about 8000 light years in diameter.

Astronomers believe a massive black hole is at the center of our Galaxy. They have detected stars orbiting the central region of the Milky Way moving as fast as 0.5 percent the speed of light. This indicates a central mass of about 2.5 million suns located in a small space. Only a black hole can generate these conditions.

RELEVANCE QUESTION: *Is it possible to know what takes place inside the event horizon of a black hole? Why or why not?*

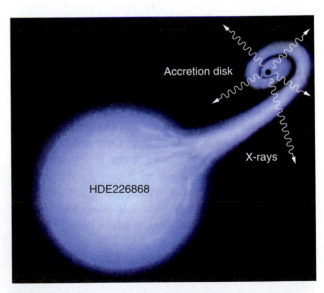

FIGURE 18.19a   Model of Cygnus X-1

The blue supergiant star HDE226868 ejects material, some of which is captured by its unseen companion (assumed to be a black hole), forming an accretion disk. X-rays generated at the inner edge of the disk have been detected by astronomers.

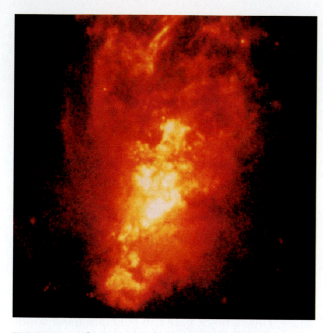

FIGURE 18.19b Spiral Galaxy M77 (NGC 1068)

A supermassive black hole is located in the center of this large spiral galaxy. M77 is some 50 million ly distant and about 8000 ly in diameter. The nucleus radiates with the intensity of 100 billion suns.

18.5 Galaxies

LEARNING GOALS

▼ Identify the classes of galaxies.

▼ State Hubble's law, and explain its use.

The Sun and its satellites occupy a very small volume of space in a very large system of stars known as the Milky Way Galaxy (Greek *galaxias*, "Milky Way"). A **galaxy** is an extremely large collection of stars bound together by gravitational attraction and occupying an enormously large volume of space. It is the fundamental component for the structure of the universe. A galaxy is classified as irregular, spiral (normal or barred), or elliptical, depending on how it appears when photographed.

Astronomer Edwin P. Hubble established this system of classifying galaxies according to their appearance in photographs. The system starts with spherical ellipticals, spreads out with increasing flatness, then branches off into a normal spiral sequence and a barred spiral sequence, with a scattering of irregulars outside the two branches (● Fig. 18.20).

The type Sa spirals have their spiral arms closely wound to the central region (● Fig. 18.21), whereas the Sb spirals have arms that spread out more from the center and the Sc spirals have very loose spiral arms (● Fig. 18.22). The S0 type links the normal spirals to the smooth ellipticals. The classification of barred spirals, which are distinguished by a broad bar that extends outward from opposite sides of the central region, follows similar unwinding of the spiral arms (● Fig. 18.23).

The irregular-type galaxies have no regular geometric shape. Examples of this type are the Small and Large Magellanic Clouds, which can be seen easily with the unaided eye from the Southern Hemisphere. They are named in honor of Magellan, who reported seeing them on his famous voyage around the world (● Fig. 18.24).

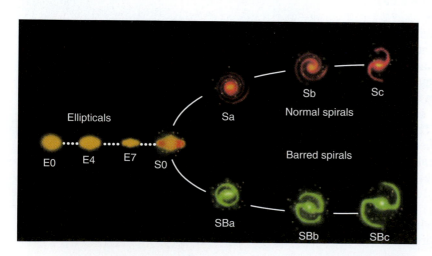

FIGURE 18.20 Hubble's Classification of Galaxies

This diagram illustrates the placement of elliptical, normal spiral, and barred spiral galaxies plus their subtypes. The diagram does not show the evolutionary sequence of galaxies.

FIGURE 18.21   The Sombrero Galaxy, M104, Type Sab Spiral

The dark band across the galaxy's center is composed of dust and gas.

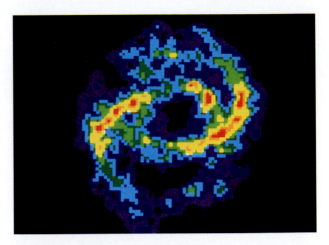

FIGURE 18.23   Barred Spiral Galaxy

In a given volume of space there are more elliptical galaxies than spiral galaxies, and the irregular types account for only about 3% of all galaxies. The elliptical galaxies are made up of older stars and are usually dimmer than the spirals. The spiral galaxies account for over 75% of the brighter galaxies observed.

Carl Seyfert, an American astronomer at the Mount Wilson Observatory in California, reported in 1944 that a few spiral galaxies exhibit very bright centers and their spectra show broad emission lines that indi-

cate that hot gas is present and expanding at very rapid rates. This evidence indicates that violent activity is taking place in the central core of the galaxies. Large amounts of energy are being released from these central cores, and the best theory of the source of this energy is the gravitational collapse of an enormous amount of matter. Over 100 of these galaxies

FIGURE 18.24   The Large Magellanic Cloud

This irregular galaxy (the closest galaxy to our Milky Way) is only 160,000 ly away. The huge bright region at the left, which is 800 ly in diameter, is the Tarantula Nebula.

FIGURE 18.22   Spiral Galaxy NGC 2997, Type Sc

FIGURE 18.25   The Seyfert Galaxy NGC 4151

Notice the very bright nucleus. If the galaxy were at a very extreme distance, only the bright nucleus would be visible. Perhaps quasars are the tremendously energetic sources in the nuclei of Seyfert galaxies.

have now been reported and are known as *Seyfert galaxies* (● Figure 18.25).

Our own Galaxy, the **Milky Way,** is believed to be a Hubble-type Sb spiral. It contains some 100 billion (10^{11}) stars and is thought to have an appearance similar to that of the Great Galaxy in Andromeda (● Fig. 18.26). Our Galaxy (● Fig. 18.27) is about 10^5 ly in diameter and has a thickness in the Sun's region of approximately 2×10^3 ly. It is rotating eastward, or counterclockwise, as viewed from the North Celestial

FIGURE 18.26   The Spiral Galaxy M31 in Andromeda

M31 is a type Sb galaxy 2.25 million ly from Earth. Two elliptical galaxies, NGC 205 (lower right) and NGC 221 (M32), are also shown.

Pole. The period of rotation, which is not the same for all regions of the galaxy, is more than 2×10^8 y for the region that contains our solar system. Our solar system is located in the plane of rotation between the Perseus and Sagittarius spiral arms of the Milky Way Galaxy. It is about 3.0×10^4 ly from the galactic center and is moving at a rate of approximately 150 mi/s in the direction of the constellation Lyra.

The galactic equator, a great circle positioned halfway between the galactic poles, is inclined about 62° from the celestial equator (● Fig. 18.28). The North Galactic Pole has a right ascension of 12 h 40 min and a declination of +28°. The South Galactic Pole has

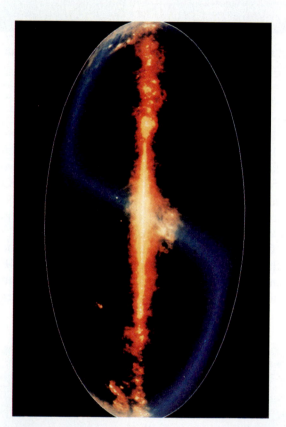

FIGURE 18.27   Infrared Image of the Milky Way Galaxy

The plane of the Milky Way Galaxy lies vertically across the middle of the image, with the galactic center at the center. (See Figure 18.28 for the correct orientation to the celestial equator.) The image is dominated by the thermal emission from interstellar dust in the Milky Way. The structured, warmer emission from interplanetary dust appears in blue. Two neighboring galaxies, the Large and Small Magellanic Clouds, are also in the image (left of the plane, halfway between the center and the lower edge of the galactic plane).

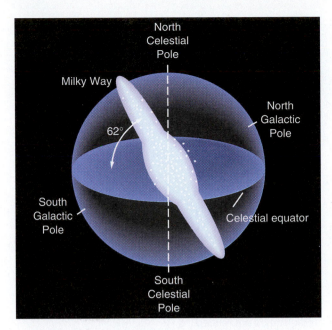

FIGURE 18.28 **Diagram Illustrating the Inclination of the Plane of Revolution of the Milky Way Galaxy to the Celestial Equator**

a right ascension of 0 h 40 min and a declination of −28°.

In the immediate neighborhood of the Milky Way and confined to an ellipsoidal volume of space—some 9×10^5 pc for the major axis and about 8×10^5 pc for the minor axis—is located a small group of galaxies known as the **Local Group.** This group has at least 28 members, most of which are dwarf ellipticals. Others are believed to exist that have not yet been detected because of their low magnitude. Our Milky Way is a member located near one end of the major axis. Messier 31 (Great Galaxy in Andromeda; see Fig. 18.26) is also a member and is positioned near the opposite end of the major axis. Messier 31 (M31) can be seen with the unaided eye from the United States. The Andromeda constellation is about 20° south of the constellation Cassiopeia.

The galaxies of the Local Group seem to be moving with random motions. The two Magellanic Clouds are moving away from our galaxy, and several others, including M31, are moving toward our galaxy.

The galaxies astronomers photograph throughout the vast volume of the universe are lumped together in **clusters** that vary in size and number. The clusters, classified as regular or irregular, range in size from 3 to 15 million ly in diameter. Some, such as our Local Group, contain a few galaxies, and others contain thousands.

The galactic clusters group together into what are called **superclusters,** that is, clusters of clusters. These superclusters have diameters as large as 300 million ly and masses equal to or greater than 10^{15} solar masses. Our Local Group and adjoining groups and clusters such as the Virgo cluster form what is called the *Local Supercluster.*

What is the origin of galaxies? Did they form from huge clouds of gas and dust? How did the universe, which seems to have been extremely uniform and smooth (energy and mass were evenly distributed) in the beginning, evolve into the galaxies, clusters, and superclusters we observe today? These questions remain unanswered. Just how and when galaxies came into existence and how they evolved remain problems that astronomers are working to solve.

Many astronomers believe that most of the matter that makes up the universe is not detected by any part of the electromagnetic spectrum. The unobserved matter has been incorrectly referred to as *missing matter,* but this term has been replaced by the term *dark matter.* No one knows the composition of this nonluminous matter, which may constitute as much as 95 percent of the total mass of the universe. Astronomers originated the concept to explain why a cluster of galaxies exists as a gravitationally bound system. The observed mass of a typical cluster is not sufficient to hold the cluster together as a unit. Therefore, matter that is undetected must be present throughout the cluster. Recent experiments seem to indicate that the neutrino (see the chapter's first Highlight on page 470) has a very small mass. Because the universe has innumerable neutrinos, they could be the dark matter astronomers are searching for.

When Edwin Hubble began looking at spectrum shifts of galaxies, he found nothing surprising at first. In the Local Group, the shifts were small: Some were blue, and some were red. But as he looked at galaxies farther and farther away, he found only redshifts. In fact, the farther away the galaxy, the larger is the redshift. From these measurements the distances to remote galaxies can be determined by converting the observed redshift of the galaxy to radial velocity and then plotting the logarithm of the velocity against the apparent magnitude (● Fig. 18.29). Hubble's discovery, now known as **Hubble's law,** can be written as

$$v = Hd \qquad \textbf{18.3}$$

where v = recessional velocity of the galaxy

d = distance away from the galaxy

H = Hubble's constant

The Hubble constant is believed to have a value of 50 to 100 km/s per million pc. If H is 50 km/s per million pc, the observed galaxy is moving away from our Sun at a speed of 50 km/s for every 1 million pc the galaxy is from our Sun. Hubble's law is extremely important; it gives astronomers vital information about the structure of the universe.

Astronomers and other scientists determine the structure of the universe by detecting and analyzing electromagnetic waves that come from the stars and galaxies. From the collected data, they contemplate the way in which these objects are distributed throughout the vast volume of the universe. Scientists estimate that at least 10^9 (1 billion) galaxies are within range of astronomers' optical telescopes and can be photographed.

Photographs of distant stars, galaxies, or galactic clusters are photographs of the way the objects appeared at the moment the light radiated from them. A galaxy that is observed (photographed) at a distance of 1 billion ly is being seen the way it appeared 1 billion years ago. In other words, we are taking a photograph of the galaxy's past, not its present. When we photograph the Sun, we photograph its surface as it existed about 8 min ago. That is the time it takes for the light to reach Earth.

Even if they were inclined to do so, astronomers would not have the time to photograph the tremendous volume of the universe to obtain photographs of all galaxies. Instead, the astronomers' model of the structure of the universe is based on a sampling of different regions. Using the 60-in. and 100-in. reflecting telescopes on Mount Wilson, Hubble obtained photographs of more than 1200 sample regions of space, counted some 44,000 galaxies, and estimated that it would be possible to photograph 100 million galaxies. After correcting his data for such things as the obstruction presented by the Milky Way and interstellar dust, he concluded that when observation of the universe is made over a large volume, the distribution of galaxies is *isotropic* and *homogeneous;* that is, when we observe a large volume of space, we observe as

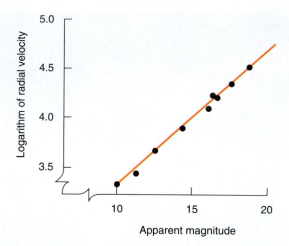

FIGURE 18.29 Hubble's Law

This graph shows the relationship (Hubble's law) between the logarithm of the radial velocity of a galaxy and its distance from the Milky Way. Recall that apparent magnitude and distance are related.

many galaxies in one direction as in any other, and the observations are the same at all distances. This concept of the uniformity of the universe is known as the **cosmological principle** and is the basic assumption for most theories of cosmology (see Section 18.7).

But recent data concerning the structuring of galaxies indicate that the universe may be quite different from what we presently believe, and the cosmological principle may not be true. Astronomers, using Hubble's law, have measured the recessional velocity of approximately 6000 nearby galaxies. They used the measured recessional velocities to calculate the distance to each galaxy. All the galaxies were within 500 million ly of Earth. The distance data plus the right ascension and declination for each galaxy were put into a computer that produced an illustration showing the location and relative positions of the 6000 galaxies.

The illustration showed the galaxies structured in volumes about 400 million ly long, 200 million ly high, and up to 15 million ly thick. This grouping of galaxies, the largest known structure observed in the universe thus far, was called the "Great Wall" of galaxies by the research astronomers (● Fig. 18.30). Nearly every galaxy in the collected data belongs to either a two-dimensional sheet or a one-dimensional thread that is millions of light-years in length. The data also revealed large volumes of space, millions of light-

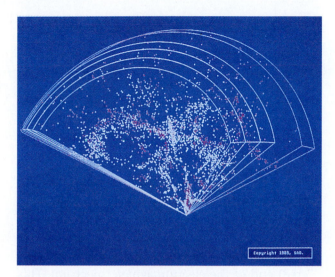

FIGURE 18.30 "Great Wall" of Galaxies
Almost 6000 galaxies are illustrated in this map. The galaxies extend horizontally some 400 million ly, vertically some 200 million ly, and 15 million ly in depth. At the bottom center is the Coma cluster, located at an average distance of 350 million ly from us. The galaxies are in the general direction of right ascension 12 h and declination +15°.

years across, that contained very few galaxies. In other words, the two-dimensional sheets were separated by large volumes of almost empty regions of space.

The 60-in. telescope at the Mount Hopkins Observatory near Tucson, Arizona, was used to collect the data. The data were taken on only four pie-shaped volumes of space. Each layer or volume, starting just above the horizon, was 6° high and slightly less than 180° wide. The other three volumes were taken above the first layer.

The data embrace an extremely small volume of space compared with the entire universe. Therefore, additional data on galaxies that are located at greater distances and in other directions of the universe will have to be obtained before conclusive decisions can be made concerning the structure of the universe.

RELEVANCE QUESTION: *Galaxies are the basic building blocks for the universe. We live in a spiral-type galaxy called the Milky Way, which is composed of 100 billion stars. If the night sky is clear of clouds and light pollution, you can observe a region of the Milky Way with the un-aided eye. Does the Milky Way, our home Galaxy, have different positions (orientations) in the sky at the same time of night throughout the quarter or semester of this physical science course? Justify your answer.*

18.6 Quasars

LEARNING GOALS

▼ List the physical characteristics of quasars.

▼ Identify the most distant objects in the known universe.

In 1963, radio astronomers, using new radio telescopes with high resolution (that is, with the ability to distinguish the separation of two points), began detecting extremely strong radio signals from sources having small angular dimensions. The sources were named **quasars**—a shortened term for "quasistellar radio sources."

Hundreds of quasars have now been detected (● Fig. 18.31), and they have two important characteristics. First, quasars have extremely large redshifts. Second, quasars emit tremendous amounts of electromagnetic radiation (energy) at all wavelengths, and most of the radiation varies in intensity over a range of a few days to years.

Quasars are very small, appear to be blue, and have absolute magnitudes as high as −25. Although not conclusive, the best evidence points to the theory that the "powerhouse" of quasar energy is a supermassive black hole at the central region of a distant galaxy.

Quasars are considered to be the most distant objects in the universe. The quasar PC1247 + 3406 is the most distant and brightest (that is, having the highest absolute magnitude) object in the known universe. Its redshift of 4.9 indicates a distance of about 13.5 billion ly from Earth. The calculated value for the distance depends on the assumed value for the Hubble constant. A value of 75 km/s/10^6 pc was assumed for this distance. The closest known quasar is 600 million ly away in the radio galaxy Cygnus A.

Recent evidence indicates that quasars are somehow related to the centers of galaxies. As mentioned in Section 18.5, Seyfert galaxies have very bright centers, and the broad emission lines present in their spectra indicate that they contain gas that is

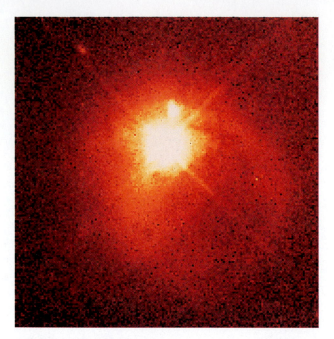

FIGURE 18.31   Quasar 3C273

One of the first quasars, 3C273, was detected in 1963. The Hubble Space Telescope obtained this recent image showing the clearest picture ever taken of the powerful jet of subatomic particles ejected from the quasar at almost the speed of light.

extremely hot. Because quasars are very distant objects, perhaps they are the bright nuclei of spiral galaxies. The spiral arms of a galaxy would not be visible at a great distance from us. The origin and evolution of quasars and galaxies are under intense study by astronomers.

RELEVANCE QUESTION: Quasars are the brightest and most distant objects in the known universe. Does this statement provide anything meaningful concerning your mental outlook on life in the universe?

18.7 Cosmology

LEARNING GOALS

▼ Describe the known structure of the universe at present.

▼ Identify theories concerning the evolution of the universe.

Cosmology is the study of the structure and evolution of the entire universe. We humans observe the universe, and our minds synthesize the observed phenomena. Concepts are then conceived and developed based on the observations. When we construct such concepts, we must remember that any concept of the universe that originates in the human mind is the way the universe is *conceived,* which is not necessarily the way it actually *is.*

The concept presently accepted by most astronomers is called the *Big Bang.* It has received broad acceptance because experimental evidence supports the concept in three major areas:

1. Astronomers observe galaxies that show a shift in their spectrum lines toward the low-frequency (red) end of the electromagnetic spectrum. This redshift is known as the **cosmological redshift.** (The cosmological redshift is not a Doppler shift, but a lengthening in the wavelength due to the expansion of the universe. The redshift is a result of the changing size of the universe and is not related to velocity, although astronomers refer to the cosmological redshift with reference to velocity.)
2. They detect a cosmic background radiation coming from space in all directions, commonly referred to as the *3-K cosmic microwave background.*
3. They observe a mass ratio of hydrogen to helium of 3 to 1 in stars and interstellar matter, as predicted by the Big Bang concept.

The redshift indicates an expanding universe.* To explain this observation, astronomers conceived the idea that the universe began with a violent event called the **Big Bang** and that we are now seeing the universe expanding from that event.

Because the event came before the expansion, we can calculate the time that has elapsed since the event. Astronomers observe galaxies receding from each other and assume that they have been receding from each other since the event. To determine the

*The geometry and mathematics of an expanding universe were first conceived and formulated by Alexander Friedmann, Russian mathematician, in 1922. Five years later, Georges Lemaitre, Belgian Catholic priest and cosmologist, developed a new cosmology theory of an expanding universe and postulated an explosive beginning from a Primeval Atom. Lemaitre, having heard Hubble lecture at Harvard University, knew of Hubble's evidence concerning the cosmological redshift of galaxies.

maximum time (t_{max}), let d represent the distance to the most remote galaxy that we observe, and assume that the rate of receding has always been equivalent to what we presently observe. The recessional velocity (v_r) is given by Hubble's law, that is, $v_r = Hd$, where H is Hubble's constant. We know, by definition, that velocity is distance divided by time ($v = d/t$).

EXAMPLE 18.4

Calculating the Age of the Universe

Calculate the maximum age of the universe in years. Use the minimum value for Hubble's constant.

$$H = \frac{50 \text{ km/s}}{10^6 \text{ pc}}$$

SOLUTION

Step 1

Given:

Hubble's law ($v = Hd$) and Hubble's constant

$$\left(\frac{50 \text{ km/s}}{10^6 \text{ pc}}\right)$$

Step 2

Wanted:

The age of the universe in years. Since Hubble's law does not include time, substitute a definition of velocity ($v = d/t$) for v.

$$v = Hd = \frac{d}{t}$$

Step 3

Cancel the distance d, and solve the equation for t.

$$H = \frac{1}{t}$$

or $t = 1/H$

Step 4

Substitute in the minimum value for H.

$$t = \frac{1}{\dfrac{50 \text{ km/s}}{10^6 \text{ pc}}}$$

Step 5

Convert 10^6 pc to kilometers in order to cancel kilometers.

$$1 \text{ pc} = 3.086 \times 10^{13} \text{ km}$$
$$10^6 \text{ pc} = 3.086 \times 10^{19} \text{ km}$$

Step 6

Substitute 3.086×10^{19} km for 10^6 pc and cancel kilometers.

$$t = \frac{1}{\dfrac{50 \text{ km/s}}{3.086 \times 10^{19} \text{ km}}}$$

Step 7

Solve for t by inverting the denominator and multiplying.

$$t = \frac{3.086 \times 10^{19} \text{ s}}{50} = 6.17 \times 10^{17} \text{ s}$$

Step 8

Convert 6.18×10^{17} s to years.

$$1 \text{ y} = 3.16 \times 10^7 \text{ s}$$
$$t = \frac{6.17 \times 10^{17} \text{ s}}{1} \times \frac{1 \text{ y}}{3.16 \times 10^7 \text{ s}}$$
$$t = 1.95 \times 10^{10} \text{ y}$$
$$t = 19.5 \text{ billion y}$$

Since the minimum value for Hubble's constant was used in the above calculation, the answer represents the maximum age of the universe. Exercise 15 at the end of the chapter asks you to calculate the minimum age of the universe.

CONFIDENCE EXERCISE 18.4

Calculate the age of the universe in years. Use a value of 85 km/s/10^6 pc for Hubble's constant. This value has recently been determined by observing cepheid variable stars in the Virgo cluster of galaxies.

As early as 1948, scientists predicted that residual radiation from the early universe should be present here and now, and the radiation should be at radio wavelengths. The radiation also should resemble the radiation from a black body at a temperature of a few degrees above absolute zero. Because the radiation from the early universe was everywhere at once, it should fill the entire universe and be isotropic; that is, the radiation should be the same in any direction that we observe.

This radiation was first detected in 1965 by Arno Penzias and Robert Wilson at Bell Telephone Laboratories in New Jersey. They were testing a new microwave horn antenna used to make measurements of the absolute intensity of microwave radiation coming from certain regions of the Milky Way. After debugging and accounting for known static sources in their equipment, an unknown source of static remained. The static was received equally from space at all times from every direction. After consulting with scientists at Princeton University who were designing and building equipment to detect the dying glow of the Big Bang, Penzias and Wilson concluded that the static (radiation) was the residual radiation from the early universe.

Today, the dying glow is called the 3-K **cosmic background radiation,** and its presence is considered evidence of an ancient, extremely hot universe. This residual radiation is the greatly redshifted radiation of the extremely hot universe that existed about 1 million years after the Big Bang.

Will the universe continue to expand forever? Presently, astronomers do not know the answer to this question, but the key to our understanding lies in determining the average density of matter in the universe. If the average density of matter is great enough, then space has a positive curvature and we live in a closed universe. In other words, gravity in the universe is great enough to stop the expansion, and eventually all matter will collapse in what is called the *Big Crunch.* If the average density of matter is too small, then space has a negative curvature and we live in an open universe that will continue to expand forever. Present data for the value of the average density of matter are not precise enough to determine whether space has a positive or a negative curvature, but it seems to indicate that we are living in a nearly flat universe.

The Big Bang theory of cosmology has problems, four of which will be mentioned here. First, the theory is based on the cosmological principle, which states that the universe is assumed to be homogeneous and isotropic. Thus proponents of the Big Bang base their theory on a universe that is uniform or smooth throughout. To support this view, which is contrary to the observation of great sheets of galaxies some 200 Mpc across, 50 Mpc high, and 5 Mpc thick, they assume that a volume of space (a few hundred Mpc on a side) placed anywhere in the universe would have approximately the same composition. Proponents of the theory assume the galaxies, the cluster of galaxies, and the huge superclusters as points in their calculations. They also assume that no structures exist that are larger than those presently observed. Obviously, opponents of the Big Bang theory believe these views are false.

Second, there is the horizon problem. As mentioned earlier, the cosmic background radiation, recently measured by NASA's Cosmic Background Explorer at 2.735 K, is the same from all directions of the universe. The horizon distance is the maximum distance light can have traveled since the Big Bang. Since nothing can travel faster than the speed of light, objects separated by a greater distance cannot be in contact with one another. Thus how can objects that we observe on opposite sides of the universe have exactly the same temperature? Proponents of the Big Bang do not have an acceptable answer.

The third problem that challenges astronomers is the age paradox. Cosmologists report the age of the universe between 8 and 12 billion years. Stellar astronomers report that the oldest stars in the known universe are between 14 and 16 billion years old. Both ages cannot be correct. The difference in values is related directly to reliable distance indicators, which are difficult to obtain. Cosmologists and stellar astronomers use different distance indicators. This gives different values for distance measurements. Until the problem is resolved, how the universe formed and evolved will remain a mystery.

Fourth, there is the flatness problem. As mentioned above, the observed density of the universe has a value that does not support an open or closed universe but one near a value that indicates the universe is flat. A "flat" universe is one that neither expands forever nor collapses back to a singularity. The Big Bang theory provides no acceptable reason why the density of the universe is so close to a value that indicates that the universe will exist forever rather than a value that indicates an open or a closed universe. (See the chapter's second Highlight.)

FIGURE 18.32 Hubble Space Telescope

This photo shows the Hubble Space Telescope being released from the payload bay of the space shuttle *Discovery* on April 25, 1990. The telescope's 2.4-m (7.8-ft) mirror is covered. The solar array panels are shown fully deployed.

The Hubble Space Telescope (● Fig. 18.32) is providing new information about the solar system and the universe beyond. The first object it observed was the planet Pluto. Figure 15.26 shows, for the first time, Pluto and its satellite Charon as two distinct objects. After the telescope was placed in orbit on April 25, 1990, astronomers and engineers discovered that images would not focus precisely to give excellent photographs. The primary mirror was not ground perfectly, causing spherical aberration. The defect was corrected in December 1993 when a servicing crew of astronauts repaired the telescope's blurry vision. At the same time, the astronauts replaced the solar arrays and the guidance system.

The search for knowledge about the universe continues on several fronts. The world's largest optical telescope—the $94 million Keck, on Mauna Kea in Hawaii—was placed in operation in 1991, and its twin, also in Hawaii, was placed in operation in 1996.

The 33-ft mirror of the Keck telescope (● Fig. 18.33) is made up of 36 hexagonal segments, each with its own support system, as well as sensors that detect misalignment between adjacent segments. The segments operate together as a single unit, and they are computer-controlled for perfect alignment. The huge mirror has four times the light-gathering power of the 200-in. reflector on Palomar Mountain.

FIGURE 18.33 The 400-Inch-Diameter Keck Telescope Under Construction

The photo shows the outer circle of 18 hexagonal mirror segments. The completed telescope has two more circles of mirrors—one inner circle of 6 mirrors and 12 more placed between the inner 6 and the outer 18. An opening at the center of the 36 mirror segments, the size of one segment, is for the light collected by the mirrors to be reflected to the recording instruments. This arrangement of 36 mirrors acting as one huge paraboloid is possible because computer technology keeps the segments in alignment. See the text for more details.

HIGHLIGHT: Other Cosmologies

In the late 1940s, Fred Hoyle, British astrophysicist and cosmologist, along with two collaborators, Thomas Gold and Hermann Bondi, formulated the *Steady-State model* of the universe. The model is based on a generalization of the cosmological principle called the *perfect cosmological principle*, which states that on a large scale the universe is the same for all observers at all places and for all time. The steady-state universe has a constant density, is infinite, had no beginning, and exists forever.

The major flaws in the model are its failure to provide an adequate explanation for a steady-state (constant-density) universe that appears to be expanding or an acceptable explanation of the microwave background radiation. Recently, one of Hoyle's collaborators indicated that there are other explanations for the observed redshifts of galaxies, which indicate an expanding universe. Incidentally, it was Hoyle, during an interview on a British radio broadcast, who coined the name Big Bang for the cataclysmic explosive event that other cosmologists put forth as the birth of the universe.

Plasma cosmology is a theoretical concept proposed by Hannes Alfvén, Swedish Nobel Laureate, in the 1970s. According to Alfvén, the universe has always existed, and its structure and evolution are controlled not solely by gravity as stated in the Big Bang model, but just as much by electric currents and magnetic fields. Using computer simulations, which neglect gravitational forces, supporters of plasma cosmology show that the rotation of plasma in a magnetic field can initiate star formation. When electric currents are extremely large (10^{16} amperes), the formation of galaxies, the basic structural units for the universe, will take place.

Any model or theory of cosmology must address the three observed facts stated in Section 18.7. Plasma cosmologists agree that the observed abundance of helium (about 25%) in the universe, and the cosmic background radiation, can be explained by the birth of massive stars in the formation of galaxies. Supporters of Alfvén's cosmology indicate that in thermonuclear reactions of massive stars, when part of the stars' hydrogen is transformed into helium, 25% of the gases produced are helium. The stars eventually explode into supernovae releasing the helium, which in turn is available in the formation of smaller stars.

The energy radiated by massive stars is absorbed by interstellar dust, which in turn emits the microwave background. This microwave background is smoothed out by the absorption and re-emission of the radiation due to intergalactic magnetic fields provided by powerful jets emitted from galactic nuclei. The radiation is scattered in all directions and exists as a radiation fog.

In reference to the cosmological redshift, Alfvén and his collaborators suggest one possible explanation. What astronomers observe is an expansion that began some 10 to 20 billion years ago in our part of the universe, but not a Big Bang that created matter, space, and time.

Another thought concerning the redshift is the idea of "tired light." Some scientists hypothesize that light loses energy as it travels through space, causing the redshift.

Paul A. M. Dirac, British theoretical physicist and Nobel Laureate, suggested in 1938 that all space (objects and the space between) is expanding. No real expansion is occurring, since the density remains constant. Thus distant galaxies only appear to be expanding.

The Big Bang, Steady-State, and Plasma cosmologies are rival concepts or models of the universe. The validity of a model must be supported by experimental evidence and valid predictions. A model cannot be accepted until these two tests are satisfied. Presently, none of the models of the universe is acceptable to all cosmologists.

A radio telescope under construction at Green Bank, West Virginia, will be the world's largest, fully steerable radio telescope. The antenna, with a reflecting surface of 2.3 acres, plus electronic and recording equipment, is scheduled to be operational in 1999.

The Very Large Array of radio telescopes in central New Mexico (see Fig. 18.17) is laid out with three arms, each 13 mi long. The Very Large Array consists of 27 radio antennas, 9 on each arm, that, with the aid of a computer, act as a single antenna and can produce a high-resolution radio map of the radiation coming from a celestial object.

Other regions of the electromagnetic spectrum are also being used to collect data from celestial objects. The Gamma Ray Observatory was placed in orbit in 1991, and the Advanced X-ray Facility was launched in January 1999.

Scientists supported by many nations will continue to send their improved instruments into orbit around Earth, and they will send more advanced probes throughout the solar system and beyond to collect data that will increase our knowledge of the universe in which we humans exist.

RELEVANCE QUESTION: *The hypothesis of a flat Earth has long been rejected. The circular disks of the Sun and Moon probably had some influence on rejecting the hypothesis and suggesting a spherical shape for planet Earth. What are your thoughts concerning the size and shape of the Universe—finite or infinite? Flat, spherical, or something else?*

Important Terms

Sun (18.1)	right ascension	supernova	clusters
photosphere	celestial prime meridian	nucleosynthesis	superclusters
plasma	light-year	neutron star	Hubble's law
chromosphere	magnitude (18.3)	pulsars	cosmological principle
corona	absolute magnitude	gravitational	quasars (18.6)
solar wind	H-R diagram	collapse (18.4)	cosmology (18.7)
heliosphere	main sequence	singularity	cosmological redshift
sunspots	red giants	event horizon	Big Bang
prominences	white dwarfs	Schwarzschild radius	cosmic background
neutrino	red dwarfs	black hole	radiation
proton-proton chain	binary star	galaxy (18.5)	
celestial sphere (18.2)	cepheid variables	Milky Way	
declination	nova	Local Group	

Important Equations

Distance to a Star: $d = \dfrac{1}{p}$

Hubble's Law: $v = Hd$

Radius of the Event Horizon of a Black Hole: $R = \dfrac{2GM}{c^2}$

Review Questions

18.1 The Sun

1. Which of the following is a feature of the Sun?
   - (a) chromosphere
   - (b) photosphere
   - (c) prominences
   - (d) corona
   - (e) all of the above

2. The Sun's energy comes from
   - (a) the fission of hydrogen to form helium.
   - (b) the fusion of protons to form nuclei of helium.
   - (c) the fusion of hydrogen to form carbon nuclei.
   - (d) none of the above.

3. Most of the light from the Sun comes from
   - (a) the chromosphere.
   - (b) the corona.
   - (c) the photosphere.
   - (d) none of the above.

4. The period of the sunspot cycle is approximately _____ years.
   - (a) 11
   - (b) 15
   - (c) 24
   - (d) 30

5. What is the temperature of the Sun (a) at its center and (b) at its surface?

6. (a) What are sunspots?
   (b) Are they cyclical? Explain.

7. Name the visible surface of the Sun.

8. What is the solar wind?

9. When is the Sun's corona visible?

10. What makes the Sun radiate energy?

11. What are neutrinos and photons, and how are they different?

18.2 The Celestial Sphere

12. Which of the following does *not* refer to the celestial sphere?
    (a) declination
    (b) parsec
    (c) right ascension
    (d) vernal equinox

13. Which of the following is *not* a unit of distance measurement to a celestial object?
    (a) astronomical unit
    (b) light-year
    (c) arc second
    (d) parsec

14. East longitude of a place on Earth's surface is comparable with the ———— of a star on the celestial sphere.
    (a) declination
    (b) right ascension
    (c) meridian
    (d) hour angle

15. The angular measure in degrees north or south of the celestial equator is called
    (a) latitude.
    (b) declination.
    (c) right ascension.
    (d) longitude.

16. Name the three celestial coordinates.

17. Define the celestial equator.

18. What is the celestial prime meridian?

19. Define each term: (a) light-year, (b) parsec, and (c) astronomical unit.

20. Which is a larger unit of distance, the light-year or the parsec? How much larger is one compared to the other?

21. (a) What are the units used to measure the annual parallax of a star?
    (b) State the relationship between the annual parallax and the distance to a star in parsecs.

22. The vernal equinox is a point on the celestial sphere that is between which two constellations?

23. What are asterisms? Give an example.

18.3 Stars

24. Which of the following does *not* refer to the name of a type of star?
    (a) red giant
    (b) cepheid variable
    (c) white dwarf
    (d) Seyfert
    (e) red dwarf

25. The Hertzsprung-Russell (H-R) diagram is a plot of the absolute magnitude of stars against their
    (a) apparent magnitude.
    (b) surface temperature.
    (c) distance.
    (d) brightness.

26. The color of a relatively cool star is
    (a) red.
    (b) yellow.
    (c) white.
    (d) blue.

27. The Sun will eventually become a
    (a) brown dwarf.
    (b) supernova.
    (c) nova.
    (d) white dwarf.

28. (a) What two major elements make up a star?
    (b) Give the relative abundance of each element.

29. What is an H-R diagram?

30. On an H-R diagram, give the position of each of the following: (a) the Sun, (b) the main sequence, (c) giants, (d) cepheid variables, (e) white dwarfs, and (f) supergiants.

31. (a) What is the closest star to the Sun?
    (b) Give its distance from the Sun in light-years.
    (c) Can the star be seen from the United States? Explain.

32. Which has the higher surface temperature, a white or a yellow star?

33. What is a binary star system? Are most stars part of a binary system?

34. What are cepheid variables? Why are they important to astronomers?

35. State the order in which the following possible stages of a star occur: hydrogen burning, planetary nebula, white dwarf, gravitational accretion, and red giant.

36. The line spectrum of a star shows a shift toward the red end of the spectrum. What does this indicate concerning the motion of the star?

37. What is (a) a neutron star and (b) a pulsar?

18.4 Gravitational Collapse and Black Holes

38. Which of the following does *not* refer to a black hole?
    (a) singularity
    (b) Schwarzschild radius
    (c) nucleosynthesis
    (d) event horizon

39. The singularity of a nonrotating black hole is surrounded by an invisible spherical boundary known as the
    (a) Schwarzschild radius.
    (b) gravitational boundary.
    (c) event horizon.
    (d) black hole boundary.

40. (a) What is a black hole?
    (b) How many black holes have astronomers detected?

41. What is a singularity?

42. What factors determine the radial distance of the event horizon?

43. What is the minimum mass that a star must have for it to undergo gravitational collapse and form a black hole?

44. Describe a method for detecting black holes.

45. How does the radial distance of the event horizon of a black hole vary with respect to (a) the mass of a collapsing star and (b) the speed of light?

18.5 Galaxies

46. Galaxies are
    (a) extremely large collections of stars.
    (b) classified as elliptical, normal spiral, barred spiral, or irregular.
    (c) found in clusters.
    (d) found in clusters of clusters called superclusters.
    (e) all of the above.

47. Hubble's law
    (a) relates the observed recessional velocity of a galaxy and its distance from us.
    (b) indicates that the universe is expanding.
    (c) is sometimes referred to as the law of redshifts.
    (d) all of the above.

48. Which of the following is *not* a classification of galaxies?
    (a) ellipticals   (c) regulars
    (b) spirals   (d) irregulars

49. The physical property used to classify galaxies is
    (a) the way they rotate.
    (b) how they appear when photographed.
    (c) according to their mass.
    (d) according to their size.

50. State the structure and dimensions of the Milky Way.

51. (a) Name the small group of galaxies of which the Milky Way is a member.
    (b) How many known members are in the group?
    (c) Which member of the group can be seen with the unaided eye from the United States?

52. Distinguish between isotropic and homogeneous distribution of galaxies.

53. What is dark matter? What is its significance?

54. What is the "Great Wall" of galaxies? Give its dimensions.

55. (a) State the cosmological principle.
    (b) Is there any reason to believe that it may not be true?

18.6 Quasars

56. Quasars
    (a) exhibit very large redshifts.
    (b) are believed to be the most distant objects from us.
    (c) exhibit extremely powerful sources of energy.
    (d) are believed to be galaxies with active galactic nuclei.
    (e) all of the above.

57. Quasars
    (a) are very large objects.
    (b) were first detected by radio telescopes.
    (c) exhibit blue shifts in their spectrum.
    (d) all of the above.

58. The word *quasars* is a shortened term for
    (a) quasistellar star radio sources.
    (b) quasistellar radio sources.
    (c) quasistellar radiation sources.
    (d) quasistellar infrared sources.

59. (a) What is the apparent color of quasars?
    (b) What is the maximum value of their absolute magnitude?

60. (a) What information is there to support the belief that quasars are the most distant objects observed by astronomers?
    (b) In what respect do quasars resemble stars?
    (c) In what respect do they differ from stars?

18.7 Cosmology

61. Which of the following is true concerning the Big Bang model of cosmology?
    (a) The universe began as a singularity of high density and temperature.
    (b) The model has received broad acceptance because of the observed cosmological redshift, the 3-K cosmic microwave background, and the three-to-one ratio of hydrogen to helium in stars.
    (c) The model predicts an open universe.
    (d) The model provides information about the universe back to the very beginning.
    (e) Answers a and b are true.

62. The cosmological principle
    (a) is based on the assumption that the universe is homogeneous.
    (b) implies that the universe has no edge.
    (c) is based on the assumption that the universe is isotropic.
    (d) implies that the universe has no center.
    (e) all of the above.

63. Describe the Big Bang model of the universe.

64. State three experimental facts that support the Big Bang model.

65. List three problems cosmologists who support the Big Bang model need to resolve.

66. What property of the universe can be determined by taking the reciprocal of Hubble's constant?

67. What is the Steady-State model of the universe?

68. Why is the universe thought to be expanding?

69. Name the Nobel Laureate who conceived the theoretical concept of Plasma cosmology.

70. How does Plasma cosmology differ from Big Bang cosmology?

Applying Your Knowledge

1. Assuming Hubble's law to be true, what is the maximum distance from which a galaxy can be measured?

2. Is the Big Bang a one-of-a-kind event, or is the Big Bang only a small expansion in a larger universe?

3. How have your personal views of the universe changed after completing the astronomy chapters?

Exercises

18.1 The Sun

1. The star Sirius has a mass of 4.3×10^{30} kg. Calculate the time, in years, it would take to convert 12% of its hydrogen into helium and energy, if 90% of its mass is hydrogen. Assume that 2.0×10^{12} kg of hydrogen are converted each second.   Answer: 7.3×10^9 y

2. A certain star has a mass of 5.0×10^{31} kg. Calculate the time, in years, it would take to convert 15% of its hydrogen into helium and energy, if 92% of its mass is hydrogen. Assume that 3.0×10^{13} kg of hydrogen are converted each second.

18.2 The Celestial Sphere

3. Find the distance in parsecs to the star Altair. The star is observed to show an annual parallax of 0.20 arcsec.
   Answer: 5.1 pc

4. The star Vega is observed to show an annual parallax of 0.125 arcsec. Determine the distance in light-years to Vega.

5. Calculate the number of miles in a light-year, using 1.86×10^5 mi/s as the speed of light.   Answer: 5.87×10^{12} mi

6. Calculate the number of meters in a light-year, using 3.00×10^8 m/s as the speed of light.

7. The great spiral galaxy in Andromeda (M31) can be seen with the unaided eye. M31 is 680 kiloparsecs from Earth. How far away is the galaxy in light-years?
   Answer: 2.2×10^6 ly

8. How many kilometers away is Alpha Centauri, which is at a distance of about 4.3 ly?

18.3 Stars

9. How many times brighter is a cepheid variable of absolute magnitude -3 than a white dwarf of absolute magnitude $+7$?   Answer: 10,000 times

10. How much brighter is a star of absolute magnitude $+1$ than a star of absolute magnitude (a) $+2$ and (b) $+6$?

18.4 Gravitational Collapse and Black Holes

11. Determine the radial distance of the event horizon of a black hole formed from the gravitational collapse of a star having a mass of 15×10^{30} kg.   Answer: 22 km

12. Calculate the distance between the singularity and the event horizon for a gravitationally collapsed star having a mass of 2.0×10^{32} kg.

18.5 Galaxies

13. Suppose our universe contained 100 billion galaxies with 100 billion stars in each. How many stars would there be?   Answer: 1×10^{22}

14. It takes the Sun 2×10^8 y to go around the center of the galaxy. How many times has it (and our solar system) gone around the galaxy during its life of 5 billion years?

18.7 Cosmology

15. Determine the minimum age of the universe, using a value for Hubble's constant of 100 km/s per 10^6 pc.
    Answer: 9.78×10^9 y

16. Hubble's constant is estimated to be between 50 and 100 km/s per 10^6 pc. Use the average of these two values and determine the age of the universe.

Solutions to Confidence Exercises

18.1 Calculate 85% of the star's mass.

4.0×10^{30} kg $\times 0.85 = 3.4 \times 10^{30}$ kg of hydrogen

Calculate 15% of the star's hydrogen.

3.4×10^{30} kg $\times 0.15 = 5.1 \times 10^{29}$ kg

Calculate the number of years required to convert the 15%.

$$\text{number of years} = \frac{\text{total amount of hydrogen converted}}{\text{rate/s} \times \text{number of seconds in one year}}$$

$$\text{number of years} = \frac{5.1 \times 10^{29} \text{ kg}}{5.0 \times 10^{12} \text{ kg/s} \times 3.16 \times 10^{7} \text{ s/y}}$$

$$= 3.2 \times 10^{9} \text{ y}$$

18.2 Calculate distance (d) using Eq. 18.1.

$$d = \frac{1}{p} = \frac{1}{0.376} = 2.66 \text{ pc}$$

18.3 Calculate the Schwarzschild radius (R) using Eq. 18.3.

$$R = \frac{2GM}{c^2} = \frac{2(6.67 \times 10^{-11} \text{ kg} \times \text{m/s}^2 \times \text{m}^2/\text{kg}^2)(8.0 \times 10^{32} \text{ kg})}{(3.0 \times 10^{8})^{2} \text{ m}^2/\text{s}^2}$$

$R = 1.2 \times 10^{6}$ m

18.4 Calculate the age of the universe using Hubble's law.

$$v = Hd = \frac{d}{t}$$

Cancel the distance d and solve the equation for t.

$$t = \frac{1}{H} = \frac{1}{\dfrac{85 \text{ km/s}}{10^{6} \text{ pc}}}$$

Convert 10^{6} pc to km in order to cancel the kilometers.

$$1 \text{ pc} = 3.086 \times 10^{13} \text{ km}$$
$$10^{6} \text{ pc} = 3.086 \times 10^{19} \text{ km}$$

Substitute 3.1×10^{19} km for 10^{6} pc and cancel kilometers.

$$t = \frac{1}{\dfrac{85 \text{ km/s}}{3.1 \times 10^{19} \text{ km}}}$$

Solve for t by inverting the denominator and multiplying.

$$t = \frac{3.1 \times 10^{19} \text{ s}}{85} = 3.6 \times 10^{17} \text{ s}$$

Convert 3.6×10^{17} s to years.

$$t = \frac{3.6 \times 10^{17} \text{ s}}{1} \times \frac{1 \text{ y}}{3.16 \times 10^{7} \text{ s}}$$

$$t = 1.1 \times 10^{10} \text{ y} = 11 \text{ billion y}$$

Answers to Multiple-Choice Review Questions

1. e	3. c	12. b	14. b	24. d	26. a	38. c	46. e	48. c
2. b	4. a	13. c	15. b	25. b	27. d	39. c	47. d	49. b

56. e	58. b	62. e
57. b	61. e	

THE ATMOSPHERE

19

...this most excellent canopy, the air.

William Shakespeare (1564–1616)

*E*arth science is a collective term that involves all aspects of our planet—land, sea, and air, and even its history. In this and the next chapter, we will be concerned with the composition and phenomena of the atmosphere.

Our atmosphere (from the Greek *atmos,* "vapor," and *sphaira,* "sphere") is the gaseous shell, or envelope, of air that surrounds Earth. Just as certain sea creatures live at the bottom of the ocean, we humans live at the bottom of this vast atmospheric sea of gases.

In recent years, the study of the atmosphere has expanded because of advances in technology. Every aspect of the atmosphere, from the ground to outer space, is now investigated in what is called **atmospheric science.** An older term, **meteorology** (from the Greek *meteora,* "the air"), is now more commonly applied to the study of the lower atmosphere. The continuously changing conditions of the lower atmosphere are what we call *weather.* Lower atmospheric conditions are monitored daily, and changing patterns are studied to help predict future conditions. Meteorologists give you the weather forecasts on TV.

With rising environmental concerns, the study of the atmosphere has grown. Because the air we breathe is such an integral part of our environment,

Photo: Our atmosphere as viewed from space (with the Moon in the background).

a knowledge of the atmosphere's constituents, properties, and workings is needed to understand, appreciate, and hopefully solve current environmental problems. We hear concerns about the greenhouse effect, holes in the ozone layer, and global warming. In this chapter we will look at normal atmospheric conditions, as well as these topics of concern. ■

19.1 Composition and Structure

LEARNING GOALS

▼ Identify the composition of air.

▼ Describe how the atmosphere is divided into regions.

The air of the atmosphere is a mixture of many gases. In addition, the air holds many suspended liquid droplets and solid particles. However, only two gases comprise about 99% of the volume of air near Earth. From ● Fig. 19.1 and Table 19.1, we see that this *air is primarily composed of nitrogen (78%) and oxygen (21%), with nitrogen being about four times as abundant as oxygen. Notice that atmospheric nitrogen and oxygen are diatomic (two-atom) molecules, N_2 and O_2. The other main constituents are argon (Ar, 0.9%) and carbon dioxide (CO_2, 0.03%).*

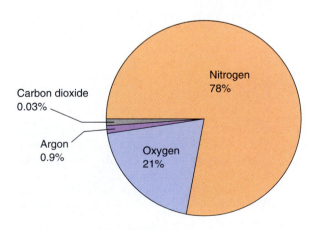

FIGURE 19.1   Composition of Dry Air

A graphic representation of the volume composition of the major constituents of air. (CO_2 percentage shown larger than scale for clarity.)

TABLE 19.1   Composition of Air

Nitrogen	N_2	78%	(by volume)
Oxygen	O_2	21%	
Argon	Ar	0.9%	
Carbon dioxide	CO_2	0.03%	
Others (traces)		*Others (variable)*	
Neon	Ne	Water vapor (H_2O) 0–4%	
Helium	He	Carbon monoxide (CO)	
Methane	CH_4	Ammonia (NH_3)	
Nitrous oxide	N_2O	Solid particles—dust,	
Hydrogen	H_2	pollen, etc.	

Minute quantities of many other gases are found in the atmosphere, along with particulate matter. Some of these gases, especially water vapor and carbon monoxide, vary in concentration, depending on conditions and locality. The amount of water vapor in the air depends to a great extent on temperature, as will be discussed later. Carbon monoxide (CO) is a product of incomplete combustion (Section 13.2). A sample of air taken near a busy freeway would contain a concentration of CO considerably higher than that normally found in the atmosphere.

In general, however, the relative amounts of the major constituents of the atmosphere remain fairly constant. Nitrogen, oxygen, and carbon dioxide are involved in the life processes of plants and animals. These gases are continuously taken from the air and replenished as by-products of these various processes. Nitrogen is taken in by some plants and released during organic decay. Animals inhale oxygen and exhale carbon dioxide, while plants convert carbon dioxide to oxygen.

Plants produce oxygen by **photosynthesis,** the process by which CO_2 and H_2O are converted into sugars (needed for plant life) and O_2, using energy from the Sun (● Fig. 19.2). Oxygen is expelled as a by-product. The key to photosynthesis is the ability of *chlorophyll*, the green pigment in plants, to convert sunlight into chemical energy. Billions of tons of CO_2 are removed from the atmosphere annually and replaced with billions of tons of oxygen. Over half the photosynthesis takes place in the oceans, which contain many forms of green plants.

FIGURE 19.2   Photosynthesis

Energy from the Sun is necessary in the photosynthesis process whereby plants produce sugars and oxygen from water and carbon dioxide.

Earth's atmosphere evolved into its present condition over millions of years. The atmospheres of other planets evolved in different fashions and contain different constituents (Chapter 15). The planet Mercury, however, has virtually no atmosphere, nor does our Moon.

The gravitational attraction between Earth and the atmosphere is greater near the planet's surface. As a result, the density of air is greatest near Earth's surface and decreases with increasing altitude. Because of this gravitational attraction, over half the mass of the atmosphere lies below an altitude of 11 km (7 mi), and almost 99% lies below an altitude of 30 km (19 mi). At a height of 320 km (200 mi) above Earth, the density of the atmosphere is such that a gas molecule may travel a distance of 1.6 km (1 mi) before encountering another gas molecule.

There is no clearly defined upper limit of Earth's atmosphere. It simply becomes more and more tenuous and merges into the interplanetary gases, which may be thought of as part of the extensive "atmosphere" of the Sun.

To distinguish different regions of the atmosphere, we look for physical variations that occur with altitude. Such changes in physical properties can be used to define vertical divisions. Atmospheric density decreases continuously with altitude, so this provides no distinction. A couple of atmospheric properties that show vertical variations are (1) temperature and (2) ozone and ion concentrations.

Temperature

In measuring the temperature of the atmosphere versus altitude, we find that distinctions do occur. These distinctions lead to major divisions of the atmosphere based on temperature variations. Near Earth's surface, the temperature of the atmosphere decreases with increasing altitude at an average rate of about $6\frac{1}{2}$ C°/km (or $3\frac{1}{2}$ F°/1000 ft) up to about 16 km (10 mi). This region is called the **troposphere** (from the Greek *tropism,* "to change," ● Fig. 19.3).

The troposphere contains over 80% of the atmospheric mass and virtually all the clouds and water vapor. There is continual mixing and a great deal of change in this region. The atmospheric conditions of the lower troposphere are referred to collectively as weather.

Changes in the weather reflect the local variations of the atmosphere near Earth's surface. The lower troposphere was the only region of the atmosphere in-

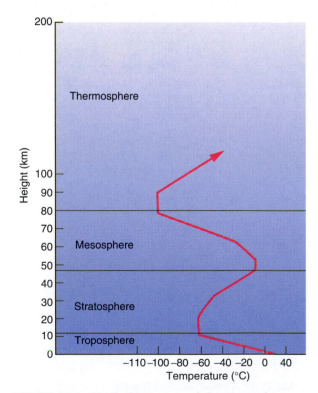

FIGURE 19.3   Vertical Structure of the Atmosphere

Divisions of the atmosphere are based on variations in physical properties such as temperature, as shown here.

vestigated before the twentieth century. At the top of the troposphere, the temperature falls to −45° to −50°C (about −50° to −60°F).

Above the troposphere the temperature of the atmosphere increases nonuniformly up to an altitude of about 50 km (30 mi). (See Fig. 19.3.) This region of the atmosphere, from approximately 16 to 50 km (10 to 30 mi) in altitude, is called the **stratosphere** (from the Greek *stratum,* "covering layer"). Together, the troposphere and stratosphere account for about 99.9% of the atmospheric mass.

The temperature of the atmosphere then decreases rather uniformly with altitude to a value of about −95°C (−140°F) at an altitude of 80 km (50 mi). This region, between about 50 and 80 km (30 to 50 mi) in altitude, is called the **mesosphere** (from the Greek *meso,* "middle").

Above the mesosphere, the thin atmosphere is heated intensely by the Sun's rays, and the temperature climbs to over 1000°C (about 1800°F). This region, extending to the outer reaches of the atmosphere, is called the **thermosphere** (from the Greek *therme,* "heat"). The temperature of the thermosphere varies considerably with solar activity.

Ozone and Ion Concentrations

The atmosphere also may be divided into two parts based on regions of concentrations of ozone and ions, with the ozone region lying below the ion region. **Ozone (O_3)** is formed by the dissociation of molecular oxygen and the combining of atomic oxygen with molecular oxygen:

$$O_2 + energy \longrightarrow O + O$$

$$O + O_2 \longrightarrow O_3$$

At high altitudes, energetic ultraviolet (uv) radiation from the Sun provides the energy necessary to dissociate the molecular oxygen. Oxygen, however, is less abundant at higher altitudes, so the production and concentration of ozone depend on the appropriate balance of uv radiation and oxygen molecules. The optimum conditions occur at an altitude of about 30 km (20 mi), and in this region is found the central concentration of the ozone layer, as illustrated in ● Fig. 19.4.

The ozone concentration becomes less with increasing altitude up to an altitude of about 70 km (45

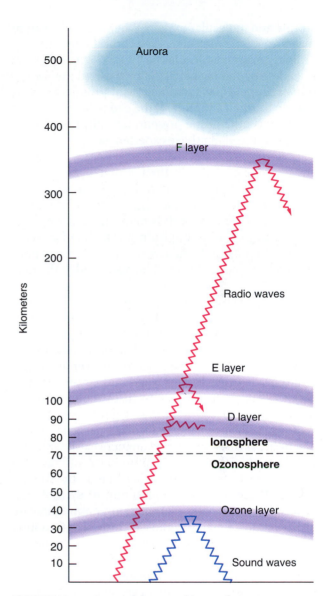

FIGURE 19.4   Ozonosphere and Ionosphere
These atmospheric regions are based on ozone and ion concentrations. A warm-air layer occurs in the ozonosphere because ozone absorbs ultraviolet radiation. The warm-air layer reflects sound waves and was first investigated in this manner. The upper ion layers reflect radio waves.

mi). The region of the atmosphere below this is referred to as the **ozonosphere.**

The ozone layer is a broad band of gas that extends through nearly all the stratosphere. Ozone is unstable in the presence of sunlight, and it dissociates into atomic and molecular oxygen. When an oxygen atom meets an ozone molecule, they combine to form two ordinary oxygen molecules ($O + O_3 \longrightarrow O_2 + O_2$) and thus destroy the ozone. This process and the formation of ozone go on simultaneously, producing a balance in the concentration of the ozone layer.

There is little ozone present naturally near Earth's surface, but you may have experienced ozone when it is formed by electrical sparking discharges. The gas is easily detected by a distinct pungent smell from which it derives its name (Greek *ozein,* "to smell"). In some areas—for example, Los Angeles—ozone is classified as a pollutant. It is found in relatively high concentrations resulting from photochemical reactions of air pollutants. Such reactions give rise to photochemical smog, which will be discussed in the next chapter.

The ozone layer in the stratosphere acts as an umbrella that shields life from harmful ultraviolet radiation from the Sun by absorbing most of the short wavelengths of this radiation. The portion of the uv radiation that gets through the ozone layer burns and tans our skin in the summer (and may cause skin cancer). Were it not for the ozone absorption, we would be badly burned and find the sunlight intolerable.

Because the ozone layer absorbs energetic ultraviolet radiation, one can expect an increase in temperature in the ozonosphere. A comparison of Figs. 19.3 and 19.4 shows that the ozone layer lies in the stratosphere. Hence the ozone absorption of uv radiation provides an explanation for the temperature increase in the stratosphere, as opposed to the continuously decreasing temperature versus altitude in the neighboring troposphere and mesosphere.

In the upper atmosphere, energetic particles from the Sun cause the ionization of gas molecules. For example,

$$N_2 + energy \longrightarrow N_2^+ + e^-$$

The electrically charged ions and electrons are trapped in Earth's magnetic field and form ionic layers in the upper region of the atmosphere called the **ionosphere.**

Variations in the ion density with altitude give rise to the labeling of three regions or layers—D, E, and F (Fig. 19.4). The D layer strongly absorbs radio waves below a certain frequency. Radio waves with frequencies above this value pass through the D layer but are reflected by the E and F layers, up to a limiting fre-

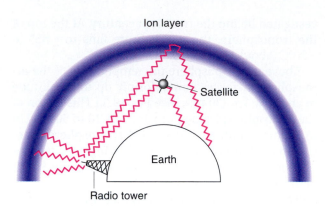

FIGURE 19.5 Global Radio Transmission
Radio waves travel in straight lines and are reflected around the curvature of Earth by ion layers. Ionic disturbances from solar activity may affect ion density and disrupt communications. Transmission by satellite is not as severely affected by ion disturbances.

quency. Thus the ionosphere provides global radio communications by the reflection of waves from ionic layers, as illustrated in ● Fig. 19.5. (Today, satellites relay many radio and television communications.)

Solar disturbances, which produce a shower of incoming energetic particles, are also associated with the beautiful displays of light in the upper atmosphere of the polar regions. In the Northern Hemisphere, these are called *northern lights,* or *aurora borealis* (Latin, *aurora,* "dawn," and *boreas,* "northern wind," ● Fig. 19.6). The Southern Hemisphere tends to be forgotten by people living north of the equator. However, light displays of equal beauty occur in the southern polar atmosphere and are called *aurora australis* (Latin, *auster,* "southern wind").

In general, the ions and electrons trapped in Earth's magnetic field are deflected toward Earth's magnetic poles (our polar regions), over which the majority of the auroras occur. However, auroras are sometimes observed at lower latitudes. The emission of light is believed to be associated with the recombination of ions and electrons. Energy is needed to ionize; and on recombining, energy is emitted in the form of visible light, or radiation.

RELEVANCE QUESTION: *On a cloud-covered day, the Sun is not exposed and you feel cool, yet you might get a severe sunburn or tan. Why is this?*

FIGURE 19.6 Aurora
The aurora borealis, or northern lights, as seen in Alberta, Canada.

19.2 Atmospheric Energy Content

LEARNING GOALS

▼ Describe how insolation is distributed in the atmosphere.

▼ Explain why the sky is blue and sunsets are red.

▼ Describe the greenhouse effect and its impact on Earth's temperature.

The Sun is by far the most important source of energy for Earth and its atmosphere. At an average distance of 93 million miles from the Sun, Earth intercepts only a small portion of the vast amount of solar energy emitted. This energy traverses space in the form of radiation, and the portion incident on Earth's atmosphere is called **insolation** (standing for *in*coming *so*lar radiation).

Because Earth's axis is tilted $23\frac{1}{2}°$ with respect to a normal (line perpendicular) to the plane of its orbit about the Sun, the insolation is not evenly distributed over Earth's surface. This tilt, coupled with Earth's revolution around the Sun, gives rise to the seasons (Chapter 16).

The solar radiation (energy) output fluctuates, but Earth receives a relatively constant average intensity at the top of the atmosphere. However, only 50% or less of the insolation reaches Earth's surface, depending on atmospheric conditions.

In considering the energy content of the atmosphere, one might think that it comes directly from insolation. However, surprisingly enough, most of the direct heating of the atmosphere comes not from the Sun but from Earth. To understand this, we need to examine the distribution and disposal of the insolation (● Fig. 19.7).

About 33% of the insolation received is returned to space as a result of reflection by clouds, scattering by particles in the atmosphere, and reflection from terrestrial surfaces such as water, ice, and ground. The reflectivity, or the amount of light a body reflects, is known as its *albedo* (from the Latin *albus,* "white"). This is expressed as a fraction or percentage of solar radiation received.

For example, Earth has an albedo of 0.33, or 33%. In other words, it reflects about one-third of the incident sunlight. The brightness of Earth as viewed from space depends on the amount of sunlight reflected; clouds play an important role (● Fig. 19.8). In comparison, the Moon with its dark surface and no

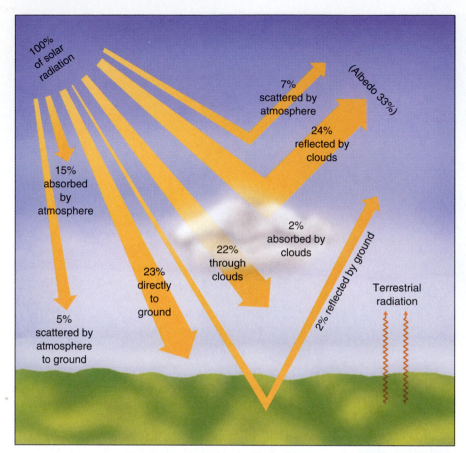

FIGURE 19.7   Insolation Distribution

An illustration of how the incoming solar radiation is distributed. The percentages vary somewhat, depending on atmospheric conditions.

atmosphere has an albedo of only 0.07; that is, it reflects only 7% of the insolation. The relatively large size and large albedo make Earth a more impressive sight than the Moon when viewed from space.

In the atmosphere, scattering of insolation occurs from gas molecules of the air, dust particles, water droplets, and so on. (*Scattering* is the absorption of incident light and its reradiation in all directions.) As shown in Fig. 19.7, some of the scattered radiation is dispersed back into space, and some is sent toward Earth's surface.

An important type of scattering involving a common atmospheric phenomenon is **Rayleigh scattering,** named after Lord Rayleigh (1842–1919), the British physicist who developed the theory. Lord Rayleigh showed that the amount of scattering by

FIGURE 19.8   Planet Earth

The brightness of Earth, as seen from space, depends on the amount of sunlight reflected, and clouds play an important role.

particles of a given molecular size was proportional to $1/\lambda^4$, where λ is the wavelength of the incident light; that is, the longer the wavelength, the less is the scattering (the denominator λ^4 becomes very large). It is this scattering that gives rise to the color of the sky. (See the chapter's first Highlight.)

About 15% of the insolation is absorbed directly by the atmosphere. Most of this absorption is accomplished by ozone, which removes the ultraviolet radiation, and by water vapor, which absorbs strongly in the infrared region of the spectrum. A major portion of the solar spectrum lies in the narrow visible region. The atmosphere is practically transparent to (absorbs little of) visible radiation, so most of the visible light reaches Earth's surface.

After reflection, scattering, and direct absorption, about 50% of the total incoming solar radiation reaches Earth's surface. This radiation goes into terrestrial surface heating, primarily through the absorption of visible radiation. As noted previously, the atmosphere—in particular, the troposphere—derives most of its energy content directly from Earth. This absorption of energy is accomplished in three main ways, listed here in order of decreasing contribution:

1. Absorption of terrestrial radiation
2. Latent heat of condensation
3. Conduction from Earth's surface

Absorption of Terrestrial Radiation

Earth, like any warm body, radiates energy that may be subsequently absorbed by the atmosphere. The wavelength of the radiation emitted by Earth depends on its temperature. From the wavelength relationship of the energy of a photon and the temperature of the emitting source, it can be shown that

$$\lambda \propto \frac{1}{T}$$

In other words, the wavelength of radiation emitted by a source is inversely proportional to its temperature.

Earth's temperature is such that it radiates energy primarily in the long-wavelength infrared region. Water vapor and carbon dioxide (CO_2) are the primary absorbers of infrared radiation in the atmosphere, with water vapor being the more important. We refer to these gases as being *selective absorbers* because they absorb certain wavelengths and transmit others.

This gives rise to the so-called **greenhouse effect,** about which you now hear a great deal. (See the chapter's second Highlight.)

Because the wavelength of the radiation is dependent on the temperature of the source, infrared photographs taken from satellites are used to study temperature variations in Earth's crust arising from such things as underground rivers and volcanic or geothermal activity (● Fig. 19.9).

Latent Heat of Condensation

Approximately 70% of Earth's surface is covered by water. Consequently, a great deal of evaporation occurs because of the insolation reaching the surface. You will recall from Chapter 5 that the latent heat of vaporization for water is 540 kcal/kg; that is, 540 kcal of heat energy is required to change 1 kg of water into the gaseous state. This is at the boiling temperature, but it is also a good approximation for evaporation.

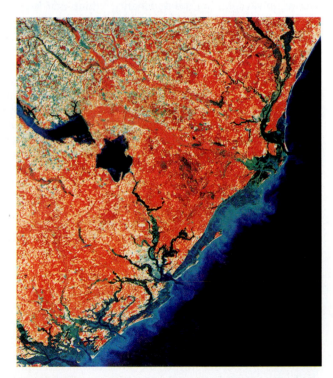

FIGURE 19.9   Infrared Analysis

An infrared photo taken from a satellite shows the variations in temperature of Earth's surface. Shown here is most of the coast of South Carolina, from Myrtle Beach (near upper right corner) south almost to Parris Island. Inland lakes and rivers are clearly visible.

HIGHLIGHT: Blue Skies and Red Sunsets

The gas molecules of the air account for most of the scattering in the visible region of the spectrum. In the visible spectrum the wavelength increases from violet to red and scattering is greater for shorter wavelengths. The violet end is therefore scattered more than the red end. (The colors of the visible spectrum—and the rainbow—may be remembered with the help of the name ROY G. BIV—red, orange, yellow, green, blue, indigo, and violet.)

As the sunlight passes through the atmosphere, the blue end of the spectrum is preferentially scattered. Some of this scattered light reaches Earth, where we see it as blue skylight (Fig. 1).

Keep in mind that all colors are present in skylight, but the dominant wavelength or color lies in the blue. You may have noticed that the skylight is more blue directly overhead or high in the sky and less blue toward the horizon, becoming white just above the horizon. You see these effects because there are fewer scatterers along a path through the atmosphere overhead than toward the horizon, and multiple scattering along the horizon path mixes the colors to give the white appearance. If Earth had no atmosphere, the sky would appear black, except in the vicinity of the Sun.

Because Rayleigh scattering is greater the shorter the wavelength, you might be wondering why the sky isn't violet, since this color has the shortest wavelength in the visible spectrum. Violet light is scattered, but the eye is more sensitive to blue light than to violet light; also, sunlight contains more blue light than violet light. The greatest color component is yellow-green, and the distribution generally decreases toward the ends of the spectrum.

The scattering of sunlight by the atmospheric gases *and* small particles gives rise to red sunsets. One might think that because sunlight travels a greater distance through the atmosphere to an observer at sunset, most of the shorter wavelengths would be scattered from the sunlight and only light in the red end of the spectrum would reach the observer. However, the dominant color of this light, were it due solely to molecular scattering, would be orange. Hence additional scattering by small particles in the atmosphere must shift the light from the setting (or rising) Sun toward the red. Foreign particles (natural or pollutants) in the atmosphere are not necessary to give a blue sky and may even detract from it. Yet they are necessary for deep red sunsets and sunrises.

The beauty of red sunrises and sunsets is often made more spectacular by layers of pink-colored clouds. The cloud color is due to the reflection of red light.

Larger particles of dust, smoke, haze, and those from air pollution in the atmosphere may preferentially scatter long wavelengths. These scattered wavelengths, along with the scattered blue light due to Rayleigh scattering, can cause the sky to have a milky blue appearance—white being the presence of all colors. Hence the blueness of the sky gives an indication of atmospheric purity. Cloud droplets and raindrops scatter even longer wavelengths. This fact is used in the principle of weather radar, which is an important means of weather monitoring, as we will learn in Chapter 20.

Thus, with condensation, a large amount of energy is transferred to the atmosphere in the form of latent heat. This energy is released during the formation of clouds, fog, rain, dew, and so on.

Conduction from Earth's Surface

A comparatively smaller, but significant, amount of heat energy is transferred to the atmosphere by conduction from Earth's surface. Because the air is a relatively poor conductor of heat, this process is restricted to the layer of air in direct contact with Earth's surface. The heated air is then transferred aloft by convection. As a result, the temperature of the air tends to be greater near the surface of Earth and decreases gradually with altitude.

19.3 Atmospheric Measurements and Observations

LEARNING GOALS

▼ Identify some important atmospheric measurements and the instruments used to make them.

▼ Demonstrate how the relative humidity may be found from psychrometric readings.

▼ Distinguish between conventional and Doppler radar.

Measurements of the atmosphere's properties and characteristics are important in its study. These properties are measured daily, and records have been

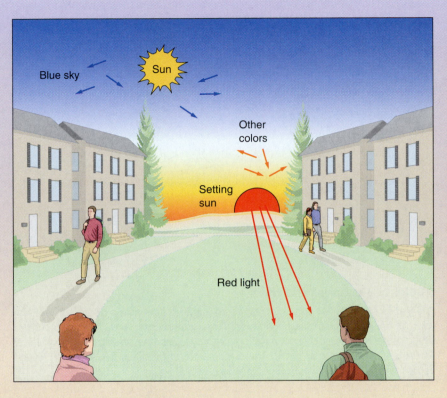

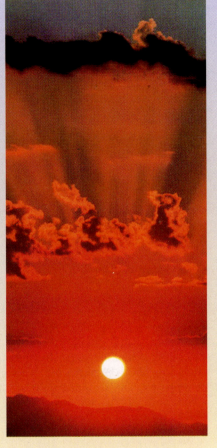

FIGURE 1 Rayleigh Scattering

(*Above*) The preferential scattering by air molecules causes the sky to be blue, which, in turn, together with the scattering caused by small particles, produces red sunsets. See text for description. (*Right*) Scattering in action. Blue sky and red sunset.

compiled over many years. Meteorologists study such records with the hope of observing cycles and trends in atmospheric behavior to better understand and predict its changes.

We observe the daily atmospheric readings to obtain a qualitative picture of the conditions for that day in the region and around the country. Fundamental atmospheric measurements include (1) *temperature,* (2) *pressure,* (3) *humidity,* (4) *wind speed and direction,* and (5) *precipitation.*

Temperature

Having discussed temperature measurements in Chapter 5, we will not dwell on this property. Keep in mind that it is the *air temperature* that is being mea-

sured. Heat transfer to a thermometer by radiation (sunlight) may give rise to a higher temperature. A truer air temperature in the summer is often expressed as being so many degrees "in the shade," implying that when making a temperature measurement, the thermometer should not be exposed to the direct rays of the Sun.

Pressure

Pressure is defined as the force per unit area ($p = F/A$). At the bottom of the atmosphere, we experience the resulting weight of the gases above us. Because we experience this weight before and after birth as a part of our natural environment, little thought is given to the

HIGHLIGHT: The Greenhouse Effect

The absorption of terrestrial radiation, primarily by water vapor and CO_2, adds to the energy content of the atmosphere. This heat-retaining process of such gases is referred to as the *greenhouse effect*, because of a similar effect that occurs in greenhouses. The absorption and transmission properties of regular glass are similar to those of the atmospheric gases—in general, visible radiation is transmitted, and infrared radiation is absorbed (Fig. 1).

We have all observed the warming effect of sunlight passing through glass—for example, in a closed car on a sunny, but cold, day. In a greenhouse, the objects inside become warm and reradiate long-wavelength infrared radiation, which is absorbed and reradiated by the glass. Thus the air inside a greenhouse heats up and is quite warm on a sunny day, even in the winter.

Actually, in this case the maintained warmth is primarily due to the glass enclosure, which prevents the escape of warm air. The temperature of the greenhouse in the summer is controlled by painting the glass panels white so as to reflect the sunlight and

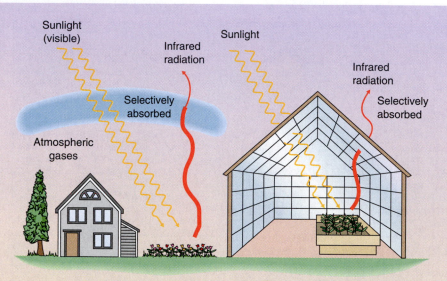

FIGURE 1 The Greenhouse Effect
The gases of the lower atmosphere transmit most of the visible portion of the sunlight, as does the glass of a greenhouse. The warmed Earth emits infrared radiation, which is selectively absorbed by atmospheric gases, the absorption spectrum of which is similar to that of glass. The absorbed energy heats the atmosphere and helps maintain Earth's average temperature.

by opening windows to allow the hot air to escape.

The greenhouse effect in the atmosphere is quite noticeable at night, particularly on cloudy nights. With a cloud and water vapor cover to absorb the terrestrial radiation, the night air is relatively warm. Without this insulating

---

fact that every square inch of our bodies sustains an average weight of 14.7 lb at sea level; that is, a pressure of 14.7 lb/in^2. We refer to 14.7 lb/in^2 as being *one standard atmosphere of pressure*.

One of the first investigations of atmospheric pressure was initiated by Galileo. In attempting to pipe water to elevated heights by evacuating air from a tube, he found that it was impossible to sustain a column of water taller than about 10 m (33 ft). Evangelista Torricelli (1608–1647), who was Galileo's successor as professor of mathematics in Florence, pointed out the difficulty through the invention of a device that showed the height of a liquid column in a tube to be dependent on the atmospheric pressure.

A glass tube filled with mercury (Hg) was inverted into a pool of mercury. Although some mercury ran out of the tube, a column of mercury 76 cm (30 in.) high was left in the tube, as illustrated in ● Fig. 19.10 on page 512. Such a device, called a **barometer** (from the Greek *baros,* "weight"), is still used to measure atmospheric pressure.

A modern version of a mercury barometer is shown in ● Fig. 19.11 on page 512. Because the column of mercury has weight, some force must hold up the column. The only available force is that of the atmospheric pressure on the surface of the mercury pool. So the greater the height (h), the greater is the pressure (p), or $p \propto h$. The equation relationship is $p = \rho g h$,

effect, the night is usually "cold and clear" because the energy from the daytime insolation is quickly lost.

Despite the daily and seasonal gain and loss of heat, the *average* temperature of Earth has remained fairly constant. Thus Earth must lose or reradiate as much energy as it receives. If it did not, the continual gain of energy would cause Earth's average temperature to rise. The selective absorption of atmospheric gases provides a thermostatic, or heat-regulating, process for the planet.

To illustrate, suppose that Earth's temperature were such that it emitted radiation with wavelengths that were absorbed by the atmospheric gases. As a result of this absorption, the lower atmosphere would become warmer and effectively hold in the heat, thus insulating Earth. With additional insolation, Earth would become warmer and its temperature would rise.

However, according to the previously mentioned wavelength-temperature relationship, the greater the temperature, the shorter the wavelength of the emitted radiation, and so the wavelength of the terrestrial radiation would be shifted to a shorter wavelength.

The wavelength would eventually be shifted to a "window" in the absorption spectrum where little or no absorption would take place and the terrestrial radiation would pass through the atmosphere into space. Thus Earth would lose energy, and its temperature would decrease. But with a temperature decrease, the terrestrial radiation would return to a longer wavelength, which would be absorbed by the atmosphere. Thus we have a turning on and off, so to speak, similar to the action of a thermostat.

Averaged over that total spectrum, the selective absorption of atmospheric gases plays an important role in maintaining Earth's average temperature. Recall from Chapter 15 that a relatively large amount of atmospheric CO_2 and the resulting greenhouse effect keep the surface of Venus very hot (480°C).

An interesting sidelight to the greenhouse effect concerns the transmission of radiation through glass. In an actual glass greenhouse, the visible portion of the insolation is transmitted through the glass, but the short-wavelength ultraviolet transmission depends on the iron content of the glass. Ordinary window glass does not efficiently transmit wavelengths below 310 nm, which is in the ultraviolet region.

Hence, although a considerable warming effect occurs through glass from visible radiation, one cannot receive much suntan or sunburn through ordinary window glass. (Ultraviolet radiation near 300 nm and below causes the skin to burn or tan. We hear more and more about protecting our skin and eyes from harmful ultraviolet rays these days.)

The effect of human activities on the greenhouse effect will be discussed in the following chapter.

---

where ρ is the density of the liquid and g is the acceleration due to gravity. Thus, the less dense the liquid, the greater the height of the column for a given pressure.

Mercury, with a relatively large density of 13.6 g/cm^3, gives a smaller and more manageable column than, say, water, with a density of 1.00 g/cm^3. In fact, a water column would be 13.6 times taller than a mercury column for a given pressure. As was noted, one atmosphere supports a mercury column 76 cm tall, so a barometer with a water column would have a column of 76 cm × 13.6 = 1034 cm, which is a little over 10 m. Hence the atmosphere does not have sufficient pressure to support a column of water much over 10 m, just as Galileo found.

The standard units of pressure ($p = F/A$) are N/m^2 or lb/in^2 in the SI and British system, respectively. However, these units are not used for the barometric readings given on radio and TV weather reports. Instead, the readings are expressed in length units, that is, so many "inches" (of mercury). For weather phenomena, we are primarily interested in changes of pressure, and since $p \propto h$, the variation in column heights will give this directly.

Because of Torricelli's barometric work, a pressure unit has been named after him. A height of one millimeter of mercury is called a *torr* (1 mm Hg = 1 torr), so one atmosphere of pressure is 76 cm = 760 mm = 760 torr. Summarizing the atmospheric pressure units,

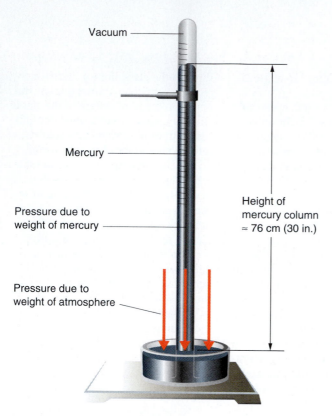

Vacuum

Mercury

Pressure due to
weight of mercury

Pressure due to
weight of atmosphere

Height of
mercury column
≈ 76 cm (30 in.)

FIGURE 19.10   Mercury Barometer

An illustration of the principle of the barometer. The external
air pressure, due to the weight of the atmosphere on the surface of the pool of mercury, supports the mercury column in
the inverted tube. The height of the column depends on the
atmospheric pressure and provides a means of measuring it.
Normally, the column will be about 76 cm (30 in.) in height.
(From Stoker, H. Stephen, *General, Organic, and Biological
Chemistry.* Copyright © 1998 by Houghton Mifflin Company.
Used with permission.)

$$\begin{aligned}
1 \text{ atmosphere (atm)} &= 76 \text{ cm Hg} = 760 \text{ mm Hg} \\
&= 760 \text{ torr} \\
&= 30 \text{ in. Hg} \\
&= 14.7 \text{ lb/in}^2
\end{aligned}$$

In the metric system, one atmosphere has a pressure
of 1.013×10^5 N/m^2, or about 10^5 N/m^2. Meteorologists use yet another unit called the *millibar* (mb). A
bar is defined to be 10^5 N/m^2, and since 1 bar = 1000
mb, we have 1 atm ≈ 1 bar = 1000 mb. With a thousand units, finer changes in atmospheric pressure can
be measured.

FIGURE 19.11   Mercury Barometer

A typical mercury barometer as used in the laboratory for
measuring atmospheric pressure. The mercury pool is seen
at the bottom, and the height of the column is measured
using the scale on the mounting board.

Because mercury vapor is toxic and a column of
mercury 30 inches tall is awkward to handle, another
type of barometer, called the *aneroid* ("without fluid")
barometer, is commonly used. This is a mechanical

device having a metal diaphragm that is sensitive to pressure, much like a drum head. A pointer is used to indicate the pressure changes on the diaphragm (● Fig. 19.12).

Aneroid barometers with dial faces are common around the home and are usually inlaid in wood, along with dial thermometers, and used as decorative wall displays.

Atmospheric pressure quickly becomes evident to us when sudden pressure changes occur. A relatively small change in altitude will cause our ears to "pop," because the pressure in the inner ear does not equalize quickly, which puts pressure on the eardrum. When the pressure equalizes (swallowing helps), the ears pop.

Airplanes are equipped with pressurized cabins that maintain the normal atmospheric pressure on the passengers' bodies. The internal pressure of the body is accustomed to an external pressure of 14.7 lb/in². Should this pressure be reduced, the excess internal pressure may be evidenced in the form of a nosebleed. Also, altitude and pressure have an effect on the boiling points of liquids, as discussed earlier, in Chapter 5.

Humidity

Humidity is a measure of the moisture, or water vapor, in the air. It affects our comfort and indirectly our ambition and state of mind. In the summer, many homes use dehumidifiers to remove moisture from the air. In the winter, humidifiers or exposed pans of water strategically placed allow water to evaporate into the air.

Humidity can be expressed in several ways. *Absolute humidity* is simply the amount of water vapor in a given volume of air. In the United States, it is commonly measured in grains per cubic foot. The grain (gr) is a small weight unit, with 1 lb = 7000 gr. Medicines are sometimes measured in grains—for example, a 5-gr aspirin tablet. An average value of humidity is around 4.5 gr/ft³.

The most common method of expressing the water vapor content of the air is in terms of relative humidity. **Relative humidity** *is the ratio of the actual moisture content and the maximum moisture capacity of a volume of air at a given temperature. This ratio is commonly expressed as a percentage:*

$$(\%) \; RH = \frac{AC}{MC} \; (\times \; 100\%) \qquad \textbf{19.1}$$

where *RH* is the relative humidity, *AC* is the actual moisture content of the air, and *MC* is the maximum moisture capacity of the air. The actual moisture content is just the absolute humidity, or the amount of water vapor in a given volume of air. The maximum moisture capacity is the maximum amount of water vapor that the volume of air can hold *at a given temperature.*

Relative humidity is essentially a measure of how "full" of moisture a volume of air is at a given temperature. For example, if the relative humidity is 0.50, or 50%, then a volume of air is "half full," or contains half as much water as it is capable of holding at that temperature.

To better understand how the water vapor content of air varies with temperature, consider an analogy with a saltwater solution. Just as a given amount of water at a certain temperature can dissolve only so much salt, a volume of air at a given temperature can

FIGURE 19.12   Aneroid Barometer

Atmospheric pressure changes on a sensitive metal diaphragm are reflected on the dial face of the barometer. For reasons discussed in Section 20.2, fair weather is generally associated with high barometric pressure, and rainy weather with low barometric pressure. (What is the barometric pressure reading on the barometer in the photo?)

only hold so much water vapor. When the maximum amount of salt is dissolved in solution, we say the solution is *saturated* (Section 11.1). This condition is analogous to a volume of air having its maximum moisture capacity.

The addition of more salt to a saturated solution results in salt on the bottom of the container. However, more salt may be put into solution if the water is heated and the temperature raised. Similarly, when air is heated, it can hold more water vapor at a higher temperature; that is, warm air has a greater capacity for water vapor than does cold air.

Conversely, if the temperature of a nearly saturated salt solution is lowered, the solution will become saturated at a certain temperature. Any additional lowering of the temperature will cause the salt to crystallize and come out of solution, because otherwise the solution would be oversaturated. Analogously, if the temperature of a sample of air is lowered, it will become saturated at a certain temperature.

The temperature to which a sample of air must be cooled to become saturated is called the **dew point** (temperature). Hence, at the dew point, the relative humidity is 100%. (Why?) Cooling below this point causes oversaturation and may result in condensation and loss of moisture in the form of precipitation.

Humidity may be measured by several means. The most common method uses the **psychrometer.** This instrument consists of two thermometers, one of which measures the air temperature, while the other has its bulb surrounded by a cloth wick that keeps it wet. These thermometers are referred to as the *dry bulb* and the *wet bulb,* respectively. They may be simply mounted, as shown in ● Fig. 19.13.

The dry bulb measures the air temperature, while the wet bulb has a lower reading that is a function of the amount of moisture in the air. This reading occurs because of the evaporation of water from the wick around the wet bulb. The evaporation removes latent heat from the thermometer bulb. If the humidity is high and the air contains a lot of water vapor, little water is evaporated and the wet bulb is only slightly cooled. Consequently, the temperature of the wet bulb is only slightly lower than the temperature of the dry bulb; that is, the wet-bulb reading is only slightly "depressed."

If, however, the humidity is low, a great deal of evaporation will occur, accompanied by considerable cooling, and the wet-bulb reading will be considerably depressed. Hence the temperature difference of

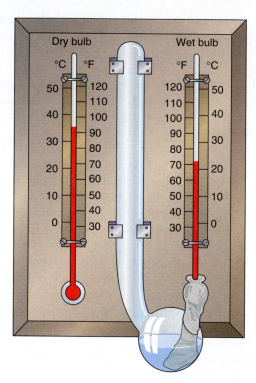

FIGURE 19.13 Psychrometer and Relative Humidity
The dry bulb of a psychrometer records the air temperature, which is greater than that of the wet bulb (because of evaporation). The lower the humidity, the greater is the evaporation and the greater is the difference in the two temperature readings. The temperature difference between the thermometers is then inversely proportional to the humidity. Thus the psychrometer provides a means for measuring relative humidity.

the thermometers, or depression of the wet-bulb reading, is a measure of relative humidity.

Using the air (dry-bulb) temperature and the wet-bulb depression, the relative humidity, the maximum moisture capacity, and the dew point can be read directly from the tables in Appendix VIII. An example of how this is done follows.

EXAMPLE 19.1

Using Psychrometric Tables

The dry bulb and wet bulb of a psychrometer have respective readings of 80°F and 73°F. Find the following for the air: (a) relative humidity, (b) maximum moisture capacity, (c) actual moisture content, and (d) dew point.

SOLUTION

Given:

80°F (dry-bulb reading)
73°F (wet-bulb reading)

Find:

(a) *RH* (relative humidity)

(b) *MC* (maximum capacity)

(c) *AC* (actual content)

(d) dew point (temperature)

First, we find the wet-bulb depression—that is, the difference between the dry-bulb and wet-bulb readings:

$$\Delta T = 80°F - 73°F = 7\ F°$$

(a) Then, using Table VIII.1 in Appendix VIII to determine the relative humidity, we find the dry-bulb temperature in the first column and then locate the wet-bulb depression in the top row of the table. Move down the column under the wet-bulb depression to the row that corresponds to the dry-bulb temperature reading. The intersection of the row and column gives the value of the relative humidity. (It may be helpful to move one finger down and another one across to help find the intersection number.) The relative humidity is 72% for this depression.

(b) The maximum moisture content (*MC*) is read directly from the table. Find the dry-bulb temperature in the first column; the *MC* for that temperature is given in the adjacent column. In this case, $MC = 10.9\ gr/ft^3$.

(c) Knowing the *RH* and *MC*, we can find the actual moisture content from Eq. 19.1, $RH = AC/MC$. Rearranging,

$$AC = RH \times MC = 0.72 \times 10.9\ gr/ft^3 = 7.8\ gr/ft^3$$

Note that the relative humidity is used in decimal form (not percent).

(d) The dew point is found using Table VIII.2 in the same manner as the relative humidity was found in part (a). The intersection value of the appropriate row and column is 70°F.

Hence, if the air is cooled to 70°F, it will be saturated and the relative humidity will be 100%, with $AC = MC$. Note from Table VIII.1 that the *MC* for 70°F is $7.8\ gr/ft^3$, which is the value found in part (c).

The value of the actual content at 80°F should correspond to the maximum capacity at the dew point (70°F).

CONFIDENCE EXERCISE 19.1

On a particular day, the dry-bulb and wet-bulb readings of a psychrometer are 70°F and 66°F, respectively. (a) What is the relative humidity, and (b) how many degrees would the air temperature have to be lowered for precipitation to be likely?

Wind Speed and Direction

Wind speed is measured with an **anemometer.** This instrument consists of three or four cups attached to a rod that is free to rotate, much like a pinwheel. The cups catch the wind, and the greater the wind speed, the faster the anemometer rotates (● Fig. 19.14).

A **wind vane** indicates the direction from which the wind is blowing. This instrument is simply a free-rotating indicator that, because of its shape, lines up

FIGURE 19.14   Anemometer and Wind Vanes

An array of two anemometers and two wind vanes is shown. The anemometer cups catch the wind to measure its speed. The shapes of the wind vanes cause them to point in the direction from which the wind is blowing.

with the wind and points the wind direction (Fig. 19.14). Wind direction is reported as the *direction from which the wind is coming.*

Precipitation

The major forms of precipitation are rain and snow. Rainfall is measured by a **rain gauge.** This device simply may be an open container with vertical sides marked in inches and placed outside. After a rainfall, the rain gauge is read, and the amount of precipitation is reported as so many inches.* The assumption is that this much rainfall is distributed relatively evenly over the surrounding area.

If precipitation is in the form of snow, the depth of the snow (where not drifted) is reported in inches. The actual amount of water received depends on the density of the snow. To obtain this reading, a rain gauge is sprayed with a chemical that melts the snow, and the actual amount of water is recorded. More elaborate rain-measuring instruments automatically measure and record rainfall and snowfall.

Weather Observations

Technology has extended our means of weather observations well beyond our direct visual capabilities. We will consider two of these: radar and satellites.

Radar (*ra*dio *d*etecting *a*nd *r*anging) is used to detect and monitor precipitation, especially that of severe storms. It operates on the principle of reflected electromagnetic waves. Radar installations are located mainly in the tornado belt of the Midwestern states and along the Atlantic and Gulf coasts, where hurricanes are probable. Additional radar information is obtained from air traffic control systems at many airports.

A more advanced radar system, called **Doppler radar,** has been developed, and a network of these new radars has been deployed. Like conventional radar, Doppler radar measures the distribution and intensity of precipitation over a broad area. However, Doppler radar also has the ability to measure wind speeds. It is based on the Doppler effect (Chapter 6), the same principle used in police radar to measure the speeds of automobiles. Doppler radar scans are commonly seen on TV weather reports.

Radar waves are reflected from raindrops in storms. The direction of a storm's wind-driven rain, and hence a wind "field" of the storm region, can be mapped. This map provides strong clues, or signatures, of developing tornadoes. Conventional radar can detect the hooked signature of a tornado only after the storm is well developed (Fig. 19.15).

Doppler radar can penetrate a storm and monitor its wind speeds. With a wind-field map, a developing tornado signature can be detected much earlier. Using Doppler radar, forecasters are able to predict tornadoes 20 min before they touch down, as compared with just over 2 min for conventional radar. Doppler radar will save many lives with this increased advance warning time.

Doppler radars are being installed at major airports for another use. Several plane crashes and near crashes have been attributed to dangerous downward wind bursts known as *wind shear*. These wind bursts generally result from high-speed downdrafts in the turbulence of thunderstorms but can occur in clear air when rain evaporates high above the ground. The downdraft spreads out when it hits the ground and forms an inward circular pattern. A plane entering the pattern experiences an unexpected upward headwind that lifts the plane. The pilot often cuts speed and lowers the plane's nose to compensate.

Further into the circular pattern, the wind quickly turns downward, and an airplane can suddenly lose altitude and possibly crash when near the ground as on landing. Since Doppler radar can detect the wind speed and the direction of raindrops in clouds, as well as the motions of dust and other objects floating in the air, it can provide an early warning of wind shear conditions.

Probably the greatest progress in general weather observation came with the advent of the weather satellite. Before satellites, weather observations were unavailable for more than 80% of the globe. The first weather picture was sent back from space in 1961. The first fully operational weather satellite system was in place by 1966. These early pole-orbiting (traveling from pole to pole) satellites, at altitudes of several hundred miles, monitored only a limited area below their orbital paths. It took almost three orbits to photograph the entire conterminous United States.

Today, a fleet of GOESs (*G*eostationary *O*rbiting *E*nvironmental *S*atellites), which orbit at fixed points, provides an almost continuous picture of weather patterns all over the globe. At an altitude of about 36,800 km (23,000 mi), the GOES orbiters have the

---

*In the United States, meteorologists commonly report precipitation in inches. Only two other countries, Brunei and the Union of Myanmar (formerly known as Burma), use this system; all others report precipitation in centimeters.

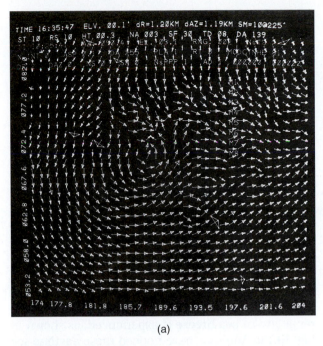

(a)

(b)

FIGURE 19.15   Radar

(a) A Doppler radar display showing wind-field patterns. Note the circular rotation. (b) A conventional radar scan showing the characteristic hooked signature of a tornado.

same orbital period as Earth, and hence are "stationary" over a particular location.

At this altitude, the GOESs can send back pictures of large portions of Earth's surface. Geographic boundaries and grids are prepared by computer and electronically combined with the picture signal so that the areas of particular weather disturbances can be easily identified (● Fig. 19.16).

With satellite photographs, meteorologists have a panoramic view of weather conditions. The dominant feature of such photographs is, of course, the cloud coverage. However, with the aid of radar, which uses wavelengths that pick up only precipitation, the storm areas are easily differentiated from regular cloud coverage.

RELEVANCE QUESTION: *Which of the atmospheric measurements discussed in this section do you commonly see on TV weather reports?*

FIGURE 19.16   Satellite Image

A GOES weather picture for the United States. A hurricane is clearly visible off the North Carolina coast, producing heavy rains to the south. A band of frontal clouds stretches to the north of the hurricane, and a few clouds appear over the Rockies. Most of the country is clear, with the exception of a band of frontal clouds over the central part of the country, producing cold rain.

19.4 Air Motion

If the air in the troposphere were static, there would be little change in the local atmospheric conditions that constitute weather. Air motion is important in many processes, even biological ones, such as carrying scents and distributing pollen. As air moves into a region, it brings with it the temperature and humidity that are mementos of its travels.

Wind is the horizontal movement of air, or air motion along Earth's surface. Vertical air motions are referred to as updrafts and downdrafts, or collectively as **air currents.** Winds and air currents require the air to be in motion. But what causes the air to move? As in all dynamic situations, forces are necessary to produce motion and changes in motion. The gases of the atmosphere are subject to two primary forces: (1) gravity and (2) pressure differences due to temperature variations.

The force of gravity is vertically downward and acts on each gas molecule. Although this force is often overruled by forces in other directions, the downward gravity component is ever present and accounts for the greater density of air near Earth's surface.

Because the air is a mixture of gases, its behavior is governed by the gas laws discussed in Chapter 5 and by other physical principles. The pressure of a gas is directly proportional to its Kelvin temperature ($p \propto T$), so if there is temperature variation, there will be a pressure difference ($\Delta p \propto \Delta T$). Recall that pressure is the force per unit area, so a pressure difference corresponds to an unbalanced force. When there is a pressure difference, the air moves from a high- to a low-pressure region.

The pressures of a region may be mapped by taking barometric readings at different locations. A line drawn through the locations (points) of equal pressure is called an **isobar.** Because all points on an isobar are of equal pressure, there will be no air movement along an isobar. The wind direction will be at right angles to the isobar in the direction of the low-pressure region, as illustrated in ● Fig. 19.17. (This is

an idealized situation. Other forces, which will be discussed shortly, may cause deflections.)

Recall from Chapter 5 that the pressure *and* volume of a gas are directly proportional to its Kelvin temperature ($pV \propto T$). A change in temperature, then, causes a change in the pressure and/or volume of a gas. With a change in volume, there is also a change in density ($\rho = m/V$). For example, if the air is heated and expands, the air density becomes less. As a result of this relationship, localized heating sets up air motion in a **convection cycle,** which gives rise to *thermal circulation* (● Fig. 19.18).

Thermal circulations due to geologic features give rise to local winds. Land areas heat up more quickly during the day than do water areas, and the warm, buoyant (less dense) air over the land rises. As air flows horizontally into this region, the rising air cools and falls, and a convection cycle is set up. As a result, during the day when the land is warmer than the water, a lake or **sea breeze** is experienced, as shown in Fig. 19.18a. You may have noticed these daytime sea breezes at an ocean beach. Notice that *a wind is named after the direction from which it comes.* For example, a wind blowing in from the sea is called a *sea breeze.* Similarly, a wind blowing *from* north *to* south is called a *north wind.*

At night, the land loses its heat more quickly than the water does, and so the air over the water is warmer. The convection cycle is then reversed, and at night a

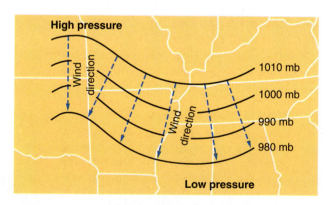

FIGURE 19.17　Isobars

Isobars are lines drawn through locations having equal atmospheric pressures. In the absence of other forces, the air motion, or wind direction, is perpendicular to the isobars from a region of greater pressure (greater mb values) to a region of lower pressure (lower mb values).

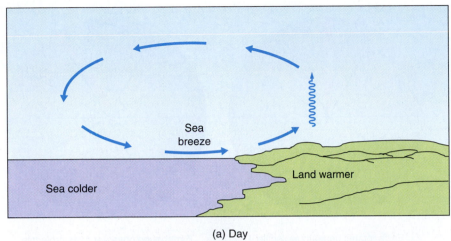

FIGURE 19.18   Daily Convection Cycles over Land and Water

(a) During the day, the land surface heats up more quickly than a large body of water. This sets up a convection cycle in which the surface winds are from the water—a sea breeze. (b) At night, the land cools more quickly than the water. The convection cycle is then reversed, with surface winds coming from the land—a land breeze.

Sea breeze

Sea colder

Land warmer

(a) Day

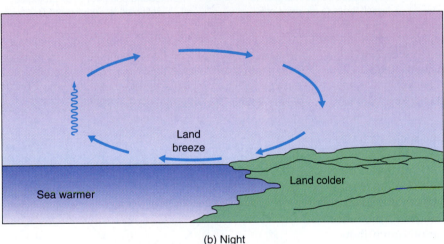

Land breeze

Sea warmer

Land colder

(b) Night

land breeze blows (Fig. 19.18b). Sea and land breezes are sometimes referred to as *onshore* and *offshore winds,* respectively.

Once the air has been set into motion, velocity-dependent forces act. These secondary forces are (1) the Coriolis force and (2) friction.

The **Coriolis force,** named after the French engineer who first described it, results because an observer on Earth is in a rotating frame of reference. This force is sometimes referred to as a *pseudoforce,* or false force, because it is introduced to account for the effect of Earth's rotation.

Humans tend to be egocentric. We commonly consider ourselves to be motionless, although we are on a rotating Earth that has a surface speed of about 1600 km/h (1000 mi/h) near the equator. Rotating with Earth, we are in an accelerating reference frame. (How

is it accelerating?) Newton's laws of motion apply to nonaccelerating reference frames and may be used for ordinary motions on Earth without correction. However, for high speeds or huge masses, such as the atmospheric gases, the correction for Earth's rotation becomes important.

To help understand this effect, imagine a high-speed projectile being fired from the North Pole southward along a meridian (● Fig. 19.19). While the projectile travels southward, Earth rotates beneath it, and hence it lands to the west of the original meridian. But to an observer at the North Pole looking southward along the meridian, it appears that the projectile is deflected to the right. By Newton's laws, this deflection requires a force, and we call it the Coriolis force, even though no such "force" really exists. Hence the Coriolis force is a pseudoforce that is

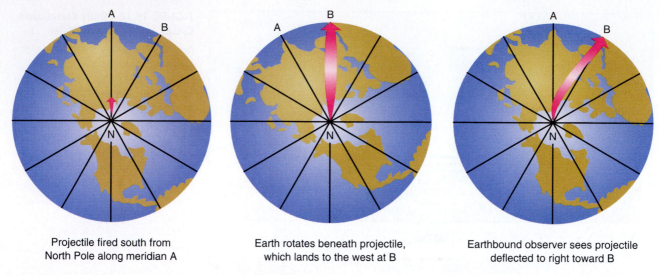

Projectile fired south from
North Pole along meridian A

Earth rotates beneath projectile,
which lands to the west at B

Earthbound observer sees projectile
deflected to right toward B

FIGURE 19.19 The Coriolis Force

(*Left*) Imagine someone firing a projectile from the North Pole toward position A. (*Center*) Earth turns while the projectile is in flight, and the projectile lands to the right of A, at location B. This is the situation you would see if viewing Earth from space over the North Pole. (*Right*) For an observer on Earth, if the projectile lands at B, it must have been deflected by some force, as required by Newton's laws. The Coriolis force was invented to account for this apparent deflection.

invented so that the effect will be consistent with the laws of motion.

In general, projectiles or particles moving in the Northern Hemisphere are apparently deflected to the right; and by similar reasoning, objects in the Southern Hemisphere appear to be deflected to the left. Hence we say that, because of the Coriolis force, moving objects are deflected to the right in the Northern Hemisphere and to the left in the Southern Hemisphere, as observed in the direction of motion. The Coriolis force is at a right angle to the direction of motion of an object, and its magnitude varies with latitude; it is zero at the equator and increases toward the poles.

Consider this effect on wind motion. Initially, air moves toward a low-pressure region (a "low") and away from a high-pressure region (a "high"). Because of the Coriolis force, the wind is deflected, and in the Northern Hemisphere the wind tends to rotate counterclockwise around a low and clockwise around a high, as viewed from above (● Fig. 19.20). These disturbances are referred to as *cyclones* and *anticyclones,* respectively. Water motion or currents in the oceans are also affected by the Coriolis force.

Friction, or **drag,** also can cause the retardation or deflection of air movements. Moving air molecules experience frictional interactions (collisions) among

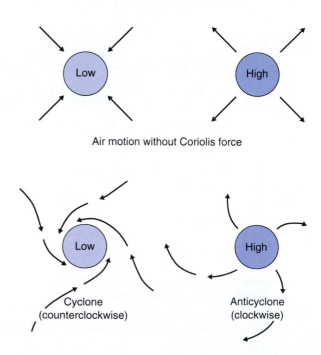

Air motion without Coriolis force

Cyclone
(counterclockwise)

Anticyclone
(clockwise)

FIGURE 19.20 Effects of the Coriolis Force on Air Motion

In the Northern Hemisphere, the Coriolis deflection to the right produces counterclockwise air motion around a low and clockwise rotation around a high (as viewed from above).

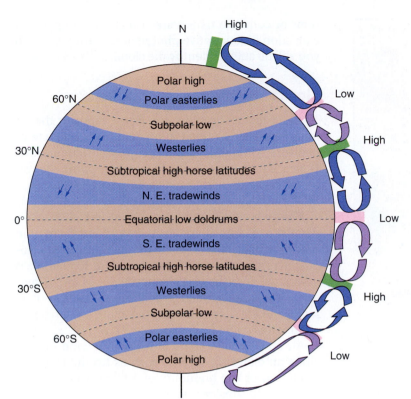

FIGURE 19.21   Earth's General Circulation Structure

For rather complicated reasons, Earth's general circulation pattern has six large convection cycles or cells. The conterminous United States lies in the Westerlies wind zone.

themselves and with terrestrial surfaces. The opposing frictional force along a surface is in the opposite direction of the air motion. Thus winds moving into a cyclonic disturbance may be deflected differently because of the sum of the forces.

Air motion changes locally with altitude, geographic features, and the seasons. However, the air near Earth's surface does possess a general circulation pattern. Because of the Coriolis force, land and sea variations, and other complicated factors, the hemispheric circulation is broken up into six general convection cycles, or pressure cells. Earth's general circulation structure is shown in ● Fig. 19.21.

Many local variations occur within the cells, which shift seasonally in latitude because of variations in insolation. However, the prevailing winds of this semipermanent circulation structure are important in influencing general weather movement around the world.

The conterminous United States lies generally between the latitudes of 30°N and 50°N.* Note in Fig. 19.21 that this is in the Westerlies wind zone. As a result, our weather conditions generally move from west to east across the country.

You commonly hear about another high-altitude wind on TV weather reports, particularly in the winter. In the upper troposphere, there are fast-moving "rivers" of air called **jet streams.** They were first noted in the 1930s but did not receive much attention until World War II, when high-flying aircraft encountered jet streams.

Several jet streams meander like rivers around each hemisphere. The behavior of jet streams is variable and not well understood. The so-called polar jet stream moves from west to east across the United States (● Fig. 19.22). It varies in altitude and latitude with the seasons, reaching lower latitudes in the winter. This jetstream is believed to have an influencing effect on the severity of our winters.

RELEVANCE QUESTION: *From what direction do changes in your weather generally come, and why?*

Conterminous means contiguous or adjoining, so the conterminous United States refers to the original 48 states. This is different from the continental United States. Why?

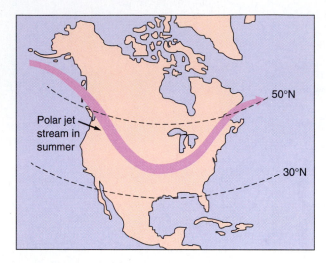

FIGURE 19.22 Jet Stream
A typical jet-stream pattern across the United States.

19.5 Clouds

LEARNING GOALS

▼ Explain how clouds are described and classified.

▼ Describe how clouds are formed.

Cloud Classification

Clouds are both a common sight and an important atmospheric consideration. **Clouds** are buoyant masses of visible water droplets or ice crystals. The size, shape, and behavior of clouds are useful keys to the weather.

Clouds are classified according to their *shape, appearance,* and *altitude*. There are four basic root names: *cirrus,* meaning "curl" and referring to wispy, fibrous forms; *cumulus,* meaning "heap" and referring to billowy, round forms; *stratus,* meaning "layer" and referring to stratified or layered forms; and *nimbus,* referring to a cloud from which precipitation is occurring or threatens to occur. These root forms are then combined to describe the types of clouds and precipitation potential.

When classified according to height, clouds are separated into four families: (1) *high clouds,* (2) *middle clouds,* (3) *low clouds,* and (4) *clouds of vertical development.* These families are listed in the Spotlight on page 523, along with the approximate heights and cloud types belonging to each family. Brief descrip-

tions or common names are also given, along with a collection of illustrative photographs. Study these so you will be able to identify the clouds.

Cloud Formation

To be visible as droplets, the water vapor in the air must condense. Condensation requires a certain temperature—namely, the dew point temperature. Hence, if moist air is cooled to the dew point, the water vapor contained therein will generally condense into fine droplets and form a cloud.

The air is continuously in motion, and when an air mass moves into a cooler region, cloud formation may take place. Because the temperature of the troposphere decreases with height, cloud formation is associated with the vertical movement of air. In general, clouds are formed in vertical air motion (currents) and are shaped and moved about by horizontal air motions (winds).

Air may rise as a result of heating or wind motion along an elevating surface, such as up the side of a mountain or the boundary of a front (Chapter 20). Here we will consider cloud formation resulting from the rising of warm, buoyant air (clouds of vertical development). As the warm air ascends, it becomes cooler, because it expands and because its internal energy is used to do the work of expansion against the surrounding stationary air. With less energy, it becomes cooler.

The temperature of the air in the troposphere decreases with altitude, and the rate of this temperature decrease with height is called the **lapse rate.** The normal lapse rate in stationary air in the troposphere is about $6\frac{1}{2}$ C°/km (or $3\frac{1}{2}$ F°/1000 ft).

Because energy is used in the expansion of a warm air mass, the rising air column has a greater lapse rate than does the surrounding air. Thus rising air cools more quickly. When the rising air mass cools to the same temperature as the surrounding stationary air, their densities become equal. The rising air mass then loses its buoyancy and is said to be in a *stable condition*. A heated air mass rises until stability is reached, and this portion of the atmosphere is referred to as a *stable layer*—that is, a layer of air of uniform temperature and density.

Clouds are formed when water vapor in the rising air condenses into droplets and can be seen. If the rising air reaches its dew point before becoming stable, condensation occurs at that height, and the rising air

A SPOTLIGHT ON: Cloud Families and Types

High Clouds (above 6 km). All composed of ice crystals.

Cirrus: Wispy and curling. Known as "artist's brushes" or "mare's tails" (a).

Cirrocumulus: Layered patches. Known as "mackerel scales" (a).

Cirrostratus: Thin veil of ice crystals. Scattering from crystals gives rise to solar and lunar halos (b).

Middle Clouds (1.8–6 km). All names have the prefix *alto*.

Altostratus: Layered forms of varying thickness. May hide the Sun or Moon and cast a shadow (c).

Altocumulus: Woolly patches or rolled, flattened layers (d).

(a) Cirrus and cirrocumulus clouds. Artist's-brush cirrus are to the left, and mackerel-scale cirrocumulus are to the right.

(b) Cirrostratus clouds. The clouds cover the sky and are evidenced by a lunar halo.

(c) Altostratus clouds. Thick, gray altostratus clouds hide the Sun.

(d) Altocumulus clouds. These clouds are often rolled and arranged in flattened layers by moving air.

Low Clouds (ground level–1.8 km).

Stratus: Thin layers of water droplets. May appear dark; common in winter. Fog may be thought of as low-lying stratus clouds (e).

Stratocumulus: Long layers of cottonlike masses, sometimes with a wavy appearance (f).

Nimbostratus: Dark, low clouds given to precipitation (g).

Advection fog: Forms when moist air moving over a colder surface is cooled below dew point. Advection fogs "roll in" (h).

Radiation fog: Condensation in stationary air overlying a surface that cools. Typically occurs in valleys, and is called a "valley fog" (i).

Clouds with Vertical Development (5–18 km). Formed by updrafts.

Cumulus Billowy: Commonly seen on a clear day (j).

Cumulonimbus: Darkened cumulus cloud, referred to as a "thunderhead" (k).

(e) Stratus clouds. Low-lying stratus clouds are sometimes called high fogs.

(f) Stratocumulus clouds. Appear as long layers of cottonlike masses.

(g) Nimbostratus clouds. Dark nimbostratus clouds are given to precipitation.

(h) Advection fog. Formed in moist air moving over a cool surface, an advection fog rolls in around San Francisco Bay's Golden Gate Bridge.

(i) Radiation fog. Commonly formed overnight in valleys when radiative heat loss cools the ground.
The nearby air is then cooled, leading to condensation and fog.

(k) Cumulonimbus cloud. This huge cloud of vertical development has a dark nimbus lower portion, from which precipitation is occurring or threatening to occur. Such dark clouds are sometimes called thunderheads.

(j) Cumulus clouds. These billowy, white clouds are commonly seen on a clear day.

Family	Types	Illustration
High Clouds (above 6 km)	Cirrus (Ci)	(a)
	Cirrocumulus (Cc)	(a)
	Cirrostratus (Cs)	(b)
Middle Clouds (1.8–6 km)	Altostratus (As)	(c)
	Altocumulus (Ac)	(d)
Low Clouds (ground level–1.8 km)	Stratus (St)	(e)
	Stratocumulus (Sc)	(f)
	Nimbostratus (Ns)	(g)
Clouds with Vertical Development (5–18 km; see Section 21.5)	Cumulus (Cu)	(j)
	Cumulonimbus (Cb)	(k)

FIGURE 19.23   Vertical Cloud Development
The cloud begins to form at the elevation at which the rising air reaches its dew point and condensation occurs. The vertical development continues until the rising air stabilizes. Air motion in the upper regions gives this cloud an anvil shape.

carries the condensed droplets upward, forming a cloud.

The height at which condensation occurs is the height of the base of the cloud. The vertical distance between the level where condensation begins and the level where stability is reached is the height, or thickness, of the cloud. For purposes of this discussion, we assume that no precipitation occurs. Once formed, the wind shapes the clouds, and they may break up into smaller forms. An example of wind action is shown in ● Fig. 19.23.

RELEVANCE QUESTION: *What types of clouds can you observe in the sky today? If there aren't any, explain why.*

Important Terms

atmospheric science	ozonosphere	psychrometer	convection cycle
meteorology	ionosphere	anemometer	sea breeze
photosynthesis (19.1)	insolation (19.2)	wind vane	land breeze
troposphere	Rayleigh scattering	rain gauge	Coriolis force
weather	greenhouse effect	radar	friction (drag)
stratosphere	barometer (19.3)	Doppler radar	jet streams
mesosphere	humidity	wind (19.4)	clouds (19.5)
thermosphere	relative humidity	air currents	lapse rate
ozone (O_3)	dew point	isobar	

Important Equation

Relative Humidity: (%) $RH = \dfrac{AC}{MC}$ ($\times$ 100%)

Review Questions

19.1 Composition and Structure

1. Which is the third most abundant gas in the atmosphere?
   (a) oxygen   (c) nitrogen
   (b) carbon dioxide (d) argon

2. In what region does the ozone layer lie?
   (a) thermosphere   (c) stratosphere
   (b) troposphere   (d) mesosphere

3. What is the difference between atmospheric science and meteorology?

4. Other than moisture, what are the three major components of the air you breathe?

5. Humans inhale oxygen and exhale carbon dioxide. With our population, wouldn't this reduce the atmospheric oxygen level over a period of time? Explain.

6. Why is the plant pigment chlorophyll so important?

7. Describe how the temperature of the atmosphere varies in each of the following regions:
   (a) the mesosphere   (c) the troposphere
   (b) the stratosphere   (d) the thermosphere

8. Of what importance is the atmospheric ozone layer?

9. What is believed to cause the displays of lights called auroras?

19.2 Atmospheric Energy Content

10. What regulates Earth's average temperature?
    (a) Rayleigh scattering   (c) atmospheric pressure
    (b) the greenhouse effect   (d) photosynthesis

11. Approximately what percentage of insolation reaches Earth's surface?
    (a) 33%   (b) 40%   (c) 50%   (d) 75%

12. What does the term *insolation* stand for?

13. From what source does the atmosphere receive most of its *direct* heating, and how is the overall heating accomplished?

14. Why is the sky blue?

15. In terms of Rayleigh scattering, why is it advantageous to have amber fog lights and red tail lights on cars?

16. Explain what is meant by the *greenhouse effect.*

17. How does the selective absorption of atmospheric gases provide a thermostatic effect for Earth?

19.3 Atmospheric Measurements and Observations

18. With what instrument is pressure measured?
    (a) an anemometer   (c) a wind vane
    (b) a barometer   (d) a psychrometer

19. With what instrument is humidity measured?
    (a) an anemometer   (c) a wind vane
    (b) a barometer   (d) a psychrometer

20. What is the principle of the liquid barometer? What is the height of a mercury barometer column for one atmosphere of pressure?

21. Explain the operation of a skydiver's altimeter.

22. Why does water condense on the outside of a glass containing an iced drink?

23. Explain the principle of the psychrometer.

24. Which way, relative to the wind direction, does a wind vane point and why?

25. At small airports, a wind sock, a tapered bag pivoted on a pole, acts as a wind vane and gives some indication of the wind speed. Explain its operation.

26. What information does Doppler radar give that conventional radar cannot?

19.4 Air Motion

27. Near a large body of water, what wind is the predominant wind during the day?
    (a) a sea breeze   (c) an updraft
    (b) a north wind   (d) a jet stream

28. What is the direction of rotation around a low in the Northern Hemisphere as viewed from above?
    (a) clockwise
    (b) counterclockwise
    (c) sometimes clockwise, sometimes counterclockwise

29. What are the primary and secondary forces of air motion, and what is the basic distinction between these forces?

30. Distinguish any differences between cyclones and anticyclones in the Northern and Southern Hemispheres, and explain.

31. What is the general wind direction for the conterminous United States and why?

32. Generally speaking, on which side of town would it be best to build a house in the United States so as to avoid smoke and other air pollutants generated in the town?

33. Should the prevailing wind direction be of any consideration in the heating plan and insulation of a house?

19.5 Clouds

34. What is the cloud root name meaning "heap"?
    (a) stratus   (c) nimbus
    (b) cirrus   (d) cumulus

35. The altostratus cloud is a member of which family?
    (a) high clouds   (c) low clouds
    (b) middle clouds   (d) vertical development

36. Name the cloud family for each of the following:
    (a) nimbostratus   (d) stratus
    (b) cirrostratus   (e) cumulonimbus
    (c) altostratus

37. Name the cloud type associated with each of the following:
    (a) mackerel sky   (c) the hazy shade of winter
    (b) solar or lunar halo   (d) thunderhead

38. What conditions are necessary for cloud formation?

39. What happens if the dew point of a rising air mass is not reached before stability?

Applying Your Knowledge

1. The maximum insolation is received daily around noon. Why, then, is the hottest part of the day around 2:00 or 3:00 P.M.?

2. (a) Why does the land lose heat more quickly at night than a body of water?

(b) Deserts are very hot during the day, and cold at night. Why is there such a large nocturnal temperature drop?

3. What are the circulation patterns around cyclones and anticyclones in the Southern Hemisphere?

Exercises

19.1 Composition and Structure

1. Express the approximate thicknesses of the (a) stratosphere, (b) mesosphere, and (c) thermosphere in terms of the thickness of the troposphere (at the equator).
   Answer: (a) 2.1   (b) 1.9   (c) 7.5

2. On a vertical scale of altitude in km and mi above sea level, compare the heights of the following (the heights not listed can be found in the chapter):
   (a) the top of Pike's Peak—14,000 ft
   (b) the top of Mt. Everest—29,000 ft
   (c) commercial airline flight—35,000 ft
   (d) supersonic transport flight (SST)—65,000 ft
   (e) communications satellite—400 mi
   (f) the E and F ion layers
   (g) aurora displays
   (h) syncom satellite—23,000 mi (satellite with period synchronous to Earth's rotation, so it stays over one location)
   (*Note:* Conversion factors are found on the inside back cover.)

3. If the air temperature is 70°F at sea level, what is the temperature at the top of Pike's Peak (14,000-ft elevation)? (*Hint:* The temperature decreases rather uniformly in the troposphere.)   *Answer:* 21°F

4. If the air temperature is 20°C at sea level, what is the temperature outside a jet aircraft flying at an altitude of 10,000 m?

19.3 Atmospheric Measurements and Observations

5. On a day when the air temperature is 75°F, the wet-bulb reading of a psychrometer is 68°F. Find each of the following: (a) relative humidity, (b) dew point, (c) maximum moisture capacity of the air, and (d) actual moisture content of the air.
   Answer: (a) 70%   (b) 64°F   (c) 9.4 gr/ft^3 (d) 6.6 gr/ft^3

6. A psychrometer has a dry-bulb reading of 95°F and a wet-bulb reading of 90°F. Find each of the quantities asked for in Exercise 5.

7. On a very hot day with an air temperature of 105°F, the wet-bulb thermometer of a psychrometer records 102°F. (a) What is the actual moisture content of the air? (b) How many degrees would the air temperature have to be lowered for the relative humidity to be 100%?
   Answer: (a) 21 gr/ft^3 (b) 4 F°

8. On a winter day, a psychrometer has a dry-bulb reading of 35°F and a wet-bulb reading of 29°F. (a) What is the actual moisture content of the air? (b) Would the water in the wick of the wet bulb freeze? Explain.

9. The dry-bulb and wet-bulb thermometers of a psychrometer read 75°F and 65°F, respectively. What are (a) the actual moisture content of the air, and (b) the temperature at which the relative humidity would be 100%?
   Answer: (a) 5.5 gr/ft^3 (b) 59°F

10. On a day when the air temperature is 80°F, the relative humidity is measured to be 79%. How many degrees would the air temperature have to be lowered for the relative humidity to be 100%?

Solution to Confidence Exercise _____

19.1 With a depression of $\Delta T = 70°F - 66°F = 4\,F°$ (from the Appendix VIII tables), (a) $RH = 81\%$, and (b) with a dew point $= 64°F$, the air temperature ($70°F$) would have to be lowered $6\,F°$ for the relative humidity to be 100% and precipitation likely.

Answers to Multiple-Choice Review Questions _____

1. d   10. b   18. b   27. a   34. d
2. c   11. c   19. d   28. b   35. b

ATMOSPHERIC EFFECTS

20

And pleas'd the Almighty's orders to perform
Rides in the whirlwind and directs the storm.

Joseph Addison (1672–1719)

On the local level, we are concerned with the daily weather conditions. Will it rain? Will it snow? How warm will it be? How should I dress? The weather changes because of atmospheric dynamics.

The air we now breathe may have been far out over the Pacific Ocean a week ago. As air moves into a region, it brings with it the temperature and humidity of previous locations. Cold, dry, arctic air may cause a sudden drop in the temperature of the regions in its path. Warm, moist air from the Gulf of Mexico may bring heat and humidity to some regions and make the summer seem unbearable. Ocean cycles affect the weather, as evidenced by the 1998 El Niño.

Thus moving air transports the physical characteristics that influence the weather and produce changes. A large mass of air can influence the region for a considerable period of time or have only a brief effect. The movement of air masses depends a great deal on Earth's air circulation structure and seasonal variations.

When air masses meet, variations of their properties may trigger storms along their common boundary. Thus the types of storms depend on the properties of the air masses involved. Also, variations within a single air mass can

Photo: Lightning—a most spectacular atmospheric effect.

give rise to storms locally. Storms can be violent and sometimes destructive. They remind us of the vast amount of energy contained in the atmosphere and also of its capability. As will be learned in this chapter, the variations of our weather are closely associated with air masses and their movements and interactions.

An unfortunate topic of atmospheric science is pollution. Various pollutants are being released into the atmosphere, affecting our health, living conditions, and environment.

Climate also may be affected by pollution. For example, a few years ago we heard a lot about *acid rain.* Government restrictions have helped reduce this problem, but what about the "ozone hole" and CFCs? These are relatively recent terms that you will learn about in this chapter. ■

20.1 Condensation and Precipitation

LEARNING GOALS

▼ Explain how precipitation is formed.

▼ Distinguish among the various types of precipitation.

In the preceding chapter's section on cloud formation, it was stated that condensation occurs when the dew point is reached. We assumed that all the essentials for condensation were present. However, it is quite possible for an air mass containing water vapor to be cooled below the dew point without condensation occurring. In this state, the air mass is said to be *supersaturated,* or *supercooled.*

How, then, are visible droplets of water formed? You might think that the collision and coalescing of water molecules would form a droplet. But this event would require the collision of millions of molecules. Moreover, only after a small droplet has reached a critical size will it have sufficient binding force to retain additional molecules. The probability of a droplet forming by this process is quite remote.

Instead, water droplets form around microscopic foreign particles, called *hygroscopic nuclei,* already present in the air. These particles are in the form of dust, combustion residue (smoke and soot), salt from seawater evaporation, and so forth. Because foreign particles initiate the formation of droplets that eventually fall as precipitation, condensation provides a mechanism for cleansing the atmosphere.

Liquid water may be cooled below the freezing point without the formation of ice if it does not contain the proper type of foreign particles to act as ice nuclei. For many years scientists believed that ice nuclei could be just about anything, such as dust. However, research has shown that "clean" dust—that is, dust without biological materials from plants or bacteria—will not act as ice nuclei. This discovery is important because precipitation involves ice crystals, as will be discussed shortly.

Because cooling and condensation occur in updrafts, the formed droplets are readily suspended in the air as a cloud. For precipitation, larger droplets or drops must form. This condition may be brought about by two processes: (1) coalescence and/or (2) the Bergeron process.

Coalescence

Coalescence is the formation of drops by the collision of droplets, the result being that larger droplets grow at the expense of smaller ones. The efficiency of this process depends on the variation in the size of the droplets.

Raindrops vary in size, reaching a maximum diameter of approximately 7 mm. A drop 1 mm in diameter would require the coalescing of a million droplets of 10-μm diameter but only 1000 droplets of 100-μm diameter. Thus we see that having larger droplets greatly enhances the coalescence process.

Bergeron Process

The **Bergeron process,** named after the Swedish meteorologist who suggested it, is probably the more important process for the initiation of precipitation. This process involves clouds that contain ice crystals in their upper portions and have become supercooled in their lower portions (● Fig. 20.1).

Mixing or agitation within such a cloud allows the ice crystals to come into contact with the supercooled vapor. Acting as nuclei, the ice crystals grow larger from the vapor condensing on them. The ice crystals melt into large droplets in the warmer lower portion

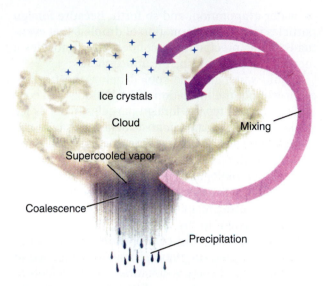

FIGURE 20.1 The Bergeron Process

The essence of the Bergeron process is the mixing of ice crystals and supercooled vapor, which produces water droplets and initiates precipitation.

of the cloud and coalesce to fall as precipitation. Air currents are the normal mixing agents.

Note that there are three essentials in the Bergeron process: (1) ice crystals, (2) supercooled vapor, and (3) mixing. *Rainmaking* is based on the essentials of the Bergeron process.

The early rainmakers were mostly charlatans. With much ceremony they would beat on drums or fire cannons and rockets into the air. Explosives may have supplied the agitation or mixing for rainmaking, assuming that the other two essentials of the Bergeron process were present.

However, modern rainmakers use a different approach. There are usually enough air currents present for mixing, but the ice-crystal nuclei may be lacking. To correct this, they "seed" clouds with silver iodide crystals or dry-ice pellets (solid CO_2).

The silver iodide crystals have a structure similar to that of ice and provide a substitute for ice crystals. Silver iodide crystals are produced by a burning process. The burning may be done on the ground, with the iodide crystals being carried aloft by the rising warm air, or the burner may be attached to an airplane and the process carried out in a cloud to be seeded.

Dry-ice pellets are seeded into a cloud from an airplane. The pellets do not act as nuclei but serve another purpose. The temperature of solid dry ice is −79°C (−110°F), and it quickly sublimes—that is, goes directly from the solid to the gaseous phase. Rapid cooling associated with the sublimation triggers the conversion of supercooled cloud droplets into ice crystals. Precipitation may then occur if this part of the Bergeron process has been absent.

Also, the latent heat released from the ice-crystal formation is available to set up convection cycles for mixing. Seeding is receiving increasing attention in initiating the precipitation of fog, which frequently hinders airport operations.

Types of Precipitation

The types of precipitation depend on atmospheric conditions and can occur in the form of rain, snow, sleet, hail, dew, or fog. *Rain* is the most common form of precipitation in the lower and middle latitudes. The formation of large water drops that fall as rain has been described previously.

If the dew point is below 0°C, the water vapor freezes on condensing, and the ice crystals that result fall as *snow*. In cold regions, these ice crystals may fall individually. In warmer regions, the ice crystals become stuck together, forming a snowflake that may be as much as 2–3 cm across. Because ice crystallizes in a hexagonal (six-sided) pattern, snowflakes are hexagonal (see Fig. 2 on page 98).

Frozen rain, or pellets of ice in the form of *sleet*, occurs when rain falls through a cold surface layer of air and freezes or, more likely, when the ice pellets fall directly from the cloud without melting before striking the ground. Large pellets of ice, or *hail*, result from successive vertical descents and ascents in vigorous convection cycles associated with thunderstorms. Additional condensation on successive cycles into supercooled regions that are below freezing may produce layered-structure hailstones the size of golf balls and baseballs (● Fig. 20.2). When cut in two, the layers of ice can be observed, much like the rings in tree growth.

Dew is formed by atmospheric water vapor condensing on various surfaces. The land cools quickly at night, and the temperature may fall below the dew point. Water vapor then condenses on available surfaces such as blades of grass, giving rise to the "early morning dew."

FIGURE 20.2   Hailstones
The successive vertical ascents of ice pellets into super-cooled air and regions of condensation produce large, layered "stones" of ice. The layered structure can be seen in these cross sections.

If the dew point is below freezing, the water vapor condenses in the form of ice crystals as *frost*. Frost is *not* frozen dew but results from the direct change of water vapor into ice (the reverse of sublimation, called *deposition*).

Interestingly, research has shown that frost is a result of bacteria-seeded ice formation. Without two common types of bacteria on leaf surfaces, water will not freeze at 0°C but can be supercooled to −6° to −8°C. These bacteria exist on plants, fruit trees, and so on and serve as nuclei for frost formation.

With frost damage to crops and fruits exceeding $1 billion annually, scientists are exploring techniques to prevent the formation of bacteria-seeded frost. One method involves the development of genetically engineered bacteria, which are altered such that they can no longer trigger ice formation.

Researchers believe that a protein on the surface of the bacterium acts as the seed for the formation of frost ice crystals. By genetically removing the gene that serves as the blueprint for this protein, it is hoped that "frost-free" bacteria can be made.

RELEVANCE QUESTION: By late morning or early afternoon the morning dew you observe is gone. Where does it go?

20.2   Air Masses

LEARNING GOALS

▼ Define *air masses,* and tell how they are classified.

▼ Identify fronts and their effects on local weather.

As we know, the weather changes with time. However, we often experience several days of relatively uniform weather conditions. Our general weather conditions depend in large part on vast air masses that move across the country.

When a large body of air takes on physical characteristics that distinguish it from the surrounding air, it is referred to as an **air mass.** The main distinguishing characteristics are *temperature* and *moisture content.*

A mass of air remaining for some time over a particular region, such as a large body of land or water, takes on the physical characteristics of the surface of the region. The region from which an air mass derives its characteristics is called its **source region.**

An air mass eventually moves from its source region, bringing its characteristics to regions in its path and bringing changes in the weather. As an air mass travels, its properties may become modified because of local variations. For example, if Canadian polar air masses did not become warmer as they travel southward, Florida would experience some extremely cold temperatures.

Whether an air mass is termed *cold* or *warm* is relative to the surface over which it moves. Quite logically, if an air mass is warmer than the land surface, it is referred to as a *warm air mass.* If the air is colder than the surface, it is called a *cold air mass.* Remember, though, that these terms are relative. The *warm* and *cold* prefixes do not always imply warm and cold weather. A "warm" air mass in winter may not raise the temperature above freezing.

Air masses are classified according to the surface and general latitude of their source regions:

Surface	*Latitude*
Maritime (m)	Arctic (A)
Continental (c)	Polar (P)
	Tropical (T)
	Equatorial (E)

TABLE 20.1 Air Masses That Affect the Weather of the United States

Classification	Symbol	Source Region
Maritime arctic	mA	Arctic regions
Continental arctic	cA	Greenland
Maritime polar	mP	Northern Atlantic and Pacific oceans
Continental polar	cP	Alaska and Canada
Maritime tropical	mT	Caribbean Sea, Gulf of Mexico, and Pacific Ocean
Continental tropical	cT	Northern Mexico, southwestern United States

The surface of the source region, abbreviated by a small letter, gives an indication of the moisture content of an air mass. Forming over a body of water (maritime), an air mass would naturally be expected to have a greater moisture content than one forming over land (continental).

The general latitude of a source region, abbreviated by a capital letter, gives an indication of the temperature of an air mass. For example, "mT" designates a maritime tropical air mass, which would be expected to be a warm, moist one. The air masses that affect the weather in the United States are listed in

Table 20.1, along with their source regions, and are illustrated in ● Fig. 20.3.

The movement of air masses is influenced to a great extent by Earth's general circulation patterns (Section 19.4). Because the conterminous United States lies predominantly in the westerlies zone, the general movement of air masses, and hence the weather, is from west to east across the country. Global circulation zones vary to some extent in latitude with the seasons, and the polar easterlies may also move air masses into the eastern United States during the winter.

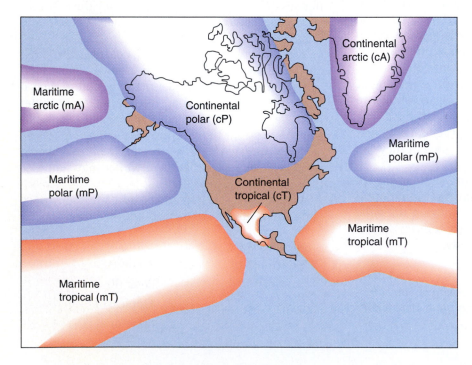

FIGURE 20.3 Air-Mass Source Regions

The map shows the source regions for the air masses of North America.

The boundary between two air masses is called a **front.** A *warm front* is the boundary of an advancing warm air mass over a colder surface, and a *cold front* is the boundary of a cold air mass moving over a warmer surface. These boundaries, called *frontal zones,* may vary in width from a few miles to a wide zone of over 160 km (100 mi). It is along fronts, which divide air masses of different physical characteristics, that drastic changes in weather occur. Turbulent weather and storms usually characterize a front.

The degree and rate of weather change depend on the difference in temperatures of the air masses and the degree of vertical slope of a front. A cold front moving into a warmer region causes the lighter, warm air to be displaced upward over the front (Fig. 20.4). The lighter air of the advancing warm front cannot displace the heavier, colder air as readily, and generally it moves slowly up and over the colder air.

Heavier, colder air is associated with high pressure, and this downward divergent air flow in a high-pressure region generally gives a cold front a greater speed than a warm front. A cold front may have an average speed of 30–40 km/h (20–25 mi/h), whereas a warm front averages about 15–25 km/h (10–15 mi/h).

Cold fronts have sharper vertical boundaries than warm fronts, and warm air is displaced upward faster by an advancing cold front. As a result, cold fronts are accompanied by more violent or sudden changes in weather. The sudden decrease in temperature is often described as a "cold snap." Dark altocumulus clouds often mark a cold front's approach (Fig. 20.4). The sudden cooling and the rising warm air may set off rainstorm or snowstorm activity along the front.

A warm front also may be characterized by precipitation and storms. Because the approach of a warm front is more gradual, it is usually heralded by a period of lowering clouds. Cirrus and mackerel scale (cirrocumulus) clouds drift ahead of the front, followed by alto clouds. As the front approaches, cumulus or cumulonimbus clouds resulting from the rising air produce precipitation and storms. Most precipitation occurs before the front passes.

The graphic symbol for a cold front is

and for a warm front

The side of the line with the symbol indicates the direction of advance.

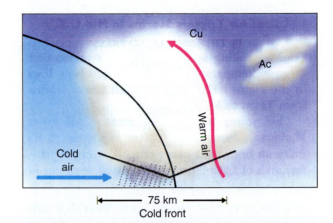

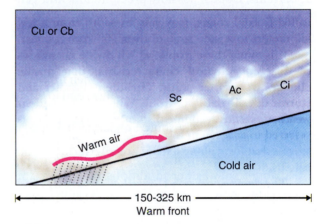

FIGURE 20.4 Side Views of Cold and Warm Fronts

Notice in the upper diagram the sharp, steep boundary that is characteristic of cold fronts. The boundary of a warm front, as shown in the lower diagram, is less steep. As a result, different cloud types are associated with the approach of the two types of fronts. (See the table on page 525 for cloud abbreviations.)

As a faster-moving cold front advances, it may overtake a warm air mass and push it upward. The boundary between these two air masses is called an *occluded front* and is indicated by

that is, the cold front occludes, or cuts off, the warm air from the ground along the occluded front. When a cold front advances under a warm front, a *cold front occlusion* results. When a warm front advances up and over a cold front, the air ahead is colder than the advancing air, and the occluded front is referred to as *warm front occlusion.*

Sometimes fronts traveling in opposite directions meet. The opposing fronts may balance each other so that no movement occurs. This case is referred to as a *stationary front* and is indicated by

Air masses and fronts move across the country bringing changes in weather. Dynamic situations give rise to cyclonic disturbances around low-pressure and high-pressure regions. As learned in Chapter 19, these disturbances are called *cyclones* and *anticyclones,* respectively (see Fig. 19.20).

As a low or cyclone moves, it carries with it rising air currents, clouds, possibly precipitation, and generally bad weather. Hence lows or cyclones are usually associated with poor weather, while highs or anticyclones are usually associated with good weather. The lack of rising air and cloud formation in highs gives clear skies and fair weather. Because of their influence on the weather, the movements of highs and lows are closely observed.

We usually think of a source region as being relatively hot or cold. But significant variations can occur within a source region at a particular latitude, giving rise to abnormal weather conditions. A classic example, El Niño, is discussed in the chapter's first Highlight.

RELEVANCE QUESTION: *What classifications of air masses generally affect your weather?*

20.3 Storms

LEARNING GOALS

▼ Identify various types of local and tropical storms.

▼ Describe the aspects of lightning safety and tornado safety.

Storms are atmospheric disturbances that may develop locally within a single air mass or may be due to frontal activity along the boundary of air masses. Several types of storms, distinguished by their intensity and violence, will be considered. These will be divided generally into local storms and tropical storms.

Local Storms

There are several types of local storms. A heavy downpour is commonly referred to as a *rainstorm*. Storms with rainfalls of 1 to 3 inches per hour are not uncommon.

A *thunderstorm* is a rainstorm distinguished by thunder and lightning, and sometimes hail. The **lightning** associated with a thunderstorm is a discharge of electrical energy. In the turmoil of a thundercloud or "thunderhead," there is a separation of charge associated with the breaking up and movement of water droplets. This gives rise to an electric potential. When this is of sufficient magnitude, lightning occurs.

Lightning can take place entirely within a cloud (intracloud or cloud discharges), between two clouds (cloud-to-cloud discharges), between a cloud and Earth (cloud-to-ground or ground discharges), or between a cloud and the surrounding air (air discharges). (See ● Fig. 20.5.)

Lightning has reportedly even occurred in clear air, apparently giving rise to the expression "a bolt from the blue." When lightning occurs below the horizon or behind clouds, it often illuminates the clouds with flickering flashes. This commonly occurs on a still summer night and is known as *heat lightning*.

Although the most frequently occurring form of lightning is the intracloud discharge, of greatest concern is lightning between a cloud and Earth. The shorter the distance from a cloud to the ground, the more easily the electric discharge takes place. For this reason, lightning often strikes trees and tall buildings. It is inadvisable, therefore, to take shelter from a thunderstorm under a tree. A person in the vicinity of a lightning strike may experience an electric shock that causes breathing to fail. In such a case, mouth-to-mouth resuscitation or some other form of artificial respiration should be given immediately and the person kept warm as a treatment for shock. (See the following discussion of lightning safety.)

Lightning Safety

If you are outside during a thunderstorm and feel an electrical charge, as evidenced by hair standing on end or skin tingling, what should you do? *Fall to the ground fast!* **Lightning may be about to strike.**

Statistics show that lightning kills, on average, **200 people a year in the United States and injures another 550. Most deaths and injuries occur at home. Indoor casualties occur most frequently when people are talking on the telephone, working in the kitchen, doing laundry, or watching TV. During severe lightning activity, the following safety rules are recommended:**

FIGURE 20.5 Lightning

Lightning discharges can occur between a cloud and Earth, between clouds, and within a cloud.

Stay indoors away from open windows, fireplaces, and electrical conductors such as sinks and stoves.

Avoid using the telephone. Lightning may strike the telephone lines outside.

Do not use electrical plug-in equipment such as radios, TVs, and lamps.

Should you be caught outside, seek shelter in a building. If no buildings are available, seek protection in a ditch or ravine. Getting wet is a lot better than being struck by lightning.

A lightning stroke's sudden release of energy explosively heats the air, producing the compressions we hear as **thunder.** When heard at a distance of about 100 m (330 ft) or less from the discharge channel, thunder consists of one loud bang, or "clap." When heard at a distance of 1 km (0.62 mi) from the discharge channel, thunder generally consists of a rumbling sound punctuated by several large claps. In general, thunder cannot be heard at distances of more than 25 km (16 mi) from the discharge channel.

Because lightning strokes generally occur near the storm center, the resulting thunder provides a method of approximating the distance to the storm. Light travels at approximately 300,000 km/s (186,000 mi/s), and so the lightning flash is seen instantaneously. Sound, however, travels at approximately $\frac{1}{3}$ km/s ($\frac{1}{5}$ mi/s), so a time lapse occurs between seeing the lightning flash and hearing the thunder.

This phenomenon also can be observed by watching someone at a distance fire a gun or hit a baseball. The report of the gun or the "crack" of the bat is heard after the smoke or flash from the gun is observed or the baseball is well on its way.

By counting the seconds between seeing the lightning and hearing the thunder (by saying, "one-thousand-one, one-thousand-two," etc.), you can estimate your distance from the lightning stroke or the storm.

EXAMPLE 20.1

Estimating the Distance of a Thunderstorm

Suppose some campers notice an approaching thunderstorm in the distance. Lightning is seen, and the thunder is heard 5.0 s later. Approximately how far away is the storm in (a) km and (b) mi?

SOLUTION

We know that the approximate or average speed of sound is $\bar{v} = \frac{1}{3}$ km/s $= \frac{1}{5}$ mi/s. Then the distance (d) the sound travels in a time (t) is given by Eq. 2.1 ($d = \bar{v}t$).

(a) Using the metric speed,

$$d = \bar{v}t = (\tfrac{1}{3} \text{ km/s})(5.0 \text{ s}) = 1.6 \text{ km}$$

(b) You could convert the distance in (a) to miles (and, in fact, you may recognize the conversion right away), but let's compute it.

$$d = \bar{v}t = (\tfrac{1}{5} \text{ mi/s})(5.0 \text{ s}) = 1.0 \text{ mi}$$

(Recall that 1 mi = 1.6 km.)

CONFIDENCE EXERCISE 20.1

If thunder is heard 3 s after seeing the flash of a lightning stroke, approximately how far away, in km and mi, is the lightning?

If the temperature of Earth's surface is below 0°C and raindrops do not freeze before striking the ground, the rain will freeze on striking cold surface objects. Such an **ice storm** builds up a layer of ice on objects exposed to the freezing rain. The ice layer may build up to over half an inch in thickness, depending on the magnitude of the rainfall. Viewed in sunlight, the ice glaze produces beautiful winter scenes with the ice-coated landscape glistening in the Sun. However, damage to trees and power lines often detracts from the beauty.

HIGHLIGHT: El Niño and La Niña

El Niño is an occasional disruption of the ocean-atmosphere system in the tropical Pacific Ocean that has important weather consequences for many parts of the world. Originally recognized by fishermen off the Pacific coast of South America, the appearance at irregular intervals of unusually warm water near the beginning of the year was named *El Niño,* meaning "the little boy" or "the Christ child" in Spanish. The latter implied the tendency of El Niño to arrive at Christmastime.

Under normal conditions, the Pacific trade winds blow generally from east to west near the equator (see Fig. 19.21) and pile up warm surface water in the western Pacific Ocean near Indonesia and the Australian continent. As a result of this warm water, the atmosphere is heated, and conditions favorable for convection and precipitation occur (Fig. 1).

This surface flow results in cooler water temperatures off the coast of South America due to the upwelling of cold water from deeper levels. The cold water is nutrient-rich and supports an abundance of fish and, thereby, a fishing industry. With an atmospheric circulation as shown in Fig. 1, the Pacific coast of South America is relatively dry because of little precipitation. (Why?)

At irregular intervals, typically every 3 to 5 years, the normal trade winds relax, and the warm pool of water in the western Pacific is free to move back eastward along the equator toward the South American continent (El Niño conditions). The displacement of warm

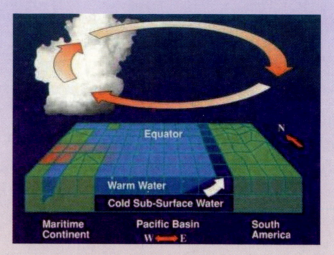

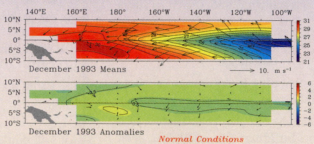

FIGURE 1   Normal Pacific Conditions.   The trade winds generally blow from east to west, causing warm surface water currents. These conditions give rise to precipitation in the western Pacific and cooler water temperatures in the eastern Pacific along the coast of South America.

water affects the atmospheric convection cycles, as shown in Fig. 2.

The convection and precipitation shift to the central and eastern Pacific and usually result in heavier than normal rains over northern Peru and Ecuador. The mechanism for precipitation for Indonesia and Australia is no longer there, and this area often will experience drought during an El Niño.

Twenty-three El Niños have occurred since 1900, and the cause of this relatively frequent event, which lasts about 18 months, is not known.

Snow is made up of ice crystals that fall from ice clouds. A *snowstorm* is an appreciable accumulation of snow. What may be considered a severe snowstorm in some regions may be thought of as a light snowfall in areas where snow is more prevalent.

When a snowstorm is accompanied by high winds and low temperatures, the storm is referred to as a *blizzard.* The winds whip the fallen snow into blinding swirls. Visibility may be reduced to a few inches. For this reason, a blizzard is often called a *blinding snowstorm.*

The swirling snow causes a loss of one's direction, and people have gotten lost only a few feet from their homes. The wind may blow the snow across level ter-

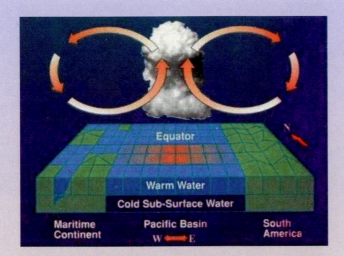

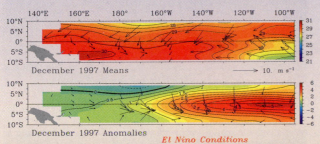

December 1997 Means

December 1997 Anomalies

El Nino Conditions

FIGURE 2   El Niño Conditions. At irregular intervals, the normal trade winds relax and warm surface water moves back eastward, causing changes in weather patterns.

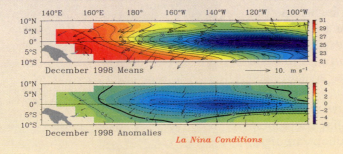

December 1998 Means

December 1998 Anomalies

La Nina Conditions

The 1982–83 El Niño was one of the worst, being responsible for the loss of as many as 2000 lives and the displacement of thousands from their homes. Indonesia and Australia suffered droughts and bush fires. Peru was hit with heavy rainfall—11 feet in areas where 6 inches was normal. The warm water spread along the coast to the United States and the West Coast was drenched with rain. Sharks were observed off the Oregon coast due to the unseasonably warm sea temperatures. Also, more hurricanes occurred in the Pacific than in the Atlantic.

Another severe El Niño occurred in 1998, with much the same effects. Perhaps you remember this one. The normal weather patterns were disrupted around the country.

At various times, the reverse conditions of El Niño are found. That is, the eastern Pacific surface waters are colder than usual, and cold water extends farther westward than usual (Fig. 3). This condition is called *La Niña* ("the little girl" in Spanish) and occurs after some, but not all, El Niño years. Strong La Niñas cause unusual weather conditions around the country. For example, during a La Niña year, winter temperatures are usually warmer than normal in the southeastern United States and cooler than normal in the Northwest.

FIGURE 3   La Niña Conditions.
At various times, cold water extends farther westward than usual, which affects coastal weather patterns.

rain, forming huge drifts against some obstructing object. Drifting is common on the flat prairies of the western United States.

The **tornado** is the most violent of storms. Although it may have less *total* energy than some other storms, the concentration of its energy in a relatively small region gives the tornado its violent distinction.

Characterized by a whirling, funnel-shaped cloud that hangs from a dark cloud mass, the tornado is commonly referred to as a *twister* (● Fig. 20.6).

Tornadoes occur around the world, but are most prevalent in the United States and Australia. In the United States, most tornadoes occur in the Deep South and in the broad, relatively flat basin between the Rockies and the Appalachians. But no state is immune. The peak months of tornado activity are April, May, and June, with southern states usually hit hardest in winter and spring, and northern states in spring and summer. However, tornadoes have occurred in every month at all times of day and night. A typical time of occurrence is between 3:00 and 7:00 P.M. on an unseasonably warm, sultry spring afternoon.

Most tornadoes travel from southwest to northeast, but the direction of travel can be erratic and may change suddenly. They usually travel at an average speed of 48 km/h (30 mi/h). The wind speed of a major tornado may vary from 160–480 km/h (100–300 mi/h). The wind speed of the devastating 1999 Oklahoma tornado was measured by Doppler radar to be 502 km/h (312 mi/h), the highest ever recorded.

Because of many variables, the complete mechanism of tornado formation is not known. One essential component, however, is rising air, which occurs in thunderstorm formation and in the collision of cold and warm air masses.

As the ascending air cools, clouds are formed that are swept to the outer portions of the cyclonic motion and outline its funnel form. Because clouds form at certain heights (Section 19.5), the outlined funnel may appear well above the ground. Under the right conditions, a full-fledged tornado develops. The winds increase and the air pressure near the center of the vortex is reduced as the air swirls upward. When the funnel is well developed, it may "touch down" or be seen extending up from the ground as a result of dust and debris picked up by the swirling winds.

The system for a tornado alert has two phases. A **tornado watch** is issued when atmospheric conditions indicate that tornadoes may form. A **tornado warning** is issued when a tornado has actually been sighted or indicated on radar.

The similarity between the terms *watch* and *warning* can be confusing. Remember that you should *watch* for a tornado when the conditions are right, and when you are given a *warning,* the situation is dangerous and critical—no more watching.

Care should be taken after a tornado has passed. There may be downed power lines, escaping natural gas, and dangerous debris. (See the following discussion of tornado safety.)

Tornado Safety

Knowing what to do in the event of a tornado is critically important. If a tornado is sighted, if the ominous roar of one is heard at night, or if a tornado

(a)

(b)

FIGURE 20.6 Tornado Destruction

(a) A tornado funnel touching down with a town in its path. (b) The high-speed winds of a tornado can push in the windward wall of a house, lift off the roof, and push the other walls outward. Imagine all of the debris that was flying around during the tornado.

warning is issued for your particular locality, *seek shelter fast!*

The basement of a home or building is one of the safest places to seek shelter.

Avoid **chimneys and windows, because there is great danger from flying glass and debris.**

Get under a sturdy piece of furniture, such as an overturned couch, or into a stairwell or closet, and *cover your head*.

In a home or building without a basement, seek the lowest level in the central portion of the structure and the shelter of a closet or hallway.

If you live in a mobile home, evacuate it. Seek shelter elsewhere.

Tropical Storms

The term *tropical storm* refers to the massive disturbances that form over tropical oceanic regions. A tropical storm becomes a **hurricane** when its wind speed exceeds 118 km/h (74 mi/h). The hurricane is known by different names in different parts of the world. For example, in southeast Asia it is called a *typhoon,* and in the Indian Ocean, a *cyclone* (● Fig. 20.7).

Regardless of the name, this type of storm is characterized by high-speed rotating winds, whose energy is spread over a large area. A hurricane may be 480–960 km (300–600 mi) in diameter and have wind speeds of 118–320 km/h (74–200 mi/h).

Hurricanes form over tropical oceanic regions where the Sun heats huge masses of moist air, and an ascending spiral motion results. When the moisture of the rising air condenses, the latent heat provides additional energy and more air rises up the column. This latent heat is the chief source of a hurricane's energy and is readily available from the condensation of the evaporated moisture from its source region.

Unlike the tornado, a hurricane gains energy from its source region. As more and more air rises, the hurricane grows, with clouds and increasing winds that blow in a large spiral around a relatively calm, low-

FIGURE 20.7   Tropical Storm Regions of the World

Tropical storms are known by different names in different parts of the world. The average tropical storm activities by month are shown in the graphs.

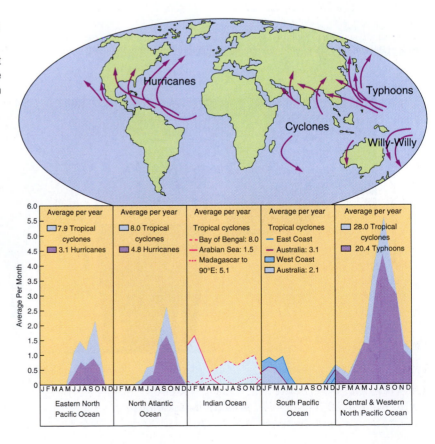

pressure center—the *eye* of the hurricane (● Fig. 20.8). The eye may be 32–48 km (20–30 mi) wide, and ships sailing into this area have found that it is usually calm and clear, with no indication of the surrounding storm. The air pressure is reduced 6–8% near the eye. Hurricanes move rather slowly, at a few miles per hour.

Covering broad areas, hurricanes can be particularly destructive. Hurricane winds do much damage, but oddly enough, drowning is the greatest cause of hurricane deaths. As the eye of a hurricane comes

ashore or *makes landfall,* a great dome of water called a **storm surge,** often over 80 km (50 mi) wide, comes sweeping across the coastline. It brings huge waves and storm tides that may reach 5 m (17 ft) or more above normal (● Fig. 20.9). The storm surge comes suddenly, often flooding coastal lowlands. Nine out of ten casualties are caused by the storm surge.

The torrential rains that accompany the hurricane commonly produce flooding as the storm moves inland. As its winds diminish, floods constitute a hurricane's greatest threat.

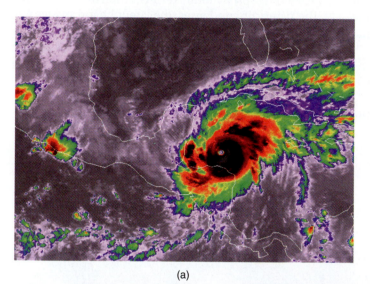

(a)

FIGURE 20.8 Hurricane Paths and Eyes

(a) A computer-enhanced infrared image of Hurricane Mitch bearing down on Central America in October 1998. As may be seen from Fig. 20.7, hurricanes forming in the Atlantic Ocean generally travel northwest until they are blown eastward by the prevailing westerlies. As a result, hurricanes can strike coastal areas around the Gulf of Mexico and the southeastern United States, both coming and going. (b) A radar profile of a hurricane. Note the generally clear eye about 35 to 55 miles from the radar station and the vertical buildup of clouds near the sides of the eye.

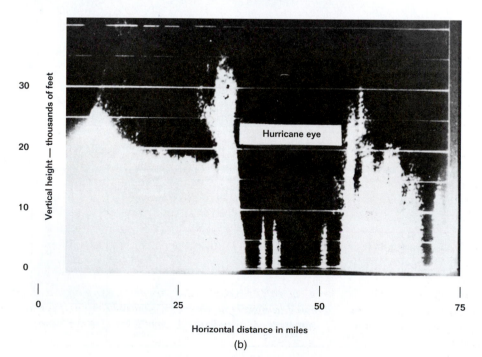

(b)

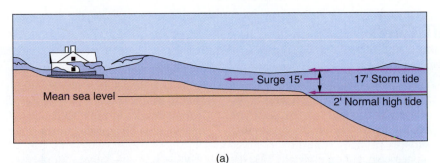

(a)

(b)

FIGURE 20.9   Hurricane Storm Surge and Damage

(a) The drawing illustrates a storm surge coming ashore at high tide. (b) An aerial view of housing destruction after a hurricane.

A terrible cyclone came out of the Indian Ocean in the spring of 1991 and struck Bangladesh with a 20-foot wall of water and winds of 145 mi/h. The flooding and devastation in the low-lying regions where the storm made landfall were enormous. Over 125,000 people lost their lives.

Once cut off from the warm ocean, the storm dies, starved for moisture and heat energy and dragged apart by friction as it moves over the land. Even though a hurricane weakens rapidly as it moves inland, the remnants of the storm can bring 6–12 inches or more of rain for hundreds of miles.

The breeding grounds of the hurricanes that affect the United States are in the Atlantic Ocean southeast of the Caribbean Sea. As hurricanes form, they move westward with the trade winds, usually making landfall along the Gulf and south Atlantic coasts. During the hurricane season, the area of their formation is constantly monitored by satellite. When a tropical storm is detected, radar-equipped airplanes, or "hurricane hunters," track the storms and make local measurements to help predict its path.

Like that for a tornado, the hurricane alerting system has two phases. A **hurricane watch** is issued for coastal areas when there is a threat of hurricane conditions within 24 to 36 hours. A **hurricane warning** indicates that hurricane conditions are expected within 24 hours (winds of 74 mi/h or greater, or dangerously high water and rough seas).

As you are probably aware, tropical storms and hurricanes are given names. How this is done is described in the chapter's second Highlight.

RELEVANCE QUESTION: *In what year was the latest tornado watch or warning issued for where you live? What should you do if a "watch" is issued? If a "warning" is issued?*

HIGHLIGHT: Naming Hurricanes

For several hundred years, many hurricanes in the West Indies were named after the saint's day on which they occurred. In 1953 the National Weather Service began to use women's names for tropical storms and hurricanes. This practice began in World War II when military personnel named typhoons in the western Pacific after their wives and girlfriends, using alphabetical order. The first hurricane of the season received a name beginning with A, the second one was given a name beginning with B, and so on.

The practice of naming hurricanes solely after women was changed in 1979 when men's names were included in the lists. A six-year list of names for Atlantic storms is given in Table 1. A similar list is available for Pacific storms. Names beginning with the letters Q, U, X, Y, and Z are excluded because of their scarcity. The names have an international flavor because hurricanes affect other nations and are tracked by the public and by weather services in many countries. The lists are recycled. For example, the 1999 list will be used again in 2005.

In 1992 Hurricane Andrew was a very destructive storm. It made landfall in Florida, causing millions of dollars worth of damage. Notice that Andrew is not in the 2004 list but has been replaced with Alex. If a hurricane is particularly destructive, its name is replaced so as not to cause confusion with another destructive storm of the same name in the next six-year cycle. Similarly, the 1989 Hurricane Hugo that hit the South Carolina coast had its name replaced.

More recently, Hurricane Mitch was particularly devastating, causing thousands of deaths and vast destruction in Central America in 1998. Mitch's name will probably be replaced in the 2004 list.

What was the largest number of tropical storms that occurred in one year? At the time of this writing, the busiest year was 1933, which holds the record with 21 tropical storms, 10 of which became hurricanes. Names for these storms would have consumed a year's list, but tropical storms were not named at the time, only numbered. The year for the largest number of hurricanes was 1969, with 12 hurricanes resulting from 18 tropical storms (third busiest year). More recently, the second busiest year with 19 tropical storms was 1995, of which 11 became hurricanes.

TABLE 1   The Six-Year List of Names for Atlantic Storms*

1999	2000	2001	2002	2003	2004
Arlene	Alberto	Allison	Arthur	Ana	Alex
Bret	Beryl	Barry	Bertha	Bill	Bonnie
Cindy	Chris	Chantal	Cristobol	Claudette	Charley
Dennis	Debby	Dean	Dolly	Danny	Danielle
Emily	Ernesto	Erin	Edouard	Erika	Earl
Floyd	Florence	Felix	Fay	Fabian	Frances
Gert	Gordon	Gabrielle	Gustav	Grace	Georges[†]
Harvey	Helene	Humberto	Hanna	Henri	Hermine
Irene	Isaac	Iris	Isidore	Isabel	Ivan
Jose	Joyce	Jerry	Josephine	Juan	Jeanne
Katrina	Keith	Karen	Kyle	Kate	Karl
Lenny	Leslie	Lorenzo	Lili	Larry	Lisa
Maria	Michael	Michelle	Marco	Mindy	Mitch[†]
Nate	Nadine	Noel	Nana	Nicholas	Nicole
Ophelia	Oscar	Olga	Omar	Odette	Otto
Philippe	Patty	Pablo	Paloma	Peter	Paula
Rita	Rafael	Rebekah	Rene	Rose	Richard
Stan	Sandy	Sebastien	Sally	Sam	Shary
Tammy	Tony	Tanya	Teddy	Teresa	Tomas
Vince	Valerie	Van	Vicky	Victor	Virginia
Wilma	William	Wendy	Wilfred	Wanda	Walter

*Names of particular individuals have not been chosen for inclusion in the list of hurricane names.

[†]Names of severe 1998 hurricanes; under consideration for replacement as of this writing.

20.4 **Atmospheric Pollution**

▼ Identify the major atmospheric pollutants.

▼ Explain some pollutant effects, such as smog and acid rain.

At the beginning of Chapter 19, a quote was presented from Shakespeare about the atmosphere: ". . . this most excellent canopy, the air. . . ." This was taken out of context and there's more: ". . . this most excellent canopy, the air, look you, this brave o'erhanging firmament, this majestical roof fretted with golden fire, why, it appears no other thing to me but a foul and pestilent congregation of vapours" (Shakespeare, *Hamlet*).

Even Shakespeare in his day made reference to atmospheric pollution—an unfortunate topic of earth science. By **pollution,** we mean any atypical contributions to the environment resulting from the activities of human beings. Of course, gases and particulate matter are spewn into the air from volcanic eruptions and lightning-initiated forest fires, but these are natural phenomena over which we have little control.

Air pollution results primarily from the products of combustion and industrial processes that are released into the atmosphere. It has long been a common practice to vent these wastes, and the resulting problems are not new, particularly in areas of population concentrations.

Smoke and soot from the burning of coal plagued England over 700 years ago. London recorded air pollution problems in the late 1200s, and particularly smoky types of coal were taxed and even banned. The problem was not alleviated, and in the middle 1600s, King Charles II was prompted to commission one of the outstanding scholars of the day, Sir John Evelyn, to make a study of the situation. The degree of London's air pollution at that time is described in the following passage from his report, *Fumifugium* (a Latin term, generally meaning *On Dispelling of Smoke*).

. . . the inhabitants breathe nothing but impure thick mist, accompanied with a fuliginous and filthy vapor, corrupting the lungs. Coughs and consumption rage more in this one city (London) than in the whole world. When in all other places the aer is most serene and pure, it is here eclipsed with such a cloud . . . as the sun itself is hardly about to penetrate. The traveler, at miles distance, sooner smells than sees the City.

However, the Industrial Revolution was about to begin, and Sir John's report was ignored and so gathered dust (and soot).

As a result of such air pollution, London has experienced several disasters involving the loss of life. Thick fogs are quite common in this island nation, and the combination of smoke and fog forms a particularly noxious mixture known as **smog,** a contraction of *smoke-fog.*

The presence of fog indicates that the temperature of the air near the ground is at the dew point, and with the release of latent heat, there is a possibility of a **temperature inversion.** As learned in Chapter 19, the atmospheric temperature decreases with increasing altitude in the troposphere (with a lapse rate of about $6\frac{1}{2}$ C°/km). As a result, hot combustive gases generally rise. However, under certain conditions, such as rapid radiative cooling near the ground surface, the temperature may locally *increase* with increasing altitude. The lapse rate is then said to be *inverted,* giving rise to a temperature inversion (● Fig. 20.10).

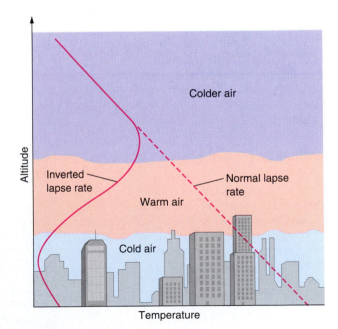

FIGURE 20.10 Temperature Inversion

Normally, the lapse rate near Earth decreases uniformly with increasing altitude. However, radiative cooling of the ground can cause the lapse rate to become inverted, and the temperature then increases with increasing altitude (usually below 1 mile). A similar condition may occur from the subsidence of a high-pressure air mass.

Most common are radiation and subsidence inversions. *Radiation temperature inversions,* which are associated with Earth's radiative heat loss, occur daily. The ground is heated by insolation during the day, and at night it cools by radiating heat back into the atmosphere (see Section 19.2). If it is a clear night, the land surface and the air near it cool quickly. The air some distance above the surface, however, remains relatively warm, thus giving rise to a temperature inversion. *Radiation fogs* provide common evidence of this cooling effect in valleys.

Subsidence temperature inversions occur when a high-pressure air mass moves over a region and becomes stationary. As the dense air settles, it becomes compressed and heated. If the temperature of the descending air exceeds that of the air below it, then the lapse rate is inverted, similar to that shown in Fig. 20.10.

With a temperature inversion, emitted gases and smoke do not rise and are held near the ground. Continued combustion causes the air to become polluted, creating particularly hazardous conditions for people with heart and lung ailments. Smog episodes in various parts of the world have contributed to numerous deaths (● Fig. 20.11).

The major source of air pollution is the combustion of *fossil fuels*—namely, coal, gas, and oil. More accurately, air pollution results from the *incomplete* combustion of *impure* fuels. Technically, combustion (burning) is the chemical combination of cer-tain substances with oxygen. Fossil fuels are the remains of plant and animal life and are composed chiefly of hydrocarbons (compounds of hydrogen and carbon).

If a fuel is pure and combustion is complete, the products, CO_2 and H_2O, are not usually considered pollutants, because they are a natural part of atmospheric cycles. For example, if carbon (as in coal) or methane, CH_4 (as in natural gas), is burned completely, the reactions are

$$C + O_2 \longrightarrow CO_2$$

$$CH_4 + 2\,O_2 \longrightarrow CO_2 + 2\,H_2O$$

However, if fuel combustion is incomplete, the products may include carbon (soot), various hydrocarbons, and carbon monoxide (CO). *Carbon monoxide* results from the incomplete combustion (oxidation) of carbon; that is,

$$2\,C + O_2 \longrightarrow 2\,CO$$

But even increased concentrations of CO_2 can affect our environment. Carbon dioxide combines with water to form carbonic acid, a mild acid many of us drink in the form of carbonated beverages (carbonated water):

$$CO_2 + H_2O \longrightarrow H_2CO_3$$

Carbonic acid is a natural agent of chemical weathering in geologic processes. But as a product of air pollution, increased concentrations may also aid in the

FIGURE 20.11 Smog Episode Waiting to Happen

Shown here is a picture of a 1940s steel mill on the Monongahela River near Pittsburgh, Pennsylvania. A similar scene in Donora, 20 miles down the river from Pittsburgh, gave rise to a 5-day smog episode in which hundreds of people became ill and at least 20 died.

corrosion of metals and react with certain materials, causing decomposition (● Fig. 20.12).

There is concern that an increase in the CO_2 content of the atmosphere may cause a change in global climate through the greenhouse effect. This concern will be considered later in the chapter.

Oddly enough, some by-products of complete combustion contribute to air pollution. For example, **nitrogen oxides (NO$_x$),** are formed when combustion temperatures are high enough to cause a reaction of the nitrogen and oxygen of the air. This reaction typically occurs when combustion is nearly complete, a condition that produces high temperatures, or when combustion takes place at high pressure, for example, in the cylinders of automobile engines.

These oxides, normally NO (nitric oxide) and NO_2 (nitrogen dioxide), can combine with water vapor in the air to form nitric acid (HNO_3), which is very corrosive. Also, this acid contributes to acid rain, which will be discussed shortly. Nitrogen dioxide (NO_2) has a pungent, sweet odor and is yellow-brown in color. During peak rush-hour traffic in some large cities, it is evident as a whiskey-brown haze.

Nitrogen oxides also can cause lung irritation and are key substances in the chemical reactions producing the so-called *Los Angeles smog,* because it was first identified in Los Angeles. This is not the classic London smoke-fog variety but a smog called **photochemical smog** that results from the chemical reactions of

hydrocarbons with oxygen in the air and other pollutants in the presence of sunlight. The sunlight supplies the energy for chemical reactions that take place in the air.

Over 15 million people live in the Los Angeles area, which has the form of a basin with the Pacific Ocean to the west and mountains to the east. This topography makes air pollution and temperature inversions a particularly hazardous combination. A temperature inversion essentially puts a "lid" on the city, which then becomes engulfed in its own fumes and exhaustive wastes.

Los Angeles has more than its share of temperature inversions, which may occur as frequently as 320 days per year. These inversions, a generous amount of air pollution, and an abundance of sunshine set the stage for the production of photochemical smog (● Fig. 20.13).

In comparison with the smoke-fogs of London, photochemical smog contains many more dangerous contaminants. These include organic compounds, some of which may be *carcinogens* (substances that cause cancer).

One of the best indicators of photochemical reactions, and a pollutant itself, is **ozone** (O_3), which is found in relatively large quantities in photochemically polluted air. In Los Angeles and elsewhere, air pollution warnings of various degrees are given on the basis of ozone concentrations in the air.

FIGURE 20.12   Decomposition
The damage done to this statue in New York City by pollution and acid rain is evident (*left*). The statue looked quite different sixty years earlier (*right*).

FIGURE 20.13 Photochemical Smog Scene in Los Angeles
Los Angeles has over 300 temperature inversions a year, giving rise to smog conditions.

Fuel Impurities

Fuel impurities occur in a variety of forms. Probably the most common impurity in fossil fuels, and the most critical to air pollution, is *sulfur.* Sulfur is present in various fossil fuels in different concentrations. A low-sulfur fuel has less than 1% sulfur content, and a high-sulfur fuel, greater than 2%. When fuels containing sulfur are burned, the sulfur combines with oxygen to form sulfur oxides (SO_x), the most common of which is **sulfur dioxide (SO_2):**

$$S + O_2 \longrightarrow SO_2$$

A majority of SO_2 emissions come from the burning of coal and an appreciable amount from the burning of fuel oils. Coal and oil are the major fuels used in electrical generation. Almost half the SO_2 pollution in the United States occurs in seven northeast industrial states.

Sulfur dioxide in the presence of oxygen and water can react chemically to produce sulfurous and sulfuric acids. Sulfurous acid (H_2SO_3) is mildly corrosive and is used as an industrial bleaching agent. Sulfuric acid (H_2SO_4), a very corrosive acid, is a widely used industrial chemical. In the atmosphere, these sulfur compounds can cause considerable damage to practically all forms of life and property. Anyone familiar with sulfuric acid, the electrolyte used in car batteries, can appreciate its undesirability as an air pollutant.

The sulfur pollution problem has received considerable attention because of the occurrence of **acid rain.** Rain is normally slightly acidic as a result of carbon dioxide combining with water vapor to form carbonic acid. However, sulfur oxide and nitrogen oxide pollutants cause precipitation from contaminated clouds to be even more acidic, giving rise to acid rain (and also acid snow, sleet, fog, and hail; ● Fig. 20.14).

The problem is most serious in New York, New England, and Canada, where pollution emissions from the industrialized areas in the midwestern United States are carried by the general weather patterns. Other areas are not immune. Acid rain now occurs in the Southeast, and acid fogs are observed on the West Coast.

The government has imposed limits on the levels of sulfur emissions, and the effects of acid rain have been reduced somewhat. However, before the regulations, rainfall with a pH of 1.4 had been recorded in the northeastern United States. This value surpasses the pH of lemon juice (pH 2.2). Canada had monthly rainfalls with an average pH of 3.5, which is as acidic as tomato juice (pH 3.5). The yearly average pH of the rain in these affected regions was about 4.2 to 4.4. Recall that a neutral solution has a pH of 7.0. (See Chapter 13 for a discussion of pH.)

In addition to acid rain, there are acid snows. Over the course of a winter, acid precipitations build up in snowpacks. During the spring thaw and resulting runoff, the sudden release of these acids gives an "acid shock" to streams and lakes.

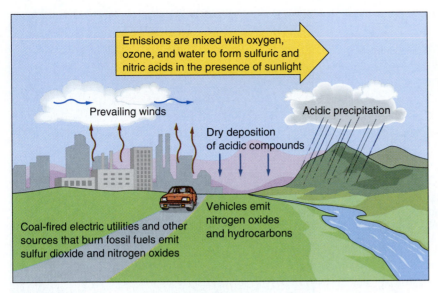

(a)

FIGURE 20.14   Acid Rain Formation

(a) Sulfur dioxide and nitrogen oxide emissions react with water vapor in the atmosphere to form acid compounds. The acids are deposited in rain or snow and also may join dry airborne particles and fall to Earth as dry deposition. (b) All rain is slightly acidic (slightly below 7.0 pH, because of natural CO_2 in the atmosphere). However, rain with a pH below 5.6 is considered acid rain. The map shows the pH values of acid rain in different parts of the country at its peak.

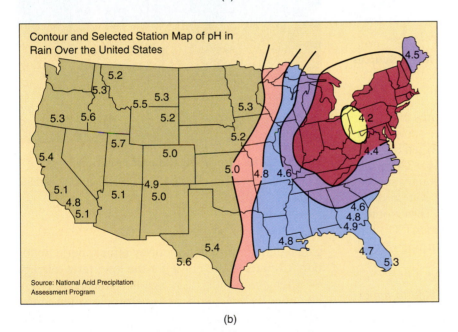

(b)

Acid precipitations lower the pH of lakes, which threatens aquatic plant and animal life. Most fish species die at a pH of 4.5 to 5.0. As a result, many lakes in the northeastern United States and Canada are "dead" or in jeopardy. Natural buffers in area soils tend to neutralize the acidity, so waterways and lakes in an area don't necessarily match the pH of the rain. However, the neutralizing capability in some regions is being taxed, and the effects of acid rain still pose a problem.

Thus air pollution can be quite insidious and can consist of a great deal more than the common particulate matter (smoke, soot, and fly ash) that blackens the outside of buildings. Pollutants may be in the form of mists and aerosols. Some pollutants are metals, such as lead and arsenic. Approximately 100 atmospheric pollutants have been identified, 20 of which are metals that come primarily from industrial processes.

● Figure 20.15 shows the sources and relative magnitudes of the various atmospheric pollutants. As can

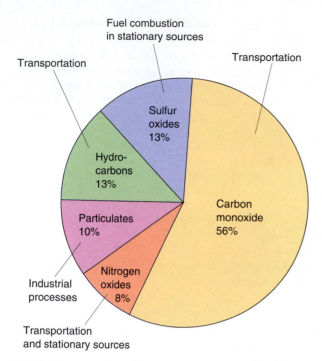

FIGURE 20.15 Air Pollutants and Their Major Sources

FIGURE 20.16 Transportation Air Pollution

Vast amounts of air pollution come from cars, trucks, and aircraft.

be seen, transportation is the major source of total air pollution. The United States is a mobile society, with over 100 million registered vehicles powered by internal combustion engines (● Fig. 20.16). The other major sources of air pollution are stationary sources and industrial processes. *Stationary source* refers mainly to electrical generation facilities. These sources account for the majority of the sulfur oxide (SO_x) pollution, resulting primarily from the burning of coal, which always has some sulfur content.

RELEVANCE QUESTION: *Do you contribute to atmospheric pollution? Explain.*

20.5 **Pollution and Climate**

LEARNING GOALS

▼ Define *climate,* and identify climatic changes.

▼ Explain the possible effects of atmospheric pollutants on climate.

Although not immediately obvious, it is generally believed that there are (and will be) changes in the global climate brought about by atmospheric pollution. **Climate** is the name for the long-term average weather conditions of a region. Some regions are identified by their climates. For example, when someone mentions Florida or California, one usually thinks of a warm climate, and Arizona is known for its dryness and low humidity. Because of such favorable conditions, the climate of a region often attracts people to live there, and thus the distribution of population (and pollution) is affected.

Dramatic changes in climate have occurred throughout Earth's history. Probably the most familiar is that of the Ice Age, when glacial ice sheets advanced southward over the world's northern continents. The most recent ice age ended some 10,000 years ago, after glaciers came as far south as the midwestern conterminous United States.

Evidence from ocean sediment cores support the theory that dramatic changes in global climate result from subtle but regular variations in Earth's orbit around the Sun. Earth's closest approach to the Sun occurs at different times of the year in a cycle of 23,000 years. Earth and the Sun are now closest in January; in 10,000 years they will be closest in July. The result will be cooler summer temperatures, less snow melting, and a growth of the polar ice caps. Such a change could slowly lead to a new ice age.

Climatic fluctuations are continually occurring on a smaller scale also. For example, in the past several decades there has been a noticeable southward shift of world climate, which has produced drought conditions in some areas. The question being asked today is whether or not air pollution may be responsible for some of the observed climate variations. For example, from 1880 to 1940, the average annual temperature of Earth's surface increased by about 0.6 C° (1.1 F°). Since 1940, the average temperature has decreased by about 0.3 C° (0.55 F°).

Associated with this lowering temperature has been a shift in the frost and ice boundaries, a weakening of zonal wind circulation, and marked variations in the world's rainfall pattern. However, this pattern appears to be changing, possibly as a result of the depletion of the ozone layer, as will be discussed shortly.

Global climate is sensitive to atmospheric contributions that affect the radiation balance of the atmosphere. These contributions include the concentration of CO_2 and other "greenhouse" gases, the particulate concentration, and the extent of cloud cover, all of which affect Earth's albedo (the fraction of insolation reflected back into space).

Air pollution and other human activities do contribute to changes in climatic conditions. Scientists are now trying to understand climate changes by using various models. These models of the workings of Earth's atmosphere and oceans allow scientists to compare theories, using historical data on climate changes; however, specific data are scant.

Particulate pollution could contribute to changes in Earth's thermal balance by decreasing the transparency of the atmosphere to insolation. We know this effect occurs from observing the results of volcanic eruptions, whose particulate matter causes changes in the albedo.

In 1991, Mount Pinatubo in the Philippines erupted. Debris was sent over 15 mi into the atmosphere, and over a foot of volcanic ash piled up in surrounding regions (● Fig. 20.17). Measurements indicate that Pinatubo was probably the largest volcanic eruption of the century, belching out tons of debris and sulfur dioxide (SO_2). The particulate matter caused beautiful sunrises and sunsets around the world during the following year. The sulfur dioxide gas reacts with oxygen and water to form tiny droplets, or aerosols, of sulfuric acid, which may stay aloft for several years before falling back to Earth.

Computer models of atmospheric chemistry suggest that the acid aerosols could cause a thinning of Earth's protective ozone layer, thus allowing more ultraviolet radiation to reach the ground. Other environmental concerns about the ozone layer are discussed in the chapter's third Highlight.

Supersonic transport (SST) aircraft operating in the lower stratosphere also have been cause for concern because of particulate and gaseous emissions

FIGURE 20.17 Volcanic Ash from Mount Pinatubo

Military personnel had to evacuate Clark Air Force Base in the Philippines when Mount Pinatubo erupted in June 1991.

HIGHLIGHT: The Ozone Hole

In 1974, scientists in California warned that chlorofluorocarbon (CFC) gases might seriously damage the ozone layer through depletion. Observations generally supported this prediction, and in 1978 the United States put a ban on the use of these gases as propellants in aerosol spray cans.

Even so, millions of tons of CFCs continue to leach into the atmosphere each year, primarily from refrigerants and spray propellants manufactured in other countries. The release of CFCs from car air conditioners is the single largest source of emissions. CFCs are also used in the manufacture of plastic foams.*

The major CFCs are CFC-11 ($CFCl_3$) and CFC-12 (CF_2Cl_2). When released, these gases slowly rise into the stratosphere, a process that takes 20 to 30 years. In the stratosphere the CFC molecules are broken apart by ultraviolet radiation, with the release of reactive chlorine atoms. These atoms in turn react with and destroy ozone molecules in the repeating cycle:

$$CFC \longrightarrow Cl$$

$$Cl + O_3 \longrightarrow ClO + O_2$$
(ozone)

$$ClO + O \longrightarrow Cl + O_2$$

Notice that the chlorine atom is again available for reaction after the process.

These atoms may remain in the atmosphere for a year or two. During this time, a single Cl atom may destroy as many as 100,000 ozone molecules.

Measurements indicate that the concentrations of CFCs in the atmosphere have more than doubled in the past 10 years. Worldwide ozone levels have declined an estimated 3% to 7% over the past few decades. Part of this decline is the result of normal fluctuations. But in 1985, scientists announced the discovery of an ozone "hole" over Antarctica (Fig. 1). Measurements have since revealed losses of greater than 50% in the total hole column, and greater than 95% at altitudes of 9 to 12 miles.

Investigations have shown that this polar hole in the ozone layer opens up annually during the southern springtime months of September and October. It is thought that the seasonal hole began forming in the late 1970s because of increasing concentrations of ozone-destroying chlorine pollutants in the stratosphere.

Ice crystals in high-altitude clouds in the lower stratosphere at the end of the polar winter are thought to provide surfaces on which chemical reactions take place. Such clouds and ozone holes are not evident at lower altitudes. Satellite measurements taken in October 1991 showed that the atmospheric concentration of ozone over the South Pole had dwindled to a low level. These low levels were observed for another 2 years and probably were a result of the airborne aerosols from the 1991 Mt. Pinatubo eruption. The aerosols increase chlorine's effectiveness of destroying the ozone.

The worldwide depletion of the ultraviolet-absorbing ozone layer will have some significant effects. Experts estimate that the number of cases of skin cancer will increase by 60% and that there will be many additional cases of cataracts. In 1994, the National Weather Service began issuing a "uv index" forecast for many large cities. On a scale of 0 to 15, the index gives a relative indication of the amount of uv light that will be received at Earth's surface at noontime the next day; and the greater the number, the greater the amount of uv. The scale is based on upper-atmosphere ozone levels and clouds.

Crops and climate will also be affected. Sea levels may rise somewhere between 1 and 4 m (3 to 12 ft) by the year 2100 as a result of global warming and melting of the polar icecaps. Also, CFCs are greenhouse gases and can contribute to the warming in this manner.

International concern over ozone depletion prompted meetings and conventions, and in 1987, some 24 na-

---

(● Fig. 20.18). In the troposphere, precipitation processes act to "wash" out particulate matter and gaseous pollutants, but there is no snow or rain washout mechanism in the stratosphere. Also, the stratosphere is a region of high chemical activity, and chemical pollutants, such as NO_x and hydrocarbons, could possibly give rise to climate-changing reactions.

There is also a temperature-increase pollution aspect. Vast amounts of CO_2 are expelled into the atmosphere as a result of the combustion of fossil fuels. As discussed in Section 19.2, CO_2 and water vapor play important roles in Earth's energy balance, because of the greenhouse effect. An increase in the atmospheric concentration of CO_2 could alter the amount of radiation absorbed from the planet's surface and produce an increase in Earth's average temperature.

During the nineteenth century, the atmospheric CO_2 content increased by about 10%, as determined

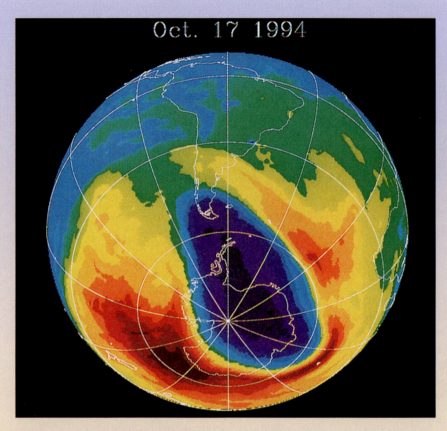

FIGURE 1   The Ozone Hole. A computer map of the South Pole region showing the total stratospheric ozone concentrations. The ozone hole, or region of minimal concentration, is shown in purple.

However, some new measurements showed worse damage to the ozone layer than was originally expected. Reacting to scientific assessments, the protocol parties decided to completely end production of halons by the beginning of 1994 and of CFCs by the beginning of 1996 in industrialized nations. (Developing countries would have a 10-year grace period.) Over 160 nations have now signed the treaty.

With only recycled and stockpiled CFCs available for use after January 1, 1996, considerable effort was made to develop replacements. These are available and used in new refrigerators and air conditioners. Such substitutes must be chemically nonreactive with ozone or be destroyed by lower-atmosphere processes before reaching the ozone layer. Most of the substitutes are not as efficient as CFCs for refrigeration purposes, and as a result, there is a greater energy cost for the same cooling.

Even so, more efficient compounds will no doubt be developed, and this is a small price to pay to protect a life-sustaining part of our atmosphere.

*Other substances that cause ozone depletion are *halons,* a group of halogen compounds used as fire-extinguishing agents, and methyl bromide, which is used as a fumigation agent and also results from biomass burning.

tions signed the *Montreal Protocol.* The agreement was to reduce the production of CFCs by half by 1998. The protocol was amended in 1990, and more than 90 nations agreed to phase out CFCs by the year 2000.

from old records. During the twentieth century, because of larger populations and more combustion, the increase has been even greater. It appears that Earth will be a warmer place in the twenty-first century, and climate changes will affect agriculture, water resources, and sea level.

Recent studies predict that the atmospheric CO_2 content will double by the year 2065, with an accompanying temperature increase of 1.5 to 4.5 C°. Note that Venus has a rich CO_2 atmosphere—200,000 times more than Earth—and a surface temperature of 470°C (900°F), hot enough to melt lead. Of course, Venus is closer to the Sun than Earth is. Depletion of the ozone layer, as discussed in the Highlight, also would result in a temperature increase.

Possible effects of increasing temperature, or global warming, include the melting of the polar ice caps, which would cause a rise in the sea level and

FIGURE 20.18   Stratospheric Pollution

The supersonic transport produces a great deal of noise pollution during takeoff and landing. It also releases pollutants in the lower stratosphere where there are no weather conditions to wash them back to Earth.

would flood coastal areas. Also, there would be drier summers in the middle latitudes. In the United States, there would be dry farmlands in the South and longer growing seasons in the North. Many other possibilities are not well understood, including the accelerated release of greenhouse gases, such as methane (CH_4), from swamps and bogs because of a warming climate.

Thus air pollution may contribute to a variety of atmospheric effects, both local and global. The effects of pollution on local climatic conditions are clearer. For example, many cities receive more rainfall than the surrounding countrysides. However, there are too few data to understand or accurately predict global climatic effects.

We are inclined to believe that increased atmospheric CO_2 concentrations and ozone-layer depletions would give rise to an increase in global temperature. Such a temperature increase would cause more water evaporation and an increase in relative humidity. Particulate pollution could then give rise to increased cloud formation, which would increase the albedo and cause a decrease in global temperature.

Certainly, little comfort can be found in this speculative, counterbalancing pollution cycle. There are too many unanswered questions on the effects of pollution, which has occurred over a relatively short time, and on the natural interacting cycles of the atmosphere and biosphere, which have taken millions of years to become established.

RELEVANCE QUESTION: *How might you personally be affected by depletion of the ozone layer and by a stronger greenhouse effect? Give at least one answer for each.*

Important Terms

coalescence (20.1)	thunder	storm surge	nitrogen oxides (NO_x)
Bergeron process	ice storm	hurricane watch	photochemical smog
air mass (20.2)	tornado	hurricane warning	ozone
source region	tornado watch	pollution (20.4)	sulfur dioxide (SO_2)
front	tornado warning	smog	acid rain
lightning (20.3)	hurricane	temperature inversion	climate (20.5)

Review Questions

20.1 Condensation and Precipitation

1. When the temperature of the air is below the dew point without precipitation, it is said to be what?
   (a) stable   (c) sublimed
   (b) supercooled   (d) coalesced

2. Which of the following is *not* essential to the Bergeron process?
   (a) silver iodide   (c) supercooled vapor
   (b) mixing   (d) ice crystals

3. What are the principles and methods of modern rain-making?

4. (a) Is frost frozen dew? Explain.
   (b) How are large hailstones formed?

20.2 Air Masses

5. Which of the following air masses would be expected to be cold and dry?
   (a) cP   (c) cT
   (b) mA   (d) mE

6. What is a cold front advancing under a warm front called?
   (a) warm front   (c) cold-warm front
   (b) stationary front   (d) warm front occlusion

7. How are air masses classified? Explain the relationship between air-mass characteristics and source regions.

8. (a) What air masses affect the weather in the conterminous United States?
   (b) How about Hawaii and Alaska?

9. What is a front? List the meteorological symbols for four types of fronts.

10. Describe the weather associated with warm and cold fronts. What is the significance of the sharpness of their vertical boundaries?

20.3 Storms

11. What is the critical alert for a tornado?
    (a) tornado alert   (c) tornado watch
    (b) tornado warning   (d) tornado prediction

12. The greatest number of hurricane casualties is caused by which of the following?
    (a) high winds   (c) flying debris
    (b) low pressure   (d) storm surge

13. How do conditions for lightning discharges occur?

14. What type of first aid should be given to someone suffering from the shock of a lightning stroke?

15. An ice storm is likely to result along what type of front? Explain.

16. What is the most violent of storms and why?

17. Describe the formation and characteristics of a tornado.

18. What is the major source of energy for a tropical storm? When does a tropical storm become a hurricane?

19. Distinguish between a hurricane watch and a hurricane warning.

20. What months are the peak season for (a) hurricanes and (b) tornadoes?

20.4 Atmospheric Pollution

21. A subsidence temperature inversion is caused by which of the following?
    (a) a high-pressure air mass
    (b) acid rain
    (c) radiative cooling
    (d) subcritical air pressure

22. Which is a major source of air pollution?
    (a) nuclear electrical generation
    (b) incomplete combustion
    (c) temperature inversions
    (d) acid rain

23. Define *air pollution*.

24. Is air pollution a relatively new problem? Explain.

25. What gives rise to most of our air pollution?

26. What are the products of complete combustion? of incomplete combustion?

27. Are nitrogen oxides products of complete or incomplete combustion? Explain.

28. Distinguish between classical smog and photochemical smog. What is one of the best indicators of the latter?

29. What is the major fossil-fuel impurity?

30. What are the causes and effects of acid rain? In which areas is acid rain a major problem and why?

31. Name the major sources of the following pollutants.
    (a) carbon monoxide   (d) nitrogen oxides
    (b) sulfur dioxide   (e) ozone
    (c) particulate matter

20.5 Pollution and Climate

32. A change in Earth's albedo could result from which of the following?
    (a) nitrogen oxides   (c) photochemical smog
    (b) acid rain   (d) particulate matter

33. Major concern about global warming arises from increased concentrations of which of the following?
    (a) sulfur oxides   (c) greenhouse gases
    (b) nitrogen oxides   (d) photochemical smog

34. How has Earth's average temperature varied over the past 100 years?

35. What are the possible effects of increased atmospheric CO_2 concentrations?

36. What are some possible explanations for changes in Earth's average temperature?

37. What effects could CFCs have on Earth's climate?

38. What is the concern about air pollution in the stratosphere?

Applying Your Knowledge

1. Why do household barometers often have descriptive adjectives such as *rain* and *fair* on their faces, along with the direct pressure readings? (See Fig. 19.12.)

2. How could CO_2 pollution be decreased while our energy needs were still being met?

3. Assuming that the name has not been retired, what will be the name of the third Atlantic tropical storm in 2007?

Exercises

20.2 Air Masses

1. Locate the source regions for the following air masses that affect the conterminous United States:
    (a) cA
    (b) mP
    (c) cT
    (d) mT
    (e) cP
    > Answer: (a) Greenland   (b) northern Atlantic and Pacific Oceans   (c) Mexico   (d) middle Atlantic and Pacific Oceans   (d) Canada

2. What would be the classifications of the air masses forming over the following source regions?
    (a) Sahara Desert
    (b) Antarctic Ocean
    (c) Greenland
    (d) Mid-Pacific Ocean
    (e) Siberia

3. On average, how far does (a) a cold front and (b) a warm front travel in 24 h?
    > Answer: (a) 840 km (521 mi)   (b) 480 km (298 mi)

4. How long does it take (a) a cold front and (b) a warm front to travel from west to east across your home state?

5. If thunder is heard 4.0 s after observing a lightning flash, approximately how far away in kilometers is the storm?
    > Answer: 1.3 km

6. While picnicking on a summer day, you hear thunder 11 s after seeing a lightning flash from an approaching storm. Approximately how far away in miles is the storm?

Solution to Confidence Exercise

20.1 $d = \bar{v}t = \frac{1}{3}\,\text{km/s} \times 3\,\text{s} = 1\,\text{km}$
        $= \frac{1}{5}\,\text{mi/s} \times 3\,\text{s} = 0.6\,\text{mi}$

Answers to Multiple-Choice Review Questions

1. b   5. a   11. b   21. a   32. d
2. a   6. d   12. d   22. b   33. c

MINERALS AND ROCKS

21

*Touch the earth, love the earth, honour the earth,
her plains, her valleys, her hills and her seas; rest
your spirit in her solitary places.*

Henry Beston (1888–1968)

Historically, **geology** refers to the study of planet Earth—its composition, structure, and history. Today, this definition is expanded to include the study of the Moon and other planets.

In this chapter and the following three, we introduce and discuss the concepts of geology necessary to comprehend the physical nature of the planet on which we live. The study begins with the outer layer of the planet. This outer layer, called the *crust,* is a thin shell with a thickness of only 4.8 to 48 km. The crust is composed of minerals and rocks, and this first chapter on geology begins with a discussion of minerals and is followed by a discussion of rocks and the processes that form them. In the chapters that follow we concentrate on the structural geology of the planet, the interactions of its crust and internal processes, and the processes that wear away and level Earth's surface. We conclude our study with the methods of geologic dating and interpreting Earth's history. ■

Photo: All three types of rocks that characterize Earth's outer layer, the crust, can be found in the Grand Canyon of the Colorado River.

21.1 **Minerals**

LEARNING GOALS

▼ Define *minerals,* identify their physical properties, and list their general uses.

▼ Describe the methods of mineral identification.

A **mineral** is a naturally occurring, crystalline, inorganic substance (element or compound) that possesses a fairly definite chemical composition and a distinctive set of physical properties. Everywhere around us we see minerals or the products of minerals. Some are quite valuable, but others are essentially worthless. For example, precious stones such as diamonds, emeralds, and rubies are valuable minerals. The chapter Highlight (page 562) deals with some of these. Also, we speak of a nation's mineral wealth when referring to natural raw materials such as ores containing iron, gold, silver, and copper. However, the minerals of common rock, such as sandstone, have little monetary value.

The term *mineral* also has taken on popular meanings. For example, foods are said to contain vitamins and "minerals." In this case *mineral* refers to compounds in food that contain elements required in small quantities by the human body, such as Fe, Na, I, Mn, Mg, and Cu. The chemical compositions of these compounds may be the same as those of naturally occurring minerals that make up Earth's crust, or they actually may be the naturally occurring minerals themselves. The names of minerals, like those of chemical elements, have historical connotations and may reflect the names of localities or people.

Minerals are composed, for the most part, of eight elements. These elements, along with their relative abundance in Earth's crust, are shown in Fig. 11.9, on page 271. Two of these eight elements, oxygen and silicon, make up about 75% of the crust. More than 2000 minerals have been found in Earth's crust; approximately 20 of them are common, and fewer than 10 account for over 90% of the crust by mass.

A **rock** is a solid, cohesive, natural aggregate of one or more minerals. Rocks are composed of 90% or more, by volume, of oxygen atoms. Therefore, the oxygen atom (or ion) is the dominating influence controlling the number of possible element combinations in the formation of minerals.

The Silicates

Most rock-forming minerals are composed mainly of oxygen and silicon. The fundamental silicon-oxygen compound is silicon dioxide, or silica, which has the formula SiO_2. Quartz, a hard and brittle solid, is an example (● Fig. 21.1). Because carbon and silicon are in the same chemical group, one might think that SiO_2 would be a molecular gas similar to carbon dioxide, CO_2. But the bonding of the two compounds is very different. Rather than being composed of SiO_2 molecules, the silicon-oxygen structure of quartz is based on a network of SiO_4 tetrahedra (● Fig. 21.2a) with shared oxygen atoms (Fig. 21.2b).

In silica, the oxygen-to-silicon ratio is 2 to 1; however, the oxygen-to-silicon ratios in the **silicates** are greater than 2 to 1 and can vary significantly. The variation occurs because the silicon-oxygen tetrahedra may exist as separate independent units or may share oxygen atoms at corners, edges, or faces in many different ways. Thus the structures of the silicate minerals are determined by the way the SiO_4 tetrahedra are arranged.

In addition to oxygen and silicon, most rocks contain aluminum and at least one more element. Feldspars are the most abundant minerals in Earth's crust. There are two main types:

1. Plagioclase feldspar, which contains oxygen, silicon, aluminum, and calcium or sodium

FIGURE 21.1   Quartz

Approximately 95% of continental crust rocks are composed of three types of minerals: quartz (shown above), the plagioclase feldspars, and the potassium feldspars.

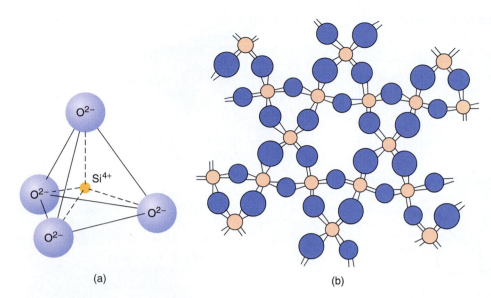

FIGURE 21.2 (a) The Silicon-Oxygen (Covalently Bonded) Tetrahedron and (b) The Structure of Quartz (SiO_2)

(a) This is the fundamental building block for all silicate minerals. (b) The structure is based on interlocking SiO_4 tetrahedra, in which each oxygen atom (blue) is shared by two silicon atoms (buff).

2. Potassium (orthoclase) feldspar, composed of oxygen, silicon, aluminum, and potassium

Some silicate structures are illustrated in ● Fig. 21.3, which shows photographs of typical minerals. Olivine is a mineral with a single independent SiO_4^{4-} tetrahedron that is combined with either Mg^{2+} or Fe^{2+}. These metal ions can substitute for each other, depending on the conditions of mineral formation.

The silicon-oxygen tetrahedra can form single- and double-chain structures, as shown in Fig. 21.3. Examples of minerals with these structures are pyroxene and hornblende. In the pyroxene chain, each tetrahedron shares two oxygens, whereas in the double hornblende chain, half the tetrahedra share two oxygens and the other half share three oxygens. These structures have various metallic ion components. The two-dimensional sheet structure of tetrahedra shown in Fig. 21.3 is the structure of mica. The three-dimensional silicate structures are too complex to be shown in a simple illustration. Figure 21.2b gives some idea of this complexity.

Nonsilicate Minerals

Nonsilicates constitute less than 10% of the mass of Earth's crust. They include valuable native elements, such as gold and silver, and gemstones, such as diamonds and sapphires. Their ores are sources of useful metals, such as iron, copper, nickel, and tin. The most common nonsilicate groups are the carbonates, the oxides, and the sulfides.

Carbonate minerals are formed when the carbonate ion (CO_3^{2-}) bonds with other ions. For example, carbonate ions commonly bond with calcium ions (Ca^{2+}) to form the mineral calcite ($CaCO_3$). Calcite is a soft mineral, and like all carbonates, it dissolves readily in acidic water (see Section 13.3). In Chapter 23 you will learn how this property of calcite, a key component of limestone, contributes to the formation of limestone caves.

Oxide minerals are produced when oxygen ions bond with metallic ions. Oxide ores, such as the iron ore, hematite (Fe_2O_3), the tin ore, cassiterite (SnO_2), and the uranium ore, uraninite (UO_2), are sources of valuable metals.

Ions of sulfur (S^{2-}) bond with various positive ions to produce the *sulfide minerals*. Sulfides such as chalcopyrite ($CuFeS_2$), which contains copper and iron, and galena (PbS), which contains lead, are valuable metal ores.

Identifying Minerals

A classification of minerals based on the physical and chemical properties of substances is advantageous because it distinguishes between different forms of minerals composed of the same element or compound. For example, graphite—a soft, black, slippery substance commonly used as a lubricant—and diamond are both composed of carbon. However, because of different crystalline structures, their properties are quite different. (Graphite mixed with other

FIGURE 21.3   Molecular Structure of Several Common Silicate Minerals

Silicate structure	Arrangement of tetrahedra (top view)	Typical mineral
Single independent tetrahedron		Olivine
Single chain	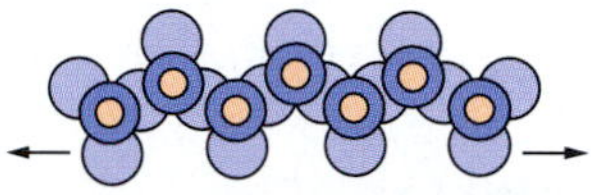	Pyroxene
Double chain	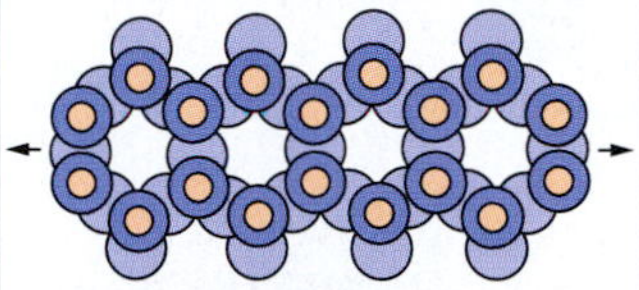	Hornblende
Continuous sheet	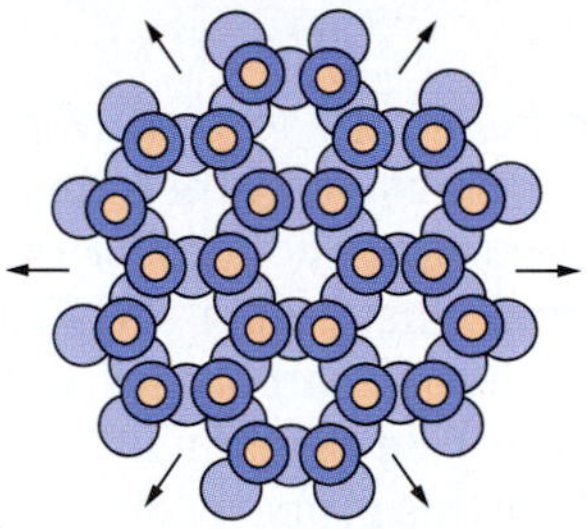	Mica
Three-dimensional network	Too complex to be shown by simple two-dimensional sketch	Quartz and feldspar

TABLE 21.1 Mohs' Scale of Hardness

1. Talc	6. Orthoclase feldspar
2. Gypsum	7. Quartz
3. Calcite	8. Topaz
4. Fluorite	9. Corundum
5. Apatite	10. Diamond

substances to obtain various degrees of hardness is the "lead" in lead pencils.)

Minerals can be identified by chemical analysis, but most of these methods are detailed and costly and are not available to the average person. More commonly, the distinctive physical properties are used as the key to mineral identification. These properties are well known to all serious rock and mineral collectors. Some of the main ones used to identify minerals are described in the following paragraphs.

Crystal form is the size and shape assumed by the crystal faces when a crystal has time and space to grow. It represents the results of the interaction of the atomic structure with the environment in which it grows. Many minerals have such characteristic crystal forms that they often can be identified by this property alone (see Fig. 21.1).

All crystalline substances crystallize in one of seven major geometric patterns. When a mineral grows in unrestricted space, it develops the external shape of its crystal form. However, during the growth of most crystals, the space is restricted, resulting in an intergrown mass that does not exhibit its crystal form. The crystalline forms are studied in detail by means of X-ray analysis.

Hardness is a comparative property that refers to the ability of a mineral to resist scratching. The degrees of hardness are represented on *Mohs' scale of hardness,* which runs from 1 to 10, soft to hard. This arbitrary scale is expressed by the 10 minerals listed in Table 21.1. Talc is the softest, and diamond is the hardest. A particular mineral on the scale is harder than (can scratch) all those with lower numbers. Using these minerals as standards, one finds the following on the hardness scale: fingernail, 2.5; penny, 3; window glass or knife blade, 5.5; steel file, 6.5.

Cleavage refers to the tendency of some minerals to break along definite smooth planes. The mineral may exhibit distinct cleavage along one or more planes, or it may exhibit indistinct cleavage or no cleavage. The degree of cleavage is a clue to the identification of the mineral (● Fig. 21.4).

Fracture refers to the way in which a mineral breaks. It may break into splinters, rough irregularly surfaced pieces, or shell-shaped forms known as conchoidal ("kon-KOID-ul") fractures (see ● Fig. 21.5).

Color is the property of reflecting light of one or more wavelengths. Although the color of a mineral may be impressive, it is not a reliable property for identifying the mineral because the presence of small amounts of impurities may cause drastic changes in the color. Some minerals, such as quartz, come in a variety of colors (● Fig. 21.6a). Azurite (Fig. 21.6b) is one of the few minerals that come in only one color.

Streak refers to the color of the powder of a mineral. A mineral may exhibit an appearance of several colors, but it will always show the same color streak. A mineral rubbed (streaked) across the surface of an unglazed porcelain tile thereby will be powdered and will show its true color.

Luster is the appearance of the mineral's surface in reflected light. Mineral surfaces appear to have a metallic or nonmetallic luster (● Fig. 21.7, on page 564). A metallic luster has the appearance of polished metal; a nonmetallic appearance may be of various lusters, such as vitreous (topaz), adamantine (diamond), pearly (opal), greasy (talc), and earthy/dull (clay).

FIGURE 21.4 Sheet-type Cleavage
This type of cleavage is common in micas.

HIGHLIGHT: Rare Beauties: Gems

A *gem* is any mineral or other precious or semiprecious stone valued for its beauty. Most gems have a crystalline structure, and when they are shaped and polished, their appearance is enhanced immensely. Their beauty depends on the special characteristics of brilliancy, color, prismatic fire, luster, optical effects, and durability.

The brilliancy of a gem is a function of its index of refraction (see Section 7.2). Diamond, which is composed only of carbon atoms bonded into a single giant molecule by single bonds, has one of the highest refractive indexes (zircon, $ZrSiO_4$, is slightly higher) of the well-known gemstones. Brilliancy also depends on the gem's transparency, the angles of the facets, and the polish (degree of smoothness) of its surfaces.

Chemical impurities contribute various colors to gems. The crystalline form of aluminum oxide (Al_2O_3) occurs in nature as the mineral corundum. Pure aluminum oxide is colorless, but minute amounts of chemical impurities create a variety of colors. For example, in sapphires (a variety of corundum), a trace of chromium creates pink, iron creates green and yellow, chromium plus iron creates orange, and titanium plus iron creates blue. Corundum is called ruby when sufficient amounts of chromium are present to create a deep red. As another example, the green of emerald (a gem variety of the mineral beryl, $Be_3Al_2Si_6O_{18}$) is due to the presence of chromium oxide.

Radiation in the visible region of the electromagnetic spectrum is composed of many wavelengths, and we see the radiation as white light. When white light passes into a gemstone, all wavelengths are bent at different angles. These changes produce flashes of different colors emerging from the gemstone. This property of several transparent gems is called *dispersion,* or prismatic fire (see Section 7.2).

Special effects are created in rubies, sapphires, and a few other gems due to the property of double refraction. These gems resolve a single beam of light into two beams, which are absorbed unequally and emerge as different colors. The six-pointed star effect seen in some rubies and sapphires is due to the reflection of light from microscopic, needle-shaped rutile (titanium oxide, TiO_2) crystals that intersect at 60° angles.

A gem must be durable to be used as an ornament. It must possess the ability to resist abrasion, cleavage, and fracture. Thus it must rank high on the

---

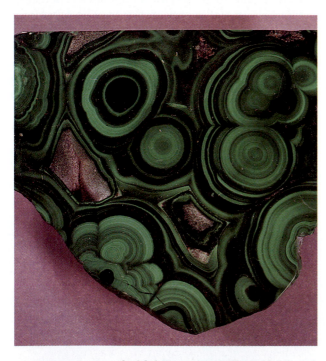

FIGURE 21.5 Conchoidal Fracture Displayed in a Sample of the Copper Mineral Named Malachite

Specific gravity is the ratio of a mineral sample's weight to the weight of an equal volume of water. Each mineral has a characteristic specific gravity. For example, fluorite (CaF_2) has a specific gravity of 3.2, whereas galena (PbS) has a value of 7.6.

RELEVANCE QUESTION: *Minerals are all around you. Prove this by identifying an item on your person or in your room that is made distinctively of one or more mineral components.*

21.2 Rocks

LEARNING GOALS

▼ Describe the three classifications of rocks.

▼ Identify how rocks are formed.

▼ Explain the rock cycle.

Rock is a natural and substantial part of Earth's crust. When we look at a mountain cliff, we see rock rather

hardness scale. Diamond and jade are very durable gems, but opal (hydrated silicon dioxide) and zircon are rather fragile.

The value of a gem depends on its beauty, rarity, size, fashion (current style), and durability. The most valued gems are the very best quality diamonds, emeralds, and rubies. However, high-quality sapphires, opals, and pearls are more expensive than lower grades of diamonds, emeralds, or rubies.

Size is a major determinant of value not only because people tend to perceive bigger as better but also because smaller gems are more common than larger gems of the same type. The enormous sapphire in Fig. 1 is large enough to merit display in the Smithsonian Institution. The size of a gem is determined primarily by its weight. The

FIGURE 1

The Logan Sapphire, a 423-carat, blue sapphire, is the largest one on public display. (Smithsonian Institution)

weight is measured in carats (ct). One carat is equivalent to 200 mg, or approximately 0.007 oz.

The major sources of diamonds are the Republic of South Africa, southwest Africa, and Tanzania. Colombia is the major source of emeralds. The finest jade, rubies, and sapphires come from Myanmar (formerly Burma). Opal is found in many countries, particularly Australia. Cambodia is a major source of zircon. The finest turquoise is found in the western United States.

Quality synthetic gems can be made with corundum. A method of fusing fine alumina in a very hot flame has been perfected, and by adding the appropriate chemical compound to give the desired color, synthetic rubies and sapphires of large size and fine quality can be made.

(a)

(b)

FIGURE 21.6   Quartz Crystals (a) and Azurite Crystals (b)

Like many minerals, quartz comes in a variety of colors. Azurite, a copper mineral, is one of the few minerals that comes in only one distinctive color.

than individual minerals. The Colorado River has carved the Grand Canyon through layers of rock, and the continents and ocean basins are composed of rock. Rocks are classified into three major categories, based on the way they originated.

Igneous rocks are formed by the solidification of magma. **Magma** is molten rock material that originates far beneath Earth's surface. Magma that reaches Earth's surface by way of a volcanic eruption, for example, is called **lava.** Thus igneous rocks may be formed deep inside the crust or at Earth's surface.

Sedimentary rocks, which are formed at Earth's surface, are aggregates of

1. Rock fragments derived from the wearing away of older rocks
2. Minerals precipitated from a solution
3. Altered remains of plants or animals

Metamorphic rocks are formed by the alteration of preexisting rock due to the effects of pressure, temperature, or the gain or loss of chemical components. *Metamorphism* occurs well below the surface but above the depths at which rock is melted.

The Rock Cycle

Most eighteenth-century scientists believed Earth had been structured by catastrophic events, and all rocks on its surface had been deposited by a great flood that took place in the recent past. This doctrine, supported by Holy Scripture, was known as *catastrophism.*

Geology became a true science early in the nineteenth century when scientists first came to understand that the geologic processes that today form and change rocks on Earth's surface and within its interior are the same processes that have been at work throughout the very long history of Earth. Thus ancient rocks were formed in the same way as modern rocks and can be interpreted in the same manner. This basic concept, known as **uniformitarianism,** is the very foundation on which the science of geology rests; that is, the present is the key to the past. James Hutton (1726–1797), a Scottish physician, gentleman farmer, and part-time geologist, is credited with developing the concept (● Fig. 21.8).

Hutton recognized that rocks are continuously being formed, broken down, and re-formed as a result of igneous, sedimentary, and metamorphic processes. His model, called the **rock cycle,** reflects the manner in which internal heat, solar energy, water, and gravity act on and transform the materials of Earth's crust (● Fig. 21.9). Hutton realized that the processes of the rock cycle require a great deal of time. It also was apparent to him that much of the rock material seen at Earth's surface today has been through the cycle many times over. Thus he hypothesized that Earth

FIGURE 21.7   Metallic and Nonmetallic Lusters

The pyrite sample (*left*) displays metallic luster. The limonite sample (*right*) displays an earthy or dull luster. Both are iron minerals.

must be very old—far older than his fellow eighteenth-century scientists had ever imagined and far, far older than the age calculated by biblical scholars. Along with Hutton's principle of uniformitarianism, the concept of the rock cycle and the recognition of Earth's great age remain central themes of modern geology.

RELEVANCE QUESTION: *Where in your local environment can you find evidence of the rock cycle at work?*

21.3 Igneous Rocks

LEARNING GOALS

▼ Describe how igneous rocks are formed.

▼ Explain the classification of common igneous rocks.

As stated previously, igneous rock forms when molten material from far beneath Earth's surface cools and solidifies. Accumulations of solid particles thrown from a volcano during an explosive eruption are also

FIGURE 21.8   Portrait of James Hutton

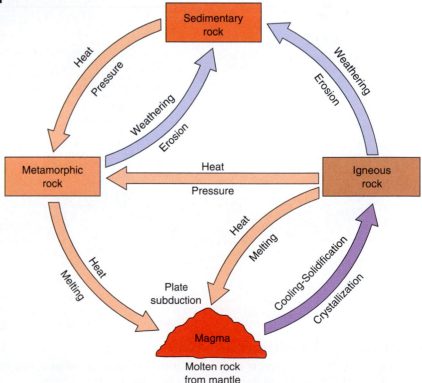

FIGURE 21.9   An Illustration of the Rock Cycle

This diagram shows the relationship of different types of rocks and the various geologic processes that transform one rock type to another.

considered igneous rocks. The molten material is known as *magma* as long as it is beneath Earth's surface but becomes *lava* if it flows on the surface. The term *lava* may be a bit confusing because it can be used to mean either the hot molten material or the resulting igneous rock.

Assuming that Earth was originally molten, the very first rocks of the continents and ocean basins must have been igneous. Geologic processes, however, long ago obscured any recognizable remnant of these most ancient materials. Nevertheless, igneous activity has continued so actively throughout the history of Earth that igneous rocks are by far the most abundant type; they are estimated to constitute as much as 80% of Earth's crust.

The eruption of a volcano is a spectacular geologic phenomenon. Any igneous rock that cools from molten lava is described as *extrusive rock*. Few people realize that the vast majority of magma (molten rock) never finds its way to the surface but cools to solid rock somewhere within Earth's interior as *intrusive rock*. We see intrusive rock only where erosion has

stripped away the overlying rock or where movement of Earth's crust has brought it to the surface.

Igneous Rocks and Plate Tectonics

The occurrence of volcanic activity is for the most part unpredictable. New volcanoes may be formed unexpectedly, and existing volcanoes lying dormant suddenly may erupt with practically no warning. However, the locations of eruptions and potential eruptions are known.

The map in ● Fig. 21.10 shows where most of the active volcanoes of the world are located. Volcanoes are so numerous along the margins of the Pacific Ocean that the region has been dubbed the "Ring of Fire." The theory of **plate tectonics** explains not only the existence of the Ring of Fire but the locations of volcanoes the world over. We will explore plate tectonics in more depth in the next chapter, but for the purpose of discussing volcanoes and other igneous activity, we will briefly summarize the theory here.

According to the theory, Earth's thin, hard outer shell, called the **lithosphere,** is divided into a number

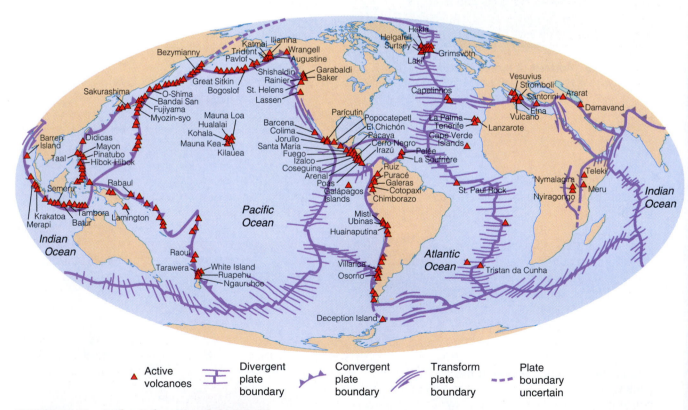

FIGURE 21.10   Active Volcanoes of the World

This map shows the three types of plate boundaries: divergent, convergent, and transform. Notice that the large majority of active volcanoes lie along these boundaries.

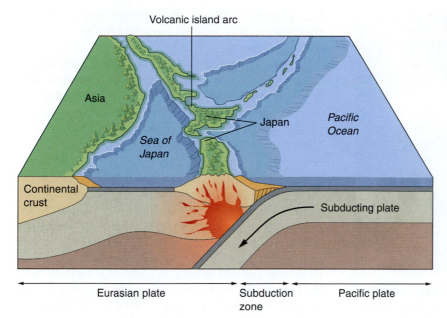

Volcanic island arc

Asia

Japan

Pacific Ocean

Sea of Japan

Continental crust

Subducting plate

Eurasian plate

Subduction zone

Pacific plate

FIGURE 21.11   Subduction of Oceanic Lithosphere Forms a Volcanic Island Arc

This model shows the tectonic activity in the region of the islands of Japan. For the bigger picture, look for this region on the western edge of the Pacific Ocean in the world map in Fig. 21.10.

of large and small fragments called *plates* (see Fig. 21.10). The rigid lithospheric plates rest on a layer of rock called the **asthenosphere.** Kept in a semimolten state by heat radiating out from Earth's interior, the asthenosphere churns in slow motion like a giant pot of boiling pudding. And like the skin that forms atop the pudding, the plates of the lithosphere are set in motion by the churning asthenosphere below. Because the plates are rigid, their movements generate friction and pressure changes along their boundaries. It is along plate boundaries that we find energy released in the form of volcanoes and earthquakes.

As fragments of the rigid outer skin of a spherical Earth, lithospheric plates can move in only three ways: They can push against one another, move away from one another, or slide by one another. Thus the designations **convergent boundary,** where plates are pushing against one another; **divergent boundary,** where they are moving apart; and **transform boundary,** where plates slide past each other. Referring again to Fig. 21.10, note that the large majority of the world's volcanoes occur where plates collide against one another at convergent boundaries.

The divergent boundaries of the world are found primarily along the ocean floor rather than on the continents. Thus most of the igneous activity at divergent boundaries occurs under water as magma wells up into the rift created by the separating plates. When the upwelling magma cools, it forms new oceanic crust. As the plates continue to move apart, magma continues to well up, cool, and move aside. The entire oceanic crust has been formed in this manner and consists of the extrusive igneous rock called *basalt.*

When colliding plate margins both consist of oceanic crust, one buckles and ever so slowly slides beneath the other. During this process of **subduction,** rock just above the subducting plate margin melts, and the magma rises to the surface to form volcanic islands (● Fig. 21.11). Various segments of the Pacific Ring of Fire, such as Indonesia, the Philippines, Japan, and the Aleutians, are volcanic island arcs.

Where the oceanic crust of one plate collides with a continent on a neighboring plate, the oceanic crust subducts, and the rising magma forms a volcanic mountain chain along the margin of the continent. The Andes Mountains of western South America and the Cascade Mountains of western North America are volcanic mountain chains. Much of the magma generated at subduction zones does not rise to the surface but cools inside the crust to form intrusive igneous rock. The giant granitic monoliths of western North America, such as those seen in Yosemite National Park, California (● Fig. 21.12), are intrusive rocks.

Igneous Rock Texture and Composition

Intrusive and extrusive igneous rocks are classified by their *texture,* or physical appearance, and their mineral composition. The texture of an igneous rock is

FIGURE 21.12   El Capitan, Yosemite Valley, California
This pluton is part of the enormous Sierra Nevada batholith.

dictated primarily by the size of its mineral grains. The mineral crystals of the basalt sample shown in ● Fig. 21.13 are tiny, resulting in a fine-grained texture, whereas the mineral grains of the granite sample are large enough to be visible to the unaided eye. Thus granite is a coarse-grained rock (see Fig. 21.13).

Grain size is determined primarily by the rate at which molten rock cools. An igneous body must cool slowly if its mineral grains are to grow large. Rapid cooling invariably yields small grains. The basalt sample in the figure, for example, was most likely brought to the surface as molten lava. Lava exposed to the cool atmosphere loses heat quickly and, therefore, develops only small grains. Globs of lava shot into the air during an explosive eruption cool so quickly that the

lava solidifies without crystallizing. The result is a rock, such as the obsidian sample in ● Fig. 21.14, that has a glassy texture.

Magma deep within Earth loses heat very slowly because the cover of rock that overlies it is a very poor conductor of heat. Igneous rocks formed at these great depths are almost invariably coarse grained. Table 21.2 shows how the location of cooling affects the rate of cooling, which in turn affects the texture of an igneous rock.

Igneous rocks can be roughly divided according to their composition into those rich in silica (SiO_2) and those relatively low in silica. Rocks such as granite are rich in silica and contain minerals with abundant silicon, sodium, and potassium. These minerals are mostly light in color. Rocks such as basalt are low in silica and are rich in iron, magnesium, and calcium. These minerals make low-silica rocks both darker and denser than silica-rich rocks. This density difference explains why the high-silica granitic continents stand higher than the low-silica basaltic ocean basins. Most volcanic islands and many continental mountain chains are built of andesite, an extrusive igneous rock of medium-silica content named after the Andes Mountains of South America, where they were first discovered. See Table 21.3 for the extrusive and intrusive forms of high-, medium-, and low-silica content igneous rocks.

RELEVANCE QUESTION: *Basalt and granite are the most common rocks of Earth's crust. From what you have learned here about the texture and composition of igneous rocks, can you think of some reasons why polished slabs of granite are often used as a decorative facade on large buildings, whereas basalt is not commonly used for such a purpose?*

FIGURE 21.13   Granite (left) and Basalt (right)

FIGURE 21.14   Obsidian

TABLE 21.2   Effects of Cooling on Igneous Rock Textures

Type of Igneous Rock	Location of Cooling	Rate of Cooling	Texture	Example
Intrusive	At depth	Slow	Coarse	Granite
Extrusive	At surface	Rapid	Fine	Basalt
Extrusive	In the air or water	Very rapid	Glassy	Obsidian

21.4 Igneous Activity

LEARNING GOALS

▼ Identify the characteristics of plutonic bodies.

▼ Describe the different types of volcanoes, and identify their eruption characteristics.

Plutons

Intrusive igneous rocks, formed below the surface of Earth by solidification of magma, are known as **plutons.** Plutons are classified according to the size and shape of the intrusive bodies and according to their relation to the surrounding rock that they penetrate. A pluton is *discordant* if it cuts across the grain of the surrounding rock and *concordant* if it is parallel to the grain. The most important discordant igneous rock body by far is a **batholith,** whose most impressive characteristic is its enormous size. It must, by definition, be exposed in an area of at least 103 km^2, but many are vastly larger than that. For example, the Coast Range Batholith in western Canada is more than 1600 km long and in places more than 160 km wide.

All surface exposures indicate that batholiths grow larger with depth, but the nature of their bottoms remains uncertain, because no canyons or mines have penetrated that deep. The invasion of a molten batholith into the rocks of Earth's crust occurs at great depth under high temperature and pressure and is an intimate part of the complex process of mountain building. Batholiths are exposed at Earth's surface by uplift or where mountain chains have been deeply scarred by erosion. ● Figure 21.15 illustrates their shape and relationship to the intruded rock.

Dikes are discordant plutons formed from magma that has filled fractures that are vertical or nearly so, and their shape therefore is tabular—thin in one dimension and extensive in the other two, as illustrated in Fig. 21.15. The sizes of dikes are just as variable as the sizes of the fractures they fill. Dikes are quite common and have been recognized in many different kinds of geologic environments.

A *sill,* illustrated in Fig. 21.15, has the same shape as a dike but is concordant rather than discordant. A *laccolith,* also shown in Fig. 21.15, is a concordant pluton formed from a blister-like intrusion that has pushed up the overlying rock layers.

Products of Volcanic Eruptions

What comes out of a volcano? People who know that volcanoes produce lava may not be aware that lava is but one of three products of volcanic eruptions.

1. The expulsion of gas is the most widespread general characteristic of all volcanoes. A volcano may

TABLE 21.3   Common Igneous Rocks Organized by Composition

Silica Content	Other Major Elements	Extrusive Rocks	Intrusive Rocks	Primarily Found
High (>70%)	Al, K, Na	Rhyolite	Granite	Continental crust
Medium (60%)	Al, Ca, Na, Fe, Mg	Andesite	Diorite	Volcanic island arcs, some continental mountain chains
Low (40–50%)	Mg, Fe, Al, Ca	Basalt	Gabbro	Oceanic crust

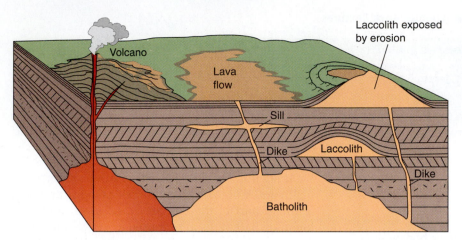

Laccolith exposed
by erosion

FIGURE 21.15 An Illustration of Plutonic Bodies

Magma solidifying within Earth forms intrusive igneous bodies. The batholith is the largest intrusive body. Sills and laccoliths are concordant bodies that lie parallel to existing rock formations. Dikes, which are discordant bodies, cut across existing rock formations.

expel gas in its earliest infancy, at the height of its activity, and during its final dying moments when all other signs of life are gone. Steam may account for as much as 90% of the gases.

2. Volcanoes may erupt lava in variable quantities under various conditions. Some volcanoes produce vast amounts of very fluid lava, which flows easily and quietly with little explosive violence. If much gas escapes at the same time, minor explosions may occur that hurl incandescent lava into the air (● Fig. 21.16). These lava fountains, impressive enough by day, create magnificent fireworks at night. Clots of lava sometimes harden in midair and strike the ground as spindle-shaped volcanic bombs of various sizes. Other volcanoes erupt relatively small volumes of lava so stiff and viscous that it can barely flow at all.

3. Some volcanoes spew enormous volumes of solids that can range in size from the finest dust to huge boulders. Such particles are known collectively as **tephra** and include not only fragments of rock but also gas-laden material that the volcano ejects in molten form but that hits the ground as a solid. Such debris is blasted into the air by violently explosive volcanoes.

Eruptive Style

From the preceding description of the products of volcanoes, it is obvious that there are two basic styles of volcanic eruptions: *peaceful* and *explosive.* In which manner a volcano erupts—peacefully or explosively—depends by and large on the **viscosity** of its magma. Low viscosity allows a magma to flow fluidly and, therefore, peacefully. Highly viscous magma

is thick and stiff and moves only when subjected to great force. Magma viscosity, in turn, depends on the temperature and silica content of the magma. High temperature and low silica content result in low viscosity; low temperature and high silica content result in high viscosity (Table 21.4).

FIGURE 21.16 A Volcanic Eruption

The force of gas escaping from the underground reservoir of magma throws molten lava high into the air over the crater of Hawaii's Kilauea volcano.

TABLE 21.4   Magma Viscosity and Eruptive Style

Magma	Temperature	Silica Content	Style of Eruption	Tectonic Setting
Rhyolitic	Low	High (>70%)	Violent	Oceanic/continental subduction zones
Andesitic	Low to moderate	Medium (60%)	Moderately violent to violent	Oceanic/continental subduction zones, volcanic island arcs
Basaltic	High	Low (40–50%)	Peaceful	Divergent boundaries, hot spots

In peaceful eruptions, lava flows fluidly out of the volcanic vent and is accompanied by the occasional fireworks caused by the expulsion of gases. Many visitors to Hawaii experience the thrill of observing such eruptions from the safety of observation points a few hundred meters from spewing vents. Peaceful volcanic eruptions involve basaltic magma. Because it originates deep beneath the crust, basaltic magma is hot. It flows fluidly because of its high temperature and low silica content.

Basaltic eruptions occur in two settings: along divergent plate boundaries, where it wells up from be-neath the lithosphere to fill the rift between spreading plates, and in plate interiors, where the plate is riding over a **hot spot.** A hot spot is the surface expression of magma that originates in the asthenosphere and deeper. Hot-spot magma rises as a plume and pierces the overlying lithospheric plate, punching out volcanic holes as the plate rides over it. For the past 70 million years, each of the volcanic islands of Hawaii and the Emperor Seamount chain have been formed in turn as the Pacific plate rides over one such hot spot (● Fig. 21.17).

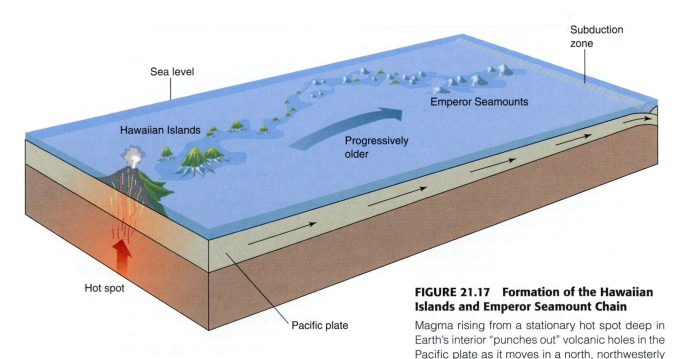

FIGURE 21.17   Formation of the Hawaiian Islands and Emperor Seamount Chain

Magma rising from a stationary hot spot deep in Earth's interior "punches out" volcanic holes in the Pacific plate as it moves in a north, northwesterly direction over the hot spot. For the bigger picture, locate the Hawaiian Islands on the world map in Fig. 21.10.

Volcanoes that erupt explosively are another matter altogether. Unpredictable in both timing and strength, explosive eruptions can cause sudden, massive devastation. For example, early in 1980, Mount Saint Helens, a moderate-sized volcano located in southwestern Washington about 60 km from Portland, Oregon, showed signs that it was going to erupt. The immediate area was evacuated, and geologists set up stations around the volcano to monitor its activity. For months the scientists watched and waited. Then, on May 18, 1980, it exploded. At 8:32 A.M., volcanologist David Johnston was standing at what he believed to be a safe distance of 8 kilometers from the summit of the volcano. He radioed to his base camp, "Vancouver! Vancouver! This is it!" Then the man, his jeep, his trailer, all the trees, and every living thing for hundreds of square kilometers around the mountain were blown away.

Explosive eruptions occur in subduction zones (for example, along the Pacific's Ring of Fire). Because subduction-zone magmas do not originate as deep inside Earth as basaltic magmas, they are cooler in temperature. They are also higher in silica content. For these reasons, they are highly viscous, and instead of flowing out, they tend to form a resistant plug in the volcanic vent. Gases that accompany fluid magmas are able to escape the vent gradually, but the gases that accompany viscous magmas are kept bottled up under high pressure beneath the plugged vent. When the plug rises and the pressure is lowered, the gases explode suddenly and violently, blowing the plug. In the case of Mount Saint Helens, much of the mountain itself was blown to pieces, propelling the debris great distances.

Volcanic Structures

Let us examine the various types of volcanic structures. Some lava reaches the surface through long fractures in the surface rocks. More commonly, however, it pours out of central vents or volcanic cones that previous eruptions have constructed.

Fissure eruptions, which issue from long fractures, have been very rare in human history. Until recently, the latest such volcanic event affected Iceland in 1783. On January 23, 1973, however, a fissure eruption began that threatened to inundate the small island of Heimaey just off the south shore of Iceland. This eruption was of great scientific interest, but it also was a matter of grave concern to residents of Iceland, whose economy depends heavily on the fishing industry centered on Heimaey.

Fissure eruptions have poured remarkably large volumes of basalt onto Earth's surface at various times in the geologic past. Virtually every continent has its own extensive area of *flood basalts,* as these thick accumulations are commonly called. The North American example, which is by no means the world's largest, is the great Columbia Plateau, a conspicuous landform over several of our northwestern states. The Columbia Plateau basalt covers an area of 576,000 km^2 to an average depth of 150 m. Calculate the total volume of this lava, and you will realize its enormous size. The ocean basins are floored with even greater floods of basalt, estimated by some to be as much as 5 km thick.

The basaltic lavas that came from fissure eruptions were so extremely fluid that they flowed many miles over Earth's surface without constructing volcanoes. Individual lava flows were not thick, however, because they extended for many miles over the surface. As one lava flow followed another over millions of years, the entire landscape was eventually drowned in a sea of solid basalt, with perhaps here and there a high peak forming an island in the black and desolate ocean of rock. The Blue Mountains of Oregon are examples of such islands within the Columbia Plateau.

Not quite as hot and fluid as the ancient basalts that built the Columbia Plateau, the frequently repeated flows of modern-day basalts form a gently sloping, low-profile **shield volcano.** The classic example of a shield volcano is Mauna Loa in the Hawaiian chain. In fact, Mauna Loa is a huge volcanic mountain, the largest single mountain on Earth in sheer bulk. Although not as tall as Mt. Everest, which is 8800 m above sea level, Mauna Loa rises 4600 m from the ocean floor to sea level and protrudes an additional 4150 m above sea level for a total height of about 8750 m. Its bulk comes from the fact that this partially submerged mountain has a base almost 160 km in diameter.

Volcanic eruptions of both lava and tephra form a more steeply sloping, layered composite cone that is called a **stratovolcano,** or *composite volcano.* The lava of stratovolcanoes has a relatively high viscosity, and eruptions are more violent and generally less frequent than those of shield volcanoes. Many stratovolcanoes have an accumulation of material up to 1800 to 2400 m above their base and have a characteristic symmetrical profile. Mount Saint Helens is a strato-

FIGURE 21.18   Mount Fuji in Japan
A dormant stratovolcano.

volcano. Dormant stratovolcanoes include Mt. Fuji in Japan (● Fig. 21.18) and Mts. Shasta, Hood, and Rainier in the Cascade Mountains.

A volcanic eruption also may consist primarily of tephra. In this case, steeply sloped **cinder cones** are formed, which rarely exceed 300 m in height. A cinder cone is shown in ● Fig. 21.19 rising above a previous lava flow.

Activity is usually not confined to the region of the central vent of a volcano. Fractures may split the cone, with volcanic material emitted along the flanks of the cone. Also, material and gases may emerge from small auxiliary vents, forming small cones on the slope of the main central vent. A funnel-shaped depression called a *volcanic crater* exists near the summit of most volcanoes, from which material and gases are ejected (● Fig. 21.20).

Many volcanoes are marked by a much larger depression called a **caldera.** These roughly circular, steep-walled depressions may be up to several miles in diameter. Calderas result primarily from the collapse of the chamber at the volcano's summit from which lava and ash were emitted. The weight of the ejected material on the partially empty chamber causes its roof to collapse, much like the collapse of a snow-laden roof of a building. Crater Lake on top of Mount Mazama in Oregon occupies the caldera formed by the collapse of the volcanic chamber of this once-active stratovolcano (● Fig. 21.21).

Historic Eruptions

Few volcanoes in written history can match the eruption of Tambora in violence and in the volume of solids hurled into the atmosphere. Tambora, on the island of Soembawa just east of Java, erupted with a thunderous roar from April 10 to 20, 1815. By the time the eruption was finished, there was a hole where previously a high mountain had stood. The 145 km^3

FIGURE 21.19   Cinder-cone Volcano Izalco in El Salvador

FIGURE 21.20 A Volcanic Crater
This volcanic crater in Costa Rica is filled with water colored yellow by suspended sulfur.

of solid debris thrown into the air so darkened the sky that for several days total darkness reigned for hundreds of miles around. Enough fine volcanic dust was circulated around Earth that the Sun's rays were partially blocked, and the year 1816 was uncommonly cold.

In 1902, geologists were reminded of the unusually destructive behavior of certain explosive volcanoes. After several months of violent, threatening activity, Mt. Pelée, at the northern end of Martinique in the Caribbean Sea, blasted out incandescent, cloudlike mixtures of superheated gas and tephra that swept

FIGURE 21.21 Crater Lake
Crater Lake on top of Mount Mazama in Oregon occupies a caldera about 10 km in diameter.

down the side of the volcano and over the nearby city of St. Pierre. In a matter of seconds, virtually all life and property were destroyed. Although nobody knows the exact loss of life because the city was crowded with refugees seeking shelter from Mt. Pelée, estimates are that between 28,000 and 40,000 people met almost instant death. Among the casualties was the governor of the island, who had come to St. Pierre to assure the inhabitants that they had nothing to fear from the nearby volcano. One of only two survivors was Auguste Ciparis, a convicted murderer, who was awaiting execution in a dungeon where he was shielded from the fiery blast.

In 1943, Paricutin erupted in a Mexican farmer's cornfield. In the first year, the volcanic cone rose to a height of over 300 m. After about 10 years of subsiding activity, Paricutin became dormant, with a cone height of over 400 m.

In 1963, volcano Surtsey (after *Surtr,* the subterranean god of fire in Icelandic mythology) boiled up from the ocean floor off the coast of Iceland. Having sufficient lava to form a barrier against the sea, Surtsey is now a permanent volcanic island that may be found on the world map (● Fig. 21.22).

Earlier in this chapter we described a massive volcanic eruption that devastated a huge area of southwestern Washington State. Prior to 1980, Mount Saint Helens was a placid, snow-capped, dormant volcano,

one of many in the Cascade Mountains in Oregon and Washington. Its last period of activity had been between 1800 and 1857.

In March 1980, a series of minor earthquakes occurred near Mount Saint Helens, and the first eruptions took place. The massive May 18 eruption devastated an area of more than 400 km^2 and left more than 60 people dead or missing. A column of ash rose to an altitude of more than 20 km. Ash blanketed nearby cities, and a light dusting fell as far away as 1450 km to the east. Huge mudflows from the ash caused flooding and silting in the rivers near the volcano, destroying and damaging many homes.

Mount Saint Helens became relatively quiet in 1981. However, it has shown minor activity in the years since, and geologists expect that intermittent activity may continue for years, and even decades, if the behavior of the volcano follows the pattern of its eruptions in the 1800s.

Mount Pinatubo in the Philippines began erupting in June 1991, throwing hot ash, rock, water, and gases into the atmosphere to a height of over 24 km. Hundreds of people were killed, and thousands were left homeless. Thousands of acres of cropland including rice, sugarcane, corn, and fish farms were covered by mud and ash, rendering the land unusable. The U.S. Department of Defense evacuated all personnel from Clark Air Force Base and gave up the base because

FIGURE 21.22 The Volcanic Island Surtsey

This volcanic island was formed in 1963 off the southern coast of Iceland.

of repair costs and its closeness to the still-smoking volcano (see Fig. 20.17).

Throughout human history, volcanoes have been viewed with awe and fear. Yet, in many areas of the world, large populations continue to live in the danger zones around volcanoes. Take southern Italy, for example. An A.D. 79 eruption of Mt. Vesuvius buried 2000 residents of the city of Pompeii under a thick, burning blanket of ash. Between then and 1944, Vesuvius has erupted some 83 times, yet the population of the region continued to grow, and the city of Naples, located only a few kilometers from the volcano, became southern Italy's largest metropolis. The reason is based on a simple economic fact: Volcanic ash forms rich, fertile soil. The region around Naples is one of the world's most agriculturally productive. However, as you will learn in Chapter 23, eruptions are not the only volcanic hazards. The soil formed from volcanic ash is rich, but it is also fine and loose and forms a gooey mass of mud when wet. In early 1998, record rainfall in southern Italy resulted in the burial of 200 people in the region around Naples, this time under 4-meter-thick rivers of mud that flowed downslope faster than people could get out of the way.

RELEVANCE QUESTION: *How far is your hometown from the nearest potential volcanic hazard?*

21.5 Sedimentary Rocks

LEARNING GOALS

▼ Describe the origin and classification of sedimentary rocks.

▼ Identify common features in sedimentary rocks.

As discussed earlier, James Hutton's model of the rock cycle was derived from his observations that rocks are continuously being formed, broken down, and reformed as a result of igneous, sedimentary, and metamorphic processes. His conceptualization began with an understanding of sedimentary processes.

In his field trips around Scotland, Hutton observed the everyday effects of rain and wind beating down on rocks and soil. He saw mud, sand, and rock fragments, large and small, carried by streams from the Scottish highlands to the sea, and he saw these parti-cles deposited as **sediment** along rivers, in lakes, and on the seafloor.

Hutton also observed rocks in the highlands that were made of cemented sand and rock fragments and concluded that they were *sedimentary rocks*—rocks whose components had been derived from the wearing away of older rocks. But how did sediments deposited on the seafloor end up as highland rocks?

Hutton knew that sediments are deposited on the seafloor in horizontal layers, or **strata,** and that each successive layer is deposited on top of the previous layer. This simple principle, called *superposition,* was established long before Hutton. He realized that as layer upon layer was piled on top, the older layers were compacted by the weight above them and converted to rock. Subsequently, he concluded, the rocks were uplifted by powerful forces from within Earth to form a mountain range. But even as the new mountains were rising, rain and wind were working to wear them down, and streams were transporting their sediments to the sea.

Sediments and sedimentary rocks make up only 5% of Earth's crust, but their importance is entirely out of proportion to their limited abundance. One reason is that they cover about 75% of the continents and even more of the ocean basins. In a sense, they are a bit like clothing; they cover only the surface, but they are conspicuous. The North American continental interior is composed of a foundation of igneous and metamorphic rocks with a sedimentary veneer only a few kilometers thick. Residents of many of the interior states of the United States may never have seen any rocks other than sedimentary ones. Our modern way of life owes much to sedimentary rocks, because they contain abundant petroleum, coal, metal deposits, and many of the materials essential to the construction industry.

Sedimentary rocks, with their many varieties, commonly form fabulous landscapes. For example, the beauty and variety of the Colorado River's Grand Canyon with its staircase-like, brightly colored slopes and cliffs of sedimentary rock are famous around the world (see the photo at the beginning of this chapter).

The Origins of Sedimentary Rocks

The transformation of sediment into a sedimentary rock is a process called **lithification.** During this process, the loose, solid particles (sediment) are com-

FIGURE 21.23   Detrital Sedimentary Rocks

Note the differences in grain size. Shale (*top left*) is very fine-grained, sandstone (*above*) is coarse, and conglomerate (*bottom left*) is very coarse, with large pebble-sized grains.

pacted by the weight of overlying material and eventually cemented together. Common cementing agents are silica, calcium carbonate ($CaCO_3$), and iron oxides, which are dissolved in groundwater that permeates the sediment. An example is the formation of shale. When fine-grained mud is subjected to pressure from overlying rock material, water is driven off, and the clay minerals begin to compact (consolidate). As groundwater moves through the compacted sediment, materials dissolved in the water precipitate around the individual small mud particles, and cementation of the particles occurs. A sedimentary rock results.

Sediments are classified according to the source of their constituents into two main groups: *detrital* and *chemical*. **Detrital sediments** are composed of solid fragments, or *detritus*, derived from preexisting rock (● Fig. 21.23). Detrital rocks are classified on the basis of the size of their components (Table 21.5). *Shale* is made from the very fine particles that comprise mud. *Sandstone* consists of sand-sized grains in a silica or calcium carbonate cement. *Conglomerate* is made from large, rounded fragments of varying sizes embedded in silica, calcium carbonate, or iron oxide cement. Rock of the same composition but with angular fragments is called *breccia*.

TABLE 21.5   Classification of Detrital Sedimentary Rocks

Sediment	Grain Size (mm)	Rock Name
Gravel	>2	Conglomerate (rounded pebbles)
		Breccia (angular pebbles)
Sand	$\frac{1}{16}$–2	Sandstone
Mud	$<\frac{1}{16}$	Shale

Chemical sediments are composed of minerals that were transported to the sea in solution. Table 21.6 shows that there are two types of chemical sedimentary rocks: *organic* and *inorganic*. Organic chemical sedimentary rocks are composed of minerals that were transported in solution but were subsequently acted on by marine organisms. *Organic limestone,* for example, consists of calcite ($CaCO_3$). The $CaCO_3$ dis-solved in seawater is extracted by microscopic marine organisms and converted to calcium carbonate skeletal and shell matter. When these organisms die, they fall to the seafloor, where their hard parts collect and eventually lithify.

Coal is classified as an organic chemical sedimentary rock. Although it does not meet the requirement of having been derived from minerals transported in solution, it is included under this heading because it is derived from the lithified remains of plant matter.

Inorganic chemical rocks are formed when the evaporation of water containing dissolved materials leaves behind a residue of chemical sediment, such as sodium chloride (halite or rock salt) and calcium sulfate (gypsum). Another example of inorganic chemical sedimentary rock is cave dripstone, which is formed primarily by calcium carbonate precipitated from dripping water. Dripstone takes on a variety of forms, but most common are the icicle-shaped *stalactites* and cone-shaped *stalagmites* (● Fig. 21.24). Note that stalactites extend down from the ceiling (*c* for ceiling), whereas stalagmites protrude up from the ground (*g* for ground).

TABLE 21.6   Types of Chemical Sedimentary Rocks

Organic Sedimentary Rocks	
Rock Name	**Characteristics**
Bituminous coal	Compacted plant remains
Organic limestone	Compacted or cemented calcareous plant or animal remains
Coquina	Porous aggregate of shell fragments

Inorganic Sedimentary Rocks	
Rock Name	**Characteristics**
Inorganic limestone	Compacted, cemented, calcareous material
Gypsum	Evaporite. Calcium sulfate
Halite	Evaporite. Sodium chloride (rock salt)
Dripstone	Calcium carbonate precipitated from dripping water

FIGURE 21.24   Dripstone
Stalactites and stalagmites in a limestone cavern.

Sedimentary Characteristics and Structures

Geologists gain much information from sedimentary rocks because their characteristics tell so much about their origin and history. Color, rounding, sorting, bedding, fossil content, ripple marks, mud cracks, footprints, and even raindrop prints are common characteristics of sedimentary rocks that indicate the conditions under which the rocks formed.

The color of a sedimentary rock is especially conspicuous in drier parts of the world where little soil or vegetation mantles the surface. The bright, variegated colors of the Painted Desert in Arizona draw crowds of tourists, but even rocks with dull colors are conspicuous in a dry landscape (● Fig. 21.25). Gray, the most common sedimentary rock color, usually reflects the rock's origin in shallow, well-aerated marine water. Rocks deposited above sea level in the presence of abundant oxygen are usually colored by iron oxides and are red-brown or yellow-brown—colors that are so striking in many of our western states. Those few

sedimentary rocks that are dark gray to black contain carbon from the organic matter that accumulated in the stagnant water where the sediment was deposited.

The larger detrital sediments tend to be worn as the currents and waves that move them grind the particles against the stream bottom and against each other. Angular fragments within the rocks inform geologists that these sediments were not carried far and that they were dropped quickly. The rounder grains of quartz sandstone tell of their long journey downstream and of the many hours they were shifted and rolled by the waves and currents of the sea.

A current or wave that can keep small grains in motion may be unable to move larger grains. Because wind and moving water transport some particles more easily than others, the particles tend to become sorted or separated according to size. The greater the distance of transportation, the more effectively the grains are sorted. In a lake or ocean, the largest particles come to rest near shore, but the finer grains are transported into the quieter water farther from shore, where they eventually settle to the bottom.

Bedding, or stratification, is the layering that develops at the time the sediment is deposited. The bedding may be conspicuous, as shown in ● Fig. 21.26, or

FIGURE 21.25 Arizona's Painted Desert

FIGURE 21.26 Bedding (Stratification) in Limestone

FIGURE 21.27 Cross-bedding in Sandstone

RELEVANCE QUESTION: *Imagine that you are standing on a pearly white, sandy beach. If the sand on this beach is caught up in the rock cycle, what transformations may it go through, and where might it be found in the geologic future?*

FIGURE 21.28 Two Common Types of Fossils

Some fossils, such as the dinosaur bones (*top*), may consist of the actual remains of the organism. However, most fossils, such as the fish cast (*bottom*), consist only of the impression left in the rock by the organism after the remains decomposed or dissolved.

vague. It may be in either thick or very thin layers. Because most sediment comes to rest on a level surface, most bedding is horizontal. In many environments, however, sediments accumulate in tilted layers known as *cross-bedding*. Much research and many pages have been devoted to the geometry and origin of the many kinds of cross-bedding. In every case, the surface is sloping where the sediments come to rest. Cross-bedding is especially common where a river empties into a lake, where sediments fill in depressions scoured by floods along river channels, or where wind drapes sand down the flanks of a dune (● Fig. 21.27). By recognizing the type of cross-bedding, the geologist makes an observation that helps to unravel the origin and history of the rock.

The most distinctive and most interesting characteristic of a sedimentary rock is the fossils it sometimes contains (● Fig. 21.28). Although a mystery to those who lived several centuries ago, a *fossil* is now known to be the remains or traces of a prehistoric organism. Geologists have collected so many fossils and have so refined the techniques of examining them that they can determine the relative age of a sedimentary rock from its fossil content. We will discuss the role of fossils in geologic age determinations in Chapter 24.

21.6 **Metamorphic Rocks**

▼ Describe the origin of metamorphic rocks.

▼ Identify the types of metamorphisms.

▼ List and describe some common metamorphic rocks.

We learned earlier that magmas that cool to form igneous rocks are created by high temperatures deep inside Earth's interior. We also learned that sedimentary rocks are formed when sediments are buried and lithified just below Earth's surface. Metamorphic rocks are created by the conditions that exist in a zone in Earth's interior that is below the region where rocks are lithified and above the region where they are melted.

About 15% of Earth's crust is metamorphic rock. *Metamorphism* is the process by which the structure and mineral content of a rock are changed while the rock remains in a solid state. All igneous or sedimentary rocks can be metamorphosed, and any metamorphic rock can be subjected to further metamorphism. The agents of metamorphism are heat and pressure. Chemically active fluids, when present, also influence the process. The resulting metamorphic rock depends on these influences as well as on the composition of the *parent rock* being metamorphosed (Table 21.7).

Both the texture and the mineral composition can change as a rock is metamorphosed. For example, in the metamorphism of organic limestone to *marble,* the texture of the rock is changed, but not the mineral content (Fig. 21.29). Organic limestone is made of fossils and a cement matrix, both of which are composed of small calcite crystals. When the limestone undergoes temperature and pressure changes, the calcite crystals grow large. All traces of fossiliferous limestone texture are obliterated in the marble, but the calcite content remains the same.

As a general rule, if a parent rock contains only one mineral (calcite, in the case of limestone), the metamorphic rock will be composed of that mineral alone. But if the parent rock contains several minerals,

TABLE 21.7   Classification of Some Common Metamorphic Rocks

Foliated Rocks			
Parent Rock	**Metamorphic Rock**	**Key Minerals**	**Characteristics**
Shale	Slate	Clay, quartz, mica, chlorite	Fined-grained, slaty cleavage
Shale, basalt, slate	Schist	Chlorite, plagioclase, mica, garnet	Coarse-grained, well-foliated
Shale, granite, slate, schist	Gneiss	Plagioclase, garnet, kyanite, sillimanite	Coarse-grained, light- and dark-colored bands

Nonfoliated Rocks			
Parent Rock	**Metamorphic Rock**	**Key Minerals**	**Characteristics**
Limestone	Marble	Calcite	Coarse, interlocking calcite grains
Sandstone	Quartzite	Quartz	Fine to coarse, interlocking quartz grains
Shale, basalt, or any fine-grained rock	Hornfels	Mica, quartz	Fine-grained, variable composition

FIGURE 21.29 **Fossiliferous Limestone (*above*) and Marble (*right*)**

When fossiliferous limestone is metamorphosed into marble, the texture of the rock is radically changed, but the calcite mineral content of the rock remains the same.

metamorphism will create new and different minerals. For example, shale is commonly composed of clay, quartz, mica, and chlorite. ● Figure 21.30 shows that as shale undergoes increasing pressure and temperature changes, it metamorphoses to *slate*, then to *schist*, and then to *gneiss* (pronounced "nice"). As it progresses through each stage, both its texture and its composition change.

Although it is always difficult to make clear-cut separations, we recognize several kinds of metamorphism: contact, shear, and regional.

Contact metamorphism is change brought about primarily by heat, with very little pressure involved. Such a change commonly occurs in shallow bedrock when it is subjected to the heat of a molten body of magma moving up from greater depths. Contact metamorphism is most obvious at such a shallow depth because bedrock near Earth's surface is normally cool, and the effects of great temperature changes are, therefore, quite pronounced. The rock immediately next to the molten magma experiences intense metamorphism and may be coarse-grained, but it grades out into finer-grained rock that has been less severely transformed by the heat. This dark, fine-grained rock, containing recrystallized minerals with random orientation, is known as *hornfels*.

Rocks changed more by pressure than temperature are said to have undergone the effects of **shear metamorphism,** which is most common in active fault zones, where one rock unit slides past another. Mechanical deformation shatters the grains or changes their shapes plastically. Recrystallization accompanies the more intense forms of shear metamorphism, but the significant physical effects are more obvious.

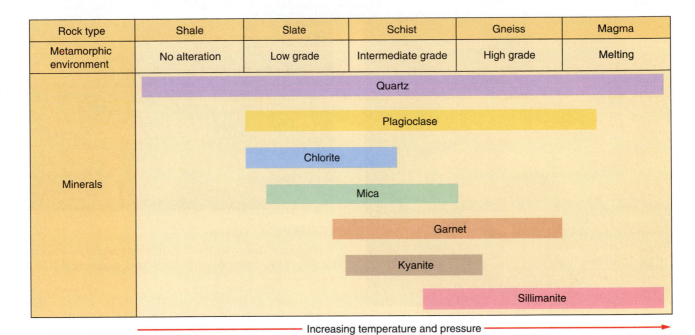

Rock type	Shale	Slate	Schist	Gneiss	Magma
Metamorphic environment	No alteration	Low grade	Intermediate grade	High grade	Melting

Minerals

Quartz

Plagioclase

Chlorite

Mica

Garnet

Kyanite

Sillimanite

Increasing temperature and pressure

FIGURE 21.30   Effects of Increasing Temperature and Pressure in Metamorphism

As sedimentary shale is subjected to increasingly intense temperatures and pressures, it becomes metamorphic slate, then schist, then gneiss, and finally the rock melts and becomes magma.

Most metamorphic rocks have been affected by both high temperature and high pressure and have, therefore, experienced both mechanical deformation and chemical recrystallization. **Regional metamorphism,** as this type of change is known, receives its name from the extremely large areas it affects. The widely exposed metamorphic rocks in central Canada are of this sort. Although we still have much to learn about regional metamorphism, it appears mostly to affect rocks undergoing intense deformation by mountain building.

In rocks subjected to regional metamorphism, the mineral grains are flattened, elongated, and aligned perpendicular to the direction of the compressive pressure. This parallel arrangement of mineral grains results in a pronounced layering of the rock called **foliation.** Rocks that are metamorphosed solely by the heat of contact metamorphism do not develop foliation.

The progressive metamorphism of shale is a good illustration of changes that occur as a sedimentary rock is subjected to more and more intense regional metamorphism. Shale, after relatively mild metamorphism, is transformed into slate (● Fig. 21.31), a fine-

FIGURE 21.31   Slate Bed

Slates are fine-grained, metamorphic rocks that possess a type of foliation known as slaty cleavage.

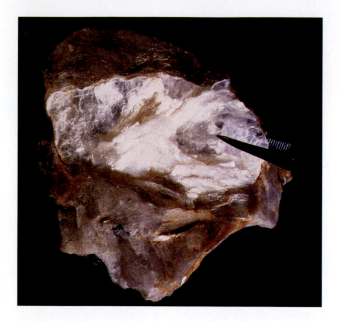

FIGURE 21.32   Schist

This mica schist is a product of intermediate-grade metamorphism.

FIGURE 21.33   Gneiss

Feldspar and quartz are the chief minerals of gneiss, a metamorphic rock, contorted under high temperature and pressure.

grained, metamorphic rock similar in many respects to shale but different fundamentally in its excellent *slaty cleavage.* A rock exhibiting slaty cleavage breaks apart easily along the planes of its thin, smooth layers. If the shale is subjected to more intense heat and pressure, it will change into schist, a foliated metamorphic rock whose grains are visible to the unaided eye (● Fig. 21.32). Very intense regional metamorphism produces gneiss, an even coarser-grained rock with rough foliation characterized by distinct banding (● Fig. 21.33). The higher grades of metamorphism, therefore, produce larger grains but rougher foliation.

Metamorphic rocks that lack foliation include such well-known examples as marble, which is formed from limestone, and quartzite, which is metamorphosed sandstone.

RELEVANCE QUESTION: *Slate has been used in various ways over the years. What are some of its uses in relation to education, building, landscaping, and recreation?*

Important Terms

geology
mineral (21.1)
rock
silicates
crystal form
hardness
cleavage
fracture
color
streak
luster
specific gravity

igneous rocks (21.2)
magma
lava
sedimentary rocks
metamorphic rocks
uniformitarianism
rock cycle
plate tectonics (21.3)
lithosphere
asthenosphere
convergent boundary
divergent boundary

transform boundary
subduction
plutons (21.4)
batholith
tephra
viscosity
hot spot
shield volcano
stratovolcano
cinder cones
caldera
sediment (21.5)

strata
lithification
detrital sediments
chemical sediments
bedding
contact
   metamorphism (21.6)
shear metamorphism
regional metamorphism
foliation

Review Questions

21.1 Minerals

1. Which of the following is a characteristic property of a mineral?
   (a) naturally occurring
   (b) crystalline
   (c) definite chemical composition
   (d) all of the above

2. The two most abundant elements in Earth's crust are
   (a) oxygen and iron.   (c) aluminum and iron.
   (b) silicon and oxygen.   (d) silicon and aluminum.

3. Feldspars
   (a) are the most abundant minerals found in Earth's crust.
   (b) are classified in two main types.
   (c) contain the elements oxygen, silicon, and aluminum.
   (d) all of the above.

4. Define *mineral.*

5. What two elements make up the majority of Earth's crust and in what percentages?

6. What is the most common mineral group of the crust?

7. Name three other important mineral groups.

8. What is the silicon-oxygen tetrahedron? Sketch its structure.

9. Name an ore of lead, of iron, and of uranium.

10. State six physical characteristics that are used to identify minerals.

21.2 Rocks

11. Rocks
    (a) are classified as igneous, sedimentary, or metamorphic.
    (b) are composed of minerals.
    (c) are classified based upon the way they originated.
    (d) form the continents and ocean basins.
    (e) all of the above.

12. The rock cycle illustrates
    (a) the interrelationships among the processes that produce the three types of rocks.
    (b) the chemical composition of basaltic rocks.
    (c) the crystallization of minerals.
    (d) all of the above.

13. Define *rock.*

14. Define *magma.*

15. What is the term for magma that reaches Earth's surface?

16. What are the three processes that produce crustal rocks?

17. Where in relation to Earth's surface does each rock-producing process occur?

21.3 Igneous Rocks

18. Igneous rocks are
    (a) the most abundant found in Earth's crust.
    (b) described as extrusive or intrusive.
    (c) classified according to their mineral composition and texture.
    (d) all of the above.

19. Grain size in igneous rocks
    (a) is determined primarily by the rate at which molten rock cools.
    (b) can be both large and small in the same rock sample.
    (c) refers to texture.
    (d) all of the above.

20. Define *igneous rock.*

21. Distinguish between magma and lava.

22. Distinguish between intrusive and extrusive rocks. State the physical characteristics of each.

23. What two major characteristics are used to classify igneous rock?

24. How does the silica content of granite compare with that of basalt?

25. Where are most of the world's active volcanoes located?

26. What tectonic process produces volcanic island arcs?

21.4 Igneous Activity

27. What is a hot spot?

28. Plutons are
    (a) intrusive igneous rocks.
    (b) solids known as tephra.
    (c) solidified lava.
    (d) explosive volcanoes.

29. Which of the following is a component of the explosive eruption of a volcano?
    (a) water vapor   (d) fine ash and dust
    (b) carbon dioxide   (e) all of the above
    (c) rock

30. Distinguish between discordant and concordant igneous rock bodies.

31. What two factors determine the viscosity of magma and lava?

32. What is the relationship between magma viscosity and the eruptive style of a volcano?

33. Distinguish among (a) shield volcanoes, (b) stratovolcanoes, and (c) cinder cones.

34. What is a fissure eruption?

35. What are calderas, and how are they formed?

21.5 Sedimentary Rocks

36. Sedimentary rocks are
    (a) the most abundant surface rocks.
    (b) the product of gradation.
    (c) a few kilometers thick.
    (d) all of the above.

37. Which of the following is *not* a common feature of sedimentary rock?
    (a) bedding   (d) discordant
    (b) stratification   (e) mud cracks
    (c) ripple marks

38. Name some of the characteristics geologists use to determine the origin of sedimentary rocks.

39. Name three sedimentary rocks, and state the sediments from which they originate.

40. What is meant by *bedding*?

41. Describe the process of lithification.

42. Distinguish between detrital and chemical sedimentary rocks.

21.6 Metamorphic rocks

43. Metamorphic rocks
    (a) are those that have been changed under the influence of great temperature and/or pressure.
    (b) make up about 30 percent of Earth's crust.
    (c) are changed physically but never chemically during metamorphism.
    (d) all of the above.

44. Which of the following is *not* a type of metamorphism?
    (a) contact   (c) regional
    (b) shear   (d) foliated

45. Metamorphic rock is classified according to its
    (a) texture.   (c) foliation.
    (b) mineral composition.   (d) all of the above.

46. Which of the following is *not* a common metamorphic rock?
    (a) slate   (c) gneiss   (e) quartzite
    (b) marble   (d) shale

47. Define *metamorphic rock.*

48. Describe three kinds of metamorphism and indicate the most essential differences among them.

49. Define *foliation.*

50. Name three metamorphic rocks and state the name of the parent rock from which each forms.

Applying Your Knowledge

1. Suppose you are vacationing in Hawaii with a friend and are sunbathing at beautiful Waikiki Beach. To impress your friend with your great knowledge of chemistry and mineralogy, explain the similarities and differences between the sand beneath your blanket and the carbonation in the beverage you are drinking.

2. While in Hawaii you want to visit Mauna Loa. Your friend, however, is afraid to go anywhere near the volcano because it might explode like Mount Saint Helens. Explain to your friend why Mauna Loa, which erupts regularly, can be expected to erupt peacefully. Then explain how geologists knew in advance that if and when Mount Saint Helen's blew, it would be a dangerously explosive eruption.

3. According to the rock-cycle concept, the content of a quartz sandstone was derived from the weathering of other, older rocks. Are there features of a sandstone sample that can tell you whether the quartz came from igneous, sedimentary, or metamorphic rocks? Explain your answer.

4. If conglomerate is classified as a sedimentary rock, why might a particular sample consist largely of pebble-sized pieces of igneous and/or metamorphic rock?

5. You are in lab class and the instructor asks you to comment on the history of a rock sample. Your first guess is that the rock is a conglomerate, because it consists primarily of smooth, pebble-sized fragments. However, the fragments are not round in shape but elongated, and they appear to be neatly arranged parallel to one another, like crayons in a box. What was happening in this rock's place of origin to cause it to look like this?

Answers to Multiple-Choice Review Questions

1. d   3. d   12. a   19. d   29. e   37. d   44. d   46. d
2. b   11. e   18. d   28. a   36. d   43. a   45. d

STRUCTURAL GEOLOGY

22

*The face of places,
and their forms decay;
And that is solid earth,
that once was sea;
Seas, in their turn,
retreating from the shore,
Make solid land,
what ocean was before.*

Ovid (43 B.C.–A.D. 18)

As we saw in Chapter 21, plate tectonics is the primary mover and shaker of Earth's outer shell, the lithosphere. In this chapter we will take a closer look at plate tectonics, beginning with its precursors, the theories of continental drift and seafloor spreading. Then we will examine three manifestations of plate movement: earthquakes, crustal deformation, and mountain building.

Earthquakes are vibrations that radiate outward in all directions when rocks grind past one another at a plate boundary. The danger of being caught in an earthquake is not in the vibrations themselves but in what the vibrations do to bring buildings and other human-made structures crashing down on us. On a more positive note, because earthquake waves travel through Earth as well as along the surface, their analysis has provided scientists with a "picture" of Earth's interior unobtainable through other means.

As churnings in the semimolten asthenosphere drive the rigid lithospheric plates against, away from, or past one another, stress is transmitted to crustal rocks, causing them to fold, stretch, or fracture. With few exceptions, mountains and other forms of crustal deformation owe their existence to movement at present or past plate boundaries. ■

Photo: Great Hanshin earthquake (1995), Kobe, Japan.

22.1 Continental Drift and Seafloor Spreading

LEARNING GOALS

▼ Explain the theory of continental drift, and list supporting evidence for it.

▼ Explain the mechanism and evidence for seafloor spreading and its relationship to continental drift.

When looking at a map of the world, we are tempted to speculate that the Atlantic coasts of Africa and North and South America could fit nicely together as though they were pieces of a jigsaw puzzle. This observation has led scientists at various times to suggest that these continents, and perhaps the other continents, were once a single, giant supercontinent that broke into fragments, which then drifted apart. However, scientists had no evidence to support the idea of such an event other than the shapes of the continents.

Continental Drift

In the early 1900s, Alfred Wegener (1880–1930), a German meteorologist and geophysicist, revived the idea of **continental drift** and brought together various geologic evidences for its support. Wegener's hypothesis gave rise to considerable controversy, and only in the last few decades has conclusive evidence been found that supports some aspects of it.

Wegener's assumption was that the continents were once part of a single giant continent, which he called **Pangaea** (from the Greek, meaning "all lands" and pronounced "pan-JEE-ah"). He proposed that this hypothetical supercontinent rifted (broke apart) about 200 million years ago and its fragments somehow drifted to their present positions and became today's continents (● Fig. 22.1).

The scientific evidence supporting Wegener's hypothesis that the continents once formed a single land mass is of several different types. Let's consider the three most prominent ones briefly.

1. *Biological evidence.* Similarities in biological species and fossils found on distant continents that are today separated by oceans strongly suggest that these land masses were at one time together as Pangaea. For example, a certain variety of garden snail is found only in the western part of Europe and the eastern part of North America. We

(a) 200 million years ago

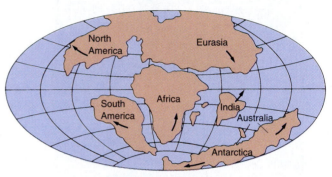

(b) 100 million years ago

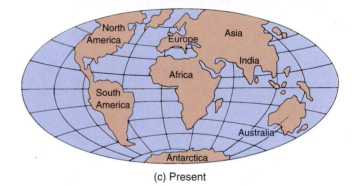

(c) Present

FIGURE 22.1   Sequence of Events in the Breakup of Pangaea

(a) 200 million years ago the giant supercontinent Pangaea rifted, and the continental fragments began to drift apart. (b) Continental positions 100 million years ago. (c) Present positions of the continents.

would not expect that the species could traverse the present-day oceans. Similarly, fossils of identical reptiles have been found in South America and Africa, and identical plant fossils have been found in South America, Africa, India, Australia, and Antarctica.

FIGURE 22.2   An Illustration of the Continuity of Continental Features

Continental jigsaw pieces cut from a printed page would show the continuity of printed lines when put back together. If the continents were once together and drifted apart, we might expect some similar "printed lines" common to the separated continents in the form of continuities of geologic features such as mountain ranges.

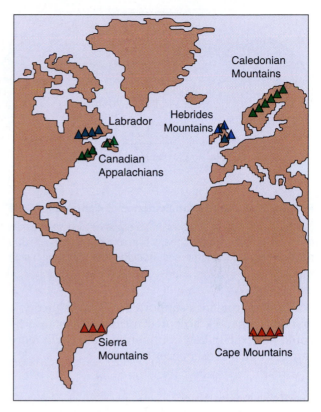

FIGURE 22.3   Continuity of Geologic Features Supporting the Theory of Continental Drift

If the continents were fitted together, various mountain ranges of similar structure and rock composition on the different continents would line up, analogous to the print in Fig. 22.2.

2. *Continuity of geologic features.* As has been noted, it was the roughly interlocking shapes of the coastlines of the African and American continents that inspired the theory of continental drift. Imagine cutting the continental shapes from a printed page like jigsaw puzzle pieces. When we put the pieces back together, we find that the continuity of the printed lines is common to the fitted pieces, as shown in ● Fig. 22.2.

If indeed the continents had rifted and drifted apart, we might expect some similar "printed lines" common to the pieces. Such evidence does occur in the form of geologic features. If the continents were put back together, the Cape Mountain Range in southern Africa would line up with the Sierra Range near Buenos Aires. These mountains are strikingly similar in geologic structure and rock composition (● Fig. 22.3). In the Northern Hemisphere the Hebrides Mountains in northern Scotland match up with similar formations in Labrador, and the Caledonian Mountains in Norway

and Sweden have a logical extension in the Canadian Appalachians.

3. *Glacial evidence.* Solid geologic evidence confirms that a glacial ice sheet covered the southern parts of South America, Africa, India, and Australia about 300 million years ago, similar to the one that covers Antarctica today. Hence a reasonable conclusion is that the southern portions of these continents were under the influence of a polar climate at this time.

Wegener's theory suggests an answer and derives support from these observations. The direction of the glacier flow is easily determined by marks of erosion on rock floors. If the continents were once grouped together, then the glaciation area was common to the various continents, as illustrated in ● Fig. 22.4.

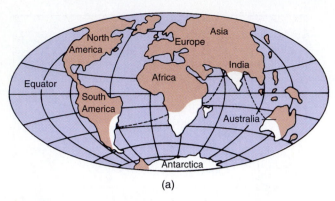

(a)

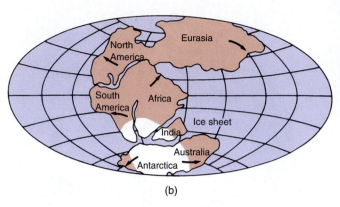

(b)

FIGURE 22.4 Glaciation Evidence of Continental Drift

(a) Geologic evidence shows that a glacial ice sheet covered parts of South America, Africa, India, and Australia some 300 million years ago, but there is no evidence of this glaciation in Europe and North America. (b) This is explained if a single ice sheet covered the southern polar region of Pangaea and the continents subsequently rifted and drifted apart.

Although evidence supported Wegener's theory, it was not generally accepted, primarily because he could not explain how continental crust could move through much denser oceanic crust.

Seafloor Spreading

The mechanism behind continental drift was suggested in 1960 by H. H. Hess, an American geologist. Geologists knew at the time that a **midocean ridge** system stretches through the major oceans of the world (● Fig. 22.5). In particular, the Mid-Atlantic Ridge runs along the center of the Atlantic Ocean between the continents. The East Pacific Rise runs along the Pacific Ocean floor just west of the North and South American continents. Geologists also knew about the deep, narrow depressions that ring the Indian and Pacific oceans. Called **deep-sea trenches,** they are the lowest places on Earth's surface.

Hess suggested a theory of **seafloor spreading,** where the seafloor spreads slowly and moves sideways away from the midocean ridges. Magma wells up into the gap and cools to form new seafloor rock (● Fig. 22.6). Thus the seafloor moves away from the ridges, conveyor-belt style, cooling and contracting as it moves. When it reaches the trenches, it descends back into the mantle.

Support for this theory has come from studies of remanent magnetism and from the determination of the ages of the rocks on each side of the midocean ridges. **Remanent magnetism** refers to the magnetism of rocks that contain the mineral magnetite. When molten material containing magnetite is extruded upward from the mantle (for example, in a volcanic eruption) and solidifies in Earth's magnetic field, it becomes magnetized. The direction of the magnetization indicates the direction of Earth's magnetic field at the time.

This solidified igneous rock is worn down by erosion. The fragments are carried away by water, and they eventually settle in bodies of water where they become layers in future sedimentary rock. In the settling process the magnetized particles, which are in

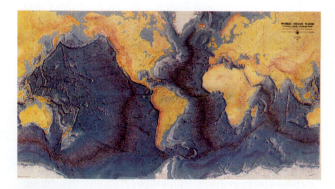

FIGURE 22.5 Map of the Ocean Basins

A system of midocean ridges extends throughout the ocean basins of the world. Notice also the deep-sea trenches that rim the Pacific and Indian oceans.

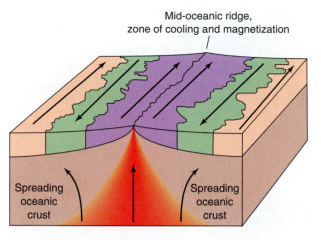

Mid-oceanic ridge,
zone of cooling and magnetization

Spreading
oceanic
crust

Spreading
oceanic
crust

Rising magma

FIGURE 22.6   Magnetic Anomalies Distributed in Symmetric, Parallel Bands on Both Sides of the Mid-Atlantic Ridge

Normal polarity is the direction of the present magnetic field at the ridge. Reverse polarity is in the opposite direction.

fact small magnets, become generally aligned with Earth's magnetic field.

Measurements of the remanent magnetism of rock on the ocean floor revealed long, narrow, symmetrical bands of **magnetic anomalies** on both sides of the Mid-Atlantic Ridge (Fig. 22.6); that is, the direction of the magnetization was reversed in adjacent parallel regions. Along with data on land rock, the magnetic anomalies indicate that Earth's magnetic field has abruptly reversed fairly frequently and regularly throughout recent geologic time. The most recent reversal was about 700,000 years ago. Why this and other reversals should occur is not known. However, the symmetry of the anomaly bands on either side of the ridge indicates movement away from the ridge at the rate of several centimeters per year and provides evidence for seafloor spreading.

These observations support the idea of seafloor spreading as a mechanism for continental drift that has culminated in the modern theory of plate tectonics. Instead of continental crust plowing through oceanic crust, however, we know today that both are carried across the upper mantle as part of a thicker layer called the *lithosphere*.

RELEVANCE QUESTION: *What similarities do you see between the methods used by geologists to solve geologic problems and those used by police detectives to solve crimes?*

22.2 Plate Tectonics

LEARNING GOALS

▼ Describe the theory of plate tectonics.

▼ List the general relative motions of plates and the resulting geologic implications.

The view of ocean basins in a process of continual self-renewal has led to the acceptance of the theory of *plate tectonics.** We now view the lithosphere not as one solid rock but as a series of solid sections or segments called *plates,* which are constantly in very slow motion and interacting with one another. The surface of the globe is segmented into about 20 plates. Some are very large, and some are small. The major plates are illustrated in ● Fig. 22.7.

As we learned in Chapter 21, the most active, restless parts of Earth's crust are located at the plate boundaries. Along the midocean ridges where one plate is pulling away from another *(divergent boundary)*, new oceanic rock is formed. Where plates are driven together *(convergent boundary)*, rock is consumed. In still other parts of Earth's surface, the abutting edges of neighboring plates slide horizontally by one another *(transform boundary)*, and rock is neither produced nor destroyed. These boundary types are illustrated in ● Fig. 22.8.

● Figure 22.9 illustrates the structure of the lithosphere and asthenosphere. The interface between these two structures is significant in terms of internal geologic processes. The asthenosphere, which lies beneath the lithosphere, is essentially solid rock, but it is so close to its melting temperature that it contains pockets of molten magma and is relatively plastic. Therefore, it is much more easily deformed than the lithosphere. The movement of plastic rock during structural adjustments takes place essentially within the asthenosphere. The lithosphere "floats" on the asthenosphere, with continents floating higher

*Tectonics (from the Greek *tekto,* "builder") is the study of Earth's general structural features and their changes.

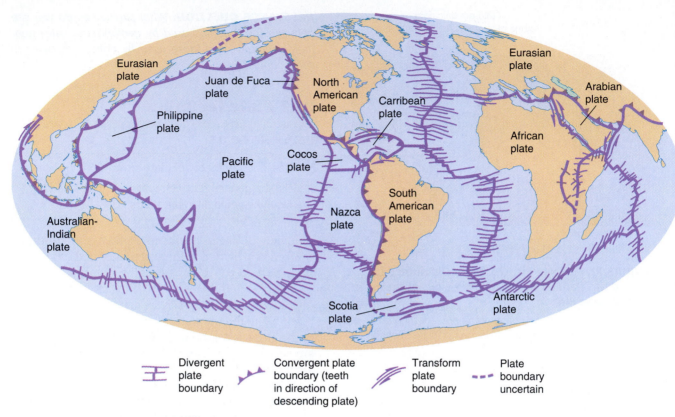

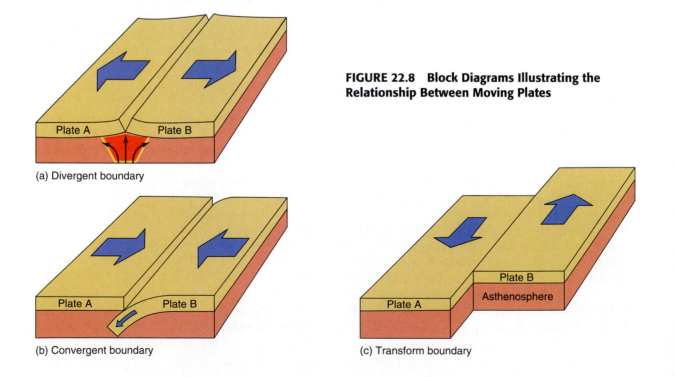

FIGURE 22.7 Map of Tectonic Plates of the World

The relative motions along the plate boundaries are indicated.

FIGURE 22.8 Block Diagrams Illustrating the Relationship Between Moving Plates

(a) Divergent boundary

(b) Convergent boundary

(c) Transform boundary

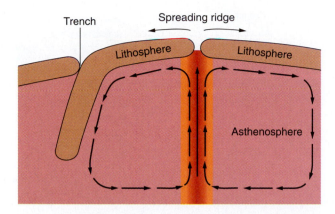

Trench   Spreading ridge

Lithosphere   Lithosphere

Asthenosphere

FIGURE 22.9   Convection Cells

Unequal distribution of temperature within Earth causes the hot, lighter material to rise and the cooler, heavier material to sink. This generates convection currents in the asthenosphere. The lithospheric plates, resting on the asthenosphere, are put in motion by the driving force of the convection cells.

because of their lower density. This state of buoyancy is called **isostasy.**

Isostatic balance can be maintained only because rock is plastic deep below the surface within the asthenosphere, where pressures and temperatures are very great. Hence, rock, though still solid, has the capacity to flow and, therefore, to adjust itself to the distribution of mass above it.

A density difference accounts for the difference in elevation between ocean basins and continental masses, but not for the differences in altitude within a continent. How can a plain barely above sea level and a towering mountain range both stand in isostatic balance if they are both composed of basically the same rock type? This arrangement is possible because the continental masses vary markedly in thickness. Just as the top of an iceberg floats higher above water than an ice cube, so a thick mass of granite stands high as a mountain, whereas a thin mass of granite reaches only slightly above sea level. The continents, though irregular in elevation, are in isostatic balance.

The forces that move the plates one against another, one away from another, or one past another are also found within the asthenosphere. The plates, segments of the lithosphere, are actually passive bodies, driven into motion by the drag of the more active asthenosphere beneath them.

Most geologists view this motion in terms of **convection cells,** with the basic source of heat provided by radioactive decay (see Fig. 22.9). The role of gravity

in the process is to drag the cooler, heavier material deeper into the Earth's interior as the hotter, less dense rock rises toward the surface, where it can lose part of its heat.

Plate Motion at Divergent Boundaries

Let's examine the role of convection cells in plate tectonics by focusing our attention first on divergent boundaries, where the plates are being driven away from one another (● Fig. 22.10). Beneath these spreading midocean ridges, convection currents lift material from the hot asthenosphere up into a region of lower pressure, where it begins to melt. The material melts only partially, however, forming a magma of basaltic composition.

As more mantle material wells up from below, the higher mantle material is shouldered to both sides and moves slowly in a horizontal direction beneath the lithospheric plates. It is the drag of the mobile asthenosphere against the bottom of the lithospheric plates that keeps the plates in motion.

A lithospheric plate cools as it moves slowly away from the hot spreading ridge, and therefore, it loses volume. As a result, the top of the plate gradually subsides and causes the oceans to grow progressively deeper away from the spreading ridges. The slope of denser, cooler rock away from the midocean ridges also contributes to the motion of a plate at convergent boundaries.

Plate Motion at Convergent Boundaries

When two plates collide, what happens next depends on whether the colliding margins are oceanic or continental crust. Only three combinations are possible: (1) oceanic-oceanic, (2) oceanic-continental, and (3) continental-continental.

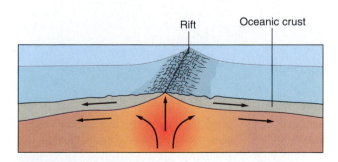

Rift   Oceanic crust

FIGURE 22.10   Spreading at the Midocean Ridge

As two plates move apart, magma rises into the rift, cools, and forms new oceanic crust.

1. *Oceanic-oceanic convergence.* Initially, when two oceanic plates collide, both begin to be subducted because each is relatively dense (3.0 g/cm^3). Eventually, one plate is subducted more than the other. The place, or zone, where the plate descends into the asthenosphere is called a **subduction zone.** The descending oceanic plate, now in contact with the asthenosphere, begins to melt. The molten material begins to rise, and a series of volcanoes develops in an arc shape on the overriding plate (● Fig. 22.11). Deep-sea trenches lie in front of the arc system, marking the places where the plates are being subducted. The deepest trench known is the Marianas Trench in the western Pacific Ocean, which is 11 km below sea level.

2. *Oceanic-continental convergence.* Whenever oceanic crust collides with lower density continental crust (2.7 g/cm^3), it is always subducted beneath the continental crust (● Fig. 22.12). A trench will develop at the point where the oceanic plate is being subducted. However, it is never as deep as the trench formed in the oceanic-oceanic convergence. The oceanic plate begins to melt as it descends into the asthenosphere. Magma then moves up into the overriding plate, causing large igneous intrusions and often volcanic mountains

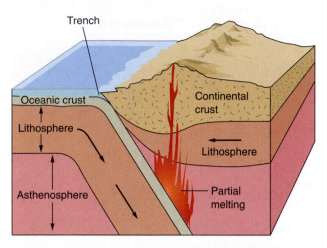

FIGURE 22.12   Diagram Illustrating the Results Produced by an Oceanic-Continental Convergence

An example is the Andes Mountains of South America.

at the surface. Examples are the Andes Mountains of South America and the Cascades of Washington and Oregon in North America.

3. *Continental-continental convergence.* Continental plates have low densities (about 2.7 g/cm^3), so when they collide with one another the boundary of rock between the continental edges is pushed and crumpled intensely to form fold-mountain belt systems (● Fig. 22.13). In this manner, continents grow in size by suturing themselves together

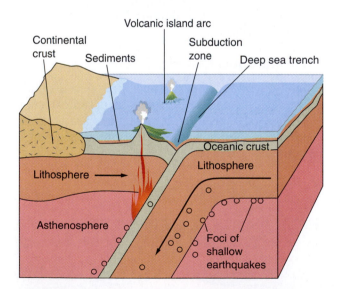

FIGURE 22.11   Diagram Illustrating the Effects Produced by the Convergence of Two Oceanic Plates

The subduction zone is the region in which one lithospheric plate plunges beneath another. Destructive earthquakes occur most often in the subduction zones.

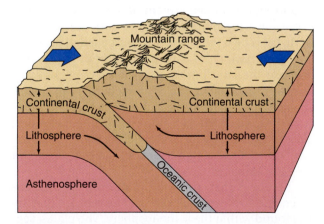

FIGURE 22.13   Diagram Illustrating a Continental-Continental Convergence

The boundary of rock between the continental edges is pushed and crumpled to form fold-mountain systems. The Appalachian Mountain Range is an example.

along fold-mountain belt systems. Examples are the Alps, Appalachians, and Himalayas. As we will see later in this chapter, geologists believe that the Himalayas formed in this manner when the Indian plate collided with the Eurasian plate.

Plate Motion at Transform Boundaries

A zone of shear, or transform, faults is a boundary at which adjacent plates slide past each other without a gain or loss in surface area (Fig. 22.8c). This zone occurs along faults that mark the plate boundaries. Movements and the resulting release of energy along these boundaries give rise to earthquakes. Examples of such fault zones are along the San Andreas Fault in California and the Anatolian Fault in Turkey. See the Spotlight feature on plate tectonics.

RELEVANCE QUESTION: *Which slow-moving plate are you riding on, and what is the direction of its movement?*

22.3 Earthquakes and Earth's Interior

LEARNING GOALS

▼ Explain the causes of earthquakes.

▼ List safety measures to be taken before, during, and after an earthquake.

▼ Describe Earth's interior structure and composition.

An **earthquake,** as those who have experienced a substantial one know, is manifested by the vibrating and sometimes violent movement of Earth's surface. Earthquakes rattle our globe perhaps a million times each year, but the vast majority are so mild that they can be detected only with sensitive instruments. A very powerful quake, however, can lay waste to a large area.

The Causes of Earthquakes

Earthquakes may be caused by explosive volcanic eruptions or even explosions caused by humans, but the large majority are associated with the movement of lithospheric plates. These movements form large crustal features called faults. In a **fault,** the rock on one side of the fracture has moved relative to the rock

on the other side of the fracture (● Fig. 22.14). The optimum place for movement is at the plate boundaries, and the major earthquake belts of the world are observed in these regions (● Fig. 22.15). Notice how the earthquake region around the Pacific Ocean is similar to that of the volcanic "Ring of Fire" (Fig. 21.10, on page 566).

Movement of neighboring plates exerts stress on the rock formations along the plate margins. Because rocks possess elastic properties, energy is stored until the stresses acting on the rocks are great enough to overcome the force of friction. Then the fault walls move suddenly, and the energy stored in the rocks is released, causing an earthquake. After a major earthquake, the rocks may continue to adjust to their new positions, causing additional vibrations called *aftershocks.*

Many large horizontal faults are associated with plate boundaries and are called *transform faults.* An example of a transform fault is the famous San Andreas Fault in California. It is the master fault of an intricate network of faults that runs along the coastal regions of California (● Fig. 22.16 on page 598). This huge fracture in Earth's crust is more than 960 km long and at least 32 km deep. Over much of its length a linear trough of narrow ridges reveals the fault's presence.

As you can see in Fig. 22.16, the San Andreas Fault system lies on the boundary of the Pacific plate and the North American plate. Movement along the

FIGURE 22.14   Faults Exposed on an Excavated Hillside

Notice that the fault planes lie at a steep angle to the bedding planes and the sides of each fault have moved in opposite directions to one another.

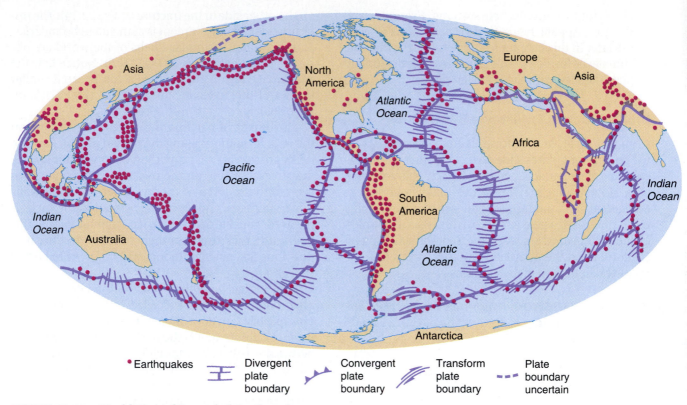

FIGURE 22.15 World Map of Recorded Earthquakes

The large majority of earthquakes occur at plate boundaries.

transform fault arises from the relative motion be-tween these plates. The Pacific plate is moving north-ward relative to the American plate at a rate of several centimeters per year. At this current rate and direc-tion, in about 10 million years Los Angeles will have moved northward to the same latitude as San Fran-cisco. In another 50 million years, the segment of con-tinental crust on the Pacific plate will have become completely separated from the continental land mass of North America.

The horizontal movement along the San Andreas Fault is cause for considerable concern in the densely populated San Francisco Bay area through which it runs. The famous San Francisco earthquake of April 18, 1906, which measured 8.3 on the Richter scale (earthquake scales are discussed shortly), resulted in the loss of approximately 700 lives and caused mil-lions of dollars of property damage. Many other milder earthquakes have occurred since then along the San Andreas and other branch faults.

In 1989, an earthquake shook the San Francisco Bay area. Measuring 7.1 on the Richter scale, this ma-

jor earthquake damaged bridges, buildings, and high-ways. In 1995, an earthquake measuring 7.2 on the Richter scale jolted western Japan. It devastated the city of Kobe, killing more than 5000 people and injur-ing over 25,000. More than 50,000 buildings were damaged (Fig. 22.17 on page 599).

Little can be done about the San Andreas Fault ex-cept to learn to live with it. Building codes in the area are strict. However, increased population and an ever-increasing need for housing have caused devel-opers to use any available land, resulting in rows of houses that straddle the trace fault of 1906 and sit on hills where earthquake-induced landslides could occur. People live in these houses with the knowl-edge of the fracture in the crust below and the hope that any future movement along the fault will occur elsewhere.

When an earthquake does occur, the point of the initial energy release or slippage is called its **focus.** A focus generally lies at considerable depth, from a few to several hundred miles. Consequently, geologists designate the location on Earth's surface directly

A SPOTLIGHT ON: Plate Tectonics—The Engine that Drives Crustal Processes

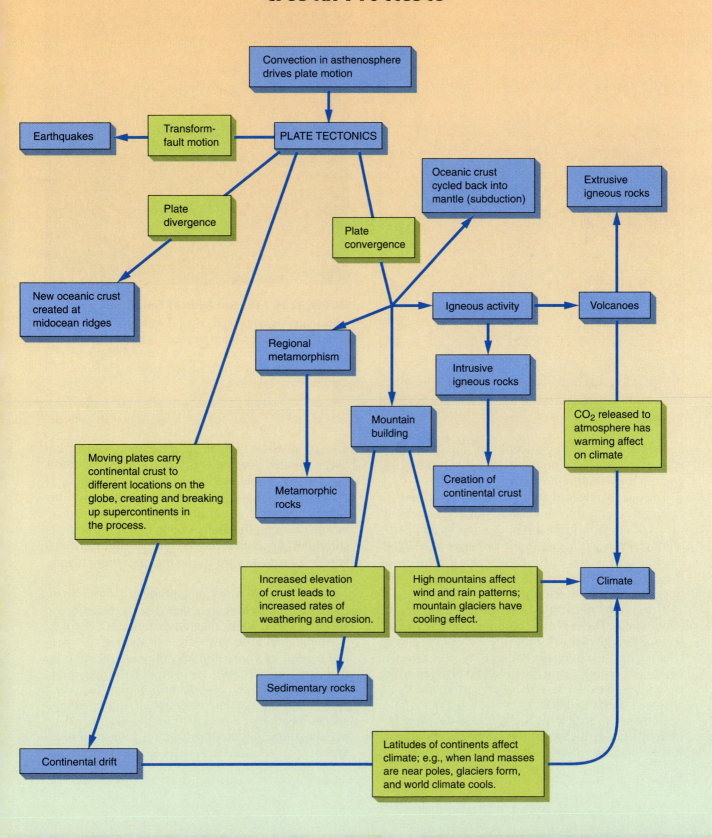

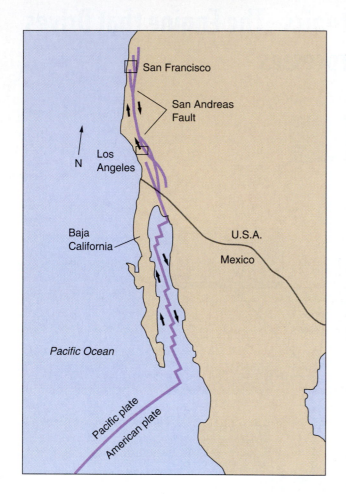

FIGURE 22.16 The San Andreas Fault

Left: A map showing the boundary, marked by the fault, between the Pacific and North American plates, which are moving in opposite directions. This is a classic example of a transform fault. *Above, right:* An aerial view of the fault on the Carizzo Plain, California.

above the focus as the **epicenter** of the quake. The epicenter is the point on the surface that receives the greatest impact from the quake.

The energy released at the focus of an earthquake propagates outwardly as **seismic waves.** The seismic waves of an earthquake are monitored by an instrument called a *seismograph,* the principle of which is illustrated in ● Fig. 22.18. The greater the energy of the earthquake, the greater is the amplitude of the traces on the recorded *seismogram.*

The severity of earthquakes is represented on different scales. The two most common are the **Richter scale,** which gives an absolute measure of the energy released by calculating the energy of seismic waves at a standard distance, and the modified **Mercalli scale,** which describes the severity of an earthquake by its observed effects (Table 22.1).

Of these two scales, the Richter scale, developed in 1935 by Charles Richter of the California Institute of Technology, is used more often, with magnitudes expressed in whole numbers and decimals, usually between 3 and 9. However, it is a logarithmic function—that is, each whole-number step represents about 31 times more energy than the preceding whole-number step. For example, an earthquake that registers a magnitude of 5.5 on the Richter scale indicates the measuring seismograph received about 31 times as much energy as from an earthquake that registers a magnitude of 4.5.

An earthquake with a magnitude of 2 is the smallest tremor felt by human beings. The largest recorded earthquakes are in the magnitude range of 8.7 to 8.9.

The Richter scale gives no indication of the damage caused by an earthquake, only its potential for damage. Damage depends not only on the magnitude of the quake but also on the location of its focus and epicenter and the environment of that region—specifically the local geologic conditions, the density of population, and the construction designs of buildings.

FIGURE 22.17 Earthquake Devastation

The photo shows the devastation caused by an earthquake measuring 7.2 on the Richter scale on January 17, 1995, in Kobe, Japan.

An earthquake can be low on the Richter scale and cause great damage, whereas another can be high on the Richter scale and cause relatively little damage. For example, in terms of loss of life, the magnitude 6.9 earthquake that struck Armenia in 1988 killed 25,000 people, but in 1992 a California earthquake of mag-

nitude 7.1—six times stronger than the Armenia quake—killed only one person. The reason is plain. Landers, California, is a tiny town in a sparsely populated region of the Mojave Desert. There are only a few buildings in the town, and most have been constructed according to strict state building codes. The

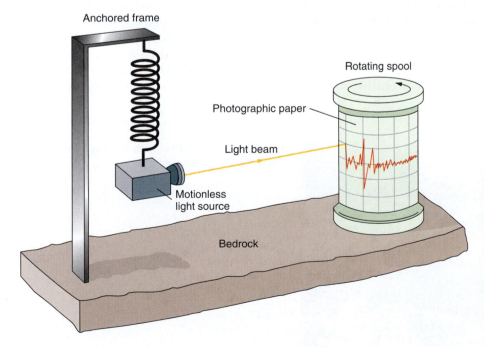

FIGURE 22.18 An Illustration of the Principle of the Seismograph

The rotating spool (anchored in bedrock) vibrates during the quake. A light beam from the relatively motionless source on a spring traces out a record, or seismogram, of the earthquake's energy on light-sensitive photographic paper.

TABLE 22.1 Modified Mercalli Scale

Scale	Description	Scale	Description
I	Not felt except by a very few under especially favorable circumstances.		nary structures; considerable in poorly built or badly designed structures.
II	Felt only by a few persons at rest, especially on upper floors of buildings.	VIII	Damage slight in specially designed structures; considerable in ordinary substantial buildings, with partial collapse; great in poorly built structures. Fall of chimneys, factory stacks, columns, monuments, and other vertically oriented features.
III	Felt quite noticeably indoors, especially on upper floors of buildings, but not generally recognized as an earthquake.		
IV	During the day, felt indoors by many, but outdoors by few. Sensation like heavy truck striking building.	IX	Damage considerable in specially designed structures. Buildings shifted off foundations. Ground cracked conspicuously.
V	Felt by nearly everyone, many awakened. Disturbances of trees, poles, and other tall objects sometimes noticed.	X	Some well-built wooden structures destroyed. Most masonry and frame structures destroyed with foundations. Ground badly cracked.
VI	Felt by all; many frightened and run outdoors. Some heavy furniture moved; a few instances of fallen plaster or damaged chimneys. Damage slight.	XI	Few, if any, masonry structures remain standing. Bridges destroyed. Broad fissures in ground.
VII	Everyone runs outdoors. Damage negligible in buildings of good design and construction; slight to moderate in well-built ordi-	XII	Damage total. Waves seen on ground surfaces. Objects thrown upward into air.

Armenian quake was centered in Spitak, an ancient, densely populated city. Thousands of poorly constructed buildings collapsed on their inhabitants almost as soon as the shaking started.

Earthquake damage may result directly from the vibrational tremors or indirectly from landslides and subsidence, as in the magnitude 8.4 earthquake that struck Anchorage, Alaska, in 1964 (● Fig. 22.19). In populated areas, a great deal of the property damage is caused by fires because of a lack of ability to fight them—a result of disrupted water mains and so on. See Table 22.2 for an abridged version of an earthquake preparedness checklist compiled by the U.S. government. The checklist offers practical steps

FIGURE 22.19 Subsidence Damage from the 1964 Alaska Earthquake

TABLE 22.2 What to Do Before, During, and After an Earthquake

Before

Check for potential fire risks, such as defective wiring and leaky gas connections. Bolt down water heaters and gas appliances.

Know where and how to shut off electricity, gas, and water at main switches and valves.

Place large and heavy objects on lower shelves. Securely fasten shelves to walls. Brace or anchor top-heavy objects.

Do not store bottled goods, glass, china, and other breakables in high places.

Securely anchor all overhead lighting fixtures.

Keep on hand:

  Portable radio

  Fire escape ladder

  Flashlight

  First aid kit

  Fire extinguisher

  Bottled water

  Telephone numbers of police, fire department, and ambulance

During

If you are outdoors, stay outdoors; if indoors, stay indoors. During earthquakes, most injuries occur as people enter or leave buildings.

If indoors, take cover under a heavy desk, table, or bench or in doorways, halls, or against inside walls. Stay away from glass. Don't use candles, matches, or other open flames either during or after the tremor. Extinguish all fires.

In a high-rise building, don't dash for exits; stairways may be broken or jammed with people. Never use an elevator.

If outdoors, move away from buildings, utility wires, and trees. Once in the open, stay there until the shaking stops. If in a moving car, drive away from underpasses and overpasses. Stop as quickly as safety permits, but stay in the vehicle.

After

Check for injuries; do not move seriously injured persons unless they are in danger of sustaining additional injury.

Be prepared for aftershocks. Stay out of severely damaged buildings; aftershocks may shake them down.

Listen to the radio for latest emergency bulletins and instructions from local authorities.

Check utilities. If you smell gas, open the windows and shut off the main gas valve. Leave the building and report gas leakage to authorities. If electrical wiring is shorting out, shut off the current at the main meter box.

Do not touch downed power lines or objects touched by downed lines.

If water pipes are damaged, shut off the supply at the main valve. Emergency water may be obtained from hot water tanks, toilet tanks, and melted ice cubes.

Check sewage lines before flushing toilets.

Source: U.S. Federal Emergency Management Agency (FEMA).

people can take to (1) be as prepared as possible in the event of an earthquake, (2) protect themselves during an earthquake, and (3) prevent further harm to themselves and minimize damage to their property during the critical hours following an earthquake. The complete checklist may be obtained from the U.S. Federal Emergency Management Agency (FEMA).

HIGHLIGHT: Deadly Tsunamis

New Guinea is an island arc located northeast of Australia and just south of the convergent plate boundary between the Australian-Indian and Pacific plates. Late in the afternoon of July 17, 1998, most of the 10,000 people who inhabited the villages along a 30-kilometer stretch of coast in northern Papua, New Guinea, were preparing for the evening meal. Suddenly, from out across the surf they heard what was later described as the roar of a jet plane taking off. As they turned to the sea, they saw the cause of the deafening noise—an enormous wall of water coming at them. They started to run, but it was too late. A 10-meter-high wave, then another, then another swept through the villages, wiping them off the map. Nearly a third of the population was lost in the disaster, most of them children.

The event was set off by an earthquake that occurred just off the coast when a portion of the seafloor along the plate boundary suddenly dropped 2 meters. The resulting vibrations sent three enormous waves of water—tsunamis—outward from the epicenter. As they neared the beach, each tsunami, one behind the other, rolled up into a breaking wave the height of a four-story building. The earthquake had been detected by seismographs nearby and around the world, but the epicenter, just off shore, was so close to the villages that there was no time to warn them. The tsunamis hit the beach within minutes of the initial shock.

Most tsunamis are generated by submarine earthquakes, and most of them originate in the subduction zones that rim the Pacific Ocean. The vibrations set off on the ocean floor are transferred to the water and radiate outward across the ocean in a series of low waves. On the open sea, a tsunami wave is difficult to detect. Though many kilometers in length, the wave may be no more than a meter-high bulge in the surface of the water. But it travels rapidly—at speeds of up to 960 km/h. When it reaches the shallow coast, friction slows the wave down at the same time as it causes it to roll up into a 5- to 30-meter-high wall of water that crashes down on the shoreline with immense force (Fig. 1).

A tsunami coming ashore is commonly and incorrectly referred to as a *tidal wave*, but these deadly waves have no relation to tides. Scientists have thus adopted the more appropriate Japanese name, *tsunami*, which means "harbor wave." The effects of a tsunami are intensified in the confined spaces of bays where most harbors are located. As a tsunami approaches the harbor and rolls up to its full height, the water that was in the harbor is drawn out and into the wave. Coastal residents of Japan know that when a harbor is suddenly emptied, it is the signal to drop everything and run to high ground, for a tsunami is only seconds away.

The coastline of Japan is especially vulnerable. A 1960 earthquake off the Chilean coast triggered a tsunami that traveled 14,500 kilometers across the Pacific Ocean to Japan, where it caused

When the energy release of a quake occurs in the vicinity of or beneath the ocean floor, huge waves called **tsunamis** are sometimes generated. See the chapter Highlight for a discussion of this potentially deadly effect of earthquakes.

Earth's Interior

Earthquakes, despite their potential for destruction, can be an aid to science. None of the remarkable scientific advances of the twentieth century revealed with certainty the composition of Earth's interior. Our ideas about it must rest on indirect evidence provided by earthquake body waves whose speed and direction reflect the types of materials they penetrate, by meteorites whose composition we believe is similar to Earth's, and by laboratory experiments performed on rocks under very high temperature and pressure.

By monitoring earthquake waves at different locations and by applying the knowledge of wave properties such as speed and refraction in various types of materials, scientists obtain information about Earth's interior structure.

Two general types of seismic waves are produced by earthquake vibrations: **surface waves,** which travel along the outer layer of Earth, and **body waves,** which travel through Earth. Surface waves cause most earthquake damage because they move along Earth's surface.

There are two types of body waves: P (for primary) or compressional waves and S (for secondary) or shear waves. The **P waves** are longitudinal compressional waves that are propagated by particles in the propagating material moving longitudinally back and forth in the same direction as the wave is traveling (see Chapter 6). In **S waves** the particles move at right

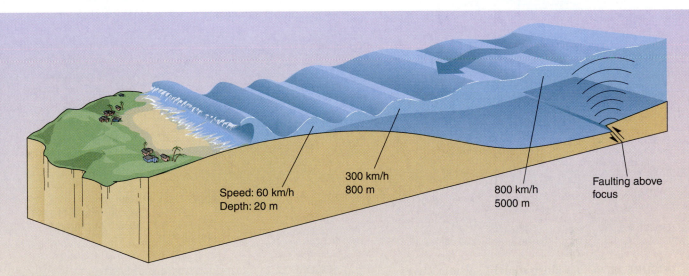

FIGURE 1 Behavior of a Tsunami

As a tsunami approaches shore, the wave is shortened in length while its height is increased. Notice that the harbor empties out just before the wave strikes.

180 deaths. Japanese history records at least 15 major tsunami disasters, some of which killed hundreds of thousands of people. The Hawaiian Islands have experienced many tsunamis that have raced across the ocean from the Alaska–Bering Strait area or other locations of earthquake activity that circumscribe the Pacific Ocean. The 1964 Alaska earthquake hit the harbor city of Anchorage with a one-two punch. As with the Papua, New Guinea, disaster, the epicenter of the Alaskan earthquake was located along the coastline. Thus the 8.4 magnitude tremors were almost immediately followed by a tsunami that devastated the harbor.

angles to the direction of wave travel and hence are transverse waves.

The P and S waves have two other important differences. S waves can travel only through solids. Liquids and gases cannot support a shear stress; that is, they have no elasticity in this direction and therefore their particles will not oscillate in the direction of a shearing force. For example, little or no resistance is felt when one shears a knife through a liquid or gas. The compressional P waves, on the other hand, can travel through any kind of material—solid, liquid, or gas. The other important difference is the speed of the waves. P waves are called primary because they always travel faster than the secondary S waves in any particular solid material and hence arrive earlier at a seismic station. It is these differences that allow seismologists to locate the focus of an earthquake and to learn about Earth's internal structure.

The speeds of the body waves depend on the density of the material, which generally increases with depth. As a result, the waves are curved, or refracted. Also, the waves are refracted when they cross the boundary, or discontinuity, between different media in the same manner that light waves are refracted (see Chapter 7). The refraction of seismic waves and the fact that S waves cannot travel through liquid media provide our present view of Earth's structure, as shown in ● Fig. 22.20.

From these indirect observations, scientists believe that Earth is made up of a series of zones, as illustrated in ● Fig. 22.21. The four layers are (1) the inner core, (2) the outer core, (3) the mantle, and (4) the crust. These layers are characterized by different composition and physical properties.

The density of the inner and outer cores suggests a metallic composition, which is believed to be chiefly

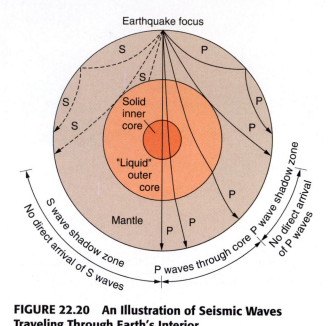

FIGURE 22.20 **An Illustration of Seismic Waves Traveling Through Earth's Interior**

Because of the S and P wave shadow zones, Earth is believed to have a liquid outer core.

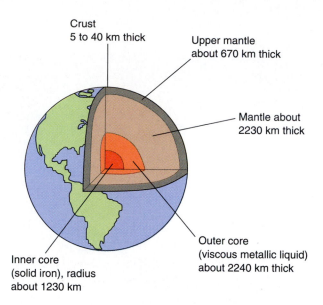

FIGURE 22.21 **Interior Structure of Earth**

iron (85%) and nickel, similar to the composition of metallic meteorites. The **outer core,** some 2240 km thick, is believed to be liquid, whereas the **inner core** is thought to be a solid ball with a radius of approximately 1230 km. Evidence for a liquid outer core and a solid inner core is found in the behavior of P and S waves. P waves slow down on entering the outer core and speed up again on entering the inner core. S waves stop altogether at the edge of the outer core. As shown in Fig. 22.20, this behavior of P and S waves creates a *shadow zone* on each side of the core.

Around the outer core is the **mantle,** which averages about 2900 km thick. The composition of the rocky mantle differs sharply from that of the metallic core, and their boundary is distinct.

Around the mantle is the thin, rocky, outer layer on which we live, called the **crust.** It ranges in thickness from about 5 to 8 km beneath the ocean basins to about 25 to 40 km under the continents.

An abrupt change in the behavior of body waves as they travel in toward Earth's interior reveals the existence of the *Mohorovicic discontinuity,* a sharply defined boundary that separates the crust from the upper mantle, the next layer down (Fig. 22.22). Both scientists and students prefer the simplified term, **Moho,** for this important boundary.

Direct observation and geophysical investigations have confirmed that the ocean basins are made of basalt and that the continents are generally granitic. Below the crust, however, we must depend on indirect information. Earth scientists are in general agreement that the upper part of the mantle is probably composed of a rock near the composition of peridotite, a very low-silica rock composed mostly of olivine and pyroxene. The lower mantle, with its much greater temperature and pressure, may contain minerals composed of magnesium silicate, magnesium oxide, and high-pressure forms of quartz.

If we view the crust and upper mantle in terms of behavior rather than composition, they can be divided somewhat differently into what we might call zones. The outer zone, which we call the *lithosphere,* extends to a depth of approximately 70 km and includes all the crust and a thin slice of the upper mantle (see Fig. 22.22). The lithosphere is rigid, brittle, and relatively resistant to deformation. Faults and earthquakes are mostly restricted to this layer. Below the lithosphere is the *asthenosphere,* which extends to a depth of roughly 700 km below Earth's surface.

Because rock within the asthenosphere is so near its melting temperature, it is plastic and mobile. For this reason, the asthenosphere plays an essential role in tectonic plate movement, as discussed earlier in this chapter.

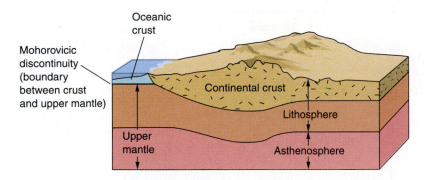

FIGURE 22.22   Lithosphere and Asthenosphere

Oceanic crust

Mohorovicic discontinuity (boundary between crust and upper mantle)

Continental crust

Lithosphere

Upper mantle

Asthenosphere

RELEVANCE QUESTION: Have you experienced an earthquake? If you have, state when and where and describe what it was like. If you have not had such an experience, ask an acquaintance who has.

22.4 Crustal Deformation and Mountain Building

LEARNING GOALS

▼ Describe structural changes in Earth's crust resulting from internal forces.

▼ Explain how mountains are formed.

The forces that build up in the vicinity of plate boundaries can buckle and fracture rocks or break them and shift their positions. Plate edges can be ruptured into huge displaced blocks or squeezed together and uplifted into great folds. In this section we will look

briefly at the ways that plate movements cause crustal deformation and create mountain ranges.

Crustal Deformation

Two major types of slow structural deformations are folding and faulting. *Folding* of Earth's crust takes place when extreme pressure is exerted horizontally or vertically. Rock can be compressed only a limited amount before it begins to buckle and fold. The folded rock layers can form an arch (**anticline**) or a trough (**syncline**), or they may fold and override an adjacent fold. ● Figures 22.23 and 22.24 illustrate the principal types of folded rocks. The folding occurs during the early stages of mountain formation.

Faulting begins with fracturing. Fractures are breaks in rock caused by stress; faults are fractures that display evidence that one side of the fracture has moved relative to the other. The stresses may be vertical and produce uplifts, they may be horizontal and produce compressions, or they may cause tension

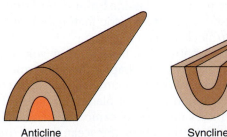
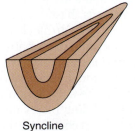

Anticline   Syncline   Overturned fold

FIGURE 22.23   Crustal Folding
Folded rock layers can occur as arches (anticlines) or troughs (synclines), or they can override another fold (overturned fold).

FIGURE 22.24   Types of Folds

Some of the more common relationships between anticlines and synclines and the landforms above them.

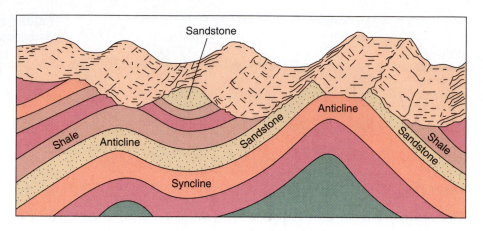

and bring about a lengthening of the crust. ● Figure 22.25 illustrates a few essential terms needed to describe fault geometry. The fault plane is the actual surface itself. The hanging wall describes the rock on the upper side of the inclined fault plane, and the footwall contains rock on the underside. The upthrown side has moved up relative to the downthrown side, as indicated by the arrows.

The three types of faults are normal, reverse, and strike-slip (● Fig. 22.26). Faults may occur in any type of rock, and they may be vertical, horizontal, or inclined at an angle. Most faults occur at an angle.

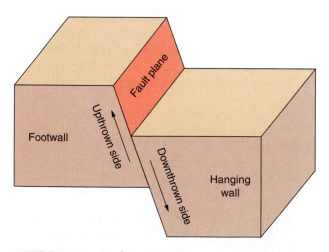

FIGURE 22.25   Fault Terminology

The surface along which the rocks move is the fault plane, and the footwall is the rock beneath it. The upthrown side moves up with respect to the downthrown side.

A **normal fault** occurs as the result of expansive forces that cause its overlying side to move downward relative to the side beneath it. In this case the stress forces are in opposite directions, and the faulting tends to pull the crust apart. A **reverse fault** occurs as the result of compressional stress forces that cause the overlying side of the fault to move upward relative to the side beneath it. A special case of reverse faulting, called **thrust faulting,** describes the faulting when the fault plane is at a small angle to the horizontal. As might be expected, this type of faulting occurs in subduction zones, in which one plate slides over a descending plate, that is, along convergent plate boundaries.

Finally, **strike-slip faulting** occurs when the stresses are parallel to the fault boundary such that the fault slip is horizontal. This type of faulting takes place along the transform boundary of two plates that *strike* and *slip* by each other without an appreciable gain or loss of surface area (Fig. 22.26c). Such faults are commonly called **transform faults.**

Mountain Building

Mountains are generally classified into three principal kinds, based on characteristic features: (1) volcanic, (2) fault-block, and (3) fold.

Volcanic mountains have been built by volcanic eruptions. As discussed in Chapter 21, most volcanoes, and hence volcanic mountains, are located above the subduction zones of plate boundaries. If the colliding plates are both oceanic, then volcanic mountain chains are formed on the ocean floor of the overlying plate. Chains of oceanic volcanic moun-

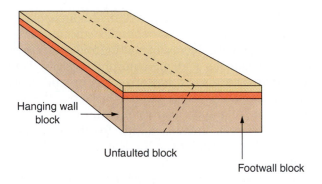

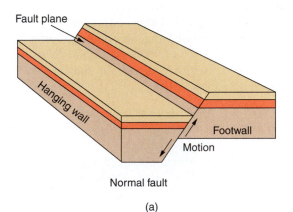

Normal fault

(a)

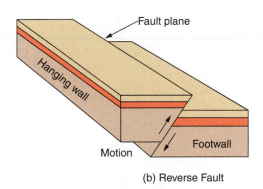

(b) Reverse Fault

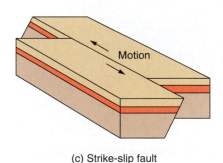

(c) Strike-slip fault

FIGURE 22.26 Block Diagrams Illustrating Three Types of Faultings

FIGURE 22.27   The Cascade Mountains in Washington and Oregon

tains are manifested above sea level by island arcs; Japan and the islands of the West Indies are good examples.

Continental volcanic mountain chains occur when the overlying plate of the subduction zone is continental crust. The Andes Mountains along the western coast of South America are an example of such a continental mountain range. The Cascade Mountains in Washington and Oregon are the only active volcanic mountains in the conterminous United States (● Fig. 22.27).

Fault-block mountains are believed to have been built by normal faulting in which giant pieces of Earth's crust were tilted and uplifted (● Fig. 22.28). These mountains evidence the great stresses within Earth's lithospheric plates. Fault-block mountains raise sharply above the surrounding plains. The Sierra Nevada range in California, the Grand Teton Mountains in Wyoming, and the Wasatch range in Utah are examples of fault-block mountains in the United States.

Fold mountains, as the name implies, are characterized by folded rock strata. Examples of fold mountains include the Alps, the Himalayas, and the Ap-

FIGURE 22.28 Fault-Block Mountains

Mount Whitney, in California, is the highest peak in the Sierra Nevada range.

FIGURE 22.29 Fold Mountains in Alberta, Canada

Note the arch (anticline) on the structure on the left.

palachian Mountains (● Fig. 22.29). Although folding is their main feature, these complex structures also contain external evidence of faulting and central evidence of igneous and metamorphic activity. Fold mountains are also characterized by exceptionally thick sedimentary strata, which indicate that the material of these mountains was once at the bottom of an ocean basin. Indeed, marine fossils have been found at high elevations in the Himalayas and other fold mountain systems.

Let's consider within the framework of plate tectonics the formation of the grandest fold mountains of all. The Himalayas are believed to have formed after the supercontinent Pangaea broke up some 200 million years ago (m.y.a.). In short, India broke away from Africa and ran into Asia, with the collision resulting in the Himalayas (● Fig. 22.30).

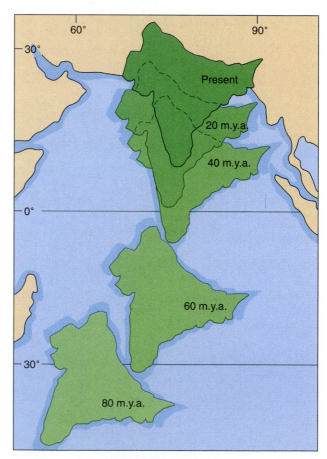

FIGURE 22.30 Collision of India with Asia

Following the breakup of Pangaea, the plate carrying the Indian continent moved northeastward toward the Eurasian plate until the Indian continent collided with Asia.

In more detail, the northward movement of the Indian plate after its separation from Africa resulted in the loss of oceanic lithosphere formerly separating India and Asia, with the eventual collision of these two continental masses. The Indian plate descended under the Eurasian plate, and the depositional basins along the Eurasian plate were folded into mountains, as illustrated in ● Fig. 22.31. When the two continental plates met, the edge of the Eurasian plate was lifted, with the spectacular result of the highest mountain range on Earth. Present-day analyses of plate movements indicate that the Indian plate is still moving northward relative to the Eurasian plate, causing the slow, continuing uplift of the Himalayas.

RELEVANCE QUESTION: *You are trying to install a 13-foot-square piece of wall-to-wall carpeting in a room that is 12 feet square in area. As you smooth out the carpet from the center, what happens to the edges? In what way is this an analogy for fold-mountain building?*

FIGURE 22.31   An Illustration of the Formation of the Himalayas

(a) As the northward-moving Indian plate descended under the Eurasian plate, the depositional basins along the Eurasian plate were folded into mountains. (b) When the two continental crusts met, the edge of the Eurasian plate was lifted, giving rise to the lofty Himalayas.

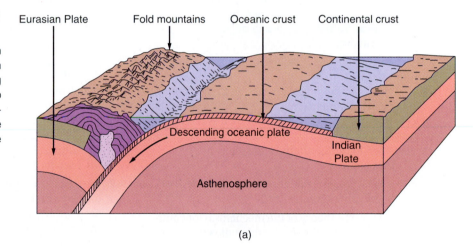

(a)

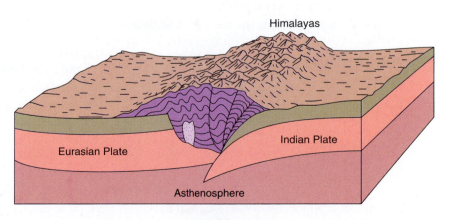

(b)

Important Terms

continental drift (22.1)
Pangaea
midocean ridge
deep-sea trenches
seafloor spreading
remanent magnetism
magnetic anomalies
isostasy (22.2)
convection cells
subduction zone

earthquake (22.3)
fault
focus
epicenter
seismic waves
Richter scale
Mercalli scale
tsunamis
surface waves
body waves

P waves
S waves
outer core
inner core
mantle
crust
Moho
anticline (22.4)
syncline
normal fault

reverse fault
thrust faulting
strike-slip faulting
transform faults
volcanic mountains
fault-block mountains
fold mountains

Review Questions

22.1 Continental Drift and Seafloor Spreading

1. Seafloor spreading
   (a) is caused by thermal convection currents.
   (b) is validated by remanent magnetism.
   (c) moves away from the midocean ridges.
   (d) all of the above.

2. Which of the following is geologic evidence supporting continental drift?
   (a) similarities in biological species and fossils found on various continents
   (b) continuity of geologic structures such as mountain ranges
   (c) glaciation in the Southern Hemisphere
   (d) all of the above

3. Describe the geologic evidence that supported Wegener's theory of continental drift.

4. What is the present theory of the mechanism for continental drift? Why was Wegener's explanation unacceptable?

5. What is meant by remanent magnetism?

6. State what evidence supports the concept of seafloor spreading.

7. What is the average rate of seafloor spreading?

22.2 Plate Tectonics

8. Which of the following is a primary cause of volcanoes, earthquakes, crustal deformation, and mountain building?
   (a) solar radiation
   (b) plate tectonics
   (c) hot spots
   (d) remanent magnetism

9. Which of the following is *not* a name for a plate boundary type?
   (a) divergent
   (b) convergent
   (c) subduction
   (d) transform

10. The basic source of energy provided for the movement of Earth's lithospheric plates is thought to be
    (a) Earth's gravitational potential energy.
    (b) radioactive decay.
    (c) residual heat from Earth's core.
    (d) solar energy.

11. What are the lithosphere and the asthenosphere?

12. What is a "plate" in the context of plate tectonics?

13. Describe the three general types of relative plate motions and the geologic results of each.

14. What is a subduction zone?

15. How are volcanic and earthquake activity explained in the theory of plate tectonics?

16. What are the names of the six major plates?

22.3 Earthquakes and Earth's Interior

17. Earthquakes
    (a) have no correlation with the "Ring of Fire."
    (b) originate in Earth's outer core.
    (c) are monitored by seismographs.
    (d) generate waves in Earth's interior called tsunamis.

18. The severity of the damage caused by an earthquake is represented on which scale?
    (a) the Richter
    (b) the Seismology
    (c) the Mercalli
    (d) the Mohs'

19. Which of the following is *not* a name labeling part of Earth's interior structure?
    (a) mantle
    (b) Mohorovicic discontinuity
    (c) Igneous discontinuity
    (d) outer core

20. The lithosphere
    (a) consists, in part, of Earth's crust.
    (b) consists, in part, of Earth's upper mantle.
    (c) is closer to Earth's surface than the asthenosphere.
    (d) all of the above.

21. What causes an earthquake?

22. What are the focus and epicenter of an earthquake?

23. Give the two general types of seismic waves and the subdivision of one of these types.

24. What is the basis of (a) the Richter scale and (b) the modified Mercalli scale?

25. Why is *tidal wave* an incorrect and misleading term for a huge ocean wave generated by an earthquake? What is the correct term?

26. What is the difference between S and P waves? How do these waves allow scientists to locate the focus of an earthquake and to investigate Earth's interior structure?

27. Explain the concept of isostasy.

22.4 Crustal Deformation and Mountain Building

28. Crustal folding of Earth's rock layers can occur as a series of arches and troughs. The arches are known as
    (a) synclines.
    (b) anticlines.
    (c) island arcs.
    (d) none of the above.

29. Which of the following is a type of fault?
    (a) normal
    (b) reverse
    (c) strike-slip
    (d) all of the above

30. The building of mountains is best explained by the
    (a) cooling and contracting of Earth's surface.
    (b) gradual accretion of material from the upper mantle.
    (c) theory of plate tectonics.
    (d) none of the above.

31. State the difference between anticlines and synclines.

32. What types of relative plate motions correspond to the different types of faulting?

33. What type of faulting is associated with a transform fault?

34. Explain each of the following terms: (a) normal faulting, (b) reverse faulting, (c) thrust faulting, and (d) strike-slip faulting.

35. What are the three principal types of mountains, and how are they distinguished? Give an example of each.

36. How is the formation of each type of mountain explained by the theory of plate tectonics?

37. Describe the history of the formation of the Himalayan Mountains.

38. Name the type of plate boundary that is most likely to produce mountain building.

Applying Your Knowledge

1. The Pacific plate is moving northwestward relative to the North American plate at the rate of about 5 centimeters per year. If this rate and direction continue, at what approximate time in the geologic future will Los Angeles become a suburb of San Francisco? (Los Angeles and San Francisco are approximately 600 km apart.)

2. How did the analysis of seismic waves help scientists determine that Earth's inner core is solid, but the outer core is liquid? Could the same technology be used to learn about the interior of the Moon? Why or why not?

3. Discuss the adage, "Earthquakes don't kill people, buildings kill people."

4. You and two friends go out on your boat and are sailing back into the harbor when you hear a radio warning that a tsunami will reach the area in 15 minutes. The first friend says, "Let's jump overboard and swim for shore." The other friend argues, "No, let's stay with the boat and head for the dock as quickly as possible." Being the captain, you decide that the best course of action is to turn around and head back out to sea. Are you correct? Why or why not?

5. Fossils of ancient marine animals and plants have been found high up in the Himalayas. How did they get there? The Hawaiian Islands are a chain of volcanic mountains rising up from the seafloor. Why would you probably *not* find marine fossils high up on the Hawaiian mountains?

Answers to Multiple-Choice Review Questions _____

1. d   8. b   10. b   18. c   20. d   29. d
2. d   9. c   17. c   19. c   28. b   30. c

SURFACE PROCESSES

23

Running water labors continually to reduce the whole of the land to the level of the sea.

Charles Lyell (1797–1875)

Nothing is permanent on Earth's surface—in the context of geologic time. Since early times, people have erected edifices and monuments as memorials to persons and peoples and as symbols of various cultures. In doing so, they used the most durable rock materials; however, even recent structures show the inevitable signs of deterioration. Buildings, statues, and tombstones eventually will crumble when exposed to the elements, and on a larger scale, nature's mountains are continually being leveled.

An integral part of the rock cycle, mentioned in Chapter 21, is the decomposition of rock by weathering. As soon as rock is exposed at Earth's surface, its destruction begins. Many of the resulting particles of rock are washed away by surface water and are transported as sediment toward the oceans in streams and rivers. Over geologic time this process constitutes a leveling of Earth by wearing away high places and transporting sediment to lower elevations—a process known as *gradation*. (See the chapter-opening quotation.)

This chapter deals with the decomposition of rock and the leveling processes that move rock particles to lower elevations under the influence of gravity. Important geologic processes take place on the ocean floor, which accounts for about 65% of Earth's surface. Therefore, waves, ocean currents, and shoreline and seafloor topography are also considered. ■

Photo: The Yellowstone River, Wyoming. The surface processes of weathering, erosion, and mass wasting work together to form the classic V-shaped stream valley.

23.1 Weathering

LEARNING GOALS

▼ Define *weathering,* and explain some weathering processes.

▼ List the chemical and physical effects associated with weathering.

Weathering is the physical disintegration and chemical decomposition of rock at or near Earth's surface. Weathering depends on a number of factors such as the type of rock, moisture, temperature, and overall climate.

Physical weathering involves the physical disintegration or fracture of rock, primarily as a result of pressure. For example, a common type of physical weathering in some regions is *frost wedging.* When rock is formed by solidification, lithification, or metamorphism, internal stresses are produced in it. One common result of these stresses is cracks or crevices in the rock, called *joints.* Joints provide an access route for water to penetrate the rock. If the water freezes, the less dense ice requires more space than the water and exerts a strong pressure on the surrounding rock, just as freezing water in a glass container usually cracks the glass. As a result, the rock may break apart, as illustrated in ● Fig. 23.1. Frost wedging is most effective in regions where freezing and thawing occur daily.

In cold upper latitudes the subsurface soil may remain frozen permanently, creating a **permafrost** layer. During a few weeks in the summer, the topsoil may thaw to a depth of a few inches to a few feet. The frozen subsurface prevents the melted water from draining. As a result, the ground surface becomes wet and spongy (● Fig. 23.2). The permafrost in Alaska caused many problems during the construction of the Alaskan oil pipeline.

Plants and animals play a relatively small role in physical weathering. Most notable is the fracture of rock by plant root systems when they invade and grow in rock crevices. Burrowing animals—earthworms in particular—loosen and bring soil to the surface. This action promotes aeration and access to moisture, which are important factors in chemical weathering. Thus one type of weathering promotes another. The activities of humans also give rise to weathering and erosion that are often unwanted, as we will discuss later in the chapter.

FIGURE 23.1   Frost Wedging
The expansive forces of freezing water in rock joints cause the rock to fracture.

In all cases of physical weathering, the disintegrated rock still has the same chemical composition. **Chemical weathering,** however, involves a chemical change in the rock's composition. Because heat and moisture are two important factors in chemical reactions, this type of weathering is most prevalent in hot, moist climates. One of the most common types of chemical weathering involves limestone, which is made up of the mineral calcite ($CaCO_3$). Rain can absorb and combine with carbon dioxide (CO_2) in the atmosphere to form a weak solution of carbonic acid:

$$H_2O + CO_2 \longrightarrow H_2CO_3$$
Carbonic acid

Also, as water moves downward through soil, it can take up even more carbon dioxide that is released by soil bacteria involved in plant decay. Recall that carbonic acid (carbonated water) is one of the weak acids in carbonated drinks.

When carbonic acid comes in contact with limestone, it reacts with the limestone to produce calcium hydrogen carbonate, or calcium bicarbonate:

$$H_2CO_3 + CaCO_3 \longrightarrow Ca(HCO_3)_2$$

Carbonic acid Limestone   Calcium
                                bicarbonate

Calcium bicarbonate dissolves readily in water and is carried away in solution. Because limestone is generally impermeable to water (and dilute carbonic acid), this type of chemical weathering acts primarily on the surfaces of limestone rock along which water flows.

Water flowing through underground limestone formations can carve out large caverns over millions of years. The cavern ceilings may collapse, causing **sinkholes** to appear on the land surface (● Fig. 23.3).

The rate of chemical weathering depends primarily on climate and the mineral content of rock, with humidity and temperature being the chief climate controls. Chemical weathering is relatively rapid in hot, humid climates, compared with polar regions. On the other hand, it is quite slow in hot, dry climates. Egyptian pyramids and statues have stood for millennia with relatively minimal weathering.

Some rock minerals are more susceptible to chemical weathering than others, as is evidenced in the

FIGURE 23.2   A "Roller-Coaster" Railroad Near Strelna, Alaska

This condition was caused by differential subsidence resulting from the thawing of topsoil over subsurface permafrost.

FIGURE 23.3   A Sinkhole in Florida

The collapse of a limestone cavern ceiling can cause a sinkhole to appear on the land surface.

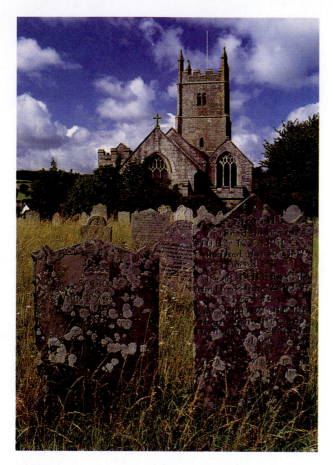

FIGURE 23.4 Weathered Tombstones in Devon, England

decomposition of tombstones (● Fig. 23.4). Marble tombstones, which consist of soluble calcite, may show a great deal of chemical weathering. Granite, due to its hardness, firmness, and durability, is the rock most suitable for tombstones.

RELEVANCE QUESTION: *What are some results of weathering on buildings, statues, and concrete walks in your community?*

23.2 Erosion

LEARNING GOAL

▼ Define *erosion,* and describe the agents of erosion.

Erosion is the wearing away of soil and rock by weathering and the downslope movement of soil and rock

fragments under the influence of gravity (mass wasting) or by the agency of streams, glaciers, wind, and waves.

Streams

Running water refers primarily to the waters of streams and rivers that erode the land surface and transport and deposit eroded materials. Rainfall that is not returned to the atmosphere by evaporation either sinks into the soil as groundwater or flows over the surface as runoff. Runoff occurs whenever rainfall exceeds the amount of water that can be immediately absorbed into the ground. It can move surprisingly large amounts of sediment, particularly from slopes without vegetation and from cultivated fields (● Fig. 23.5). Runoff usually occurs only for short distances before the water ends up in a stream.

A **stream** is defined by geologists as any flow of water occurring between well-defined banks. The term is applied to channeled flows of any size, from meter-wide mountain brooks to kilometers-wide rivers. The material transported by a stream is referred to as *stream load.* The load of a stream varies from dissolved minerals and fine particles to large rocks, and the transportation process, as well as the degree to which the stream can erode its channel, depends on the volume and swiftness (discharge) of the stream's current. A stream's load is divided into three components: dissolved load, suspended load, and bed load.

The *dissolved load* consists of dissolved water-soluble minerals that are carried along by a stream in solution. As much as 20% of the material reaching the oceans is transported in solution. Fine particles not heavy enough to sink to the bottom are carried along in suspension. The *suspended load* is quite evident after a heavy rain when the stream appears muddy. Coarse particles and rocks along or close to the bed of the stream constitute its *bed load.* These particles and rocks are rolled and bounced along by the current.

The action of the flowing water in a river causes its bed to erode. The principal landform resulting from a stream's erosive power is its V-shaped valley. The depth of an eroded valley is limited by the level of the body of water into which the river flows. The limiting level below which a stream cannot erode the land is called its *base level.* In general, the ultimate base level for rivers is sea level.

Once a stream has eroded its channel close to base level, downward erosion slows, and the stream begins to expend its energy by eroding its bed from side to

FIGURE 23.5   Runoff
This runoff is beginning to carve out a channel. If its supply of water continues, it may become a stream.

side. As one bank is eroded and then the other, the valley floor is widened. Heavy rains or spring thaw in the mountains upstream causes periodic flooding. During flooding, the water overflows its banks and deposits sediment across the widened valley floor, or **flood plain.** Due to this natural process, the flood plains of large rivers offer some of the most fertile soil on Earth.

Rich soil is the reason flood plains become heavily populated. However, the people who come to live on a flood plain want the soil but not the floods that go with it. So they build artificial levees to hold the river in its channel. See the chapter Highlight for a discussion of the problems associated with flood-prevention programs.

As a stream flows overland, it twists and turns, following the path of least resistance. A looplike bend in a river channel is called a **meander** (● Fig. 23.6). Meanders shift or migrate because of greater erosion on the outside of the curved loops. The speed of the streamflow is greater in this region than near the inside bank of the meander where sediment is deposited. When the erosion along two sharp meanders causes the stream to meet itself, it may abandon the water-filled meander, which is then known as an *oxbow lake.*

The amount of material eroded and transported by streams is enormous. The oceans receive billions of tons of sediment each year as a result of the action of running water. The Mississippi River alone discharges approximately 500 million tons of sediment yearly into the Gulf of Mexico. A river's suspended and bed loads may accumulate at its mouth and form a **delta,** such as the Nile River Delta (● Fig. 23.7). The rich sediment makes delta areas important for agriculture.

FIGURE 23.6   The Meandering Sioux River in Iowa
Notice how the stream's winding path has widened its valley.

HIGHLIGHT: The Great Mississippi Flood of 1993

In 1927, a prolonged period of heavy rainfall caused extensive flooding in the upper Mississippi River valley. More than 200 lives were lost in the disaster. Soon after, Congress enacted the first Mississippi River Flood Control Act, which provided funding for the U.S. Army Corps of Engineers to build levees, flood walls, dams, reservoirs, and pumping stations that would hold back the river and prevent it from spilling out onto its natural flood plain.

The program has been a mixed blessing for the region. On the positive side, it has contained the frequent overflowing that previously kept people from living along the banks of the river. As a result, the population along the river has grown in geometric proportions since the program began. The program has even prevented the moderate flooding that plagued communities along the valley every 10 to 20 years or so.

The downside was slower in coming but became obvious in the summer of 1993, when exceptional rainfall across the upper Mississippi valley followed an equally rainy spring. Along the 800-kilometer stretch from St. Paul, Minnesota, to St. Louis, Missouri, the river rose as high as 7 meters above normal flood stage and burst through its artificial levees with intense force, spreading as far as 13 kilometers from its channel (Fig. 1).

During this period of catastrophic flooding, which lasted for 2 months, 50 people were killed, 50,000 people were rendered homeless, and 14 million acres of land—including whole towns, farms, parts of cities, and suburbs—were inundated. Estimates of the total damage were in the neighborhood of 15 billion dollars.

In the years that followed, scientists, engineers, politicians, and Mississippi valley citizens debated these questions: (1) Did 65 years of engineering the river to prevent normal flooding actually intensify the severity of the 1993 flood? (2) What can valley communities do to prevent such disasters in the future? To better understand the issues,

let's look briefly at how basic flood-control mechanisms work for normal flooding.

When a stream overflows its channel, the coarser sediments are deposited along the river banks. As the water spreads out, the finer sediments are distributed in a thin layer across the valley floor. Repeated flooding results in the formation of *natural levees,* narrow, thick beds of coarse sediments built up on either side of the river. The basic mechanism of flood control is simply to construct *artificial levees* on top of the river's natural levees.

In the case of the Mississippi, this approach worked reasonably well for normal flooding, and the response to the occasional major floods that breached the levees was to build higher levees. In some places along the upper Mississippi, the levees were built as much as 12 meters high. As a result, when the extraordinary rains of 1993 caused the extrahigh levees to be breached, it was like a dam bursting. The roaring floodwaters attacked the

FIGURE 23.7 Delta of the Nile River

FIGURE 1   Flooding Along the Mississippi River
Albany, Illinois, was completely inundated by the Mississippi floods in the summer of 1993.

land with unprecedented volume, force, and speed. Furthermore, in the weeks that followed, the remains of the levees prevented the floodwaters from draining back into the river channel. This left the land flooded for a much longer time than it would have been in a natural flood.

Thus the widely agreed on answer to the first question was yes, the engineering did exacerbate the 1993 disaster. Nevertheless, different valley communities have arrived at different conclusions regarding the best solution to the problem. Many of the urban communities with large populations along the river and agricultural communities with valuable cropland along the river have decided that they have no choice but to build higher, thicker levees.

Other communities have decided to eliminate or greatly reduce artificial flood-retention systems and are initiating changes in their zoning laws that will prevent further development of the land along the river banks. Their strategy is to allow the river area to flood naturally by preserving the few stretches of natural flood plain that are still left and by buying back additional lands from private owners and returning them to their natural state. Regardless of their choice, all involved hope that the solution they have chosen will not be put to the test in the near future.

However, some sediment deposits are not always to our benefit. Many harbors must be dredged to remove sediment deposits to keep them navigable, and the buildup behind dams poses operational problems. Human activities that remove erosion-preventing vegetation from the land may give rise to sediment pollution in rivers and streams, and result in environmental problems.

Glaciers

Frost wedging, one of the eroding actions of ice, was mentioned previously. Many of us are familiar with the ice of winter, but parts of Earth are covered with large masses of ice year-round. These large masses are called **glaciers.** To most of us, the term *glacier* usually brings to mind the ice age. Indeed, large areas of Greenland and Antarctica are presently covered with glacial ice sheets, or *continental glaciers,* similar in size to those that covered Europe and North America during the last ice age over 10,000 years ago. But are there any glaciers in the United States today? The answer is yes. In fact, you may be surprised to learn that there are about 1100 glaciers in the western part of the conterminous United States and that 3% (44,000 km^2) of the land area of Alaska is covered by glaciers.

Glaciers are formed when, over a number of years, more snow falls than melts. As the snow accumulates and becomes deeper, it is compressed into solid ice by its own weight. When enough ice accumulates, the glacier "flows" downhill or, if it is on a flat region, out from its center. The icebergs commonly found in the North Atlantic and Antarctic oceans are huge chunks of ice that have broken off from the edges of the glacial ice sheets of Greenland and Antarctica, respectively.

Small glaciers, called *cirque* (pronounced "sirk") *glaciers,* form along mountains in hollow depressions that are protected from the Sun. A majority of the glaciers in the United States are of this variety. The

FIGURE 23.8 Jöstedalsbreen, the Largest Glacier in Europe

ice movement further erodes the land and forms an amphitheater-like depression called a *cirque*. When the ice melts, these glacier-eroded cirques often become lakes.

If snow accumulates in a valley, the valley floor may be covered with compressed glacial ice, and a *valley glacier* is formed that flows down the valley (● Fig. 23.8). At lower and warmer elevations the ice melts, and where the melting rate equals the glacier's flow rate, the glacier becomes stationary. The flow rate may be from a few inches to over 30 m per day, depending on the glacier's size and other conditions.

The erosive action of a valley glacier is not unlike that of a stream. As the ice flows, it loosens and carries away materials, or bed load, that will be ground fine by abrasion. Although much slower than a stream, a glacier can pick up huge boulders and gouge deep holes in the valley. The paths of vigorous, preexisting mountain glaciers are well marked by the deep U-shaped valleys they leave.

Glaciers, like streams, also deposit the material they carry. The general term *drift* is applied to any type of glacial sediment deposit. Material that is transported and deposited by ice, in contrast to melt water, is called *till*. Till deposits are not layered or sorted, as is the sediment carried away by the melt water of a glacier. At the end and along the sides of a glacier the till may form ridges known as **moraines** (● Fig. 23.9). The terminal moraine marks the farthest advance of the glacier. Terminal moraines give us an indication of the extent and advance of the glacial ice sheet in North America, which retreated about 10,000 years ago. These moraines lie as far south as Indiana, Ohio, and Long Island, New York.

Wind

Erosion by wind is a slow process, but wind contributes significantly to the leveling of the land surface, especially in deserts. Most people think of a desert as a hot, dry place, but to a geologist, the term **desert** is defined by lack of precipitation, not by temperature. Occurring in cold as well as torrid regions, deserts account for one-fifth of Earth's land surface. Because vegetation is sparse in deserts, the occasional rainfall wreaks erosive havoc on the land surface. However, from day to day, wind is the prime mover of the land, transporting it grain by grain from high places to low.

A region of arid and semiarid climate in western North America extends from northern Mexico to eastern Washington State and includes such well-known

FIGURE 23.9   Valley Glacier in Alaska Showing Lateral and Medial Moraines

deserts as the Sonoran Desert, the Mojave Desert, and the Great Basin.

Dust particles that are small enough are transported great distances by the wind, and larger particles are moved short distances by rolling or bouncing along the surface. This action is quite evident in areas with large quantities of loose, weathered debris. Dust storms may darken the sky and be of such intensity that visibility is reduced to almost zero (● Fig. 23.10). During the 1930s in the Midwest, drought conditions created areas known as *dust bowls,* in which layers of fertile topsoil were blown away by the wind.

FIGURE 23.10   Sandstorm in Tibet

Waves

As wind blows across the ocean, friction drags on the water surface, creating waves. The wind-driven water waves erode the land along the shorelines. This erosion is most evident where the ocean surf pounds the shore. Some coasts are rocky and jagged, which shows that only the hardest materials can withstand the unrelenting wave action over long periods of time. Along other coastlines, cliffs are formed, terraced by the eroding action of waves. Waves are discussed further in Section 23.4.

Mass Wasting

Wherever the ground slopes, debris consisting of soil and rock fragments that have been loosened by weathering—or shaken loose by an earthquake or volcanic eruption—is pulled downslope by gravity. **Mass wasting** is the general geologic term for the downslope movement of soil and rock under the direct influence of gravity. To start the debris moving, the force of gravity must overcome friction, the force that keeps the material stationary.

Both the forces of gravity and friction are influenced by the presence of water. Water adds weight to the debris, thus increasing the pull of gravity, and it acts as a lubricant, thereby lessening the effects of friction. For this reason, mass-wasting disasters occur most frequently during or after a heavy snow or rainfall.

Mass movement may be fast or slow, depending on the steepness of the slope, and the amount of material involved in a single episode may be a few pebbles or an entire mountainside. Thus the effect of an episode on the local environment and on any people living there can range from unnoticeable to devastating.

Two important types of *fast mass wasting* are landslides and mudflows. **Landslides** involve the downslope movement of large blocks of weathered materials. Spectacular landslides occur in mountainous areas when large quantities of rock break off and move rapidly down the steep slopes. This type of landslide is termed a *rockslide.*

A comparatively slower form of landslide is a **slump,** the downslope movement of an unbroken block of overburden, which leaves a curved depression on the slope (● Fig. 23.11). Slumps are commonly accompanied by debris flows that consist of a mixture of rock fragments, mud, and water flowing

FIGURE 23.11   Slump—A Common Form of Mass Wasting

downslope as a viscous liquid. Small slumps are commonly observed on the bare slopes of new road construction.

Mudflows are the movements of large masses of soil that have accumulated on steep slopes and become unstable due to the absorption of large quantities of water from melting snows and heavy rains. Vegetation hinders mass movement. Consequently, mudflows are common in hilly and mountainous regions where land is cleared for development without regard to soil conservation.

In a very recent disaster that occurred on May 5, 1998, more than 200 people died and 1500 lost their homes and all their possessions in mudflows that devastated several towns in the Sarno valley of southern Italy (● Fig. 23.12). The media dubbed the disaster a "modern-day Pompeii," referring to the volcanic eruption that buried thousands of people in volcanic ash near Naples in A.D. 79 (see Chapter 21).

Nature played a key role in triggering the 1998 disaster. Rainfall at the time was exceptional—as much fell in a week as would normally fall in a year. But the tragedy was only to a limited extent a natural disaster. The loose volcanic soils of Campania, the region surrounding Naples, are inherently perilous. During the past 50 years, 631 landslides and mudflows have hit the region, killing 3800 people. Geologists have continually warned about the construction of towns and

FIGURE 23.12 May 1998 Mudflow Disaster in Southern Italy

Rescue efforts bogged down as the mud dried and hardened.

housing in the region, declaring it a "risk zone." Nevertheless, homes, schools, and hospitals are still being built on the dangerous slopes.

Fast mass wasting is quite dramatic, but *slow mass wasting* is a more effective geologic transport process. In contrast to the rates of fast mass wasting, the rates of slow mass wasting are generally imperceptible. One important type is called **creep**—the slow particle-by-particle movement of weathered debris down a slope, taking place year after year. It cannot be seen happening, but the manifestations of creep are evident (● Fig. 23.13). Although spectacular landslides and slumps involve large quantities of mass movement, this is but a fraction of the cumulative total of mass movement by creep over a period of time.

RELEVANCE QUESTION: *What good example of erosion can be found somewhere in your community?*

FIGURE 23.13 Creep

The downhill bends in these trees are evidence of creep in the surface beneath them.

23.3 Groundwater

LEARNING GOALS

▼ Describe Earth's water resources.

▼ Explain the hydrologic cycle.

Water is often referred to as the basis of life. The human body is composed of 55% to 60% water by weight, and water is necessary to maintain our body functions. This common chemical compound is an essential part of our physical environment not only in life processes but also in other areas, such as agriculture, industry, sanitation, firefighting, and even religious ceremonies. Early civilizations developed in valleys where water was abundant, and even today the distribution of water is a critical issue. Consider how your life would be affected without an adequate water supply.

Our water supply is a reusable resource that is constantly being redistributed over Earth. Many factors enter into this redistribution, but in general it is a movement of moisture from large reservoirs of water, such as oceans and seas, to the atmosphere, to the land, and back to the sea. This gigantic cyclic process is known as the **hydrologic cycle** (● Fig. 23.14).

Moisture evaporated from the oceans moves over the continents through atmospheric processes and falls as precipitation. Some of this water evaporates and returns to the atmosphere, some of it becomes runoff, but a large part of it soaks into the soil and down into the subsurface, where it collects as **groundwater.**

Groundwater Mechanics

Earth's water supply, some 1.25×10^{18} m³, may seem inexhaustible because it is one of our most abundant natural resources. Approximately 70% of Earth's surface is covered with water. However, about 98% of that water is saltwater, and only about 2% is fresh. Most of the fresh water is frozen in the glacial ice sheets of Greenland and Antarctica. A mere 0.6% of fresh water is groundwater, yet groundwater is the source of half our drinking water and more than half the water used in agriculture and industry.

While the rate of human use of groundwater increases every year, the rate of yearly rainfall is not enough to replenish the groundwater reserves that

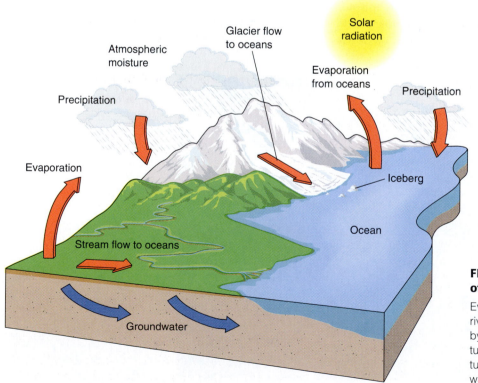

FIGURE 23.14   An Illustration of the Hydrologic Cycle

Evaporated moisture from oceans, rivers, lakes, and soil is distributed by atmospheric processes. It eventually falls as precipitation and returns to the soil and bodies of water.

took thousands of years to accumulate. In addition, groundwater supplies in many places have been contaminated by human, agricultural, and industrial wastes. The use and abuse of groundwater have become key environmental issues facing all nations of the world, including the United States, where bottled water has become a standard commodity sold in supermarkets.

The movement of groundwater is controlled by the physical properties of soil and rocks. *Porosity* is the percentage volume of unoccupied space in the total volume of a substance. The porosity of rocks and soil near Earth's surface determines the ground's capacity to store water. The most common rock for storing water is sandstone.

The factor that determines the availability of groundwater is the permeability of the subsurface soil and rock. *Permeability* is a material's capacity to transmit fluids, which is a function of the porosity of the material. Of course, loosely packed soil components, such as sand and gravel, permit greater movement of water. Clay, on the other hand, has fine openings and relatively low permeability. The average rate of movement of groundwater through rocks is about 14 m/y.

Under the influence of gravity, water percolates downward through the soil until at some level the ground becomes saturated. The upper boundary of this *zone of saturation* is called the **water table.** The unsaturated zone above the water table is called the *zone of aeration* (● Fig. 23.15).

In the zone of aeration, the pores of the soil and rocks are partially or completely filled with air. In the zone of saturation, all the voids are saturated with

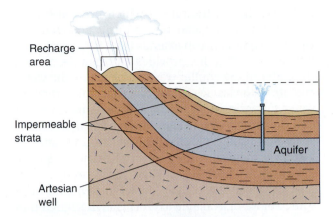

FIGURE 23.16   Confined Aquifer, Recharge Area, and Artesian Well

As long as the top of the well is below the altitude of the recharge area (dashed line), water will rise up out of the well without the aid of pumping.

groundwater, forming a reservoir from which we obtain part of our water supply by drilling wells to depths below the surface of the water table (see Fig. 23.15). Lakes, rivers, and springs occur where the water table intersects the surface. The level of the water table shows seasonal variations, and shallow wells may go dry in late summer.

A body of permeable rock through which groundwater moves is called an **aquifer** (from the Latin for "water carrier"). Sand, gravel, and loose sedimentary rock are good aquifer materials. Aquifers are found under more than half the area of the conterminous United States.

Water must be pumped out of water-table wells, but a special geometry of impermeable rock layers gives rise to what are called *artesian wells,* in which water under pressure rises to the surface without the aid of pumping. As illustrated in ● Fig. 23.16, this geometry can occur when an aquifer is sandwiched between sloping, impermeable rock strata and the higher end is exposed to the surface so that it can receive water to replenish the aquifer. The pressure of gravity can cause water to spurt or bubble onto the surface above the aquifer. The name *artesian* comes from the French province of Artois, where such wells and springs are common.

Groundwater Depletion and Contamination

Groundwater reserves are not inexhaustible. When the rate of extraction is greater than the rate of

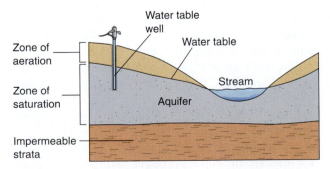

FIGURE 23.15   Zone of Aeration, Zone of Saturation (Aquifer), and Water-Table Well

A water-table well requires pumping in order to draw water up from an aquifer.

recharge, that is, the rate at which the water is replaced, an aquifer can be seriously depleted. The Ogallala aquifer, which extends from South Dakota to the Texas panhandle, provides an example. Most of the water in the aquifer was added during the wetter climate of the last ice age. Today, the aquifer provides water for agricultural irrigation in a semiarid region covering 65,000 square kilometers. More than 150,000 wells are currently tapping the Ogallala at 10 times the rate of recharge, and the zone of saturation is shrinking drastically. In 30 years' time, this essentially nonrenewable aquifer will be depleted. With no other source of water available to replace it, the agricultural economy of the region surely will fail.

Besides depletion, excessive groundwater extraction can lead to a variety of problems, one of which is land subsidence. Because the water in a confined aquifer helps to support the weight of the overlying strata, excessive extraction of water may cause the land surface to sink, or subside. Land subsidence is accelerated if the surface strata are supporting additional weight, such as buildings, roads, and other structures. Subsidence may cause water and drainage pipes to rupture, building foundations may shift, cracks can develop suddenly in roads, and roadbeds can sag.

Another problem of excessive groundwater extraction that occurs in coastal regions is saltwater contamination of wells. Because saltwater is denser than fresh water, fresh groundwater in the zone of saturation floats on saltwater that has seeped in from the ocean and penetrated the strata. As pumping lifts fresh water up to the surface through the well, the saltwater directly beneath the well rises to fill the gap (● Fig. 23.17). Thus excessive pumping brings saltwater up into the well very quickly.

Other forms of contamination also threaten the supply of usable groundwater. The quality of water is usually expressed in terms of the amount of dissolved chemical substances present as a concentration in parts per thousand (ppt), parts per million (ppm), or parts per billion (ppb). For example, if a water sample contains 1 ppm salt, it contains 1 g of salt/1 million g of water.

Also important with respect to quality are the types of chemicals water contains. Chemicals range from nontoxic to extremely toxic; concentrations range from very high to very minute values. The water quality may be very good with a fairly high concentration of calcium carbonate but very toxic with minute amounts of another substance such as methylmercury.

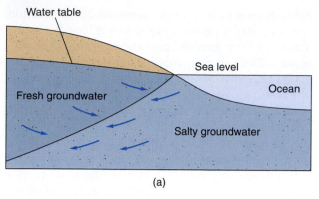

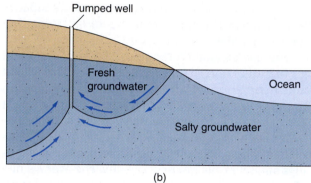

FIGURE 23.17 Saltwater Contamination of a Coastal Water-Table Well

(a) Saltwater seeps into the ground beneath the aquifer, but because it is denser, it stays beneath the fresh water. (b) Overdrawing the fresh water causes the saltwater to rise up into the well.

Our best natural source of fresh water is rainwater, but even that contains a variety of dissolved chemicals due to air pollution. Acid rain, which is caused mainly by the burning of fossil fuels, is a common pollutant these days (see Section 20.4). Rain falling on Earth's surface enters the ground where it takes on more chemicals when it reacts with the soil, rock, and organic material. Water quality is reduced further, sometimes drastically, when the groundwater is contaminated by phosphates (found in laundry detergents and fertilizers), sewage and other waste materials, pesticides (organic compounds that are used to kill insects), herbicides (organic compounds that are used to kill unwanted plant life), and industrial wastes.

Municipal water supplies are treated by chlorination to eliminate bacteria and viruses and thus increase the quality of our drinking water. This process does not remove organic compounds such as acetone, benzene, carbon tetrachloride, or chloroform that may enter the water supply by way of industrial wastes. Some of these compounds can be removed by using charcoal filters.

In some cases water containing bicarbonates (hydrogen carbonates) and chlorides is sold commercially as "mineral water." However, dissolved minerals in "hard" water have undesirable effects. *Hard water* is so called because of its high content of dissolved calcium and magnesium salts (bicarbonates, chlorides, and sulfates). Iron salts also contribute to hard water. Such dissolved minerals not only affect the taste of the water but also cause use problems. The salts combine with the organic acids in soaps to form insoluble compounds, thereby reducing the lathering and cleansing qualities of the soap. It is because of these insoluble compounds from soap that clothing does not come out "whiter white" on wash day, and also why there is a ring around the bathtub. Water softening is an active business in many parts of the country.

Because water makes up 55% to 60% of the human body weight, good quality water is extremely important for good health. Therefore, it is important for all individuals, wherever they live, to protect their fresh water supply.

RELEVANCE QUESTION: *How would your life be affected without an adequate water supply?*

23.4 Shoreline and Seafloor Topography

LEARNING GOALS

▼ List the chemical and physical properties of the oceans.

▼ Describe the major features of seafloor topography.

Waves, Currents, and Tides

The vastness of the restless oceans imparts an awe-inspiring feeling to most observers, and where the ocean meets the land, we observe the beauty of coastal regions, shaped by surface waves, that vary from broad low beaches to steep rocky cliffs. The oceans cover about 71% of Earth's surface. The five major oceans in order of decreasing size are the Pa-

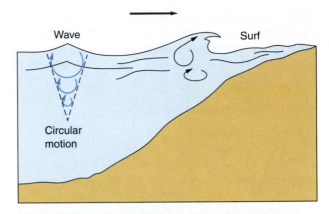

FIGURE 23.18   An Illustration of a Surface Wave Approaching the Shore

cific, Atlantic, Indian, Antarctic, and Arctic oceans. Their average depth is about 4 km; the greatest measured depth is about 11 km in the Marianas Trench in the western Pacific.

Ocean water is in constant motion. Three types of seawater movements are waves, currents, and tides. Ocean waves continually lap the shore. While the form of a wave moves directly toward the shore, the water "particles" move in more or less circular paths, as illustrated in ● Fig. 23.18. This circular pattern is the reason that debris bobs up and down as the waveforms pass under it.

As a wave approaches shallower water near the shore, the water particles experience difficulty completing their circular paths and are forced into more elliptical paths. The surface wave then grows higher and steeper. Finally, when the depth becomes too shallow, the water particles can no longer move through the bottom part of their paths, and the wave breaks, with the crest of the wave falling forward to form *surf*.

At the beach, you may have noticed that unless a piece of bobbing debris comes close enough to be caught in the surf and thrown up on shore, it appears to move steadily in a direction parallel to the shoreline. This movement is an indication of a **long-shore current** flowing along the shore. The current arises from incoming ocean waves that break at an angle to the shore. The component of water motion along the shore causes a current in that direction. Waves and their resulting long-shore currents are important agents of erosion along coastlines.

The periodic rise and fall of the tides are also quite evident at the beach. **Tides** result from the two tidal bulges that "move" around Earth daily as a result of the gravitational attractions of the Moon and Sun and the rotation of Earth (see Chapter 17). The water level may rise as much as 12 meters in some regions at high tide.

Shoreline Topography

When incoming waves attack an irregular stretch of coastline, the parts of the shoreline that jut out are subjected to the most erosion. If the land along the coast is elevated, wave action cuts into the base of the slopes below the high-tide mark. As shown in ● Fig. 23.19, when the overlying rock loses support, it collapses, leaving behind a near-vertical *wave-cut cliff*. Isolated remnants of resistant rock remain as *sea stacks* and *sea arches,* and hollowed-out portions of the wave-cut cliff form *sea caves.*

Along with the sediments transported to the sea by rivers, the fragments eroded from elevated stretches of coastline are transported by long-shore currents to bays and to calmer stretches of the coast where the land is at a lower elevation. Here we see beaches and other features of a depositional coastline.

As shown in ● Fig. 23.20, some of the more common depositional features are

1. *Pocket beaches,* which form in the low-energy wave environment between headlands.
2. *Barrier islands,* which extend more or less parallel to the mainland. New York's Fire Island is a well-known example. Between a barrier island and the mainland is a protected body of water called a *lagoon.*
3. *Spits,* which are narrow, curved projections of beach that extend into the sea, elongating the shoreline. Cape Cod, Massachusetts, is a spit.

Seafloor Topography

Scientists once thought that the surface features or topography of the ocean basins consisted of an occasional volcanic island arc on a relatively smooth sediment-covered floor. This incorrect view resulted from the lack of direct observation. The surface of the oceanic crust was not explored in great detail until after World War II. With the advent of modern technology, sounding and drilling operations revealed that the ocean floor is about as irregular as the surfaces of the continents, if not more so (● Fig. 23.21).

FIGURE 23.19   Features of Coastal Erosion

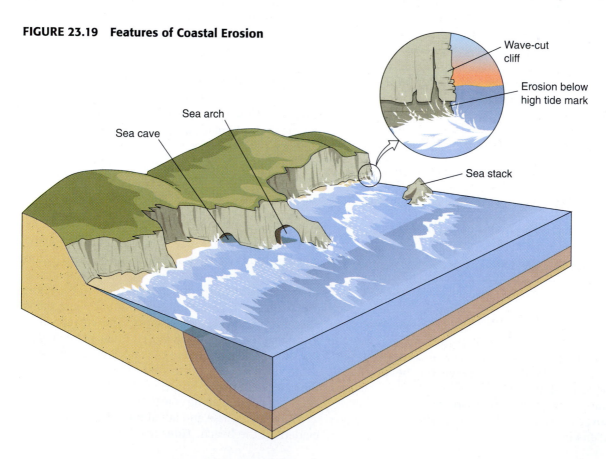

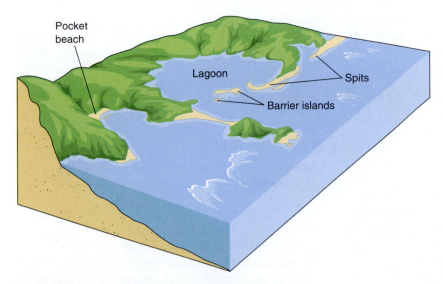

FIGURE 23.20   Features of Coastal Deposition

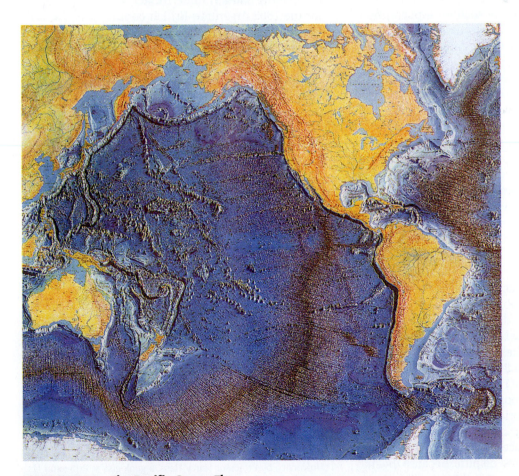

FIGURE 23.21   The Pacific Ocean Floor

Relatively recent explorations have shown the seafloor surfaces to be as irregular as the surfaces of the continents.

We now know that the seafloor has a system of midocean ridges—rocky, submarine mountain chains that mark divergent plate boundaries. Volcanic island arcs rim the Pacific and Indian oceans. Large volcanic mountains also rise from the ocean floor away from plate boundaries, marking the place where the plate has ridden over a mantle hot spot.

Many isolated, submarine, volcanic mountains also have been discovered. They are known as **seamounts** and are individual mountains that may extend to heights of over 1.6 km above the seafloor. Some seamounts have flat tops and are given the special name of *guyots** ("GEE-ohs"). Their shapes suggest that the tops were once islands that were eroded away by wave action. However, many of the guyot tops are several thousand feet below sea level. The eroded seamounts subsided and sank below sea level as the oceanic crust moved away from a spreading ridge.

Another marked feature of seafloor topography is *trenches,* which mark the locations of the deep-sea subduction zones. These trenches are as much as 240 km in width, 24,000 km or more in length, and 11 km in depth.

The huge volumes of sediment flowing into the oceans from continental regions do have an effect on seafloor topography. Distributed by ocean currents, sediment accumulates in some regions such that a layer covers and masks the irregular features of the rocky ocean floor. The resulting large, flat areas are called **abyssal plains.** Abyssal plains are most common near the continents, which supply the sediment.

Although 70% of Earth's surface is covered with water, the oceanic crust basins account for only about 65% of the surface area. Thus the continental crust makes up 35% of Earth's surface area, but since only about 30% of Earth's surface is land, 5% of the continental crust must be submerged. These shallowly submerged areas that border the continental masses are called **continental shelves** (● Fig. 23.22).

The widths of these shelves vary greatly, but average on the order of 64 to 80 km. The Pacific coast of South America has almost no continental shelf, only a relatively sharp, abrupt continental slope. However, off the north coast of Siberia, the continental shelf extends outward into the ocean for about 1280 km.

Continental shelves recently have become a point of international interest and dispute, because the majority of commercial fishing is done in the waters

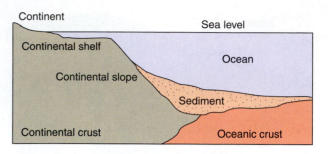

FIGURE 23.22   A Cross-Sectional Illustration of a Continental Shelf

above the shelves. Also, the shelves are the locations of oil deposits that are now being tapped by offshore drilling. As a result, many countries, including the United States, have extended their territorial claims to an offshore 320-km limit. In fact, one of the reasons for the 1982 Argentine-British conflict over the Falkland Islands was potential offshore oil deposits.

Beyond a continental shelf, the surface of the continental landmass slopes downward to the floor of the ocean basin. The **continental slopes** define the true edges of the continental landmasses. Erosion along these slopes gives rise to deep submarine canyons that extend downward toward the ocean basins; and near the edges of the ocean basins, the sediment collects in depositional basins.

Undersea exploration is a relatively new phase of scientific investigation. Indeed, a great deal more is to be learned about this vast region. As advances in technology provide more data on these previously inaccessible depths, we may expect our knowledge of Earth and its geologic processes to grow.

---

*Named in honor of Arnold Guyot, the first geologist at Princeton University, by Professor Harry Hess, a geologist at Princeton University who discovered the first flat-topped seamounts in the 1950s.

Important Terms

weathering (23.1)
physical weathering
permafrost
chemical weathering
sinkholes
erosion (23.2)
stream

flood plain
meander
delta
glaciers
moraines
desert
mass wasting

landslides
slump
creep
hydrologic cycle (23.3)
groundwater
water table
aquifer

long-shore current (23.4)
tides
seamounts
abyssal plains
continental shelves
continental slopes

Review Questions

23.1 Weathering

1. Weathering is the alteration that rocks undergo due to exposure to
   (a) air.   (c) living organisms.
   (b) water.   (d) all of the above.

2. The rate of chemical weathering of rock depends on the
   (a) climate.   (d) temperature.
   (b) mineral content.   (e) all of the above.
   (c) humidity.

3. Distinguish between physical and chemical weathering.

4. What is frost wedging?

5. How do plants and animals contribute to weathering?

6. On what factors does chemical weathering depend?

7. Describe briefly the chemical weathering process of limestone and the formation of caverns.

8. What is a sinkhole?

23.2 Erosion

9. Which of the following are agents that cause erosion?
   (a) running water   (c) wind
   (b) ice   (d) all of the above

10. By which of the following factors is a desert defined?
    (a) lack of precipitation   (c) high winds
    (b) intense heat   (d) all of the above

11. What is the downslope movement of soil and rock fragments under the influence of gravity called?
    (a) sheet erosion   (d) drift
    (b) mass wasting   (e) none of the above
    (c) meandering

12. What is erosion?

13. State and describe the three components of a stream's load.

14. Why does a river meander?

15. Distinguish between continental and valley glaciers.

16. Are there any glaciers in the United States? Explain.

17. Describe each of the following: (a) drift, (b) till, and (c) moraine.

18. What is mass wasting?

19. Explain each of the following and state whether it is a fast or slow type of mass wasting: (a) rockslide, (b) creep, (c) slump, and (d) mudflow.

23.3 Groundwater

20. The total amount of water on planet Earth is
    (a) increasing.   (c) remaining constant.
    (b) decreasing.   (d) none of the above.

21. The continuous circulation of Earth's water supply is known as the _____ cycle.
    (a) aqua   (c) hydrologic
    (b) redistribution   (d) hydrogeology

22. Describe the hydrologic cycle.

23. Explain each of the following: (a) porosity, (b) permeability, (c) the zone of aeration, (d) the zone of saturation, and (e) the water table.

24. What is an aquifer?

25. Distinguish between a water-table well and an artesian well.

26. What is the cause of "hard" water, and what are some of its effects?

23.4 Shoreline and Seafloor Topography

27. The oceans cover about what percent of Earth's surface?
    (a) 50   (b) 65   (c) 70   (d) 85

28. Which of the following is *not* a feature of coastal erosion?
    (a) spit   (d) sea cave
    (b) sea stack   (e) sea arch
    (c) wave-cut cliff

29. Name and describe three features of coastal deposition.

30. Define and explain the formation of each of the following: (a) seamounts, (b) guyots, and (c) abyssal plains.

31. What are continental shelves and continental slopes? What defines the true edges of the continental landmasses?

32. Why are the continental shelves the focus of current international interest?

Applying Your Knowledge

1. The Moon has approximately one-sixth the surface gravity of Earth and has neither an atmosphere nor surface water. Can the process we know as weathering occur on the Moon? Can mass wasting occur on the Moon? Explain.

2. State some procedures that can reduce the hazards of landslides.

3. You are on the town board of a small coastal community that depends on local water-table wells for its water supply. The board is considering plans to build a large retirement village on the outskirts of town. The village will install its own well to supply all its water needs. The added population plus the new golf course and three swimming pools will increase by 20% the total demand on the aquifer. Explain to the other board members how the retirement village well might quickly turn the fresh water in all the local wells to saltwater.

4. What are the source and purity of the water supplied to your home?

5. Apply your knowledge of plate tectonics and seafloor topography to predict the geologic future of the Hawaiian Islands.

Answers to Multiple-Choice Review Questions

1. d   9. d   11. b   21. c   28. a
2. e   10. a   20. c   27. c

GEOLOGIC TIME

24

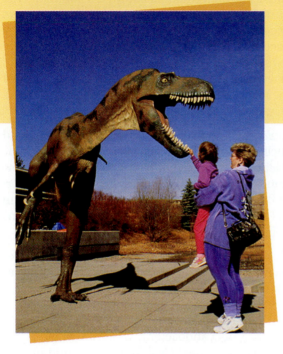

Lives of great men all remind us
We can make our lives sublime,
And, departing, leave behind us
Footprints on the sands of time.

Longfellow, Psalm of Life *(1838)*

Geology is indeed a broad topic. It studies the composition, structure, and processes of both the surface and interior of Earth. Not only do geologists study Earth as it exists today, they also look back at its long history over the span called **geologic time.**

Perhaps geology's major contribution to human knowledge is the finding that Earth is very old—about 4600 million years—and that humans (the genus *Homo*) have been around for only a minute part—about 2 million years—of that immense span of time. This recognition has an impact on our view of reality in a manner comparable with the finding in astronomy that, far from being the center and major part of the universe, Earth is more like a mere speck of dust.

In science, we have to follow where the evidence leads; what we might wish to be true must adjust to what actually seems to be correct. This perspective of our place in time and space should heighten both our sense of responsibility as the present-day custodians of Earth and our appreciation of the size and complexity of nature and what the human mind can discover and comprehend.

Photo: Two Cenozoic-era mammals admire an *Albertosaurus*, a Mesozoic-era reptile from about 70 million years ago.

This final chapter starts with a discussion of what *fossils* are, how they are formed, and what part they play in our understanding of geologic time. We then examine *relative geologic time*—determined by placing rocks, and the geologic events that they record, into chronologic order. After that, we discuss how *radiometric dating* is used to determine *absolute geologic time*—the actual ages of rocks and geologic events, including the age of Earth. Finally, we show that all this information enables construction of the absolute *geologic time scale*. The chapter Highlight discusses the event that is believed to have caused the extinction of the dinosaurs. ▪

24.1 Fossils

LEARNING GOALS

▼ Tell what fossils are and how they are formed.

▼ Recognize some types of fossil organisms.

The fossil record is vital to the understanding of geologic time. A **fossil** is any remnant or indication of prehistoric life preserved in rock. The study of fossils is called **paleontology,** an area of interest to both biologists and geologists. Evidence of ancient plants and animals can be preserved in several ways.

1. *Original remains.* Ancient insects have been preserved by the sticky tree resin in which they were trapped. The hardened resin, called *amber* and often used for jewelry, is found in Eastern Europe and the Dominican Republic (● Fig. 24.1a). The entire bodies of woolly mammoths have been found frozen in the permafrost of Alaska and Siberia. More often, only the hardest parts of organisms are preserved, such as bones. Shark teeth and the shells of shallow-water marine organisms endure well, are easily buried in sediment, and are thus common types of fossils.

2. *Replaced remains.* The hard parts (bone, shell, etc.) of a buried organism can be slowly replaced by minerals such as silica (SiO_2), calcite ($CaCO_3$), and pyrite (FeS_2) in circulating groundwater. A copy of the original plant or animal material results. Petrified wood, such as the beautiful samples from Arizona's Petrified National Forest, is a common type of replacement fossil (Fig. 24.1b). *Carbonization* occurs when plant remains are decomposed by bacteria under anaerobic (airless) conditions. The hydrogen, nitrogen, and oxygen are driven off, leaving a carbon residue that may retain many of the features of the original plant (Fig. 24.1c). In this way, coal was formed.

3. *Molds and casts of remains.* When an embedded shell or bone is dissolved completely out of a rock, it leaves a hollow depression called a *mold.* If new mineral material fills the mold, it forms a *cast* of the original shell or bone. Molds and casts can only show the original shape of the remains (Fig. 24.1d).

4. *Trace fossils.* A fossil imprint made by the movement of an animal is called a *trace fossil.* Examples are tracks, borings, and burrows (Fig. 24.1e).

The earliest evidence of ancient life is fossil blue-green algae, or cyanobacteria, which are single-celled organisms. The oldest algal fossils are found in Australia and date back to about 3500 million years ago (● Fig. 24.2). The fossil record shows that as time passed, larger and more complex life forms developed.

Not only do fossils play a major role in helping to determine the relative ages of rocks, they also can tell geologists something about past climatic conditions. For example, the finding of a coral reef in a farmer's cornfield indicates that the area was once a warm, shallow sea.

Certain microfossils, such as certain species of foraminifera, when found in rock layers have been proven to indicate the presence of nearby oil deposits (● Fig. 24.3). Companies involved in oil exploration drill deep underground and examine the cores removed, in search of these characteristic fossils. As our discussion of geologic time continues, we will become acquainted with other fossil creatures.

RELEVANCE QUESTION: *What fossils occur in your locality or close by?*

(a)

(b)

(c)

(d)

(e)

FIGURE 24.1   Some Modes of Fossil Formation

(a) *Original remains.* About 40 million years ago, this grasshopper was trapped in tree resin that later hardened to amber. (b) *Replaced remains.* About 200 million years ago, the organic matter in this log from Arizona was replaced by silica bit by bit to form petrified wood. (c) *Replaced remains.* About 350 million years ago, this seed fern was changed to carbon, but its form was preserved. (d) *Molds and casts.* This near-perfect cast of *Archaeopteryx,* one of the earliest birds (note the feather impressions), was found in Germany and dates from 145 million years ago. It had teeth, claws, and a bony tail. (Also refer to Fig. 24.18 on page 649.) (e) *Trace fossils.* About 200 million years ago in Arizona's Painted Desert, a dinosaur left these tracks in what was then a mudflat.

(a)

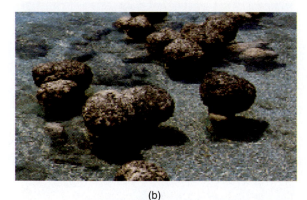

(b)

FIGURE 24.2   Fossil and Modern Algae

(a) The fossils in this rock are *stromatolites*, structures created by an early form of algae during the Archean eon about 3000 million years ago. (b) Modern stromatolites in Shark Bay, Western Australia.

24.2 Relative Geologic Time

LEARNING GOALS

▼ Explain and apply the principles for relative dating of rocks and geologic events.

▼ Name the present eon, its three eras, and its twelve periods.

Relative geologic time is obtained when rocks and the geologic events that they record are placed in chronologic order without regard to actual dates. As an analogy, you might find out that a friend of yours graduated from college, then served in the military, and then got married. You know the *relative time*— the *order* in which the events occurred. But what if you know the actual date of each of these events in

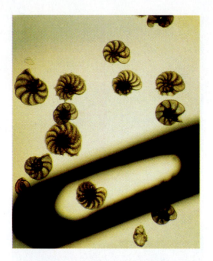

FIGURE 24.3   Foraminifera from the Mississippian Period

The needle's eye shows how tiny were these aquatic creatures named *foraminifera* but usually referred to as *forams*.

your friend's life? Then you would know what geologists call *absolute time* (Section 24.3).

The principles, or laws, used to determine the relative ages of rocks in one locality are based on common sense, and we will discuss the two major ones. The **principle of superposition** states that in a sequence of undisturbed sedimentary rocks, lavas, or ash, each layer is younger than the layer beneath it and older than the layer above it. (● Fig. 24.4). Sometimes layers are disturbed by folding or faulting, and the geologist must look for such evidence and, if found, take it into account.

The **principle of cross-cutting relationships** states that an igneous rock is younger than the rock layers it has intruded (cut into or across). The same principle applies to faults, in that a fault must be younger than any of the rocks it has affected (● Fig. 24.5).

In a given locality, sediments are not deposited continuously over time. Breaks called **unconformities** occur in the rock record. The missing layers of rocks may never have been deposited, or they may have been deposited and then eroded away after the rock surface was uplifted by a geologic event. If the area was then resubmerged, new layers may have been deposited. Unconformities represent gaps in the geologic record, and usually it is impossible to determine how much time is represented by an unconformity.

FIGURE 24.4   The Principle of Superposition

The principle of superposition tells us that the shale is older than the sandstone, which in turn is older than the mudstone in Arizona's Moenkopi formation.

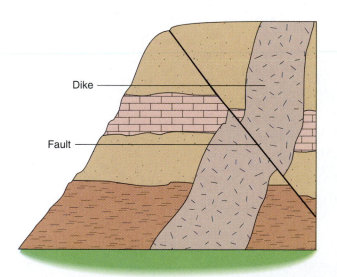

FIGURE 24.5   The Principle of Cross-Cutting Relationships

In this diagram of a road cut, the dike of igneous rock has cut across the sedimentary strata and so is younger than the strata. The fault has displaced the dike (as well as the strata) and so is younger than the dike. (From Dolgoff, Anatole, *Essentials of Physical Geology.* Copyright © 1998 by Houghton Mifflin Company. Used with permission.)

EXAMPLE 24.1

Applying Principles of Relative Dating

Using the principles of relative dating, analyze ● Fig. 24.6 and put the rocks marked 1 through 5 in sequence of youngest to oldest.

SOLUTION

The principle of superposition indicates that the topmost layer, rock 5, is the youngest, followed by rock 4, rock 3, and rock 1. The principle of cross-cutting relationships shows that rock 2 is younger than rock 1 but older than rock 3. So the correct order from youngest to oldest is 5, 4, 3, 2, 1.

CONFIDENCE EXERCISE 24.1

Figure 24.6 shows an unconformity. Where is it, and when (relatively) must it have been formed?

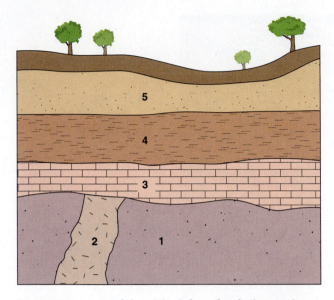

FIGURE 24.6 Applying Principles of Relative Dating

See Example 24.1.

Geologists use the principles just discussed to determine the relative ages of rocks in a specific locality. **Correlation** is the process of matching rock layers in different localities by use of index fossils or other means. If the age of rock in locality *A* is known, and rock in locality *B* is *correlated* with *A*, then the age of locality *B* rock is thereby determined to be the same age as *A*.

Certain fossils, called *index fossils*, are a major aid in correlation. **Index fossils** are those that are typical of a particular limited time segment of Earth's history, widespread, numerous, and easily identified. For example, many species of trilobites are important index fossils (● Fig. 24.7). After an index fossil has been thoroughly established, geologists know that any newly investigated rock layer in which it is found is the same age as previously known layers in which that index fossil was found. Let's look at a simple example.

FIGURE 24.7 Trilobites

The trilobites *Modicia* (large) and *Ptychagnostus* (small) are index fossils of the Cambrian period. Trilobites were early marine arthropods with hard exoskeletons, whose name refers to the "three lobes" that run lengthwise down the body. They were the first organisms with eyes, ruled the early Paleozoic seas, but were wiped out by the extinction event that ended the Paleozoic era.

EXAMPLE 24.2

Using the Process of Correlation

The Cambrian is the oldest (earliest) and the Permian is the youngest (latest) of the six geologic periods named at the left in ● Fig. 24.8. Four fossils, labeled *A* through *D*, are shown, along with their time range. (a) Which fossil would be the most useful as an index fossil? (b) If a rock layer from a certain locality contains both fossils *C* and *D*, what can be said about the period of the rock?

SOLUTION

(a) Fossil *A* would be the best index fossil because of the narrow range of time in which it lived.

(b) Rock that contains both fossils *C* and *D* would have to be from the Silurian period, because only during this period did *both* fossils live.

CONFIDENCE EXERCISE 24.2

Refer once again to Fig. 24.8. (a) Which fossil would be of the least use as an index fossil? (b) If a rock layer from a certain locality contains the index fossil *C, Phacops,* what can be said about the period of the rock? (c) From its sketch, what type of creature is *Phacops*?

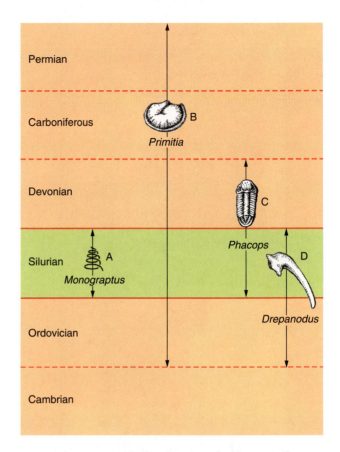

FIGURE 24.8   Correlation by Use of Index Fossils

See Example 24.2. (From Dolgoff, Anatole, *Essentials of Physical Geology.* Copyright © 1998 by Houghton Mifflin Company. Used with permission.)

By correlating rocks over large areas, the relative ages of most of the rocks on the surface of Earth have been determined. Thus geologists were able to establish a relative time scale for Earth's history. Please refer to ● Fig. 24.9 while we discuss some features of the relative geologic time scale.

The largest units of geologic time are the **eons.** We live in the *Phanerozoic* ("evident life") *eon.* The time before that is collectively called *Precambrian time,* because it immediately precedes the Cambrian period. Eons are subdivided into **eras,** and the oldest era in our eon is the Paleozoic era (the "age of ancient life"). Then comes the Mesozoic era (the "age of reptiles"), and the Cenozoic Era (the "age of mammals").

In turn, eras are divided into smaller time units called **periods.** The Paleozoic era is split into seven periods. (In Europe, the Mississippian and Pennsyl-

vanian periods are combined and called the *Carboniferous period.*) Three periods make up the Mesozoic era (you have probably heard of the middle one). Our era, the Cenozoic, has only two periods—the Tertiary and Quaternary ("qua-TUR-nay-ry"). We owe a debt to the plants that lived during the Pennsylvanian period, because much of our energy comes from the coal that was eventually formed from them (● Fig. 24.10).

In Section 24.5 and the Spotlight feature on page 647, we will put some absolute ages on these and additional geologic time divisions and briefly discuss what events cause the divisions to be made.

RELEVANCE QUESTION: *To what geologic period or periods do rocks in your locality belong?*

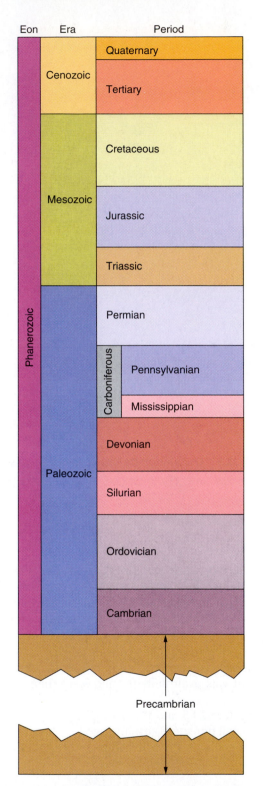

Eon	Era	Period
	Cenozoic	Quaternary
		Tertiary
	Mesozoic	Cretaceous
		Jurassic
		Triassic
Phanerozoic		Permian
	Paleozoic	Pennsylvanian / Carboniferous
		Mississippian
		Devonian
		Silurian
		Ordovician
		Cambrian

Precambrian

FIGURE 24.9 The Relative Geologic Time Scale

The time units are arranged with the oldest at the bottom and successively younger ones on top.

FIGURE 24.10   A Pennsylvanian Period Swamp

This reconstruction of a swamp of 300 million years ago shows conifers, ferns, and cycads, which were buried and slowly changed to coal in the oxygen-depleted layers.

24.3 **Radiometric Dating**

LEARNING GOALS

▼ Explain how radioactivity allows some rocks to be dated.

▼ Discuss the use of carbon dating.

The establishment of the relative geologic time scale was a significant scientific achievement but one that left geologists with a sense of frustration. They were not content to know only the *order* in which geologic events occurred, they needed to know *how long ago* in years these events occurred. That is, they wanted absolute ages in addition to relative ages. The need to measure **absolute geologic time**—the actual age of geologic events—became apparent in 1785 when James Hutton's concept of uniformitarianism (Section 21.2) indicated that Earth was very ancient, and it became crucial in 1859 when Charles Darwin's theory of organic evolution redoubled interest in the early history of our planet.

Radiometric dating, the determination of age by using radioactivity, has become geology's best tool for establishing absolute geologic time. Recall from Section 10.3 that atomic nuclei that decay of their own accord are said to be *radioactive*. A decay product, or *daughter nucleus* as it is commonly called, may be a stable nucleus. If so, the transformation reaches completion in a single step. However, the daughter nuclei of many naturally occurring radionuclides are themselves radioactive and hence undergo further decay. The radioactive series can be long and complex, as shown in Fig. 10.8 on page 241 for the decay of uranium-238 to, ultimately, stable lead-206.

Scientists usually express the rate of decay of radionuclides in terms of *half-life*—the span of time required for half the parent nuclei in a sample to decay (Section 10.3). Because the rate of decay for a given radionuclide remains constant—unaffected by temperature, pressure, and chemical environment—we can use radioactivity as a clock to measure the march of geologic time. *The older the rock, the greater is the ratio of the stable daughter to the radioactive parent* (● Fig. 24.11). For example, uranium-238 has a half-life of 4.46 billion years for its decay to lead-206. This means that for every 1000 atoms of uranium-238 present in a rock 4.46 billion years ago, only 500 atoms of uranium-238 would still be there today, and 500 atoms of lead-206 would have been formed. And 4.46 billion years in the future, only 250 atoms of uranium-238 will have survived, with 750 atoms of lead-206 formed.

Therefore, the rate of decay of a radionuclide in a rock can serve as a "clock" for dating the rock. Ideally, a radionuclide can tell the age of the rock that contains it under the following conditions:

1. No addition or subtraction of the parent or daughter has occurred over the lifetime of the rock other than that caused by radioactive decay.
2. The age of the rock does not differ too much from the half-life of the parent radionuclide.

3. None of the daughter element was present in the rock when it formed, or if it was, it is possible to tell how much.

Satisfying condition 1 is usually not difficult. However, the dating of orthoclase rocks by the decay of potassium-40 to argon-40 had to be abandoned because too much argon gas leaks from orthoclase.

Condition 2 must be satisfied because if the rock's age is too much greater than the half-life of the radionuclide, then too many half-lives go by and so little of the parent nuclide remains that its amount cannot be measured accurately. On the other hand, if the rock has an age very much less than the half-life, so little of the daughter may have formed that it might be hard to accurately measure it. Because most rocks that geologists are interested in dating have ages of hundreds of millions, or even billions, of years, only parent radionuclides of similarly long half-lives can be used. Of course, many rocks do not contain appropriate radionuclides and thus cannot be dated by radiometric methods. Radionuclides that are commonly used to date rocks are listed in Table 24.1.

Satisfying condition 3 is sometimes a problem, as can be illustrated by considering *uranium-lead dating* of rocks. (Uranium-lead dating means that uranium is the parent radionuclide and lead is the final daughter product.) Lead that comes from radioactive decay is called *radiogenic lead,* whereas lead that does *not* come from radioactive decay is termed *primordial lead.*

FIGURE 24.11 Half-Life and Radiometric Dating

As a parent radionuclide decays to a daughter, the proportion of the parent decreases (red line), while the proportion of the daughter increases (blue line). By measuring the proportion of parent to daughter in a sample, the number of half-lives since "time zero" (100% percent) can be obtained. Multiplying the number of half-lives by the half-life of the radionuclide gives the age of the rock. (From Chernicoff, Stanley, *Geology*, Second Edition. Copyright © 1999 by Houghton Mifflin Company. Used with permission.)

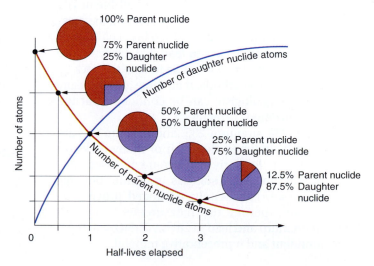

TABLE 24.1 Six Major Radionuclides Used for Radiometric Dating

Parent Nuclide	Half-Life (years)	Daughter Nuclide	Effective Range (years)	Some Materials That Can Be Dated
Rubidium-87	49 billion	Strontium-87	>10 million	Mica, microcline, whole metamorphic rock
Thorium-232	14 billion	Lead-208	>10 million	Zircon, uraninite
Uranium-238	4.46 billion	Lead-206	>10 million	Zircon, uraninite
Uranium-235	704 million	Lead-207	>10 million	Zircon, uraninite
Potassium-40	1.25 billion	Argon-40	>100 thousand	Mica, hornblende, whole volcanic rock
Carbon-14	5730	Nitrogen-14	<75 thousand	Bone, charcoal

Primordial lead always consists of 1.4% lead-204, 24.1% lead-206, 22.1% lead-206, and 52.4% lead-208. (Recall from Section 10.2 that primordial elements on Earth occur in a fixed ratio of isotopes.) If any primordial lead is in the rock being dated, not only will the lead isotopes 206, 207, and 208 be present but also 204, *which never comes from radioactive decay*. For every 1.4 g of lead-204 present, the geologist knows that 24.1 g of lead-206, 22.1 g of lead-207, and 52.4 g of lead-208 are also primordial and can subtract that amount from the total of each lead isotope to determine the amount of radiogenic lead. For example, in ● Fig. 24.12, if the rock has 1.4 g of lead-204 and 46.0 g of lead-206, then the amount of radiogenic lead-206 is 46.0 g minus 24.1 g, or 21.9 g. The 21.9-g value would be the amount of lead-206 used to help date the rock.

As stated previously, uranium-238, the most abundant isotope of uranium, has a half-life of 4.46 billion years and decays to lead-206. If uranium-238 is in the rock, then uranium-235, which has a half-life of 704 million years and decays to lead-207, also will be present. In addition, thorium-232, which has a half-life of 14 billion years and decays to lead-208, likely will be present. Thus it might be possible to date the rock by three independent methods. If all three give basically the same age, geologists can be confident that the rock has been dated correctly.

Uranium-238, uranium-235, and thorium-232 are clocks that have timed the birth of many rocks on Earth. Most of the early measurements used uranium-bearing minerals such as uraninite, but this mineral is rare. Fortunately, advanced procedures now available permit the measurement of small traces of uranium and lead in zircon ($ZrSiO_4$), a much more abundant and representative mineral.

Potassium, one of the most abundant and widespread elements, contains a very small percentage (0.012%) of radioactive potassium-40. This radionuclide is found in many rocks that do not contain measurable quantities of uranium, and so it has become one of the most useful tools in determining where a rock falls on the calendar of geologic time. Potassium-40 (half-life of 1.25 billion years) undergoes beta decay to argon-40, so this type of radiometric dating is referred to as *potassium-argon dating*. As mentioned previously, orthoclase rocks leak too much of the ar-

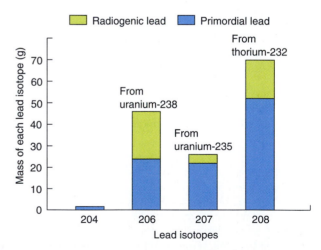

FIGURE 24.12 Primordial and Radiogenic Lead

Because lead-204 is *never* radiogenic, if any lead-204 is found in a rock, the geologist knows that primordial 206, 207, and 208 isotopes are there as well. By knowing the constant proportion of the primordial lead isotopes (blue bars), geologists can tell how much of the total amount of each isotope is actually radiogenic (the green bars).

gon gas formed, but other potassium-bearing minerals such as biotite, muscovite, and hornblende are better able to retain the argon and can be dated. Biotite from normal igneous rocks and from volcanic ash deposits has proved to be especially useful. All these minerals are susceptible to argon leakage if they have been heated. A potassium-argon date may merely reveal the last time the rock was heated rather than its true age. Therefore, geologists tend to regard potassium-argon dates as the *minimum* ages of the rocks and require that the dates be consistent with other geologic evidence before being fully accepted as the correct ages.

Rubidium-87 (half-life of 49 billion years) is found in many rocks and undergoes beta decay to strontium-87. Rubidium-87, which commonly occurs in the same minerals that contain potassium, is more abundant than potassium-40 and has the further advantage of decaying to a daughter that is not a gas. A disadvantage is that a lot of strontium-87 is primordial, for which a correction must be made. Geologists often use *rubidium-strontium dating* to compare with potassium-argon determinations from the same rock.

EXAMPLE 24.3

Using Radiometric Dating

Analysis of samples of a certain igneous rock layer shows that the ratio of uranium-235 to its daughter, radiogenic lead-207, is 1.00 to 3.00; that is, only 25.0% of its original uranium-235 (half-life of 704×10^6 years) remains. How old is the rock?

SOLUTION

To decay from 100% to 25.0% would take two half-lives, as shown in Fig. 24.11 or found by

$$100\% \longrightarrow 50\% \longrightarrow 25\%$$

Now, to find the time in years, multiply the 2.00 half-lives by the half-life of the radionuclide. Thus (2.00 half-lives)(704×10^6 years/half-life) $= 1.41 \times 10^9$ years, or 1.41 billion years.

CONFIDENCE EXERCISE 24.3

Analysis of samples of another igneous rock layer shows that the ratio of uranium-235 to its daughter,

radiogenic lead-207, is 1.00 to 7.00; that is, only 12.5% of its original uranium-235 (half-life of 704×10^6 years) remains. How old is the rock?

Carbon Dating

A dating technique developed in 1950 by Willard Libby, an American chemist, is the only radiometric method that is used to find the age of ancient, once-living remains such as charcoal, parchment, or bones. Carbon-14 (symbolized ^{14}C) is a radionuclide whose relatively short half-life of 5730 years puts a second hand on the radioactive time clock. **Carbon dating dates organic remains by measuring the amount of ^{14}C in an ancient sample and comparing it with the amount in present-day organic matter.**

Throughout history, ^{14}C has been produced in the atmosphere by the action of neutrons on atmospheric nitrogen (Fig. 24.13). The newly formed ^{14}C reacts with oxygen in the air to form radioactive carbon dioxide, ^{14}CO$_2$, which, along with ordinary ^{12}CO$_2$, is used by plants in photosynthesis. About one out of every trillion (10^{12}) carbon atoms in plants is ^{14}C. Animals that eat the plants incorporate the radioactive ^{14}C in their cells, as do animals that eat the animals that ate the plants (see Fig. 24.13).

Thus all living matter has about the same level of radioactivity due to ^{14}C—an activity of about 15.3 counts per minute per gram of total carbon. Once an organism dies, it ceases to take in ^{14}C, but the ^{14}C in its remains continues to undergo radioactive decay. Therefore, the longer the organism has been dead, the lower is the radioactivity of each gram of carbon in its remains.

The newest method of carbon dating relies on a specially designed mass spectrometer that separates and counts both the ^{14}C atoms and the ^{12}C atoms in a sample. By comparing the ratio of the isotopes in the specimen to the ratio in living matter, the time since its death can be calculated. This method uses tiny samples and can date specimens as old as 75,000 years. Beyond that age, only about 0.02% of the original ^{14}C is still undecayed and is too small to measure accurately.

Carbon dating assumes that the amount of ^{14}C in the atmosphere (and hence in the biosphere) has been the same throughout the past 75,000 years. However, because of changes in solar activity and Earth's magnetic field, it apparently has varied by as

FIGURE 24.13 Carbon Dating

An illustration of how carbon-14 forms in the atmosphere and enters the biosphere. See text for further discussion.

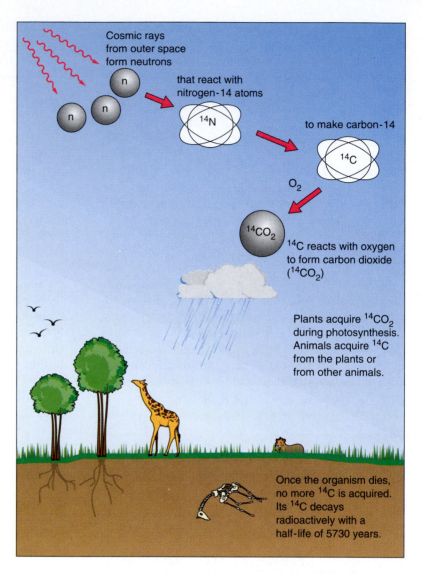

Cosmic rays from outer space form neutrons

that react with nitrogen-14 atoms

^{14}N

to make carbon-14

^{14}C

O_2

$^{14}CO_2$

^{14}C reacts with oxygen to form carbon dioxide ($^{14}CO_2$)

Plants acquire $^{14}CO_2$ during photosynthesis. Animals acquire ^{14}C from the plants or from other animals.

Once the organism dies, no more ^{14}C is acquired. Its ^{14}C decays radioactively with a half-life of 5730 years.

much as 5% above or below normal. California's bristlecone pines, which can live for as long as 5000 years, allow geologists to correct for the slight changes in the abundance of ^{14}C. By studying the ^{14}C activity of samples taken from the annual growth rings in both dead and living trees, geologists developed a calibration curve for ^{14}C dates as far back as about 5000 B.C. Therefore, carbon dating is most reliable for specimens no more than 7000 years old.

Carbon dating has been used to determine the age of organic remains such as bones, charcoal from ancient fires, the beams in pyramids, the dung of ground sloths, the Dead Sea Scrolls, and the Shroud of Turin. Such dates have a certain range of possible error asso-

ciated with them. For example, the flax from which the Shroud of Turin was woven was dated by three independent laboratories as having been grown about A.D. 1325, plus or minus 65 years, a date consistent with the shroud first surfacing in France about A.D. 1357.

24.4 The Age of Earth

LEARNING GOAL

▼ State the age of Earth and the evidence for that value.

In the middle and late 1800s, Lord Kelvin, the distinguished physicist (Section 5.1), attempted to determine Earth's absolute age from the rate of heat loss from its interior. Kelvin assumed that Earth began as a hot, molten body, which became solid as it cooled and continued to lose its residual heat from its still hot interior. From a measurement of the present rate of heat loss, Kelvin calculated that Earth became solid between 20 and 40 million years ago. This estimate, although based on actual measurements and supported by Kelvin's considerable prestige, was at odds with estimates of both geologists and biologists, who thought the present slow pace of geologic processes and organic evolution hinted at a much greater age.

However, after the discovery of radioactivity in 1896, it became evident that Kelvin's calculation was badly in error because of the incorrectness of one of his basic assumptions—that all the heat in Earth's interior was residual. We now know that most of the heat is actually from radioactive decay, which is a continuous process. Therefore, much more heat is available to flow out over a much longer time than Kelvin reckoned. Ironically, the phenomenon of radioactivity, which torpedoed his calculation, became geology's best tool for establishing absolute geologic time.

Kelvin did the best he could with the knowledge of the time, but this story brings home the importance of examining basic assumptions, of realizing that scientific results are always subject to change as new evidence accumulates, and of not getting overly concerned if scientists disagree (as do two camps of astronomers right now about the age of the universe; see Section 18.7). As research continues on a scientific problem, usually the correct answer finally emerges.

Fortunately, geologists are confident that they now have an accurate value for Earth's age—4.6 billion years old, or 4600 My (*My* means "megayears," or million years). Three major pieces of evidence support this date.

1. *The age of Earth rocks.* Using radiometric dating, it is possible to put absolute dates on many igneous and metamorphic rocks. At present, the longevity record is held by 4.3-billion-year-old zircon crystals from Australia. Other ancient deposits include 4.0-billion-year-old metamorphic and igneous rocks in Canada, 3.8-billion-year-old metamorphic rocks in Minnesota, 3.7-billion-year-old granites in southwestern Greenland (● Fig. 24.14), and 3.4-billion-year-old granites in South Africa.

FIGURE 24.14   Ancient Rocks
One of Earth's oldest known rock formations (approximately 3.8 billion years old) is found in Greenland. (The pocket knife indicates scale.)

2. *The age of meteorites.* Meteorites from the asteroid belt (Section 15.5), which presumably formed about the same time as Earth, date at 4.6 billion years. This age is given by both uranium-lead and rubidium-strontium methods.

3. *The age of Moon rocks.* The rocks from the lunar highlands are the oldest materials brought back from the Moon, and these yield dates of 4.55 billion years.

So strong evidence exists that the planets, moons, and asteroids of the solar system all formed about 4.6 billion years ago. It is unlikely that rocks formed on Earth will be found that are quite 4.6 billion years old, because the Earth's surface was probably molten for several hundred million years. Weathering and subduction due to plate tectonics undoubtedly have destroyed many ancient rocks, so we are fortunate to find as many old rock formations as we do.

One final point needs to be addressed, and we cannot improve on the statement of American geologist

Anatole Dolgoff: "Some people argue on religious grounds that the Earth is only 5000 to 10,000 years old, and geologists are often drawn into public debate as advocates of their scientific estimates. Sometimes, the position of geologists in this debate is misunderstood—for in the final analysis, geologists have no stake in how old the Earth is. They simply want to *know* how old it is! If the Earth is only 5000 years old, as some who take a literal interpretation of the Scriptures claim, so be it. However, the evidence points overwhelmingly to the contrary."*

24.5 The Geologic Time Scale

LEARNING GOALS

▼ Describe the use of igneous rock to date sedimentary rock.

▼ State the dates and events that mark the divisions of eons and eras.

Relative geologic time and absolute geologic time are combined to give the **geologic time scale** shown in the Spotlight feature. Time in millions of years, as determined by radiometric dating methods, is shown on the right side of the relative time units, along with a list of some major geologic and biological events that took place at given times.

Although radiometric dating has provided geologists with numerous essential dates for the ages of rocks, many of the dates on the geologic time scale are estimated values and are subject to minor changes as new evidence is found.

Geologists have principally used sedimentary rocks to establish the relative time scale, whereas most of their radiometric determinations have been made on igneous rocks. Even though sedimentary rocks often contain radionuclides, the sediments that form the rocks have been weathered from rocks of different, and older, ages. So the age of the sedimentary rock is not the same as the ages of its constituents.

Reasonable values for the absolute dates for sedimentary rock layers are frequently determined by relating them to igneous rocks, as shown in the following example.

*Anatole Dolgoff, *Essentials of Physical Geology,* Houghton Mifflin Company, Boston, MA, 1998, page 357.

EXAMPLE 24.4

Using Igneous Rocks to Date Sedimentary Rocks

● Figure 24.15 shows the intrusion of two igneous dikes (X and Y, radiometrically dated at 400 My and 350 My, respectively) across strata whose relative ages were determined by the fossils they contained. What can be said about the age of the Devonian stratum labeled B?

SOLUTION

Igneous dike X intruded the Silurian strata, and then its top and part of the Silurian strata were eroded. Stratum B was then deposited at the unconformity, so stratum B must be younger than 400 My (the date of X). Because of the intrusion of dike Y through stratum B, stratum B must be older than 350 My (the date of Y). So radiometric dating tells us that the Devonian stratum is between 350 My and 400 My old. Geologists say that the age of the stratum has been *bracketed*.

CONFIDENCE EXERCISE 24.4

Referring to Fig. 24.15, what can be said about the absolute age of the sedimentary rock layer A from the Mississippian period?

Space does not permit a discussion of all the information in the geologic time scale shown in the Spotlight feature, so please examine the feature closely. The Hadean eon (at the bottom of the scale) ends and the Archean eon begins about 4000 My ago, the date of the earliest known Earth rocks. The Proterozoic eon begins 2500 My ago, when the core rocks of what is now North America came together. ● Figure 24.16 shows a reconstruction of the huge supercontinent that formed during the Proterozoic and broke up during the early Paleozoic.

The Phanerozoic eon (our present eon) and the Paleozoic era began about 545 My ago when the hard-shelled marine invertebrate fossils first become abundant. At this time, an extinction event occurred, followed by a great proliferation of life forms that is sometimes referred to as the *Cambrian explosion.* Rather suddenly, oceans that had previously held nothing more complicated than burrowing worms teemed with complex animal life.

A SPOTLIGHT ON: The Geologic Time Scale

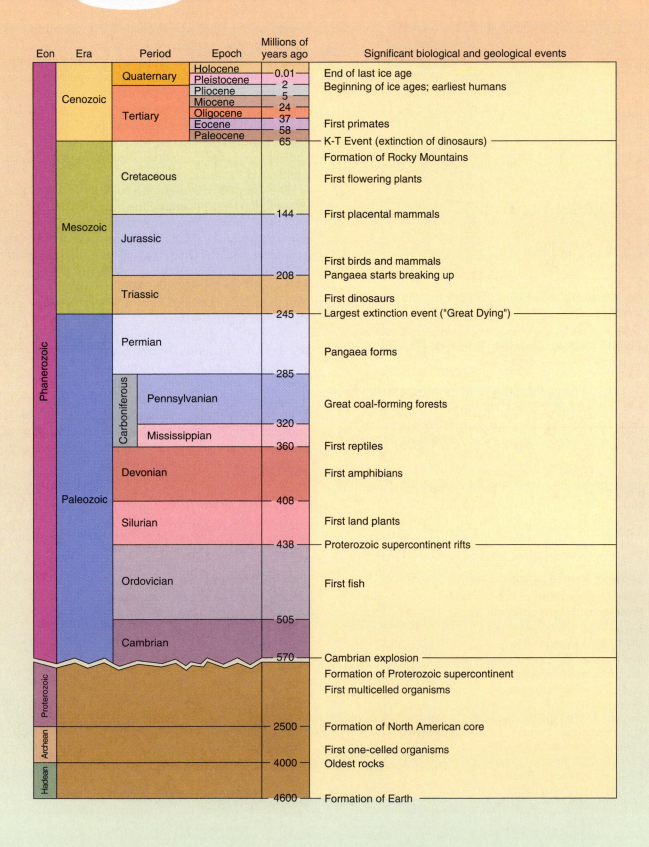

Eon	Era	Period	Epoch	Millions of years ago	Significant biological and geological events
Phanerozoic	Cenozoic	Quaternary	Holocene	0.01	End of last ice age
			Pleistocene	2	Beginning of ice ages; earliest humans
		Tertiary	Pliocene	5	
			Miocene	24	
			Oligocene	37	First primates
			Eocene	58	
			Paleocene	65	K-T Event (extinction of dinosaurs)
	Mesozoic	Cretaceous			Formation of Rocky Mountains
					First flowering plants
				144	First placental mammals
		Jurassic			
				208	First birds and mammals / Pangaea starts breaking up
		Triassic			First dinosaurs
				245	Largest extinction event ("Great Dying")
	Paleozoic	Permian			Pangaea forms
				285	
		Carboniferous — Pennsylvanian			Great coal-forming forests
				320	
		Carboniferous — Mississippian		360	First reptiles
		Devonian			First amphibians
				408	First land plants
		Silurian		438	Proterozoic supercontinent rifts
		Ordovician			First fish
				505	
		Cambrian		570	Cambrian explosion
Proterozoic					Formation of Proterozoic supercontinent / First multicelled organisms
				2500	Formation of North American core
Archean					First one-celled organisms
				4000	Oldest rocks
Hadean				4600	Formation of Earth

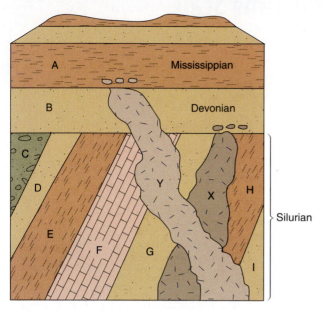

FIGURE 24.15 Using Igneous Rock to Date Sedimentary Rock

See Example 24.4. (From Dolgoff, Anatole, *Essentials of Physical Geology*. Copyright © 1998 by Houghton Mifflin Company. Used with permission.)

The Paleozoic era ended and the Mesozoic era began about 245 My ago when the most devastating extinction known to geologists occurred—the Permian event sometimes called the *Great Dying*. An estimated 90% of living species expired, including all the trilobites and most of the crinoids (● Fig. 24.17) and brachiopods (● Fig. 24.18). A recent hypothesis about the cause of this event involves a buildup of carbon dioxide, perhaps as a consequence of the formation of the supercontinent, Pangaea, during the late Paleozoic era. As discussed in Chapters 21 and 22, the conver-

gence of Earth's landmasses into one supercontinent involved a lot of oceanic-oceanic and oceanic-continental subduction, which created a huge number of carbon dioxide–spewing volcanoes.

In the Mesozoic era, Pangaea broke into the familiar continents of today (see Fig. 22.1 on page 588). The Mesozoic climate was mild. Coral grew in what is now Europe, and the poles were free of glacial ice. Dinosaurs lived on all the continents, but the numerous fossils found in the western United States and Canada indicate that these localities provided a particularly good environment for dinosaurs.

The chapter Highlight discusses the extinction event that 65 My ago ended the Mesozoic era and began the Cenozoic era, sometimes called the *age of mammals* (see the chapter-opening illustration). Our present period, the Quaternary, began about 2 My ago with the appearance of the oldest fossils of the genus *Homo*, which was preceded for about another 2 million years by other hominids, who were members of the genus *Australopithecus*.

The periods of the Cenozoic era are subdivided into **epochs,** the names of which end in *-cene*, a suffix meaning "recent." The Pleistocene epoch is also known as the *Ice Ages*. Our present epoch, the Holocene, begins with the last retreat of the glaciers from North America and Europe about 10,000 years ago. The timeline illustrated in ● Fig. 24.19 helps us to realize how long geologic time really is and how short is the part we call "recorded history."

RELEVANCE QUESTION: *If asked where you live, you might respond by giving the state, city, street, and house number. In what way might you answer the question of when you live?*

FIGURE 24.16 Approximate Configuration of the Proterozoic Supercontinent

A supercontinent formed in the late Proterozoic eon. It began to rift (break apart) in the early Paleozoic era, as shown here. (In the late Paleozoic era the pieces reassembled into the supercontinent called *Pangaea*, which itself began to rift in the middle Mesozoic era.) (From Dolgoff, Anatole, *Essentials of Physical Geology*. Copyright © 1998 by Houghton Mifflin Company. Used with permission.)

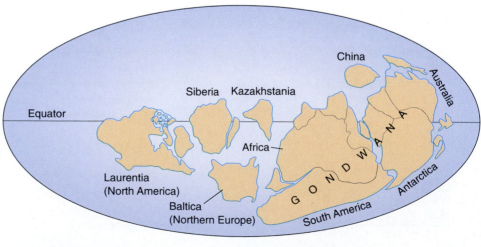

FIGURE 24.17   Crinoid Fossils of the Mississippian Period, Found in Indiana

Crinoids are marine invertebrate animals that look like plants. They were abundant (about 5000 species) during most of the Paleozoic era, and a few species (commonly called "sea lilies") exist today. Flat, circular or star-shaped segments of crinoid columns are common fossils sometimes called "Indian money."

FIGURE 24.18   Brachiopods from the Paleozoic Era

Brachiopods are hard-shelled marine invertebrate animals that were abundant during the Paleozoic era. At the left are shown *casts,* whereas *molds* are shown on the right. Unlike mussels, whose two shell valves are similar but opposite, brachiopod valves are unlike and unequal. A few species of brachiopods (called *lamp shells*) exist today.

FIGURE 24.19   Geologic Time in Perspective

A timeline drawn across the United States helps to put geologic time in perspective. Each kilometer corresponds to about 1 million years of Earth history. Only for about 6000 years have we had recorded, or written, history. (From Chernicoff, Stanley, *Geology*, Second Edition. Copyright © 1999 by Houghton Mifflin Company. Used with permission.)

HIGHLIGHT: The K-T Event: The Disappearance of Dinosaurs

What killed about 70% of all the world's plant and animal species, including all the dinosaurs, 65 million years ago (Fig. 1)? The answer began to emerge in the late 1970s, when a team of scientists from the University of California at Berkeley began investigating a thin layer of clay that marks the boundary between the Mesozoic and Cenozoic eras. The Cretaceous (symbolized K) is the latest period of the Mesozoic era, whereas the earliest period of the Cenozoic era is the Tertiary (symbolized T). Thus where the two periods meet is called the *K-T boundary*, and the event that marks the division is often referred to as the **K-T Event.**

In Cretaceous rocks are found fossils of thousands of marine invertebrate species, dinosaurs, and other organisms, none of which survived to leave fossils in Tertiary rock strata. No doubt exists that a gigantic mass extinction occurred. In the Tertiary strata are found, for the first time, fossils of modern types of plants and fossils of mammals larger than rodents.

The team, led by Walter and Luis Alvarez, was investigating the clay layer in the cliffs near Gubbio in northern Italy. First, they found that the clay layer contained a concentration of the rare element *iridium* (Ir) hundreds of times greater than in normal clay or in Earth's crust as a whole. However, the concentration closely matched the iridium content of some meteorites. Second, the clay layer contained distinctive glassy beads, called *spherules,* that were once molten droplets like those associated with meteorite impact craters on the Moon and Earth. Third, the clay also contained grains of *shocked quartz,* a type of quartz formed from meteorite impacts. Fourth, particles of soot were found, apparently from extensive fires.

The Alvarez team hypothesized that a massive meteorite struck Earth 65 million years ago, causing widespread fires and driving dust, ash, and other debris (about 100 trillion tons!) high into the stratosphere. The cloud girdled the globe, blocking out sunlight for perhaps a decade, causing acid rain, and catastrophically disrupting the food chain on land and in the oceans.

Subsequently, the same materials (iridium, spherules, shocked quartz, and soot) have been identified at all the 95 thin K-T boundary layer locations scattered throughout the world. Apparently, when the distinctive debris from the impact slowly settled back to Earth, it formed the thin layer of sediment that is now found intact at only certain localities because plate tectonic movement and erosion have wiped it out in most places.

Obviously, the Alvarez hypothesis of a "doomsday rock" needed testing and further proof. A search was launched for an impact crater that was of the appropriate date (65 million years old) and size (many miles wide) to do such damage. And it was found! Satellite images detected such a filled-in crater about 150 miles in diameter in the

FIGURE 1   A Casualty of the K-T Event
An *Albertosaurus* of the late Cretaceous greets visitors to Canada's Royal Tyrrell Museum.

Caribbean on the Yucatan peninsula of Mexico, a site called Chicxulub ("CHEEK-shoe-lube"). The rock strata at the location contain a lot of sulfur, which would have formed a deadly amount of acid rain in the fallout. A strike at that location has been likened to a cannon shell hitting the ammunition magazine of a ship—it could hardly have hit at a worse location.

Some of the latest findings indicate that the continental shelf 100 miles away was broken off by the force of the impact, causing a huge tsunami that led to the piled gravel deposits found on the shores of Texas hundreds of miles away. As research continues, the evidence grows that the dinosaurs and many other species met their demise at the hands of a 6-mile-wide visitor from space. Evidence for more than a dozen large mass extinctions is found in the geologic record, but the K-T Event is second only to the Great Dying, when about 90% of the world's species vanished at the boundary between the Paleozoic and Mesozoic eras about 245 million years ago.

Important Terms

geologic time
fossil (24.1)
paleontology
relative geologic
 time (24.2)
principle of superposition

principle of cross-cutting
 relationships
unconformities
correlation
index fossils
eons

eras
periods
absolute geologic
 time (24.3)
radiometric dating
carbon dating

geologic time scale (24.5)
epochs
K-T Event

Review Questions

1. What is meant by *geologic time*?

24.1 Fossils

2. Fossils
   (a) are any indication of prehistoric life.
   (b) help determine the relative ages of rocks.
   (c) can be tracks imprinted in rocks.
   (d) all of the above.

3. What is the name for the branch of science that specifically studies fossils?
   (a) petrology   (c) paleontology
   (b) mineralogy   (d) fossiligraphy

4. What is the name for a hollow volume that has the same shape as the plant or animal that was originally present in the rock?
   (a) mold   (b) cast   (c) trace fossil   (d) nodule

5. What is amber?

6. Why are the most abundant fossils those of marine creatures that lived in shallow water?

7. What is the name for a type of fossil formed when a mineral replaces once-living material?

8. Arizona is famous for what variety of fossil?

9. Carbonization is the process that formed what useful material?

10. Explain the difference in a mold and a cast.

11. What types of organisms are the oldest for which geologists can find evidence? About how long ago did the earliest of these apparently live?

12. How do fossils aid in oil exploration?

13. What is the name for a fossil imprint made by the movement of an animal?

24.2 Relative Geologic Time

14. What is obtained when rocks and geologic events are put into chronologic order without regard to the actual dates?
    (a) absolute geologic time
    (b) relative geologic time
    (c) a geologic formation
    (d) a correlation period

15. Into what time spans are eons next divided?
    (a) eras   (b) ages   (c) epochs   (d) periods

16. Which of these is the earliest (oldest) period in the Paleozoic era?
    (a) Triassic   (c) Cambrian
    (b) Tertiary   (d) Permian

17. State the principle of superposition, and give an example.

18. State the principle of cross-cutting relationships, and give an example.

19. What is an unconformity?

20. What is meant by the term *correlation*?

21. What is the name for fossils that can be used for correlation?

22. What four features characterize the best index fossils?

23. Define the term *relative geologic time.*

24. Name the three eras of the Phanerozoic eon in order of oldest to youngest.

25. Into how many periods is the Paleozoic era divided? Name the latest (youngest) of these periods.

26. Name the three periods of the Mesozoic era in order of oldest to youngest.

27. Name the two periods of the Cenozoic era, with the oldest first.

28. The Mississippian and Pennsylvanian periods in North America go by what name in Europe?

29. What is meant by *Precambrian time*?

24.3 Radiometric Dating

30. Radioactive _____, which gradually changes to lead, can be used to date ancient rocks.
    (a) rubidium   (c) carbon
    (b) potassium   (d) uranium

31. The maximum age of a specimen that can be carbon dated is about how many years?
    (a) 500 thousand   (c) 50 million
    (b) 4.6 billion   (d) 75 thousand

32. Radiometric dating is used to determine _____ geologic time.

33. What is the daughter product of potassium-40?

34. What is the parent radionuclide of strontium-87?

35. Carbon dating does *not* measure the ratio of carbon-14 to its daughter, nitrogen-14, but instead to what nuclide?

36. What are the three conditions for using a radionuclide in a rock to date it?

37. Distinguish between primordial lead and radiogenic lead.

38. How can a geologist tell whether all the lead in a rock is radiogenic?

24.4 The Age of Earth

39. Which scientist tried to determine Earth's age by measuring its loss of interior heat?
    (a) Darwin   (c) Hutton
    (b) Kelvin   (d) Dolgoff

40. State three pieces of evidence that Earth is about 4600 million years old.

41. About how old are the oldest rocks found on Earth? On the Moon?

24.5 The Geologic Time Scale (and Highlight)

42. Which became the dominant life form in the Cenozoic era?
    (a) dinosaurs   (c) insects
    (b) trilobites   (d) mammals

43. Fossils from Precambrian time could include which of these?
    (a) human skulls
    (b) dinosaur bones
    (c) algae, bacteria, and sea worms
    (d) leaf impressions and shark teeth

44. From the suffix of the name, which must be an epoch?
    (a) Devonian   (c) Jurassic
    (b) Miocene   (d) Archean

45. In what era were dinosaurs common?
    (a) Mesozoic   (c) Cambrian
    (b) Paleozoic   (d) Cenozoic

46. Briefly, how can the absolute age of a sedimentary rock be determined?

47. On the basis of what geologic or biologic events is Precambrian time split into three eons? Name the three eons.

48. Which era on the geologic time scale encompasses the shortest time span?

49. In what era do you live? What period? What epoch?

50. The Great Dying separates what two eras?

51. What geologic event separates the Pleistocene epoch from the Holocene epoch?

52. What biologic event is used by geologists to separate the Tertiary period from the Quaternary period?

53. What is the biologic event at the start of the Paleozoic era called?

54. What is the event called that caused the demise of the dinosaurs and many other species?

55. To what was the extinction at the end of the Mesozoic probably due? State some evidence for that explanation.

56. What is the significance of the word *Chicxulub* in geology?

57. What are the starting and ending dates of the Paleozoic era?

58. Briefly, what effect would a supercontinent such as Pangaea be expected to have on the distribution of land plants and animals of the time?

Applying Your Knowledge

1. Your uncle wants to know if the Cambrian explosion was a terrorist attack. What is your reply?

2. Your friend shows you a cow bone dug up from a field on his farm and declares the bone to be a fossil. Do you agree? Why or why not?

3. You see a comic strip in which cavemen and dinosaurs are shown coexisting. Why do you realize that artistic license is being taken?

4. At a jewelry show, you overhear a person looking at an amber brooch say that she wishes it contained an insect so that she would have a fossil. What error is she making?

5. While standing in line at the grocery store, you see a magazine headline stating that carbon dating has been used to find the age of a dinosaur bone. Why are you skeptical of that claim?

Exercises

24.2 Relative Geologic Time

1. Table 24.2 shows, in color, the range in the rock record of six different Paleozoic era fossils. Along the top of the chart is a letter for each period of the era (C for Cambrian, P for Pennsylvanian, PR for Permian, etc.).
   (a) What is the range of geologic periods for the crinoid *Platycrinites*?
   (b) To what period does rock belong that contains the brachiopod *Zygospira* and the trilobite *Phacops*?
   (c) List the fossils shown that might be found in rock of the Pennsylvanian period.
   (d) Considering only the time range, which of the fossils would be the best index fossil?
       Answer: (a) Mississippian, Pennsylvanian, Permian
                (b) Silurian   (c) *Platycrinites* and *Lingula*
                (d) *Elrathia*

2. Refer to Table 24.2.
   (a) What is the range of geologic periods for the brachiopod *Zygospira*?
   (b) To what period does rock belong that contains the crinoid *Taxocrinus* and the trilobite *Phacops*?
   (c) List the fossils shown that might be found in rock of the Silurian period.
   (d) Why could neither of the two trilobites listed be used to identify an Ordovician period rock?
   (e) Considering only the time range, which fossil would be the worst index fossil?

TABLE 24.2   Some Fossils and Their Periods

	C	O	S	D	M	P	PR
Phacops (a trilobite)			■	■			
Elrathia (a trilobite)	■						
Taxocrinus (a crinoid)				■	■		
Platycrinites (a crinoid)					■	■	■
Zygospira (a brachiopod)		■	■				
Lingula (a brachiopod)		■	■	■	■	■	■

3. Refer to ● Fig. 24.20.
   (a) Which rock stratum is younger, *A* or *B*? What geologic principle did you use?
   (b) Which is younger, the rock stratum *C* or the igneous intrusion marked *E*? What geologic principle did you use?
       Answer: (a) *A* is the younger; superposition   (b) *E* is the younger; cross-cutting relationships

4. Refer to Fig. 24.20.
   (a) Which is younger, the rock stratum *C* or the fault marked *D*? What geologic principle did you use?
   (b) Which is younger, the fault marked *D* or the igneous intrusion marked *E*? What geologic principle did you use?

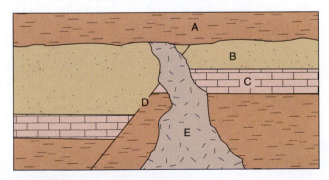

FIGURE 24.20   Relative Dating
See Exercises 3 and 4.

24.3 Radiometric Dating

5. Charcoal from an ancient campfire has a ^{14}C to ^{12}C ratio that is one-fourth that of new wood. About how old is the charcoal? The half-life of ^{14}C is 5730 years.

   Answer: About 11,460 y

6. Carbon obtained from the cloth wrappings of an Egyptian mummy is found to have one-half the ^{14}C to ^{12}C ratio of present-day carbon. About how old are the wrappings? The half-life of ^{14}C is 5730 years.

24.5 The Geologic Time Scale

7. Refer to ● Fig. 24.21, which shows a lava flow radiometrically dated at 245 My and an igneous dike dated at 210 My. What can be said about the absolute age of the sandstone layer?

   Answer: Older than 210 My but younger than 245 My

8. Refer to Fig. 24.21. What can be said about the absolute age of the shale layer and its age relative to the metamorphic rock layer?

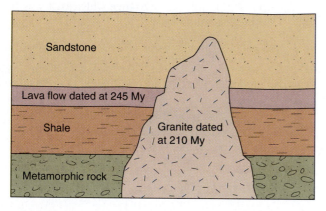

FIGURE 24.21   **Absolute Dating of Sedimentary Rocks**
See Exercises 7 and 8.

Solutions to Confidence Exercises

24.1 The unconformity is represented by the wavy border between rock 1 and rock 3, where rock 2 has been truncated by erosion. The unconformity must be older than rock 2 and younger than rock 3.

24.2 (a) Fossil *B* has such a wide time range that it would be of limited use as an index fossil.
   (b) Rock that contains fossil *C* must be from *either* the Silurian or Devonian, because these are the only two periods in which this fossil has ever been found. Additional evidence would be needed to say exactly to which of the two periods the rock belongs.
   (c) *Phacops* is a trilobite. (See Fig. 24.7.)

24.3 To decay from 100% to 12.5% would take three half-lives, as shown in Fig. 24.11 or found by

$$100\% \longrightarrow 50\% \longrightarrow 25\% \longrightarrow 12.5\%$$

Find the time in years by multiplying the 3.00 half-lives by the half-life of the radionuclide. Thus (3.00 half-lives)(704 × 10^6 years/half-life) = 2.11 × 10^9 years, or 2.11 billion years.

24.4 The principle of cross-cutting relationships tells us that *A* is younger than dike *Y*, because dike *Y* was eroded before *A* was deposited. So *A* must be younger than 350 My.

Answers to Multiple-Choice Review Questions

2. d   4. a   15. a   30. d   39. b   43. c   45. a
3. c   14. b   16. c   31. d   42. d   44. b

Appendixes

APPENDIX I The Seven Base Units of the International System of Units (SI)

1. **meter, m (length):** The meter is defined in reference to the standard unit of time. One meter is the length of the path traveled by light in a vacuum during a time interval of 1/299,792,458 of a second. That is, the speed of light is a universal constant of nature whose value is defined to be 299,792,458 meters per second.

2. **kilogram, kg (mass):** The kilogram is a cylinder of platinum-iridium alloy kept by the International Bureau of Weights and Measures in Paris. A duplicate in the custody of the National Institute of Standards and Technology serves as the mass standard for the United States. This is the only base unit still defined by an artifact.

3. **second, s (time):** The second is defined as the duration of 9,192,631,770 cycles of the radiation associated with a specified transition of the cesium-133 atom.

4. **ampere, A (electric current):** The ampere is defined as that current that, if maintained in each of two long parallel wires separated by one meter in free space, would produce a force between the two wires (due to their magnetic fields) of 2×10^{-7} newtons for each meter of length.

5. **kelvin, K (temperature):** The kelvin is defined as the fraction 1/273.16 of the thermodynamic temperature of the triple point of water. The temperature 0 K is called *absolute zero.*

6. **mole, mol (amount of substance):** The mole is the amount of substance of a system that contains as many elementary entities as there are atoms in 0.012 kilograms of carbon-12.

7. **candela, cd (luminous intensity):** The candela is defined as the luminous intensity of 1/600,000 of a square meter of a black body at the temperature of freezing platinum (2045 K).

TABLE I.1 Prefixes Representing Powers of 10*

Multiple	Prefix	Abbreviation
10^{18}	exa-	E
10^{15}	peta-	P
10^{12}	tera-	T
10^{9}	giga-	G
10^{6}	mega-	M
10^{3}	kilo-	k
10^{2}	hecto-	h
10	deka-	da
10^{-1}	deci-	d
10^{-2}	centi-	c
10^{-3}	milli-	m
10^{-6}	micro-	μ
10^{-9}	nano-	n
10^{-12}	pico-	p
10^{-15}	femto-	f
10^{-18}	atto-	a

*The most commonly used prefixes are highlighted in color.

APPENDIX II Solving Mathematical Problems in Science

Mathematics is fundamental in the physical sciences, and it is difficult to understand or appreciate these sciences without certain basic mathematical abilities. Dealing with the quantitative side of science gives the student a chance to review and gain confidence in the use of basic mathematical skills, to learn to analyze problems and reason them through, and to see the importance and power of a systematic approach to problems. We recommend the following approach to solving mathematical problems in science.

1. Using symbol notation, list what you are given and what is unknown. Include all units—not just the numbers.
2. From that information, decide the type of problem with which you are dealing, and select the appropriate equation.
3. If necessary, rearrange the equation for the unknown. (Appendix III discusses equation rearrangement.)
4. Substitute the known numbers *and their units* into the rearranged equation.
5. See if the units combine to give you the appropriate unit for the unknown. (Appendix IV discusses the analysis of units.)
6. Once the units are adjudged correct, do the math, being sure to express the answer to the proper number of significant figures and to include the unit. (Appendixes V, VI, and VII describe how to use positive and negative numbers, powers-of-10 notation, and significant figures.)
7. Evaluate the answer for reasonableness.

APPENDIX III   Equation Rearrangement

An important skill for solving mathematical problems in science is the ability to rearrange an equation for the unknown quantity. Basically, we want to get the unknown

1. into the numerator
2. positive in sign
3. to the first power (not squared, cubed, etc.)
4. alone on one side of the equals sign ($=$)

Addition or Subtraction to Both Sides of an Equation

The equation $X - 4 = 12$ states that the numbers $X - 4$ and 12 are equal. To solve for X, use the rule, whatever is added to (or subtracted from) one side can be added to (or subtracted from) the other side and equality will be maintained. We want to get X, the unknown in this case, alone on one side of the equation, so we add 4 to both sides (so that $-4 + 4 = 0$ on the left side). Example III.1 illustrates this procedure.

EXAMPLE III.1

$$X - 4 = 12$$

$$X - 4 + 4 = 12 + 4 \qquad \text{(4 added to both sides)}$$

$$X = 16$$

EXAMPLE III.2

Suppose in the scientific equation $T_K = T_C + 273$ we wish to solve for T_C. All we need to do is to get T_C

alone on one side, so we subtract 273 from each side. The procedure is

$$T_K = T_C + 273$$

$$T_K - 273 = T_C + 273 - 273 \qquad \text{(273 is subtracted from both sides)}$$

$$T_K - 273 = T_C \text{ (or } T_C = T_K - 273)$$

Shortcut

After you understand the principle of adding or subtracting the same number or symbol from each side, you may wish to use a shortcut for this type of problem. Namely, to move a number or symbol added to (or subtracted from) the unknown, take it to the other side but change its sign. See how Examples III.1 and III.2 are solved using the shortcut.

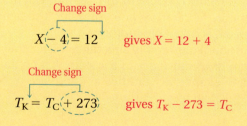

Change sign

$$X - 4 = 12 \qquad \text{gives } X = 12 + 4$$

Change sign

$$T_K = T_C + 273 \qquad \text{gives } T_K - 273 = T_C$$

PRACTICE PROBLEMS

Solve each equation for the unknown. As you proceed, check your answers against those given at the end of this appendix.

(a) $X + 9 = 11$; $X = ?$

(b) $T_f - T_i = \Delta T$; $T_f = ?$

(c) $\lambda f = E_i - E_f$; $E_i = ?$

(d) $H = \Delta E_i + W$; $W = ?$

(e) $\dfrac{\Delta E_p}{mg} = h_2 - h_1; h_2 = ?$

(f) $H = E_p - \dfrac{mv^2}{2}; E_p = ?$

Multiplication or Division to Both Sides of an Equation

For equations such as $\frac{X}{4} = 6$, the rule is, multiply (or divide) both sides by the number or symbol that will leave the unknown alone on one side of the equation. So, in the equation $\frac{X}{4} = 6$, we multiply both sides by 4, as shown in Example III.3.

EXAMPLE III.3

$$\frac{X}{4} = 6$$

$$\frac{4X}{4} = 6 \times 4 \qquad \text{(both sides multiplied by 4)}$$

$$X = 24$$

EXAMPLE III.4

In the scientific equation $F = ma$, solve for a. The unknown is already in the numerator, so all we must do is move the m by dividing both sides by m.

$$F = ma$$

$$\frac{F}{m} = \frac{ma}{m} \qquad \text{(both sides divided by } m\text{)}$$

$$\frac{F}{m} = a \left(\text{or } a = \frac{F}{m}\right)$$

EXAMPLE III.5

Suppose the unknown is in the denominator to start with; for example, solving for t in $P = \frac{W}{t}$. Multiply both sides by t to get t in the numerator, then divide both sides by P to move P to the other side. The procedure is

$$P = \frac{W}{t}$$

$$tP = \frac{tW}{t} \qquad \text{(both sides multiplied by } t\text{)}$$

$$tP = W$$

$$\frac{tP}{P} = \frac{W}{P} \qquad \text{(both sides divided by } P\text{)}$$

$$t = \frac{W}{P}$$

Shortcut

After you understand the principle of multiplying or dividing both sides of the equation by the same number or symbol, you may wish to use a shortcut to move the number or symbol. Namely, whatever is multiplying (or dividing) the whole of one side, winds up dividing (or multiplying) the whole other side. See how Examples III.3, III.4, and III.5 are solved using the shortcut.

$$\frac{X}{4} = 6 \qquad \text{gives } X = 4 \times 6 \text{ (or 24)}$$

$$F = ma \qquad \text{gives } \frac{F}{m} = a$$

$$P = \frac{W}{t} \qquad \text{gives } t = \frac{W}{P}$$

(The shortcut saves many steps when an unknown is in the denominator.)

PRACTICE PROBLEMS

Solve for the unknown. The answers are given at the end of this appendix.

(g) $7X = 21; X = ?$ (h) $\dfrac{X}{3} = 2; X = ?$

(i) $w = mg; m = ?$ (j) $\lambda = \dfrac{c}{f}; c = ?$

(k) $\lambda = \dfrac{c}{f}; f = ?$ (l) $\lambda = \dfrac{h}{mf}; m = ?$

(m) $E_k = \dfrac{mv^2}{2}; m = ?$ (n) $v^2 = \dfrac{3kT}{m}; T = ?$

(o) $pV = nRT; R = ?$

Multiple Operations to Both Sides of an Equation

To solve some problems requires the use of both the addition/subtraction and the multiplication/division rules. For example, solving $\frac{X-4}{3} = 8$ and $\frac{X}{3} - 4 = 8$ requires both rules. These two equations are similar, but not identical. We will follow a different order of application of the rules as we solve each.

A good way to proceed is to follow the principle: If the *whole side* on which the unknown is found is multiplied and/or divided by a number or symbol, get that number or symbol to the other side first. *Then,* move any numbers or symbols that are added to or subtracted from the unknown. (This is the case in the first example.) On the other hand, if only the unknown (and not the whole side) is multiplied or divided by a number or symbol, first move any number or symbol added or subtracted. (This is the case in the second example.) The successive steps for each example are shown.

Example III.7, 4 was added to both sides first, and then both sides were divided by 3. We used different strategies because in the first example the 3 was dividing the whole side that X was on, whereas in the second example it wasn't. Other strategies may be applied to problems like these, but things generally work out better if the recommended principle is used.

The rules for rearranging scientific equations are exactly the same. Before substituting numbers and units for the various letters in a scientific equation, most experts advise that the equation first be rearranged so that the unknown is positive, to the first power, and alone in the numerator on one side. It is easier and more accurate to move several letters than it is to move a bunch of numbers and units, as you would have to do if you substituted first.

EXAMPLE III.6

$$\frac{(X-4)}{3} = 8$$

$$\frac{3 \times (X-4)}{3} = 3 \times 8 \qquad \text{(both sides multiplied by 3)}$$

$$X - 4 = 24$$

$$X - 4 + 4 = 24 + 4 \qquad \text{(4 added to both sides)}$$

$$X = 28$$

EXAMPLE III.7

$$\frac{X}{3} - 4 = 8$$

$$\frac{X}{3} - 4 + 4 = 8 + 4 \qquad \text{(4 added to both sides)}$$

$$\frac{X}{3} = 12$$

$$\frac{3X}{3} = 12 \times 3 \qquad \text{(both sides multiplied by 3)}$$

$$X = 36$$

EXAMPLE III.8

How would you proceed in solving the equation $a = \frac{v_f - v_i}{t}$ for v_f? The v_i and the t need to be moved, and t should be moved first because it is dividing the whole side that the unknown, v_f, is on. The successive steps are

$$a = \frac{v_f - v_i}{t}$$

$$at = \frac{t(v_f - v_i)}{t} \qquad \text{(both sides multiplied by } t\text{)}$$

$$at = v_f - v_i$$

$$at + v_i = v_f - v_i + v_i \qquad \text{(} v_i \text{ added to both sides)}$$

$$at + v_i = v_f \text{ (or } v_f = at + v_i)$$

Shortcut

Of course, the same shortcuts can be used in these more complicated equations. Just be sure you use them in the correct order. For Example III.6, III.7, and III.8, the circled numbers show the correct order for the shortcuts:

$$\frac{(X - 4)}{3} = 8 \qquad \text{gives } X = (3 \times 8) + 4 = 28$$

Note that in Example III.6 both sides were first multiplied by 3, and then 4 was added to both sides. But in

$$\frac{X}{3} - 4 = 8$$ gives $X = 3 \times (8 + 4) = 36$

$$a = \frac{v_f - v_i}{t}$$ gives $at + v_i = v_f$

PRACTICE PROBLEMS

Solve for the unknown. The answers are given at the end of this appendix.

(p) $3X + 2 = 27$; $X = $?

(q) $\dfrac{X + 2}{4} = 6$; $X = $?

(r) $a = \dfrac{F_2 - F_1}{m}$; $F_2 = $?

(s) $H = \dfrac{B}{U} - M$; $B = $?

(t) $T_F = 1.8 T_C + 32$; $T_C = $?

(u) $H = m c_w (T_f - T_i)$; $T_f = $?

(v) $F = \dfrac{m(v - v_o)}{t}$; $v = $?

Squared or Negative Unknowns

You will encounter some equations in which the unknown is squared (raised to the second power). For example, solve for v in $a = \dfrac{v^2}{r}$. In such cases, solve for the squared form first; then, either before or after substituting the numbers and units, take the square root of both sides.

EXAMPLE III.9

$$a = \frac{v^2}{r}$$

$$ar = \frac{rv^2}{r} \quad \text{(both sides multiplied by } r\text{)}$$

$$ar = v^2$$

$$\sqrt{ar} = \sqrt{v^2} \quad \text{(the square root of both sides taken)}$$

$$\sqrt{ar} = v \text{ (or } v = \sqrt{ar}\text{)}$$

Occasionally, you may need to solve for an unknown that is negative at the start. For example, find T_i in $\Delta T = T_f - T_i$. In such cases, multiply both numerators by -1, then proceed as usual.

EXAMPLE III.10

$$\Delta T = T_f - T_i$$

$$-\Delta T = -T_f + T_i \quad \begin{array}{l}\text{(both sides} \\ \text{multiplied by } -1\text{)}\end{array}$$

$$T_f - \Delta T = -T_f + T_i + T_f \quad \begin{array}{l}(T_f \text{ added to} \\ \text{both sides)}\end{array}$$

$$T_f - \Delta T = T_i$$

PRACTICE PROBLEMS

Solve for the unknown. The answers are given at the end of this appendix.

(w) $E_k = \dfrac{m v^2}{2}$; $v = $?   (x) $a = \dfrac{v_f - v_i}{t}$; $v_i = $?

(y) $F_G = \dfrac{G m_1 m_2}{r^2}$; $r = $?

ANSWERS

(a) $X = 2$

(b) $T_f = \Delta T + T_i$

(c) $E_i = \lambda f + E_f$

(d) $W = H - \Delta E_i$

(e) $h_2 = \dfrac{\Delta E_p}{mg} + h_i$

(f) $E_p = H + \dfrac{m v^2}{2}$

(g) $X = 3$

(h) $X = 6$

(i) $m = \dfrac{w}{g}$

(j) $c = \lambda f$

(k) $f = \dfrac{c}{\lambda}$

(l) $m = \dfrac{h}{\lambda f}$

(m) $m = \dfrac{2 E_k}{v^2}$

(n) $T = \dfrac{v^2 m}{3k}$

(o) $R = \dfrac{pV}{nT}$

(p) $X = -3$

(q) $X = 22$

(r) $F_2 = ma + F_1$

(s) $B = U(H + M)$

(t) $T_C = \dfrac{T_F - 32}{1.8}$

(u) $T_f = \dfrac{H}{mc_w} + T_i$

(v) $v = \dfrac{Ft}{m} + v_o$

(w) $v = \sqrt{\dfrac{2E_k}{m}}$

(x) $v_i = v_f - at$

(y) $r = \sqrt{\dfrac{Gm_1 m_2}{F_G}}$

APPENDIX IV Analysis of Units

Measured quantities always have dimensions, or *units*; for example, 17 *grams*, 1.4 *meters*, 23°*C*, and so forth. Analyzing the units is important when dealing with scientific equations because the units can often show you whether you have rearranged the equation and put in your data correctly. For example, if the unknown for which you are solving is a distance, yet the units show you coming out with m/s² (a combination of units characteristic of an acceleration), you have done something wrong (probably rearranged the equation incorrectly). Units follow the same rules as numbers. Only a few basic situations are encountered when analyzing units, and most are illustrated in the problems given in this appendix.

You have often heard that you cannot add apples to oranges; neither can you add, say, grams to meters. When adding and subtracting, the units must be the same.

EXAMPLE IV.1 (TWO DIFFERENT CASES)

$8 \text{ mL} + 2 \text{ mL} = 10 \text{ mL}$ (that is, mL + mL gives mL)

$15 \text{ g} - 7 \text{ g} = 8 \text{ g}$ (that is, g − g = g)

When multiplying and dividing, a unit in the numerator will cancel its counterpart in the denominator. A unit multiplied by the same unit gives the unit squared.

EXAMPLE IV.2 (THREE DIFFERENT CASES)

$\dfrac{8 \text{ m}}{2 \text{ m}} = 4$ $5.0 \dfrac{\text{m}}{\text{s}^2} \times 3.0 \text{ s} = 15 \dfrac{\text{m}}{\text{s}}$

$8 \text{ m} \times 2 \text{ m} = 16 \text{ m}^2$

What about a problem where $\frac{\text{m}}{\text{s}}$ is divided by $\frac{\text{m}}{\text{s}^2}$? Recall how fractions are divided. To divide $\frac{3}{4}$ by $\frac{1}{4}$ you *invert* (turn over) the fraction in the denominator and then multiply by the inverted fraction. Units are handled the same way. Be sure to practice this trickiest part of analyzing units. Do not omit the intermediate step; this is no place to take shortcuts.

EXAMPLE IV.3 (TWO DIFFERENT CASES)

$\dfrac{\frac{3}{4}}{\frac{1}{4}} = \dfrac{3}{4} \times \dfrac{4}{1} = 3$ $\dfrac{\frac{\text{m}}{\text{s}}}{\frac{\text{m}}{\text{s}^2}} = \dfrac{\text{m}}{\text{s}} \times \dfrac{\text{s}^2}{\text{m}} = \text{s}$

PRACTICE PROBLEMS

Analyze these units. As you proceed, check your answers against those given at the end of this appendix.

(a) $\dfrac{\text{g}}{\text{mol}} \times \text{mol}$

(b) $\dfrac{\frac{\text{g}}{\text{mol}}}{\text{mol}}$

(c) $\dfrac{\text{cm}}{\text{s}} + \dfrac{\text{cm}}{\text{s}}$

(d) $\dfrac{\text{g}}{\frac{\text{g}}{\text{mol}}}$

(e) $\dfrac{\frac{\text{g}}{\text{mol}}}{\text{g}}$

(f) $\dfrac{\text{m}}{\text{s}} \times \text{m}$

(g) $\dfrac{\frac{\text{m}}{\text{s}}}{\text{m}}$

(h) $\dfrac{\text{J}}{\text{cal}} \times \text{J}$

(i) $\dfrac{\text{J}}{\text{cal}} \times \text{cal}$

(j) $\text{kg} - \text{kg}$

(k) $\dfrac{\dfrac{g}{mL}}{mL}$ (l) $kg \times \dfrac{m}{s^2}$

Chapter 1 discusses the use of conversion factors, a procedure in which the proper analysis of units is crucial. Using conversion factors, try the following practice problems. In (o) through (t), put in the units of the conversion factor first to make sure they will work out correctly, then insert the proper numbers. Check your answers.

PRACTICE PROBLEMS

(m) If 1 L = 1.06 qt, then how many $\frac{qt}{L}$ are there? How many $\frac{L}{qt}$?

(n) If there are 1.61 $\frac{km}{mi}$, then how many $\frac{mi}{km}$ are there?

(o) 15.0 m = ? yd, if 1 m = 1.09 yd

(p) 52.0 L = ? qt, if 1 L = 1.06 qt

(q) 46.0 km = ? mi, if 1 mi = 1.61 km

(r) 93.0 lb = ? kg, if 1 kg = 2.20 lb (at Earth's surface)

(s) 87 kg = ? g, if 1 kg = 1000 g

(t) 49 cm = ? m, if 1 m = 100 cm

ANSWERS

(a) g

(b) $\dfrac{g}{mol^2}$

(c) $\dfrac{cm}{s}$

(d) mol

(e) $\dfrac{1}{mol}$

(f) $\dfrac{m^2}{s}$

(g) $\dfrac{m^2}{s}$

(h) $\dfrac{J^2}{cal}$

(i) J

(j) kg

(k) $\dfrac{g}{mL^2}$

(l) $\dfrac{kg\text{-}m}{s^2}$

(m) $1.06 \dfrac{qt}{L}$; $\dfrac{1}{1.06}\dfrac{L}{qt}$

(n) $\dfrac{1}{1.61}\dfrac{mi}{km}$

(o) $15.0 \text{ m } (1.09)\dfrac{yd}{m} = 16.4 \text{ yd}$

(p) $52.0 \text{ L}(1.06)\dfrac{qt}{L} = 55.1 \text{ qt}$

(q) $46.0 \text{ km}\left(\dfrac{1}{1.61}\right)\dfrac{mi}{km} = 28.6 \text{ mi}$

(r) $93.0 \text{ lb}\left(\dfrac{1}{2.20}\right)\dfrac{kg}{lb} = 42.3 \text{ kg}$

(s) $87 \text{ kg } (1000)\dfrac{g}{kg} = 87 \times 10^3 \text{ g}$

(t) $49 \text{ cm}\left(\dfrac{1}{100}\right)\dfrac{m}{cm} = 0.49 \text{ m}$

APPENDIX V **Positive and Negative Numbers**

Multiplying and Dividing

To multiply and divide positive and negative numbers, follow these simple rules:

If *both* numbers are positive or *both* numbers are negative, the result is positive.

If one number is positive and the other is negative, the result is negative.

EXAMPLE V.1 (THREE DIFFERENT CASES)

$3 \times 4 = 12$
$-20 \div -5 = 4$
$(-3) \times 4 = -12$

PRACTICE PROBLEMS

Perform the designated operations. The answers are given at the end of this appendix.

(a) $14 \div 2$ (b) $-30 \div 6$

(c) 5×8 (d) $-7 \times (-6)$

(e) $8 \times (-2)$ (f) $-40 \div -5$

Algebraic Addition

Algebraic addition of positive and negative numbers is illustrated in Example V.2 on page A8. The numbers may be grouped and added or subtracted in any sequence without affecting the result.

EXAMPLE V.2 (SIX DIFFERENT CASES)

$$4 + 5 = 9$$
$$4 + (-5) = 4 - 5 = -1$$
$$4 - 5 = -1$$
$$4 - (-5) = 4 + 5 = 9$$
$$-4 - 5 = -9$$
$$-5 + 4 - 6 + 8 = -11 + 12 = 1$$

PRACTICE PROBLEMS

Perform the designated operations. Check your answers.

(g) $7 + 6$ (h) $-7 + 3 - 6$

(i) $-8 - 7$ (j) $14 - 5 + 8$

(k) $18 - 10$ (l) $8 - (-6)$

ANSWERS

(a) 7 (b) -5 (c) 40 (d) 42

(e) -16 (f) 8 (g) 13 (h) -10

(i) -15 (j) 17 (k) 8 (l) 14

APPENDIX VI Powers-of-10 Notation

Changing Between Decimal Form and Powers-of-10 Form

Chapter 1 discusses how to switch a number from the decimal form to the powers-of-10 form, and vice versa. Try the following practice problems.

PRACTICE PROBLEMS

Put a–d in standard powers-of-10 form, and e–h in decimal form. The answers are given at the end of this appendix.

(a) 2500 (b) 870,000

(c) 0.0000008 (d) 0.0357

(e) 6×10^4 (f) 5.6×10^3

(g) 5.6×10^{-6} (h) 7.9×10^{-2}

Changing Between Powers-of-10 Forms

When changing from one powers-of-10 form to another, the final number must be equal to the number with which you started. So, if the exponential part is made larger, the decimal part must become correspondingly smaller, and vice versa.

EXAMPLE VI.1

Change 83×10^5 to the 10^6 form.
Going from 10^5 to 10^6 is an *increase* by a factor of 10. Therefore, the decimal part must *decrease* by a factor

---

of 10. Since 83 divided by 10 is 8.3, the answer is 8.3×10^6.

Change 4.5×10^{-9} to the 10^{-10} form.
Going from 10^{-9} to 10^{-10} is a *decrease* by a factor of 10. Therefore, the decimal part must *increase* by a factor of 10. Since 4.5 multiplied by 10 is 45, the answer is 45×10^{-10}.

PRACTICE PROBLEMS

Determine the value required in place of the question mark for the equation to be true. Check your answers.

(i) $3.02 \times 10^7 = ? \times 10^6$ (j) $126 \times 10^{-3} = ? \times 10^{-2}$

(k) $896 \times 10^4 = ? \times 10^6$ (l) $32.7 \times 10^5 = 3.27 \times 10^?$

Addition and Subtraction of Powers-of-10

In addition or subtraction, the exponents of 10 must be the same value.

EXAMPLE VI.2

$$\begin{array}{r} 4.6 \times 10^{-8} \\ + 1.2 \times 10^{-8} \\ \hline 5.8 \times 10^{-8} \end{array}$$ and $$\begin{array}{r} 4.8 \times 10^{7} \\ - 2.5 \times 10^{7} \\ \hline 2.3 \times 10^{7} \end{array}$$

PRACTICE PROBLEMS

Perform the designated arithmetical operations. Check your answers.

(m) $\begin{array}{r} 4.5 \times 10^5 \\ + 3.2 \times 10^5 \\ \hline ? \end{array}$ (n) $\begin{array}{r} 5.66 \times 10^{-3} \\ - 3.24 \times 10^{-3} \\ \hline ? \end{array}$

Multiplication of Powers-of-10

In multiplication, the exponents are added.

EXAMPLE VI.3

$(2 \times 10^4)(4 \times 10^3) = 8 \times 10^7$ and

$(1.2 \times 10^{-2})(3 \times 10^6) = 3.6 \times 10^4$

PRACTICE PROBLEMS

Perform the designated arithmetical operations.

(o) $(7 \times 10^5)(3 \times 10^4) = ?$

(p) $(2 \times 10^{-3})(4 \times 10^6) = ?$

Division of Powers-of-10

In division, the exponents are subtracted.

EXAMPLE VI.4

$\dfrac{4.8 \times 10^8}{2.4 \times 10^2} = 2.0 \times 10^6$ and

$\dfrac{3.4 \times 10^{-8}}{1.7 \times 10^{-2}} = 2.0 \times 10^{-6}$

An alternative method for division is to transfer all powers-of-10 from the denominator to the numerator by changing the sign of the exponent. Then, the exponents of the powers-of-10 may be added, because they are now multiplying. The decimal parts are not transferred; they are divided in the usual manner. This method requires an additional step, but many students find it leads to the correct answer more consistently. Thus:

$$\frac{4.8 \times 10^8}{2.4 \times 10^2} = \frac{4.8 \times 10^8 \times 10^{-2}}{2.4} = 2.0 \times 10^6$$

PRACTICE PROBLEMS

Perform the designated arithmetical operations.

(q) $\dfrac{18 \times 10^7}{3 \times 10^4} = ?$ (r) $\dfrac{(3 \times 10^{17})(4 \times 10^{-8})}{6 \times 10^{-11}} = ?$

Squaring Powers-of-10

When squaring exponential numbers, multiply the exponent by 2. The decimal part is multiplied by itself.

EXAMPLE VI.5

$$(3 \times 10^4)^2 = 9 \times 10^8$$
$$(4 \times 10^{-7})^2 = 16 \times 10^{-14}$$

PRACTICE PROBLEMS

Perform the designated algebraic operations. Check your answers.

(s) $(8 \times 10^{-5})^2$ (t) $(4 \times 10^3)^2$ (u) $(3 \times 10^{-8})^2$

Finding the Square Root of Powers-of-10

To find the square root of an exponential number, follow the rule, $\sqrt{10^a} = 10^{\left(\frac{a}{2}\right)}$. Note that the exponent must be an *even* number. If it is not, change to a power-of-10 form that gives an even exponent. Find the square root of the decimal part by determining what number multiplied by itself gives that number.

EXAMPLE VI.6

$$\sqrt{9 \times 10^8} = 3 \times 10^4$$
$$\sqrt{2.5 \times 10^{-17}} = \sqrt{25 \times 10^{-18}} = 5 \times 10^{-9}$$

PRACTICE PROBLEMS

Perform the designated algebraic operations. Check your answers.

(v) $\sqrt{4 \times 10^8}$

(w) $\sqrt{16 \times 10^{-10}}$

(x) $\sqrt{78 \times 10^{-11}}$

ANSWERS

(a) 2.5×10^3

(b) 8.7×10^5

(c) 8×10^{-7}

(d) 3.57×10^{-2}

(e) 60,000

(f) 5600

(g) 0.0000056

(h) 0.079

(i) 30.2×10^6

(j) 12.6×10^{-2}

(k) 8.96×10^6

(l) 3.27×10^6

(m) 7.7×10^5

(n) 2.42×10^{-3}

(o) 21×10^9

(p) 8×10^3

(q) 6×10^3

(r) 2×10^{20}

(s) 64×10^{-10}

(t) 16×10^6

(u) 9×10^{-16}

(v) 2×10^4

(w) 4×10^{-5}

(x) 2.8×10^{-5}

APPENDIX VII   Significant Figures

In scientific work, most numbers are *measured* quantities and thus are not exact. All measured quantities are limited in significant figures (abbreviated SF) by the precision of the instrument used to make the measurement. The measurement must be recorded in such a way as to show the degree of precision to which it was made—no more, no less. Furthermore, calculations based on the measured quantities can have no more (or no less) precision than the measurements themselves. Thus the answers to the calculations must be recorded to the proper number of significant figures. To do otherwise is misleading and improper.

Counting Significant Figures

Measured Quantities

Rule 1: **Nonzero integers** are always significant (e.g., both 23.4 g and 234 g have 3 SF).

Rule 2: **Captive zeros,** those bounded on both sides by nonzero integers, are always significant (e.g., 20.05 g has 4 SF; 407 g has 3 SF).

Rule 3: **Leading zeros,** those *not bounded* on the *left* by nonzero integers, are never significant (e.g., 0.04 g has 1 SF; 0.00035 has 2 SF). Such zeros just set the decimal point; they always disappear if the number is converted to powers-of-10 notation.

Rule 4: **Trailing zeros,** those bounded *only on the left* by nonzero integers are probably not significant *unless a decimal point is shown,* in which case they are always significant. For example, 45.0 L has 3 SF but 450 L probably has only 2 SF; 21.00 kg has 4 SF but 2100 kg probably has only 2 SF; 55.20 mm has 4 SF;

151.10 cal has 5 SF; 3.0×10^4 J has 2 SF. If you wish to show *for sure* that, say, 150 m is to be interpreted as having 3 SF, change to powers-of-10 notation and show it as 1.50×10^2 m.

Exact Numbers

Rule 5: Exact numbers are those not obtained by measurement but by definition or by counting small numbers of objects. They are assumed to have an unlimited number of significant figures. For example, in the equation $C = 2\pi r$, the "2" is a defined quantity, not a measured one, so it has no effect on the number of significant figures to which the answer can be reported. In counting, say, 15 pennies, you can see that the number is exact because you cannot have 14.9 pennies or 15.13 pennies. The $\frac{9}{5}$ or 1.8 found in temperature conversion equations are exact numbers based on definitions.

PRACTICE PROBLEMS

Using Rules 1–5, determine the number of significant figures in the following measurements. As you proceed, check your answers against those given at the end of the appendix.

(a) 4853 g

(b) 36.200 km

(c) 0.088 s

(d) 30.003 J

(e) 6 dogs

(f) 74.0 m

(g) 340 cm

(h) 40 mi

(i) 8.9 L

(j) 1.30×10^2 cal

(k) 0.002710 ft

(l) 4000 mi

(m) 0.0507 mL

(n) 1.6×10^4 N

(o) The 2 in $E_k = \dfrac{mv^2}{2}$

(s) $7.43 \dfrac{\text{kg}}{\text{L}} \times 15$ L $= 111.45$ kg

(t) $5766 \dfrac{\text{m}}{\text{s}} \times 322$ s $= 1{,}856{,}652$ m

Multiplication and Division Involving Significant Figures

Rule 6: In calculations involving only multiplication and/or division of measured quantities, the answer shall have the same number of significant figures as the *fewest* possessed by any measured quantity in the calculation.

EXAMPLE VII.1

A calculator gives 4572.768 cm^3 when 130.8 cm is multiplied by 15.2 cm and then by 2.3 cm. However, this answer would be rounded and reported as 4.6×10^3 cm^3, since 2.3 cm has the fewest significant figures (2). The reasoning behind this is that the measured 2.3 cm could easily be wrong by 0.1 cm. Suppose it were really 2.4 cm; what difference would this make in the answer on the calculator? You would get 4771.584 cm^3! Comparing this to what the calculator originally gave, you see that the uncertainty in the answer is in the hundreds place; so the answer is properly reported only to the hundreds place; that is, to 2 SF. The measured quantity with the fewest significant figures will have the greatest effect on the answer because of percentage effects (a miss of 1 out of 23 is more damaging than a miss of 1 out of 1308, for example).

PRACTICE PROBLEMS

Rewrite the following calculator-given answers so that the proper number of significant figures is shown in each case. When necessary, use exponential notation to avoid ambiguity in the answer.

(p) $7.7 \dfrac{\text{m}}{\text{s}^2} \times 3.222$ s $\times 2.4423$ s $= 60.59199762$ m

(q) 0.0075 cm $\times$ 0.005 cm $\times$ 8211 cm $=$
$\qquad\qquad\qquad\qquad$ 0.3079125 cm^3

(r) 93.0067 g $\div$ 35 mL $= 2.65733428571 \dfrac{\text{g}}{\text{mL}}$

Addition and Subtraction Involving Significant Figures

Rule 7: In calculations where measured quantities are added or subtracted, the final answer can have only one "uncertain" figure, so it stops at the place on the right that any of the data first stops.

Carry the calculation one place farther, and then round up the answer if justified. (If the last figure is less than 5, drop it; if it is 5 or greater, round the preceding figure up one.)

EXAMPLE VII.2

In each of these examples, the vertical dashed line shows how far over the answer can go. Note that each answer has been calculated to one place farther than will be reported. This is so we can see if rounding is necessary. The final answers are shown below the double underline.

46.6⋮ m	38⋮ cm	5.68⋮7 × 10³ g
+ 5.7⋮2 m	− 7⋮.44 cm	+ 11.11⋮ × 10³ g
52.3⋮2 m	30⋮.6 cm	16.79⋮7 × 10³ g
52.3⋮ m	31⋮ cm	16.80⋮ × 10³ g

PRACTICE PROBLEMS

Perform the designated arithmetic operations, being careful to retain the proper number of significant figures.

(u) $\begin{array}{r} 0.0012 \text{ m} \\ + \ 1.334 \text{ m} \\ \hline ? \end{array}$

(v) $\begin{array}{r} 879 \ \text{ g} \\ - \ 79.9 \text{ g} \\ \hline ? \end{array}$

(w) $\begin{array}{r} 6.788 \text{ cm} \\ + \ 5.6 \ \ \text{ cm} \\ \hline ? \end{array}$

(x) $\begin{array}{r} 67.4 \ \text{ kg} \\ - \ 0.06 \text{ kg} \\ \hline ? \end{array}$

(y) $\begin{array}{r} 54.09 \times 10^4 \text{ g} \\ + \ 3 \ \ \ \times 10^4 \text{ g} \\ \hline ? \end{array}$

ANSWERS

(a) 4 (Rule 1)

(b) 5 (Rule 4)

(c) 2 (Rule 3)

(d) 5 (Rule 2)

(e) unlimited (Rule 5)

(f) 3 (Rule 4)

(g) 2 or 3 (Rule 4)

(h) 1 or 2 (Rule 4)

(i) 2 (Rule 1)

(j) 3 (Rule 4)

(k) 4 (Rules 3, 4)

(l) 1, 2, 3, or 4 (Rule 4)

(m) 3 (Rules 2, 3)

(n) 2 (Rule 1)

(o) unlimited (Rule 5)

(p) 61 m

(q) 0.3 cm^3

(r) $2.7 \frac{\text{g}}{\text{mL}}$

(s) $1.1 \times 10^2 \text{ kg}$

(t) $1.86 \times 10^6 \text{ m}$

(u) 1.335 m

(v) 799 g

(w) 12.4 cm

(x) 67.3 kg

(y) $57 \times 10^4 \text{ g}$

APPENDIX VIII   Psychrometric Tables (pressure: 30 in. of Hg)

TABLE VIII.1   Relative Humidity (%) and Maximum Moisture Capacity

Air Temp. (°F) (Dry Bulb)	Max. Moisture Capacity (gr/ft^3)	Degrees Depression of Wet Bulb Thermometer (F°)													
		1	2	3	4	5	6	7	8	9	10	15	20	25	30
25	1.6	87	74	62	49	37	25	13	1						
30	1.9	89	78	67	56	46	36	26	16	6					
35	2.4	91	81	72	63	54	45	36	27	19	10				
40	2.8	92	83	75	68	60	52	45	37	29	22				
45	3.4	93	86	78	71	64	57	51	44	38	31				
50	4.1	93	87	80	74	67	61	55	49	43	38	10			
55	4.8	94	88	82	76	70	65	59	54	49	43	19			
60	5.7	94	89	83	78	73	68	63	58	53	48	26	5		
65	6.8	95	90	85	80	75	70	66	61	56	52	31	12		
70	7.8	95	90	86	81	77	72	68	64	59	55	36	19	3	
75	9.4	96	91	86	82	78	74	70	66	62	58	40	24	9	
80	10.9	96	91	87	83	79	75	72	68	64	61	44	29	15	3
85	12.7	96	92	88	84	80	76	73	69	66	62	46	32	20	8
90	14.8	96	92	89	85	81	78	74	71	68	65	49	36	24	13
95	17.1	96	93	89	85	82	79	75	72	69	66	51	38	27	17
100	19.8	96	93	89	86	83	80	77	73	70	68	54	41	30	21
105	23.4	97	93	90	87	83	80	77	74	71	69	55	43	33	23
110	26.0	97	93	90	87	84	81	78	75	73	70	57	46	36	26

Note: To use the table, determine the air temperature with a dry bulb thermometer and degrees depressed on the wet bulb thermometer. Read the maximum capacity directly. Read the relative humidity (in percent) opposite and below these values.

TABLE VIII.2   Dew Point (°F)

Air Temp (°F) (Dry Bulb)	Degrees Depression of Wet Bulb Thermometer (F°)													
	1	2	3	4	5	6	7	8	9	10	15	20	25	30
25	22	19	15	10	5	−3	−15	−51						
30	27	25	21	18	14	8	2	−7	−25					
35	33	30	28	25	21	17	13	7	0	−11				
40	38	35	33	30	28	25	21	18	13	7				
45	43	41	38	36	34	31	28	25	22	18				
50	48	46	44	42	40	37	34	32	29	26	0			
55	53	51	50	48	45	43	41	38	36	33	15			
60	58	57	55	53	51	49	47	45	43	40	25	−8		
65	63	62	60	59	57	55	53	51	49	47	34	14		
70	69	67	65	64	62	61	59	57	55	53	42	26	−11	
75	74	72	71	69	68	66	64	63	61	59	49	36	15	
80	79	77	76	74	73	72	70	68	67	65	56	44	28	−7
85	84	82	81	80	78	77	75	74	72	71	62	52	39	19
90	89	87	86	85	83	82	81	79	78	76	69	59	48	32
95	94	93	91	90	89	87	86	85	83	82	74	66	56	43
100	99	98	96	95	94	93	91	90	89	87	80	72	63	52
105	104	103	101	100	99	98	96	95	94	93	86	78	70	61
110	109	108	106	105	104	103	102	100	99	98	91	84	77	68

Note: To use the table, determine the air temperature with a dry bulb thermometer and degrees depressed on the wet bulb thermometer. Find the dew point opposite and below these values.

Glossary

Aberration of starlight the apparent displacement in the direction of light coming from a star due to the orbital motion of Earth. (p. 391)

Absolute geologic time the time of past geologic events based on the radioactive decay of certain atomic nuclei. (p. 640)

Absolute magnitude the apparent magnitude that a star would have if it were placed 10 pc from Earth. (p. 476)

Abyssal plain a large flat area on the ocean floor where layers of sediment have covered the original seafloor topography. (p. 630)

Acceleration the change in velocity divided by the change in time; $a = \Delta v / \Delta t$. (p. 29).

Acceleration due to gravity usually given as the symbol g; equal to 9.80 m/s^2 or 32 ft/s^2. (pp. 31, 56)

Acid a substance that gives hydrogen ions (or hydronium ions) in water (Arrhenius definition). (p. 332)

Acid rain rain that has a relatively low pH (acid) due to air pollution. (p. 548)

Acid–base reaction the H$^+$ of the acid unites with the OH$^-$ of the base to form water, while the cation of the base combines with the anion of the acid to form a salt. (p. 334)

Acid–carbonate reaction an acid and a carbonate (or hydrogen carbonate) react to give carbon dioxide, water, and a salt. (p. 337)

Activation energy the energy necessary to start a chemical reaction; a measure of the minimum kinetic energy colliding molecules must possess in order to react. (p. 327)

Activity series a list of elements in order of relative ability of their atoms to be oxidized in solution. (p. 340)

Addition polymers those formed when molecules of an alkene monomer add to one another. (p. 369)

Air current vertical air movement. (p. 518)

Air mass a mass of air with physical characteristics that distinguish it from other air. (p. 533)

Albedo the fraction of incident sunlight reflected by a surface. (p. 389)

Alcohols organic compounds containing a hydroxyl group, —OH, attached to an alkyl group; general formula, R—OH. (p. 361)

Aliphatic hydrocarbon A carbon–hydrogen compound that contains no benzene rings. (p. 352)

Alkali metals the elements in Group 1A of the periodic table, except for hydrogen. (p. 284)

Alkaline earth metals the elements in Group 2A of the periodic table. (p. 286)

Alkanes hydrocarbons that contain only single bonds; general formula C_nH_{2n+2}. (p. 353)

Alkenes hydrocarbons that have a double bond between two carbon atoms; general molecular formula C_nH_{2n}. (p. 357)

Alkyl group a substituent that contains one less hydrogen atom than the corresponding alkane; general symbol, R. (p. 354)

Alkyl halide an alkane derivative in which one or more of the hydrogen atoms have been replaced by a halogen atom; general formula, R—X. (p. 359)

Alkynes hydrocarbons that have a triple bond between two carbon atoms; general molecular formula, C_nH_{2n-2}. (p. 358)

Allotropes two or more forms of the same element that have different bonding structures in the same physical phase. (p. 273)

Alpha decay the disintegration of a nucleus into a nucleus of another element with the emission of an alpha particle. (p. 239)

Alternating current electric current produced by constantly changing the voltage from positive to negative to positive, and so on. (p. 181)

Altitude the angle measured from the horizon to a celestial object. (p. 433)

Amides nitrogen-containing organic compounds having the general formula RCONHR′. (p. 368)

Amine a basic (alkaline) organic compound that contains nitrogen; general formula, R—NH$_2$. (p. 362)

Amino acid an organic compound that contains both an amino group and a carboxyl group. (p. 368)

Amino group a substituent of general formula —NH$_2$, which forms an amine when attached to an alkyl group. (p. 362)

Ampere the unit of electric current defined as that current which, if maintained in each of two long parallel wires separated by one meter in free space, would produce a magnetic force between the two wires of 2×10^{-7} newtons for each meter of length. (p. 173)

Amplitude   the maximum displacement of a wave from its equilibrium position. (p. 123)

Anemometer   an instrument used to measure wind speed. (p. 515)

Angular momentum mvr for a mass m going at a speed v in a circle of radius r. (p. 61)

Anions   negative ions; so-called because they move toward the anode (the positive electrode) of an electrochemical cell. (p. 301)

Annular eclipse   a solar eclipse in which the Moon blocks out all of the Sun except for a ring around the Sun's outer edge. (p. 456)

Ante meridiem   pertaining to time before 12 noon. (p. 427)

Anticline   a rock fold that forms an upward-pointing arch; the opposite of syncline. (p. 605)

Apparent solar day   the duration of one rotation of Earth on its axis with respect to the Sun. (p. 426)

Aquifer   a body of permeable rock through which groundwater moves. (p. 625)

Aromatic hydrocarbon   a carbon–hydrogen compound that contains one or more benzene rings. (p. 351)

Ascending node   the point where the Moon's path crosses the ecliptic plane going northward. (p. 456)

Asteroids   large chunks of matter that orbit the Sun (usually between Mars and Jupiter) and that are too small to be labeled as planets. (p. 407)

Asthenosphere   the rocky substratum below the lithosphere that is hot enough to be deformed and is capable of internal flow. (p. 567)

Astronomical unit   the average distance between Earth and the Sun, which is 93 million miles. (p. 383)

Astronomy   the scientific study of the universe beyond Earth's atmosphere. (p. 380)

Atmospheric science   the investigation of every aspect of the atmosphere, from the ground to the edge of outer space. (p. 500)

Atom   the smallest particle of an element that can enter into a chemical combination. (p. 205)

Atomic mass   the average mass (in *atomic mass units*, u) of an atom of the element in naturally occurring samples. (p. 237)

Atomic number   symbolized Z, it is equal to the number of protons in the nucleus of an atom of that element. (p. 235)

Autumnal equinox   the point where the Sun crosses the celestial equator from north to south, around September 21. (p. 433)

Avogadro's number 6.02×10^{23}, symbolized N_A; the number of entities in a mole. (p. 344)

Barometer   a device used to measure atmospheric pressure. (p. 510)

Base   a substance that produces hydroxide ions in water (Arrhenius definition). (p. 334)

Batholith   a large intrusive igneous rock formation that has an area of at least 40 square miles. (p. 569)

Bedding   the stratification of sedimentary rock formations. (p. 579)

Bergeron process   the process by which precipitation is formed in clouds. (p. 531)

Beta decay   the disintegration of a nucleus into a nucleus of another element, with the emission of a beta particle. (p. 240)

Big Bang   theory of the origin of the universe that states that the known universe was concentrated in a point of energy that exploded approximately 12 billion years ago. (p. 490)

Binary star   two stars in orbit about their common center of mass, bound together by gravity. (p. 477)

Black hole   an object whose gravity is so strong that the escape velocity is equal to or greater than the speed of light; thus no radiation can escape from the object. (p. 483)

Body waves   seismic waves that travel through Earth's interior. (p. 602)

British system   the system of units used in the United States, which uses the foot, pound, and second, as the standards of length, weight, and time, respectively. The system is sometimes referred to as the *fps* (foot-pound-second) *system*. (p. 4)

Btu   the amount of heat required to raise one pound of water one degree Fahrenheit at normal atmospheric pressure. (p. 96)

Caldera   a roughly circular, steep-walled depression formed as a result of the collapse of a volcanic chamber. (p. 573)

Calorie   the amount of heat necessary to raise one gram of pure liquid water one degree Celsius at normal atmospheric pressure. (p. 96)

Carbohydrates   organic compounds that contain multiple hydroxyl groups in their molecular structure. A basic component of living matter. (p. 361)

Carbon dating   a procedure used to establish the age of ancient organic remains by measuring the concentration of ^{14}C and comparing it to that of present-day organic remains. (p. 643)

Carboxyl group   a substituent of general formula –COOH, which, when attached to an alkyl group, forms a carboxylic acid. (p. 362)

Carboxylic acids   a class of organic compounds characterized by the presence of a carboxyl group; general formula, RCOOH. (p. 362)

Carcinogen   a cancer-causing agent. (p. 351)

Carnot (ideal) efficiency   the theoretical maximum efficiency for a heat engine. It can never be attained physically. (p. 113)

Cartesian coordinate system   a two-dimensional coordinate system in which two number lines are drawn perpendicular to each other, and the origin is assigned at the point of intersection. (p. 422)

Catalyst   a substance that increases the rate of reaction but is not itself consumed in the reaction. (p. 329)

Cations   positive ions; so-called because they move toward the cathode (the negative electrode) of an electrochemical cell. (p. 301)

Celestial prime meridian   an imaginary half-circle running from the North Celestial Pole to the South Celestial Pole and crossing perpendicular to the celestial equator at the point of the vernal equinox. (p. 473)

Celestial sphere   the apparent sphere of the sky on which all the stars seem to appear. (p. 472)

Celsius scale   a temperature scale with 0°C as the ice point and 100°C as the steam point. (p. 93)

Centi   the metric prefix meaning 1/100, or 0.01. (p. 8)

Centripetal acceleration   the "center-seeking" acceleration necessary for circular motion; $a = v^2/r$. (p. 35)

Centripetal force   the "center-seeking" force that causes an object to travel in a circle. (p. 49)

Cepheid variables   stars that vary in magnitude with a fixed period of between 1 and 50 days. (p. 477)

CFCs   chlorofluorocarbons, which are used in air conditioners, refrigerators, heat pumps, and so forth, and which deplete the ozone layer. (p. 359)

Chain reaction   occurs when each fission event causes at least one more fission event. (p. 250)

Chemical properties   characteristics that describe the chemical reactivity of a substance, that is, its ability to transform into another substance. (p. 322)

Chemical reaction   a change that alters the chemical composition of a substance. (p. 322)

Chemical sediments   particles composed of minerals that were transported to the sea in solution. (p. 578)

Chemical weathering   a change in a rock's composition or size due to a chemical change. (p. 614)

Chemistry   the division of physical science that studies the composition and structure of matter and the reactions by which substances are changed into other substances. (p. 265)

Chromosphere   a transparent layer of gas that rests on the photosphere in the atmosphere of the Sun. (p. 468)

Cinder cone   a volcano with a steeply sloped cinder cone formed by eruptions of tephra. (p. 573)

Cleavage   the splitting of a mineral along an internal molecular plane. (p. 561)

Climate   the long-term average weather conditions of a region of the world. (p. 550)

Cloud   a buoyant mass of visible droplets of water and ice crystals in the lower troposphere. (p. 522)

Clusters   groups of neighboring galaxies held together by mutual gravitation. The clusters, classified as regular or irregular, range in size from 3 to 15 million ly in diameter. (p. 487)

Coalescence   the combining of small droplets of water vapor to make larger drops. (p. 531)

Color (of mineral)   the visual sensation received by the light reflected from a mineral sample. (p. 561)

Combination reaction   one in which at least two reactants combine to form just one product: $A + B \rightarrow AB$. (p. 324)

Combustion reaction   the reaction of a substance with oxygen to form an oxide, along with heat and light in the form of fire. (p. 327)

Comet   a small mass of ice and dust that revolves around the Sun in a highly elliptical orbit. (p. 409)

Compound   a substance composed of two or more elements chemically combined in a definite, fixed proportion by mass. (p. 266)

Concave lens   a lens that has the shape of the inside (concave side) of a spherical section. (p. 163)

Concave mirror   a mirror shaped like the inside of a small section of a sphere. (p. 158)

Condensation polymers   large molecules constructed from smaller molecules that have two or more reactive groups. Each small molecule attaches to two others by ester or amide linkages. (p. 370)

Condensation theory   a process of solar system formation in which interstellar dust grains act as condensation nuclei. (p. 413)

Conduction (thermal)   the transfer of heat energy by molecular collisions. (p. 103)

Conjunction   the time at which a planet and the Sun are on the same meridian. (p. 386)

Conservation of angular momentum, law of the angular momentum of a system remains constant unless acted upon by an external torque. (p. 62)

Conservation of (total) energy, law of the total energy of an isolated system remains constant. (p. 76)

Conservation of linear momentum, law of the total linear momentum of a system remains constant if there are no external unbalanced forces acting on the system. (p. 60)

Conservation of mass, law of no detectable change in the total mass occurs during a chemical reaction. (p. 294)

Conservation of mechanical energy, law of in an ideal system, the sum of the kinetic and potential energies are constant: $E_k + E_p = E$ (a constant). (p. 76)

Constitutional (structural) isomers compounds that have the same molecular formula but differ in structural formula. (p. 355)

Contact metamorphism a change in rock brought about primarily by heat rather than pressure. (p. 582)

Continental drift the theory that continents move, drifting apart or together. (p. 588)

Continental shelf the moderately sloping submerged margin of a continental land mass. (p. 630)

Continental slope the seaward slope beyond the continental shelf. It extends downward to the ocean basin. (p. 630)

Convection the transfer of heat through the movement of a substance. (p. 103)

Convection cells large masses of material circulating in the semiplastic asthenosphere because of heating at the bottom and cooling from above. (p. 593)

Convection cycle the cyclic movement of matter (e.g., air) due to localized heating and convectional heat transfer. (p. 518)

Convergent boundary a region where moving plates of the lithosphere are driven together, causing one of the plates to be consumed into the mantle as it descends beneath an overriding plate. (p. 567)

Conversion factor an equivalence statement expressed as a ratio. (p. 12)

Convex lens a lens that has the shape of the outside (convex side) of a spherical section. (p. 163)

Convex mirror a mirror shaped like the outside of a small section of a sphere. (p. 158)

Coriolis force a pseudoforce arising in an accelerated reference frame on the rotating (accelerating) Earth. Projectiles are deflected to the right in the Northern Hemisphere and to the left in the Southern Hemisphere, as observed in the direction of motion. (p. 519)

Corona the Sun's outer atmosphere. (p. 468)

Correlation establishing the equivalence of rocks in separate regions; correlation by fossils is an example. (p. 638)

Cosmic background radiation the microwave radiation that fills all space and is believed to be the redshifted glow from the Big Bang. (p. 492)

Cosmological principle the hypothesis that on a large scale the galaxies in the universe are distributed equally in all directions and at all distances. The latest observations indicate this may not be true. (p. 488)

Cosmological redshift the shift toward longer wavelengths caused by the expansion of the universe. (p. 490)

Cosmology the study of the structure and evolution of the universe. (p. 490)

Coulomb the unit of electric charge equal to one ampere-second (A-s). (p. 173)

Coulomb's law the force of attraction or repulsion between two charged bodies is directly proportional to the product of the two charges and inversely proportional to the square of the distance between them. (p. 175)

Covalent bond the force of attraction caused by a pair of electrons being shared by two atoms. (p. 306)

Covalent compounds those in which the atoms share pairs of electrons to form molecules. (p. 305)

Crater (lunar) a circular depression on the surface of the Moon caused by the impact of a meteorite. (p. 445)

Creep a type of slow mass wasting involving the particle-by-particle movement of weathered debris down a slope, which takes place year after year. (p. 623)

Crescent moon the Moon viewed when less than one-quarter of the illuminated surface is facing an observer on Earth. (p. 450)

Critical mass the minimum amount of fissionable material necessary to sustain a chain reaction. (p. 250)

Cross-cutting relationships, principle of a rock or fault is younger than any rock or fault through which it cuts. (p. 636)

Crust the thin outer layer of Earth. (p. 604)

Crystal form the size and shape assumed by crystal faces of a mineral when the crystal has time and space to grow. (p. 561)

Curie temperature the temperature above which ferromagnetic materials cease to be magnetic. (p. 191)

Current the rate of flow of electric charge; $I = q/t$. (p. 173).

Cycloalkanes hydrocarbons that have the general molecular formula C_nH_{2n} and possess rings of carbon atoms, with each carbon atom bonded to a total of four carbon or hydrogen atoms. (p. 357)

Daylight Saving Time   time advanced one hour from standard time, adopted during the spring and summer months to take advantage of longer evening daylight hours and save electricity. (p. 430)

Decibel (dB)   a unit of sound level intensity; 0.01 bel. (p. 129)

Declination   the angular measure in degrees north or south of the celestial equator. (p. 473)

Decomposition reaction   one in which only one reactant is present and breaks into two (or more) products: $AB \rightarrow A + B$. (p. 324)

Deep-sea trenches   deep, narrow depressions in the ocean floor that ring the Indian and Pacific Oceans. They are the lowest places on Earth's surface. (p. 590)

Definite proportions, law of   different samples of a pure compound always contain the same elements in the same proportion by mass. (p. 295)

Delta   the accumulation of sediment formed where running water enters a large body of water such as a lake or ocean. (p. 617)

Density   a measure of the compactness of matter using a ratio of mass to volume; $\rho = m/V$. (p. 11)

Derived units   combinations of fundamental units. (p. 10)

Descending node   the point where the Moon's path crosses the ecliptic plane going southward. (p. 456)

Desert   an area on Earth's surface that has a severe lack of precipitation. (p. 620)

Detrital sediments   rocks formed from sediment composed of fragments of pre-existing rocks. (p. 577)

Dew point   the temperature at which a sample of air becomes saturated, that is, has a relative humidity of 100 percent. (p. 514)

Diffraction   the bending of waves when moving past an opening or obstacle that has a size smaller than or equal to the wavelength. (p. 153)

Direct current (dc)   electric current in which the electrons flow directionally from the negative (−) terminal toward the positive (+) terminal. (p. 181)

Dispersion   different frequencies of light refracted at slightly different angles giving rise to a spectrum. (p. 147)

Displacement   the directional straight-line distance between two points. (p. 26)

Distance   the actual path length between two points. (p. 25)

Divergent boundary   a region where plates of the lithosphere are moving away from one another. (p. 567)

Doppler effect   an apparent change in frequency resulting from the relative motion of the source and the observer. (p. 133)

Doppler radar   radar that uses the Doppler effect on water droplets in clouds to measure the wind speed and direction. (p. 516)

Double-replacement reactions   ones that take the form $AB + CD \rightarrow AD + CB$. The positive and negative components of the two compounds "change partners." (p. 338)

Dual nature of light   light sometimes behaves as waves and sometimes as particles. (p. 209)

Dual nature of matter   matter sometimes behaves as particles and sometimes as waves. (p. 224)

Earthquake   the sudden release or transfer of energy because of sudden movement resulting from stresses in Earth's lithosphere. (p. 595)

Eclipse   an occurrence in which one celestial object is partially or totally blocked from view by another. (p. 454)

Ecliptic   the apparent path of the Sun on the celestial sphere. (p. 389)

Electric charge   a fundamental property of matter that can be either positive or negative and gives rise to electric forces. (p. 173)

Electric potential energy   the potential energy that results from work done in separating electric charges. (p. 178)

Electric power   the expenditure of electrical work divided by time; $P = W/t = IV$. (p. 180)

Electromagnetic wave   a wave consisting of oscillating electric and magnetic fields. (p. 125)

Electromagnetism   the interaction of electric and magnetic effects. (p. 192)

Electron configuration   the order of electrons in the energy levels of an atom. (p. 278)

Electronegativity   a measure of the ability of an atom to attract shared electrons to itself. (p. 310)

Electrons   negatively charged particles that surround the nucleus of an atom. (pp. 173, 206, 233)

Element   a substance in which all the atoms have the same number of protons, that is, have the same atomic number, Z. (p. 235)

Elliptical paths, law of   all planets, asteroids, and comets revolve around the Sun in elliptical orbits. (p. 382)

Endothermic reaction   one that absorbs energy from the surroundings. (p. 326)

Energy   the capacity to do work. (p. 72)

Entropy   a measure of the disorder of a system. (p. 114)

Eon   the largest units of geologic time. They are divided into eras. (p. 639)

Epicenter   the point on the surface of Earth directly above the focus of an earthquake. (p. 598)

Epoch   an interval of geologic time that is a subdivision of a period. (p. 648)

Equal areas, law of   as a planet (or asteroid or comet) revolves around the Sun, an imaginary line joining the planet to the Sun sweeps out equal areas in equal periods of time. (p. 383)

Equilibrium   in chemistry, a dynamic process in which the reactants are combining to form the products at the same rate that the products are combining to form the reactants. (p. 333)

Era   an interval of geologic time made up of periods and epochs. (p. 639)

Erosion   the downslope movement of surface and near-surface materials due to gravity and the agents that cause such movements. (p. 616)

Ester   an organic compound that has the general formula RCOOR'. (p. 363)

Event horizon   the position in space at which the escape velocity from a black hole equals the speed of light. (p. 482)

Excess reactant   a starting material that is only partially used up in a chemical reaction. (p. 297)

Excited states   the energy levels above the ground state in an atom; see Ground state. (p. 214)

Exothermic reaction   one that releases energy to the surroundings. (p. 325)

Experiment   an observation of natural phenomena carried out in a controlled manner so that the results can be duplicated. (p. 3)

Fahrenheit scale   a temperature scale with 32°F as the ice point and 212°F as the steam point. (p. 93)

Fats   esters composed of the trialcohol named *glycerol*, $C_3H_5(OH)_3$, and long-chain carboxylic acids known as *fatty acids*. (p. 365)

Fault   a fracture along which a relative displacement of the sides has occurred. (p. 595)

Fault-block mountains   mountains that were built by normal faulting, in which giant pieces of Earth's crust were uplifted. (p. 607)

First-quarter phase   the Moon that is exactly 90° east of the Sun and appears as a quarter moon on the observer's meridian at 6 P.M. local solar time. (p. 450)

Fission   a process in which a large nucleus is split into two intermediate-sized nuclei, with the emission of neutrons and the conversion of mass into energy. (p. 249)

Flood plain   the land next to a river or stream that can become enundated when the river or stream overflows. (p. 617)

Fluorescence   the property of a substance, such as the mineral fluorite, of producing light while it is being acted upon by ultraviolet light. (p. 220)

Focal length   the distance from the vertex of a mirror or lens to the focal point. (p. 158)

Focus (earthquake)   the point within Earth at which the initial energy release of an earthquake occurs. (p. 596)

Fold mountains   mountains characterized by folded rock strata, with external evidence of faulting and central evidence of igneous metamorphic activity. Fold mountains are believed to be formed at convergent plate boundaries. (p. 607)

Foliation   the mineral orientation characteristic of some metamorphic rocks due to directional pressures during transformation. (p. 583)

Foot-pound   the unit of work (or energy) in the British system. (p. 71)

Force   any quantity capable of producing motion. (p. 44)

Formula mass   the sum of the atomic masses of the atoms showing in the chemical formula of the compound or element. (p. 295)

Fossil   a remnant or trace of an organism preserved from prehistoric times. (p. 634)

Foucault pendulum   any pendulum that is used to demonstrate the rotation of Earth. (p. 389)

Fracture   refers to the way some minerals break unevenly rather than cleave smoothly. (p. 561)

Frequency   the number of oscillations of a wave during one second. (p. 124)

Friction   the opposing force that exists when contact surfaces tend to slide past one another. (p. 520)

Front   the boundary between two air masses. (p. 535)

Full moon   the phase of the Moon that occurs when the Moon is 180° east of the Sun. (p. 451)

Functional group   any atom, group of atoms, or organization of bonds that determines specific properties of a molecule. (p. 359)

Fusion   a process in which smaller nuclei are fused (joined) to form larger ones, with the release of energy. (p. 254)

G   the universal gravitational constant; $G = 6.67 \times 10^{-11} \text{ N} = \text{m}^2/\text{kg}^2$. (p. 53)

Galaxy   A large-scale aggregate of stars (plus some gas and dust) held together by gravity. Galaxies have a spiral, elliptical, or irregular structure. Each contains, on average, one hundred billion solar masses. (p. 484)

Gamma decay   an event in which a nucleus emits a gamma ray and becomes a less energetic form of the same nucleus. (p. 240)

Gas   matter that has no definite volume or shape. (p. 106)

Generator   a device that converts mechanical work or energy into electrical energy. (p. 195)

Genetic effects   defects in the subsequent offspring of recipients of radiation. (p. 259)

Geocentric model   the old false theory of the solar system, which placed Earth at its center. (p. 381)

Geologic time   the time span that covers the long history of Earth. (p. 633)

Geologic time scale   a relative time scale based on the fossil contents of rock strata and the principles of superposition and cross-cutting relationships. (p. 646)

Geology   the study of planet Earth—its dynamics, composition, structure, and history. Also, the study of the chemical and physical properties of other solar system bodies. (p. 557)

Gibbous moon   the Moon viewed when more than one-quarter of its surface appears illuminated. (p. 450)

Glacier   a large ice mass that consists of recrystallized snow and which flows on a land surface under the influence of gravity. (p. 619)

Gravitational collapse   the caving in of a very massive body because of its attraction for itself. (p. 482)

Gravitational potential energy   the potential energy resulting from an object's position in a gravitational field. In other words, the stored energy that comes from doing work against gravity. (p. 74)

Great circle   any circle on the surface of a sphere whose center is at the center of the sphere. Applies especially to imaginary circles on Earth's surface that pass through both the North Pole and South Pole. (p. 424)

Greenhouse effect   the heat-retaining process of atmospheric gases, such as water vapor and CO_2, due to the selective absorption of long-wavelength terrestrial radiation. (p. 507)

Greenwich meridian   the reference meridian of longitude, which passes through the old Royal Greenwich Observatory near London. (p. 423)

Gregorian calendar   the reformed Julian calendar — our present-day calendar. (p. 438)

Ground state   the lowest energy level of an atom. (p. 214)

Groundwater   water that soaks into the soil. (p. 624)

Groups   the vertical columns in the periodic table. (p. 275)

Half-life   the time it takes for the decay of half the nuclei in a sample of a given radionuclide. (p. 244)

Halogens   the elements in Group 7A of the periodic table. (p. 286)

Hardness   a comparative property that refers to the ability of a mineral to resist scratching. Usually measured using the Mohs' scale. (p. 561)

Harmonic law   the square of the sidereal period of a planet is proportional to the cube of its semimajor axis (one-half the major axis). (p. 383)

Heat   a form of energy; energy in transit from one body to another as a result of a temperature difference. (p. 96)

Heat engine   a device that uses heat energy to perform useful work. (p. 111)

Heat pump   a device used to transfer heat from a low-temperature reservoir to a high-temperature reservoir. (p. 114)

Heisenberg's uncertainty principle   it is impossible to know simultaneously the exact velocity and position of a particle. (p. 222)

Heliocentric model   the model of the solar system that places the Sun at its center. (p. 382)

Heliosphere   the volume of space around the Sun over which the effect of the solar wind extends. (p. 468)

Hertz   one cycle per second. The SI unit of frequency. (p. 124)

Horizon   the dividing line where Earth and sky appear to meet. (p. 433)

Horsepower   a unit of power equal to 550 ft-lb/s. (p. 78)

Hot spot   a large, fixed source of heat below the lithosphere that can generate volcanoes or other geothermal phenomena as a lithospheric plate moves over it. (p. 571)

H-R diagram   a plot of the absolute magnitude versus the temperature of stars. (p. 476)

Hubble's law   the recessional speed of a distant galaxy is directly proportional to its distance away. (p. 487)

Humidity   a measure of the water vapor in the air. (p. 513)

Hurricane   a tropical storm with winds of 74 mi/h or greater. (p. 541)

Hurricane warning   an alert that hurricane conditions are expected within 24 hours. (p. 543)

Hurricane watch   an advisory alert that hurricane conditions are a definite possibility. (p. 543)

Hydrocarbons   organic compounds that contain only carbon and hydrogen. (p. 351)

Hydrogen bond   the dipole-dipole force between a hydrogen atom in one molecule and a nearby oxygen, nitrogen, or fluorine atom in the same or a neighboring molecule. (p. 315)

Hydrologic cycle   the cyclic movement of Earth's water supply from the oceans to the mountains and back again to the oceans. (p. 624)

Hydroxyl group a substituent with the formula —OH. When attached to an alkyl group, an alcohol (R—OH) is formed. (p. 361)

Hypothesis a tentative explanation of some regularity in nature, or a tentative answer of any kind. (p. 3)

Ice storm a storm with accumulations of ice as a result of the surface temperature being below the freezing point. (p. 537)

Ideal gas law relates the pressure, volume, and absolute temperature of a gas; $p_1 V_1 / T_1 = p_2 V_2 / T_2$. (p. 108)

Igneous rock rock formed by the cooling and solidification of hot, molten material. (p. 564)

Index fossil a fossil that is related to a specific span of geologic time. (p. 638)

Index of refraction the ratio of the speed of light in a vacuum to the speed of light in a medium. (p. 142)

Inertia the natural tendency of an object to remain in a state of rest or in uniform motion in a straight line. (p. 45)

Inner core the innermost region of Earth, which is solid and probably composed of about 85% iron and 15% nickel. (p. 604)

Inner transition elements the *lanthanides* and *actinides*, the two rows at the bottom of the periodic table, make up the inner transition elements. (p. 276)

Insolation the solar radiation received by Earth and its atmosphere; **in**coming **sol**ar radi**ation**. (p. 505)

Intensity (of sound wave) the rate of energy transfer through a given area, with units of watts per square meter (W/m^2). (p. 128)

Interference, constructive a superposition of waves for which the combined waveform has a greater amplitude. (p. 155)

Interference, destructive a superposition of waves for which the combined waveform has a smaller amplitude. (p. 155)

International Date Line the meridian that is 180° E or W of the prime meridian. (p. 429)

Interplanetary dust very small solid particles known as *micrometeoroids* that exist in the space between the planets. (p. 411)

Ion an atom, or chemical combination of atoms, having a net electric charge. (p. 282)

Ionic bonds electrical forces that hold the ions together in the crystal lattice of an ionic compound. (p. 302)

Ionic compounds compounds formed by an electron transfer process in which one or more atoms lose electrons and one or more other atoms gain them to form ions. (p. 299)

Ionization energy the amount of energy it takes to remove an electron from an atom. (p. 280)

Ionosphere the region of the atmosphere between about 70 km (43 mi) and several hundred kilometers in altitude. It is characterized by a high concentration of ions. (p. 504)

Irregular (diffuse) reflection reflection from a rough surface in which the reflected rays are not parallel. (p. 140)

Isobar a line on a weather map denoting locations with the same atmospheric presure. (p. 518)

Isostasy the concept that Earth's crustal material "floats" in gravitational equilibrium on a "fluid" substratum. (p. 593)

Isotopes forms of atoms of an element that have the same numbers of protons but differ in their numbers of neutrons. (p. 236)

Jet streams fast moving "rivers" of air in the upper troposphere. (p. 521)

Joule a unit of energy equivalent to 1 N-m or 1 kg-m^2/s^2. (p. 70)

Jovian planets the four outer planets — Saturn, Jupiter, Uranus, and Neptune. All have characteristics resembling Jupiter. (p. 385)

Kelvin the unit of temperature on the Kelvin (absolute) temperature scale. A kelvin is equal in magnitude to a degree Celsius. (p. 95)

Kelvin scale the "absolute" temperature scale that takes absolute zero as 0 K. (p. 95)

Kilo metric prefix that means 10^3, or one thousand. (p. 8)

Kilocalorie the amount of heat necessary to raise 1 kg of water 1 C°. (p. 96)

Kilogram the unit of mass in the mks system; one kilogram has an equivalent weight of 2.2 pounds. (p. 4)

Kilowatt-hour (kWh) a unit of energy (power × time); $P = E/t$, and $E = Pt$. (p. 79)

Kinetic energy energy of motion equal to $1/2\ mv^2$. (p. 72)

Kinetic theory a gas consists of molecules moving independently in all directions at high speeds (the higher the temperature, the higher is the average speed), colliding with each other and the walls of the container, and having a distance between molecules that is large, on average, compared with the size of the molecules themselves. (p. 107)

K-T Event the extinction episode that marks the transition from the Cretaceous period (K) to the Tertiary period (T) . (p. 650)

Land breeze a local wind from land to sea resulting from a convection cycle. (p. 519)

Landslide   a type of fast mass wasting that involves the downslope movement of large blocks of weathered material. (p. 622)

Lapse rate   the rate of temperature decrease with increasing altitude. In the troposphere the normal lapse rate is -6.5 C°/km or -3.5 F°/1000 ft. (p. 522)

Laser   an acronym for **l**ight **a**mplification by **s**timulated **e**mission of **r**adiation; it is coherent, monochromatic light. (p. 219)

Last-quarter phase   occurs at the instant the Moon is 270° east of the Sun. (p. 451)

Latent heat of fusion   the amount of heat required to change one kilogram of a substance from the solid to the liquid phase at the melting point temperature. (p. 100)

Latent heat of vaporization   the amount of heat required to change one kilogram of a substance from the liquid to the gas phase at the boiling point temperature. (p. 100)

Latitude   the angular measurement in degrees north or south of the equator for a point on the surface of Earth. (p. 423)

Lava   magma that reaches Earth's surface through a volcanic vent. (p. 564)

Law   a concise statement, in words or a mathematical equation, about a fundamental relationship or regularity of nature. (p. 3)

Length   the measurement of space in any direction. (p. 4)

Lewis structures   "electron-dot" symbols used to show valence electrons in molecules and ions of compounds. (p. 300)

Lewis symbol   the element's symbol represents the nucleus and inner electrons of an atom, and the valence electrons are shown as dots arranged around the symbol. (p. 299)

Lightning   an electric discharge in the atmosphere. (p. 536)

Light-year   the distance light travels in one year, which is 6 trillion miles. (p. 473)

Limiting reactant   a starting material that is used up completely in a chemical reaction. (p. 297)

Line absorption spectrum   a set of dark spectral lines of certain frequencies or wavelengths formed by dispersion of light that has come from an incandescent source and has then passed through a sample of cool gas. (p. 212)

Line emission spectrum   a set of bright spectral lines of certain frequencies or wavelengths formed by dispersion of light from a gas-discharge tube. Each element gives a different set of lines. (p. 212)

Linear momentum   the product of an object's mass and its velocity. (p. 59)

Linearly polarized light   the condition of transverse light waves that vibrate in only one plane. (p. 151)

Liquid   matter that has a definite volume but no definite shape. (p. 106)

Liter (L)   a metric unit of volume or capacity; 1 L = 1000 cm^3. (p. 9)

Lithification   the process of forming sedimentary rock from sediment; also called *consolidation*. (p. 576)

Lithosphere   the outermost solid portion of Earth, which includes the crust and part of the upper mantle. (p. 566)

Local Group   The cluster of galaxies that includes our own Milky Way. (p. 487)

Longitude   the angular measurement in degrees east or west of the prime meridian for a point on the surface of Earth. (p. 423)

Longitudinal wave   a wave in which the particle motion and the wave velocity are parallel to each other. (p. 123)

Long-shore current   a current along a shore due to waves that break at an angle to the shoreline. (p. 627)

Lunar eclipse   an eclipse of the Moon caused by Earth blocking the Sun's rays to the Moon. (p. 456)

Luster   the appearance of a mineral's surface in reflected light. (p. 561)

Magma   hot, molten, underground rock material. (p. 564)

Magnetic anomalies   adjacent regions of rocks with remanent magnetism of opposite polarities. That is, the directions of the magnetism are reversed. (p. 591)

Magnetic declination   the angular variation of a compass from geographic north. (p. 191)

Magnetic field   a set of imaginary lines that indicate the direction a small compass needle would point if it were placed at a particular spot. (p. 187)

Magnification factor   the ratio of the image distance and the object distance of a spherical mirror or lens that gives the magnification of an object; $M = -D_i / D_o$. (p. 161)

Magnitude (apparent)   a measure of the brightness of a star as observed from Earth. (p. 474)

Main sequence   a narrow band on the H-R diagram on which most stars fall. (p. 476)

Mantle   the interior region of Earth between the core and the crust. (p. 604)

Mass   a quantity of matter and a measure of the amount of inertia that an object possesses. (pp. 4, 45)

Mass defect   any decrease in mass during a nuclear reaction. (p. 257)

Mass number   the number of protons plus neutrons in a nucleus; the total number of nucleons. (p. 236)

Mass wasting   the downslope movement of overburden under the influence of gravity. (p. 622)

Matter (de Broglie) waves   the waves produced by moving particles. (p. 223)

Meander   the looping, ribbonlike path of a river channel that results from accumulated deposits of eroded material having diverted the stream flow. (p. 617)

Mercalli scale   a scale of earthquake severity based on the physical effects produced by an earthquake. (p. 598)

Meridians   imaginary lines along the surface of Earth running from the geographic North Pole, perpendicular to the equator, to the geographic South Pole. (p. 423)

Mesosphere   the region of Earth's atmosphere that lies between approximately 50 and 80 km (30 and 50 mi) in altitude. (p. 503)

Metal   an element whose atoms tend to lose valence electrons during chemical reactions. (p. 277)

Metamorphic rock   rock that results from a change in preexisting rock due to high pressure or temperature or both. (p. 564)

Meteor   a metallic or stony object that burns up as it passes through Earth's atmosphere and appears to be a "shooting star." (p. 408)

Meteorite   a metallic or stony object from the solar system that strikes Earth's surface. (p. 408)

Meteoroids   small, interplanetary objects in space before they encounter Earth. (p. 408)

Meteorology   the study of atmospheric phenomena. (p. 500)

Meter   the standard unit of length in the mks system. It is equal to 39.37 inches or 3.28 feet. (p. 4)

Metric prefixes   a series of prefixes, such as *centi-* and *milli-*, used in the metric system to express multiples of 10. See Appendix I for a complete listing of metric prefixes. (p. 8)

Metric system   the decimal (base-10) system of units used predominantly throughout the world. (p. 4)

Micro   prefix that means 10^{-6}, or one one-millionth. (p. 8)

Midocean ridge   a series of mountain ranges on the ocean floor, more than 84,000 km in length, extending through the North and South Atlantic, the Indian Ocean, and the South Pacific. (p. 590)

Milky Way   the name of our galaxy. (p. 486)

Milli   the metric prefix that means 10^{-3}, or one one-thousandth. (p. 8)

Mineral   any naturally occurring, inorganic, crystalline substance (element or compound) that possesses a fairly definite chemical composition and a distinctive set of physical properties. (p. 558)

Mixture   a type of matter composed of varying proportions of two or more substances that are just physically mixed, *not* chemically combined. (p. 266)

Mks system   the metric system that has the meter, kilogram, second, and coulomb as the standard units of length, mass, time, and electric charge, respectively. (p. 7)

Moho (Mohorovicic discontinuity)   the boundary between Earth's crust and mantle. (p. 604)

Mole   the quantity of a substance that contains 6.02×10^{23} formula units (the number of atoms in exactly 12 g of carbon-12). (p. 344)

Molecule   an electrically neutral particle composed of two or more atoms chemically combined. (p. 272)

Monomer   a fundamental repeating unit of a polymer. (p. 369)

Moraine   a ridge of glacial till. (p. 620)

Motion   the changing of position. (p. 25)

Motor   a device that converts electrical energy into mechanical energy. (p. 194)

Mountain range   a geologic unit or series of mountains. (p. 446)

Neap tide   moderate tides with the least variation between high and low. (p. 458)

Neutrino   a subatomic particle that has no electric charge but does possess a very small rest mass, energy, and momentum. (p. 469)

Neutron number N, the number of neutrons in the nucleus of an atom. (p. 236)

Neutron star   an extremely-high-density star composed almost entirely of neutrons. (p. 480)

Neutrons   neutral particles found in the nuclei of atoms. (p. 233)

New moon   the phase of the Moon that occurs when the Moon is on the same meridian as the Sun. (p. 450)

Newton   the unit of force in the mks system; 1 kg-m/s^2. (p. 47)

Newton's first law of motion   an object will move at a constant velocity unless acted upon by an external unbalanced force. (p. 44)

Newton's law of universal gravitation   the gravitational force between two masses, m_1 and m_2, is directly proportional to the products of the masses and inversely proportional to the square of the distance r between their centers of mass; $F = Gm_1m_2/r^2$. (p. 53)

Newton's second law of motion   the acceleration of an object is equal to the net force on the object divided by the mass of the object; $a = F/m$. (p. 47)

Newton's third law of motion   whenever one mass exerts a force upon a second mass, the second mass exerts an equal and opposite force upon the first mass. (p. 51)

Nitrogen oxides (NO$_x$)   chemical combinations of nitrogen and oxygen, such as NO and NO_2. (p. 547)

Noble gases   the elements of Group 8A of the periodic table. (p. 283)

Nonmetal   an element whose atoms tend to gain (or share) valence electrons during chemical reactions. (p. 277)

Normal fault   a nonvertical fault in which the overlying side (hanging wall) moves downward relative to the side beneath it (footwall). (p. 606)

Nova   a white dwarf star that suddenly increases dramatically in brightness for a brief period of time. (p. 477)

Nucleons   a collective term for neutrons and protons (particles in the nucleus). (p. 233)

Nucleosynthesis   the creation of the nuclei of elements inside stars. (p. 480)

Nucleus   the central core of an atom; composed of protons and neutrons. (p. 233)

Nuclide   a specific type of nucleus, characterized by specifying the atomic number and mass mumber. (p. 239)

Octet rule   In forming compounds, atoms tend to gain, lose, or share electrons to achieve electron configurations of the noble gases. (p. 299)

Ohm   the unit of resistance; equal to one volt per ampere. (p. 178)

Ohm's law   the voltage across two points is equal to the current flowing between the points times the resistance between the points; $V = IR$. (p. 179)

Oort cloud   the cloud of cometary objects believed to be orbiting the Sun at 50,000 AU and from which comets originate. (p. 410)

Opposition   the time at which a planet is on the opposite side of Earth from the Sun. (p. 388)

Organic chemistry   the study of compounds that contain carbon. (p. 349)

Outer core   the innermost region of Earth, which is composed of two parts: a solid inner core and a molten, highly viscous, "liquid" outer core. (p. 604)

Oxidation   occurs when oxygen combines with another substance (or when an atom or ion loses electrons). (p. 339)

Ozone O_3, a form of oxygen found naturally in the atmosphere in the ozonosphere. It is also a constituent of photochemical smog. (pp. 503, 547)

Ozonosphere   a region of the atmosphere, between approximately 15 and 50 km (9 and 31 mi) in altitude, characterized by ozone concentration. (p. 503)

P waves   primary (P) waves, so called because they reach a seismic station before the secondary (S) waves. P waves are longitudinal or compressional waves; that is, their particle oscillations are in the direction of propagation and are transmitted by solids, liquids, and gases. (p. 602)

Paleontology   the systematic study of fossils and prehistoric life forms. (p. 634)

Pangaea   the giant supercontinent that is believed to have existed over 200 million years ago. (p. 588)

Parallax   the apparent motion, or shift, that occurs between two fixed objects when the observer changes position. (p. 390)

Parallel circuit   a circuit in which an entering current divides proportionally among the circuit elements. (p. 182)

Parallels   imaginary lines encircling Earth parallel to the plane of the equator and representing degrees of latitude. (p. 423)

Parsec   the distance to a star when the star exhibits a parallax of one second of arc. This distance is equal to 3.26 light-years or 206,265 astronomical units. (p. 391)

Penumbra   a region of partial shadow. During an eclipse, an observer in the penumbra sees only a partial eclipse. (p. 455)

Period   In physics, the time for a complete cycle of motion. In chemistry, one of the seven horizontal rows of the periodic table. In geology, an interval of geologic time that is a subdivision of an era and is made up of epochs. (pp. 124, 275, 639)

Periodic law   the properties of elements are periodic functions of their atomic numbers. (p. 275)

Permafrost   ground that is permanently frozen. (p. 614)

pH   a measure (on a logarithmic scale) of the hydrogen ion (or hydronium ion) concentration in a solution. (p. 334)

Phases of matter   the physical forms of matter; most commonly, solid, liquid, and gas. (p. 105)

Phosphorescence   a glow of light that persists after the removal of the source of photons needed for excitation of the material's electrons. (p. 220)

Photochemical smog   air-pollution conditions resulting from the photochemical reactions of hydrocarbons with other pollutants and atmospheric oxygen in the presence of sunlight. (p. 547)

Photoelectric effect   the emission of electrons that occurs when certain metals are exposed to light. (p. 208)

Photon a "particle" of electromagnetic energy. (p. 208)

Photosphere the Sun's outer surface, visible to the eye. (p. 467)

Photosynthesis the process by which plants convert CO_2 and H_2O to sugars and oxygen. (p. 501)

Physical weathering the physical disintegration or fracture of rock, primarily as a result of pressure. (p. 614)

Plains (lunar) large, dark, flat areas on the Moon believed to be craters formed by meteorite impact that then filled with volcanic lava. (p. 446)

Plasma a high-temperature gas of electrons and protons or other nuclei. (pp. 106, 254, 467)

Plate tectonics the theory that Earth's lithosphere is made up of rigid plates that are in relative motion with respect to each other. (p. 566)

Pluton a large body of intrusive igneous rock. (p. 569)

Polar covalent bond one in which the pair of bonding electrons is unequally shared, leading to the bond having a slightly positive end and a slightly negative end. (p. 310)

Polar molecule one that has a positive end and a negative end, that is, one that has a dipole. (p. 312)

Polarization the restriction of the electric vector of a light wave to one plane. (p. 151)

Poles, law of (magnetic) like poles repel, unlike poles attract. (p. 187)

Pollution any atypical contributions to the environment resulting from the activities of humans. (p. 545)

Polyatomic ion an electrically charged combination of atoms. Table 11.6 lists the common ones. (p. 282)

Polymer a compound of very high molecular mass whose chainlike molecules are made up of repeating units called *monomers*. (p. 369)

Position the location of an object with respect to another object. (p. 25)

Post meridiem pertaining to time after 12 noon. (p. 427)

Potential energy the energy a body possesses because of its position in a force field. (p. 73)

Power work or energy per unit time. (p. 78)

Powers-of-ten notation notation in which numbers are expressed by a coefficient and a power of ten; for example, $2500 = 2.5 \times 10^3$. Also called *scientific notation*. (p. 16)

Precession the slow rotation of the axis of spin of Earth around an axis perpendicular to the ecliptic plane. The rotation is clockwise as observed from the North Celestial Pole. (p. 435)

Precipitate an insoluble solid that appears when two clear liquids (usually aqueous solutions) are mixed. (p. 338).

Pressure the force per unit area; $p = F/A$. (p. 107)

Primordial nebula a large, swirling volume of interstellar cold gas and dust that contracted under the influence of its own gravity and formed in the shape of a flattened rotating disk. (p. 413)

Principal quantum number the numbers $n = 1, 2, 3, \ldots$ used to designate the various principal energy levels that an electron may occupy in an atom. (p. 212)

Products the substances formed during a chemical reaction. (p. 322)

Prograde motion orbital or rotational motion in the forward direction. In the solar system, this is west-to-east, or counterclockwise, as viewed from above Earth's North Pole. (p. 386)

Prominence a flamelike loop or sheet of glowing gas erupting from an active region of the solar chromosphere into the corona. (p. 468)

Proteins extremely long-chain polyamides formed by the enzyme-catalyzed condensation of amino acids. (p. 368)

Proton-proton chain a series of stellar nuclear reactions in which four hydrogen nuclei (protons) combine to form one helium nucleus and release energy. (p. 470)

Protons positively charged particles in the nuclei of atoms. (pp. 173, 233)

Psychrometer an instrument used to measure relative humidity. (p. 514)

Pulsar a rapidly rotating neutron star. (p. 480)

Pure substance a type of matter (element or compound) in which all samples have fixed composition and identical properties. (p. 266)

Quantum a discrete amount. (p. 208)

Quantum mechanics the branch of physics that replaced the classical-mechanical view that everything moved according to exact laws of nature with the concept of probability. Schrödinger's equation forms the basis of quantum wave mechanics. (p. 225)

Quasar a shortened term for quasi-stellar radio source. (p. 489)

Radar an instrument that sends out electromagnetic (radio) waves, monitors the returning waves that are reflected by some object, and thereby locates the object. Radar stands for radio detecting and ranging. Radar is used to detect and monitor precipitation and severe storms. (p. 516)

Radiation the transfer of energy by means of electromagnetic waves. (p. 105)

Radioactivity the spontaneous process of a sample of a radionuclide undergoing a change by the emission of particles or rays. (p. 239)

Radiometric dating a general name for dating rocks and organic remains by measurements utilizing the rate of decay of radionuclides they contain. (p. 640)

Radionuclides types of nuclei that undergo radioactive decay. (p. 239)

Rain gauge an open, calibrated container used to measure amounts of precipitation. (p. 516)

Ray a straight line that represents the path of light. (p. 140)

Rayleigh scattering the preferential scattering of light by air molecules and particles that accounts for the blueness of the sky. The scattering is proportional to $1/\lambda^4$. (p. 506)

Rays (lunar) streaks of light-colored material extending outward from craters on the Moon. (p. 446)

Reactants the original substances in a chemical reaction. (p. 322)

Real image an image from a mirror or lens that can be brought to focus on a screen. (p. 159)

Red dwarfs small, cool, red, type M stars, whose color and size give them their name. (p. 477)

Red giant a relatively cool, very bright star that has a diameter much larger than average. (p. 476)

Redshift a Doppler effect caused when a light source, such as a galaxy, moves away from the observer and shifts the light frequencies lower or toward the red end of the electromagnetic spectrum. (p. 134)

Reduction occurs when oxygen is removed from a compound (or when an atom or ion gains electrons). (p. 339)

Reflection the change in the direction of a wave when it strikes and rebounds from a surface or the boundary between two media. (p. 140)

Reflection, law of the angle of incidence equals the angle of reflection, $\theta_i = \theta_r$, as measured relative to the normal, a line perpendicular to the reflecting surface. (p. 140)

Refraction the bending of light waves caused by a speed change as light goes from one medium to another. (p. 141)

Regional metamorphism a change in rock over a large area and brought about by both heat and pressure. (p. 588)

Regular (specular) reflection reflection from a smooth surface in which the reflected rays are parallel. (p. 140)

Relative geologic time a time scale obtained when rocks and the geologic events they record are placed in chronologic order without regard to actual dates. (p. 536)

Relative humidity the ratio of the actual moisture content of a volume of air to its maximum moisture capacity at a given temperature. (p. 513)

Remanent magnetism the magnetism retained in rocks containing ferrite minerals after solidifying in Earth's magnetic field. (p. 590)

Representative elements the A group elements in the periodic table. (p. 275)

Resistance (electrical) the opposition to the flow of electrical charge. (p. 178)

Resonance a wave effect that occurs when an object has a natural frequency that corresponds to an external frequency. (p. 134)

Retrograde motion orbital or rotational motion in the backward direction. In the solar system, this is east-to-west, or clockwise, as viewed from above Earth's North Pole. (p. 386)

Reverse fault a nonvertical fault in which the overlying side (hanging wall) moves upward relative to the side beneath it (footwall). (p. 606)

Revolution the movement of one mass around another. (p. 389)

Richter scale a scale of earthquake severity based on the amplitude or intensity of seismic waves. (p. 598)

Right ascension a coordinate for measuring the east-west positions of celestial objects. The angle is measured eastward from the vernal equinox in hours, minutes, and seconds. (p. 473)

Rill a narrow trench or valley on the Moon. (p. 446)

Roche limit the minimum distance from a planet or other object at which a second object can be held together by gravitational forces. (p. 401)

Rock a solid, cohesive natural aggregate of one or more minerals. (p. 558)

Rock cycle the cyclic changes of rock, during which the rock is created, destroyed, and metamorphosed by Earth's internal and external geologic processes. (p. 564)

Rotation the turning of a mass about an axis passing through the mass. (p. 389)

S waves secondary (S) seismic waves, so called because they reach a seismic station after the primary (P) waves. S waves are transmitted only by solids and are transverse or shear waves; that is, their particle oscillations are at right angles to the direction of propagation. (p. 602)

Salt an ionic compound that contains any cation except H^+ combined with any anion except OH^-. (p. 335)

Saturated solution one having the maximum amount of solute dissolved in the solvent at a given temperature. (p. 267)

Scalar a quantity that has a magnitude but has no direction associated with it. (p. 25)

Schwarzschild radius   the distance from the singularity to the event horizon in a nonrotating black hole. (p. 482)

Scientific method   an investigative process which holds that no concept or model of nature is valid unless the predictions are in accord with experiment. That is, all hypotheses should be based on as much relevant data as possible and then should be tested and verified. (p. 3)

Seabreeze   a local wind from the sea to land as a result of a convection cycle. (p. 518)

Seafloor spreading   the theory that the seafloor slowly spreads and moves away from midocean ridges. The spreading is believed to be due to convection cycles of subterranean molten material that cause the formation of the ridges and a surface motion in a lateral direction from the ridges. (p. 590)

Seamount   an isolated submarine volcanic structure. (p. 630)

Second   the standard unit of time. It is now defined in terms of the frequency of a certain transition in the cesium atom. (p. 7)

Sedimentary rock   rock formed from the lithification, or consolidation, of layers of sediment. (p. 564)

Sediments   mineral or organic matter deposited by water, air, or ice. (p. 576)

Seismic waves   the waves generated by the energy release of an earthquake. (p. 598)

Series circuit   a circuit in which an entering current flows individually through all the circuit elements. (p. 181)

Shear metamorphism   a change in rock brought about primarily by pressure rather than heat. (p. 582)

Shield volcano   a volcano with a low, gently sloping profile formed by a fissure eruption of low-viscosity lava. (p. 572)

SI (International System of Units)   a modernized version of the metric system that contains seven base units. (p. 7)

Sidereal day   the rotation period of Earth with respect to the vernal equinox. One sidereal day is 23 h, 56 min, 4.091 s. (p. 426)

Sidereal month   the orbital period of the Moon around Earth with respect to a star other than the Sun. (p. 449)

Sidereal period   the orbital or rotational period of any object with respect to the stars. (p. 386)

Sidereal year   the time interval for Earth to make one complete revolution around the Sun with respect to any particular star other than the Sun. (p. 432)

Significant figures   a method of estimating or expressing error in mathematical operations and measurements. (p. 14)

Silicate   any one of numerous minerals that have the oxygen and silicon tetrahedron as their basic structure. (p. 558)

Single-replacement reactions   reactions in which one element replaces another that is in a compound: $A + BC \rightarrow B + AC$. (p. 340)

Singularity   the center of a black hole. The point to which the entire mass of a star has contracted. (p. 482)

Sinkholes   a depression on the land surface where soluble rock (limestone) has been removed by groundwater. (p. 615)

Slump   a type of landslide involving the downslope movement of an unbroken block of overburden that leaves a curved depression on the slope. (p. 622)

Smog   a contraction of smoke-fog used to describe the combination of these conditions. (p. 545)

Soap   sodium salts of fatty acids that are formed, along with glycerol, when fats are treated with a base such as sodium hydroxide. (p. 367)

Solar eclipse   an eclipse of the Sun caused by the Moon blocking the Sun's rays to an observer on Earth. (p. 455)

Solar system   the Sun, nine planets and their satellites, the asteroids, comets, meteoroids, and interplanetary dust. (p. 381)

Solar wind   an outward flow of charged particles, mainly electrons and protons, from the Sun. (p. 468)

Solid   matter that has a definite volume and a definite shape. (p. 105)

Solubility   the amount of solute that will dissolve in a specified volume or mass of solvent (at a given temperature) to produce a saturated solution. (p. 268)

Solution   a mixture that is uniform throughout. Also called a *homogeneous mixture*. (p. 267)

Somatic effects   short-term and long-term effects on the health of a recipient of radiation. (p. 258)

Sound   a wave phenomenon caused by variations in pressure in a medium such as air. (p. 127)

Sound spectrum   an ordered arrangement of various frequencies or wavelengths of sound. The three main regions of the sound spectrum are the infrasonic, the audible, and the ultrasonic. (p. 128)

Source region   the region or surface from which an air mass derives its physical characteristics. (p. 533)

Specific gravity   the ratio of a substance's weight to the weight of an equal volume of water. (p. 562)

Specific heat   the amount of heat energy in kilocalories necessary to raise the temperature of one kilogram of the substance one degree Celsius. (p. 97)

Speed of light   how fast light travels. In air or a vacuum, $c = 3.00 \times 10^8$ m/s or 186,000 mi/s. (p. 126)

Speed of sound   how fast sound travels in a medium; for example, $v_s = 344$ m/s in air at room temperature. (p. 131)

Speed, average   the distance traveled divided by the time to travel that distance. (p. 25)

Speed, instantaneous   how fast an object is traveling at a particular moment or instant. (p. 26)

Spherical mirror equation   a mathematical expression relating the object distance, image distance, and focal length of a spherical mirror; $1/D_o + 1/D_i = 1/f$. (p. 161)

Spring tides   the tides of greatest variation between high and low. (p. 458)

Standard time zones   the division of the surface of Earth into 24 time zones, each containing about 15° of longitude. (p. 427)

Standard unit   a fixed and reproducible reference value used for the purpose of taking accurate measurements. (p. 4).

Standing wave   A "stationary" waveform arising from the interference of waves traveling in opposite directions. (p. 134)

Stock system   a system of nomenclature for compounds of metals that form more than one ion. A Roman numeral placed in parentheses directly after the name of the metal denotes its ionic charge in the compound being named. (p. 305)

Storm surge   the great dome of water associated with a hurricane when it makes landfall. (p. 542)

Strata   layers of rock or sediment lying on top of one another. (p. 576)

Stratosphere   the region of Earth's atmosphere approximately 16 to 50 km (10 to 30 mi) in altitude. (p. 503)

Stratovolcano   a volcano with a steeply sloping symmetric cone formed by eruption of high-viscosity lava and tephra. (p. 572)

Streak   the color of a powdered mineral on a streak plate (unglazed porcelain). (p. 561)

Stream   any flow of water occurring between well-defined banks. (p. 616)

Strike-slip faulting   movement along a horizontal transform fault in which the two sides of the fault strike and slip by each other. (p. 606)

Strong nuclear force   the short-range force of attraction that acts between two nucleons and holds the nucleus together. (p. 238)

Structural formula   a graphic representation of the way the atoms are connected to one another in a molecule. (p. 353)

Subduction   the process in which one plate is deflected downward beneath another plate into the asthenosphere. (p. 567)

Subduction zone   the place where a lithospheric plate descends beneath another and into the asthenosphere. (p. 594)

Sulfur dioxide (SO_2)   an atmospheric pollutant formed by the oxidation of sulfur; it contributes to acid rain. (p. 548)

Summer solstice   the farthest point of the Sun's declination north of the equator (for the Northern Hemisphere . (p. 423)

Sun   a star; a self-luminous sphere of gas held together by its own gravity and energized by nuclear fusion reactions in its interior. (p. 467)

Sunspots   patches of cooler, darker material on the surface of the Sun. (p. 468)

Supercluster   a large volume of space (diameter as large as 300 million ly) where matter is concentrated into galaxies; a cluster of clusters of galaxies. (p. 487)

Supernova   an exploding star. (p. 478)

Superposition, principle of   the principle that in a succession of stratified deposits the younger layers lie over the older layers. (p. 636)

Supersaturated solution   a solution that contains more than the normal maximum amount of dissolved solute at a given temperature and hence is unstable. (p. 268)

Surface wave (seismic)   a seismic wave that travels along Earth's surface or a boundary within it. (p. 602)

Syncline   a rock fold that forms a trough; the opposite of an anticline. (p. 605)

Synodic month   the orbital period of the Moon with respect to the Sun; the month of the Moon's phases. (p. 449)

Synodic period   the orbital or rotational period of an object as seen by an observer on Earth. (p. 386)

Synthetic detergents   soap substitutes. Their molecules contain a long hydrocarbon chain that is nonpolar (e.g., $C_{12}H_{25}-$) and a polar group such as sodium sulfate ($-OSO_3^- Na^+$). (p. 368)

Synthetics   materials whose molecules have no duplicates in nature. (p. 369)

System of units   a group of standard units and their combinations. The two major systems of units in use today are the metric system and the British system. (p. 4)

Temperature   a measure of the average kinetic energy of the molecules in a sample. (p. 93)

Temperature inversion   a condition characterized by an inverted lapse rate. (p. 545)

Tephra   solid material emitted by volcanoes; it ranges in size from fine dust to large boulders. (p. 570)

Terminator   the boundary line dividing day and night on the surface of a planet or moon. (p. 402)

Terrestrial planets   the inner four planets — Mercury, Venus, Earth, and Mars. All are similar to Earth in general chemical and physical properties. (p. 385)

Theory   a tested explanation of a broad segment of basic natural phenomena. (p. 3)

Thermal efficiency   the ratio of the work output and the heat input of a heat engine (usually expressed as a percent). (p. 111)

Thermodynamics, first law of   the heat energy added to a system must go into increasing the internal energy of the system, or any work done by the system, or both. The law, which is based on the conservation of energy, also states that heat energy removed from a system must produce a decrease in the internal energy of the system, or any work done on the system, or both. (p. 111)

Thermodynamics, second law of   it is impossible for heat to flow spontaneously from an object having a lower temperature to an object having a higher temperature. (p. 112)

Thermodynamics, third law of   a temperature of absolute zero can never be attained. (p. 113)

Thermosphere   the region of Earth's atmosphere extending from about 80 km (50 mi) in altitude to the outer reaches of the atmosphere. (p. 503)

Thin-lens equation   a mathematical expression relating the object distance, image distance, and focal length for thin, spherical lenses; $1/D_o + 1/D_i = 1/f$. (p. 165)

Thrust faulting   a special case of reverse faulting in which the fault plane is at a small angle to the horizontal. (p. 606)

Thunder   the sound associated with lightning that arises from the explosive release of electrical energy. (p. 537)

Tidal force   a differential gravitational force that occurs when two separate bodies interact gravitationally. The variation of the gravitational forces across the bodies tends to stretch or change the shape of each body. (p. 458)

Tides   the periodic rise and fall of the water level along the shores of large bodies of water. (p. 627)

Time   the continuous forward-flowing of events. (p. 7)

Tornado   a violent storm characterized by a funnel-shaped cloud and high winds. (p. 539)

Tornado warning   the alert issued when a tornado has actually been sighted, or indicated on radar. (p. 540)

Tornado watch   the alert issued when conditions are favorable for tornado formation. (p. 540)

Torque   a force about an axis. (p. 61)

Total internal reflection   a phenomenon in which light is totally reflected because refraction is impossible. (p. 145)

Transform boundary   a region of the lithosphere where a moving plate slides along one side of another without creating or destroying lithosphere. (p. 567)

Transform fault   a fault with a horizontal fault plane along which strike-slip faulting occurs. (p. 606)

Transformer   a device that increases or decreases the voltage or alternating current. (p. 196)

Transition elements   the B group elements in the periodic table. (p. 276)

Transverse wave   a wave in which the vibrations are perpendicular to the wave velocity. (p. 123)

Tropical year   the time interval from one vernal equinox to the next. (p. 432)

Troposphere   the region of Earth's atmosphere from the ground up to about 16 km (10 mi). (p. 502)

Tsunami   a Japanese word for a seismic sea wave — an unusually large sea wave produced by a seaquake or undersea volcanic eruption. (p. 602)

Ultrasound   sound with frequency above 20 kHz. (p. 130)

Umbra   a region of total darkness in a shadow. During an eclipse, an observer in the umbra sees a total eclipse. (p. 455)

Unconformity   a break in the geologic rock record. (p. 636)

Uniformitarianism   the principle that the same processes operate today on and within Earth as in the past. Hence, the present is considered the key to the past. (p. 564)

Universe   everything that is — all energy, matter, and space. (p. 380)

Unsaturated solution   one in which more solute can be dissolved at the same temperature. (p. 267)

Valence electrons   the electrons that are involved in bond formation, usually those in an atom's outer shell. (p. 278)

Valence shell   an atom's outer shell, which contains the valence electrons. (p. 278)

Vector   a quantity that has both magnitude and direction. (p. 25)

Velocity, average   the change in displacement divided by the change in time; $v = \Delta d/\Delta t$. (p. 26)

Velocity, instantaneous   the velocity at a particular instant of time. (p. 26)

Vernal equinox   the point where the Sun crosses the celestial equator from south to north, around March 21. (p. 433)

Virtual image   an image from a lens or mirror that cannot be brought to focus on a screen. (p. 159)

Viscosity   the internal property of a substance that offers resistance to flow. (p. 570)

Volcanic mountains   mountains that have been built by volcanic eruptions. (p. 606)

Volt   the unit of voltage equal to one joule per coulomb. (p. 178)

Voltage   the amount of work it would take to move an electric charge between two points, divided by the value of the charge, that is, work per unit charge. (p. 178)

Waning phase   the illuminated portion of the Moon is getting smaller as observed from Earth. (p. 450)

Water table   the boundary between the zone of aeration and the zone of saturation. (p. 625)

Watt   a unit of power equivalent to $1 \text{ kg-m}^2/\text{s}^3$, or 1 J/s. (p. 78)

Wave   the propagation of energy from a disturbance. (p. 121)

Wave speed   the distance a wave travels divided by the time of travel. (p. 124)

Wavelength   the distance from any point on a wave to an identical point on the adjacent wave. (p. 123)

Waxing phase   the illuminated portion of the Moon is getting larger as observed from Earth. (p. 450)

Weather   the atmospheric conditions of the lower troposphere. (p. 502)

Weathering   the physical disintegration and chemical decomposition of rock. (p. 614)

Weight   a measure of the force due to gravitational attraction ($w = mg$, on Earth's surface). (p. 48)

White dwarf   a white star that has a much smaller diameter and much higher density than average. It is believed to be the final stage of small- and average-mass stars. (p. 476)

Wind   the horizontal movement of air; air motion along Earth's surface. (p. 518)

Wind vane   a free-rotating device that, because of its shape, lines up with the wind and points the wind direction. (p. 515)

Winter solstice   the farthest point of the Sun's declination south of the equator (for the Northern Hemisphere). (p. 433)

Work   the product of a force and the parallel distance through which it acts. (p. 70)

X-Rays   high-frequency, high-energy electromagnetic radiation formed when high-speed electrons strike a metallic target. (p. 217)

Zenith   the position directly overhead for an observer on Earth. (p. 433)

Zenith angle   the angle between the zenith and the Sun at noon. (p. 433)

Photo Credits

Chapter 1: p. 1, Steve Krongard/The Image Bank p. 4, Pascal Quittemelle/Stock Boston p. 6, National Institute of Standards and Technology p. 7, Photo Researchers, Inc. p. 8, National Institute of Standards and Technology p. 9, left Richard Megna/Fundamental Photographs p. 9, middle Tom Pantages p. 9, right Aaron Haupt/Photo Researchers, Inc. p. 9 bot., Susan Van Etten/Photo Researchers, Inc. p. 11, Richard Megna/Fundamental Photographs p. 12, Dr. E. R. Degginger/Earth Scenes p. 13, left Ruth Dixon/Stock Boston p. 13, right Leonard Lessin/Peter Arnold, Inc. p. 15, Phil Degginger p. 22, Phil Degginger

Chapter 2: p. 24, Bob Daemmrich/Stock Boston p. 25, Steve Umland/The Image Works p. 33, left The Granger Collection p. 33 right, Olympia/Sipa Press p. 36 top, Farrell Grehan/Photo Researchers, Inc. p. 36 bottom, Richard Megna/Fundamental Photographs p. 38 left, Andrew D. Bernstein/Allsport p. 38 right, Focus on Sports

Chapter 3: p. 43, Henry Groskinsky/Peter Arnold, Inc. p. 45, Ken O'Donoghue p. 46 left, The Granger Collection p. 46 right, The Granger Collection p. 50, Tom Pantages p. 52 left, NASA/Johnson Space Center p. 52 right, Stephen Marks/Image Bank p. 54 both, National Institute for Highway Safety p. 57 top, Craig Melvin/SportsChrome p. 57 middle, Richard Megna/Fundamental Photographs p. 57 bot., Bob Daemmrich/Stock Boston p. 58, NASA/Johnson Space Center p. 59, NASA/Johnson Space Center p. 63 top, Manny Millan © Time, Inc./Sports Illustrated p. 63 top, Manny Millan © Time, Inc./Sports Illustrated p. 63 bottom, George Hall/Woodfin Camp & Associates p. 63 bottom, Bell Helicopter Textron, Inc.

Chapter 4: p. 69, Phil Degginger p. 74, Paul Sutton/Duomo p. 79, Bob Daemmrich/The Image Works p. 81, © Ex Rouchon/Photo Researchers, Inc. p. 83 top, Grant Heilman/Grant Heilman Photography p. 83 bottom, T. J. Florian/Rainbow p. 84, M. Attar/Sygma p. 86 left, Dr. E. R. Degginger p. 86 right, Mark C. Burnett/Stock Boston

Chapter 5: p. 92, Jeff Persons/Stock Boston p. 94 top, John Smith p. 94 bot., Jerry Wilson p. 97, Bob Daemmrich/The Image Works p. 99 left, James Carmichael/The Image Bank p. 99 right, John Banagen/The Image Bank p. 104, Dan McCoy/Rainbow p. 110 top, Tom Pantages p. 110 bot., Larry Lefever/Grant Heilman Photography p. 117, Martha Shethar

Chapter 6: p. 121, Aaron Chang/The Stock Market p. 128, John Smith p. 130, United Airlines p. 130, Steven Stone/The Picture Cube p. 130, bottom 3 photos, Henry Rachlin p. 131, Mehau Kulyk/Science Photo Library/Photo Researchers, Inc. p. 132, Bill Gallery/The Picture Cube p. 133, Jerry Wilson p. 135 top, Richard Megna/Fundamental Photographs p. 135 bot., The Bettmann Archive

Chapter 7: p. 139, Alan Mercer/Stock Boston p. 141, Coco McCoy/Rainbow p. 142, Fundamental Photographs p. 144, Phil Degginger p. 145 bot., Phillip Hayson/Photo Researchers, Inc. p. 145 top, Richard Megna/Fundamental Photographs p. 146 left, Jeff Persons/Stock Boston p. 146 right, Tom Raymond/Medichrome/Division of The Stock Shop p. 147, David Parker/Science Library/Photo Researchers, Inc. p. 147, David Parker/Science Library/Photo Researchers, Inc. p. 149, Dr. E. R. Degginger p. 151, Leonard Lessing/Peter Arnold, Inc. p. 152, Polaroid Corporation p. 153 top, Dr. E. R. Degginger p. 153 bot., Dr. E. R. Degginger p. 154, Ken Kay/Fundamental Photographs p. 155 left, Jerry Wilson p. 155 right, Jerry Wilson p. 156, Peter Aprahamian/Science Phot Library/Photo Researchers, Inc. p. 157, Educational Development Center p. 158 bot., Phil Degginger p. 158 top, Tom Pantages p. 159, Aldo Mastrocola/Lightwave p. 163, Aldo Mastrocola/Lightwave p. 164, Ken O'Donoghue p. 165, Ken O'Donoghue

Chapter 8: p. 172, Bryan Peterson/The Stock Market p. 177, Ken O'Donoghue p. 178, John Smith p. 180 left, Ken O'Donoghue p. 180 right, Ken O'Donoghue p. 184, Tom Lyle/The Stock Shop p. 186, Tom Pantages p. 187 (a&b), Ken O'Donoghue p. 188 left, Paul Silverman/Fundamental Photographs p. 188 right, Richard Megna/Fundamental Photographs p. 189 all, Richard Megna/Fundamental Photographs p. 194 both, Richard Megna/Fundamental Photographs p. 198 top, John Zoiner/Peter Arnold, Inc. p. 198 bot. left, Yoav Levy/Phototake p. 198 bot., E. R. Degginger p. 199 left, Ken O'Donoghue p. 199 right, Ferranti Electronics/A. Sternberg/Science Photo Library/Photo Researchers, Inc.

Chapter 9: p. 204, Dr. E. R. Degginger p. 206 bot., Stacy Pick/Stock Boston p. 206 top, Richard Megna/Fundamental Photographs p. 207, American Institute of Physics, Emilio Segre Visual Archives p. 208, Tom Stack & Associates p. 210, California Institute of Technology p. 218, The Burndy Library p. 219, The Mary Evans Picture Library p. 221 both, Dr. E. R. Degginger p. 222, Will & Deni McIntryre/Photo Researchers, Inc. p. 225 right, Courtesy IBM Almaden Research Center p. 225 right, David Scharf/Peter Arnold, Inc. p. 227 top left, The Mary Evans Picture Collection p. 227 bot. left, The Granger Collection p. 227 top right, The Bettmann Archive p. 227 middle right, The Bettmann Archive p. 227 bot. right, The Granger Collection

Chapter 10: p. 232, David York/Medichrome/Division of The Stock Shop p. 240, The Granger Collection p. 242 left, Brown Brothers p. 242 right, American Institute of Physics p. 247, Fermi National Laboratory Batavia, IL p. 253, Chicago Historical Society p. 255, U. S. Air Force p. 256, Alexander Tsiaras/Science Source/Photo Researchers, Inc. p. 260 right, The Bettmann Archive

Chapter 11: p. 265, Richard Megna/Fundamental Photographs p. 267, Sean Brady p. 269 bot., The Bettmann Archive p. 269 top photos, Sean Brady p. 270, T. Orban/Sygma p. 273, Paul Silverman/Fundamental Photographs p. 276, The Granger Collection p. 279, Ken O'Donoghue p. 284 bot., Hiroyuki Matsumoto/Tony Stone Images p. 284 top, Sean Brady p. 286, Sean Brady p. 287 left, Bob Daemmrich/The Image Works p. 287 right, Science Photo Library/Photo Researchers, Inc. p. 288, The Bettmann Archive

Chapter 12: p. 293, Bill Ross/West Light p. 296, David, Jacques-Louis, *Antoine Laurent Lavoisier (1743–1794) and His Wife, Marie-Anne Pierrette Paulze, (1758–1836).* The Metropolitan Museum of Art, purchase, Mr. and Mrs. Charles Wrightsman. Gift, in honor of Everett Fahy, 1977. (1977.10) p. 297, Sean Brady p. 300 (a), Richard Megna/Fundamental Photographs p. 300 (b–d), Ken O'Donoghue p. 304 both, Ken O'Donoghue p. 305, Sean Brady p. 308, David Glass/AP/Wide World p. 313, both Sean Brady p. 316, Chip Clark

Chapter 13: p. 321, Kaz Mori/The Image Bank p. 322, Sean Brady p. 325, both Sean Brady p. 327, Sean Brady p. 328, The Bettmann Archive p. 328 left, Ken O'Donoghue p. 328 right, Ken O'Donoghue p. 329, Wide World Photos p. 330 left, Richard Megna/Fundamental Photographs p. 330 top, Ken O'Donoghue p. 330 right, Richard Megna/Fundamental Photographs p. 331, Thomas Eisner and Daniel Aneshausley, Cornell University p. 332 left, Ken O'Donoghue p. 332 right, Ken O'Donoghue p. 333 bot., Runk/Schoenberger/Grant Heilman Photography p. 333 top, Greg Mancusco p. 334, Runk/Schoenberger/Grant Heilman Photography p. 335 left, Ken O'Donoghue p. 335 right, Richard Megna/Fundamental Photographs p. 336, Max A. Listgarden/Visuals Unlimited p. 337 right, Sean Brady p. 337 left, Dave Buresh/Denver Post p. 339, Ken O'Donoghue p. 340 both, W. H. Breazeale, Jr. p. 341 both, Sean Brady p. 342 left, Chip Clark p. 342 right, Ken O'Donoghue p. 343 all photos except bot. right, Sean Brady p. 343 bot. right, Ken O'Donoghue p. 344, Ken O'Donoghue

Chapter 14: p. 349, Charles Pefley/Stock Boston p. 352, W. H. Breazeale, Jr. p. 353 all, W. H. Breazeale, Jr. p. 355, AMOCO Corporation/American Petroleum Institute p. 358, Blair Seitz/Photo Researchers, Inc. p. 359, Ken Greer/Visuals Unlimited p. 360, Leland C. Clark, Jr. p. 361 top, Ken O'Donoghue p. 361 bot., Norman O. Tomalin/Bruce Coleman Inc. p. 363, *The Death of Socrates* by Jacques Louis David, the Metropolitan Museum of Art, Wolfe Fund, 1931. Catharine Lorillard Wolfe Collection (31.45) p. 364 bot., Dr. Jeremy Burgess/Science Photo Library/Photo Researchers, Inc. p. 364 top, Jay Reiter/AP/Wide World p. 366 right, Phil Degginger p. 366 left, Phil Degginger p. 367, Ted Streshinsky/Photo 20-20 p. 370 left, Bob Masini/Phototake p. 370 right, Phil Degginger p. 372, The Dupont Company p. 373 right, Dr. Jeremy Burgess/Science Photo Library/Photo Researchers, Inc. p. 373 left, Sean Brady

Chapter 15: p. 380, NASA p. 381, The Granger Collection p. 382 left, The Granger Collection p. 382 right, Erich Lessing/Art Resource p. 390, National Museum of American History/The Smithsonian Institution p. 393, NASA p. 394 left, NASA/JPL

p. 394 right, NASA p. 395 left, USGS p. 395 right, NASA/Johnson Space Center pp. 396–397 all, NASA p. 399, NASA/Johnson Space Center p. 401 right, NASA Johnson Space Center p. 401 right, NASA/Johnson Space Center p. 401 left, NASA p. 402 left, Hubble Heritage Team/NASA p. 403 left, NASA/JPL p. 403 right, NASA/Johnson Space Center p. 404, photo of Jupiter, NASA/Science Source/Photo Researchers, Inc. p. 404, all other photos, NASA p. 405, photo of Neptune, NASA/Science Source/Photo Researchers, Inc. p. 405, all other photos, NASA p. 406, NASA/JPL p. 407, NASA/Goddard Space Flight Center p. 408 left, NASA p. 408 right, Meteor Crater, Northern Arizona, USA p. 409 left, Chip Clark p. 409 right, Mt. Wilson Las Campanas Observatories, Carnegie Institution, Washington, D. C. p. 412, NASA/JPL p. 414, NASA/JPL

Chapter 16: p. 421, Carl Purcell/Photo Researchers, Inc. p. 425, National Institute of Standards, Boulder Laboratories, U. S. Dept. of Commerce

Chapter 17: p. 443, NASA p. 444, NASA p. 445 left, NASA/Lunar and Planetary Institute p. 445 right, USGS Clementine Space Team p. 447 left, NASA/Lunar and Planetary Institute p. 447 right, NASA/Johnson Space Center p. 452, Lick Observatory p. 455 left, Tom Pantages p. 455 right, Tom Pantages p. 460 top, NASA p. 460, four bottom photos, Willy Benz, Harvard-Smithsonian Center for Astrophysics

Chapter 18: p. 466, NASA p. 468, Hale Observatories p. 469, left NASA/Johnson Space Center p. 472, Anglo-Australian Observatory p. 478, Palomar Observatories p. 479 left, NASA p. 479 right, Hansen Planetarium p. 480, National Optical Astronomy Observatories p. 481, Science Photo Library/Photo Researchers, Inc. p. 483, NASA p. 484, NASA p. 485 top left, National Optical Astronomy Observatories p. 485 bot. left, Anglo-Australian Observatory p. 485 bot. right, Palomar Observatory p. 485 top right, Dr. Martin W. England/Science Photo Library/Photo Researchers, Inc. p. 486 top left, NOAA p. 486 bot. left, Hale Observatories/Photo Researchers, Inc. p. 486 bot. right, NASA p. 489, M. L. Geller/Harvard-Smithsonian Center for Astrophysical Observatory p. 490, NASA p. 493 top, NASA/Johnson Space Center p. 493 bot., Peter French/W. M. Keck Observatory

Chapter 19: p. 500, NASA p. 502, Gregory G. Dimijian/Photo Researchers, Inc. p. 505, Daryl Benson/Masterfile p. 506, NASA/Johnson Space Center p. 507, NASA/Johnson Space Center p. 509, Tom Ives p. 512, Science Kit Inc. p. 513, Dr. E. R. Degginger p. 515, Ray Nelson/Phototake Inc. p. 517 bot. left, NOAA p. 517 top left, NOAA p. 517 bot. right, NOAA p. 523 (a), Dr. E. R. Degginger p. 523 (b), Joseph Sohm/Chromosohm/The Stock Market p. 523 (c), Dr. E. R. Degginger p. 523 (d), Wayne Decker/Fundamental Photographs p. 524 (e), Martin W. Grosnick/Bruce Coleman, Inc. p. 524 (f), Dr. E. R. Degginger p. 524 (g), Dr. E. R. Degginger p. 524 (h), Phil Degginger p. 525 (i), Tom Bean p. 525 (j), Dr. E. R. Degginger p. 525 (k), Phil Degginger p. 526, Tom Bean

Chapter 20: p. 530, Gary Milburn/Tom Stack & Associates p. 533, Howard Bluestein/Photo Researchers, Inc. p. 537, Tom

Ives/The Stock Market   p. 540 top, Merilee Thomas/Tom Stack & Associates   p. 540 bot., Bob Daemmrich/Stock Boston   p. 542, NOAA/Roger Weldon   p. 543, Odyssey/Woodfin Camp & Associates   p. 546, Standard Oil Company/Carnegie Library, Pittsburgh   p. 547 left, Kristen Brochman/Fundamental Photographs   p. 547 right, NYC Photo Archive/Fundamental Photographs   p. 548, Brett Fromer/The Image Bank   p. 550, Stephanie Maze/Woodfin Camp & Associates   p. 551, Wide World Photos   p. 553, NASA   p. 554, Gordon Langsbury/Bruce Coleman, Inc.

Chapter 21:   p. 557, Stephen Trimble   p. 558, Dr. E. R. Degginger   p. 560 all, Ward's Natural Scientific Establishment Inc.   p. 561, Ward's Natural Scientific Establishment Inc.   p. 562, E. R. Degginger/Bruce Coleman Inc.   p. 563 top, Chip Clark   p. 563 (a), Paul Silverman/Fundamental Photographs   p. 563 (b), E. R. Degginger/Bruce Coleman Inc.   p. 564 (a), Lee Boltin Picture Library   p. 564 (b), Paul Silverman/Fundamental Photographs   p. 565, The Trustees of The British Museum   p. 568 left, Lee Boltin Picture Library   p. 568 middle, Lee Boltin Picture Library   p. 568 right, Lee Boltin Picture Library   p. 568 top, J. Messerschmidt/Bruce Coleman Inc.   p. 570, D. Griggs/USGS   p. 573 bot., Luis Villota/Bruce Coleman, Inc.   p. 573 top, James Montgomery/Bruce Coleman Inc.   p. 574 top, Gregory Dimijian/Photo Researchers, Inc.   p. 574 bot., C. C. Lockwood/Bruce Coleman Inc.   p. 575, Norman Tomalin/Bruce Coleman Inc.   p. 577 top left, Paul Silverman/Fundamental Photographs   p. 577 top right, Tom Bean   p. 577 bot., Tom Bean   p. 578, Runk/Schoenberger/Grant Heilman Photography   p. 579 right, Gary Withey/Bruce Coleman Inc.   p. 579 left, David Brown/Tom Stack & Associates   p. 580 left, Stephen Trimble   p. 580 bot. right, Boltin Picture Library   p. 580 top right, Kevin Schafer/Peter Arnold, Inc.   p. 582 left, Brian Parker/Tom Stack & Associates   p. 582 right, Tom Bean   p. 583, Dr. E. R. Degginger   p. 584 right, Noble Proctor/Photo Researchers, Inc.   p. 584 left, Paul Silverman/Fundamental Photographs

Chapter 22:   p. 587, Patrick Robert/Sygma   p. 590, World Ocean Floor, Bruce C. Heezen and Marie Tharp, 1977. Copyright by Marie Tharp, 1977. Reproduced by permission of Marie Tharp, 1 Washington Ave., South Nyack, NJ 10960   p. 595, Tom Bean   p. 598, Peter Turnley/Black Star   p. 599, Patrick Robert/Sygma   p. 600, USGS   p. 607, Grant Heilman/Grant Heilman Photography   p. 608 top, Chuck O'Rear/Westlight   p. 608 bot., John Montagne

Chapter 23:   p. 613, John Kieffer/Peter Arnold, Inc.   p. 614, P. Carrara/USGS   p. 615 top, L. A. Yettle/USGS   p. 615 bot., A. S. Navoy/USGS   p. 616, Charles Kennard/Stock Boston   p. 617 left, Dr. E. R. Degginger   p. 617 right, Tom Bean   p. 618, Dr. E. R. Degginger   p. 619, Larry Mayer/Gamma Liaison   p. 620, Paolo Koch/Photo Researchers, Inc.   p. 621 top, Dr. E. R. Degginger   p. 621 bot., Galen Rowell/Peter Arnold, Inc.   p. 622, J.T. McGill/USGS   p. 623 bot., Dr. E. R. Degginger   p. 623 top, Eric Vandville/Gamma Liaison   p. 629, World Ocean Floor, Bruce C. Heezen and Marie Tharp, 1 Washington Ave, South Nyack, NJ 10960

Chapter 24:   p. 633, Royal Tyrell Museum of Paleontology/Albert Community Development   p. 635 (a), Ken Lucas/Biological Photo Service   p. 635 (b), William Ferguson   p. 635 (c), Dr. E. R. Degginger   p. 635 (d), Tom Bean   p. 635 (e), Tom Bean   p. 636 top left, Breck Kent   P. 636 bot. left, William Ferguson   p. 636 right, Bruce Iverson   p. 637, Tom Bean   p. 638, J. Amos/Photo Researchers, Inc.   p. 640, Ludek Pesek/Science Photo Library/Photo Researchers, Inc.   p. 645, Robert Dymek   p. 649 left, Dr. E. R. Degginger/Bruce Coleman Inc.   p. 649 right, Walter Hodge/Peter Arnold, Inc.   p. 650, Royal Tyrrell Museum of Paleontology/Alberta Community Development

Index

Conversion Factors

Mass
1 gram $= 10^{-3}$ kg
1 kg $= 10^3$ g (equivalent weight $=$ 2.20 lb)
1 u $= 1.66 \times 10^{-24}$ g $= 1.66 \times 10^{-27}$ kg

Length
1 cm $= 10^{-2}$ m $= 0.394$ in.
1 m $= 10^{-3}$ km $= 1.09$ yd $= 3.28$ ft $= 39.4$ in.
1 km $= 10^3$ m $= 0.62$ mi
1 in. $= 2.54$ cm $= 2.54 \times 10^{-2}$ m
1 ft $= 12$ in. $= 30.48$ cm $= 0.3048$ m
1 yd $= 3$ ft $= 0.914$ m
1 mi $= 5280$ ft $= 1609$ m $= 1.609$ km
1 pc $= 3.26$ ly $= 2.05 \times 10^5$ AU
1 AU $= 1.499 \times 10^{11}$ m
1 ly $= 9.461 \times 10^{15}$ m
1 pc $= 3.086 \times 10^{16}$ m

Volume
1 m^3 $= 10^3$ L $= 264$ gal
1 L $= 10^{-3}$ m^3 $= 1.06$ qt $= 0.264$ gal
1 ft^3 $= 7.48$ gal $= 0.0283$ m^3 $= 28.3$ L
1 qt $= 2$ pt $= 0.946$ L $= 946$ mL
1 gal $= 4$ qt $= 3.785$ L

Time
1 h $= 60$ min $= 3600$ s
1 day $= 24$ h $= 1440$ min $= 8.64 \times 10^4$ s
1 year $= 365$ days $= 8.76 \times 10^3$ h $= 5.26 \times 10^5$ min $= 3.16 \times 10^7$ s

Energy
1 joule $= 0.738$ ft-lb $= 0.239$ cal $= 9.48 \times 10^{-4}$ Btu $= 6.24 \times 10^{18}$ eV
1 kcal $= 4186$ J $= 3.97$ Btu $= 0.00116$ kWh
1 Btu $= 1055$ J $= 778$ ft-lb $= 0.252$ kcal
1 cal $= 4.186$ J $= 3.97 \times 10^{-3}$ Btu $= 3.09$ ft-lb
1 ft-lb $= 1.36$ J $= 1.29 \times 10^{-3}$ Btu
1 eV $= 1.60 \times 10^{-19}$ J
1 kWh $= 3.60 \times 10^6$ J $= 3.413 \times 10^3$ Btu $= 860$ kcal

Speed
1 m/s $= 3.6$ km/h $= 3.28$ ft/s $= 2.24$ mi/h
1 km/h $= 0.278$ m/s $= 0.621$ mi/h $= 0.911$ ft/s
1 ft/s $= 0.682$ mi/h $= 0.305$ m/s $= 1.10$ km/h
1 mi/h $= 1.467$ ft/s $= 1.609$ km/h $= 0.447$ m/s
60 mi/h $= 88$ ft/s

Force
1 newton $= 0.225$ lb
1 lb $= 4.45$ N
Equivalent weight of 1 kg mass $= 2.20$ lb $= 9.80$ N

Pressure
1 atm $= 14.7$ lb/in.^{2} $= 1.013 \times 10^5$ N/m^2 $= 30$ in. Hg $= 76$ cm Hg
1 bar $= 10^5$ Pa
1 millibar $= 10^2$ Pa
1 Pa $= 1$ N/m^2 $= 10^{-2}$ millibar

Power
1 watt $= 0.738$ ft-lb/s $= 1.34 \times 10^{-3}$ hp $= 3.41$ Btu/h
1 ft-lb/s $= 1.36$ W $= 1.82 \times 10^{-3}$ hp
1 hp $= 550$ ft-lb/s $= 745.7$ watt $= 2545$ Btu/h